Studies in Systems, Decision and Control

Volume 216

The series "Studies in Systems, Decision and Control" (SSDC) covers both new developments and advances, as well as the state of the art, in the various areas of broadly perceived systems, decision making and control–quickly, up to date and with a high quality. The intent is to cover the theory, applications, and perspectives on the state of the art and future developments relevant to systems, decision making, control, complex processes and related areas, as embedded in the fields of engineering, computer science, physics, economics, social and life sciences, as well as the paradigms and methodologies behind them. The series contains monographs, textbooks, lecture notes and edited volumes in systems, decision making and control spanning the areas of Cyber-Physical Systems, Autonomous Systems, Sensor Networks, Control Systems, Energy Systems, Automotive Systems, Biological Systems, Vehicular Networking and Connected Vehicles, Aerospace Systems, Automation, Manufacturing, Smart Grids, Nonlinear Systems, Power Systems, Robotics, Social Systems, Economic Systems and other. Of particular value to both the contributors and the readership are the short publication timeframe and the world-wide distribution and exposure which enable both a wide and rapid dissemination of research output.

Indexed by SCOPUS, DBLP, WTI Frankfurt eG, zbMATH, SCImago.

All books published in the series are submitted for consideration in Web of Science.

Allam Hamdan · Haneen Mohammad Shoaib ·
Bahaaeddin Alareeni · Reem Hamdan
Editors

The Implementation of Smart Technologies for Business Success and Sustainability

During COVID-19 Crises in Developing Countries

Springer

Editors
Allam Hamdan
Ahlia University
Manama, Bahrain

Bahaaeddin Alareeni
Middle East Technical University
Northern Cyprus Campus
Kalkanli, Güzelyurt, Mersin, Turkey

Haneen Mohammad Shoaib
The University of Business and Technology
Jeddah, Saudi Arabia

Reem Hamdan
University College of Bahrain
Manama, Bahrain

ISSN 2198-4182 ISSN 2198-4190 (electronic)
Studies in Systems, Decision and Control
ISBN 978-3-031-10211-0 ISBN 978-3-031-10212-7 (eBook)
https://doi.org/10.1007/978-3-031-10212-7

This Springer imprint is published by the registered company Springer Nature Switzerland AG
The registered company address is: Gewerbestrasse 11, 6330 Cham, Switzerland

Preface

Smart technologies are defined as information, material, and social–political technologies with advanced and intelligent features; such technologies with intelligent features included big data, Internet of things, cloud computing and artificial intelligence, and other modern technologies that contributed to digital economy trends. Thus, it can examine the implications of SMEs sustainability strategy within innovative business models. However, SMEs challenging their survival during uncertainty manners among the crises, although customers and suppliers may increase their expectations through effective technology. Literature finds that organizations have to adopt sustainability strategy in their management practices, to help obtain successful business performance in profitability, while ensuring environmental stability and social sustainability. Industry 4.0 technologies are identified as the main contributor to the digitalization era. New technology delivers optimal outputs by utilization of effective resource. Therefore, smart technologies that have inventive and creative objects became critical to enterprise; recent studies show that it has led enterprises business such as SMEs to considerable investments, which many organizations over the world attempt to use innovative technologies such as IoT and AI, and these technologies have potential on sustainable business models.

This book includes 82 chapters by authors from 29 countries. Parts of the chapters of this book were based on direct personal invitations, while the other part was carefully selected from the CBF'2022. All of the chapters have been evaluated by the editorial board and reviewed based on a double-blind peer-review system by at least two reviewers.

The chapters of the book are divided into seven main parts:

I. E-Marketing, Advertisement, E-Commerce, and Business Success.
II. Artificial Intelligence, TechManagement, Women Entrepreneurship, and Social Implications.
III. Driving Innovative, Sustainability, and Creative Practices.
IV. Modern Social Media, Public Relations, and Its Effects on Society and Business.

V. Education Management, Technology, Smart Universities, and COVID-19 Impact.
VI. Sustainable Finance, Innovation, and Business Uncertain Situations.
VII. The Fourth Industrial Revolution, Tourism and Hospitality, and Healthcare Services.

The chapters of this book present a selection of high-quality research on the theoretical and practical levels, which ground the uses of smart technologies in business, health care, media, marketing, education, entrepreneurship, and other vital areas. We hope that the contribution of this book will be at the academic level and decision-makers in the various economic and executive levels.

Manama, Bahrain
Jeddah, Saudi Arabia
Güzelyurt, Turkey
Manama, Bahrain
March, 2022

Allam Hamdan
Haneen Mohammad Shoaib
Bahaaeddin Alareeni
Reem Hamdan

Contents

E-Marketing, Advertisement, E-Commerce, and Business Success

An Analytical Study of the Relationship Between Network Capability and e-Marketing to Achieve the Competitive Advantage of MSEs

Indah Fatmawati and Askar Garad

Abstract *Purpose* Two technical developments that stimulate reality perception are network capabilities and e-marketing. Both have been used in marketing situations to help businesses increase their competitiveness. From a scientific standpoint, the goal of this study is to frame the relationship between Network Capability and e-marketing to improve the competitiveness of MSEs advancements over the last five years. *Design/Methodology/Approach* One method employed in this research is to conduct a comprehensive literature review of the available international studies. For examining publications published in highly rated scientific journals indexed in Scopus, we used the meta-analysis method. We used (WordStat 8 and MAXQDA 2020) as part of the analysis process to look at articles and summarize descriptive statistics and correlation and keyword cloud. *Findings* Network Capability and e-marketing research are heavily reliant on mobile technologies. The gap between wearable gadgets and their counterparts is expected to close shortly. No analysis found on Big Data/machine e-marketing based on secondary data is a serious gap. Also, the fact that there are minor studies on the relationship between network capability and e-marketing and their impact on the competitiveness of SMEs is evident here. *Future implication* When we look to the future during the Corona pandemic, we can see that companies that adopt e-marketing and network capabilities will recover and eventually get out of the pandemic crisis faster than other companies. This is based on the ability of companies, particularly technology companies, to adapt and change, as well as invent new tools. As a result, a set of reasons and directives must be provided to enable us to understand this transformation and capitalize on it as an opportunity to change marketing methods and develop company network capabilities so that companies can continue to compete while maintaining interaction and customer loyalty. *Originality/Value* E-marketing technology and network capabilities are maturing, and additional marketing applications are emerging. For research

I. Fatmawati (✉) · A. Garad
Doctoral of Management, Postgraduate Program, Universitas Muhammadiyah Yogyakarta, Yogyakarta, Indonesia
e-mail: indahfatmawati@umy.ac.id

A. Garad
e-mail: askargarad@umy.ac.id

A. Hamdan et al. (eds.), *The Implementation of Smart Technologies for Business Success and Sustainability*, Studies in Systems, Decision and Control 216,
https://doi.org/10.1007/978-3-031-10212-7_1

to progress forward, scholars must keep pace and fill in the research gaps in both fields. The literature review study reflects this association between Network Capability, Competitiveness, and Marketing in MSEs. It also allows scholars to perform a more thorough analysis of theoretical development and collaborative work on the relationship between Network Capability, Competitiveness, and Marketing.

Keywords Network capability · Competitiveness · e-marketing · MSEs

1 Introduction

The ability to build networking is an essential asset of a company that can determine the success of its business performance. Previous studies have shown that the ability to establish a network determines the performance of superior businesses. Limited capabilities of companies, both in terms of financial capability, knowledge, technology, human resources, and access to the market, make them need cooperation with partners to create value for customers.

In the small and medium-sized businesses context, the limited size requires MSMEs to rely on the strength of external relationships to overcome various business problems they encounter [1]. The ability to manage internal and external information is known to improve the business performance of MSMEs by stimulating knowledge sharing, cost reduction, speed of innovation, business reputation, and opportunity identification [2]. Businesses can replenish each other through this network, supplement their less competitive resources, and gain opportunities and broader market access.

More companies are innovating in the network in recent decades than relying solely on research and development [3, 4]. Network Capability (NC) is an essential asset for the company's performance, not least MSMEs. In addition, developed NC can encourage knowledge transfer among different business units, create opportunities to learn and promote collaboration with each other [5], and facilitate the creation of new knowledge within the company [6, 7], all of which are critical to the success of innovation.

Networking research has been a popular topic for the past two decades. NC's important role becomes critical considering that the company will not survive and prosper only through its efforts. Each depends on the activities and performance of other companies [8, 9]. Nevertheless, there are still exciting gaps to expecting what constructs improve the effectiveness of the relationship between NC and MSME performance. This notion refers to mediating the influence of network capability on the company's business performance. Analytically, a process is alleged to occur because of the formation of the company's network capability, which affects the company's business performance [10]. One of the exciting opportunities investigated to see the effectiveness of NC on the achievement of MSME performance is in the possible role of sharing value creation. Cooperation conducted by MSMEs in the field of fashion in the creation of value and distributional is suspected to be antecedents of the creation of better business performance at a time when businesses could cooperate (network capability). This critical issue will be investigated in research over two years of implementation with a variable focus and different consequences in each year of research implementation [11].

The NC concept is derived from Resource-Based View (RBV), a critical pillar in the strategic management literature [12]. This research will use resource-based theory to explain NC constructs. Capability views were developed from RBV of competitive strategy. RBV states that companies in the same industry have different performances because they differ in resources and capabilities [13].

In the view of dynamic capability, contrary to traditional RBV, it is believed that companies need to develop new capabilities to identify opportunities and respond quickly [14, 15]. NC allows companies to gain access to various resources, identify opportunities, and provide a rapid response to the market's changing needs [16]. NC allows companies to connect their resources with other resources by building relationships [1].

Based on the above, this study aims to investigate reflecting the relationship between Network Capability, Competitiveness, Marketing on MSEs performance. We will present identifiable patterns in Network Capability, Competitiveness, Marketing and the relationship between them and their impact on improving an institution's performance. Then we will suggest emerging core topics that deserve more academic attention.

2 Methodology

This paper uses the meta-analysis method. A literature review of meta-analysis involves taking findings from the selected literature and analyzing them. Polit and Beck [17] argue that meta-analysis help in drawing conclusions and discovering aspects and relationships between findings. It is a non-statistical procedure that evaluates and analyzes results from studies and builds on previous concepts and interpretations. This research is based on articles published in scientific journals with a high evaluation and indexed by https://www.scopus.com/. The following keywords (see Fig. 2) were used to find relevant research by looking for them in the title or abstract of publications at https://scholar.google.com/ and journals with a specialty in the field of literature review for this study at https://www.scimagojr.com/. The scientific journals were chosen between 2017 and 2021 for review. According to Cronin [18] the review process consists of five distinct steps; the first step was Collection 534 articles using the keywords above. In the second step, during the Elimination of duplication in literature, 281 articles were selected, and in the third step, check journal name in SCOPUS, 353 articles were selected. In the fourth step, check title and abstract, 31 articles were selected, and in the final step, full-text analysis and 31 papers were approved in this literature review. Articles that discuss the three variables were carefully chosen in this study, namely the relationship between network capability and e-marketing to achieve the competitive advantage of MSEs.

Table 1 showing articles that were selected in this research. This literature review is a statistical review of existing scientific references about the relationship of governance with financial performance and its impact on the institution's performance. The literature review gives an overview of the level reached by science on this topic. It allows the reader to learn about similar theories, methods, and gaps present in this research.

Through the analysis process, (Wordstat 8) was used to investigate every journal and summarize the descriptive statistics, correlation, and keywords cloud. The focus of the analysis was on the results of the articles.

Figure 1 shows the distribution of articles over the years of publication. We can notice from the figure the fluctuation of focused research on network capability and e-marketing between high and low. This phenomenon means there is no clear plan by higher education institutions and other institutions specialized in governance that is interested in linking universities with economic development and directing research on this topic.

The final step in statistical analysis is the keyword cloud of the articles. Figure 2 shows the main words that were the most repeated and used to describe the relationship among variables. As we notice, marketing was the most frequent word, and the

Table 1 Articles list in review by groups

No.	Network capability and competitiveness MSEs	E-marketing and competitiveness MSEs
1	2017, Daniel Battaglia	2017, Saim Kashmiri
2	2017, Elena	2017, Sutapa
3	2017, Helen McGrath	2018, Byoungho Jin
4	2017, Thomas	2018, Daniel Quaye
5	2018, Danny	2018, Nuryakin
6	2018, Frances	2019, 2020, Saleh, Hasan
7	2019, Hsian-Ming	2019, Irina Mihailova
8	2019, Jongwoo Kim	2019, Sadiku-Dushi
9	2019, So Won Jeong	2020, Abdullah
10	2020, Drummond et al.	2020, Amirhossein Tohidi
11	2021, Helen McGrath	2020, Mohammad Alrousan
12	2021, Hong Yun Tian	2020, Mohammad Falahat
13	2021, Naili Farida	2020, Nehu
14	2021, Pedro Mendonça	2020, Umu Khouroh
15		2020, Vera Butkouskaya
16		2021, František Pollák
17		2021, Hector Cuevas
Total	14	17

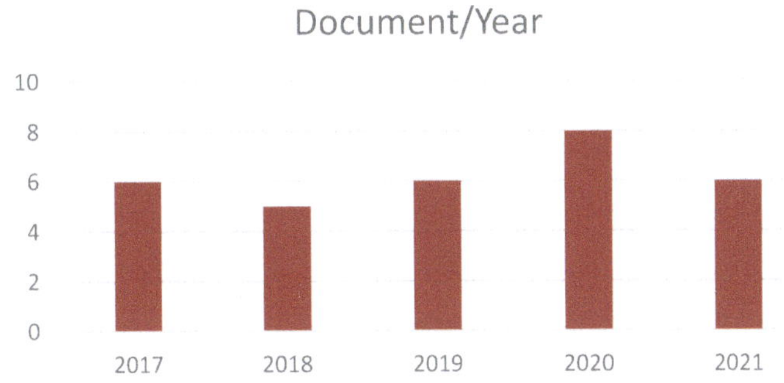

Fig. 1 Distribution of articles by year of publication

Fig. 2 Cloud distribution of keywords

second level was business, network and performance. In addition, the figure also has some related terms to describe this relationship.

We divided the articles into two groups as in Table 2, the first representing networking capability and competitiveness SMEs, and the second representing e-marketing and competitiveness SMEs.

Table 2 shows us the number of articles associated with each variable and the proportion of their distribution between the two groups. We note that all articles are 100% related to variables, except for the networking capability variable, in only 11 of the 17 articles in the second group. We also note that the e-marketing variable is present in 12 articles, in proportion to 14 articles found in the first group. Here shows us the lack of articles that examine the relationship between network capability and e-marketing and its impact on the competitiveness of SMEs.

Table 2 Distribute the articles between the two groups

	Network capability and competitiveness MSEs		E-marketing and competitiveness MSEs		Total	
MSEs	14	100.0%	17	100.0%	31	100.0%
Network capability	14	100.0%	11	64.7%	25	80.6%
E-marketing	12	85.7%	17	100.0%	29	93.5%
Competitiveness	14	100.0%	16	94.1%	30	96.8%
SUM	54	385.71	61	358.82	115	370.97
N = Documents	14	100.00	17	100.00	31	100.00

3 Discussions

3.1 Network Capability and Competitiveness MSEs

The organizations operate in ever-changing environments, interconnected in cooperative arrangements that concrete value. They seek value concretion through collaborative networks rather than in isolation. However, there is a lack of understanding of how networked organizations concrete value in the network through the innovative use of information systems (IS). Existing multi-firm studies focus primarily on dyadic relationships. Many studies have reported either increased organizational efficiencies (exploitative capability) or resulting innovations (exploitative capability) as a result of the concretion process, but rarely both aspects at the same time [19].

The competitiveness of Strategic Small Business Networks (SBSNs) is enhanced by a number of factors, including network management, resource/capabilities sharing, the depth of exchange relationships, and the quality of the partnerships formed. Self-directing and standardizing skills may be relevant to governance [20]. Centralized production can also make it easier to share resources and make the best use of available resources. The higher the strength of a partnership, the more likely it is to contribute to the forging of alliances and the exchange of information [21].

Competitiveness has an indirect impact on the efficacy of internationalization since online and offline communication mediates the relationship between the two variables [21]. For new projects to obtain and mobilize external resources and engage in interactive networked activities, network capacity development is critical. For the Industrial Procurement and Marketing (IMP) group, developing strategies and new project contexts are relatively recent research streams [22]. Moderating effects suggest that combining a few great marketing partnerships with many weak customer service relationships improves salespeople's effectiveness. When managing cross-functional relationships, salespeople benefit from network routing capabilities. Furthermore, sales managers benefit from a variety of intranet initiatives that boost the sales team's effectiveness [23].

That SME innovation boosts export potential, despite the fact that, contrary to previous research, there is no link between environmental competitiveness and export potential. Furthermore, there is a strong positive correlation between government networks and export potential, as well as a slight positive correlation between industry networks and export potential. Further research into industry sub-sectors reveals a greater link between industry networks and export potential [24].

3.2 E-Marketing and Competitiveness MSEs

SMEs were driven to increase their technology and marketing capabilities as a result of their international entrepreneurial orientation and local market competitiveness, resulting in improved performance in foreign markets. The coexistence of an internal drive (i.e. international entrepreneurial orientation) and a hostile external environment (i.e. local rivalry) can effectively foster SME capabilities, highlighting the importance of context in creating organizational capacities [25].

Market orientation has a big impact on innovation and performance. Excellence and competitive success are also influenced by innovation. Furthermore, competitive advantage has a substantial impact on the performance of a company [26].

Although an intra-industry spillover from a significant retail data breach is possible, managers can employ elements such as information technology, marketing, and corporate social responsibility to help insulate their companies from the contagion effect [27].

Product design and packaging advances, as well as advertising, retail, and pricing innovations, offer SMEs with a long-term competitive advantage. New product designs and packages, as well as creative retail channels, are the primary drivers of long-term market advantage. Integrating marketing competency (marketing resources and capabilities) with innovative marketing activities improves competitive advantage by a small margin. Physical resources may provide a competitive edge, but combining them with dynamic marketing capabilities ensures competitive sustainability in a competitive market.

According to Nuryakin et al. [10], marketing capability has a negligible impact on marketing performance. Marketing competence had a major impact on competitive advantage, while market orientation had an impact on marketing performance. Product innovation was also influenced by market orientation. Product innovation has a big impact on a company's competitive edge. Product innovation has a major impact on marketing performance, and competitive advantage has a significant impact on marketing performance.

4 Conclusion and Recommendation

This study aimed to determine the relationship between network capability and e-marketing and the impact of attaining this on MSEs competitiveness. The studies' findings reveal a strong and positive link between network capability, e-marketing, and each of the three competitive strategies: cost leadership, excellence, and focus. It also demonstrated the influence of network and marketing capabilities on competitive advantage across all four dimensions: (quality, speed, flexibility, cost). In addition, the three competitive strategies (cost leadership, excellence, and focus) impact the competitive advantage in its four dimensions: (quality, speed, flexibility, cost).

Firms that deploy network capabilities will benefit from lower information uncertainty and increased capacity to develop international network capabilities. Furthermore, when a firm has a high level of international strategic orientation and international network capabilities, Network Capability leads to indirect international market growth. Overall, e-marketing capabilities improve the firm's ability to generate other internal capabilities, which has a positive impact on the firm's international market growth and competitiveness MSEs.

On the other hand, e-marketing capabilities have a direct impact on worldwide business network interactions. Furthermore, the marketing process is inextricably linked to economic success. Overall, network competence and entrepreneurial marketing are crucial aspects that have a substantial impact on SMEs' competitive advantage. Also, networking skills, entrepreneurial marketing, and competitive advantage are all crucial characteristics that have a substantial impact on SMEs' marketing performance. For SMEs to be successful and competitive in the long run, networking capability, entrepreneurial marketing, competitive advantage, and marketing performance must all be established.

Based on above, the e-marketing and Network Capability have been shown to positively enhance internationalization and competitiveness MSEs, but scant empirical testing limits our understanding of the explicit impact on firm internationalization.

When we look to the future during the Corona pandemic, we can see that companies that adopt e-marketing and network capabilities will recover and eventually get out of the pandemic crisis faster than other companies. This is based on the ability of companies, particularly technology companies, to adapt and change, as well as invent new tools. As a result, a set of reasons and directives must be provided to enable us to understand this transformation and capitalize on it as an opportunity to change marketing methods and develop company network capabilities so that companies can continue to compete while maintaining interaction and customer loyalty.

The study makes several recommendations, the most essential of which is to development promotional programs to attract buyers from internal and external countries. It's also crucial to learn from successful worldwide experiences that can aid in developing new environmentally friendly services. The study also suggests that MSEs concentrate on their capabilities and resources to lower the costs of producing high-quality services and the need to expand their use of current technology to promote their products or services.

References

1. Solano Acosta, A., Herrero Crespo, Á., Collado Agudo, J.: Effect of market orientation, network capability and entrepreneurial orientation on international performance of small and medium enterprises (SMEs). Int. Bus. Rev. **27**(6), 1128–1140 (2018). https://doi.org/10.1016/j.ibusrev.2018.04.004
2. Bharadwaj, S.G., Varadarajan, P.R., Fahy, J.: Sustainable competitive advantage in service industries: a conceptual model and research propositions. J. Mark. **57**(4), 83 (1993). https://doi.org/10.2307/1252221
3. Cenamor, J., Parida, V., Wincent, J.: How entrepreneurial SMEs compete through digital platforms: the roles of digital platform capability, network capability and ambidexterity. J. Bus. Res. **100**, 196–206 (2019). https://doi.org/10.1016/j.jbusres.2019.03.035
4. Chang, C.C.: Exploring IT entrepreneurs' dynamic capabilities using Q-technique. Ind. Manag. Data Syst. **112**(8), 1201–1216 (2012). https://doi.org/10.1108/02635571211264627
5. Clarke, A., Fuller, M.: Collaborative strategic management: strategy formulation and implementation by multi-organizational cross-sector social partnerships. J. Bus. Ethics **94**(Suppl. 1), 85–101 (2010). https://doi.org/10.1007/s10551-011-0781-5
6. Jaakkola, M., Möller, K., Parvinen, P., Evanschitzky, H., Mühlbacher, H.: Strategic marketing and business performance: a study in three European 'engineering countries.' Ind. Mark. Manag. **39**(8), 1300–1310 (2010). https://doi.org/10.1016/j.indmarman.2010.06.005
7. Nandakumar, M.K., Ghobadian, A., O'Regan, N.: Business-level strategy and performance: The moderating effects of environment and structure. Manag. Decis. **48**(6), 907–939 (2010). https://doi.org/10.1108/00251741011053460
8. Ngugi, I.K., Johnsen, R.E., Erde´lyi, P.: Relational capabilities for value co-creation and innovation in SMEs. J. Small Bus. Enterp. Dev. **12**(2), 260–278 (2010). https://doi.org/10.1108/14626001011041256
9. Garad, A., Al-Ansi, A.M., Qamari, I.N.: The role of e-learning infrastructure and cognitive competence in distance learning effectiveness during the covid-19 pandemic. Cakrawala Pendidik. **40**(1), 81–91 (2021). https://doi.org/10.21831/cp.v40i1.33474
10. Nuryakin, N., Aryanto, V.D.W., Setiawan, M.B.: Mediating effect of value creation in the relationship between relational capabilities on business performance. Contad. Adm. **63**(1), 1–21 (2018). https://doi.org/10.22201/fca.24488410e.2018.1178
11. Constantinescu-Dobra, A.: The Internet marketing and the SMEs. A comparative analysis of dentistry strategies for online and printed advertising. Mark. Inf. Decis. **5**, 75–88 (2012) [Online]. Available: https://www.ceeol.com/search/article-detail?id=170129
12. Garad, A., Rahmawati, A., Pratolo, S.: The impact of board directors, audit committee and ownership on financial performance and firms value. Univ. J. Account. Financ. **9**(5), 982–994 (2021). https://doi.org/10.13189/ujaf.2021.090509
13. Kumar, K., Boesso, G., Favotto, F., Menini, A.: Strategic orientation, innovation patterns and performances of SMEs and large companies. J. Small Bus. Enterp. Dev. **19**(1), 132–145 (2012). https://doi.org/10.1108/14626001211196442
14. Eshlaghy, A.T., Maatofi, A.: Learning orientation, innovation and performance: evidence from small-sized business firms in Iran. Eur. J. Soc. Sci. **19**(1), 114–122 (2011)
15. Jämsä, P., Tähtinen, J., Ryan, A., Pallari, M.: Sustainable SMEs network utilization: the case of food enterprises. J. Small Bus. Enterp. Dev. **18**(1), 141–156 (2011). https://doi.org/10.1108/14626001111106479
16. Holtzman, Y.: Innovation in research and development: tool of strategic growth. J. Manag. Dev. **27**(10), 1037–1052 (2008). https://doi.org/10.1108/02621710810916295
17. Polit, D.F., Beck, C.T.: The content validity index: are you sure you know what's being reported? critique and recommendations. Res. Nurs. Health **29**, 489–497 (2006). https://doi.org/10.1002/nur
18. Cronin, P., Ryan, F., Coughlan, M.: Undertaking a literature review: a step-by-step approach. Br. J. Nurs. **34**(8–9), 1045–1055 (2008). https://doi.org/10.1016/0965-5425(94)90007-8

19. Battaglia, D., Pohlmann, C., Moura, J.B., Borchardt, M., Pereira, G.M.: Leveraging elements to small business strategic networks: the perception of business partners. Int. J. Bus. Excell. **12**(1), 83–101 (2017). https://doi.org/10.1504/IJBEX.2017.083340
20. Kirichenko, E.: Capacity-Building in Cluster Networks: An Analysis of the Cosmetics Clusters International Network. Universität Linz (2017)
21. McGrath, H., O'Toole, T., Marino, L., Sutton-Brady, C.: A relational lifecycle model of the emergence of network capability in new ventures. Int. Small Bus. J. Res. Entrep. **36**(5), 521–545 (2017). https://doi.org/10.1177/0266242617738571
22. Claro, D.P., Ramos, C.: Sales intrafirm networks and the performance impact of sales cross-functional collaboration with marketing and customer service. J. Pers. Sell. Sales Manag. **38**(2), 172–190 (2018). https://doi.org/10.1080/08853134.2018.1437353
23. Francis, S.: India's electronics manufacturing sector getting the diagnosis right. Econ. Polit. Wkly. **53**(34), 112–117 (2018). https://doi.org/10.1002/tie.21972
24. Sutapa, S., Mulyana, M., Wasitowati, W.: The role of market orientation, creativity and innovation in creating competitive advantages and creative industry performance. J. Din. Manaj. **8**(2), 152–166 (2017). https://doi.org/10.15294/jdm.v1i1.12756
25. Kashmiri, S., Nicol, C.D., Hsu, L.: Birds of a feather: intra-industry spillover of the target customer data breach and the shielding role of IT, marketing, and CSR. J. Acad. Mark. Sci. **45**(2), 208–228 (2017). https://doi.org/10.1007/s11747-016-0486-5
26. Quaye, D., Mensah, I.: Marketing innovation and sustainable competitive advantage of manufacturing SMEs in Ghana. Manag. Decis. **57**(7), 1535–1553 (2019). https://doi.org/10.1108/MD-08-2017-0784
27. Khouroh, U., Sudiro, A., Rahayu, M., Indrawati, N.K.: The mediating effect of entrepreneurial marketing in the relationship between environmental turbulence and dynamic capability with sustainable competitive advantage: an empirical study in Indonesian MSMEs. Manag. Sci. Lett. **10**, 709–720 (2020). https://doi.org/10.5267/j.msl.2019.9.007

Applying Marketing Intelligence System in Improving Marketing Performance for Jordanian Corporations During COVID19 Pandemic

Nafez Nimer, Mahmoud Allan, and Tareq Nael Hashem

Abstract The research looks at how a marketing intelligence system might help Jordanian businesses improve their marketing effectiveness in the wake of the COVID19 Pandemic. The study's sample size includes 234 Jordanian marketing managers. A stratified random sample of 146 Jordanian firms was chosen, and 118 valid questionnaires were obtained from marketing managers for statistical analysis, yielding an 80.8% response rate. According to the findings, the marketing intelligence system is crucial in influencing the company's marketing performance, and its simple and practical design increases the possibility of benefiting from it in improving the company's marketing performance. The researchers recommend allocating a separate department for marketing intelligence systems in order to maximize the system's benefit in terms of improving Jordanian corporations' marketing performance, as well as involving the department's director in various strategic decisions affecting the company.

Keywords Marketing intelligence system · Marketing performance · Jordanian corporations

1 Introduction

Knowing your competition is crucial, but many businesses do not do a good job of acquiring information about them in a methodical way. If you work for a company, you may keep a lot of important information in your head. Everything is kept in giant

N. Nimer · M. Allan
Marketing Department, Business Faculty, Al-Zaytoonah University of Jordan, Amman, Jordan
e-mail: n.ali@zuj.edu.jo

M. Allan
e-mail: mahmoud.allan@zuj.edu.jo

T. N. Hashem (✉)
Marketing Department, Isra University, Amman, Jordan
e-mail: tareq.hashem@iu.edu.jo

A. Hamdan et al. (eds.), *The Implementation of Smart Technologies for Business Success and Sustainability*, Studies in Systems, Decision and Control 216,
https://doi.org/10.1007/978-3-031-10212-7_2

paper folders, such as corporate accounts, as well as scattered papers that maintain sales data, customer information, and so on around the company. If you don't spend a lot of time looking for it, you may not find it. As a result, managers often base their decisions on emotions rather than facts. This is because the time spent hunting for information often costs businesses money [11].

Managers must have the proper information in order to make less risky and uncertain strategic planning decisions. The value of information is decided only by the decisions and actions that people take in response to their knowledge [18]. Marketing "intelligence" has been proved to be useful as a practical guide and analytical tool for managers looking to fill gaps in their competitors' knowledge created by flaws in their strategy. People think of intelligence as a tool for gathering information rather than as information in and of itself [3]. Keep an eye on your company and its surroundings, acquire information that might be useful to your business, be alerted to a threat or capitalize on a good chance, and support you in preparing for and making decisions. This is what the marketing intelligence system is for.

This study's importance might be summed up as a substantial scientific addition to the field of marketing intelligence systems. In addition to highlighting the need of giving marketing information through the marketing intelligence system in order to improve Jordanian corporations marketing performance. This study is also being carried out in light of the economic fluctuations caused by the COVID19 Pandemic, as the need for Jordanian companies in various sectors to use the information provided by the marketing intelligence system in designing their marketing strategies and developing their marketing performance to help them face the increasing intensity of competition in the local market grows. It is becoming more important for firms to use marketing intelligence systems to record and create connections with their consumers, as well as to increase their competitiveness in a variety of industries. Many Jordanian firms have been criticized for their lack of interest in marketing intelligence systems after the COVID19 Pandemic. Marketing effectiveness may be improved and a company's competitive advantage can be achieved by making better use of resources that are occasionally accessible. This research investigates how the marketing intelligence system may assist in improving the marketing performance of Jordanian corporations in the aftermath of the COVID19 Pandemic.

2 Theoretical Framework

2.1 Marketing Intelligence System

A marketing intelligence system can be defined as "a system that collects data from the environment serving management by keeping it informed of the activities of its competitors, customers and other elements that affect its marketing operations" [14].

It can also be defined as: The ability to obtain appropriate information for marketing managers, executives, and management teams in order to make decisions under conditions of certainty, ambiguity, and risk is critical [10]. Smith [19] also defined marketing intelligence as "the process of acquiring data in the sphere of business that helps company managers to make faster decisions". The marketing intelligence system, in our opinion, may be described as a collection of methods and sources that managers utilize to gather daily information on key changes in the marketing environment.

Criteria for Marketing Intelligence System

Many authors have emphasized the strategic value of competition intelligence. They stressed the need of two fundamental characteristics for a marketing intelligence system:

- **Data on the competition**: Data must be gathered and processed in ways that allow it to be shown in a fashion that supports strategic decision-making [13].
- **The degree to which competitive strategic information is suitable**: It relies on how well the information gathered, evaluated, and disseminated is incorporated into the organization's strategic decision-making process.

This is deemed required since many businesses acquire all accessible information on all rivals, resulting in an influx of unstructured data that is difficult to handle and so becomes worthless or of little use [8].

The following are the major elements that the organization should identify for each rival, according to Kotler [12]:

- **Objectives**: this section highlights the competitor's profitability and market share goals.
- **Strategies**: The competitor's strategy is decided by whether it focuses on cheap pricing, good quality, better service, or reduced costs, as well as if the competitor's activities are long or short-term.
- **Strengths and Weaknesses**: The company compares the strengths and weaknesses of its competitors.
- **Forms of response (reaction)**: when the company evaluates how rivals will respond to its activities in terms of pricing (increasing or lowering) and promotional mix components.

2.2 *Dimensions of the Marketing Intelligence System*

Various aspects of the marketing intelligence system are included in the system, including the following dimensions:

- **Customer understanding** is one of the elements that marketing intelligence systems are concerned with. By understanding the nature and desire of customers, along with their purchasing motives, identifying their behavior and communicating it in the form of information to decision-makers in the marketing department, marketing intelligence systems can work on taking them into consideration in the design of their products, provided that they include an appropriate marketing mix [3]. In order to differentiate between organizations, it is necessary to collect realistic information and data about the customer and his purchasing motives. When an organization is able to collect realistic information and data about the customer and his purchasing motives, it is able to retain the customer more by providing him with what he actually wants and achieving his satisfaction [7].

- **Market understanding** is one of the elements that a marketing intelligence system is concerned with, as it is concerned with understanding the nature of market and its variables, which are represented by factors such as share of the market, rate of growth, and market size, and delivering it in the form of information to decision makers in the marketing department [3].
- **Product intelligence**: It is a contemporary of the marketing intelligence system, which is considered the tool used to meet the needs and desires of customers, and is concerned with determining the quality of the product that the company or organization will offer and how to develop it [13].
- **Competitor Intelligence**: A marketing intelligence system element that is focused with watching the activity of competitors in a given market and collecting information about them on a continual basis [3].

2.3 Marketing Performance

Organizations strive to achieve the objectives set forth in their marketing strategy. They attempt to do so in order to enhance their regulatory market share and achieve long-term growth in the target market. As a consequence, corporations attempt to keep constant control over their performance. The enforcement of this law is accomplished via a number of methods. An organization's strategic performance is being evaluated in order to expand market share and achieve growth. They also want to modify their plans in order to take into account the most current development [9, 17]. It is possible to define marketing performance as the extent to which a company's marketing objectives are fulfilled [21].

Given the above, it is clear that gaining a competitive advantage over one's competitors is essential for every firm. This point should be stressed: all businesses attempt to develop marketing plans and strategies in order to get a competitive advantage and enhance their marketing performance [1].

In order to evaluate marketing performance, the following factors are considered: the current volume of sales for a product line, the market share held by the company, the percentage of sales distributed through intermediaries, consumer awareness of the company's advertising campaigns, consumer and intermediary perceptions of the organization's image, the level of satisfaction with consumer needs and satisfaction with the organization's products, and the percentage of marketing cost incurred. The failure rate of new items has a significant impact on total costs [5].

According to Aggarwal and Gupta [2], traditional financial measures of marketing performance (profit, sales, and cash flow) have been broadened to include a variety of non-financial factors (such as market share, quality, customer satisfaction, loyalty, and brand equity), as well as input (marketing audit, implementation) and output (marketing audit, efficiency/effectiveness) measures.

3 Research Hypotheses

Following an assessment of the research's theoretical literature, we submit the following study hypotheses:

- There is an impact of Marketing Intelligence System in improving Marketing Performance for Jordanian Corporations during COVID19 Pandemic.
- There is an impact of Customer Understanding in improving Marketing Performance for Jordanian Corporations during COVID19 Pandemic.
- There is an impact of Market Understanding in improving Marketing Performance for Jordanian Corporations during COVID19 Pandemic.
- There is an impact of product intelligence in improving Marketing Performance for Jordanian Corporations during COVID19 Pandemic.
- There is an impact of Competitor Intelligence in improving Marketing Performance for Jordanian Corporations during COVID19 Pandemic.

4 Research Model

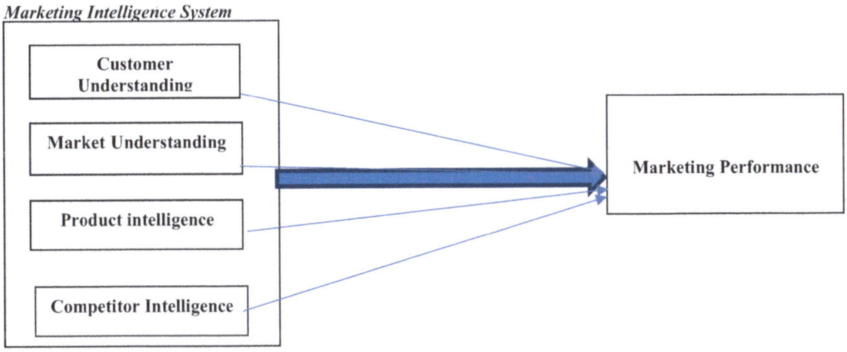

Source Research model was developed based on [3, 15]

5 Methodology

The researchers used a descriptive analytical technique to conduct this study in order to investigate the function of the marketing intelligence system in boosting marketing performance in Jordanian corporations in the aftermath of the COVID19 Pandemic.

5.1 The Study's Population and Sample

The study population consists of 234 Jordanian marketing managers, who are distributed as follows (https://www.sdc.com.jo):—Banking and Financial Services (16)—Insurance and Insurance (22)—Sector of Services (140)—The Industrial Sector (56).

A stratified random sample was drawn from 146 Jordanian corporations (www.surveysystem.com), which were divided into four economic sectors based on their proportion of the study population, and 118 valid questionnaires were retrieved for statistical analysis from marketing managers, resulting in an 80.8% response rate, which is a good percentage for scientific research purposes.

5.2 Tool of the Study

The questionnaire (the study tool) was divided into two sections: the first contained personal information about the participants (gender, age, educational level, and number of years of experience), and the second contained paragraphs that measured both the independent and dependent variables of the study.

5.3 Statistical Analysis

Descriptive statistics were employed to extract frequencies and percentages to define the characteristics of the study sample, as it was determined that the majority of the study sample members are men, as shown by their 62.5% share of the research sample. We also observe that the majority of the research sample members are young individuals, with the age group (30–50 years) accounting for 77% of the sample. We also observe that the majority of the research sample members had university degrees, with 75.6% of the study sample being bachelor's degree holders. As it turns out, the majority of the research sample members had more than 10 years of practical experience (66.9%).

Also, mean and standard deviation of the study variables were extracted, as they are summarized from the following Table 1.

Table 1 Descriptive statistics

	N	Minimum	Maximum	Mean	Std. deviation
Customer understanding	146	1.00	5.00	3.2616	0.83759
Market understanding	146	1.00	5.00	3.4233	1.08825
Product intelligence	146	1.00	5.00	3.0164	1.05347
Competitor intelligence	146	1.40	5.00	3.6397	0.95545
Marketing performance	146	1.00	5.00	3.1712	0.99986

Because the means of the aforementioned variables are larger than the crucial number (3), the data in the preceding table indicates that people have favorable opinions about them. Additionally, it is discovered that Market Understanding has the most favorable attitudes, while product intelligence has the lowest good attitudes.

5.4 *Validity and Reliability Tests*

Factor analysis was done to assess the questionnaire's validity, and the factor loading of each item is more than 0.40. This represents the questionnaire's validity. Furthermore, as shown in Table 2, Cronbach alpha values are higher than the approved 0.60%, indicating that the questionnaire is reliable [4].

5.5 *Collinearity Statistics Test*

The values of VIF and Tolerance in the preceding table are within the acceptable range, which indicates that Tolerance values are larger than 0.10 and VIF values are fewer than 10. This indicates that there is no multicollinearity among the independent variables [6] (Table 3).

5.6 *Hypothesis Testing*

- **There is an impact of Marketing Intelligence System in improving Marketing Performance for Jordanian Corporations during COVID19 Pandemic** (Table 4).

When the Multiple Regression test is applied to test the above hypothesis, it is discovered that $F = 75.919$ is significant at the 0.05 level, and $r = 0.826$ demonstrates a high degree of correlation between the variables. Furthermore, the independent factors explain 68.3% of the variation in the dependent variable. That is, there is an

Table 2 Validity and reliability tests

Customer understanding	Factor loading	Cronbach alpha (α)
v1	0.758	0.931
v2	0.824	
v3	0.795	
v4	0.813	
v5	0.749	
Market understanding	Factor loading	Cronbach alpha (α)
v6	0.891	0.941
v7	0.899	
v8	0.871	
v9	0.741	
v10	0.869	
Product intelligence	Factor loading	Cronbach alpha (α)
v11	0.807	0.94
v12	0.840	
v13	0.821	
v14	0.860	
v15	0.714	
Competitor intelligence	Factor loading	Cronbach alpha (α)
v16	0.941	0.93
v17	0.936	
v18	0.905	
v19	0.936	
v20	0.925	
Marketing performance	Factor loading	Cronbach alpha (α)
v21	0.916	0.934
v22	0.909	
v23	0.939	
v24	0.950	
v25	0.933	

Table 3 Collinearity statistics test

Model	Collinearity statistics	
	Tolerance	VIF
Customer understanding	0.619	1.616
Market understanding	0.303	3.298
Product intelligence	0.477	2.097
Competitor intelligence	0.324	3.082

Table 4 Hypothesis testing

ANOVA

Model		Sum of squares	df	Mean square	F	Sig.	R	R square
1	Regression	98.995	4	24.749	75.919	0.000	0.826	0.683
	Residual	45.964	141	0.326				
	Total	144.959	145					

Coefficients

Model		Unstandardized coefficients		Standardized coefficients	t	Sig.
		B	Std. error	Beta		
1	(Constant)	−0.203	0.227		−0.894	0.373
	Customer understanding	0.363	0.072	0.304	5.050	0.000
	Market understanding	0.592	0.059	0.645	10.117	0.000
	Product intelligence	0.370	0.065	0.390	5.676	0.000
	Competitor intelligence	0.225	0.087	0.215	2.587	0.011

impact of Marketing Intelligence System in improving Marketing Performance for Jordanian Corporations during COVID19 Pandemic.

Furthermore, the coefficient table demonstrates that the t value for each variable is significant at the 0.05 level. This means:

- There is an impact of Customer Understanding in improving Marketing Performance for Jordanian Corporations during COVID19 Pandemic.
- There is an impact of Market Understanding in improving Marketing Performance for Jordanian Corporations during COVID19 Pandemic.
- There is an impact of product intelligence in improving Marketing Performance for Jordanian Corporations during COVID19 Pandemic.
- There is an impact of Competitor Intelligence in improving Marketing Performance for Jordanian Corporations during COVID19 Pandemic.

Furthermore, it has been discovered that "Market Understanding" has the greatest influence on enhancing Marketing Performance when beta = 0.645. While "Competitor intelligence" has the least influence on enhancing Marketing Performance since beta = 0.215.

6 Conclusion

The purpose of this research was to ascertain the marketing intelligence system's contribution to the improvement of marketing performance in Jordanian corporations in the aftermath of the COVID19 Pandemic. The research focused on marketing managers of Jordanian corporations. It has been determined that marketing intelligence systems are used on an average basis in the firms studied.

Additionally, it was discovered that the marketing intelligence system is critical in affecting the company's marketing performance, and its simple and practical design enhances the likelihood of benefitting from it in increasing the company's marketing performance.

Similarly, the findings of [20] support this conclusion (information about customers, competitors, and suppliers has the greatest direct impact on the perception of the overall quality of marketing intelligence and, given its strong connections to other categories of marketing intelligence, it may also serve as a mediating factor). Additionally, this conclusion is supported by [16] findings that marketing intelligence and information systematization in market contexts are "critical for the rise of competitiveness".

In light of the study's findings, the researchers propose that organizations stay current with advancements in the area of marketing intelligence systems in order to maximize their effectiveness in using these systems to boost marketing performance. So that this system does not become a burden for these businesses and diminishes their gain from it. Additionally, it is necessary to allocate a separate department for marketing intelligence systems in order to maximize the system's benefit in terms of improving the marketing performance of Jordanian corporations, as well as the necessity of involving the department's director in various strategic decisions affecting the company. Additionally, it is necessary to train employees to use the marketing intelligence system in a manner consistent with the demands of their jobs, while also focusing on diversifying the methods for collecting data and information from the surrounding environment and addressing the companies' apparent weaknesses in this area.

References

1. Adams, G.L., Lamont, T.: Knowledge management systems and developing sustainable competitive advantage. J. Knowl. Manag. **7**(2), 142–154 (2003)
2. Aggarwal, N., Gupta, M.: Marketing performance measures: Current status in Indian companies. Decision **33**(1), 47–74 (2006)
3. Crowley, E.A.: Market intelligence versus market research. In: Mukerjee, H.S. (ed.) Industrial Marketing, 1st edn., pp. 169–172. Excel Books, Delhi, India (2009)
4. Gujarati, D.N., Porter, D.C.: Basic Econometrics, 5th edn. McGraw Hill Inc., New York (2009)
5. Haidar, M.: The Information Systems—An Approach Towards Achieving a Competitive Advantage. Al-Dar Al-Jame'eya Publishing House, Alexandria (2002)

6. Hair, J., Black, W., Babin, B., Anderson, R.: Multivariate Data Analysis, 8th edn. Cengage Learning, Pearson Prentice Hall, NJ (2018)
7. Hashem, T.: The role of customer information system IN improving marketing performance of Jordanian commercial banks. Int. J. Manag. Appl. Sci. **5**(11), 81–85 (2019)
8. Hashem, T.: The role of marketing information systems in improving marketing performance for Jordanian corporations. Ph.D. thesis. Amman Arab University, Jordan (2006)
9. Hashem, T., Al Maani, A.: The impact of strategic fit on the marketing performance of the industrial corporations in Jordan. Int. J. Bus. Manag. **14**(2), 115–124 (2019)
10. Igbaekemen, G.O.: Marketing intelligence as a strategic tool for competitive edge. Br. J. Mark. Stud. **2**(5), 17–34 (2014)
11. Johnson, G., Scholes, K., Whittington, R.: Exploring Corporate Strategy: Text and Cases, 8th edn. Prentice Hall, Harlow (2008)
12. Kotler, P.: Kotler on Marketing. Simon & Schuster Inc., New York (1999)
13. Ladipo, P.K.A., Awoniyi, M.A., Arebi, I.T.: The influence of marketing intelligence on business competitive advantage (a study of Diamond Bank Plc). J. Compet. **9**(1), 51–71 (2017). https://doi.org/10.7441/joc.2017.01.04
14. Mcleod, R., Schell, G.: Management Information Systems. Prentice Hall, New Jersey (2004)
15. Norfazlina, G., Akma, S., Nurul Adrina, S., Noorizan, M.M.: Customer information system satisfaction and task productivity: the moderating effect of training. Proc. Econ. Financ. **37**, 7–12 (2016)
16. Queiroz, J., Oliveira, B.: Benefits of the marketing information system in the clothing retail business. J. Inf. Syst. Technol. Manag. **11**(1), 153–168 (2014)
17. Rifai, F., Obeidi, M., Al-Maqousi, K.: The impact of business entrepreneurship on organizational performance: an empirical study at Jordanian telecommunication companies. Acad. Entrepr. J. **28**(1), 1–10 (2022)
18. Sharma, B.: Marketing strategy, contextual factors and performance—an investigation of their relationship. Mark. Intell. Plann. **22**(2) (2004) [Online]. Available: file://A:emeraldinsight.com
19. Smith, O.: The power of marketing intelligence [Online]. http://www.smetimes.in/smetimes/in-depth/2011/May/31/the-power-of-marketing-intelligence354533.html (2011)
20. Venter, P., Rensburg, M.: The relationship between marketing intelligence and strategic marketing. UNISA Grad. Sch. Bus. Leadersh. **17**(4), 440–456 (2014)
21. Wheelen, L., Hunger, J.: Strategic Management and Business Policy, 11th edn. Pearson Education Inc., Upper Saddle River, New Jersey (2008)

The Moderator Effect of COVID Pandemic on the Relationship Between Online Shopping Services and Online Consumer Purchasing Behavior. A Case of Jordan

Zakaria Azzam

Abstract This study aims to identify the moderator effect of Covid Pandemic on the relationship between Online Shopping Services and Online Consumer Purchasing Behavior in Jordan. For this purpose the researcher determined certain variables related to Online Shopping Services which are (speed and response, quality of service, availability of e_application, risks perception, usage flexibility, Privacy and Trust, customer experience), and on the other hand variables related to (Online Consumer Purchasing Behavior) to explore the relationship between them. Later on a questionnaire is designed, and was distributed on google form to 520 respondents who deal with Online Shopping Services in Amman city the capital of Jordan. Out of which (42) questionnaires were missed which represents (8%). And finally the sample size represented by (478) respondents, which represents (92%) of the distributed questionnaires. Suitable statistical analytical method is adopted which suits the nature of the present study using through SPSS program which is used to analyze the data collected from the respondents. The main conclusion of this paper is that there is statistically significant relationship between Online Shopping Services and Online Consumer Purchasing Behavior in Jordan of various factor (speed and response, quality of service, availability of application, risks perception, usage flexibility, Privacy and Trust, customer experience), on the other hand, Online Shopping Services have a high contribution on Online Consumer Purchasing Behavior by about (68%). The main recommendations of this study is that marketers of online shopping services should review continuously the significance of online shopping services, in order to judge customer satisfaction and face the development taking place in during the Pandemic, and enhance the availability of e_application which should be easy to use and have full safety issues in order to motivate customers to use such applications.

Keywords Covid pandemic · Online shopping services · Consumer purchasing behavior · Jordan

---

Z. Azzam (✉)
Marketing Department, Zarqa University, Zarqa, Jordan
e-mail: zazzam@zu.edu.jo

A. Hamdan et al. (eds.), *The Implementation of Smart Technologies for Business Success and Sustainability*, Studies in Systems, Decision and Control 216,
https://doi.org/10.1007/978-3-031-10212-7_3

1 Introduction

The development taking place in the field technology during Corona Pandemic has radically changed how different marketers and services providers interact with their target markets. The services sectors such as restaurants, food suppliers, banks, clothing shops and insurance companies which are at the main sectors that depend on online shopping services has been forces to change their way of providing their products specifically during Corona Pandemic to customer [1].

The concept of online services refers to "electronic delivery of different services to customers through electronic communication channels, as many services providers try to distinguish themselves in a highly competitive sector in order to adapt their services to the ever-changing customer needs and to change the traditional functions of the company to become electronically provided and offered", which may results in benefits to the marketers and customers such as reducing the expenditures, saving time and efforts of both the parties involved [2, 3].

Marketers of online services to survive with the intense competition during Corona Pandemic era, they must respond to all patterns of changes in social life style that has been taken place through interacting with customers in order to motivate and attract them to increase their purchases, which fit the needs of the customers attitudes during the pandemic which was totally different. Therefore, it was necessary for services providers to change the way customer get products through, from the classical one to electronical way. Which encourages customers to shift to online shopping services [1, 3–6]. There is also an increasing number of Jordanian companies looking for innovative ways to provide their online services, such as restaurants, food suppliers, banks, clothing shops and insurance, in order to make these services more convenient and efficient for customers, it has become important for marketers to ensure that customers are aware about the quality of online services, its speed and response, risks perception, ease of use, privacy and trust, e_aplication availability, in order to judge service and customer's satisfaction for online services [1, 3–7]. Generally speaking customers are hesitant to adopt online shopping services due of many reasons that induced customers to accept the classical way of getting online shopping services, but with the changes taking place during and post Corona Pandemic era in Jordan, the adoption of online shopping services has become a must, which motivated all marketers and producers to adopt online services which became a shape of Online Consumer Purchasing Behavior during and post Corona Pandemic era [3].

2 Research Problem

All online services providers are keen to motivate their customers to go for online shopping services in different ways through providing suitable, easy, and comfortable online shopping services to their clients during and post Corona Pandemic which can allow them to get their products electronically in convenient way which ensures a lot of savings represented by efforts, cost and time to the customers who deals with online shopping services [8–10].

On the other hand, the choice for some customers using online shopping services may result in increasing customer doubts and fear, complexity, risks, and difficulties in accessing the application electronically due to lack of experience and distrust in services providers due to bad experiences, which forces customer to reject the idea of to using online services some times. Therefore, online services adoption may not be the best choice for those customers as an Online Consumer Purchasing Behavior during and post Corona Pandemic era which make burden on marketers to convince them to adopt such applications and change their online behavior [3].

For some consumers who are hesitant to adopt online shopping services, for some reasons represented by fear of using the system, application technical problem and, risks associated with application, distrust in the e_application and difficulty in accessing these services, therefore, customers may refuse to use online shopping services and this make burden on online services providers to search for different ways to convince the target market to adopt it [3].

The current study is an attempt to analyze the moderator effect of Covid Pandemic on the relationship between Online Shopping Services and Online Consumer Purchasing Behavior in Jordan.

This paper aims to answer the following:

What is the Effect of COVID Pandemic as a moderator variable on relationship between online shopping services and online consumer purchasing behavior in Jordan?

3 Research Objectives

This paper is an attempt to achieve the following:

1. To identify the Relationship Between Online Shopping Services elements (represented by speed and response trust and safety, quality of services, risk perception, Availability of e_application, and finally ease of use) and Online Consumer Purchasing Behavior in Jordan.
2. To explore the Effect of COVID Pandemic as a Moderator variable on The Relationship Between Online Shopping Services and Online Consumer Purchasing Behavior in Jordan.
3. To provide an in depth empirical investigation of the issue of the competitive ability concept of Online services providers during Covid Pandemic in Jordan.

4 Research Hypotheses

The researcher supposes the following main hypotheses based on the previous studies reviewed and literature related to the variables selected carefully [1, 3, 11–16]:

H0: There is no statistical significant Relationship at the significance level ($\alpha \leq 0.05$) of The Relationship Between Online Shopping Services and Online Consumer Purchasing Behavior in Jordan.
H.02: There is no statistically significant effect at ($\alpha \leq 0.05$) of Corona Pandemic as a moderating variable on the relationship between Online Shopping Services and Online Consumer Purchasing Behavior in Jordan.

5 Literature Review

Literature related to online consumer behavior indicated that there are many issues that may affect consumers adoption of online shopping services reviewed represented [1, 3, 11–16] summarize the main elements of online shopping services which may affecting online consumer behavior during Corona Pandemic which are as follows.

5.1 Expected Benefit

Expected benefit which can be identified as "the level to which a customer believes that a certain application would encourages him for online shopping services adoption". The expected benefit has a potential impact consumers' on the adoption of online shopping services for certain products desired by him [16].

The consumer' adoption and usage of online shopping services relies on specific variables, including the customers' expectations regarding the advantages he may receive after dealing electronically with the application of online shopping services such as reducing the time necessary to complete the dealings, effort minimizing, cost savings …etc., the customer will benefit from adopting online shopping those services [17].

On the other side, if the customer assured that using online shopping services, would give him some advantages definitely he is going to adopt those online shopping services [17].

5.2 Risk Perception

Risks Perception as another element can be expressed as "the possible negative consequences in comparison with the desired outcome from online services". The risks of insufficient security for the application relate to the lack of adequate security in the application, where information is identified by hackers which allow them to misuse the private information of the identified customer. The risks that others may misuse the application [3, 17].

5.3 *Availability of Application*

Availability of e_application that which should be Easy and simple to be used by the customers. The Availability of e_application electronically used by the customer is another important issue to be taken into consideration by online services providers, the more simple, streamlined and clear the application the higher the adoption of online shopping behavior of the customer. Usage smoothness of any application "is the degree to which a person believes that using an application electronically or information technology would be free of effort".

5.4 *Usage Smoothness of Application*

Usage smoothness is one of the main factors to predict consumer behavior related to online shopping services adoption. Online shopping services applications provided by marketers should be simple and easy to use, which gives the customer the chance to eliminate all the technical and non technical obstacles and problems while using this application which will have a good positive consequences on convincing the consumers to adopt and use online shopping services, specifically during Corona Pandemic [1–3, 15, 17–19].

5.5 *Trust and Safety*

Trust can be defined as the expectations that a customer has about the quality, traits, of others. Trust factor is one of the most important issues of customers for users, specifically online shopping customers, who assured that trust has a good impact on the acceptance of online shopping services. Trust as a matter influencing customers' adoption and using online shopping services, may motivate them to deal with services providers online s, at all the times and all places which make them ready to order the products electronically [19, 20].

Fraud is the most significant danger to the trust and security of online shopping services adoption for consumers, specifically online hackers which occurs during dealing electronically with some services providers online, which may be misused by others [19, 20].

Safety on the other hand is identified as a parameter that expresses the level of feeling cool and calm when using online shopping g services provided by marketers, compared with the fear that results of receiving the service from services provider [21].

5.6 Services Quality

Services is defined as "the ability of the services to meet the expectations of the customers through providing an excellent level of that service". Literature related to quality identified many elements of quality that can achieve the desired expectations to the customer such as tangibility, reliability, accuracy, credibility, responsiveness, safety and confidence. Online services need such dimensions to parallel with the service in order to encourage customer to adopt online shopping services [22].

6 Online Shopping Services and Online Consumer Purchasing Behavior

Online Shopping services are the methods and procedures used in getting products from local marketers and other channels online through e_applications of certain services providers electronically which ensure getting needed product delivered to a specific customer or their doorstep at any place the are available in. Technology used in different sectors has shifted from selling classically to online shopping services for certain products that make customers to find desired products, and to give their place to ask for them electronically from marketers websites for those products. Technological development has allowed customers to increase the use of online shopping services encouraged several online shopping ordering needed products online. This change in customer's life style attracted marketers to develop the way they provide their products to those customers in a way better than the competitors with high level of credibility and in order to, to increase their sales, profit and market share which foster customer relationships, [1, 2, 4, 15, 17]. COVID-19 generated a relative change in consumers' pattern of dealing with online shopping services which resulted in consumption behavioral change [7, 10, 17–19, 23]. Technology adopted during Corona Pandemic saves a lot of costs and efforts of consumers which has realized a positive influences on the national economy of the whole world and this motivated marketers and consumers to adopt Online shopping services [17–19, 23]. The demand of customers for online shopping services has increased during Corona Pandemic in which technology has been as an attracting factor for customers to adopt online shopping behavior where such services satisfy the wants and desires individuals by providing fast and suitable services to them during Corona Pandemic particularly conveniently [25–28].

7 Research Model

The researcher supposes the research model based on literature review and previous studies [1–3, 5, 6, 10, 11, 15, 17–19, 23, 24] as shown in Fig. 1.

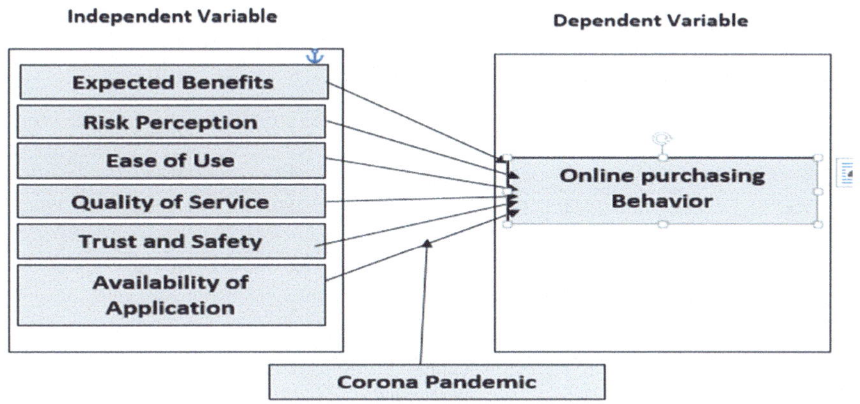

Fig. 1 Research Model

8 Methodology

The method used in this study is t descriptive as it describes the situation it accurately and clarifies it characteristics through collecting, analyzing, and interpretation the data. In order to identify the Moderator Effect of COVID Pandemic on the Relationship Between Online Shopping Services and Online Consumer Purchasing Behavior in Jordan. The descriptive analytical methodology which reaches meaningful generalizations that enable to enrich the knowledge about that phenomenon, in the study [20]. The researcher designed one questionnaire for The Moderator Effect of COVID Pandemic on the Relationship Between Online Shopping Services Online Consumer Purchasing Behavior in Jordan which was distributed electronically to a purposive non probability sample. SPSS version 24 was used to analyze the data through descriptive analysis, such as means, standard deviations, and frequencies, also analyze the hypotheses. Reliability is also calculated in order to assess the questionnaire by using both the internal consistency measures of Cronbach's alpha. Validity also assessed.

8.1 Population and Sample Size

The sample size of this study which represents the population are the customers of online shopping services, who deals with food delivery, medicines, clothing, insurance, financial etc. The population size, should not be less than (386) participants in order to represent the population, for more accuracy, the researcher distributed (520) questionnaires on online shopping services consumers who adopted e_Online shopping services for ordering products in Amman City the Capital of Jordan by using online (Google form). The researcher depended on purposive non probability

Table 1 Reliability of the independent and dependent variables (Cronbach's alpha)

Number	Variables	Statements	Cronbach's alpha
1	Expected benefit	6	0.91
2	Risk perception	5	0.801
3	Ease of use	4	0.882
4	Quality of services	4	0.798
5	Trust and safety	6	0.778
6	Availability of e_application	4	0.799
7	Online purchasing behavior	5	0.813
	Total	34	0.881

sample for distributing the questionnaires. About (42) Questionnaires were unreturned, which represents about (8%) of all questionnaires. The questionnaires were valid for actual analysis, which composited (478) as the sample size.

8.2 Research Instrument Reliability

Five-point Likert scale, which starts from range one strongly agrees until range five strongly disagree adopted in research instrument. Cronbach's alpha is designed as a measure of reliability. The normal range of Cronbach's coefficient alpha value between 0.0 and +1.0, and the higher values reflects a higher degree of reliability. The lower limit for Cronbach's Alpha value is (0.70) although according to [11]. The Cronbach's coefficient alpha was calculated for each field of the questionnaire. Table 1 shows the reliability of the independent and dependent variables (Cronbach's Alpha).

9 Results and Discussion

H0: There is no statistical significant Relationship at the significance level ($\alpha \leq 0.05$) of The Relationship Between Online Shopping Services and Online Consumer Purchasing Behavior in Jordan. Regression's several methods are used to examine the hypothesis at the significance level ($\alpha \leq 0.05$). Table 2 shows the results.

Table 2 shows dependent variable (Online Purchasing Behavior) and the set of the independent variables factor (Perceived Benefit, speed and response, privacy and safety, quality of services, risk perception, and finally ease of use), are significant

Table 2 Results of multiple regressions of the main hypothesis

Dependent variable	R	R2	F	DF	Sig.	Independent variable	B	T	Sig.
Online purchasing behavior	0.716	0.512	75.89	6	0.00	Perceived benefit	0.289	9.788	0.000
				478		Risk perception	0.210	8.32	0.001
						Ease of use	0.214	12.99	0.000
				472					
						Services quality	0.197	6.481	0.002
						Confidence and safety	0.121	11.45	0.004

because F significant was (0.000) which is less than (0.05), and the calculated F value is (75.89), which is more than F value Table 2. Therefore, we rejected the null hypothesis and accepted the substitute hypothesis, which states that: There is statistical significant Relationship at the significance level ($\alpha \leq 0.05$) of The Relationship Between Online Shopping Services and Online Consumer Purchasing Behavior in Jordan. The relationship between the dependent and independent variables is strong and positive. It is more than (0.5), R = 0.716. Also, the $R^2 = 0.512$, which means that the independent variables contribute to the dependent variable in a percentage of 51.2%. Since the values of the calculated t values of the variables (Perceived benefit 9.788, Risk perception 8.32, ease of use 12.99, service quality 6.481 and safety and confidence 11.45), are more than the t value table (1.986). Also, the t significant is less than (0.05). It means that there is a statistically significant relationship for Perceived benefit, Risk perception, ease of use, service quality and safety and confidence on online consumer purchasing behavior).

H.02: There is no statistically significant effect at ($\alpha \leq 0.05$) of Corona Pandemic as a moderating variable on the relationship between Online Shopping Services and Online Consumer Purchasing. Based on the result showed in Table 3, Corona Pandemic level has statistically significant role at ($\alpha \leq 0.05$) on as a moderating variable on the relationship between Online Shopping Services and Online Consumer Purchasing which is supported by the Frequency and percentage for the participants according to "Does Corona Pandemic Encourages you to Purchase Online" variable (n = 478) results shows that the percentage for "Does Corona Pandemic Encourages you to Purchase Online" with 89.33% for (yes), but the lowest percentage was (10.67%) for (no).

Table 3 One-way ANOVA analysis was used to test second main hypothesis

Variables		Sum of square	DF	F	Sig.
Online shopping services	Between group	10.33	5	14.22	0.002
	Within group	112.47	473		
	Total	122.8	478		
Online consumer purchasing behavior	Between group	6.64	5	7.89	0.000
	Within group	147.12	473		
	Total	153.76	478		

10 Recommendations

1. Marketers should review continuously the significance of online shopping services, to ensure customer satisfaction and keep pace with the continuous development in the field of technology and communication systems, and work to educate customers about the importance of using online shopping services and demonstrate their importance and advantages in the speed of dealing with and meeting their needs.
2. Online shopping Services providers should ensure that customers have high trust in adopting online shopping services on an ongoing basis and receive customer comments about any challenges they face, expand the study of the obstacles that affect customers' use of electronic banking services.
3. Online Services providers should develop e_applications that is easy to use in order to motivate consumers to adopt online shopping services.
4. Further and future related articles using other variables and factors to find out the extent of their effect of Corona Pandemic on online shopping services and Online purchasing behavior.
5. The findings of this paper would help marketers and producers of online shopping services to develop the design and shape of their sites as well as develop their promotional tools in order to convince customer to change their buying behavior towards online shopping.

References

1. Hashem, T.N.: Examining the influence of COVID 19 pandemic in changing customers? orientation towards E-shopping. Mod. Appl. Sci. (2020)
2. Ali, S., Khaled, N., Javed, H., Islam, D.: Consumer adoption of online food delivery services in Pakistan. J. Innov. Technol. Mark. Complexity **7**(1) (2020)
3. Azzam, Z., Ali, M.W., Alshiehk, M., Khanfar, I., Ismail, A.R.: Factors affecting the use of electronic banking services provided by Jordanian commercial banks from customers' point of view. In: Proceedings of the International Arab Conference on Information Technology (ACIT, 2021). Qaboos University, Omman (2021)
4. Omori, Y.: Online consumption during the COVID-19 crisis: evidence from Japan (2020)

5. Richards, T.J., Rickard, B.: COVID-19 impact on fruit and vegetable markets. Can. J. Agric. (2020)
6. Rahman, S.U., Bhatti, A., Mohamed, R., Ayoup, H.: The moderating role of trust and commitment between consumer purchase intention and online shopping behavior in the context of Pakistan. J. Glob. Entrep. Res. **9**, 43 (2019)
7. Wei, L.H., Osman, M.A., Zakaria, N., Bo, T.: Adoption of e-commerce online shopping in Malaysia. In: Proceedings of the IEEE 7th International Conference on E-Business Engineering, Shanghai, China, pp. 140–143, 10–12 Nov 2010
8. Chai, L.T., Yat, D.N.C.: Online food delivery services: making food delivery the new normal. J. Mark. Adv. Pract. **1**, 62–77 (2019)
9. Ganapathi, P., Abu-Shanab, E.A.: Customer satisfaction with online food ordering portals in Qatar. Int. J. E-Serv. Mob. Appl. **12**, 57–79 (2020)
10. Richards, T.J., Rickard, B.: COVID-19 impact on fruit and vegetable markets. Can. J. Agric. Econ. **68**, 189–194 (2020)
11. Al-Adayleh, R., Al-Smeirat, M., Mobaideen, H.: Factors affecting the use of electronic banking services via the Internet from the viewpoint of customers—a field study on the South Region—Jordan Jordanian. J. Bus. Adm. **12**(3) (2016)
12. Anouze, A.L.M., Alamro, A.S.: Factors affecting intention to use e-banking in Jordan. Int. J. Bank Mark. **38**(1), 86–112 (2019). Emerald Publishing Limited, 0265-2323. https://doi.org/10.1108/IJBM-10-2018-0271
13. Mohammad, A.A.S.: Customers' electronic loyalty of banks working in Jordan: the effect of electronic customer relationship management. Int. J. Sci. Technol. Res. **8**(12) (2019)
14. Almestarihi, R.: Factors affecting Jordanian electronic banking services. Manag. Sci. Lett. **10**(4), 915–922 (2019). https://doi.org/10.5267/j.msl.2019.10.004
15. Nguyen, H.V., Tran, H.X., Van Huy, L., Nguyen, X.N., Do, M.T., Nguyen, N.: Online book shopping in Vietnam: the impact of the COVID-19 pandemic situation. Publ. Res. Q. (2020)
16. Rawash, H.N., Masa'd, F., Enaizan, O., Eanizan, B., Adaileh, M.J., Saleh, A.M., Almestarihi, R.: Factors affecting Jordanian electronic banking services. Manag. Sci. Lett. **10**(4), 915–922 (2019). https://doi.org/10.5267/j.msl..10.004
17. Azzam, Z.A.: Managers' perspective towards perceived risks associated with technology based self services a case of-Jordan banks. Inst. Interdisc. Bus. Res. **4**(11), 46–64 (2013)
18. Wold, S.: How Covid-19 has changed shopper behavior. Xeim-Mark. Week. (2020)
19. Yeo, V.C.S., Goh, S.K., Rezaei, S.: Consumer experiences, attitude and behavioral intention toward online food delivery (OFD) services. J. Retail. Consum. Serv. **35**, 150–162 (2017)
20. Yoona, H., Occena, L.: Impacts of customers perceptions on Internet banking use with a smart phone. J. Comput. Inf. Syst. **54**(3), 1–9 (2014)
21. Al-Damour, H.H., Helouz, F.M.: The impact of banks service quality on the loyalty of customers to Jordanian banks studies. Adm. Sci. 39(1) (2012)
22. Ali, N., Allan, M., Azzam, Z.: The impact of total quality management dimensions on achieving competitive advantage: managerial perspective of the quality department staff at five star hotels. Case of Jordan. Int. J. Sci. Technol. Res **9**(02) (2020)
23. Rahman, S.U., Bhatti, A., Mohamed, R., Ayoup, H.: The moderating role of trust and commitment between consumer purchase intention and online shopping behavior in the context of Pakistan. J. Glob. Entrep. Res. **9** (2019)
24. Nickle, A.: How the coronavirus is affecting the produce industry. Packer (2020)
25. Featherman, M.S., Miyazaki, A.D., Sprott, D.E.: Reducing Online (2010)
26. Moriarty, B.: 5 reasons why your restaurant needs an online ordering system. Ind. Trends Top. (2016)
27. Nickle, A.: How the coronavirus is affecting the produce industry
28. Richard, E., Mandari, E.: Factors influencing use of mobile banking services: the case of Ilala district in Tanzania. Oper. Res. Soc. Eastern Africa **7**(1), 42–54 (2017)

E-Marketing and Customers' Bank Loyalty Enhancement: Jordanians' Perspectives

Mohammad Habes, Mahmoud Alghizzawi, Mokhtar Elareshi, Abdulkrim Ziani, Mohammad Qudah, and Maryam Mohammed Al Hammadi

Abstract Customer care and satisfaction are on-demand and sought by any business, especially in increasing online activities and interactions, and purchases. Providing such services simultaneously is a crucial and essential matter to get customers' loyalty. This paper examines how banks in Jordan rely on E-marketing to build long-lasting relationships with customers and the main factors assigned to understand their bank loyalty. Theoretically, based on the TAM paradigm and online survey of n = 330 respondents, the study found that all the E-marketing factors (e.g., social media features, word of mouth, informativeness) and the two TAM factors (mainly PEU and PU) reported having a positive impact on Jordanians' bank loyalty, suggesting that these factors played a vital role in evaluating the bank services. The possible techniques about the online environment that can practically improve the commitment and loyalty among the banking sector in Jordan are also discussed.

Keywords E-marketing · Social media · Loyalty · Banking · Customers · Jordan

---

M. Habes · M. Qudah · M. M. Al Hammadi
Department of Radio and Television, Yarmouk University, Irbid 21163, Jordan
e-mail: mohammad.habes@yu.edu.jo

M. Qudah
e-mail: mohammad.qudah@yu.edu.jo

M. Alghizzawi (✉)
Universiti Sultan Zainal Abidin (UniSZA), Kuala Terengganu, Malaysia
e-mail: dr.alghzawi87@gmail.com

M. Elareshi
Communication and Media College, Al Ain University, 64141 Al Ain, UAE
e-mail: mokhtar.elareshi@aau.ac.ae

A. Ziani
Mass Communication College, Umm Al Quwain University, 536, Umm Al Quwain, UAE
e-mail: drabdul.karim@uaqu.ac.ae

School of Business Management, UUM College of Business, Sintok, Malaysia

A. Hamdan et al. (eds.), *The Implementation of Smart Technologies for Business Success and Sustainability*, Studies in Systems, Decision and Control 216,
https://doi.org/10.1007/978-3-031-10212-7_4

1 Introduction

There is intense competition between companies in marketing in promoting their quality products/services. This has led many to look for an alternative to promote and market their stuff, including an online environment (e-marketing). Strategically, they have increased their appearance on the Internet and focused on online customers' care and satisfaction, not to mention the desire to meet their endless needs [1]. Online promotion is crucial for any company, including banks [2, 3] as they need to build trust and long-lasting relationships with their clients and gain their loyalty and satisfaction. In Jordan, it is estimated that there are 6.84 million active internet users, with 6.30 million social media users increasing by 11% between 2020 and 2021. This growth in an online environment has impacted Jordan's growing use of mobile and social media. Therefore, social media has become a central revenue-generating platform for marketers and advertisers. However, they still lack trust and reliability, mainly if ordinal people use them. Due to this quick and effective communication, customers especially possess their opinion about the brand and the company [4]. This study aimed to gain insights into the role of e-marketing in the banking sector by discussing the possibility of digital technologies improving the loyalty among customers of the banking sector in Jordan.

2 Literature Review

2.1 Social Media as a Marketing Tool

Social media becomes an influential platform in every industry, including the banking sector. For example, customers are easily able to meet their needs and desires through the Internet and social media platforms. The growth rate of any firm/institution could be linked to its appearance on social media platforms. Companies' growth largely depends on managing their marketing strategies in online and offline environments [1]. Social media marketing refers to online publications that use modern technologies such as websites. It is also linked to the two parties' conversation, communication, and involvement. For Veland et al. [5], social media marketing is a multichannel media outlet that can influence users' interest in making purchases. This part gives knowledge of contemporary advancements in web-based media showcasing with unique reference to the development of online innovation and its impact upon different advertising activities as social media marketing [6]. Küpper et al. [7] stated that the prevalence and far-reaching utilization of online media like Facebook, Twitter, and LinkedIn had furnished organizations with stages to team up and convey for all intents and purposes. Other studies have confirmed the power of social media marketing on shopping decisions [5], brand development, promotion, and publicity [8, 9].

2.2 E-Marketing

Digital marketing, Internet marketing or electronic marketing (e-marketing) is the component of marketing based on the Internet, and online digital technologies use desktops, mobile phones etc., to connect with whomever we go. It can also comprise e-mail marketing, e-newsletters, viral e-mail, and banner ads [10]. Thus, e-marketing is about planning and executing the conception, distribution, promotion of products and services using a networked environment such as the Internet to facilitate exchanges and satisfy customer needs and demands. Prensa and Suryawardani [10] indicated that e-marketing refers to marketing through electronic modern communication technology such as the Internet [7, 11]. Based on the Internet and websites, the future e-marketing is hugely growing, and it is one of the backbones of any national economy. For example, people who live in big cities are expected to use the Internet as a lifestyle, including in Jordan. It offers them the opportunities to introduce products and market them [1].

2.3 Customer Loyalty

Customer loyalty to any organization influences its future. The banking sector is one of the touchiest areas in customer loyalty [12]. Taylor et al. [12] recommend that banking sector value and trust are reliably the main precursors to client steadfastness's social and attitudinal types. Ngo and Nguyen [13] revealed that examination uncovers that administration quality and consumer loyalty are significant predecessors of client dependability and consumer loyalty intervenes in the impacts of administration quality on client faithfulness. These discoveries propose that there are non-straight connections between the three developments and stress the need to treat client reliability on the board as a cycle that incorporates many components interfacing with one another.

2.4 Social Media Features

Social media platforms arguably have an impact on customers' attitudes and loyalty. It allows users to exchange content and feedback about any company by displaying videos, photos, and conversations. This, in turn, affects both other customers' perspectives and the company itself [14]. Social media features such as a Watch, collaborative stories in groups, free sound collection (in Facebook), stories highlights (in Instagram), broadcast live in split-screen, sponsored filter and lens link (in Snapchat), 280-character story (in Twitter). Theoretically, social media platforms' perceived ease of use and perceived usefulness can provide sufficient incentives for customers to accept banking services. Sharma and Bhatt [15] examined the impact of social media on e-commerce and marketing and found a positive impact of social media features on respondents, as they bring them the perceived usefulness and ease of use when viewing corporate services and completing purchases. Accordingly, it is suggested:

H1: Social media features have a positive role in perceived usefulness.
H2: Social media features positively impact perceived ease of use.

2.5 *Word of Mouth*

Word of mouth is the transmission of information and talking between individuals about specific goods, products and services positively or negatively [11]. With the development of social media and increased communication between online users, word of mouth is often seen as an influential and transmitted faster and more effective platform [16, 17]. It is about sharing the experience with others regarding some goods, products and services. This stage usually comes before buying, so the work searches for information to ensure the quality of the service [18]. Customers consider word of mouth as a circulated among themselves that is more credible than promotional operations. In addition to that, the website's features are consistent with the perceived benefit of customers and ease of use, which enhances word of mouth. This can be linked to the perceived usefulness and ease of use [17]. Accordingly, it is suggested:

H3: Word of mouth has a positive role in perceived usefulness.
H4: Word of mouth has positively impacts perceived ease of use.

2.6 *Informativeness*

Information is one of the primary motivators for customers' response to marketing ads via social media platforms. It is a significant key to driving customers to the intention of buying [19]. It is a concept that shows the abundance of information about a particular good or service, which is a sensitive matter for customers [20]. Perceived ease of use and usefulness are linked to the abundance of information available through social media platforms. Kim et al. [21] examined the technology acceptance model and the impact of digital advertising. They revealed that the abundance of information through digital advertising through social media platforms and smartphones positively affects the ease of use and the perceived usefulness of the information features available through digital advertisements. Accordingly, it is suggested:

H5: Informativeness has a positive role in perceived usefulness.
H6: Informativeness has a positive role in ease of use.

2.7 *Technology Acceptance Model*

Davis [22] proposed the TAM paradigm is considered a fundamental model in technological adoption. The model suggested two main structures; perceived ease of use, which is the individual's feeling that using a technological system will be easy and uncomplicated. Perceived usefulness is for the individual using the technology to feel that their work has become more accessible [23]. Thus, he has the desired benefit that completing his work has become more smooth [24]. Hence, the two structures influence customers [17] toward banking. Mulia et al. [25, 26] examined the acceptance model of technology in banks. They found a positive relationship between the perceived usefulness and ease of use of electronic banking services in achieving loyalty among customers. They emphasized the effect of loyalty among customers and confirmed the existence of a relationship positive among themselves. Accordingly, it is suggested:

H7: Perceived ease of use has a positive role in perceived usefulness.
H8: Perceived ease of use has a positive role in customer loyalty.
H9: Perceived usefulness has a positive role in customer loyalty.

3 Methods

3.1 *Participants*

The primary purpose of this research is to examine and assess customer loyalty toward the Jordan of Bank. It specifically focuses on factors that measure customers' loyalty to the Bank of Jordan through e-marketing factors (e.g., social media features, word of mouth, informativeness) and TAM factors (PEU and PU) and its loyalty enhancement. The study instrument was designed based on previous related studies using an online survey method during October and November 2021. The population targeted in this research was the customers of Bank of Jordan, Amman, Jordan.

3.2 *Instrument of Measurement*

The scale measures were adapted from previously validated research. The questionnaire survey included measures for e-marketing's impact on enhancing customers' loyalty. The survey was tested for its internal reliability and validity before distribution (Table 1). The pre-structured questionnaire had two parts: First part invited respondents to evaluate their demographic details such as gender, age, education level, residence and marital status.

The second part was used to measure the relationship between e-marketing factors and customers' bank loyalty. A pilot study was used for the mentioned factors. The

Table 1 Validity and reliability of measurement scales

Constructs	Items	Mean	(SD)	Factor loading	CA	AVE
Social media features (SMF) (M = 3.56)	SMF1	3.60	(0.924)	0.676	0.825	0.498
	SMF2	3.62	(0.916)	0.685		
	SMF3	3.45	(0.885)	0.597		
	SMF4	3.59	(0.843)	0.645		
Word of mouth (WM) (M = 3.58)	WM1	3.73	(0.976)	0.698	0.768	0.400
	WM2	3.34	(0.875)	0.672		
	WM3	3.64	(0.776)	0.652		
	WM4	3.61	(0.792)	0.690		
Informativeness (INF) (M = 3.40)	INF1	3.41	(0.836)	0.681	0.691	0.41
	INF2	3.37	(0.838)	0.585		
	INF3	3.43	(0.897)	0.755		
Perceived ease of use (PEU) (M = 3.26)	PEU1	3.31	(0.916)	0.690	0.844	0.541
	PEU2	3.33	(0.953)	0.672		
	PEU3	3.20	(0.957)	0.671		
	PEU4	3.20	(0.920)	0.660		
Perceived usefulness (PU) (M = 3.54)	PU1	3.71	(0.777)	0.685	0.737	0.41
	PU2	3.56	(0.801)	0.586		
	PU3	3.42	(0.815)	0.570		
	PU4	3.48	(0.769)	0.602		
Customer loyalty (CL) (M = 3.47)	CL1	3.42	(0.906)	0.560	0.717	0.51
	CL2	3.51	(0.773)	0.655		
	CL3	3.50	(0.804)	0.607		

exploratory factor analysis was performed using the principal component analysis extraction method with Varimax rotation and Kaiser Normalization (Eigenvalue > 1). Table 1 reported the results of validity and reliability of the measurement scales, as well as the results of Cronbach's alpha evaluating the internal reliability, which was around 0.70 suggesting the value of internal reliability is met, and the outcomes of AVE values to ensure the validity of the construct, and they exceed the threshold value of 0.50. The results indicated that the research constructs for measuring e-marketing in enhancing customers' bank loyalty were reliable, and the sample size is sufficient.

4 Results

The descriptive statistics included the mean scores of constructs and their standard deviations (SD). The results indicated that the highest mean score with the positive impact of e-marketing on customer loyalty was social media features (M = 3.56), followed by word of mouth (M = 3.58), perceived usefulness (M = 3.54), customers' loyalty (M = 3.47), informativeness (M = 3.40) and perceived ease of use (M = 3.26).

4.1 Relationships Between Customers' Loyalty and E-Marketing with TAM

Furthermore, the bivariate analysis (Pearson correlation) was carried out to examine the relationship between the e-marketing factors and ATM factors in enhancing customers' loyalty. The findings of the Pearson correlation stated the nature and strength of the correlation between the constructs. It revealed that strong positive pairwise correlation between the WM and SMF ($r = 0.670$), INF and SMF ($r = 0.641$), INF and WM ($r = 0.666$), PEU and SMF ($r = 0.555$), PEU and WM ($r = 0.591$), PEU and INF ($r = 0.683$), PU and SMF ($r = 0.536$), PU and WM ($r = 0.624$), PU and PEU ($r = 0.572$), CL and SMF ($r = 0.460$), and CL and WM ($r = 0.570$), CL and INF ($r = 0.584$), CL and PEU ($r = 0.522$), and CL and PU ($r = 0.674$). These findings have supported the stated hypothesis (**H1**) by indicating a significant positive relationship between social media features and perceived usefulness ($r = 0.54$) and perceived ease of use (**H2**) ($r = 0.56$).

Social media features were expected to become a vital part of marketing, including e-marketing, by providing different options/functions that allow customers to communicate and interact with the services provider, e.g., the Jordan Bank. The data analysis also indicated that word of mouth has a positive role on respondents' perceived usefulness ($r = 0.61$) and perceived ease of use ($r = 0.60$), supporting (**H3** and **H4**). This is an exciting finding that would help understand the effectiveness of word of mouth in the online environment.

The analyzed data confirmed the positive relationship between informativeness and perceived usefulness ($r = 0.62$) and perceived ease of use ($r = 0.68$), meaning that the more information given to customers, the more they perceived it as usefulness and ease to use (**H5** and **H6**). The findings supported our **H7** that assumed the relationship between perceived ease of use and perceived effectiveness ($r = 0.57$). This positive relationship indicated that perceived ease of use impacts respondents' consideration of e-marketing usefulness in enhancing further customers' loyalty, supporting our **H8** and **H9**. For the moderating variables, the results of the data analysis showed no significant relationship between all the independent variables (demographic features) and the dependent variables (SMF, WM, INF, PEU, PU, CL). Unlike the other demographic characteristics, the variable social status showed a significant negative relationship with perceived ease of use ($r = -0.120$). In contrast, no important relationships were reported by the DI and the other IV.

Based on the results of Pearson correlation analysis, a linear regression analysis was performed to assess whether e-marketing factors (EMF) and TAM factors

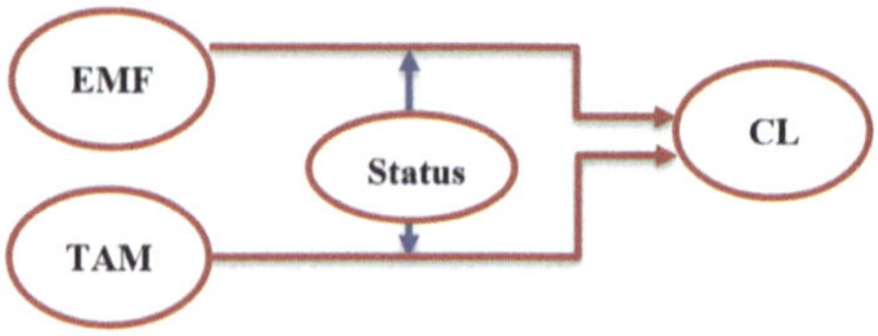

Fig. 1 Structural equation modelling for EMF and TAM

predict customers' loyalty. Before running the linear regression analysis, factor analysis through the principal component analysis with Varimax rotation was run to the eleven SMF, WM and INF items to derive one factor for subsequent multivariate data analysis, linear regression. The factoring criteria were: a factor Eigenvalue ≤ 1, a minimum primary loading ≤ 0.50 on the element and each item has a loading of ≤ 0.50.

Further statistics were applied; for example, the reliability of the measurement for EMR factor was tested by using a standard reliability test of Cronbach's alpha (0.889) and Bartlett's test of Sphericity [(1495.468) ($P = 0.000$) followed by the KMO value, which was 0.889 ($P = 0.000$), Eigenvalue (3.909). The same criteria were applied to eight PEU and PU items to derive the TAM factor. The reliability of the measurement for TMA factor was tested by using a common reliability test of Cronbach's alpha (0.851) and Bartlett's test of Sphericity [(936.249) ($P = 0.000$) followed by the KMO value, which was 0.879 ($P = 0.000$), Eigenvalue (3.058).

For CL items, the reliability of the measurement for CL factor was tested by using a common reliability test of Cronbach's alpha (0.717) and Bartlett's test of Sphericity [(211.432) ($P = 0.000$) followed by the KMO value, which was 0.648 ($P = 0.000$), Eigenvalue (1.317). a KMO value of more than 0.60 is acceptable for a measure of factorability. Secondly, gender, age, education, and residence were excluded from this analysis. They lacked a significant relationship with e-marketing, meaning that social status was the only variable included in the study (Fig. 1). The linear regression tests were based on the partial least squares structural equation modelling (PLS-SEM).

The results of the regression suggested that EMF explained 35% of the variance, $R^2 = 0.35$, $F(1{,}328) = 179.774$, $p = 0.000$ (Table 2). EMF significantly predicted customers' loyalty enhancement, $\beta = 0.595$, $t = 13.408$, $p = 0.000$. The results of the regression suggested that TAM explained 40% of the variance, $R^2 = 0.40$, $F(1{,}328) = 217.167$, $p = 0.000$. TAM significantly also predicted customers' loyalty enhancement, $\beta = 0.631$, $t = 14.737$, $p = 0.000$.

The third predicted variable was status; the regression analysis revealed social status has a significant impact on customers' loyalty, and the results explained only 12% of variance, $R^2 = 0.012$, $F(1{,}328) = 3.872$, $p = 0.050$. However, the prediction of social status (single or married) on customers' loyalty seemed very weak and, $\beta = -0.662$, $t = -1.968$, $p = 0.050$. This means that respondents' status did not influence the relationship between EMF and CL. Given the results of all the variations reported here, it seemed that other variables could predict more how to enhance customers' loyalty.

Table 2 Path analysis of structural equation modelling

Variable	R^2	β	t	p-value
Constant (CL)				0.000
EMF	0.354	0.595	13.408	0.000
TMA	0.398	0.631	14.737	0.000
Status	0.012	-0.662	-1.968	0.050

Note Dependent variable = CL. Independent variable = EMF, TAM, status

5 Discussions and Conclusion

This research examines and assesses customer loyalty toward the Jordan of Bank. It focuses on measuring customers' loyalty through e-marketing factors. Besides, the two TAM factors and their loyalty to build a lasting relationship, especially in the banking sector. A total of 330 responses were gathered and reported using a purposive convenience sampling approach and further complex statistical analysis to test our hypotheses. This research is underpinned by the extent to which e-marketing and TAM factors influence customers' loyalty toward the bank of Jordan, Amman, Jordan. The findings of this research appear to be consistent with several prior studies on the significance of EMF and TAM variables.

Concerning EMF, the findings of the reported research confirmed that social media features, word of mouth, and informativeness impact customers' loyalty toward the bank of Jordan, especially about enhancing such relationships. For example, such results indicated that these factors are an essential part of any e-marketing tools in the digital environment. These factors can also be used to enhance and promote Jordan's bank services. This is perhaps because the Internet facilities (e.g., through social media features websites) have proliferated in Jordan, and Jordanian customers are one click away from buying all their desires and need products and services.

Our findings support the assumption that e-marketing is now part of any e-business to achieve marketing goals, primarily through social media platforms and word of mouth. Such media have continued to establish new ways for businesses to collaborate. They effectively communicate with their online customers through word of mouth, which effectively supports customers and businesses. Now information through SMF and WM can help promote and meet customers' needs. Concerning TAM factors, not surprisingly, all the TAM factors appear to influence customers' loyalty toward the bank of Jordan. These findings are consistent with those that highlighted the importance of medium (e.g., social media) to be ease of use and perceived usefulness in banking.

Although all our hypotheses were supported, the data analyzed of some independent variables showed no significant effect on this relationship. For example, gender and status variables seemed to impact the customers' bank loyalty. It is not clear why certain independent variables had no significant results. To sum up, social media has proven to be an effective platform for communication and interaction between companies and customers. It plays a vital role in e-marketing by connecting products and services with customers, especially helping customers' choices. The findings reported provided a structural framework for understanding an emergent field, e-marketing, especially in the new media era and its integration and interaction between customers and products and services.

References

1. Krypton, A., Tantuah, N., Satria, H.W.: The effect of social media e-marketing towards consumers shopping behavior. J. Vokasi Indones. **6**(2), 1–7 (2018). https://doi.org/10.7454/JVI.V6I2.118
2. Elbasir, M., Elareshi, M., Habes, M.: The influence of trust, security and reliability of multimedia payment on the adoption of EPS in Libya. Multicult. Educ. **6**(5) (2020)
3. Salloum, S.A., Al-Emran, M., Habes, M., Alghizzawi, M., Ghani, M.A., Shaalan, K.: What impacts the acceptance of E-learning through social media? an empirical study. Rec. Adv. Technol. Accept. Model. Theor., 419–431 (2021)
4. Singh, S. Social Media Marketing for Dummies (2010)
5. Veland, R., Amir, D., Samije, S.-D.: Social media channels: the factors that influence the behavioural intention of customers. Int. J. Bus. Glob. **12**(3), 297–314 (2014). https://doi.org/10.1504/IJBG.2014.060214
6. Plume, C.J., Dwivedi, Y.K., Slade, E.L.: The new marketing environment. In: Social Media in the Marketing Context (2017)
7. Küpper, T., Lehmkuhl, T., Wieneke, A., Jung, R.: Technology Use of Social Media Within Customer Relationship Management: An Organizational Perspective (2015)
8. Wilson, L.: 30-Minute Social Media Marketing (SMM) Actions, pp. 57–71. Emerald Publishing Ltd. (2019). https://doi.org/10.1108/978-1-83867-078-820191005
9. Permatasari, A., Laydi, F.: The effects of social media advertising on consumer purchase intention: a case study of Indonesian family start-up enterprises. Int. J. Technol. Transf. Commer. **16**(2), 172 (2018). https://doi.org/10.1504/IJTTC.2018.097422
10. Priansa, D.J., Suryawardani, B.: Effects of E-marketing and social media marketing on E-commerce shopping decisions. J. Manaj. Indones. **20**(1), 76–82 (2020). https://doi.org/10.25124/jmi.v20i1.2800
11. Fox, G., Longart, P.: Electronic word-of-mouth: successful communication strategies for restaurants. Tour. Hosp. Manag. **22**(2), 211–223 (2016). https://doi.org/10.20867/thm.22.2.5
12. Taylor, S.A., Goodwin, S., Celuch, K.: The importance of brand equity to customer loyalty. J. Prod. Brand Manag. (2004). https://doi.org/10.1108/10610420410546934
13. Ngo, M.V., Nguyen, H.H.: The relationship between service quality, customer satisfaction and customer loyalty: an investigation in vietnamese retail banking sector. J. Compet. (2016). https://doi.org/10.7441/joc.2016.02.08
14. Ebrahim, R.S.: The role of trust in understanding the impact of social media marketing on brand equity and brand loyalty. J. Relatsh. Mark. **19**(4), 287–308 (2020). https://doi.org/10.1080/15332667.2019.1705742
15. Sharma, B.K., Bhatt, V.K.: Impact of social media on consumer buying behavior—a descriptive study on Tam model. i-Manager's J. Manag. **13**(1), 34 (2018)
16. Minazzi, R.: The digitization of word-of-mouth. In: Social Media Marketing in Tourism and Hospitality, pp. 21–45. Springer (2015)
17. Mehrad, D., Mohammadi, S.: Word of mouth impact on the adoption of mobile banking in Iran. Telemat. Inform. **34**(7), 1351–1363 (2017)
18. Ku, Y.C., Wei, C.P., Hsiao, H.W.: To whom should i listen? finding reputable reviewers in opinion-sharing communities. Decis. Support Syst. (2012). https://doi.org/10.1016/j.dss.2012.03.003

19. Lee, J., Hong, I.B.: Predicting positive user responses to social media advertising: the roles of emotional appeal, informativeness, and creativity. Int. J. Inf. Manage. **36**(3), 360–373 (2016)
20. Wu, R., Wang, G., Yan, L.: The effects of online store informativeness and entertainment on consumers' approach behaviors: empirical evidence from China. Asia Pac. J. Mark. Logist. (2019)
21. Kim, Y.B., Joo, H.C., Lee, B.G.: How to forecast behavioral effects on mobile advertising in the smart environment using the technology acceptance model and web advertising effect model. KSII Trans. Internet Inf. Syst. **10**(10), 4997–5013 (2016)
22. Davis, F.D.: Perceived usefulness, perceived ease of use, and user acceptance of information technology. MIS Q., 319–340 (1989)
23. Salloum, S.A., Al-Emran, M., Habes, M., Alghizzawi, M., Ghani, M.A., Shaalan, K.: Understanding the impact of social media practices on E-learning systems acceptance. In: International Conference on Advanced Intelligent Systems and Informatics, pp. 360–369 (2019)
24. Davis, F.D., Bagozzi, R.P., Warshaw, P.R.: User acceptance of computer technology: a comparison of two theoretical models. Manage. Sci. **35**(8), 982–1003 (1989)
25. Mulia, D., Usman, H., Parwanto, N.B.: The role of customer intimacy in increasing Islamic bank customer loyalty in using e-banking and m-banking. J. Islam. Mark. (2020)
26. Salloum, S.A., AlAhbabi, N.M.N., Habes, M., Aburayya, A., Akour, I.: Predicting the intention to use social media sites: a hybrid SEM-machine learning approach. Adv. Mach. Learn. Technol. Appl. Proc. AMLTA **2021**, 324–334 (2021)

The Effect of the Pharmaceutical Marketing Mix Elements During Corona Virus Pandemic on the Decision of the Prescription for Foreign-Made Drugs

Iyad A. A. Khanfar, Mustafa Said Al-Shaikh, and Mohammad Al-Abed

Abstract The study aimed to identify the effect of the pharmaceutical marketing mix elements during the corona virus pandemic on the decision to prescribe foreign-made medicines. They were based on the descriptive analytical method, using the questionnaire to collect study data. The study tool was distributed to (114) specialist physicians in Zarqa governorate in an intentional inspection manner after ensuring its validity for application. The study found a set of results and the most important are: there is an effect at the level ($\alpha \leq 0.05$) for the elements of the pharmaceutical marketing mix (product price, distribution, and promotion) combined and individually on the decision for prescribed foreign-made medicines for specialist doctors in Zarqa governorate. The study recommended that Jordanian pharmaceutical companies should take attention to developing marketing strategies. That include all elements of the pharmaceutical marketing mix, (pharmaceutical product, drug price, drug distribution, and drug promotion). Therefore enhancing its competition with foreign companies producing medicines, as this elements directly impact the prescription decision to consider demographic variables for physicians when preparing marketing plans, because of its impact on the prescription decision.

Keywords Pharmaceutical marketing mix · Pharmaceutical industry · Foreign drug · Jordan

I. A. A. Khanfar (✉) · M. S. Al-Shaikh
Zarqa University, Zarqa, Jordan
e-mail: i.khanfar@zu.edu.jo

M. S. Al-Shaikh
e-mail: malshaikh@zu.edu.jo

M. Al-Abed
Ministry of Health, Amman, Jordan

A. Hamdan et al. (eds.), *The Implementation of Smart Technologies for Business Success and Sustainability*, Studies in Systems, Decision and Control 216,
https://doi.org/10.1007/978-3-031-10212-7_5

1 Introduction

The corona virus pandemic has reduced demand in most business sectors, except the pharmaceutical sector, which witnessed a rebound. The use of virus prevention products was widely promoted. The necessity of using medical goods was unprecedented for all individuals. with the obligation to exchange medical services and ideas between physicians, and protocols for the use of medicine through personal sales have been established to convince governments with the information to use those protocols and buying them to citizens, and stimulating sales through a contract non-faceted conferences, to drive and encourage the demand for some products [1].

Foreign pharmaceutical companies direct their marketing mix towards doctors in order to influence their decision to prescribe, The targeted sample is chosen carefully through which a larger market share can be achieved; and it's done by targeting specialist doctors in regions with high population density and less public health services, where the population of Zarqa ranges between (1300–1400) thousand With a rate of 14% of the total population, there are 71% of health insured people in the governorate, and a percentage less than the average of the health insured in the Kingdom 87%, and the proportion of the governorate's population of health centers is (31,000) compared to (16.5) thousand in the Kingdom [2].

Therefore, specialist doctors in Zarqa Governorate are more reliable to marketing campaigns of foreign pharmaceutical companies compared to other governorates. That is to say, specialist doctors in the health sector occupy a particular position.

This category of physicians is usually targeted by drug-producing companies with a specific marketing mix.

Namely, to influence the prescription decision, the report of health ministry indicated that the number of specialist doctors in Zarqa governorate are only 153, [3], this what makes a conflict in the minds of the specialist doctors in the prescription decision between the local medical product and the foreign medical product. This happens because of the increased level of competition between local pharmaceutical companies and foreign ones. presenting them with a marketing mix specifically targeting them. In addition to what the Zarqa governorate has a feature down ratio of the health-ensured population compared to their percentage in the Kingdom that increases the number of patients they have, what constitutes a marketing opportunity for drug producing companies, therefore, the study problem emerged, which can be formulated with the main study question: What is the effect of the components of the pharmaceutical marketing mix corona virus pandemic on the decision to prescribe foreign-made medicines for specialist doctors in Zarqa Governorate?

2 Procedural and Operational Definitions

Pharmaceutical Marketing Mix: Marketing mix is "a set of tactics for marketing tools that include—product, price, promotion, and distribution, it is managed by the company to obtain the desired response in the target markets [4]. **The researchers define it** as all the activities carried out by foreign companies producing medicines such as pricing, distribution, promotion, and products, which affect the decision to prescribe foreign-made medicines for specialist doctors in Zarqa Governorate.

Pharmaceutical Product: It is "a physical product or service for a patient who wants to buy it, and it is classified into two types, tangible products. Moreover, the intangible (services) that users buy" [5]. **The researchers define it as the recommendations contained in the prescription that use** specific drugs for recovery, which is one of the components of the drug marketing mix.

Pharmaceutical Price: the amount of money to be paid for a product or service, or the exchange value that helps patients receive a product or service for a certain amount" [4]. The **researchers defines it** as the value the patient pays to obtain the treatment that a physician has prescribed, and is one component of the drug marketing mix.

Pharmaceutical Distribution: It is "all measures that aim to ensure the flow of products (goods, services) at the appropriate time and place for patients in the target market [6]. The researchers define it as all activities the company that produces the drug to deliver it to the consumer (patient) in the right place and time.

Pharmaceutical Promotion: It is "a marketing activity that involves a persuasive communication process, through which a good, service, idea or behavioral pattern is introduced to influence the minds of a particular audience to their behavioral response to what it promotes" [7]. The **researchers define it** as all Activities carried out by drug-producing companies that aim to influence the decision of the doctor and the patient (the consumer) and entice them towards its products.

Medical Recommendation Decision: defined as "an order that is written to give the appropriate treatment to the patient" [8]. It is also **defined by the researchers** as a group of medicines that specialist doctors in Zarqa governorate prescribe to patients.

3 Theoretical Framework and Hypothesis Development

Most the morals in the field of marketing indicated that the elements of the marketing mix of physical goods are (price, promotion, distribution, and product); below is a detailed explanation of each one of them:

Pharmaceutical Product: The product can be defined as "a set of benefits that the consumer obtains to satisfy his needs, and these benefits include, material benefits such as the characteristics that make up the product and the psychological benefits obtained by the product's acquisition, such as social status" [9, 10]. As for the marketing of pharmaceutical products, they are the products that help the person to get rid of diseases or eliminate them. A medicinal product has a set of tangible or intangible properties, which have been designed in specialized pharmaceutical organizations to satisfy the needs of the end consumer in this case the patient and his desire for treatment and recovery.

Pharmaceutical Price: the price is one of the most important strategic decisions of the organization, on which the organization's success may be based or not. It affects not only the rest of the marketing mix elements, but also the organization's overall performance, in addition to the contribution of particular decisions. Price increases the profitability of the organization, it is an essential and influential element in attracting new consumers to the organization for some products, but when pricing pharmaceutical products, as the pharmaceutical price represents the value of what one party (seller) takes from the other party (buyer) the patient in the process interchangeability, and the price decision is the most crucial element of the pharmaceutical marketing mixture because it is the only one that represents revenues for the organization, unlike other elements.

Pharmaceutical Distribution: companies seek to find a kind of interconnection with all members of the distribution chain, which helps them distribute their products and achieve its objectives; these members are called distribution outlets, which are a group of organizational units that has a way is to market goods, and it means distribution, as [11] sees pharmaceutical distribution is the process of delivering products (goods and services) to the end consumer or industrial buyer, as for the pharmaceutical industry, [12] defines drug distribution as "all the procedures that aim to ensure the flow of products at the appropriate time and place for consumers in the target market, which includes distribution: major sales outlets such as pharmaceutical companies producing medicine, drug stores, retailers, private pharmacies in private and government hospitals, and from these outlets the medicine reaches the pharmacists and from them to the patient".

Pharmaceutical Promotion: definitions of promotion depend on the coordination of efforts undertaken by the organization in establishing information outlets to facilitate the sale of drugs. There are several definitions of promotion, but we will adopt the most comprehensive definition, which is the next definition which states that promotion is "the activity undertaken within the framework of any marketing effort that involves a process of persuasive communication" [4, 13]. As for pharmaceutical marketing, it is all efforts aimed at providing consumers, whether they are individuals or organizations, with basic information about products to persuade and motivate them to buy.

The Decision to Prescribe a Foreign Drug: Is defined as a complex activity that involves the interaction of some factors that vary from medical to psychological and social factors that are important criteria in the decision of physicians to prescribe medications. The impact of these psychological and social factors varies across the different demographic characteristics of physicians [14].

The study of the affecting factors on the doctor's behavior prescribing medications is a highly complex matter. Because it depends on many aspects and the interaction between these aspects, as most of the manners and previous studies indicated that the factors affecting physicians' behavior when choosing a particular medicinal product are divided into product-related factors, patient, company-related factors, and physician-related factors.

Factors related to the doctor are considered among the most important ones, such as the essential/primary specialization: knowing the specialty and the type of practice, a marketing pharmaceutical persons can consist in counting the number of physicians who are specialized in a particular specialty, and then knowing the methods of prescribing the drug by each doctor in his specialty category, also the drug marketing persons can develop the promotional plan, and the promoters and advertisers can be familiar with the type of medicines that specialist doctors are accustomed to prescribing demographic factors such as age and gender: demographics influence the doctor's behavior when giving a prescription. For example, female doctors have a strong interest in cosmetic and cosmetic medicine skincare, even if their specialty does not go with this field also the doctor at this age has more experience and management of the patient's condition. Therefore, age should be at least an important factor in making promotional decisions, place of graduation, co-workers, reading magazines and medical books., geographical area, attendance at medical conferences, type, and activity status [15, 16].

4 Study Hypotheses

Based on the study problem and its objectives, the researchers proposed the following hypotheses:

H01: **The first main hypothesis: there is no statistically significant effect at the level of significance ($\alpha < 0.05$) for the marketing mix elements through Covid-19 on the decision to prescribe foreign medicines for specialist doctors Zarqa Governorate**. From the first main hypothesis, we derive the following sub-hypotheses:

1-H01: There is no statistically significant effect at the significance level ($\alpha \leq 0.05$) for the medicinal products for foreign medicines for specialist doctors in Zarqa governorate.

2-H01: There is no statistically significant effect at the level of significance ($\alpha \leq 0.05$) for drug pricing under COVID-19 on the prescription decision medicines for foreign medicines for specialist doctors in the Zarqa governorate.

3-H01: There is no statistically significant effect at the level of significance ($\alpha \leq 0.05$) for drug distribution under COVID-19 on the prescription decision medicines for foreign medicines for specialist doctors in the Zarqa governorate.

4-H01: There is no statistically significant effect at the level of significance ($\alpha \leq 0.05$) for drug promotion under COVID-19 on the prescription decision medicines for foreign medicines for specialist doctors in the Zarqa governorate.

H02: The second main hypothesis: There are no statistically significant differences at the level of significance ($\alpha \leq 0.05$) on the prescription decision for foreign manufactured medicines for specialist doctors through COVID-19 in Zarqa governorate according to different demographic factors.

5 Study Methodology

This study follows the descriptive approach, as this approach fits with the nature of the study, and this is what became clear to the researchers after perusal based on previous studies, the descriptive approach is defined as "a method of analysis based on sufficient and accurate information about a phenomenon" or a specific topic over a known period or periods in order to obtain practical results that have been interpreted in an objective and consistent manner with the actual data of the phenomenon" [17–19].

5.1 Study Population and Sample

The study population consists of all specialist doctors working in Zarqa governorate during this study's preparation period. The study, whose number is estimated at (153), is a practicing physician according to the statistics of the ministry of health for the year (2020), and given the nature of the study population. And its geographical spread, the researchers found that the most appropriate sampling method is the intentional sampling method, and based on [20] equation (110), Physician and the sample size required to adequately represent the study population.

5.2 Study Tool and Description

In order to achieve the study objectives of the, and after referring to previous studies, a tool was formulated to measure the effect of marketing mix on the decision to prescribe foreign medicines, and the study tool consisted of five main axes, in addition to the axis of personal information as in Table 1.

5.3 Controlling Study Tool

To verify the validity and stability of the study tool, the apparent and internal construct validity were measured, and the alpha coefficient was calculated. Cronbach for each axis of the study tool, and the total sum of the tool is as follows.

Table 1 The study tool in its initial form

Subject	Sequence	Indicator
Personal information	4 articles	Multiple choice
Pharmaceutical product	1–5	1–5
Pharmaceutical price	6–9	1–5
Pharmaceutical promotion	10–16	1–5
Pharmaceutical distribution	17–20	1–5
Recommendations	21–24	1–5
Total	**28**	

5.4 *Virtual Sincerity*

The study tool was presented to a group of arbitrators with experience and expertise in marketing. They were asked to express their opinion on the comprehensiveness of the paragraphs, their affiliation with the field, the appropriateness of the linguistic formulation, the clarity of the paragraphs, and the addition or deletion of the paragraphs or modify as they see fit.

5.5 *The Sincerity of the Internal Construction*

The validity of the study tool was verified by using the validity of the internal construction, where the tool was applied to a sample.

a survey of (20) doctors were randomly selected from within the study community and from outside its sample, to calculate the correlation coefficient between a score for each phrase and the total score for the tool, and the results were as in Tables 2 and 3.

Table 2 shows that all the expressions that measure the axis of the marketing mix were directly related to the total of the expressions of the axis to which it belongs. The expression (14) was significant at the level of significance ($\alpha < 0.05$), and the expressions Mean of which), ($\alpha < 0.01$), significant level at significant significance was (15, 16, 17, 18, 19, 20, 1, 2, 3, 4, 5, 6, 7, 8, 9, 10, 11, 12, 13) The phrases measure what they are set to measure in this axis in each axis.

All statements that measure the axis (recommendation to buy foreign medicines) were directly related to the total number of statements. The axis, and all of them were significant at the level of significance ($\alpha < 0.01$), which means that the expressions measure what they were set to measure in this axis, whereas [21] confirmed that the stability coefficient of the internal consistency of the tool for the variables of any study should be greater than 0.06. This was confirmed by the above results for all study variables.

Table 2 Internal consistency validity of the study tool (marketing mix)

Variable	#	Sentence	Correlation
Product	1	Pharmaceutical companies are keen to provide attractive designs for the pharmaceutical product	0.895**
	2	Pharmaceutical companies rely on safe manufacturing techniques for the pharmaceutical product, which motivates me to recommend the purchase of their products	0.795**
	3	Good pharmaceutical products are characterized by low side effects	0.809**
	4	A drug that is internationally recognized as an incentive to recommend its purchase	0.794**
	5	The information attached to the drug motivates me to recommend its purchase	0.822**
Price	6	Good drug companies offer price quotes	0.925**
	7	The price, in general, motivates me to recommend it	0.670**
	8	The competitive prices offered by pharmaceutical companies make me recommend buying their medicines	0.898**
	9	The convenience of the price for the effectiveness of the drug makes me recommend it	0.781**
Promotion	10	The credibility of advertising for the drug motivates the recommendation to buy it	0.718**
	11	The good relationship with the sales representative of the pharmaceutical company motivates the doctor to recommend the purchase of its products	0.824**
	12	Conferences, scientific lectures, seminars, and other events held by companies motivate the doctor to buy their products	0.785**
	13	The reputation of the drug company motivates the purchase of its products	0.668**
	14	Companies that give gifts and souvenirs to doctors motivate me to recommend their drugs	0.446*
	15	Be sure to recommend prescription drugs for which I see promotions	0.734**
	16	Be sure to recommend the prescription of foreign medicines about which I get information through various promotional means	0.851**
Distribution	17	Constant availability of the drug is an important condition for recommending its purchase	0.788**
	18	I recommend buying medicines from companies that supply their products in all distribution outlets	0.720**
	19	The prompt replacement of damaged samples motivates me in my decision to recommend the purchase of foreign medicines	0.730**
	20	The flawless delivery of foreign medicines encourages me to recommend them to buy	0.851**

Table 3 The validity of the internal consistency of the study tool (the decision to recommend the purchase of foreign medicines)

#	Sentence	Correlation
1	Pharmaceutical products that are characterized by low side effects influence my decision to recommend its purchase	0.853**
2	The constant availability of the drug affects my decision to recommend it	0.849**
3	The competitive prices offered by pharmaceutical companies affect my recommendation to buy their medicines	0.901**
4	Companies that hold conferences, scientific lectures, symposia and other events influence my decision to recommend the purchase of their medicines	0.791**

5.6 *Study Tool Stability*

To ensure the stability of the study tool, the researchers used Cronbach's alpha coefficient for each axis of the study tool. And the total sum of the tool, the results were as in Table 4.

We note from Table 4 that all axes of the study tool have achieved a stability coefficient greater than 0.07 as indicated by [22]. To indicate the tool's suitability to achieve the study's objectives and its validity for application, and after the researchers made sure of the validity and reliability of the study tool, they applied it to the target study population.

6 Hypothesis Testing

H01: There is no statistically significant effect at the level of significance ($\alpha < 0.05$) for the drug marketing mix elements on the prescription decision. For foreign-manufactured medicines for physicians of specialization in Zarqa governorate, the researchers used the multiple linear regression test to test this hypothesis, and Table 5 illustrating the test results.

Table 4 The stability of the study tool

Subject	#	Cronbach alpha value
Pharmaceutical marketing mix	20	0.0834
pharmaceutical product	5	0.782
Pharmaceutical price	4	0.741
Pharmaceutical distribution	7	0.722
Pharmaceutical promotion	4	0.742
Recommending the purchase of foreign medicines	4	0.0756
overall stability	24	0.0845

Table 5 Multiple linear regression analysis results

Independent variable	Dependent variable	R	R^2	F	Ind.
Marketing mix elements	Prescription decision for foreign drugs	0.736	0.541	32.149	0

We notice from Table 5 that the value of (F) reached (32.149), and its significance level is (0. 00), which is less than the significance level (0.05), which means rejecting the null hypothesis that "there is no statistically significant effect at the level ($\alpha \le 0.05$) for the pharmaceutical marketing mix elements on the decision to prescribe foreign-made medicines for specialist doctors in Zarqa governorate" and accept the alternative hypothesis: There is a statistically significant effect at the $\alpha < 0.05$ level of the drug marketing mix elements on the decision to prescribe foreign drugs made for specialist doctors in Zarqa governorate.

6.1 The First Sub-hypothesis Test

1-H01: There is no statistically significant effect at the level of significance ($\alpha < 0.05$) for the product as one of the marketing mix elements on the decision of prescription for foreign medicines for specialist doctors in Zarqa governorate. In testing this hypothesis, the researchers relied on the test Simple linear regression, as in Table 6.

We note from Table 6 that the value of (T) amounted to (8.154) and its level of significance is (0.00), which is less than (0.05), which means that there is an effect with statistical significance of the product as one of the elements of the marketing mix on the decision to prescribe foreign medicines for specialist doctors in Zarqa governorate, we reject the null hypothesis that "there is no statistically significant effect at the level of significance ($\alpha < 0.05$) for the product as one of the elements of the marketing mix on the decision to prescribe foreign medicines for specialist doctors in Zarqa governorate, and we accept the following alternative hypothesis: There is a statistically significant effect at the level of significance ($\alpha < 0.05$) for the product as one of the elements of the marketing mix on the prescription decision for foreign medicines for specialist doctors in Zarqa governorate.

Table 6 Simple linear regression for the first sub-hypothesis

Dependent variable	R	R^2	Independent variable	B	T	Ind.
Prescription decision for foreign drugs	0.61	0.372	Product	0.732	8.154	0

Table 7 Simple linear regression for the second sub-hypothesis

Dependent variable	R	R^2	Independent variable	B	T	Ind.
Prescription decision for foreign drugs	0.621	0.385	Price	0.542	7.743	0

6.2 *The Second Sub-hypothesis Test*

2-H01: There is no statistically significant effect at the level of significance ($\alpha < 0.05$) for the price as one of the marketing mix elements on the decision of prescription for foreign medicines for specialist doctors in Zarqa governorate. In testing this hypothesis, the researchers relied on the test simple linear regression, as in Table 7.

We note from Table 7 that the value of (T) is (7.743) and its level of significance is (0.00), which is less than (0.05), which means that there is an effect with statistical significance of price as one of the elements of the marketing mix on the decision to prescribe foreign medicines for specialist doctors in the governorate in blue, we reject the null hypothesis that "there is no statistically significant effect at the level of significance ($\alpha < 0.05$) for the price as one of the elements the marketing mix on the decision to prescribe foreign medicines for specialist doctors in Zarqa governorate, and we accept the following alternative hypothesis: There is a statistically significant effect at a significant level ($\alpha < 0.05$) for the price as one of the marketing mix elements on the prescription decision for foreign medicines for specialist doctors in Zarqa governorate.

6.3 *The Third Sub-hypothesis Test*

3-H01: There is no statistically significant effect at the level of significance ($\alpha < 0.05$) for distribution as one of the marketing mix elements on the decision of prescription for foreign medicines for specialist doctors in Zarqa governorate. In testing this hypothesis, the researchers relied on the test simple linear regression, as in Table 8.

We note from Table 8 that the value of (T) reached (9.589) and its level of significance is (0.00), which is less than (0.05), which means that there is an effect with statistical significance of distribution as one of the elements of the marketing mix on the decision to prescribe foreign medicines for specialist doctors in the governorate in blue, we reject the null hypothesis that "there is no statistically significant effect at

Table 8 Simple linear regression for the third sub-hypothesis

Dependent variable	R	R^2	Independent variable	B	T	Ind.
Prescription decision for foreign drugs	0.671	0.451	Distribution	0.749	9.589	0

Table 9 Simple linear regression for the fourth sub-hypothesis

Dependent variable	R	R^2	Independent variable	B	T	Ind.
Prescription decision for foreign drugs	0.614	0.377	Promotion	0.704	8.237	0

the level of significance (α < 0.05) for the distribution as one of the elements of the marketing mix on the decision to prescribe foreign medicines for specialist doctors in Zarqa governorate, and we accept the hypothesis for the following alternative: There is a statistically significant effect at a significant level (α < 0.05) of the distribution as one of the marketing mix elements on the prescription decision for foreign-made medicines for specialist doctors in Zarqa governorate.

There is also a direct relationship between the distribution and the decision to prescribe foreign medicines. The value of the correlation between them was (0.671), which explains the distribution amounted to 1.45% of the changes that occurred in the decision to prescribe foreign medicines.

6.4 *Fourth Sub-hypothesis Test*

4-H01: There is no statistically significant effect at the level of significance (α < 0.05) for promotion as one of the marketing mix elements on the decision to prescribe foreign medicines for specialist doctors in Zarqa governorate. In testing this hypothesis, the researchers relied on a simple linear regression test, as shown in Table 9.

We note from Table 9 that the value of (T) reached (8.237) and its level of significance is (0.00), which is less than (0.05), which means that there is an effect with Statistical significance of the promotional mix as one of the elements of the marketing mix on the decision to prescribe foreign medicines for specialist doctors In Zarqa governorate, we reject the null hypothesis that "there is no statistically significant effect at the level of significance (α < 0.05) for the marketing mix. The promotional campaign is one of the marketing mix elements on the decision to prescribe foreign medicines for specialist doctors in Zarqa governorate. We accept the following alternative hypothesis: There is a statistically significant effect at a significant level (α < 0.05) of promotion as one of the marketing mix elements on the prescription decision for foreign medicines for specialist doctors in Zarqa governorate.

6.5 *Testing the Second Main Hypothesis*

H02: According to demographic factors, there are no statistically significant differences at the level of significance (α < 0.05) on the decision to prescribe foreign

Table 10 One-way ANOVA for the variable of specialization

Contrast source	Sum	Degree	Square mean	F	Ind.
Between groups	4.881	13	0.375	0.875	0.582
Inside groups	42.932	100	0.429		
Sum	47.814	113			

Table 11 (ANOVA way one) for the variable number of years of experience

Contrast source	Sum	Degree	Square mean	F	Ind.
Between groups	2.007	4	0.502	1.194	0.318
Inside groups	45.807	109	0.42		
Sum	47.814	113			

medicines for doctors specializing in Zarqa governorate. In testing this hypothesis, the researchers relied on the one-way ANOVA test for variable j (specialization, number of years of experience), as in Tables 10 and 11.

Through Table 10, we conclude that the value of (F) reached 0.875, and its level of significance is (0.582), which is greater than 0.05, and therefore we accept the null hypothesis: There are no statistically significant differences at the level of significance ($\alpha < 0.05$) on the decision to prescribe foreign medicines for physicians specialization in Zarqa governorate, according to their medical specialization.

Through Table 11, we conclude that the value of (F) reached 1.194, and its level of significance is (0.318), which is greater than 0.05, and therefore we accept the null hypothesis that **there are no** statistically significant differences at the level of significance ($\alpha < 0.05$) on the decision to prescribe Foreign drugs for physicians of specialization in Zarqa governorate, according to the number of years of their experience.

7 Results Discussion

The results indicate an effect of the dimensions of the independent variable marketing mix on the dependent variable in the prescription decision medical prescriptions for foreign medicines, as these dimensions explained (1.54) % of the changes that occurred in the decision to prescribe foreign medicines. This result agrees with what the study of [7] indicated that there is an effect of the marketing mix on customer satisfaction, a study [23] that the marketing mix significantly affects the purchasing decision, and with [12] study of the presence of an effect of the elements of the pharmacological marketing mix on the purchasing decision, a study by [24]. The elements of the marketing mix, collectively and individually, are: (product, price, promotion, and distribution) influence drug preferences, and with a study [25], there is a general effect of perceiving physicians for the prescription drug marketing mix strategies, [26] study that pharmaceutical marketing strategies is linked to the behavior of doctors and the study of [27] that multinational pharmaceutical companies are responsible for the transformation in marketing strategies, The study also found that there were no statistically significant differences at the level of significance at ($\alpha < 0.05$).

The decision to prescribe foreign medicines for specialist doctors in Zarqa governorate, according to different demographic factors (medical specialization, and the number of years of experience). This result is what distinguishes the current study from other previous studies, as it used the demographics factors (experience and specialization) as modifying factors for the relationship between the marketing mix and the decision to prescribe a drug product, which have not been determined Its impact on any of the previous studies. However, some of these studies found an impact of demographic factors, such as the study of [14, 23, 24, 28–30].

8 Study Recommendations

According to the findings of the study, it recommends that Jordanian pharmaceutical companies should pay attention to developing marketing strategies that include all elements of the pharmaceutical marketing mix (pharmaceutical product, pharmaceutical price, pharmaceutical distribution, and pharmaceutical promotion) to promote its competition with foreign companies producing medicines, as these elements have a direct impact on the decision to prescribe, And Jordanian medicine companies must develop integrated strategic plans capable of diversifying their products, and integrating vertically and horizontally as much as possible to control, and to get rid of the threats facing the Jordanian pharmaceutical sector by following robust distribution strategies.

Jordanian pharmaceutical companies work to enhance their competitiveness by setting different prices for local pharmaceutical products compared to the foreign product, With the need to pay attention to promotion as an important and influencing factor in the doctors' decision regarding the prescription, especially the means of stimulating sales, and taking into consideration the demographic variables of doctors when preparing the marketing plans, because of their impact on the decision of the prescription.

References

1. Abdel-Malik, H.: The global corona crisis and the opportunity for growth in the pharmaceutical industry business. Shuaa J. Econ. Stud. **5**(1) (2021)
2. Ministry of Planning and International Cooperation, Annual Report (2017)
3. Annual Statistical Report: Directorate of Institutional Development and Quality Control. Jordanian Ministry of Health (2020)

4. Kotler, P., Armstrong, G.: Principles of Marketing, 11th edn. Prentice Hall, New Jersey (2014)
5. Singh, M.: Marketing mix of 4P's for competitive advantage. IOSR J. Bus. Manag. **3**(6) (2012)
6. Kotler, P., Armstrong, G.: Principles of Marketing. Pearson Education, New York (2019)
7. Amora, A., Eram, S.: The effect of the marketing mix on customer satisfaction. Unpublished master's thesis. University Center of Abdel Hafeez Boulasouf Mila, Al-Jeria (2009)
8. Kumar, A., Jain, S., Dangi, I., Chowdary, S., Choubitker, O., Kumar, K.: Ideal drug prescription writing. World J. Pharm. Pharm. Sci. **8**(3) (2019)
9. Datta, A., Dave, D.: Effects of physician-directed pharmaceutical promotion on prescription behavior: longitudinal evidence. Health Econ. (2016)
10. Elyan, M.: Foundations of Contemporary Marketing. Dar Safaa for Publishing and Distribution, Amman (2009)
11. Al-Bakri, T.: Communication and Promotion. Dar Al-Hamid Publication, Amman (2008)
12. Khanfar, I.A.-I.: The effect of pharmacological mix components on locally made forces (applied study in the city of Amman). J. Manag. Dev. Res. Stud. **9** (2016)
13. Tawanda, N., Elizabeth, G., Frank, G.: Preferred sources of information, factors influencing the attitudes and prescribing behavior of medical doctors in the Western Suburbs of Harare, Zimbabwe. Int. J. Health Med. Inf. **2** (2013)
14. Azhar, A.: The role of personal selling in choosing pharmaceutical products from the point of view of doctors. Unpublished master's thesis. Faculty of Economics, Tartous University, Syria (2018)
15. Al Alaq, B.: The Promotion and Commercial Advertisement. Al Yazori Publication, Amman (2007)
16. ILO Report: Skills for Trade and Economic Diversity (2015)
17. G/egziabher, S.: Prescription drug promotion and prescribing behavior of physicians in case of Addis Ababa green licensed private hospitals. Master thesis. Addis Ababa University School of Commerce Department of Marketing Management (2017)
18. Gonul, F.F., Carter, F., Petrova, E., Srinivasan, K.: Promotion of prescription drugs and its impact on physicians' choice behavior. J. Mark. **56** (2001)
19. Obeidat, T., Adas, A.-R., Kayed, A.-H.: Scientific Research: Its Concept, Tools and Methods. Dar Al-Fikr Publishers and Distributors, Amman (2020)
20. Mason, R.L., Gunst, R.F., Hess, J.L.: NetLibrary: Statistical Design and Analysis of Experiments: With Applications to Engineering and Science, 2nd edn. Wiley (2003)
21. Hair, J.F., Hult, G.T.M., Ringle, C.M., Sarstedt, M.: A Primer on Partial Least Squares Structural Equation Modeling (PLS-SEM). Sage, Thousand Oaks, CA (2017)
22. Taghza, M.: The logical structure of Cronbach's alpha coefficient and the extent of its Rudder in estimating domes in the light of the assumptions of measurement models. King Saud Univ. J. **21**(3) (2012)
23. Ajman, H.: The capital governorate of the effect of the elements of the mental mix on the decision to purchase medicines from the point of view of pharmacists in Damascus. Unpublished master's thesis. Faculty of Administrative Sciences, Syrian Allafit Ardeyah University (2016)
24. Obeidat, M., Ghadeer, H.: The effect of industrial mix components on Jordanian pharmacists' preference for local medicine compared to its foreign counterpart. Darsat J. Adm. Sci. **38**(1) (2011)
25. Hailu, A., Workneh, B., Kahissa, M.: Influence of Pharmaceutical Marketing Mix Strategies on Physicians' Prescribing Behaviors in Public and Private Hospitals Dessie, Ethiopia: A Mixed Study Design. BMC Public Health (2021)
26. Khazzaka, M.: Pharmaceutical marketing strategies' influence on physicians' prescribing pattern in Lebanon: ethics, gifts, and samples. MC Health Serv. Res. (2019)
27. Kalotra, A.: Marketing strategies of different pharmaceutical companies. J. Drug Deliv. Ther. **4**(2) (2014)
28. Al-Abbasi, R., Zahwani, R.: The effect of promotional mix elements on over-the-counter drug products for small pharmaceutical enterprises from the pharmacist's point of view. AFal J. Econ. Stud. **5** (2018)

29. Davari, M.E.K., Bereket, M.T.: Factors influencing prescribing decisions of physicians: a review. Ethiop. J. Sci. **28**(6) (2018)
30. Masoud, I., Kanaan, A.: The effect of rewards, audit on burdens in Jordanian companies. Int. J. Pro-Econ. **9**(1) (2020)

Web Sites

31. https://gis.moh.gov.jo/MOHHealthMap/Home.html
32. https://mop.gov.jo/AR/List/0172019

The Impact of Choosing Distribution Channels Factors During the Coronavirus Pandemic on the Wholesaler's Satisfaction

Iyad A. A. Khanfar, Iyad Dalbah, Abdul Aziz Atteyh, and Mustafa Said Al-Shaikh

Abstract The study aimed to clarify the impact of choosing distribution channel factors during the Coronavirus pandemic and the wholesaler's satisfaction: internal control as a mediating variable (A case study of the Petra Investment and Trade Company/Jordan). The study population consisted (450) of wholesalers using the comprehensive survey method, (370) questionnaires were retrieved electronically, with a percentage of (82.2%), the dimensions of distribution channels were (efficiency in performance, speed, cost, length of the distribution channel). The study reached a set of results which are: there is a statistically significant impact of the factors of choosing distribution channels during Corona pandemic (performance efficiency, speed, cost, length of the distribution channel) on the internal control, there is an intermediary role of internal control over the relationship between the factors of choosing distribution channels (efficiency in performance, speed, cost, length of the distribution channel) on the satisfaction of wholesalers in the Petra Investment and Trade Company. The study recommended that it is necessary for the Petra Investment and Trade Company to continue to pay attention to the factors affecting the satisfaction of wholesalers (performance efficiency, speed, cost, and length of the distribution channel). This should happen through conducting training courses and establishing updated electronic systems that contribute to raising the effectiveness of the distribution channel. In addition, increasing the utilize in the internal control system of Petra Investment and Trade Company for its significant

I. A. A. Khanfar (✉) · M. S. Al-Shaikh
Zarqa University, Zarqa, Jordan
e-mail: i.khanfar@zu.edu.jo

M. S. Al-Shaikh
e-mail: malshaikh@zu.edu.jo

I. Dalbah
Arab American University, Jenin, Palestine

A. A. Atteyh
Petra Group, Amman, Jordan
e-mail: a.ribhy@petragroup.com.jo

A. Hamdan et al. (eds.), *The Implementation of Smart Technologies for Business Success and Sustainability*, Studies in Systems, Decision and Control 216,
https://doi.org/10.1007/978-3-031-10212-7_6

role in influencing the satisfaction of wholesalers. It has to happen through continuous follow-up of wholesalers' business, updating information and establishing a system for exchanging the information with them to Raise the level of loyalty and manage the relationship more efficiently.

Keywords Distribution channels · Internal control · Satisfaction of wholesalers · Petra Investment and Trade Company/Jordan

1 Introduction

Organizations place their relations with the intermediaries of the distribution channel as one of their strategic objectives because their success in the delivery of their goods depends on the type of relationship with intermediaries from wholesalers and retailers, and because of their direct connection with wholesalers, the satisfaction of wholesalers is one of their strategic goals that must be achieved and measured periodically [5].

Wholesale trade includes selling products and services to companies, factories, and organizations that buy these products to resell them or use them to produce other products or services. Examples of wholesale trade are a bank's purchase of a new computer for use in data processing, a school buying a projector for use in class, or a clothing store buying ready-made dresses for resale, another example or wholesalers those where products and goods are distributed to them from investment and trading organizations to reach the final consumer [19].

In order to ensure the success of their distribution channel, the organizations work to choose the appropriate factors of their internal and external environment, and these choices come after a reasonable investigation of their strengths, weaknesses, opportunities, and potential threats provided by the external environment. Therefore, some factors may affect the organization's distribution channel, [21] point out the necessity of evaluating the distribution channel to ensure its validity, considering several factors such as control, flexibility, and legal restrictions. Therefore, it required an influential and tight internal control factor in order to reassure the senior management of the efficient use of its resources; thus, it is one of the most critical sectors that work in developing economic work, therefore, a tight internal control system must be available, because internal control has a prominent and vital role in promoting performance and reducing risks, so it must be taken care of, developed and supplied with qualified employees with experience and high efficiency [15].

Andjelkovic and Radosavljevic [6] stresses the need to pay special attention to the distribution channel length while ensuring its efficiency and performance speed, so the distribution channel's length does not prejudice its effectiveness without any additional costs. These four factors must be considered when designing the distribution channel, especially for food commodities, as [8] confirms that: measuring the efficiency of the distribution channel continuously for food companies contributes to the ability to remedy the problems they face before its increase and hinder their work.

The organizations ensure the effectiveness of their distribution channel in terms of channel length, efficiency, and speed, while maintaining an appropriate cost size during the Corona pandemic by establishing internal control systems, as [13] indicates that the speed and efficiency of the distribution channel increase the volume of customer satisfaction and loyalty.

2 Study Problem

Based on the above, there is a need to investigate the impact of choosing distribution channels factors during the Coronavirus pandemic and them on the wholesaler's satisfaction; the internal control as a mediating variable, the problem of the study can be formulated by the following central question:

What is the impact of choosing distribution channel factors during the Coronavirus pandemic and the wholesaler's satisfaction, considering internal control as an intermediary variable?

3 Operational and Procedural Definitions

The efficiency of distribution: "delivery of products in the right place, in the right form, at the right time, and at the lowest possible cost", [14] and the researchers define it procedurally as all the activities undertaken by the Petra Investment and Trade Company in order to deliver goods to wholesalers during the Corona pandemic, at the lowest possible cost and at the right time and place, measured in terms (1–5) of the study tool.

Speed in distribution: "the economic and practical solution to provide the product at the right time through intermediaries due to the long distance between the producer and the consumer" [11], and the researchers procedurally define it as the time of product delivery from Investment and Trade Company stores to wholesalers during the Corona pandemic, and It is measured in terms (6–10) of the study tool.

Cost in performance: "the means that the organization uses to reduce the number of expenses that spent to deliver goods to the final consumer using intermediaries" [11], and the researchers define it procedurally as the sum of expenses incurred by Petra Investment and Trade Company in order to deliver goods to wholesalers during the Corona pandemic, and it is measured in terms (11–15) as the study tool.

The length of the distribution channel: "The number of agencies that mediate the producer and the consumer, this division is usually due to economic reasons" [1], the researchers define it procedurally as the number of organizations that mediate the relationship between the Petra Investment and Trade Company and the final consumer during a pandemic of Corona, measured in terms (16–20) as the study tool.

Internal control: "unit, device or entity which has the responsibility to set the red lines or prohibited practices to being crossed in the organization" [2], the researchers define them procedurally as all regulations and instructions which the Petra Investment and Trade Company follows intending to control the flow of goods through distribution channels during the Corona pandemic to wholesalers, and it is measured in terms (21–25) in the study tool.

Customer satisfaction: "all efforts that the organization planned to meet the needs and requirements of customers, responding to all their inquiries, satisfying their needs in time according to their characteristics, and within the specified terms, to ensures the continuity of the relationship" [9], the researchers defines it procedurally as all efforts made by the Petra Investment and Trade Company during the Corona pandemic to building positive relations with the wholesalers, it is measured in terms (26–30) as the study tool.

Wholesalers: "merchants sell goods in large quantities to the retailer. It is carried out by repacking and selling it in smaller quantities at a higher price" [5]. The researchers define them procedurally as the segment of external customers that Petra Investment and Trading Company deals with, who buy in large quantities and perform the distribution function for retailers during the Corona pandemic.

4 Theoretical Framework

Distribution: defined as "the set of activities and processes that allow the product to be reached to the consumer or user after its production, allows to directing the products from the manufacturer to the place of their sale, and placing the products within reach of consumers". Distribution involves a set of essential functions. Among them is the transportation of products from the place of manufacture to the place of sale to the final consumers, storage of products to the time the consumer needs it, provides a set of commercial functions represented in three aspects: coordination, information provision, and service provision [20].

From the researcher's point of view, distribution is a path through which goods pass from producers to consumers, and distribution needs a successful plan to distribute products promptly to consumers to achieve the desired benefit from effective distribution.

Satisfaction: it is a feeling of happiness or disappointment, resulting from comparing the perceived performance of the product with the expected performance; if the performance is less than the expectations, the result is unsatisfactory, but if it matches the expectations, the customer is satisfied or pleased [17], several factors increase the satisfaction of wholesalers, such as responding to complaints, because meeting the needs of consumers is one of the primary duties of wholesalers, giving wholesalers significant discounts based on the volume of purchase, and adding free corners for the company producing in wholesalers stores to attracting the consumers, continuous communication With wholesalers, preparing the market needs of goods, so that the desires of consumers are met, and on the other hand, the company's profit increases [12].

Internal control: is defined as one of the "basic duties of any organization, whether civil organizations or governmental organizations, as it is an integral part of evaluating the unit performance and directing its operations reliably and effectively. The responsibility of establishing a sound system of internal control and maintaining it is one of the administrations tasks [4]. From the researcher's point of view, the concept of internal control came to preserve the funds and assets of companies, from fraud or fraud attempts. It also to provide appropriate information with accuracy and reliability. In addition, to maintain continuity in business operations and financial sustainability. Therefore, to maintain business objectives in the short, medium, and long term for all types of a commodity that organizations produce. The system would be also needed to increase profit and attract more customers. Based on that, Petra Company has an internal control system where all the operations are monitored, which starts from importing products and raw materials, manufacturing and selling processes, and ending with post-sale control and feedback processes.

5 Study Hypothesis

Based on the study problem and its objectives, the researchers proposed the following hypothesis:

H01: There is no statistically significant impact at the level ($\alpha \leq 0.05$) for the factors of choosing distribution channels (performance efficiency, speed, cost, and length of the distribution channel) on the satisfaction of wholesalers for the Petra Investment and Trade Company/Jordan.

H02: There is no statistically significant impact at the level ($\alpha \leq 0.05$) for the factors of choosing distribution channels (performance efficiency, speed, cost, and length of the distribution channel) on the internal control of the Petra Investment and Trade Company/Jordan.

H03: There is no statistically significant impact at the level ($\alpha \leq 0.05$) for internal control on the satisfaction of wholesalers for the Petra Investment and Trade Company/Jordan.

H04: There is no statistically significant impact at the level ($\alpha \leq 0.05$) for the factors of choosing distribution channels (performance efficiency, speed, cost, and length of the distribution channel) on the satisfaction of wholesalers in the presence of internal control as a mediating variable for the Petra Investment and Trade Company/Jordan.

6 Study Methodology

This study follows the descriptive approach, as this approach fits with the study nature; the descriptive approach is defined as "a method of analysis based on sufficient and accurate information about a specific phenomenon or topic over a known period or periods, in order to obtain practical results that have been interpreted in an objective manner consistent with the actual data of the phenomenon" [16].

7 Study Population

The study population represents all wholesalers dealing with the Petra Investment and Trade Company/Jordan, where they are, during the period of this study (450) merchants listed in the company's database. The researchers used the comprehensive survey method due to the small size of the study population, 450 questionnaires were distributed, while the number of valid retrieved questionnaires for statistical analysis are (370), and this represents (82.2%) of the distributed questionnaires, which is a statistically acceptable percentage [10].

8 Study Tools

The researchers used the questionnaire to collect the data, which was designed and distributed to the study population. The questionnaire consisted of four main parts: the first part includes demographic variables (gender, educational qualification, job status, number of years of experience, age), the second part consisted of (4) dimensions that measure the independent variable represented (choice of distribution channels) divided into (5) items that measure the first dimension (performance efficiency), (5) items that measure the second dimension (speed), (5) Items that measure the third dimension (cost) and (5) items that measure the fourth dimension (the length of the distribution channel). The third part of the questionnaire consisted of a dimension that measures the mediating variable (internal control), the last part consisted of a dimension that measures the dependent variable (the satisfaction of wholesalers), and Table 1 shows the sections of the questionnaire.

9 Adjusting the Study Tool

In order to ensure the validity of the study tool, the research ensured its face validity; it was presented to a group of professors with experience and specialization in

Table 1 Description of the study tool

Dimension	Items
Performance efficiency	5
Speed	5
Cost	5
Distribution channel length	5
Total of independent variable items	20
The mediating variable: internal control	5
Dependent variable: wholesaler satisfaction	5
Total of questionnaire items	30

marketing in some Jordanian universities, some modifications were made, reformulated according to their observations and recommendations. Therefore the questionnaire took its final form that was distributed to the pilot study sample whose size is (20) were chosen randomly, to measure the internal validity and reliability for each questionnaire item the total correlation score resulted as in Tables 2, 3 and 4.

By Table 2, we can conclude that; all the Items that measure the distribution channel selection variables had a direct relationship with the sum of the items of the scope to which they belong, the Items (1, 2, 3, 5, 7, 8, 9, 12, 14, 16, 17, 19) were Significant at the level of significance ($\alpha \leq 0.01$), the Items (4, 7, 10, 13, 15, 18, 20) were significant at the level of significance ($\alpha \leq 0.05$), which means that the Items measuring what they were set to measure in this scope [22].

By Table 3, we can conclude that; all the Items that measure the scope of the mediating variable (internal control), were directly related to the sum of the Items scope to which they belong, the items (2 and 4) were significant at the level of significance ($\alpha \leq 0.01$), the item (1) was significant at the level of significance ($\alpha \leq 0.05$), which means that items measure what they were set to measure in this scope.

By Table 4, we can conclude that; all the Items that measure the scope of the dependent variable (wholesaler's satisfaction), were directly related to the sum of the Items scope to which they belong, the item (1, 3) was significant at the level of significance ($\alpha \leq 0.05$), which means that items measure what they were set to measure in this scope.

As the questionnaire in its final form, the reliability coefficient was extracted through the Cronbach Alpha values for study tool scopes, and Table 5 shows the values of the reliability coefficients for them:

Table 5 shows that Cronbach's alpha coefficient for all the study scopes was more significant than 70%, and all reliability coefficients are high and acceptable for the study, where the reliability coefficient (Cronbach's alpha) is considered acceptable if it exceeds (70%), [18].

Table 2 The internal consistency of the study tool (distribution channel selection factors)

Variable	#	Item	Correlation
Performance efficiency	1	The company employees can perform their roles in a correct manner	0.895
	2	The company employees are adequately qualified to perform their roles in a specified manner	0.795
	3	The high management depends on determining the employees' behavior to evaluate them	0.809
	4	The company employees' have cognitive skills that contribute to improving their performance	0.794
	5	The company employees' have practical and theoretical knowledge that contributes to improving their performance	0.822
Speed	6	The employees can perform tasks on time	0.925
	7	High management can encourage and lead the employees to work at the right time	0.670
	8	The company seeks to reach a quick response to the questions and concerns of consumers	0.898
	9	The high management is interested in moving forward to achieve competitive excellence	0.781
	10	The high management is keen to deal with obstacles and overcome them on time	0.650
Cost	11	The company's management placed a suitable system for operations purposes and provided the necessary information	0.718
	12	The company's management placed a suitable financial politics for operations at reasonable costs	0.824
	13	The benefits derived from the financial system outweigh its costs	0.785
	14	To ascertain the cost accurately, a good plan for gathering the overhead expenditures	0.668
	15	There is a clear delineation of authority and responsibility if the cost system operates successfully	0.446
Distribution channel length	16	The company relies on the distribution channel as a marketing element that saves many costs	0.788
	17	A company's marketing strategies depends on the distribution channel as a changeable variable in the marketing planning process; to ensure its success	0.720
	18	To achieve effectiveness in the distribution channels, there is a continuous improvement in the company's marketing efforts	0.730
	19	The distribution channels have specific objectives to ensure its activities' direction efficiently and effectively	0.851

(continued)

Table 2 (continued)

Variable	#	Item	Correlation
	20	There is effective management of distribution channels, which leads to achieving a competitive advantage	0.753

Table 3 The internal consistency of the study tool (internal control)

#	Item	Correlation
1	The company provides a well-developed control system based on detecting errors and deviations to address them	0.833
2	The company has a set of standards of the quality system at the lowest possible cost	0.869
3	The high management provides an internal control system that achieves the desired goals efficiently and effectively	0.701
4	The company's control system can obtain accurate information about the company's performance	0.791
5	The control system is keen to ascertain the information sources	0.654

Table 4 The internal consistency of the study tool (wholesaler's satisfaction)

#	Item	Correlation
1	The company controls the wholesaler's activities as they are an essential part of distribution channels	0.754
2	The needs of wholesalers are met constantly and on time	0.744
3	All wholesalers' inquiries are answered to achieve effective communication with them	0.656
4	The company is making significant efforts to achieve the satisfaction of wholesalers; to build a long term relationship with them	0.545
5	There are continuous follow-up procedures for wholesalers to prevent their acquisition by other competing companies	0.653

Table 5 Reliability coefficients for the study tool Cronbach alpha

Scope	# of items	Cronbach alpha (%)
Performance efficiency	5	80.3
Speed	5	80.1
Cost	5	80.7
Distribution channel length	5	80.5
Independent variable: choosing distribution channels factors during the Corona pandemic	20	83.7
The mediating variable: internal control	5	81.9
Dependent variable: wholesaler satisfaction	5	82.9
All the items	30	84.3

10 Test of the Hypothesis

H01: There is no significant impact at the level of significance ($\alpha \geq 0.05$) for the factors of choosing distribution channels dimensions (efficiency in performance, speed, cost, length of the distribution channel) on the satisfaction of wholesalers in Petra Investment and Trade Company.

Table 6 shows the results of testing the first hypothesis, which represents the effect of the factors of choosing distribution channels with their dimensions (efficiency in performance, speed, cost, length of the distribution channel) on the satisfaction of wholesalers in Petra Investment and Trade Company, the above table indicates that; the value of the coefficient of determination (R^2) amounted to (75.4%), which means that the distribution channel selection factors explained (75.4%) of the change in the satisfaction of wholesalers in the Petra Investment and Trade Company, and the significance value of the variance test (F) was (0.000) at the level of Significance less than (0.01), which means that there is a positive and statistically significant effect of the factors of choosing distribution channels dimensions (efficiency in performance, speed, cost, length of the distribution channel) on the satisfaction of wholesalers in Petra Investment and Trade Company, and therefore the null hypothesis was rejected and accepted The alternative hypothesis states: "**There is a statistical impact at the level of significance ($\alpha \geq 0.05$) for the factors of choosing distribution channels dimensions (efficiency in performance, speed, cost, length of the distribution channel) on the satisfaction of wholesalers in the Petra Investment and Trade Company**".

Table 6 also shows that there is no statistical impact at the significance level ($\alpha \geq 0.05$) of the factor (efficiency in performance) on the satisfaction of wholesalers in the Petra Investment and Trade Company.

H02: There is no statistically significant impact at the level of significance ($\alpha \leq 0.05$) for the factors of choosing distribution channels dimensions (performance efficiency, speed, cost, and length of the distribution channel) on the internal control of the Petra Investment and Trade Company/Jordan.

Table 7 shows the results of the second hypothesis test, which represents the impact of the factors for choosing distribution channels (performance efficiency,

Table 6 H01test

Dependent variable	R	R^2	F	*P* value	Independent variable	B	t	*P* value
Wholesaler satisfaction	0.869	0.754	280.197	0.00	Performance efficiency	0.043	0.7340	0.4630
					Speed	0.194	2.865	0.0040
					Cost	0.331	5.175	0.0000
					Distribution channel length	0.328	7.243	0.0000

Table 7 H02 test

Dependent variable	R	R^2	F	*P* value	Independent variable	B	t	*P* value
Internal control	0.89	0.792	347.785	0.00	Performance efficiency	−0.90	−1.600	0.109
					Speed	0.186	2.838	0.005
					Cost	0.193	3.120	0.002
					Distribution channel length	0.619	14.189	0.000

speed, cost, length of the distribution channel) on the internal control of Petra Investment and Trade Company, it indicates that the value of the coefficient of determination (R^2) It reached (79.2%), which means that the distribution channel selection factors explained (79.2%) of the change in the internal control in Petra Investment and Trade Company, the significance value of the variance test (F) was (0.000) at a lower significance level from (0.01) which means that; there is a positive statistically significant impact of the factors of choosing distribution channels dimensions (efficiency in performance, speed, cost, length of the distribution channel) on the internal control of Petra Investment and Trade Company, therefore the null hypothesis was rejected and accepted The alternative hypothesis states: "**There is a statistical impact at the level of significance ($\alpha \geq 0.05$) for the factors of choosing distribution channels dimensions (efficiency in performance, speed, cost, length of the distribution channel) on the internal control of the Petra Investment and Trade Company**". Table 7 also shows that there is no statistical effect at the level of significance ($\alpha \geq 0.05$) for the factor (efficiency in performance) on the internal control of the Petra Investment and Trade Company.

H03: There is no statistically significant impact at the level of morality ($\alpha \leq 0.05$) of internal control on the satisfaction of wholesalers of the Petra Investment and Trade Company/Jordan.

Table 8 shows the results of testing the third hypothesis, which represents the impact of internal control on the satisfaction of wholesalers in Petra Investment and Trade Company, the table indicates that; the value of the coefficient of determination (R^2) reached (74.3%), which means that the internal control has explained a (74.3%) of changes in the satisfaction of wholesalers in the Petra Investment and Trade Company, the significance value of the variance test (F) reached (0.000) at a level of significance less than (1%), which means that there is a positive a statistical significance for internal control On the satisfaction of wholesalers in the Petra Investment and Trade Company, accordingly, the null hypothesis was rejected and the alternative hypothesis was accepted, which states: "**There is a statistical impact at the level of significance ($\alpha \leq 0.05$ of internal control on the satisfaction of wholesalers in the Petra Investment and Trade Company**".

Table 8 H03 test

Dependent variable	R	R^2	Independent variable	B	t	*P* value
Wholesaler satisfaction	0.862	0.743	Internal control	0.862	32.658	0.00

H04: There is no statistically significant impact at the level of significance ($\alpha \leq 0.05$) for the factors of choosing distribution channels dimensions (efficiency in performance, speed, cost, length of the distribution channel) on the satisfaction of wholesalers at the presence of internal control as a mediating variable for the Petra Company Investment and Trade/Jordan.

To test the mediating role of the internal control variable between the factors of choosing distribution channels dimensions (performance efficiency, speed, cost, length of the distribution channel) and the satisfaction of wholesalers, the researchers adopted the steps identified by [7], which are:

1. There is a significant relationship between the independent variable and the mediating variable.
2. There is a significant relationship between the independent variable and the dependent variable.
3. There is a significant relationship between the mediating variable and the dependent variable in the regression analysis equation with the independent variable as a control variable.
4. The third condition includes two terms:

   If the value of the impact factor (B) in the third equation decreased from its value in the second equation, there is a partial mediating role between the independent and dependent variables.

   If the impact coefficient (Zero-effect) value in the third equation is more than its value in the second equation, this means there is a complete mediating role between the independent and dependent variables.

The results showed the existence of a significant relationship between the factors of choosing distribution channels dimensions (efficiency in performance, speed, cost, length of the distribution channel) and the mediating variable (internal control), the existence of a significant relationship between the independent variable factors of choosing distribution channels dimensions (efficiency in performance, speed, cost, length of the distribution channel) and the dependent variable, wholesaler satisfaction.

Table 9 shows the results of testing the fourth hypothesis, which represents the impact of distribution channel selection factors on the satisfaction of wholesalers in the presence of internal control as a mediating variable in Petra Investment and Trade Company. The table indicates that; the value of the coefficient of determination (R^2) has reached (80.0%), which means the distribution channel selection factors have explained (80.0%) of the changes in wholesaler satisfaction in the presence of internal control as a mediating variable in Petra Investment and Trade Company. The significance value of the variance test (F) was (0.000) at the level Less than (1%),

Table 9 H04 test

Dependent variable	R	R^2	F	*P* value	Dependent variable	B	T	*P* value
Wholesaler satisfaction	0.895	0.8	735.953	0.00	Choosing distribution channels factors	0.492	10.234	0.00
					Internal control	0.445	10.345	0.00

which means that there is a positive statistically significant impact of the factors of choosing distribution channels on the satisfaction of wholesalers in the presence of the internal control of the Petra Investment and Trade Company.

Therefore, the null hypothesis was rejected and the alternative hypothesis was accepted, which states: '**There is a statistically significant impact at the level ($\alpha \leq 0.05$) for the factors of choosing distribution channels dimensions (efficiency in performance, speed, cost, length of the distribution channel) for the satisfaction of wholesalers at the presence of internal control as a mediating variable for the Petra Investment and Trade Company/Jordan**'.

11 Discussion

The study concluded that; there is a significant effect at the level ($\alpha \leq 0.05$) for the factors of choosing distribution channels dimensions (performance efficiency, speed, cost, length of the distribution channel) on the satisfaction of wholesalers in Petra Investment and Trade Company, and this result agrees with the findings of [6] study of the importance of distribution channel selection factors in responding to customer requests, distribution channels can provide a competitive advantage for all partners of the supply chain, through a short delivery time and higher product availability. Statistical significance at the level ($\alpha \leq 0.05$) for the internal control on the satisfaction of wholesalers for the Petra Investment and Trade Company /Jordan. There is of great importance for the company to monitor all the company's operations on an ongoing basis through an effective internal control system within the organizational context, to maintain operational financial sustainability to achieve profitability and continuity, in addition to maintaining product or service efficiency within the framework of certain standards by evaluating these applications and the final outputs.

The importance of internal control emerges from the importance of controlling the organization operations and financial positions, which are carried out through internal control reports, which leads to raising the confidence of wholesalers in the distribution channel efficiency without affecting its speed and cost.

There is an impact of factors of choosing distribution channels dimensions (performance, speed, cost, length of the distribution channel) on the internal control in Petra Investment and Trade Company. This result is consistent with what was pointed out by [14, 15] that the internal control system helps in achieving price competition.

Attention to quality, providing the best products and developing them. In addition to expanding the provision of services and products, which helps in supporting the organization competitiveness, facing challenges, improving productivity, maintaining the organization's position in the market and ensuring the continuity of its operations and progress. The study results indicate to a mediating role to the internal control between the relation of factors of choosing distribution channels dimensions (performance efficiency, speed, cost, length of the distribution channel) and the satisfaction of wholesalers in the Petra Investment and Trade Company, these results agree with what was indicated by [3] there is a major role for the internal control system in reducing the risks facing companies because it contributes to raising the quality of services provided, preserving the company's assets, increasing profits, in light of economic openness, independence of decision-making and economic freedom of customers; which leads to the creation of intense competition among companies based on services quality that offered, the ability to provide after-sales services; Thus, the company's ability to protect its assets and attract more customers; Leads to an increase in the competitive advantage.

12 Recommendations

The necessity of the Petra Investment and Trade Company continuing to pay Attention to the factors affecting the satisfaction of wholesalers (performance efficiency, speed, cost and length of the distribution channel) through conducting training courses and establishing updated electronic systems that contribute to raising the effectiveness of the distribution channel, increasing the utilize in the internal control system of the Petra Investment and Trade Company for its great role in influencing the satisfaction of wholesalers, continuous follow-up of wholesalers' business, updating information and establishing a system for exchanging the information with them to raise the level of loyalty and manage the relationship for more efficiently.

References

1. Al-Khelaifi, B.A.-D.: Distribution strategy in the economic enterprise. Unpublished master's thesis. Mentouri University-Constantine, Faculty of Economics and Management Sciences, Algeria (2005)
2. Al-Mashhadi, A.S.J.: Review on development of the internal control system. Technology **31**(35) (2021)
3. AlSharif, B.M.M., Al-Slehat, Z.A.F.: The effect of internal control on the competitive advantage of the bank. Int. J. Bus. Manag. **14**(9), 1–91 (2021)
4. Al-Taweel, Z.S.: Internal control and reporting of failures and the lack of scientific methodology for reporting—the case of Al-Bireh municipality—Palestine. J. Legal Econ. Adm. Sci. 5(13):41–24 (2021)

5. Altayeb, B.: The role of distribution strategies in activating sales activation in the economic institution. Unpublished master's thesis. University of M'sila, Faculty of Economic and Commercial Sciences, Algeria (2014)
6. Andjelkovic, A., Radosavljevic, M.: The length of the distribution channel as a factor of its efficiency. Strateg. Manag. **25**(2), 3–17 (2020)
7. Baron, R.M., Kenny, D.A.: The moderator–mediator variable distinction in social psychological research: conceptual, strategic, and statistical considerations. J. Pers. Soc. Psychol. **51**(6) (1986)
8. Chadoka, P., Murairwa, S.: A framework for measuring the efficiency of C&P enterprise's distribution channels in Harare, Zimbabwe. Amity J. Mark. **3**(1), 1–12 (2018)
9. Daoud, G.Q., Rahman, A., Hisham, H.: The effect of evaluating the performance of members of distribution channels on customer satisfaction—field research. J. Adm. Res. **19**(74) (2013)
10. Duncan, D.N.: The adequacy of response rates to online and paper surveys: what can be done? Assess. Eval. High. Educ. **33**(3) (2008)
11. El-Deeb, H.: Pricing and Distribution. Syrian Virtual University Publications, Syria (2020)
12. Hasibuan, C.F.: The measurement of customer satisfaction towards the service quality at Xyz wholesale by using fuzzy service quality method. In: IOP Conference Series: Materials Science and Engineering, vol. 909(1). IOP Publishing (2020)
13. Khater, S.A.: The impact of the dimensions of marketing flexibility in light of the Corona pandemic on the behavior of customer citizenship: the mediating role of the mental image of the organization (applied study). Sci. J. Trade Financ. **40**(3) (2020)
14. Kotler, P., Armstrong, G.: Principles of Marketing. Pearson Education, New York (2019)
15. Nasir, M., Simsek, S., Cornelsen, E., Ragothaman, S., Dag, A.: Developing a decision support system to detect material weaknesses in internal control. Decis. Supp. Syst. 113631 (2021)
16. Obeidat, T., Adas, A.R., Kayed, A.H.: Scientific Research, its Concept, Tools and Methods, 9th edn., Dar Al-Fikr Publishers and Distributors, Amman (2020)
17. Putra, A., Tarigan, Z.J.H., Siagian, H.: Influence of information quality on retailer satisfaction through supply chain flexibility and supplier relationship management in the retail industry. J. Tek. Ind. **22**(2), 93–102 (2021)
18. Shkurkin, D.V., Krasyuk, I.A., Krymov, S.M., Kazantseva, I.G., Zakharenko, G.N: Sales policy and sales marketing system. Int. J. Appl. Bus. Econ. **15**(12), 203–213 (2017)
19. Shpak, N., Kuzmin, O., Dvulit, Z., Onysenko, T., Sroka, W.: Digitalization of the marketing activities of enterprises: case study. Information **11**(2), 109 (2020)
20. Tartar, A., Wagdy, S.: Factors affecting decision-making for the design of distribution channels in the economic enterprise. In: Among the Activities of the International Forum: Decision-making in the Economic Institution, Faculty of Economics, Larbi Tebessi University, Algeria (2019)
21. Trippi, R., Settle, R.: A nonparametric coefficient of internal consistency. Multivar. Behav. Res. **11**(4), 419–424 (1976)

Web Sites

22. http://www.petragroup.com.jo/en-us/ourcapabilities.aspx

The Impact of Salesperson Skills on the Purchasing Decision in Electronics Showroom During the Coronavirus (COVID-19) Pandemic in Zarqa City

Khaled Tawfeq Al-Assaf and Ayed Moh'd Al Muala

Abstract This study focused on knowing the effect of the skills of sales representatives on the purchasing decision in electronics showrooms during the pandemic of the Corona virus in Zarqa city. A refereed questionnaire from experts in the same field was used to know the opinions of the study sample. The sample size that was valid for analysis was 370, and the response rate was 82%. It was distributed to a number of salespeople in electronics showrooms during the coronavirus pandemic in Zarqa. The results showed that all hypotheses have an important and positive impact on the purchasing decision in the electronics showrooms during the Corona Virus pandemic in Zarqa city (H1, H2, H3, and H4). Therefore, the study recommends that electronics showrooms should have everything that leads to improving the skills of the sales representative.

Keywords Salesperson skills · Purchase decision · Electronics showrooms · Jordan

1 Introduction

Organizations are competing with each other to sell their products in light of the pandemic, creating a sustainable competitive advantage in the market through training salesmen on modern marketing strategies, directing and encouraging customers to buy. Nowadays, he Coronavirus (COVID-19) pandemic appeared and spread all over the world, where salespeople started to increasingly dependent on technology to make their sales process. A successful sales process depends on the salesperson's ability to effectively choose the preferred communication platform for the customer [1]. And, taking into account the increasing range of sales techniques available to salespeople, becausethey must understand how these technologies help them take advantage of it and how to use it optimally, which lead them to achieve their selling goals [2]. Within the technology development in the sales

K. T. Al-Assaf · A. M. Al Muala (✉)
Faculty of Economics and Administrative Sciences, Zarqa University, Zarqa, Jordan
e-mail: amuala@zu.edu.jo

A. Hamdan et al. (eds.), *The Implementation of Smart Technologies for Business Success and Sustainability*, Studies in Systems, Decision and Control 216,
https://doi.org/10.1007/978-3-031-10212-7_7

force, the success of sales management in organizations relied on the sales power of the organization as they are linking between the organization and consumers [3]. Where the personal selling process involves a persuasive communication process that targets the purchasing behavior of the consumer through personal and direct contact, and this contributes to giving direct feedback, which contributes to developing the sales dialogue skill more quickly, which increases the chances of persuasion and inducements of answers by the final consumer [4]. Hence, the importance of the salesman is highlighted in influencing the purchasing decision of the consumer in Zarqa city towards the purchase of electrical appliances, knowing that most of the electricalappliances are considered from those market commodities that need the skill in convincing customers. Meanwhile, the spread of the Coronavirus (COVID-19) pandemic caused a change in the purchasing patterns in the markets, besides the increasing competition between organizations that deal with shopping goods in general, and electrical appliances in particular, in which the great role of the salesman appears, which affects the consumer's decision, and thussupportsincreasing the number of these organizations' customers; whether an agent or an intermediary, particularly in the showrooms.

2 Literature Review

The findings of [5] study shed the light on the need to broaden the focus on the use of social media to become a concept that understands social selling as a broader application of digital marketing principles at the level of the salesperson. Whereas, social selling benefits from the social and digital channels in order to understand and connect with influencers, as well as expected and existing customers, also engage them in relevant customer buying journey touch points for the purpose of building valuable business relationships. While, the most prominent results of [6] study, that direct selling is the main pillar and the best and most effective method in distributing products, which leads to increased sales. Provided that the scientific method is followed in selecting and qualifying salesmen, and recommended the necessity for organizations to pay attention toward rehabilitation and training of the salespeople, therefore, achieves a balance between satisfying the needs of the consumer on the one hand, and the needs of the salespeople on the other hand, which leads to achieving the interest of the organization, and working to address some shortcomings of some organizations that suffer from a weakness in their sales, notably, this highlight the importanceof the salespeople role.

Furthermore, [7] study also found that there are results concerningthe impact of advertising in each of these stages. It can be said that it is the only factor that moves the individual from one stage to another, as a matter of fact; advertising affects the consumer's behavior effectively. Meanwhile, it is worth mentioning here that [2] study, concluded that the level of personal selling in organizations working in the field of electrical industries is high, as it plays a prominent role in achieving high and distinguished performance, and emphasizing the importance of defining the salespeople skills and tasks in all its dimensions, whether communication skills, personal skills, scientific background, credibility, confidence or experience. Additionally, the study recommended the importance of adopting the concept of personal selling by the organization's management, because the weakness and shortcoming in adopting the knowledge of this activity lead to the wasting of many competitive advantageopportunities for the organization.

Also, [8] study results indicate that there are two types of social network characteristics, namely; relationship centralization and positional centralization. Thus, contribute greatly to the performance of a salesperson. Furthermore, this finding indicates that researchers should not assume that all centralities represent similar underlying network properties. So, in light of these results, there are several impacts on managers and researchers as well as trends.

As a consequence, the researcher reviewed [9] study. Where, the research relied on evaluating the sales force and mostly on measures that reflect the performance of salesmen, such as sales volume, revenue, and evaluations, as well as the manager evaluating and managing for the sales force. Nevertheless, organizations that care about sales must emphasize the need to adapt to the ever-changing market to achieve maximum performance. Additionally, the researchers suggest looking to develop a scale to assess and demonstrate the impacts of training type andincentive type on a salesperson's future value for profit. It can be noted that trade shows are of great importance in the marketing of electrical appliances, whereas [10] study recommended working on a review of a developed research agenda to help upgrade the trade marketing literature.

In the same context, but from the other side, a study was conducted on the salesmen participatingexperiences in virtual trade shows (VTS), where the study aimed to expand the understanding of the main factors and challenges of participating in virtual trade shows (VTS) as well as the virtual marketing capabilities required for organizations concerned with trade shows, and among the most prominent results that the study found is the main motives for participating in these programs are to increase sales revenue, reduce costs, reach new or different markets, and build brand credibility. Thereby, the results also indicate that the future of these programs will likely depend on the technological development of VTS and the way in which they are incorporated into the existing technological and strategical processes of exhibitors and visitors [5].

3 Hypotheses of Study

The study adopted four hypotheses, as shown below:

H1 Conversation positively and significantly affects on purchase decision.
H2 Communication positively and significantly affects on purchase decision.
H3 Experience positively and significantly affects on purchase decision.
H4 Responsibility positively and significantly affects on purchase decision.

4 Research Methodology

4.1 Population and Sample

The researcher used and designed the descriptive analytical research, where the size of the study population consisted of sales representatives in the city of Amman and the sample size that targeted 450. The study population consists of a sales representative in the city of Amman. About 450 questionnaires were distributed to sales representatives in Amman. Only 370 questionnaires obtained had correct answers and were used to analyze the data in this research. According to [8], 370 answers is an acceptable number for researchers to move forward with data analysis. The researcher designed questionnaires, analysis method, and results.

4.2 Questionnaires Design

The questionnaires were pre-tested with 30 of sales representatives in Amman city. Furthermore, the purpose of the research was explained to sales representatives in Amman' to get feedback regarding questionnaire questions, a number of participants suggested that a few words in the questions somewhat not clear. Except for these comments, the result of pre-test questionnaire shows that, questions are realistic, accurate, and easy to understand. After the pre-test, unclear words and sentences were modified. The appropriate sample was used in the study, as it is considered most popular for sample design, the research in the social sciences as it provides researchers with an acceptable database for the use of statistical techniques. The questionnaire consists of three parts. The first part deals with the demographic factors of the respondents: gender, age, education, income, and marital status. The second part of the questionnaires is about the independent variable- conversation, communication, experience and responsibility. The third part was about the dependent factor—the purchase decision. In addition, Likert scale used to measure independent and dependent variables, this scale is widely used in both marketing [11]. However, many researchers have argued that using a five-point format is as good as any other [12].

5 Results

5.1 Reliability Test

Many studies focuses on calculating the reliability of the variables, and they all agree with the lower limit of 0.60 [13]. Table 1, lists Cronbach's alpha (alpha coefficient)

Table 1 Results of reliability

Variable	Items number	Alpha coefficient
Conversation	5	0.78
Communication	6	0.74
Experience	4	0.80
Responsibility		
Purchasing decision	4	0.89

for each variable. We note that all variables show a high value of reliability. Table 1 shows that there are four independent variables and one dependent variable of the purchase decision.

5.2 *Demographic Variables*

According to the results of the descriptive research in Table 2, we note that the largest number of respondents were males (50.8%), while females (49.2%). As for the age group, it ranges between 26 and 35 years (53.8%). Regarding to education, the percentage of respondents who hold the majority of education level were a bachelor's degree (67.8%).

Table 2 Summary of respondent characteristics

Characteristics	Title	Frequencies	Percentage (%)
Gender	Male	188	50.8
	Female	182	49.2
Age	Less than 25 years old	9	2.45
	26–35 years old	199	53.8
	36–45 years old	134	36.2
	More than 46 years old	28	7.6
Education	Secondary school	9	2.4
	Bachelor	251	67.8
	Master	103	27.8
	Ph.D.	7	1.95

Table 3 Hypothesis results

Model		Unstandardized coefficients		Standardized coefficients	T	Sig.
		B	Std. error	Beta		
1	(Constant)	0.801	0.162		5.080	0.000
	Conversation	0.111	0.052	0.110	2.105	0.036
	Communication	0.271	0.059	0.114	3.920	0.000
	Experience	0.374	0.053	0.180	3.550	0.000
	Responsibility	0.220	0.053	0.115	2.995	0.000

5.3 Hypothesis Test

The results of the study, as shown in Table 3, showed that the hypotheses values of the skills of the salesperson significantly affect the purchasing decision of the total sample. Thus, the results confirm the hypotheses of the study (H1, H2, H3, H4) (2), and the data indicate that the results of the hypothesis are importantly related to the purchase decision.

6 Conclusion

The study confirmed the impact of the salesperson's skills (conversation, communication, experience and responsibility) on the purchase decision in the electronics fair during the spread of the Corona virus pandemic in Zarqa city, after data collection and analysis using the SPSS program, the statistical analysis showed that all skills have an impact on the purchase decision. Furthermore, all independent variables fulfilled the study objectives. According to these results, the owners of electronics showrooms in Zarqa city should pay more attention to the communication dimension to improve the skills of sales representatives in order to enhance the purchasing decision.

The result indicates there is statistical significant impact for salesperson's skills on purchase decision. In addition, the salesperson's skills have the highest contribution on purchase decision in the electronics fair during the spread of the Corona virus pandemic in Zarqa city.

References

1. Serviss, E., Ledet, A., Henderson, J.: Communication alignment in sales: the role of emotional intelligence. In: 2020 Society for Marketing Advances Conference. Auburn University (2021)

2. Itani, O.S., Krush, M.T., Agnihotri, R., Trainor, K.J.: Social media and customer relationship management technologies: influencing buyer-seller information exchanges. Ind. Mark. Manage. **90**, 264–275 (2020)
3. Al-Sumaida'i, M., Youssef, R.O.: Sales Department, 1st edn. Dar Al Masirah Publishing (2010)
4. Mualla, N., Tawfiq, R.: The Origins of Marketing, an Analytical Introduction, 3rd edn. Wael Publishing House (2005)
5. Ancillai, C., Terho, H., Cardinali, S., Pascucci, F.: Advancing social media driven sales research: establishing conceptual foundations for B-to-B social selling. Ind. Mark. Manage. **82**, 293–308 (2019)
6. Moza, M.R.: "The effect of the salesman in marketing", is studying a case at a company called "Pasgianos Soft Drinks Co. Ltd.". Master's thesis. Sudan University of Science and Technology (2004)
7. Gottlieb, U., Bianchi, C.: Virtual trade shows: exhibitors' perspectives on virtual marketing capability requirements. Electron. Commer. Res. Appl. **21**, 17–26 (2017)
8. Sekaran, U.: Research Methods for Business: A Skill-Building Approach, 4th edn. Wiley (2003)
9. Kumar, V., Sunder, S., Leone, R.P.: Measuring and managing a salesperson's future value to the firm. J. Mark. Res. **51**(5), 591–608 (2014)
10. Tafesse, W., Skallerud, K.: A systematic review of the trade show marketing literature: 1980–2014. Ind. Mark. Manage. **63**, 18–30 (2017)
11. Burns, A., Bush, R.: Marketing Research: Online Research Applications, 4th edn. Prentice Hall, New Jersey (2002)
12. Churchill, G., Iacobucci, D.: Marketing Research: Methodological Foundations, 9th edn. Thomson South-Western, Ohio (2004)
13. Hair, J.F., Black, W.C., Babin, B.J., Anderson, R.E., Tatham, R.L.: Multivariate Data Analysis, 6th edn. Prentice Hall (2006)

Artificial Intelligence, TechManagement, Women Entrepreneurship, and Social Implications

FinTech Adoption in Jordan: Extending UTAUT2 with eWOM and COVID-19 Perceived Risk

Ali Aljaafreh, Nour Qatawneh, Rehan Awad, Hassan Alamro, and Sulaiman Ma'aitah

Abstract One of the most significant problems that communities face during times of crisis is gaining access to financial services. Over the last few decades, rapid developments in financial technology have significantly changed both people's lives and business practices especially in the emergency circumstances. This research examines the role of FinTech applications in building resilience during the COVID-19 pandemic. The research attempts to empirically examines factors affecting Jordanian users' intention to use FinTech applications by developing a conceptual model as an extension of UTAUT2 with: COVID-19 perceived risk and eWOM as predicted constructs, both of which the study expect to increase FinTech adoption among Jordanians. As a result, the current paper develops a model for FinTech adoption in Jordan.

Keywords FinTech · COVID-19 perceived risk · eWOM · UTAUT2 · COVID-19

1 Introduction

Since the beginning of twenty-first century, technology has become a compulsory component in our businesses and lives. The rapid revolution in Information Technology (IT), the need for accuracy, and other environmental conditions such as

A. Aljaafreh (✉) · N. Qatawneh · R. Awad · H. Alamro · S. Ma'aitah
Mutah University, Al-Karak, Jordan
e-mail: Ali.jaafreh@mutah.edu.jo

N. Qatawneh
e-mail: nour_qatawneh@mutah.edu.jo

R. Awad
e-mail: rehan.essa@mutah.edu.jo

H. Alamro
e-mail: hassan.a@mutah.edu.jo

S. Ma'aitah
e-mail: suliman_m98@mutah.edu.jo

A. Hamdan et al. (eds.), *The Implementation of Smart Technologies for Business Success and Sustainability*, Studies in Systems, Decision and Control 216,
https://doi.org/10.1007/978-3-031-10212-7_8

COVID-19 have accelerate adopting IT in different sectors [1]. Banking and financial sector has been also affected by the expansion of IT in developing or creating new financial products and services known as FinTech [2].

The concept FinTech is becoming well known in recent years and defined as" a new financial industry that applies technology to improve financial activities" [3]. However, during COVID pandemic, FinTech has exploded and considered as a reliable and secure way to perform financial transactions [4]. The pandemic forced the governments around the world applied lockdown period in which aggressively affected the economy [5]. The World Health Organization (WHO) has reported that the COVID-19 is dangerous and causes death to elderly people. In Jordan, the government followed the world protocol that deal with COVID-19 and the country faced lockdown periods during 2020–2021 [6].

Literature showed many factors that affected the adoption of different forms of FinTech. For instance, [1] discussed the factors that affect the adoption of e-Banking and reported that initial trust, national culture are crucial influencers. In addition, eWallet adoption has been investigated also and the author explained that the adoption process is affected by COVID-19 perceived risk [7]. Moreover, other researchers focused on the influence of electronic reviews and comments (also known as electronic word of mouth) about mobile banking on adopting this type of FinTech [2, 8].

Despite the fact that the literature has thoroughly discussed FinTech adoption in developing countries [9–11]; the current paper attempts to investigate the issue in Jordan by develop a model that combine different factors based on the literature and the context.

The rest of the current paper is structured as follow.

2 Literature Review and Theoretical Framework

2.1 FinTech in Jordan During COVID-19 Pandemic

Financial institutions observed stark variations in user behavior, where demand on digital channels increased. Moreover, the crisis has put the weaknesses of the financial sector under the spotlight, where services which had lacked behind on digitalization, such as credit, were amongst the hardest hit. Users who were brought inside the financial inclusion umbrella during the COVID 19 crisis, on the other hand, claimed experiencing problems using various services. To counteract this, a number of institutions have increased their offerings and overall market integration in an effort to remove the obstacles to substantial financial services access.

Financial institutions emphasized the expansion of digital capabilities over branch expansion in the light of the COVID-19 pandemic, advancing the financial sector's digital transformation process. During the crisis, Jordan Payments and Clearing

Company (JoPACC) played a critical role in assuring the transmission of financial education materials by focusing on the primary issues and challenges faced by clients. During the pandemic, JoPACC also launched a new payment system that enabled instant payment methods. JoPACC today owns five payment systems: ECC, ACH, JoMoPay, eFAWATEERcom, and the Instant Payment System (CliQ), which was launched during Phase 3 of the epidemic (beginning June 4th). These systems were subjected to varying degrees of change and demonstrated varying levels of resilience [12].

Jordan's financial sector is governed by the Central Bank of Jordan (CBJ), which was established in October 1964 [13]. In response to the revolution in payment services and the rapid transformation to the use of technology in the recent decade, CBJ established a firm, Jordan Payments and Clearing Company (JoMoPay), which was officially launched on April 1, 2014 for payments and financial services that are offered through mobile wallets [14]. JoMoPay is one of well-developed financial system that is made up of a variety of different stakeholders.

Reports reveal a substantial increase in FinTech usage in Jordan since the start of COVID-19. For example, according to a report, the total number of mobile wallets in January 2020 was 642,356, but by December 2021, the number of users had tripled to 1,759,799 [15]. According to the same organization, the total rise in wallets in 2021 was 462,769, implying that 654,674 consumers adopted FinTech in 2020 [16]. Despite the fact that Jordan has curbed the spread of COVID-19, the crisis has had a tremendous impact on Jordanian households [17]. In response to these challenges, Jordan began implementing processes to support its economy during the crisis early on, including encouraging citizens to use FinTech applications to complete their financial activities during the lockdown and even later.

Human behavior of adopting new technologies have been widely investigated in several discipline such as information system (IS), sociology, and psychology. During a period of three decades, scientists have created their own theoretical models [18–20] to address the main factors that affect the adoption process, however; they did not consider the contributions of others' work [21]. Accordingly, in 2003, a group of researchers decided to create a unified model based on thoroughly review for previous theoretical models [18–20, 22–26] and named it the unified theory of acceptance and use of technology (UTAUT) model [27]. The model consists of four main constructs that are hypothesized to have an effect on users' intention to adopt technology in organizational contexts: performance expectancy, effort expectancy, social influence, and facilitating conditions. In addition, the model also hypothesized relationships for four moderators: age, gender, experience, and voluntariness of use. However, in 2012, Venkatesh and his colleagues have recommended some additions to UTAUT to include other factors such as habit, hedonic motivation, and price value [28].

Despite the fact that UTAUT2 model is widely used and have high variance value 70% in explaining user behavior toward adopting new technology, the current context of COVID-19 and the increased literacy of IT have created new attention to investigate factors such as COVID-19 perceived risk and electronic word of mouth (eWOM).

2.2 COVID-19 Perceived Risk (Disease Risk)

Perceived risk of COVID-19 was defined as the individuals' risk perception, both cognitive and emotional, regarding the pandemic. Risk dimension can be divided into two categories: cognitive and disease risk [29]. Few of research, examined "disease risk" as a factor influencing people's willingness to accept digital payments. According to [7], the emergence of COVID-19 had a negative impact on Indonesian and Malaysian customers' willingness to use physical money. However, it resulted in a greater use of e-wallets in their financial transactions. In the same vein, mobile payment is viewed as a disease-prevention strategy that lowers the risk of catching the virus [30]. Despite the fact that the majority of studies found a negative impact of perceived risk on intention [31, 32], a recent study investigating the impact of COVID-19 perceived risk on intention to use e-Wallet confirmed that COVID-19 perceived risk has a positive and significant impact [7].

2.3 eWOM

Electronic word of mouth (eWOM) communication can be defined as a positive or negative statement made through the Internet about a product, company, or new technology that is accessible to a group of people and institutions either through the mobile Internet and device [33–35] stated that eWOM has a greater influence when a product is eaten rather than sought, implying that the impact of eWOM grows stronger after a service encounter. Consumers can utilize eWOM communicate and exchange information about a product or service by using online discussion and forum sites, social networking sites, and consumer review sites [36]. Because the identity of the reviewer cannot be revealed, eWOM eliminates the unpleasantness associated with bias information transmission among friends and family, as opposed to traditional WOM communication [37]. Several researchers discovered that eWOM can influence IT adoption both directly and indirectly by influencing attitude [38], along with purchase intention. However, the use of online banking services may also be influenced by electronic word of mouth.

3 Materials and Method

The study's key contribution is to develop a theoretical model as an extension of UTAUT2 by including two additional constructs: COV19 perceived risks and electronic word of mouth (eWOM). The aim of the presented framework is to investigate factors influencing FinTech behavioral intention and use behavior among Jordanians. All the constructs presented in Fig. 1 act as key direct predictors for intension

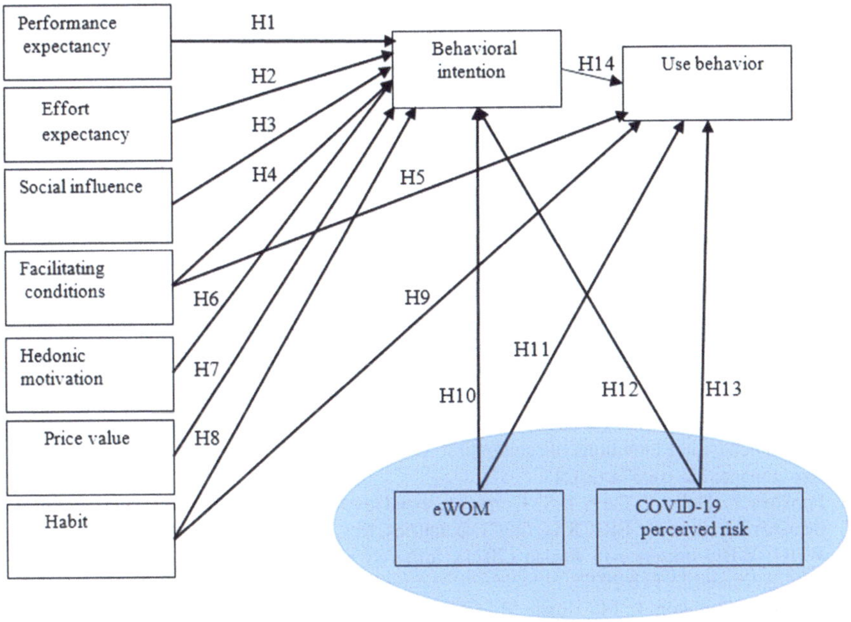

Fig. 1 The proposed model

behavior. The relationships between the constructs are postulated based on a thorough literature review, which demonstrates their integration into an understandable research model, as seen in Fig. 1's blue-shaded parts.

4 Results and Discussion

As presented in the proposed model, each of (effort expectancy, performance expectancy, social influence, facilitating conditions, hedonic motivation, price value, and habits) are constructs which contextually taken from UTAUT2, whereas eWOM and COV-19 perceived risk are new constructs added to the model.

5 Conclusion

According to the literature, several factors influencing users' behavior to use FinTech applications. These factors include (namely: performance expectancy, effort expectancy, social influence, hedonic motivation, price value, habit, facilitating condition). Based on the presented study, it is important to consider both of (eWOM and COV-19 perceived risk) that are expected to significantly affect users' behavior to use FinTech applications in Jordan.

References

1. Aljaafreh, A., Al-Hujran, O., Al-Ani, A., Al-Debei, M.M., Al-Dmour, N.: Investigating the role of online initial trust in explaining the adoption intention of internet banking services. Int. J. Bus. Inf. Syst. **36**(4), 474–505 (2021)
2. Abubakar, A.M., Ilkan, M., Al-Tal, R.M., Eluwole, K.K.: eWOM, revisit intention, destination trust and gender. J. Hosp. Tour. Manag. **31**, 220–227 (2017)
3. Schueffel, P.: Taming the beast: a scientific definition of fintech. J. Innov. Manag. **4**(4), 32–54 (2016)
4. Daragmeh, A., Lentner, C., Sági, J.: FinTech payments in the era of COVID-19: factors influencing behavioral intentions of 'generation X' in Hungary to use mobile payment. J. Behav. Exp. Financ. **32**, 100574 (2021)
5. Popkova, E., DeLo, P., Sergi, B.S.: Corporate social responsibility amid social distancing during the COVID-19 crisis: BRICS vs. OECD countries. Res. Int. Bus. Financ. **55**, 101315 (2021)
6. WHO: WHO presence in Jordan (2021). http://www.emro.who.int/jor/who-presence-in-jordan/. Accessed 21 Feb 2022
7. Aji, H.M., Berakon, I., Md Husin, M.: COVID-19 and e-wallet usage intention: a multigroup analysis between Indonesia and Malaysia. Cogent Bus. Manag. **7**(1) (2020). https://doi.org/10.1080/23311975.2020.1804181
8. Petrovic, M., Rajin, D., Milenković, D., Marić, D.: The influence of eWOM on the use of mobile banking. Ekon. Preduz. **69**, 95–104 (2021). https://doi.org/10.5937/EKOPRE2102095P
9. Hasan, R. Ashfaq, M., Shao, L.: Evaluating drivers of Fintech adoption in the Netherlands. Glob. Bus. Rev., 09721509211027402 (2021)
10. Morgan, P.J.: Fintech, financial literacy, and financial education. Routledge Handb. Financ. Lit., 239–258 (2022)
11. Singh S., Sahni, M.M., Kovid, R.K.: What drives FinTech adoption? a multi-method evaluation using an adapted technology acceptance model. Manag. Decis. (2020)
12. JoMoPay: About JoPACC (2022). https://www.jopacc.com/EN/Pages/About_JoPACC. Accessed 25 Feb 2022
13. CBJ: Central Bank of Jordan: a brief outlook (2022). https://www.cbj.gov.jo/Pages/viewpage.aspx?pageID=174. Accessed 25 Feb 2022
14. CBJ: National switching for mobile payment (2022). https://www.cbj.gov.jo/Pages/viewpage.aspx?pageID=321. Accessed 25 Feb 2022
15. JoMoPay: Mobile wallet payments made in Jordan (2020) [Online]. Available: https://www.jopacc.com/ebv4.0/root_storage/en/eb_list_page/jomopay_report_jan._2020.pdf
16. JoMoPay: JoMoPay report, December 2021 (2021) [Online]. Available: https://www.jopacc.com/ebv4.0/root_storage/en/eb_list_page/jomopay_transactions_-_december_2021-0.pdf
17. Bashayreh, A., Wadi, R.M.A.: The effect of Fintech on banks' performance: Jordan case. In: International Conference on Business and Technology, pp. 812–821 (2020)
18. Davis, F.D.: Perceived usefulness, perceived ease of use, and user acceptance of information technology. MIS Q. 319–340 (1989)
19. Fishbein, M., Ajzen, I.: Belief, attitude, intention, and behavior: an introduction to theory and research. Philos. Rhetor. **10**(2) (1977)
20. Rogersvenka, E.M.: Diffusion of Innovations. The Free Press, New York, NY, USA (1995)
21. Xie, J. Ye, L., Huang, W., Ye, M.: Understanding FinTech platform adoption: impacts of perceived value and perceived risk. J. Theor. Appl. Electron. Commer. Res. **16**(5), 1893–1911 (2021) [Online]. Available: http://10.0.13.62/jtaer16050106

22. Ajzen, I.: The theory of planned behavior. Organ. Behav. Hum. Decis. Process. **50**(2), 179–211 (1991)
23. Heffernan, C.J.: Social foundations of thought and action: a social cognitive theory, Albert Bandura Englewood Cliffs, New Jersey: Prentice Hall, 1986, xiii+ 617 pp. Hardback. US $39.50. Behav. Chang. **5**(1), 37–38 (1988)
24. Davis, F.D., Bagozzi, R.P., Warshaw, P.R.: Extrinsic and intrinsic motivation to use computers in the workplace 1. J. Appl. Soc. Psychol. **22**(14), 1111–1132 (1992)
25. Taylor, S., Todd, P.: Assessing IT usage: the role of prior experience. MIS Q., 561–570 (1995)
26. Thompson, R.L., Higgins, C.A., Howell, J.M.: Personal computing: toward a conceptual model of utilization. MIS Q., 125–143 (1991)
27. Venkatesh, V., Morris, M.G., Davis, G.B., Davis, F.D.: User acceptance of information technology: toward a unified view. MIS Q., 425–478 (2003)
28. Venkatesh, V., Thong, J.Y.L., Xu, X.: Consumer acceptance and use of information technology: extending the unified theory of acceptance and use of technology. MIS Q., 157–178 (2012)
29. Yıldırım, M., Güler, A.: Factor analysis of the COVID-19 perceived risk scale: a preliminary study. Death Stud., 1–8 (2020). https://doi.org/10.1080/07481187.2020.1784311
30. Sreelakshmi, C.C., Prathap, S.K.: Continuance adoption of mobile-based payments in Covid-19 context: an integrated framework of health belief model and expectation confirmation model. Int. J. Pervasive Comput. Commun. (2020)
31. Marafon, D.L., Basso, K., Espartel, L.B., de Barcellos, M.D., Rech, E.: Perceived risk and intention to use internet banking: the effects of self-confidence and risk acceptance. Int. J. Bank Mark. (2018)
32. Kassim, N.M., Ramayah, T.: Perceived risk factors influence on intention to continue using Internet banking among Malaysians. Glob. Bus. Rev. **16**(3), 393–414 (2015)
33. Kudeshia, C., Kumar, A.: Social eWOM: does it affect the brand attitude and purchase intention of brands? Manag. Res. Rev. (2017)
34. Bronner, F., De Hoog, R.: Vacationers and eWOM: who posts, and why, where, and what? J. Travel Res. **50**(1), 15–26 (2011)
35. Park, S.Y.: An analysis of the technology acceptance model in understanding university students' behavioral intention to use e-learning. J. Educ. Technol. Soc. **12**(3), 150–162 (2009)
36. Wu, J., Gaytán, E.A.A.: The role of online seller reviews and product price on buyers' willingness-to-pay: a risk perspective. Eur. J. Inf. Syst. **22**(4), 416–433 (2013)
37. Etikan, I., Musa, S.A., Alkassim, R.S.: Comparison of convenience sampling and purposive sampling. Am. J. Theor. Appl. Stat. **5**(1), 1–4 (2016)
38. Al-Ja'afreh, A.L.I., Al-Adaileh, R.: The impact of electronic word of mouth on consumers purchasing intention. J. Theor. Appl. Inf. Technol. **98**(02) (2020)
39. Yıldırım, M., Güler, A.: COVID-19 severity, self-efficacy, knowledge, preventive behaviors, and mental health in Turkey. Death Stud. **1**(8) (2020)
40. Hennig-Thurau, T., Gwinner, K.P., Walsh, G., Gremler, D.D.: Electronic word-of-mouth via consumer-opinion platforms: what motivates consumers to articulate themselves on the internet? J. Interact. Mark. **18**(1), 38–52 (2004)

Assessing the Influences of Need of Achievement and Confidence on the Women-Owned-Micro Enterprises' Success in Malaysia

Mohd Nor Hakimin Yusoff, Mohd Rafi Yaacob, and Zainon Che Mat

Abstract This study examined factors that influence women entrepreneurs' success. Based on psychological traits theory, two constructs were identified; the need of achievement and self-confidence. This study used a quantitative approach, altogether involved 300 micro-sized women entrepreneurs throughout Malaysia. The data collected subsequently analyzed using SEM-PLS. According to the study's findings, both variables showed a strong association with the success of women entrepreneurs. These findings not only reinforce previous findings on the vital role of personality in women entrepreneurship activities but also emphasizes the roles of the government to spur and stimulate the above-mentioned factors through relevant training programs.

Keywords Women entrepreneur · Success · Need of achievement · Self-confidence · Micro enterprise

1 Introduction

Micro-Small enterprises are well-known as the backbones for most countries and women entrepreneurs contribute significantly to the nation' wealth. In Malaysia, women owned about 19.7% of the total SMEs with an equity more than 51%. Women entrepreneurship is recognized as one of key activities to promote socio-economic stability. The involvement of more women in entrepreneurship will have a greater impact on the economic growth. Previous studies revealed that women's

M. N. H. Yusoff · M. R. Yaacob (✉)
Universiti Malaysia Kelantan, Kota Bharu, Malaysia
e-mail: rafi@umk.edu.my

M. N. H. Yusoff
e-mail: hakimin@umk.edu.my

Z. C. Mat
Universiti Kuala Lumpur, Kuala Lumpur, Malaysia
e-mail: yanniez_01@yahoo.com.my

A. Hamdan et al. (eds.), *The Implementation of Smart Technologies for Business Success and Sustainability*, Studies in Systems, Decision and Control 216,
https://doi.org/10.1007/978-3-031-10212-7_9

involvement in entrepreneurship, amongst others were driven by financial freedom, independence, loss of employment and business opportunities. Despite of desire to success, they faced myriad of challenges and constraints which negatively impacted their success [1]. They were lacking in entrepreneurial aptitude toward success due to limited abilities and capabilities [2]. It was rather disappointing when numerous training programs failed to enhance their management skills and personality [3]. Consequently, personality become a major issue in women entrepreneurs' success. One reason why women entrepreneurs hardly benefited from training was due to lack of focus. Most of them were heavily burdened with family matters and on the same time having businesses to look after [4]. These issues inevitably limit their focus which finally failed to give full commitment in business activities and training [5].

Comparatively, women entrepreneurs are lagging behind compared to male entrepreneurs. This issue is very worrying as it would demotivate others from venturing into entrepreneurship. This is reflected when the current trend of women's entrepreneurial involvement showed no or a negative increment in women involvement in entrepreneurship. According to the GEM report (2016), the women entrepreneurship activity in Malaysia was at 88.9% which was slightly higher than men (77.8%). However, its experience a slow growth which is 13.6% (2004) to 13.9% (2007) down to 13.1% in 2008, up to 19.2% in 2010, and slightly up to 20.3% in 2015.

Previous studies posited that apart from family commitment, the success of women entrepreneurs was hindered by the level work-family conflict, gender discrimination, financial constraints, lack of infrastructural support, lack of skill and personality-based constraints which refers to absence of tendency to take risks and ability to manage businesses. In another development, Kuratko and Hodgetts [6] argued that the significant factors that influenced entrepreneurs' success were personality characteristics, living conditions, and environmental factors. On another point, Ismail and Nasir [7] studied on Malaysian's entrepreneurs revealed that the psychological motives such as self-satisfaction, independence and supportive environmental factors affected their success in businesses. According to Ruf et al. [8] owner–managers' characters were the key factors that influenced firms' behavior. In a further study by Fatimah et al. [9], in order for rural women entrepreneurs in Malaysia to succeed, they needed to have self-confidence and determination as well as vision.

Grounding on the Psychological Traits theory (PTT) this study aimed to further investigate the effect of personality traits that referred to the internalities of entrepreneurs such as needs of achievement (henceforth NA) and confidence to the success of women entrepreneurs. This study is deemed paramount and timely to close the gap as studies on the influence of personality on entrepreneurial success are still lacking, especially amongst women entrepreneurs in developing countries including Malaysia [10–12]. Furthermore, the findings of the study are instructive to the government and its relevant agencies pertaining to entrepreneurship.

2 Literature Review

2.1 Women Entrepreneur and Success

Simply speaking an entrepreneur is an individual who has the ability to create an on-going business activity [13, 14]. Lavoie describes a woman entrepreneur as a leader in any small business that is responsible for the first attempt to launch a new business venture, a financial risk controller, and an administrative coordinator, who holds social responsibilities, and who also manages day-to-day business activity. While an entrepreneur is a color blind, where both genders; male and female can be one, due to a plethora of reasons including opportunities, there are many male entrepreneurs compared with female entrepreneurs. On a broader perspective, Anwar and Rashid [15] posit that woman entrepreneur is a woman that uses her knowledge, experience, and resources to develop or create new business opportunities, either informally, that operates at home without formal registration of the business, or formally, via business registration which includes hiring workers and has premises. Similarly, if a woman or/and a group of women starts, organizes, and operates own business, they are considered to be an entrepreneur [16]. Women entrepreneurs like their male counterparts are also known as people who venture into entrepreneurship, create new entrepreneurial values through their time spent in the business effort by assuming the accompanying financial risk, psychological risk, and social risks, as well as getting the rewards in terms of financial growth, individual satisfaction, and also independence [17].

Lately, women entrepreneurs have made an increasing contribution to the nation's economic growth [18]. Globally, one third of all businesses in developed countries are owned by women who are self-employed, and yet are capable of growth and sustainability. Generally, women define success as when they accomplish their set of targets, or through personality motivation, realize an achievement with external criterion [19]. Women entrepreneurs were driven to attain their aspirations of independence and financial freedom [20]. In a recent study conducted by Shagufta et al. (to be published) in Kashmir Valley, India showed women entrepreneurs were better off than women non-entrepreneurs in terms of economic, social, political and psychological indicators of empowerment.

2.2 The Context

In Malaysia, women entrepreneurs are mostly involved in selling foods, retailing and wholesale businesses, as this particular sector does not require much skilled labor, or a high capital investment, it is not complex and is a low risk involvement [21]. At the outset, the women involvement in entrepreneurial activities was to support family financial burden by generating extra income beside spouse [22]. Majority of them operates as micro and small enterprises. Generally, small enterprises were dominated by male entrepreneurs; however, the number of women entrepreneurs has experienced a positive growth in line with favorable support from the government. Various support programs involving large financial allocations are provided to ensure that the entrepreneurial activities undertaken are viable and sustainable.

The increasing involvement of women in the field of entrepreneurship has made the role of women increasingly significant in the country's economic development. Therefore, this study aimed to fill the literature gap in relating with the success of women from various dimensions. It examines women's success factors, grounding on the theoretical field of traits that were under research and success was measured on the achievement of personal missions. It is suffice to say this approach can meet the lack of research among developing countries including Malaysia [12].

2.3 *Need of Achievement*

Need of achievement (NA) refers the desire to succeed or the desire to achieve something. It can also be said to be a boost to one's success or a boost to avoid failure [23]. The need for achievement motivation involves the process of natural urging that occurs with individuals within the goal of achieving satisfaction and success in their activities. A person with high Need of achievement inclines towards positive choice to do something that can meet his or her needs. McClelland [23], described high need of achievement motivated individuals to take personal responsibility, setting goals and achieving them through their efforts. They tend to seize greater opportunities for success. People who have accomplished a lot have a strong desire to take on new tasks as well as a strong sense of independence [24]. As a result, those who have a strong desire to succeed are more likely to prepare ahead of time [25]. Indeed, entrepreneurs who have a strong Need for Achievement more often than not outperformed those who had a low Need for Achievement [26, 27]. Therefore, this study hypothesized that;

> H1: Need of achievement significantly affects the success of women entrepreneurs.

Self-confidence

Self-confidence is defined as an individual's belief his own abilities to perform any task and is one of the key drivers to succeed in business [28]. An entrepreneur with high self-confidence is the individual who is willing to and not afraid of making their own decisions [29, 30]. In dealing with business pressures, a lack of confidence was also a major problem. In the same vein [31] noted that the entrepreneurs with high confidence levels would be more efficient in problem-solving and decision making. This view is supported by Galanaski and Giourka [32], who suggested that entrepreneurs with high confidence level were committed to achieving their goals to succeed. In short, self-confidence remains vital for entrepreneurs' success, therefore, this study hypothesized that;

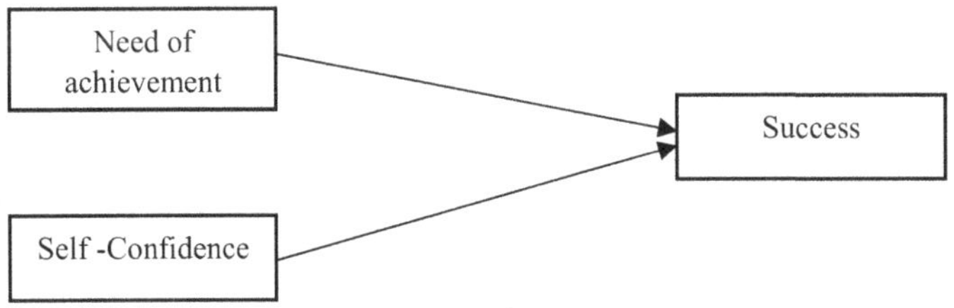

Fig. 1 Theoretical framework

H2: Self-confidence significantly affects the success of women entrepreneurs.

3 Methodology

This study adopted a cross sectional approach, using quantitative method where data were obtained from a randomly selected sample using survey questions. A total of 300 samples were selected, consisting of women entrepreneurs under the guidance and monitoring of the entrepreneurship development agencies in Malaysia. A total of 600 questions were distributed, which exceeded the total samples to address the low response rate of between 15 and 25%. A total of 300 respondents returned the completed survey questions which deemed useable for the purpose of the study.

This study adapted research questions from the previous researchers by making modifications to suit the needs of local environment. The NA measurement was adapted from Zeffane [33] using 5 measurement items. While to measure self-confidence, the measurement items were adapted from Benzing et al. [34] who also used 5 items in their study. The measurement items for success were also adapted from Fisher et al. [35] which used a subjective approach with 4 measurement items. A total of 15 measurement items with a five-point Likert scale method ranging from 1—strongly disagree to 5—strongly agree were used in measuring the variables studied. The method of adaptation from previous studies used in this study is because it is higher validity and reliability as it had been tested in previous studies. As far as data analyses were concerned, this study used the SEM-PLS instrument for dealing with complex models with a large number of constructs and indicators (Fig. 1).

4 Finding

4.1 Demographics

The unit of analysis of this study was a woman entrepreneur who owned a micro -sized business. Five demographic characteristics were used to describe the samples; namely marital status, age, level of education, age of business and business activity.

Table 1 Respondent background

		N	%
Status	Single	75	23.7
	Married	228	72.2
	Divorced	13	4.1
Age	25 and below	51	16.1
	26–35	163	51.6
	36–45	75	23.7
	46–55	21	6.6
	56–65	6	1.9
Education level	Primary	46	14.6
	Secondary	195	61.7
	Tertiary	75	23.7
Tenure	3–4 years	171	54.1
	4–5	106	33.5
	5–7	32	10.1
	More than 7 years	7	2.2
Activity	Services	275	87.03
	Manufacturing	41	12.97

Table 1 shows that the majority of the entrepreneurs involved were already married. The largest group is comprised of those aged 26–35 years (51.6%). The majority of respondents (61.7%) graduated from secondary school and only 23.7% were educated in institutions of higher learning. Meanwhile, in terms of business age, 54% of respondents had been in business for 3–4 years. 33.5% between 4 and 5 years. The overwhelming majority (87%) of respondents are involved in the services sector.

4.2 Results of Measurement Model

Under the model measurement evaluator four tests were performed to assess the reliability and validity of the constructs of the measurement items. These conducted to assess (i) indicator reliability, (ii) internal consistency [18] (iii) convergent validity and (iv) discriminant validity. The results of all four tests are discussed in the subsequent sections.

Indicator Reliability and Internal Consistency-Loading and Cross-Loading
The reliability of item measurement was assessed by looking at the loading value of each item. According to Hair et al. [36] item measurement is reliable when the loading value exceeds 0.4 While the internal consistency of item measurement is acceptable when the loading score is above 0.6. The results of this test are shown in Table 2 where all the loading items measurement values exceeded 0.6.

Table 2 Construct validity

	Self-confidence	Entrepreneurial success	N of ach
Con1	0.612	0.211	0.275
Con2	0.752	0.198	0.280
Con3	0.749	0.213	0.256
Con4	0.744	0.149	0.186
Con5	0.734	0.171	0.270
Con6	0.784	0.274	0.320
Con7	0.696	0.269	0.240
ES1	0.313	0.824	0.480
ES2	0.242	0.826	0.503
ES3	0.232	0.755	0.414
ES4	0.235	0.824	0.458
ES5	0.167	0.754	0.371
ES6	0.254	0.806	0.499
NaC1	0.277	0.490	0.765
NaC2	0.272	0.428	0.737
NaC3	0.281	0.441	0.796
NaC4	0.245	0.422	0.795
NaC5	0.292	0.476	0.846
NaC6	0.292	0.424	0.763
NaC7	0.337	0.431	0.737

Factor loading, average variance extracted (AVE), and composite reliability (CR) were used to assess convergent validity. According to the results, all criteria were met since the item loadings were greater than 0.6, the AVE greater than 0.5, and the CR values greater than 0.7 (Table 3).

Discriminant Validity

Fornell-Lacker criterion analysis was used to assess discriminatory validity.

Result shows the established AVE exceeded the square correlation value, confirming the validity with no multicollinearity issues (Fig. 2; Tables 4 and 5).

5 Discussion and Conclusion

This study merely based on the Psychology Traits Theory that posit personality's factor can influence individual actions. Two personality indicators were tested in this study; namely the NA and Self-confidence on women entrepreneurs' success. The results of the study successfully supported the both hypotheses where NA and

Table 3 Convergent validity

Construct	Label	Std loading	CR	AVE
Need of achievement	NaC1	0.765	0.915	0.605
	NaC2	0.737		
	NaC3	0.796		
	NaC4	0.795		
	NaC5	0.846		
	NaC6	0.763		
	NaC7	0.737		
Self-confidence	Conf1	0.612	0.886	0.527
	Conf2	0.752		
	Conf3	0.749		
	Conf4	0.744		
	Conf5	0.734		
	Conf6	0.784		
	Conf7	0.696		
Entrepreneur success	Ent1	0.824	0.914	0.638
	EnS2	0.826		
	EnS3	0.755		
	EnS4	0.824		
	EnS5	0.754		
	EnS6	0.806		

self-confidence significantly influenced the women entrepreneurs' success. Both personality traits were significant factors in the study of the success of women entrepreneurs in the context of Malaysia. These findings are consistent with previous studies conducted elsewhere by Khan et al. [37] and Anggadwita et al. [38]. The high NA in women entrepreneurs needs to be nurtured so that the chances of success in entrepreneurial activities are likely higher. Instilling this spirit on women entrepreneurs should require a careful planning, before entering into the world of entrepreneurship. Among the things that can be done is to prepare a business plan prior or as early as the establishment of the business. Having a comprehensive business plan not only gives a clear direction about business, it also helps entrepreneurs to bounce back in the event of failure because through its entrepreneurs can objectively and analytically analyze of their mistakes. At the same time, it can also avoid creating unrealistic objectives or targets that are impossible to achieve, obviously failure to reach the target will have a negative impact on motivation. Apart from that, external parties also play a significant role in pushing NA to the maximum level. As one of the motivational elements, high NA can be achieved if external supports are provided by relevant stakeholders to women entrepreneurs to overcome the problems they

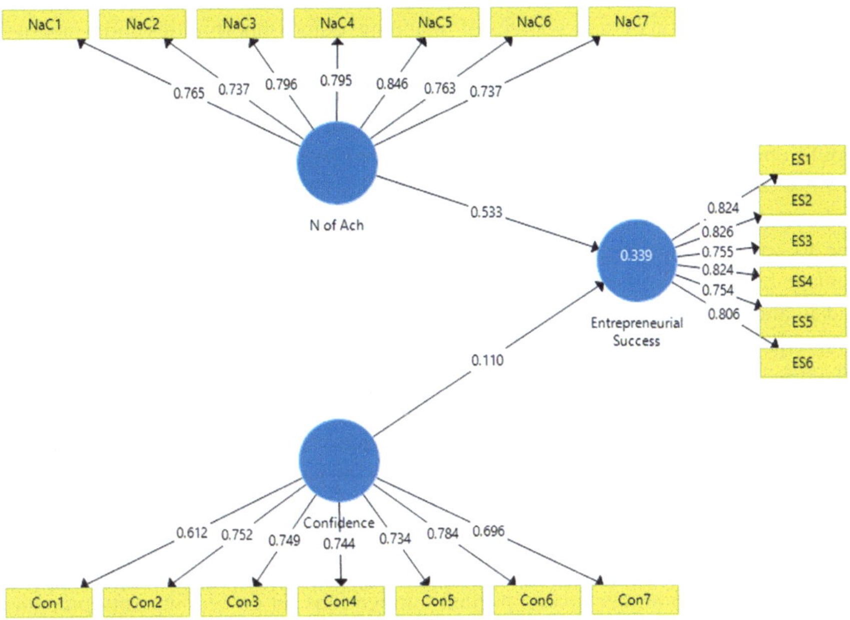

Fig. 2 Path analysis

Table 4 The Fornell–Larcker criterion

	Confidence	Entrepreneurial success	Need of ach
Self-confidence	0.726		
Entrepreneurial success	0.305	0.799	
N of ach	0.366	0.573	0.778

Table 5 Hypotheses result

Hypothesis	Regression path	Original sample (O)	Std. deviation	*t*-value	*p*-value	Hypotheses supported
H1	Self-confidence → entrepreneurial success	0.110	0.049	2.250	0.025	Yes
H2	N of ach → entrepreneurial success	0.533	0.048	11.055	0.000	Yes

faced. The provision of easy sources of financing, equal opportunities and a positive perception of their capabilities especially among investors are paramount important.

Furthermore, this study a found self-confidence had a significant relationship with the success of women entrepreneurs, indicating a high level of self-confidence would open up high opportunities for successful women entrepreneurs compared to those who do not. Generally, previous studies showed lack of self-confidence was one of the main problems of women entrepreneurs. Amongst others, this is due to society's negative perception and pessimistic view of women's ability to manage business. Women were seen as weak objects in terms of management skills and financial capabilities; therefore, they would not be able to compete with men counterparts in the marketplace. This connotation has weakened their morale in turn affecting their self-confidence. Therefore, the support of the community and the government through entrepreneurship development agencies is very important to increase the self -confidence of women entrepreneurs. Business support as discussed above is one of the inevitable and smart steps that the government can take to increase management capacity and strong funds [39]. At the same time campaign programs can lead to a paradigm shift of society's thinking towards women's involvement in entrepreneurship.

Overall, these findings have important implications for developing entrepreneurship development scheme, specializing for the real needs of women entrepreneurs. The identification of two vitals variables would enhance the effectiveness of programs. Theoretically, this combination of findings provides a clearer and better understanding on the role of personality traits on women entrepreneurs' success, by and large in developing country. The contribution of this study has been to confirm the significant roles of NA and self-confidence on the entrepreneurs' success. These findings explain that strong desired personnel in women entrepreneurs play an important role in regulating their behaviors that lead to success. Suggesting the Theory of Need of Achievement foundation confirms and validates the argument that women entrepreneurial personality traits and self-confidence significantly affect business success.

Like other studies, this study is not without its limitations. First was limited by the absence of large sample size which include women entrepreneurs across types and sizes of businesses. The present study focused on women entrepreneurs in micro-sized enterprises whom may demographically distinct to small and medium size enterprises as well. Further studies with larger samples need to be carried out to validate and confirm the result of this study. Albeit with all the said limitations, the samples in this study are sufficient to represent all women where the majority of women entrepreneurs are composed of those who also hold the title of wife and mother and aged in the range of 26–35 years. More than half of them also do not have a high level of education. These data reflect the demographics of the majority of women micro -entrepreneurs. Their levels of NA were high. Resultantly, a healthy phenomenon in the government's efforts to encourage the involvement of more women in entrepreneurship. While this study is an eye opener, many relevant questions that needs to be resolved are: when does the spirit of high need of achievement begin? What was before getting involved in entrepreneurship or after venturing into it?

References

1. Isa, F.M., et al.: Malaysian women entrepreneurs: some emerging issues and challenges of entering global market. Int. J. Acad. Res. Bus. Soc. Sci. **8**(12), 10 (2018)
2. Kuschel, K., et al.: Funding challenges of Latin American women start-up founders in the technology industry. Cross Cult. Strateg. Manag. **24**(2), 310–331 (2017)
3. Imeraj, E., Gruda, S.: Challenges of women entrepreneurs in Albania. Mediterr. J. Soc. Sci. **4**(11) (2013)
4. Pareek, P., Bagrecha, C.: A thematic analysis of the challenges and work-life balance of women entrepreneurs working in small-scale industries. Vision **21**(4), 461–472 (2017)
5. De Clercq, D., Brieger, S.A.: When discrimination is worse, autonomy is key: how women entrepreneurs leverage job autonomy resources to find work–life balance. J. Bus. Ethics (2021)
6. Kuratko, F.D., Hodgetts, M.R.: Entrepreneurship Theory, Process, Practice. Thomson South-Western, Quebec (2007)
7. Ismail, N.N.H.M., Nasir, M.K.M., Rahman, R.S.A.R.A.: Factors that influence women to be involved in entrepreneurship: a case study in Malaysia. Creat. Educ. 12(04), 11 (2021)
8. Ruf, P.J., et al.: The pillars of family firm performance? how owner manager's values impact SEW and performance. Acad. Manag. Proc. **2021**(1), 13514 (2021)
9. Fatimah, H., Aznarahayu, R., Nasina, M.D.: Rural women entrepreneurs in Malaysia: what drives their success? Int. J. Bus. Manag. **9**(4) (2014)
10. Murugesan, R., Jayavelu, R.: The influence of big five personality traits and self-efficacy on entrepreneurial intention: the role of gender. J. Entrep. Innov. Emerg. Econ. **3**(1), 41–61 (2017)
11. Shastri, R.K., Sinha, A.: The socio-cultural and economic effect on the development of women entrepreneurs (with special reference to India). Asian J. Bus. Manag. **2**(2), 5 (2010)
12. Yadav, V., Unni, J.: Women entrepreneurship: research review and future directions. J. Glob. Entrep. Res. **6**(1), 12 (2016)
13. Cunningham, J.: Defining entrepreneurship. J. Small Bus. Manage. **13**, 21 (2011)
14. Yusoff, M.N.H., et al.: The role of government financial support programmes, risk-taking propensity, and self-confidence on propensity in business ventures. Sustainability **13**(1), 380 (2021)
15. Anwar, M.U., Rashid, A.G.: Female entrepreneurs: a review of the literature and proposed conceptual framework. In: Proceedings of 2nd International Conference on Business Management and Marketing (2011)
16. Sharma, P.: Women entrepreneurship development in India. Glob. J. Manag. Bus. Stud. **3**(4), 6 (2013)
17. Hisrich, D.R., Peters, P.M., Shepherd, A.D.: Entrepreneurship, 9th edn. McGraw-Hill Irwin, New York (2013)
18. Kamberidou, I.: Women entrepreneurs: we cannot have change unless we have men in the room. J. Innov. Entrep., 18 (2013)
19. Aldrich, H.E., Cliff, J.E.: The pervasive effects of family on entrepreneurship: toward a family embeddedness perspective. J. Bus. Ventur. **18**(5), 23 (2003)
20. Orhan, M., Scott, D.: Why women enter into entrepreneurship: an explanatory model. Women Manag. Rev. **16**(5), 232–247 (2001)
21. Hassan, F., Ramli, A., Desa, N.M.: Rural women entrepreneurs in Malaysia: what drives their success? (2014)
22. Jennings, J.E., Brush, C.G.: Research on women entrepreneurs: challenges to (and from) the broader entrepreneurship literature? Acad. Manag. Ann. **7**(1), 663–715 (2013)

23. McClelland, D.C.: N achievement and entrepreneurship: a longitudinal study. J. Pers. Soc. Psychol. **1**(4), 389–392 (1965)
24. Handrito Radityo, P., Slabbinck, H., Vanderstraeten, J.: Enjoying or refraining from risk? The impact of implicit need for achievement and risk perception on SME internationalization. Cross Cult. Strateg. Manag. **27**(3), 317–342 (2020)
25. Isaga, N.: The relationship of personality to cognitive characteristics and SME performance in Tanzania. J. Small Bus. Enterp. Dev. **25**(4), 667–686 (2018)
26. Thapa, A.: Determinants of microenterprise performance in Nepal. Small Bus. Econ. **45**(3), 581–594 (2015)
27. Yusoff, M.N.H., et al.: Measuring and comparing functional business skills and knowledge among Asnaf community in Malaysia. Econ. Soc. **11**(2), 229–247 (2018)
28. Kirkwood, J.: Is a lack of self-confidence hindering women entrepreneurs? Int. J. Gend. Entrep. **1**(2), 118–133 (2009)
29. Martins, I., Monsalve, J.P.P., Martinez, A.V.: Self-confidence and fear of failure among university students and their relationship with entrepreneurial orientation. Acad. Rev. Latinoam. Adm. **31**(3), 471–485 (2018)
30. Hung, C.M., Katsioloudes, M.I.: Cultural context and the Vietnamese-American entrepreneurial experience. J. Trans. Manag. Dev. **7**(2), 37–46 (2002)
31. Yusoff, M.N., et al.: Psychological traits and intention to use E-commerce among rural micro-entrepreneurs in Malaysia. J. Theoret. Appl. Electr. Commer. Res. **16**(5) (2021)
32. Galanaski, K., Giourka, P.: Entrepreneurial path: decoupling the complexity of entrepreneurial process. Int. J. Entrep. Behav. Res., 317–335 (2017)
33. Zeffane, R.: Need for achievement, personality and entrerpreneurial potential: a study of young adults in the United Arab Emirates. J. Enterp. Cult. **21**(01), 75–105 (2013)
34. Benzing, C., Chu, H.M., Kara, O.: Entrepreneurs in Turkey: a factor analysis of motivations, success factors, and problems. J. Small Bus. Manage. **47**(1), 58–91 (2009)
35. Fisher, R., Maritz, A., Lobo, A.: Evaluating entrepreneurs' perception of success. Int. J. Entrep. Behav. Res. **20**(5), 478–492 (2014)
36. Hair, F.J., et al.: Multivariate Data Analysis: A Global Perspective. Prentice Hall, Pearson Education, Inc., New Jersey (2006)
37. Khan, R.U., et al.: Factors affecting women entrepreneurs' success: a study of small- and medium-sized enterprises in emerging market of Pakistan. J. Innov. Entrep. **10**(1), 11 (2021)
38. Anggadwita, G., et al.: Socio-cultural environments and emerging economy entrepreneurship. J. Entrep. Emerg. Econ. **9**(1), 85–96 (2017)
39. Yusoff, M.N.H., Zainol, F.A.: The public business support services: does it really works? Asian Soc. Sci. **10**(14) (2014)

Critical Success Factors of Outsourced Public Practice Accountants as Business Advisors Among SMEs in Malaysia

Emy Filzah Binti Zulkifli and Mohd Nor Hakimin Bin Yusoff

Abstract Public practice accountants (PPA) play an important role as business advisors in supporting SMEs to sustain on business performance. Furthermore, SMEs outsourcing for business advisors to lower costs, increase efficiency, obtain expertise and keep sustainable competitive advantages. The purpose of the present study is to analyze the critical success factors (CSF) of SMEs' intention to outsource PPA advisory services from the Decomposed Theory of Planned Behavior (DTPB) perspectives. A total of 78 responses from SMEs' owners were collected through emails and online questionnaires. In this study, the population was comprised of SMEs in the East Coast of Peninsular Malaysia. Meanwhile, the criteria to determine the sample were based on the definition of SME in Malaysia. The researchers positively figured out that awareness, perception of benefits, and growth intention acted as CSF that affected SMEs' decision to outsource PPA as business advisors. Meanwhile, the researchers found that the influence of peers, entrepreneurial association and professional fees cost are not crucial factors with moderate scores toward intention-behavior of outsourcing PPA advisory services. This present study is expected to provide a better understanding of SMEs' needs and contribute new insights for CSF of outsourcing advisory service from PPA. In addition, this study also provides guidelines to implement CSF to build SME owners' resilience and enable them to manage their business activities effectively.

Keywords Business advisors · Critical success factors · Decomposed theory of planned behavior · Public practice accountants · SMEs

E. F. B. Zulkifli (✉) · M. N. H. Bin Yusoff
Faculty of Entrepreneurship and Business, Universiti Malaysia Kelantan (UMK), Kota Bharu, Malaysia
e-mail: a20d0143f@siswa.umk.edu.my

M. N. H. Bin Yusoff
e-mail: hakimin@umk.edu.my

A. Hamdan et al. (eds.), *The Implementation of Smart Technologies for Business Success and Sustainability*, Studies in Systems, Decision and Control 216,
https://doi.org/10.1007/978-3-031-10212-7_10

1 Introduction

Small and Medium Enterprises (SMEs) are recognized as the main support driving the country's economy by representing 97.2% of business establishments. SMEs make a significant contribution to the growth of the Malaysian economy as a developing country through the job creation of 7.3 million workers, which is 48% of the total employment in Malaysia and more than one-third of the total Gross Domestic Product [1]. Based on statistics released by the [1], the contribution of SMEs to total Gross Domestic Product (GDP) increased to 38.9% in 2019 compared to 38.3% recorded in 2018. However, due to the COVID-19 health crisis in 2020, a decrease of 7.3–38.2% in 2020 and a depreciation of RM512.8 billion from RM553.5 billion in the previous year.

For that reason, all SMEs owners need to develop their business performance. As well known, SMEs are plagued by various weaknesses due to limited capital, constraints in obtaining skilled manpower and lack of a strategic and systematic management system. This is agreed by some researchers who point out that among the causes of SMEs not being able to operate effectively are due to internal resource constraints such as highly skilled human resources and limited capital [2]. This is consistent with [3] the result found that SMEs are tough to manage and control their internal system well in order to prepare financial statements. Besides difficulties in preparing financial statements, SMEs owners also lack sufficient resources on financial or non-financial performances. If this problem constantly occurs, it is not impossible that SMEs have potency continuance under risk and are forced to face major challenges in maintaining a competitive business environment.

In order to overcome this problem, some studies propose that SMEs need to outsource public practice accountant advisory services to implement the accounting function [4, 5]. Previous studies mention that SMEs need to outsource advisory services to change their traditional routine tasks to specific accounting functions [6]. Outsourced PPA advisory service is an opportunity for SMEs to sustain business growth and competitive advantage. In addition, outsourced PPA advisory services also address the problems of financial management, strategic management and business operations [7] as well as assist in the management and financial accounting matters [5]. For these reasons, outsourced PPA advisory service is the most popular among SMEs [8] and can be professional advisors to confirm that SMEs endure sustainability.

Therefore, based on the view above, the researchers are urged to investigate the CSF of outsourced PPA as business advisors among SMEs in Malaysia as not many studies have been carried out on this topic in Malaysia. The CSF is a method that aids business owners to specify and identify the most relevant and crucial factors determining an organization's success. This research is a continuity from a previous study in Indonesia [9], which used the Theory of Planned Behavior (TPB) as a basic theory to prove the influence of the competence of accountant external, the size of the business and the competitiveness of the industry toward SMEs' intention to use the services from external accountants. Considering the situation of SMEs in using advisory services from PPA, there are still fewer studies that use TBP as the basis of their research. This is supported in research by [10] regarding the role of external accountants in offering advisory services using the Agency Theory.

In short, this research serves the six new CSFs based on the Decomposed Theory of Planned Behavior (DTPB) to fill the research gap and would be useful to generate new findings and insights. Furthermore, this study will analyses the influences of awareness, perceived benefits, the influence of peers and entrepreneurial association, professional fees and growth intention towards intention to outsource advisory services offered by PPA among SME owners in the three East Coast of Peninsular states of Malaysia: Kelantan, Terengganu, and Pahang.

2 Literature Review

2.1 Public Practice Accountant as Business Advisors

Member in public practice is a term used to describe an accounting firm run by a chartered accountant in a sole proprietorship or partnership business [11]. Accountants engaged in public practice must have a valid practice certificate issued by the MIA. In addition, PPA are professional accountants who are trained and skilled to meet clients' diverse demands [4]. In public practice agencies, certified accountants provide professional services to clients for a fee. Their clients typically come from public sector organizations, large commercial companies, or private individuals. Chartered accountants can work in the nonprofit, industrial, commercial and public sectors in reporting, financial management, procurement, or treasury management roles.

PPA advisory services are defined as the provision of professional services in assisting or advising clients in aspects of business management [12]. The advisory services offered to SMEs are closely linked to business objectives. According to [10], business advisory services from PPA are an emerging service among SMEs. Previously, these services were only used by large firms but are now the main source of choice among SMEs [13]. Even advisory services have been widely used, and it is believed that PPA can be a professional business advisor among SMEs [14]. This is in line with the findings of a study by [15], which has proven that 86% of SMEs use business advisory services from PPA compared to other external sources.

2.2 *Factors of Outsourcing PPA as Business Advisors by SMEs*

In the Malaysian context, the findings from the [16] survey have identified twelve factors that influence the decision of SMEs in outsourcing PPA services (Table 1). All these factors can be classified as quality factors against PPA. In summary, the three factors that have the highest percentage influencing the decision of SMEs to purchase business advisory services from PPA are high knowledge and technical, responsive nature, and knowing the best needs for the business. In comparison, the branding factor is rated the lowest. This may be because SMEs are not aware of any branding or advertising activities on the services offered by PPA.

A summary of the survey conducted reveals many factors identified in helping SMEs' owners in deciding to use business advisory services from PPA. By leveraging the advantages and qualities of the outsourced advisory services, this clearly gives the impression that advisory services are very important in overcoming the various constraints faced by SMEs. However, these factors still cannot be explained conclusively and clearly. Therefore, it provides space for this study to identify the determining factors that influence the Behaviour of the use of business advisory services from PPA among SMEs in Malaysia.

3 Methodology

In developing the above research questions, this research designed a set of survey questionnaires to certify most of the issues concerning advisory services by PPA.

Table 1 Factors of outsourcing PPA as business advisors

Rankings	Factors	(%)
1	Strengths in technical science	75.8
2	Responsive	72.4
3	Business needs	69.6
4	Competitive price	67.7
5	Punctuality	65.0
6	Skills and competencies	63.2
7	Best quality service	61.3
8	Trustworthiness	54.8
9	Relationships	39.5
10	Capabilities in networking	33.1
11	Creative and innovative	28.8
12	Branding	11.2

Source [16]

The final form is developed for SMEs' owners consisting of the CSF, which affects their intentions-behavior to outsource PPA advisory services. The questionnaire was divided into various sections, starting with respondents' general information and then focusing on dimensions of CSFs: Attitudes (awareness, perceived benefits), Subjective Norms (influence of peers, entrepreneurial association), and Perceived Behavioral Control (professional fees and growth intention). The questionnaires were distributed individually to the respondents by emails and WhatsApp (self-administered), and the primary data was collected using online survey questionnaires. A total of 384 questionnaires were distributed, and primary data was successfully collected from 78 SMEs' owners located in the three East Coast of Peninsular states of Malaysia: Kelantan, Terengganu, and Pahang. The constructs in this research utilized the five-point Likert type scale (1 = strongly disagree, and 5 = strongly disagree). Thus, the analysis data had been performed using Statistical Package for the Social Sciences (SPSS) version 23 for descriptive analysis to identify the means and Cronbach Alpha of each instrument.

The results of empirical studies are divided into demographics of respondents, reliability and descriptive analysis. The questionnaire response rate was 20.8%, where out of 384 distributed questionnaires, 80 questionnaires were returned to the researchers. From 80 questionnaires, 78 (97.5%) were processable and the remaining 2 (2.5%) questionnaires could not be processed. Even though the number and rate of samples are low, it is still effective because, according to [17], the response rate should not be considered the only measure of the validity and quality of research in segregation from other important factors.

4 Results and Analysis

4.1 Respondent Profile

The data on respondent profiles in this research were classified based on gender, education level, type of industry, business activities, location of business and status, age of business, number of full-time employees, company financial sources, and annual sales. The details are summarized in Table 2.

Based on Table 2, from 78 collected research data, it could be seen that most respondents were female SME owners (52.6%), and 47.4% were male. Next, most respondents graduated with first degree (47.74%), and the lowest is Ph.D. level (1.3%). Furthermore, the profile of respondents by industry showed that 10 owners (12.8%) were from the manufacturing industry, 32 (41.0%) were service industry, and the remaining 36 (46.2%) were from other industries. Findings by business activities showed that 20.5% (16) from wholesale and retail, 12.8% (10) from food and beverage, 10.3% (8) from textile & clothing, 3.8% (3) from furniture/wood products, 2.6% (2) from transport and storage, 16.7% (13) from printing activities and 33.3% (26) from other business activities. The profile of respondents by business

Table 2 Respondent profile

Respondent classification	N	(%)
Gender		
Male	37	47.4
Female	41	52.6
Total	**78**	**100**
Education		
Certificate/PMR	0	0.0
SPM	7	9.0
STPM	7	9.0
Diploma	17	21.8
First degree	37	47.4
Master	9	11.5
PHD	1	1.3
Total	**78**	**100**
Type of industry		
Manufacturing	10	12.8
Service	36	41.0
Others	32	46.2
Total	**78**	**100**
Business activities		
Wholesalers and retail	16	20.5
Foods and drinks	10	12.8
Textiles and clothing	8	10.3
Furniture/wood products	3	3.8
Transportation and storage	2	2.6
Printing	13	16.7
Plastic and rubber	0	0
Others	26	33.3
Total	**78**	**100**
Location of business		
Kelantan	26	33.3
Terengganu	30	38.5
Pahang	22	28.2
Total	**78**	**100**
Location status		
City	47	60.3
Rural	31	39.7
Total	**78**	**100**

(continued)

Table 2 (continued)

Respondent classification	N	(%)
Age of business		
Less than 5 years	17	21.8
6–20 years	55	70.5
More than 20 years	6	7.7
Total	**78**	**100**
Number of full-time employees		
Less than 5 employees	41	52.6
5 to <75 employees	33	42.3
75 to <200 employees	2	2.6
Others	2	2.6
Total	**78**	**100**
Company financial sources		
Own financial	30	38.5
Family members	11	14.1
Government agencies	4	5.1
Bank loan	28	35.9
Others	5	6.4
Total	**78**	**100**
Annual sales		
<RM300,000	38	48.7
RM300,000–RM3,000,000	26	33.3
RM3,000,000–RM15,000,000	10	12.8
RM15,000,000–RM20,000,000	3	3.8
>RM20,000,000	1	1.3
Total	**78**	**100**

location showed a total of 26 (33.3%) from Kelantan, a total of 30 (38.5%) from Terengganu and the remaining 22 (28.2%) from Pahang. Moreover, according to location status, the profile of respondents showed that 47 (60.3%) were from urban areas, while another 31 (39.7%) were from rural areas.

According to business age, it reveals that the business age group that is most shown by the respondents is 6–20 years which is 70.5% (55). Followed by the business age group less than 5 years which is 21.8% (17), and over 20 years old, which is 7.7% (6). Furthermore, in terms of the number of full-time employees, it shows that most respondents have less than 5 employees with a total of 41 (52.6%). They were followed by 5 to <75 employees with a total of 33 (42.3%) and lastly the number of employees between 75 to $\leq$200 employees and exceeding 200 employees with a total of 2 (2.6%). The findings according to business financial sources found that 38.5% (30) were from their own finances, 14.1% (11) from family members, 5.1% (4) from sourced government agencies, 35.9% (28) by bank loans, and the rest 6.4% (5) came from other financial sources. Finally, related to the profile of respondents according to annual sales showed a total of 38 (48.7%) obtained annual sales <RM300,000, a total of 26 (33.3%) gained between RM300,000–RM3,000,000, 10 (12.8%) for annual sales between RM3,000,000–RM15,000,000, 3 (3.8%) attained annual sales between RM15,000,000–RM20,000,000 and 1 (1.3%) obtained annual sales >RM20,000,000.

4.2 Reliability Analysis (Cronbach Alpha)

The questionnaire was tested using the Cronbach Alpha method to see the value of its reliability. According to [18], the questionnaire instruments tested through the Cronbach Alpha method, and attained a value of at least 0.7 indicate that the questionnaire instrument is consistent in its reliability. All Cronbach Alpha values are as Table 3.

Table 3 shows that the range of reliability for the dimensions of attitudes, subjective norms, and perceived behavioral control towards outsourced business advisory services from PPA among SMEs' owners is 0.768–0.951. The overall Cronbach's alpha of attitudes, subjective norms, perceived behavioral control, and intentions toward the behavior of outsourced PPA advisory services were 0.851 (10 items), 0.949 (10 items), 0.816 (10 items), and 0.791 (5 items). Thus, Cronbach's Alpha table above shows that the instruments used have high reliability and are acceptable.

Table 3 Cronbach alpha

Factors	Alpha Cronbach	Number of items
Attitudes	**0.851**	**10**
Perceived benefits	0.785	5
Awareness	0.824	5
Subjective norms	**0.949**	**10**
Peers	0.900	5
Entrepreneurial association	0.951	5
Perceived behavioral control	**0.816**	**10**
Professional fees	0.768	5
Growth intention	0.850	5
Intentions towards behavior of outsourced PPA as business advisors	**0.791**	**5**

4.3 *Descriptive Analysis*

Attitudes Towards Intentions-Behavior of Outsourcing PPA as Business Advisors by SMEs

Table 4 shows that the dimension of attitude, perceived benefits (mean = 4.27, SD = 0.436) and awareness (mean = 4.11, SD = 0.510) are at a high level. Overall, the attitude score towards outsourcing business advisory services from PPA (mean = 4.19, SD = 0.413) among SMEs owners is high.

Subjective Norms Towards Intentions-Behavior of Outsourcing PPA as Business Advisors by SMEs

Table 5 shows that the dimensions of subjective norms from the factor of peers (mean = 3.61, SD = 0.695) and entrepreneurial associations (mean = 3.29, SD = 0.846) are at a moderate level. Generally, the subjective norm score on the outsourced business advisory services from PPA (mean = 3.45, SD = 0.726) among SMEs' owners is moderate.

Perceived Behavioral Control Towards Intentions-Behavior of Outsourcing PPA as Business Advisors by SMEs

Table 6 shows that the perceived Behavioural control dimension, professional fees (mean = 3.49, SD = 0.591), is moderate. While the growth orientation (mean = 4.20, SD = 0.549) is at a high level. Overall, the perceived Behavioural control score on outsourced business advisory services from PPA (mean = 3.85, SD = 0.466) among SMEs' owner is at a high level.

Intentions-Behavior of Outsourcing PPA as Business Advisors by SMEs

Table 7 shows the mean and SD scores for the item of Intention-Behaviour of outsourcing advisory services from PPA. Overall, it shows that the level of intention score (mean = 4.41, SP = 0.446) among SME owners in three states in the East Coast of Peninsular Malaysia is to high.

Table 4 Attitudes towards intentions-behaviour of outsourcing PPA

	Mean	SD	Level
Perceived benefits	4.27	0.436	High
Awareness	4.11	0.510	High
Total (attitudes)	4.19	0.413	High

Table 5 Subjective norms towards intentions-behavior of outsourcing PPA

	Mean	SD	Level
Peers	3.61	0.695	Moderate
Entrepreneurial association	3.29	0.846	Moderate
Total (subjective norms)	3.45	0.726	Moderate

Table 6 Perceived behavioral control towards intentions-behavior of outsourcing PPA

	Mean	SD	Level
Professional fees	3.49	0.591	Moderate
Growth intention	4.20	0.549	High
Total (perceived behavioural control)	3.85	0.466	High

Table 7 Intentions-behaviour of outsourcing PPA

	Mean	SD	Level
Intention of outsourcing PPA	4.41	0.446	High
Total	4.41	0.446	High

5 Discussion

This research extends to fill the theoretical gap by using the DTPB model, which has never been used in any previous study to identify antecedent's variables of outsourcing public practice accountant as a business advisor among SMEs. DTPB framework has originally contributed new six CSF, namely awareness, perception of benefits, entrepreneurial peers and association, professional fees and growth intention that influences outsourcing advice services provided by PPA as an added value and the most useful resource. This study investigates whether CSF derives SMEs' intention-behavior from outsourcing advisory services from PPA. Empirical study approves that SMEs having a greater need to outsource business advice from PPA show greater performance.

Based on the results above, the most CSF of outsourced PPA advisory services among SMEs' owners are awareness, perception of benefits and growth orientation where all these factors are grounded by Models of DTPB. These results are strongly supported by a previous study by [9], who conducted research on proven attitude toward Behavior, subjective norms, and perceived behavioral control among SMEs owners in the City of Bengkulu, Indonesia. Respondents agreed that the dimension of attitudes (awareness and perception of benefits) and perceived behavioral control (growth intention) are the most crucial factors among SMEs in Malaysia who are intent to outsource advisory services by PPA. Furthermore, SMEs owners have obtained the benefits and consciousness that outsourcing PPA advisory services is a good decision to help them perform their accounting activities for the survival of their business activities. The finding has significant implications in help them to solve internal and external issues such as strategic market planning, communication, competition and sustainable competitive advantage.

Moreover, this study provides empirical evidence consistent with the Theory of Planned Behavior (TPB) that remarks the attitude toward a Behavior is one of the significant factors of the intention to perform a Behavior [19]. In line with that, these results proved that the attitude toward Behavior and perceived Behavioral control have an effect on the awareness, perception of benefit, and growth intention when

SMEs' owners outsourced the advisory services from PPA and all these findings were supported by recent studies, namely [20–22].

6 Conclusion

Ultimately, these research findings confirms the implications for SMEs to achieve this critical success factors (awareness, perception of benefits and growth orientation). They would have high responsiveness and responsibilities to perceive a performance benefit from outsourcing PPA business advisory services and proved this CSF was valid for SMEs industries. It is because all the CSF have become a significant element and give further support and incentive to the accounting profession to continue promoting the expertise as business advisors. Implementing advisory services from PPA can assist the quality of management and add value to SMEs with their business capability. The results of this study should be considered as a potential limitation. First, because the findings do not examine the relationship between SMEs and PPA. Next, this study cannot be show that SMEs outsourced business advice have impact on actual business performance.

However, explaining the Intention-Behavior outsourced PPA as a business advisor requires further research; the results may be taken as input and guidelines to discover new insights on business advisory support. Agenda for future research, the authors need to fill gaps by looking at the relationship between theory, practices, barriers and the accounting profession. In short, the relationship between CSF of outsourcing PPA advisory services might measure on actual SMEs performance and the results should be compared with other countries. Based on this result, it could be seen that the decision to outsource PPA advisory services was a significant factor in improving SMEs' performance. Earlier research also suggested that PPA as an external accountant became their first choice for SMEs to increase their business performance in a long term [5]. Subsequently, PPA requires extensive encouragement by extending their advisory services to support for the business's strategy and activities to attain an economic scale, which in turn enhanced SMEs' accomplishment.

Acknowledgements The authors gratefully acknowledge to the Ministry of Education Malaysia (MOE) for the scholarships and funding of my master's degree.

References

1. Department of Statistics Malaysia: Pertumbuhan ekonomi dan sumbangan peratusan mengikut negeri, 2019. http://dspace.unimap.edu.my/xmlui/handle/123456789/69039 (2020). Last accessed 27 Feb 2022
2. De Bruyckere, S., Verplancke, F., Everaert, P., Sarens, G., Coppens, C.: The importance of mutual understanding between external accountants and owner-managers of SMEs. Aust. Account. Rev. **30**(1), 4–21 (2020). https://doi.org/10.1111/auar.12251

3. Tazilah, M.D.A.K., Majid, M., Suffari, N.F.: Effects of outsourcing internal audit functions among small & medium enterprises. Int. J. Bus. Technol. Manag. **1**(1), 28–34 (2019)
4. Carey, P.J.: External accountants' business advice and SME performance. Pac. Account. Rev. **27**(2), 166–188 (2015). https://doi.org/10.1108/PAR-04-2013-0020
5. Ganesan, Y., Haron, H., Amran, A., Pitchay, A.A.: It is the external accountant matter in business performance of small and medium enterprises? the role of non-audit services. Int. Acad. J. Account. Financ. Manag. **5**(4), 36–48 (2018)
6. Cullinan, C.P., Zheng, X.: Accounting outsourcing and audit lag. Manag. Audit. J. **32**(3), 276–294 (2017). https://doi.org/10.1108/MAJ-03-2016-1349
7. Kamyabi, Y., Devi, S.: The impact of accounting outsourcing on Iranian SME performance: transaction cost economics and resource-based perspectives. World Appl. Sci. J. **15**(2), 244–252 (2011)
8. Devi, S.S., Samujh, H.R.: Accountants as Providers of Support and Advice to SMEs in Malaysia. ACCA Global Homepage. https://www.accaglobal.com/content/dam/acca/global/PDF-technical/small-business/rr-118-001.pdf (2010). Last accessed 27 Feb 2022
9. Irwansyah, I., Hatta, M., Al Masriki, G.F.: The business owners SMEs sector of intention to use the services by external accountants. J. Audit. Financ. Forensic Account. **6**(2), 57–70 (2019). https://doi.org/10.21107/jaffa.v6i2.4934
10. Carey, P.J.: The provision of business advice to SMEs by external accountants. Manag. Audit. J. **31**(3), 290–313 (2016). https://doi.org/10.1108/MAJ-12-2014-1131
11. Malaysian Institute of Accountants: #Nation building: driving sustainability through trust and collaboration—integrated report 2020. https://www.mia.org.my/v2/downloads/resources/publications/integrated_annual_report/2020/MIA_Integrated_Annual_Report_2020.pdf (2020). Last accessed 27 Feb 2022
12. Oosthuizen, A.: Predicting the benefits that small business owners obtain from their accounting practitioners. Doctoral thesis. University of Pretoria. University of Pretoria Institutional Repository. http://hdl.handle.net/2263/67864 (2018)
13. Ibrahimova-Guluzada, A.: Outsourcing of financial accounting function research. Master's thesis. Kaunas University of Technology (2017)
14. Blackburn, R., Carey, P., Tanewski, G.: Business advice by accountants to SMEs: relationships and trust. Qual. Res. Account. Manag. **15**(3), 358–384 (2018). https://doi.org/10.1108/QRAM-04-2017-0022
15. Banham, H., He, Y.: Accounting professionals and small and medium enterprises (SMEs). J. Bus. Econ. Res. **12**(3), 1–6 (2014)
16. Malaysian Institute of Accountants: A report on current usage pattern of business and professional services among SMEs in the manufacturing and distributive trade sectors. https://www.mia.org.my/v2/downloads/ppt/research_publications/2009/04/15/MIA_SME_Survey_Full_Report.pdf (2008). Last accessed 27 Feb 2022
17. Mellahi, K., Harris, L.C.: Response rate in business and management research: an overview of current practice and suggestions for future directions. Br. J. Manag. **27**(2), 426–437 (2016). https://doi.org/10.1111/1467-8551.12154
18. Sekaran, U., Bougie, R.: Research Methods for Business: A Skill Building Approach, 7th edn. Wiley, New York (2016)
19. Ajzen, I.: The theory of planned behaviour. Organ. Behav. Hum. Decis. Process. **50**(2), 179–211 (1991). https://doi.org/10.1016/0749-5978(91)90020-T
20. Chang, Y.C., Enkhjargal, U., Huang, C.I., Lin, W.L., Ho, C.M.: Factors affecting the Internet banking adoption. J. Ekon. Malays. **54**(3), 117–131 (2021). https://doi.org/10.17576/JEM-2020-5403-09
21. Shafiai, M.H.M., Ali, S.A., Yaacob, S.E.: Kesan sikap, faktor halangan dan manfaat tanggapan pekebun kecil sawit persendirian kepada niat menggunakan mikro kredit Islam. Islamiyyat Int. J. Islam. Stud. **42**(1), 3–14 (2020)
22. Nasir, S.Z., Wahid, H., Noor, M.A.M.: Pembayaran zakat pendapatan melebihi kadar kewajipan zakat: Kajian dalam kalangan kakitangan kerajaan di Malaysia. Islamiyyat Int. J. Islam. Stud. **43**(1), 65–73 (2017)

The Impact of Economic Climate and R&D Transfer on Entrepreneurship Capacity During COVID-19 Pandemic: A Qualitative Study

Abrar AL Alawi, Abdullah Al Shukaili, Fadi Abdelfattah, and Khalid Abed Dahleez

Abstract Entrepreneurship capability is a relatively new topic that has piqued the interest of academics, practitioners, and policymakers because of its critical role in economic development in today's dynamic environment. As a result, this research will examine the impact of the economic climate and research and development (R&D) transfers on the Sultanate of Oman's entrepreneurial capacity. The investigation was motivated by the dearth of available literature and empirical evidence in Oman. The current study analyzed data from the Oman Global Entrepreneurship Monitor (GEM) 2020, compiled by national experts from various industries. This research followed a qualitative approach, conducting in-depth interviews with thirty-six professionals from nine industries who collectively reflect Oman's entrepreneurial ecosystem. The data was then analyzed and descriptively interpreted. The most significant findings indicate that factors affecting the economic climate and the function of R&D transfer affected entrepreneurial capacity during the COVID-19 pandemic. This will enable the current study to provide evidence to aid policymakers in developing nations, notably Oman, in strengthening entrepreneurial capacity and identify areas for the empirical investigation to develop efficient ecosystems to support entrepreneurs. This study recommends that government and non-government stakeholders develop strategies for a post-COVID-19 entrepreneurial revival. This is likely one of the first qualitative assessments of the COVID-19 pandemic's likely impact on entrepreneurship. As a result, the practical implications and solutions are examined in the context

A. AL Alawi (✉) · A. Al Shukaili
Entrepreneurship Center, University of Nizwa, Birkat Al Mawz, Oman
e-mail: abraralalawi@unizwa.edu.om

A. Al Shukaili
e-mail: a.alshukaili@unizwa.edu.om

K. A. Dahleez
College of Business Administration, A'Sharqiyah University, Ibra, Oman
e-mail: khalid.dahleez@asu.edu.om

F. Abdelfattah
Management Department, Modern College of Business and Science, Muscat, Oman
e-mail: fadi_fattah@yahoo.com

A. Hamdan et al. (eds.), *The Implementation of Smart Technologies for Business Success and Sustainability*, Studies in Systems, Decision and Control 216,
https://doi.org/10.1007/978-3-031-10212-7_11

of modern economic, social, and environmental challenges. Additionally, it recommends relevant research areas and makes recommendations on how to empower entrepreneurs to overcome the aforementioned obstacles.

Keywords Economic climate · Entrepreneurship · COVID-19 · R&D · Oman

1 Introduction

Over the world, economic prosperity and stability pour in the industrialized Small and Medium Enterprises (SMEs) [1, 2]. As the industrial structure characterized by diversification facilitate a better environment for innovation that supports the development of SMEs, constituting social force and guideline for future development plans [3–7] this growth will dramatically result in reduced poverty and enhanced public well-being. Despite the numerous natural, human, and physical resources worldwide, employment opportunities prevailed limited [8–11]. However, the entrepreneurship sector constitutes multiple generators for innovative ideas, employment opportunities, socio-economic improvement, wealth, and overall prosperity for developed and under-developed countries [12–14]. Nevertheless, the adequate capacity for entrepreneurship remains lacking, consistent with the inadequate training and learning programs in entrepreneurship management [15–17]. Moreover, startup entrepreneurs require more extraordinary educational qualifications and economic climates than an established business. As such, multiple researchers claimed that research and development (R&D) in the industrialized field of entrepreneurship drive entrepreneurial activity [2, 18–20].

Since unemployment increasingly concerns the public and decision-makers, raising the entrepreneurship capacity is imperative [12, 21]. Existing studies demonstrate a significant association between the prevailing economic climate and the capacity for entrepreneurship [20, 22–24]. Furthermore, changing and unpredictable economic climate during the wide outbreak of COVID-19 represents a challenge influencing the extent of entrepreneurship capacity [25, 26–29]. Some existing literature argued that R&D transfer programs boost entrepreneurial capacity fundamentally. This is explained by stakeholders claiming the broad need of R&D in most of the business activities and processes, considering it as the means to overcome the business issues and challenges [5, 7, 9, 11, 18]. However, Global Entrepreneurship Monitor (GEM) discussed the main challenges and opportunities confronting businesses and entrepreneurs in Oman. Based on the Oman GEM report, the COVID-19 pandemic drives the development of business activities towards advanced technological dependence [30]. Governmental policies and programs in Oman outline supportive strategies represented in Oman vision 2040 in order to generate innovative business activities and increases employment opportunities [31–34].

Based on the above debate, we aim to explore the national expert's perception of the influence of economic climate and R&D transfer on the extent of capacity for entrepreneurship during the pandemic. This study has a prominent role in adjusting the transformation policies and strategies towards enhancing entrepreneurship capacity to support the economy and reduce the unemployment rate. Further, prior research studies in developing countries ignored determining the factors that result in enhanced entrepreneurship capacity, which the current paper strives to fill this knowledge gap through a qualitative research approach.

2 Literature Review

2.1 Economic Climate

The dynamic economic climate provides an attractive environment for the business in the marketplace [26, 27, 29]. Economic climate differs significantly along with the geographical and political conditions. However, the continuous social, environmental, and technological changes respond to the economic conditions, affecting both organizations and businesses [6, 35, 36]. More precisely, focusing the effort on technological environment improvement prompts the economic climate influence, especially on the society. As a result, this formulates efficient predictors of the national economic growth and related productivity [12, 22, 37, 38]. From another point of view, the prevailing economic climate postulates diverse political changes to support business activities and maintain competitive economic circumstances [6, 14, 23, 24, 26]. In order to strengthen the economic business activities, growth suggests a practical impact on the economic climate, which enhances the research and development transfer [18, 20, 24]. COVID-19 pandemic generated several challenges and opportunities under the changing economic climate. Hence, several entrepreneurs utilize the emerging opportunities to expand their business capabilities [8, 12].

2.2 R&D Transfer and Economic Climate

The efficient development of any country worldwide is attached to improved research and technology [5, 7, 11]. Several researchers argue that the current economic climate requires an efficient strategic plan to ascertain the transfer of research and development (R&D). Considering that R&D provides realistic and updated solutions for the emerged issues [9, 10, 24, 39].

Previous studies confirm a direct link between economic climate and R&D transfer. Moreover, development in the marketplace is supported and enabled by the characteristics of the economic climate, including the stability and taxes system [5, 7, 11]. Economic prosperity sought as reflection emphasized the human capital, R&D transfer, and the spread of entrepreneurial culture [1, 2, 15, 40]. Meanwhile, the spread of COVID-19 pandemic suggests improvements in the local and international markets, increasingly intense competition, and the prevailed socio-economic climate impose changes keen to offer development plans based on the research and qualitative alteration purposes [6, 8, 19, 35, 38, 41].

2.3 Economic Climate and Capacity for Entrepreneurship

The global economic improvement and the increasingly competing socio-economic climate force entrepreneurs to explore new opportunities that advance entrepreneurial ventures' performance [1, 5, 16]. Interestingly, the capacity for entrepreneurship sought as the impact of the economic climate constituted from multiple aspects, including government regulations and policies that organize facilitate marketplace environment activities [10, 22, 28, 35, 38, 42]. The newly emerged entrepreneurs in the dynamic economy require enhanced capabilities and training that constitute dramatic support for the induced entrepreneurial capacity [16, 17, 39]. Coelho et al. [16] debate the entrepreneurial capacity in educational capabilities that substantially play a significant role in qualifying the students with the skills and knowledge that advances their ability to create a novel entrepreneurial venture. Meanwhile, COVID-19 spread forced multiple economic conditions such as limited resources, budget, and investment affected by the lockdown policies, and precautionary measures impacted the overall capacity for entrepreneurship [11, 15, 17, 20, 28, 43].

2.4 R&D Transfer and Capacity for Entrepreneurship

Entrepreneurial capacity is considered one of the drivers for recognizing an emerging entrepreneurial opportunity that boots their potentiality towards starting a new entrepreneurial venture [11, 15, 17, 43]. The higher capacity for entrepreneurship, the more individuals are involved in initiating new business or benefits from their previous entrepreneurial experiences to expand their business activities [5, 39, 43, 44]. COVID-19 pandemic spread imposed new policies and regulations that affected the entrepreneur's management of their business and individual's entrance to the entrepreneurship market. In this respect, R&D transfer plays a significant role in directing the entrepreneurial efforts towards expanded capacity for entrepreneurship [8, 24, 28, 43, 45]. Investment in R&D programs formulates a competitive domain, mainly to recover the economy. Therefore, generating innovative business ideas of new products, processes, or technology will result in strengthening the economy and widen the competitive capacity for entrepreneurship in the face of various local and global challenges [10, 15, 18, 20, 39, 45].

3 Method

3.1 Approach

Interviewing techniques are well-known and often used in qualitative research [46]. A personal interview is conducted between the interviewer and the study subjects (interviewee). Both spend a few minutes debating a particular subject [47]. This study used semi-structured interviews that were generalized to a broader group based on their concerns and relevance [48]. In-depth interviews aid in developing survey instruments, selecting survey items, and the content validity assessment of survey items [49]. Because it does not need any special abilities, contact between researcher and participant enables the researcher to comprehend the participant's perspective, thoughts, and experiences on a given subject [45].

The approach of in-depth interviewing presents various difficulties. To begin with, ambiguity in language—when an interviewer asks reviewees questions, there is a possibility that one will not receive a comprehensive response owing to ambiguity and a lack of comprehension about the interviewers' unique issue [50, 51]. Second, interviews can go wrong; a probability of miscommunication between interviewer and interviewee might result in the interview being canceled [51, 52]. Third, there is a risk that the interviewer may encounter responsible participants rather than non-elite perspectives, which may result in prejudice and a lack of grasp of specific themes [51, 53, 54, 55].

There are numerous strategies for overcoming these obstacles. Due to the uncertainty of the language, interview questions were asked in Arabic using a structured format to avoid various interpretations. Thus, it aided in collecting precise and pertinent responses [51, 56, 57]. Controlling interviews required the interviewer to be "*capable of responding by deviating from the topic, rephrasing the question, or, in certain situations, halting or terminating the interview*" [58]. To avoid elitist bias, relevant participants were chosen from numerous industries that shaped entrepreneurs who owned or managed small and medium-sized businesses during the COVID-19 epidemic.

3.2 Data Collection

The study analyzed qualitative data from the 2020 Global Entrepreneurship Monitor (GEM). Individual in-depth interviews were undertaken to ascertain the effect of the entrepreneurial capacity predictor. Thirty-six national experts from several industries in Oman were interviewed in a series of semi-structured interviews. The interview's primary objective is to elicit information on the factors that influence entrepreneurial potential. To preserve privacy and confidentiality, individual interviews were performed in private locations. The interviews began with a brief introduction of the researcher and a description of the study's goal. Three elements compose the interview guide: an introduction, a consent form, and open-ended questions.

The participants consented to volunteer; all interviews were recorded by writing with their approval and kept anonymous by assigning each interview a code. The interview questions centered on the COVID-19 epidemic and its profound effect on business. The researcher subsequently compiled and transcribed interview notes. Before beginning the transcribing operations, a summary was created.

4 Findings and Discussion

The interviewee of this study were national experts consisting of 75% male and 25% female. Most of the interviewed experts aged 40–45 years old hold a Master's or Ph.D. degree. Moreover, the experts were selected from nine different industries and organizational contexts, including finance, governmental institutions, and industrial. The characteristics of the interviewee are presented in Table 1.

Since this research follows the qualitative research method, descriptive analysis was performed. This study discusses the effect of the economic climate and R&D transfer on entrepreneurial capacity.

Table 1 National expert's characteristics

Demographic		No	%
Gender	Male	27	75.0
	Female	9	25.0
Age	25–30	5	13.9
	31–35	6	16.7
	36–40	9	25.1
	40–45	12	33.3
	46–50	4	11.2
Education	Secondary	2	5.6
	Diploma/bachelor	16	44.4
	Master/Ph.D.	18	50.0
Organization type	Finance	4	11.1
	Governmental policies	4	11.1
	Governmental programs	4	11.1
	Education and training	4	11.1
	Research and development	4	11.1
	Commercial infrastructure	4	11.1
	Industrial/business	4	11.1
	Physical infrastructure	4	11.1
	Cultural/social	4	11.1

Referring to the **economic climate Theme**, the interviewees express the following statements:

- *"In Oman, the government has adopted effective measures for new and growing firms to adjust to the economic reality caused by the COVID-19 pandemic."*
- *"In Oman, the government has adopted effective measures to avoid the massive loss of new and growing firms due to the COVID-19 pandemic."*
- *"In Oman, the government has acted to protect workers and customers of new and growing firms from COVID-19 during the pandemic."*
- *"As a result of the COVID-19 pandemic in Oman, the government has substantially increased the digital or online delivery of regulations for new and growing firms."*

It can conclude that most interviewees assured that there are ineffective measures pursued by the government that seeks to adapt to the fact that economic conditions affected the new and growing firms by the current propagation of COVID-19. On the other hand, the interviewees state that there are effective parameters the Omani government utilized to shrink the considerable loss experienced by the new and growing firms due to the pandemic crisis. However, in the Oman context, the government responded to the conquered economic climate amid COVID-19 spread by taking several actions to protect the customers and employees of new and growing firms. Most of the interviewees declared that "*the regulations stated by the government to the new and growing firms are declared and conveyed through digital means because of the current pandemic crisis.*" In the same vein, the obtained results are explained by [20], who concluded that the challenges and opportunities of running a business venture are mainly driven by the economic climate, especially in terms of long-term investment and budget capacity.

Referring to the **R&D transfer Theme**, the interviewees express the following statements:

- *"In Oman, new technology science and other knowledge are efficiently transferred from universities and public research centers to new and growing firms."*
- *"In Oman, new and growing firms have just as much access to new research and technology as large, established firms."*
- *"In Oman, new and growing firms can afford the latest technology."*
- *"In Oman, there are adequate government subsidies for new and growing firms to acquire new technology."*
- *"In Oman, the science and technology base efficiently supports the creation of world-class new technology-based ventures in at least one area."*
- *"In Oman, there is good support available for engineers and scientists to have their ideas commercialized through new and growing firms."*

The majority of the interviewees express a lack of efficient transfer of knowledge, new technology, and science from the universities and centers of public research to the firms that are either new or growing. Most experts state that the access to the new technology and research for the growing companies is not similar to that of the established large companies in Oman. Simultaneously, the latest emerged technologies cannot be afforded by the growing and new firms as elucidated by the national experts. However, most respondents stated that governmental subsidies in Oman are inadequately available for the new and growing firms to acquire the updated new technology. As per the national expert's illustration, efficient support was provided from the base of technology and science in Oman to develop new business ventures based on the newest technology in at least one business aspect during the COVID-19 outbreak.

Further, about half of interviewees state that the scientists and engineers' new ideas in Oman acquire good support through ideas commercialization by firms that are new or growing, while on the other side, the remaining thinks the opposite, especially during the spread of COVID-19. These findings suggest that R&D programs shape and strengthen entrepreneurship capacity in line with previous studies [2, 16, 38, 39]. The following are the respondent's statements:

Referring to the **Entrepreneurship Capacity Theme**, the interviewees express the following statements: Moreover, the interviewees' experts illustrate that the spread of the COVID-19 pandemic affected entrepreneurship capacity. The majority of the interviewees interpret that working from home due to the forced circumstances of the COVID-19 prevalence was inadequately promoted, especially in new and growing firms in Oman [25]. The national experts ascertain that many new and growing firms recognize abundant emerging opportunities due to the spread of the COVID-19 pandemic. Meanwhile, most of the respondents confirmed that the pandemic crisis had not leveraged the level of new, growing, and established firm cooperation between firms in Oman. The COVID-19 spread impact on enhanced new, growing, and established Omani firm's global collaboration significantly in the context of proposals, social activities, and challenges. Previous research work is consistent with the result stating that running a new business venture requires increasing capacity for entrepreneurship that encourages both new and existing entrepreneurs to continue and expand their business activities [2, 7, 11, 15, 43].

5 Conclusion and Recommendations

This study aimed to determine the effect of the economic climate and R&D transfers on the Sultanate of Oman's entrepreneurial ability during the COVID-19 epidemic. The study's findings suggested that entrepreneurial activity has been negatively impacted in Oman due to COVID-19. The most noteworthy findings suggest that during the COVID-19 pandemic, variables impacting the economic climate and the role of R&D transfer influenced entrepreneurial potential. There is a need to assist entrepreneurs in developing their resilience in the face of the COVID-19 issue by digitalizing their business models. Additionally, this study discovered a gap in emergency planning and capacity-building trends for SMEs, which resulted in considerable harm to SMEs during the pandemic caused by the coronavirus. Thus, our findings corroborate those of Shafi et al. [59], who discovered that over 83 percent of SMEs were unprepared for the COVID-19 epidemic and lacked a plan or strategy for dealing with an unanticipated circumstance, the Coronavirus pandemic. SMEs lacked the expertise and capability to adjust their business strategies to a climate of social isolation and lockdown. This pandemic may provide an opportunity to rethink the objectives of SMEs by including capacity building for unforeseen events such as the COVID-19 pandemic. The research used a qualitative technique, with in-depth interviews conducted with thirty-six individuals from nine industries representing Oman's entrepreneurial ecosystem of entrepreneurs and managers of small and medium-sized businesses. The study's conclusions cannot be extrapolated to other geographic or corporate contexts. It would be preferable if this inquiry explored other locations and issues to be more comprehensive.

Due to the COVID-19 epidemic, developing inclusive and long-term economic recovery solutions is necessary. According to Oman's Vision 2040 [3], diversifying the country's economy away from oil and focusing on a diverse range of objectives would help alleviate the burden on Oman's economy and counteract the negative economic effect. On the other hand, the Oman government's efforts and swift actions have aided in the virus's containment and assistance for small and medium-sized businesses [60]. Several government initiatives are aimed at building an inclusive and sustainable economic strategy. As the current pandemic has highlighted the fragility of our economic, health, and disaster preparation systems, it is vital to strengthen social protection because it acts as an automatic stabilizer, protecting the most vulnerable persons.

Like other developing countries, Oman must prioritize prevention overtreatment, and the time has come to foresee and prepare for future shocks. SME owners should utilize the COVID-19 experience to establish a proactive plan for managing their operations. Additionally, SMEs should adhere to the country's recovery strategy to ensure the seamless functioning of their businesses.

6 Limitations and Suggestions for Future Research

Due to the government-imposed shutdown to prevent the spread of coronavirus during COVID-19, recruitment and availability of research volunteers were hampered. Additionally, there were limitations due to a paucity of scientific research undertaken in an Omani environment during data collection for this study since the coronavirus situation was still developing. Due to the scarcity of data on entrepreneurial aptitude and the fact that not all of it was publicly accessible, designing and implementing research projects was hard. The inquiry was prompted by Oman's paucity of available literature and empirical evidence. However, empirical research would have an advantage in quantifying and generalizing findings related to these issues. An empirical study examining and quantifying the elements affecting the supply chain, sales, and income of micro, small, and medium-sized firms might be a future research option for determining the impact of the COVID-19 pandemic in developed and developing nations.

References

1. Dębicka, A., Olejniczak, K., Skąpska, J.: Enterprises' perception and practice of humane entrepreneurship. J. Small Bus. Enterp. Dev. **29**(1), 127–146 (2022). https://doi.org/10.1108/JSBED-01-2021-0028
2. Oyolola, F., Otonne, A.: Entrepreneurship, capacity development, and youth employment generation: a study of selected Sub-Saharan Africa Countries. Munich Personal RePEc Archive 99156, 4 (2020). https://ideas.repec.org/p/pra/mprapa/99156.html
3. Al Alawi, A., Fattah, F.A., Dulal, M.: Financial performance analysis of firms: a focus on oil and gas industry sustainable practices in Oman. In: Artificial intelligence for sustainable finance and sustainable technology (AMAM Al-Sartawi, pp. 83–96). Springer, Cham (2022). https://doi.org/10.1007/978-3-030-93464-4_9
4. Fritsch, M.: Related Variety, Unrelated Variety, and Regional Growth: The Role of Absorptive Capacity and Entrepreneurship (2016)
5. Jiménez-Barrionuevo, M.A.M., Molina, L.M., García-Morales, V.J.: Combined influence of absorptive capacity and corporate entrepreneurship on performance. Sustain. (Switz.), **11**(11) (2019). https://doi.org/10.3390/su11113034
6. Shilomboleni, H.: Political economy challenges for climate-smart agriculture in Africa. Agric. Hum. Values **37**(4), 1195–1206 (2020). https://doi.org/10.1007/s10460-020-10126-5
7. Towers, N., Santoso, A.S., Sulkowski, N., Jameson, J.: Entrepreneurial capacity-building in HEIs for embedding entrepreneurship and enterprise creation—a tripartite approach. Int. J. Retail Distrib. Manag. **48**(8), 881–899 (2020). https://doi.org/10.1108/IJRDM-06-2019-0185
8. Juergensen, J., Guimón, J., Narula, R.: European SMEs amidst the COVID-19 crisis: assessing impact and policy responses. J. Ind. Bus. Econ. **47**(3), 499–510 (2020). https://doi.org/10.1007/s40812-020-00169-4
9. Madzikanda, B., Li, C., Dabuo, F.T.: What determines the geography of entrepreneurship? A comparative study between Sub-Saharan Africa and South-East Asia. J. Entrepreneurship Innov. Emerg. Econ. **7**(2), 246–262 (2021). https://doi.org/10.1177/23939575211023629
10. Orobia, L.A., Tusiime, I., Mwesigwa, R., Ssekiziyivu, B.: Entrepreneurial framework conditions and business sustainability among the youth and women entrepreneurs. Asia Pac. J. Innov. Entrepreneurship **14**(1), 60–75 (2020). https://doi.org/10.1108/apjie-07-2019-0059
11. Rehman, N., Razaq, S., Farooq, A., Zohaib, N.M., Nazri, M.: Information technology and firm performance: mediation role of absorptive capacity and corporate entrepreneurship in manufacturing SMEs. Technol. Anal. Strateg. Manag. **32**(9), 1049–1065 (2020). https://doi.org/10.1080/09537325.2020.1740192
12. Goschin, Z., Antonia, M., Tigau, H.: Entrepreneurship recovery in Romania after the great recession. A dynamic spatial panel approach. Sustainability (Switzerland), **13**(19), 1–12 (2021). https://doi.org/10.3390/su131910702
13. Kangogo, D., Dentoni, D., Bijman, J.: Determinants of farm resilience to climate change: the role of farmer entrepreneurship and value chain collaborations. Sustain. (Switz.) **12**(3) . https://doi.org/10.3390/su12030868
14. To, C.K.M., Chau, K.P., Kan, C.W.: The logic of innovative value proposition: a schema for characterizing and predicting business model evolution. J. Bus. Res. **112**(October), 502–520 (2020). https://doi.org/10.1016/j.jbusres.2019.10.023
15. Akinbola, O.A., Sanni, S.A., Akinbola, O.S.: Appraisal of entrepreneurship capacity programs and internationalization of small and medium enterprises in Nigeria. Acta Univ Danubius Oeconomica **14**(6), 72–89 (2018). http://ezproxy.umgc.edu/login?https://search.ebscohost.com/login.aspx?direct=true&db=bth&AN=133379990&site=eds-live&scope=site
16. Coelho, F.J.M., Marques, C., Loureiro, A., Ratten, V.: Evaluation of the impact of an entrepreneurship training program in Recife, Brazil. J. Entrepreneurship Emerg. Econ. **10**(3), 472–488 (2018). https://doi.org/10.1108/JEEE-10-2017-0077
17. Shafique, I., Kalyar, M.N.: Linking transformational leadership, absorptive capacity, and corporate entrepreneurship. Adm. Sci. **8**(2) (2018). https://doi.org/10.3390/admsci8020009

18. Herrington, M., Coduras, A.: The national entrepreneurship framework conditions in sub-Saharan Africa: a comparative study of GEM data/National Expert Surveys for South Africa, Angola, Mozambique, and Madagascar. J. Glob. Entrep. Res. **9**(1), 1–24 (2019). https://doi.org/10.1186/s40497-019-0183-1
19. Kah, S., O'Brien, S., Kok, S., Gallagher, E.: Entrepreneurial motivations, opportunities, and challenges: an international perspective. J. Afr. Bus. **23**(2), 1–20 (2020). https://doi.org/10.1080/15228916.2020.1838835
20. Yoo, J., Kim, J.: The effects of entrepreneurial orientation and environmental uncertainty on Korean Technology Firms' R & D investment. J. Open Innov. Technol. Market Complex. **5**(2) (2019). https://doi.org/10.3390/joitmc5020029
21. Bakry, D., Khalifa, R., Dabab, M.: The effectiveness of entrepreneurship programs to reduce unemployment in developing countries: the case of Saudi Arabia (2019)
22. Antonites, A., Schoeman, W.J., van Deventer, W.F.J.: A critical analysis of social innovation: a qualitative exploration of a religious organisation. HTS Teologiese Stud./Theological Stud. **75**(4), 1–12 (2019). https://doi.org/10.4102/hts.v75i4.5282
23. Beynon, M.J.: Cronfa-Swansea University Open Access Repository Entrepreneurial climate and self-perceptions about entrepreneurship: a country comparison using fsQCA with dual outcomes Cardiff Business School, Cardiff University, Cardiff (2018)
24. Zinecker, M., Skalicka, M., Balcerzak, A.P., Pietrzak, M.B.: (2021). Identifying the impact of external environment on business angel activity. Econ. Res.-Ekonomska Istrazivanja 1–23. https://doi.org/10.1080/1331677X.2021.1888140
25. Abdel Fattah, A.M., Al Halbusi, H., Dahleez, K.A., Saleh, A.S., Al Wahibi, G.: An investigation of components deriving entrepreneurial activities in Duqm City : a special economic zone in Oman. Int. J. Entrepreneurship **25**(4), 1–14 (2021)
26. Ferreira, J.J.M., Fernandes, C., Ratten, V.: The effects of technology transfers and institutional factors on economic growth: evidence from Europe and Oceania. J. Technol. Transfer **44**(5), 1505–1528 (2019). https://doi.org/10.1007/s10961-019-09730-3
27. Liu, B., Xia, L., Han, L., Shen, J., Dan, E., Sun, H.: Research on the construction of Hangzhou power economic climate index model. In: Proceedings of 2021 IEEE 4th international electrical and energy conference, CIEEC 2021 (2021). https://doi.org/10.1109/CIEEC50170.2021.9510352
28. Phillipson, J., Gorton, M., Turner, R., Shucksmith, M., Aitken-McDermott, K., Areal, F., Cowie, P., Hubbard, C., Maioli, S., McAreavey, R., Souza-Monteiro, D., Newbery, R., Panzone, L., Rowe, F., Shortall, S.: The COVID-19 pandemic and its implications for rural economies. Sustain. (Switz.) **12**(10), 1–9 (2020). https://doi.org/10.3390/SU12103973
29. Tong, D., Cheng, J., Liu, Y., Yu, S., Yan, L., Hong, C., Qin, Y., Zhao, H., Zheng, Y., Geng, G., Li, M., Liu, F., Zhang, Y., Zheng, B., Clarke, L., Zhang, Q.: Dynamic projection of anthropogenic emissions in China: Methodology and 2015–2050 emission pathways under a range of socio-economic, climate policy, and pollution control scenarios. Atmos. Chem. Phys. **20**(9), 5729–5757 (2020). https://doi.org/10.5194/acp-20-5729-2020
30. GEM. GEM 2020/2021 Global Report (2021)
31. Al Shukaili, A., Cucculelli, M., González-Pernía, J.L., Legazkue, I.P.: Entrepreneurship policy to overcome barriers to new firm growth in a developing economy: evidence from Oman. Int. J. Entrep. Small Bus. **35**(4), 511–537 (2018). https://doi.org/10.1504/IJESB.2018.096173
32. Al Shukaili, A.M., Al Kindi, K., Kassim, N.M., Ahmed, Z., Al Hosni, K.: Can government financial support enhance job creations: insights from Oman. J. Sci. Technol. Policy Manag. (2022). https://doi.org/10.1108/JSTPM-07-2021-0100
33. Hilal, B., Abdul Rahim, N.F., Iranmanesh, M.: The conceptual framework of SMEs financial success in sultanate of Oman. J. Governance Integrity **3**(2), 17–36 (2020). https://doi.org/10.15282/jgi.3.2.2020.5307
34. Khan, F.R., Krishnamurthy, J.: Future-proofing of tourism entrepreneurship in Oman: challenges and prospects. J. Work Appl. Manag. **8**(1), 79–94 (2016). https://doi.org/10.1108/jwam-06-2016-0008

35. Hatmanu, M., Cautisanu, C., Ifrim, M.: The impact of interest rate, exchange rate and European business climate on economic growth in Romania: an ARDL approach with structural breaks. Sustain. (Switz.) **12**(7) (2020). https://doi.org/10.3390/su12072798
36. Mukti, I.Y., Reza Firdausy, D., Aldea, A., Prambudia, Y., Van Sinderen, M., Iacob, M.E.: Architecting a service-oriented rural platform : improving rural economic climate through participation in the digital business ecosystem. In: Proceedings of 2021 IEEE 25th International Enterprise Distributed Object Computing Conference, EDOC 2021, pp. 92–103 (2021). https://doi.org/10.1109/EDOC52215.2021.00013
37. Chang, T., Deng, X., Hwang, B.G.: Investigating political risk paths in international high-speed railway projects: the case of Chinese international contractors. Sustainability (Switzerland), 11(15) (2019). https://doi.org/10.3390/su11154157
38. Miragaia, D., Ferreira, J., Carvalho, A., Ratten, V.: Interactions between financial efficiency and sports performance: data for a sustainable entrepreneurial approach of European professional football clubs. J. Entrepreneurship Publ. Policy **8**(1), 84–102 (2019). https://doi.org/10.1108/JEPP-D-18-00060
39. Trundle, R.C.: Empowering entrepreneurial capacity: training, innovation and business ethics. Eurasian Bus. Rev. **66**(3), 297–309 (2019). https://doi.org/10.5840/thought199166326
40. Cristo-Andrade, S., Ferreira, J.J.: Knowledge spillovers and strategic entrepreneurship: what researches and approaches? Int. Entrepreneurship Manag. J. **16**(1), 263–286 (2020). https://doi.org/10.1007/s11365-018-0541-4
41. García-Sánchez, E., García-Morales, V.J., Martín-Rojas, R.: Analysis of the influence of the environment, stakeholder integration capability, absorptive capacity, and technological skills on organizational performance through corporate entrepreneurship. Int. Entrepreneurship Manag. J. **14**(2), 345–377 (2018). https://doi.org/10.1007/s11365-017-0436-9
42. Alazaiza, M.Y., AbdelFattah, F., Al Maskari, T., Bashir, M.J., Nassani, D.E., Albahnasawi, A., Abushammala, M.F.M., Hamad, R.J.: Effect Of COVID-19 pandemic on food purchasing and waste generation during the lockdown period in the sultanate of Oman. Glob.l NEST J. **24**(1) (2022). https://doi.org/10.30955/gnj.004157
43. Clarysse, B., Tartari, V., Salter, A.: The impact of entrepreneurial capacity, experience, and organizational support on academic entrepreneurship. Res. Policy **40**(8), 1084–1093 (2011). https://doi.org/10.1016/j.respol.2011.05.010
44. Rahman, M.S., AbdelFattah, F.A., Bag, S., Gani, M.O.: Survival strategies of SMEs amidst the COVID-19 pandemic: application of SEM and fsQCA. J. Bus. Ind. Mark. ahead-of-p(ahead-of-print) (2022). https://doi.org/10.1108/JBIM-12-2020-0564
45. Silverman, D. (2020). *Qualitative research.* SAGE.
46. Bryman, A.: Business research methods. Oxford University (2016)
47. Nasar, A., Akram, M., Safdar, M.R., Akbar, M.S.: A qualitative assessment of entrepreneurship amidst COVID-19 pandemic in Pakistan. Asia Pac. Manag. Rev. (2021). https://doi.org/10.1016/j.apmrv.2021.08.001
48. Suthanthiran, K. (2011). A cost-effective solution for the healthcare crisis in the United States and around the world
49. Creswell, J.W., Creswell, J.: Research design: qualitative, quantitative, and mixed methods approaches. Sage Publications (2017)
50. Craig, C., Douglas, S.: International marketing research, 3rd edn. Wiley (2005)
51. Creswell, J.W.: Research design: qualitative, quantitative, and mixed methods approaches, 2nd edn. Sage Publications (2003)
52. Krosnick, J.A..: Question and questionnaire design. In: The Palgrave handbook of survey research. Palgrave Macmillan, Cham. pp. 263–313 (2018)
53. Brush, C.G., de Bruin, A., Welter, F.: A gender-aware framework for women's entrepreneurship. Int. J. Gend. Entrep. **1**(1), 8–24 (2009). https://doi.org/10.1108/17566260910942318
54. Krishnan, V.: A comparison of principal components analysis and factor analysis for uncovering the early development instrument (EDI) Domains. (2011, July)
55. Shigeta, K., Koizumi, M., Sakai, H., Tsuji, Y., Inaba, R., Hiraoka, N.: A survey of the awareness, offering, and adoption of OERs and MOOCs in Japan. Open Praxis **9**(2), 195 (2017). https://doi.org/10.5944/openpraxis.9.2.568

56. Maronick, T.J.: An empirical investigation of consumer perceptions of "made in USA" claims. Int. Mark. Rev. **12**(3), 15–30 (1995). https://doi.org/10.1108/02651339510091735
57. Teece, D., Pisano, G., Shuen, A.: Dynamic capabilities and strategic management. Strateg. Manag. J. **18**(7), 509–533 (1997)
58. Doody, O., Noonan, M.: Preparing and conducting interviews to collect data. Nurse Res. **20**(5), 28–32 (2013). https://doi.org/10.7748/nr2013.05.20.5.28.e327
59. Shafi, M., Liu, J., Ren, W.: Impact of COVID-19 pandemic on micro, small, and medium-sized enterprises operating in Pakistan. Research in Globalization, 2, 100018 (2020).
60. Hassan, M.S., Al Halbusi, H., Najem, A., Razali, A., Abdel Fattah, F.A.M., Williams, K.A.: Risk perception, self-efficacy, trust in government, and the moderating role of perceived social media content during the COVID-19 pandemic. Changing Soc. Personalities **5**(1), 9. https://doi.org/10.15826/csp.2021.5.1.120

The Response of Islamic Banks Face the Covid-19 Pandemic and the Role of Financial Technology

Mohamed Bechir Chenguel

Abstract The Covid-19 pandemic is not yet over, and its impact has affected global economic and financial sectors. This represents a challenge for Islamic finance considered as an alternative finance especially after the advent of the 2009 global financial crisis. Islamic finance certainly has an important market share in crowd-funding and also in microfinance, and its role in the financing of SMEs all over the world is not negligible, and this financial crisis should therefore pose different challenges to Islamic finance. It requires a set of instruments specific to its category, and also needs new technologies (FinTech), strategies and financial services to cope with the possible phenomenon. Our work seeks to present how Islamic finance, with its specific financing instruments, can fight against the disastrous effects of this health crisis, in particular the instruments of Zakat, funding by Qardh-Al-Hassan and fundraising like the Sukuks and finally the importance of using a set of technologies, called Fintech, to build an alternative and sustainable financial system after Covid-19. Indeed, this health crisis should present different challenges to Islamic finance, which requires FinTech, in its new strategies and its financial services. Our Paper is structured as following, we start with a review of the literature, in a second part we present the covid-19 crisis and Islamic finance, the roles of Islamic products in the face of covid-19, and in a last part, the role of artificial intelligence in islamic finance.

Keywords Covid-19 · Islamic finance · Alternative funding · Islamic banks · Fintech

1 Introduction

The Covid-19 pandemic is an unprecedented global shock involving supply and demand disruptions in an interconnected global economy. Indeed, infections have reduced labor supply and productivity, and lockdowns, business closures and social

M. B. Chenguel (✉)
Kairouan University, Kairouan, Tunisia
e-mail: bechir.chenguel@gmail.com

A. Hamdan et al. (eds.), *The Implementation of Smart Technologies for Business Success and Sustainability*, Studies in Systems, Decision and Control 216,
https://doi.org/10.1007/978-3-031-10212-7_12

distancing have resulted in supply disruptions, and layoffs and loss of income as well. that the deteriorating economic outlook has reduced household consumption and business investment. Beyond its obvious impact on human health and safety, the spread of COVID-19 has thus generated an unprecedented social, political and economic impact. A series of measures taken by governments to prevent the spread of the virus shows that a global economic recession is inevitable, and economies must prepare for one of the worst crises they have experienced since the Great Depression of the 1930s [1]. The first studies and analyzes, however, on the economic impact of the epidemic suggest a decrease in GDP ranging from 3 to 6% depending on the country and the evolution of the spread of the virus in the world [2]. The economic and financial situation of all countries has been seriously affected by the spread of the Covid-19 crisis. Global financial and regulatory authorities have put in place several preventive measures and actions to deal with this crisis. The financial and banking sector was one of the sectors most affected by the health crisis, and mainly, the classic and Islamic banks which were heavily affected [3].

The main objective of this work is to determine whether certain specific Islamic financial products can constitute a solution in times of crisis. Our contribution is to focus on certain Islamic financing instruments and their role to revive and exit the recession, by offering adequate financing tools. And we also talked about new innovations in artificial intelligence and fintech and their growth following the covid-19 crisis. Our work will be organized as follows: first, we will focus on the harmful effects of the Covid-19 health crisis, the impact on the economic and financial situation. Second, we will try to present the impact of the health crisis on Islamic financial instruments, and at the end we will take a closer look at the different fintech instruments that have emerged and are considered as a solution to get out of this health crisis.

2 Literature Review

The Covid-19 health crisis once again shows the dysfunctions of conventional finance based on profit maximization, and highlights the need to turn to more ethical finance capable of helping those in need. Islamic finance has certain tools like Qardh-Al Hasan, Sadaqa and Zakat which allow liquidity to be injected quickly into the market [4]. Currently the whole world is suffering from the covid19 pandemic with its new variants, and several researchers are trying to see and design new steps to follow and new methods to adopt to put the world economy back on the right track, thus researchers are trying to demonstrate the importance of Islamic finance in overcoming this crisis and the role that FinTech can play in achieving this goal, Kabir et al. [5]. The fundamental principles of Islamic finance such as the prohibition of the "Riba" interest rate and the prohibition of products resulting from conventional financial engineering (options, swaps and toxic assets ...), allow Islamic banks to protect oneself from crises, and to better resist the 2008 financial crisis; Taleb Bendiab [6], but the current Covid-19 health crisis is completely different. And Islamic finance has many assets that will allow it to contribute significantly to the recovery of economic activity gradually, indeed, this type of finance is based solely on real activity, and on the principle of profit and loss sharing, considering money only as a medium of exchange [4, 7]. Social financial instruments offered by Islamic finance like Quardh-Al-Hasan, Zakat, Social Sukuk and Waqf can be used during the Covid-19 crisis, to assist and help companies in difficulty [8]. In times of crisis, the principle of social justice and the equitable distribution of income are very important [9] and these principles are the basic principles of Islamic finance. With the health crisis we have financial markets in distress, factories, restaurants and hotels, and stores closed, with a considerable slowdown in transport and international travel activities, the COVID-19 health crisis has hit the global economy hard, and the spread of the coronavirus has created both a supply shock and a demand shock.

And this situation has generated a real and growing risk of recession [10]. Global financial authorities like the IMF are aware of the severity of the global economic recession caused by the coronavirus [11]. The scenario remains pessimistic, the economy could fall by 8% globally this year, and nearly 5% in emerging and developing economies, while the global recovery is limited to just over 1% in 2021 [12]. And the outlook for global growth in 2021 will not be good [11]. Nevertheless, in a world where finance is globalized, we know that financial shocks contribute to cyclical fluctuations which can generate financial crises which can in turn have negative repercussions on the banking system: solvency crises with defaults on corporate and household debts. To counteract the negative effects of the crisis, the major central banks of the G20 countries have announced major monetary support measures. The US Federal Reserve and the European Central Bank have opted for a cut in key rates, massive asset and sovereign debt redemptions to support the bond market and avoid a growth in risk premiums on the debt of the most weakened states by crisis [13]. Baldwin and Tomiura [14] believe that the Covid-19 pandemic would exert both health and economic contagion effects, they also note that this virus constitutes a supply and demand shock, affecting international trade in goods and services. According to the IMF [15], in its report on the Covid-19 health crisis in Africa, found that the health crisis is accompanied by an economic crisis and this involves decreases in production and demand, significant drop in commodity prices, especially oil, which have fallen by around 50% since the start of 2020.

The world economy is now facing an economic shock, the nature of which needs to be analyzed to find appropriate measures. These practices have led to a recession of the planetary economy, a disruption in the global production chain which has generated speculation in the financial markets, a decrease in foreign exchange reserves, a deterioration in the value of goods, a decrease in foreign trade and an instability and depreciation of the currency. According to Claudiu [16] the Covid-19 shock is as much monetary as financial, in fact, the Covid-19 pandemic creates fear and uncertainty, affecting the global economy and thus amplifying the volatility of financial markets. The recession induced by the Covid-19 pandemic will cause solvency problems for some businesses and households and will put pressure on banks' balance sheets, causing growth in the share of non-performing loans and thus causing bank failures [13].

3 Islamic Finance COVID-19 Crisis

The global pandemic (COVID-19) has crippled the global economy, putting companies and people in a difficult situation. Oxfam (2020) found that stopping economic activities has caused more than 500 million people in poverty. According to Haider et al. [4], the only way that can save economies is where the rich and governments around the world come together and join forces to help poor governments. With the same principle of solidarity, Islamic finance, already with some of its financial instruments, can provide support to the community. This happens through its religious instruments like Zakat and Sadaqa and non-profit contracts aimed at bringing aid to the needy.

3.1 Islamic Finance Tools Used in Covid19 Crisis Period:

3.1.1 Zakat

Zakat is one of the five pillars of Islam, which consists of paying the poor 2.5% of assets owned for a year. The Koran has clearly defined the eligible beneficiaries of Zakât. These include the poor, needy, to those who are heavily in debt, and travelers in distress. There is thus a general consensus that the main objective of zakat is the eradication of poverty, that is to say that it is addressed primarily to the needy and the poor [17]. This good practice makes it possible to fight against income inequality and to bring a certain socio-economic justice in society in order to fight poverty. Zakt is therefore well regarded as short-term emergency assistance for those most affected by the Covid-19 crisis. With the current pandemic, the poor and workers operating in the informal sector urgently need funds to enable them to meet their basic needs. Zakat can be set up as a dedicated fund for dealing with COVID-19 with a goal of helping people and even SMEs that have been shut down due to this pandemic. By making these instruments of Islamic banks institutional, the authorities can take advantage of these significant funds that can be deployed in the fight against the adverse effects of this health crisis. Moreover, the reach of Zakat remains limited to some categories fixed by religion, while other people, who are not eligible for this alms, also need help in this period. We can cite as an example the managers who have contracted micro credits and have suffered very serious economic consequences.

3.1.2 Qard Al-Hassan

Qard Al-Hassan refers to the interest-free loan. It is a mechanism that is part of the aid tools and means of mutual aid accessible to all categories of people who need help in this period of crisis and who do not necessarily fall into the category of people benefiting from the Zakat, it can be seen as a short-term rescue fund allowing clients of Islamic financial institutions to raise funds quickly and at zero cost [18]. The Al Qard Al-Hassan loan, unlike the previous instrument Zakat, can be granted to anyone, regardless of their situation, rich or poor. And this specificity of versatility makes the use of this financial instrument greater than that of other instruments such as Zakat. Thus, entrepreneurs who have suffered the devastating effects of the health crisis can use this fundraising tool to relaunch their activities. Widiyanto et al. [19] conducted a study on this instrument and proved the effectiveness of Al Qard Al-Hassan in strengthening the economic activities of the poor and needy. According to Iqbal and Shafiq [11] this mechanism can be used to help poor entrepreneurs to restore their business which is suffering from this crisis. And this Islamic financial tool is used effectively, it will become an excellent means of supporting growth in times of crisis. Unfortunately in practice, this instrument is largely ignored by Islamic banks because of its non-profit nature, but its commercialization and usefulness in times of crisis is one of the strengths of Islamic finance that can help entrepreneurs restore their previous position as quickly as possible. This instrument of support for entrepreneurs is limited to the period of the COVID-19 health crisis, while the integration of Al Qard Al-Hassan in the portfolio of Islamic financial institutions could provide invaluable help during the support of actors economic in the event of crises specific to companies or particular sectors [20].

3.1.3 Sukuks

Are the newest Islamic finance products. The Sukuk have been the subject of several recent research including that of Ogunbado [21] which has focused on this Islamic financial instrument and its interests in financing the development of basic infrastructure. Sukuks in general have been used to finance large state projects like building roads or dams. But it was in 2014 that the first Sukuk were intended to promote vaccination (Vaccine Sukuk), this initiative was favorably received by the markets, and several global financial institutions appreciated this type of financing and recognized the impact. positive of this sukuk market. Following this and the success of this financing operation, the World Bank and the International Finance Facility for Immunization, jointly decided to reiterate the issuance of this type of Sukuk in 2015. According to the International Financial Facility for Vaccination, the two fundraising operations in 2014 and 2015 encouraged the launch of the third fundraising operation in 2019 with the issuance of "Vaccine Sukuk" and this in collaboration with the Group of the Islamic Development Bank (ISBD). Noordin et al. [21] and Badeeu et al. [22] have shown through the study that the issuance of sukuk is a viable financing instrument for health care providers to develop the health sector. However, researchers associate the use of Islamic financial instruments in financing public health with the existence of political will, government support and an appropriate Islamic financial ecosystem. These fundraising experiences thanks to sukuk can be an encouraging source for authorities and governments wishing to revive their key sectors weakened by the Covid-19 crisis, namely: the health sector, education as well as the economic and financial sector Abdullah and Keshminder (2020).

3.2 The Role of Islamic FinTech During the Covid19 Period:

The word fintech is the abbreviation of two terms namely "finance and technology. It is a sector resulting from the use of digital platforms and artificial intelligence in financial services, as opposed to traditional financial services based on direct contact; Lacasse et al. (2016). FinTech is seen as an industry that uses new digital technologies to provide innovative solutions to improve financial services [23]. Thus new financial technologies have contributed to the establishment of new business models. These innovations in the banking sector have affected payment systems, and also financial regulation. And in recent years, the evolution of technological innovations has grown tremendously. These latest technological innovations have affected all banking activities in the front office and in the back office, the operations of collection, disbursement, borrowing, the risk management department, and even the financial advice of clients have been robotic. Fintech has turned the world of finance upside down with its technological innovations. The use of blockchain and artificial intelligence has already revolutionized the world of finance [17]. Currently finance is facilitated by the process of technological innovations where financial services are provided through artificial intelligence, and robots that have replaced financial advisers [24]. Along the same lines, Using Fintech in the delivery of Islamic services such as Waqf, Crowdfunding and Zakat ... will make financial services to clients easy, transparent and efficient Todorof (2018). Currently, the COVID-19 pandemic has affected all countries and has had a devastating impact on the global economy. And this leads to drawing attention to the potential of Islamic finance and the solutions it can provide in times of crisis. Thus we observed an acceleration of the digital transformation process, already underway before the covid-19 pandemic. Islamic finance and its instruments is open to all innovations as long as it does not violate the principles of Sharia law. Islamic fintech companies will thus offer better Islamic financial services, saving time and at a lower cost [25]. With the first months of the pandemic, the number of fintech users increased dramatically in the first months of 2020 [26]. Islamic Fintech will play a very important role in promoting the Islamic finance sector, since Islamic Fintech is flexible, attractive, efficient and leads to financial inclusion and innovation [24, 27]. The success of Islamic finance and the role it will play in the post-Covid-19 period will be primarily determined by the use and adoption of Fintech over the next two months [4]. Fintech and the transformation of its financial services by Islamic banks will follow the great behavioral changes of people and society, thus the confinement of people, and the imposed distancing will impact people even after the pandemic [28]. The period of health crisis has lasted, and this has led people to use fintech-based payments such as card payment, online, through apps instead of cash. This method of payment without going out, and without using cash is convenient and very efficient. Fintech companies have offered Islamic banks new products and services, which present new challenges. Indeed, replacing traditional products, services and business models with more efficient digital technologies is not so easy [29]. The use of fintech through blockchain technology, big data and machine learning in Islamic finance will certainly bring more efficiency and transparency for clients and also for regulators of Islamic Financial Institutions [30–33]. And thus we can say that the Covid-19 crisis presented an opportunity for Islamic FinTech to develop and provide an alternative financial system. Islamic FinTech will provide practical, innovative financial services while respecting the basic principles of Sharia law, and therefore ensure financial and banking services accessible to all and also participate in poverty reduction [31].

3.2.1 Islamic Fintech: New Financial Products

The covid-19 health crisis has engendered an economic crisis, pushing companies to many challenges at different levels: a level of logistics, transforming workplaces and transforming the supply chain, and this has impacted corporate finances and cash flow. These changes will push for the creation of new solutions and new opportunities to improve the performance of Islamic Financial Institutions, and this by offering new innovative financial services based on FinTech.

The Islamic Crowdfunding:

Islamic crowdfunding can be defined as a combination of crowdfunding and the financial technology of crowdfunding with Islamic financial contracts. More and more start-ups offer financial products in accordance with Islamic law through platforms through which any user finances the project of a new business (Start up) or even of an existing business through loans or donations [34]. Based on new communication technologies, collaborative and participatory financing makes it possible to overcome the main challenges of entrepreneurs. Indeed, depending on the context, a crowdfunding platform can be qualified as "Islamic", "participatory" if it respects the rules established by the Shari'ah namely sharing and loss and profit, the elimination of the interest rate on loans, speculation and the non-involvement of prohibited industries in transactions such as the sale of alcohol, pork etc..... The transactions of the platform must be simply in accordance with the principles of Islamic finance [17]. The impact of this social funding can be seen as a complementary source to government funding. The platform acts as an intermediary between entrepreneurs with Shari'ah-compliant ideas and investors who subscribe to these projects. Compared to conventional crowdfunding there are four forms of Shari'ah-compliant platforms [35]:

- Donation-based crowdfunding platforms: these are platforms that collect sadaqa funds or "zakat" funds intended for social projects or initiatives.
- Platforms based on rewarded donation models: it consists in collecting small amounts which will contribute to the success of projects resulting from the platform, in exchange for rewards to donors, and the reward comes from the funded project.

- Platforms based on equity participation: shareholders invest through Islamic contracts for the financing of projects through the platform.

Platforms based on loan models: lenders grant loans to start-ups or projects in the platform.

These loans are made through "Murabaha" or "Ijara" contracts. Thus a FinTech initiative, through crowdfunding can provide borrowers with an easy, faster, cheaper and affordable way to obtain financing directly from the public. Crowdfunding has become the primary source of funding for rapidly growing startups. It has the potential to become the biggest challenger in the traditional media of finance. Crowdfunding and P2P loans will undoubtedly be the future of financing. EthisCrowd is an Islamic crowdfunding platform investing in entrepreneurial, commercial and real estate activities in emerging Asia. Based in Singapore, with a presence in Indonesia, Malaysia and Australia, the company finances the construction of affordable and commercial housing, primarily in Indonesia, through private and institutional investors, as well as Islamic banks. PayZakat is the newest Islamic fintech platform that uses artificial intelligence (AI) -based chatbots and entirely new digital tools to help users make payments from Zakat (a mandatory wealth tax), Sadaqah (a payment from alms not compulsory) and other charitable payments. User can select a country and a specific charity in that country to help.

Smarts Contracts

Smart contracts are used to facilitate the transfer of digital assets on the basis of previously agreed terms and conditions. One of the fundamental goals of Islamic finance is to bring transparency and efficiency in its transactions and smart contracts help to achieve this goal. The digital smart contract is definitely the future of finance and it will replace the traditional paper contract because it helps bring more reliability and transparency without endangering its legitimacy [36]. Islamic smart banking: The smart world is very often used today for smart cities, smart phones and smart banking. Smartness is fundamentally achieved by adopting innovative technologies such as automation, cloud IoT, etc. Smart Banking means a smarter, technology-based approach to customer service in the provision of financial services. Smart Banking is a new concept of Islamic banking where customer engagement and satisfaction is taken to another level. Example: Qitmeer is the first new generation Islamic banking platform based on BlockDAG, which will serve the clients of Islamic banks for ethical investments according to the Quran and Sunnah [30–32].

Islamic Crypto-Currency

The blockchain is the decentralized distributed ledger for recording the financial transaction. Blockchain technology can be applied in financial services such as Zakat, Qardh-Al-Hasan, P2p and Mudharaba loans, etc., and also for the creation of a new

Islamic virtual currency. Blockchain technology will bring more transparency and efficiency Khan (2020). A cryptocurrency is a digital asset based and secured by blockchain cryptography technology. At present, and given the expansion of virtual currencies in the financial world, a large number of Muslim investors wish to invest in Bitcoin but are reluctant due to respect for Sharia law. They must therefore be able to invest in a less volatile Islamic cryptocurrency [37]. It is possible to issue Islamic cryptocurrency if the volatility issue is resolved [38]. Thus the creation of an Islamic cryptocurrency can be a popular Islamic finance tool during and after the health crisis to attract new investors and mobilize resources. Indeed, there are cryptocurrencies that are backed by gold and this is Sharia compliant. And already some virtual currencies already exist like Monnaie X8, OneGram …

3.2.2 Artificial Intelligence and Islamic Finance

Artificial intelligence, along with blockchain, is one of the technologies enabling banking and finance professions to accelerate their digital transformation; the banking and financial sectors can be proud of being among the most advanced in the use of machine learning technologies. Between current new laws, security threats from fraud and increasing consumer pressure for better customer service, traditional banking practices quickly became obsolete.

To develop, Artificial intelligence is based on two main pillars: the progress made in machine learning, which is mainly based on the selection of the most relevant data, and the development of IT and technological tools which are its vectors [25]. Machine learning consists in providing computer systems with means comparable to those of human beings, such as reasoning, learning and self-improvement, without pre-programming the machine. To this end, machine learning programs analyze a large amount of data, from which they identify human trends and behaviors; they offer an answer to the growing data challenge, which has continued to grow. To meet these challenges, banks have turned to digital transformation to maintain their competitive advantage on all fronts, with early adopters being the most rewarded. Analysts and experts estimate that artificial intelligence will save the banking industry around $1 trillion by 2030. The role of artificial intelligence is becoming increasingly important in the study of banking and finance. Banks and financial services organizations use robotic advisory services to provide financial services to clients. This will change the experience of financial services, Robotics advisory services for Islamic banks include Chatbot for financial advice and Zakat calculation [25].

4 Conclusion

For about half a century, several crises of different kinds have followed one another: financial, economic and health crises (2008 subprime crisis 2009 global financial crisis, Ebola, SARS, H1N1, etc.), But the scale of this Covid-19 crisis has severely impacted the global economy, affecting jobs, health, businesses and industries. Certainly the banking sector is one of the main players in the transmission of the crisis to all sectors, indeed with the containment and closure of borders, the drop in demand caused by the Covid-19, caused the outbreak of all types of risks, namely solvency risk, since companies were unable to meet their commitments and pay off their bank loans. And to reduce their costs, companies were forced to lay off their employees and workers, which generated a recession. The monetary policies applied by the countries to guard against the crisis, and even the preventive measures were insufficient. And the various financial institutions that will have to take the shocks of this type of crisis into account in their risk management have not been successful. The various monetary and budgetary authorities reacted by injecting liquidity into the financial system and granting targeted loans to various SMEs, households and vulnerable people in order to limit the impact of the crisis on the financial system.

In order to study the role that Islamic finance can play in countering the harmful effects of the current health crisis. We first presented the strengths available to the Islamic financial sector, highlighting some Islamic financial instruments that can help alleviate the adverse effects of this health crisis, and explaining the role of Islamic finance in the period post-covid-19. Indeed we tried to present and reveal that the instruments of Islamic finance can to cope with the economic crisis suffered by the COVID-19 health crisis, and how these instruments can offer solutions and leave the state crisis for companies and individuals.

Islamic finance supports social development by distributing and sharing a portion of excess wealth through the quard al hassan and sukuk who can support economically distressed communities by providing them with cash flows to encourage consumption. It should be mentioned that the efficiency of Islamic financial instruments in the fight against the harmful effects of the covid-19 crisis, is conditioned by the existence of a strong will of the Islamic banking sector, government support and existence of an appropriate ecosystem. The second aspect evoked in this research work is the use of fintech tools in Islamic finance and how these fintechs can offer solutions during this health crisis. In fact fintech is seen as a fundamental innovation, because the automation of operations and their registration allows to create entirely new products and services more for customers and for companies, it can also create models of urban development oriented. On decentralization and customer data. Islamic fintech can help Islamic finance thrive in the post-covid era. Either through the creation of Islamic crypto currency, or the use of Islamic smart contracts.

As in all research work, our work has limits. In fact, given the novelty of the topic dealt with concerning Islamic finance in the face of the covid-19 crisis, we currently do not have figures on the use of these financial instruments during this health crisis. And we must pay special attention to the importance of adopting fintech in the Islamic financial sector, and take a closer look at the solutions offered during this crisis. We have presented during our research work only theoretical aspects, and we must test them empirically in future research.

References

1. Rhee, C.: Pandémie de COVID-19: plus faible croissance dans la région AsiePacifique depuis les années 1960. Disponible à (2020). https://www.imf.org/fr/News/Articles/2020/04/15/blog-covid-19-pandemic-and-the-asiapacific-lowest-growth-since-the-1960s.consulté le 10/06/2020
2. Fernandes, N.: Economic effects of coronavirus outbreak (COVID-19) on the world economy (2020, March 22). Disponible sur SSRN https://ssrn.com/abstract=3557504
3. Raza Rabbani, M., Asad Mohd. Ali, M., Rahiman, H. U., Atif, M., Zulfikar, Z. and Naseem, Y.: The response of Islamic financial service to the COVID-19 pandemic: the open social innovation of the financial system. J. Open Innov. Technol. Mark. Complex. **7**(1), 85 (2021)
4. Haider, M., Khan, S., Rabbani, M. R. and Thallasinos, Y.E.: An artificial intelligence and NLP based Islamic FinTech model combining Zakat and Qardh-Al-Hasan for countering the adverse impact of COVID 19 on SMEs and individuals. Int. J. Econ. Bus. Adm. **8**(2), 348–361 (2020). (Haider Syed et al. 2020)
5. Hassan, M. K., Muneeza, A. and Sarea, A. M.: Impact of COVID-19 on Islamic social finance. In: COVID-19 and Islamic Social Finance, pp. 9–23. Routledge (2021)
6. Taleb Bendiab, I.A.: Évaluation de l'efficience, la performance et le risque des banques conventionnelles et con-conventionnelles: essais empiriques [Doctoral dissertation, Université Paris-Saclay (ComUE)] (2018)
7. Bureshaid, N., Lu, K., Sarea, A.: Adoption of FinTech services in the banking industry. Stud. Comput. Intell. **954**, 125–138 (2021)
8. Mohamed, H.Y., Hamdan, A., Karolak, M., Razzaque, A., Alareeni, B.: FinTech in Bahrain: the role of FinTech in empowering women. Lecture Notes in Networks and Systems, 194 LNNS, pp. 757–766 (2021)
9. Smolo, E., Mirakhor, A.: The GFC and its implications for the Islamic financial industry. Int. J. Islam. Middle East. Financ. Manag. **3**, 372–385 (2010)
10. Guterres, A.: Secrétaire général de l'ONU, «Briefing to the Security Council on Global Governance Post-covid-19». Etats-unis, New York (2020)
11. Iqbal, Z., Shafiq, B.: Islamic finance and the role of Qard-al-Hassan (Benevolent Loans) in enhancing inclusion: a case study of AKHUWAT. ACRN Oxford J. Financ Risk Perspect. **4**(4), 23–40 (2015). (K Georgieva 2020)
12. World Bank Group (2020) Global Economic Prospects. flagship report 30 edition
13. Beck, T. «Finance in the times of coronavirus». In: Baldwin, R., di Mauro, B.W. (eds.) Economics in the Time of COVID-19, pp. 73–77 (2020). ISBN 978-1-912179-28-2 (Braquet 2020)
14. Baldwin, R., Tomiura, E.: Université de Kinshasa, Nature du choc de la Covid-19 d'un point de vue économique, p. 14 (2020)
15. IMF (2020) «La crise de la COVID-19 menace la stabilité financière" rapport 2020
16. Claudiu, A.: Coronavirus and financial volatility: 40 days of fasting and fear, p. 14. Poitiers France (2020)
17. Hasan, R., Hassan, M.K., Aliyu, S.: Fintech and Islamic Finance: literature review and research Agenda. Int. J. Islamic Econ. Finance (IJIEF) **3**(1), 75–94 (2020). (Hassan and Ashraf 2010, IFSB 2017)
18. Afkar, T.: Influence analysis of Mudharabah financing and Qardh financing to the profitability of Islamic Banking in Indonesia. Asian J. Innov. Entrepreneurship **2**(3), 340–351 (2017)
19. Widiyanto, W., Mutamimah, S., Hendar, H.: Effectiveness of Qard al-hasan financing as a poverty alleviation model. Econ. J. Emerg. Markets **3**(1), 27–42 (2011)
20. Brgchou, K., Hamimida, M.: «La finance islamique face aux effets économiques de la crise sanitaire «COVID-19» Recherches et Applications en Finance Islamique Researches and Applications in Islamic Finance, vol 5, Numéro 1, février 2021 (2021). ISSN 9052-0224
21. Noordin, K.A.: Islamic Finance: Is Cryptocurrency Halal? The Edge, Malaysia (2018). (Ogunbado 2019)

22. Badeeu, F.N., Nafiz, A.R., Muneeza, A.: Developing regional healthcare facilities in Maldives through Mudharabah perpetual sukuk. Int. J. Manag. Appl. Res. **6**(2), 81–96 (2019)
23. Razzaque, A., Cummings, R.T., Karolak, M., Hamdan, A.: The propensity to use FinTech: input from bankers in the Kingdom of Bahrain. J. Inf. Knowl. Manag. **19**(1), 2040025 (2020)
24. Khan, S., Rabbani, M.R.: Artificial intelligence and NLP based Chabot as Islamic banking and finance expert. Int. J. Inf. Retrieval Res. (IJIRR) (2020)
25. Khan, S., Rabbani, M.R.: Artificial intelligence and NLP based Chatbot as Islamic banking and finance expert. In 2020 international conference on Computational Linguistics and Natural Language Processing (CLNLP 2020), Seoul, South Korea on July 20–22, 2020 (2020)
26. Fu, J., Mishra, M.: The global impact of COVID-19 on Fintech adoption. Swiss Finan. Inst. Res. Paper (20–38) (2020)
27. Khan, S.N., Rabbani, M.R.: Artificial intelligence and NLP based Chatbot as Islamic banking and finance expert. 2020 International Conference on Computational Linguistics and Natural Language Processing (CLNLP 2020), Seoul, South Korea on July 20–22.
28. Wijayanti, D.M., Pradipta, H.: Sharia FinTech: Positive Innovation in Consumer Perspective. Paper presented at the International Seminar Academic Network on Competition Policy, Bali (2017)
29. Khan, S., Hassan, M.K., Rabbani, M.R.: An artificial intelligence-based Islamic FinTech model on Qardh-Al-Hasan for COVID 19 affected SMEs. In Islamic Perspective for Sustainable Financial System. Istanbul University Press, Istanbul, Turkey (2021a)
30. Rabbani, M.R.: The competitive structure and strategic positioning of commercial banks in Saudi Arabia. Int. J. Emerg. Technol. **11**(3), 43–46 (2020)
31. Rabbani, M.R.: The competitive structure and strategic positioning of commercial banks in Saudi Arabia. Int. J. Emerg. Technol. **11**(3), 43–46 (2020). (Vasenska et al. 2021)
32. Rabbani, M.R.: Financial literacy among the Private University Students in Kingdom of Bahrain. Int. J. Sci. Technol. Res. **9**(3) (2020)
33. Sun, H., Rabbani, M.R., Sial, M.S., Yu, S., Filipe, J. A., Cherian. J.: Identifying big data's opportunities, challenges, and implications in finance. Mathematics **8**(10) (2020)
34. Alazemi, M., Al Omari, A.M.: The application level of institutional governance in Islamic Institutions and Banks in Kuwait. Int. J. Bus. Ethics Governance **3**(3), 85–101 (2020). https://doi.org/10.51325/ijbeg.v3i3.58
35. Marzban, S., Asutay, M., Boseli, A.: Shariah-Compliant Crowd Funding: An Efficient Framework for Entrepreneurship Development in Islamic Countries. Havard Islamic Finance Forum 2014 (2014)
36. Rahim, S.R.M., Mohamad, Z.Z., Bakar, J.A., Mohsin, F.H., Isa, N.M.: Artificial intelligence, smart contract and islamic finance. Asian Soc Sci **14**(2), 145 (2018)
37. Baker, S.R., Bloom, N., Davis, S.J., Terry, S.J.: COVID-induced economic uncertainty (No. w26983). National Bureau of Economic Research (2020)
38. Alzubaidi, I.B., Abdullah, A.: Developing a digital currency from an Islamic perspective: case of blockchain technology. Int. Bus. Res. **10**, 1–9 (2017)

Factors Affecting the Adoption of Social Media Among Jordanian SMEs Based on UTAUT Mode

Sultan Alshourah, Manal Altawalbeh, Ibrahim Jodeh, and Munner Hasan

Abstract The growing adoption of Social Media (SM) and its capability to help small and medium-sized businesses (SMEs) target large audiences simultaneously has transformed how small and medium businesses perform their routine operations. This study objectives to explore the dimensions that inspiration the SM adoption in SMEs Jordanian to enhance their financial and marketing performance and business technology use using the approach of Unified Theory of Acceptance and Use of Technology approach (UTAUT). Expected performance, Expected effort, Social impact, and Facilitation conditions as dimensions of the UTAUT Model, it's generally agreed upon by previous studies and scholars, and not yet, the interest of managers, researchers, and policymakers to investigate (UTAUT) model in developing countries specifically. To achieve this objective, quantitative research has been approved. A questionnaire was sent to 202 participants randomly designated SMEs in Amman and, 162 usable questionnaires were obtained, and data was tested with confirmatory factor analysis, and regression was used. The results of this research discovered that the (UTAUT) model had fitted to display 65% of the variance in social media behavioral managers. After the analysis, the results of the study indicated that all dimensions mentioned previously had significant and positive effects on Behavioral Intention (BI) to use Social Media in Jordanian SMEs. Based on these results, practical and theoretical implications are presented to researchers, SMEs' managers, and owners. Output, create employment opportunities and improve global competitiveness.

Keywords Jordan · SMEs · UTAUT · Social media

S. Alshourah (✉) · I. Jodeh · M. Hasan
Zarka University, Zarqa, Jordan
e-mail: salshourah@zu.edu.jo

M. Altawalbeh
Middle East University, Zarqa, Jordan

A. Hamdan et al. (eds.), *The Implementation of Smart Technologies for Business Success and Sustainability*, Studies in Systems, Decision and Control 216,
https://doi.org/10.1007/978-3-031-10212-7_13

1 Introduction

SMEs play the largest role in economic growth in every country in the world. It plays a pivotal role in the economy Jordanian; they absorb nearly 60% of the workforce in Jordanian society. Studies indicate that they constitute more than 90% of the total organizations operating in various sectors [1]. In Addition, it contributes to 50% of Jordan's GDP [2]. Under environment dynamics and competitive conditions, business characteristics are constantly changing. Thus, SMEs managers need tools of marketing that are not only active but also efficient to enhance their performance of the financial and marketing and enable SMEs to customer's engagement within multiple [3, 4]. Therefore, Social Media is a common tool used by SMEs because they can use customer feedback to SMEs and easy technical requirements, and also low cost [1].

Jordanian ranks worldwide from more than 50 countries using social networking platforms. The number of Internet users in Jordan has about 8 million users and the number of Jordanian accounts on various social networks about 10.9 million calculations, 96% of them from the youth generation they are on daily, while the number of users of social networking is 2.7 million users, 25% From the population of 10 million [5]. Thus, Social Media has had an enormous impact on Jordanian society, and it has become a remarkable promotional tool to SME businesses in Jordan [2]. Despite this, social media applications mostly in the context of SMEs in Jordan are still substandard or weak [1, 6]. According to [1] study, most Jordanian SMEs use traditional media for their strategies of marketing, but it cannot depend on it, and they have had to adapt and explore how to take advantage of the new interactive social media. A similar study, [2] results showed that the adoption of digital marketing (social media) is still weak marketing processes or is still in its early stages of adoption in small and medium companies. Besides, the SMEs majorities have limited their usage to social media marketing while email marking is a common tool for digital marketing. Finally, a study by [7] revealed that SM applications were generally used for marketing, communication, information for the improvement of business reputation, and both fake identities and security issues are the main limitations.

Although previous studies are looking at the factors that influence social media adoption in SMEs [2, 8–10]. There is still needed to gain a further understanding of the manager's behavioral adoption of social media as a business platform and to provide evidence of empirical for the literature on adoption of social media in SMEs [3, 4]. Also, to identify the social media adoption factors success in SMEs [3, 11]. Additionally, Most of the studies had conducted in countries of developed [12], and only a little number of studies have been elaborated in countries of developing [1]. Based on some previous studies, however, confirm that SMEs need to understand how to identify and effectively address the dimensions affecting the adoption process to ensure that the promised benefits are achieved [10]. To gap fill in previous studies local and international, this study objective is to recognize undertaking factors influencing the social media adoption media by SME managers using the UTAUT approach—a unified theory of technology acceptance and use (UTAUT) and validation (UTAUT) model. Therefore, according to UTAUT dimensions, this study answers the following questions of research.

- What is the effect of the Expected performance on SM adoption in SMEs Jordanian?
- What is the effect of the Expected effort on SM adoption in SMEs adoption in SMEs Jordanian?
- What is the effect of the Social Impact on SM adoption in SMEs adoption in SMEs Jordanian?
- What is the effect of the facilitating conditions on SM adoption in SMEs Jordanian?

In other words, the (UTAUT) model clarifies what factors SMEs use and adoption of social media in Jordan. Therefore, the (UTAUT) model review is predicted to help practitioners and researchers understand the current state of research during COVID-19 and support future research on this topic. Additionally, it will help Jordanian SMEs managers adopt and explore how to take advantage of their contemporary and interactive SM as part of their e-marketing strategies in order to become more relevant to their customers and suppliers so that they can increase their financial and marketing performance during COVID-19.

2 Literature Review and (UTAUT)

UTAUT model was revealed by Venkatesh et al. [13], which has been used widely in studies of different sciences and different settings. Venkatesh et al. [14], argued that the UTAUT integrates eight well-known models/theories in diverse disciplines. However, the integration of eight theories in terms of empirical similarities and conceptual differences was the main reason for unifying these theories is to reach a unified view of user acceptance of information technology. The eight theories used include logical action theory, TAM, the theory of planned behavior, TAM and TPB together, and the diffusion of innovation theory, social cognitive theory, the motivational model, and the personal computer use model. The research showed that the UTAUT model was capable. Moreover, The UTAUT model has been highly praised by Waehama et al. [15] due to its ability to explain at least 70% of technology acceptance behavior in contrast to other models, which can only describe more than 40%, as well as its function in demonstrating new incoming technology acceptance. According to [16] argued that few studies have focused on the intentions of SMEs to SM adoption to market their goods using the UTAUT model. Additionally, they add to this research gap that needs further investigation.

The idea addressed by the UTUAT model is to determine the user's intentions towards the use of innovations and the resulting use behavior through four main pillars of acceptance of use, which are the performance of expected, expected effort, social impact, and facilitation conditions [11, 13, 14]. Meanwhile, elements such as gender, age, experience, and voluntarism of use were brought into the mediation of the four constructs and their influence on intent to use and behavior [13] performance of expected refers to "the degree to which the user believes that using a new system or using new technology will help to achieve tasks with excellent performance" [13]. Besides that, Expected effort refers to "the degree to which a technology is easy to use and can have a positive impact on its use of intended" [11]. Social impact refers to "the degree to which users adopt and believe in new technology or system" [17] and the facilitation condition refers to "the level at which an individual believes that there is an institutional and technical infrastructure to assist in the use of the system" [11, 18].

Recently, to estimate and predict the intent to use technology, UTAUT is considered to be the healthiest and most suitable technology model for adaptation in SMEs [11, 19]. Furthermore, few studies focused on the factors affecting social media adoption and the behavioral intention of SMEs managers in developing countries. However, several authors have conducted to UTAUT model to understand Behavioral Intention to use Social Media (please see [11, 14, 20–22]. Among these studies, a semi-structured interviews after implementation and adoption of Face book business page study by Mandal and McQueen [22] found that the conditions of facilitating and social impact insignificant the behavioral and adoption intentions of social media by micro business owners. While the expectancy of performance and expected effort played an important role, user behavior was dominated by owner characteristics and codification efforts. Dahnil et al. [20], found that technological, environmental, and organizational (TOE) elements positive affect adoption of social media on the part of SMEs in the 250 SMEs in the city of Irbid, north Jordan. Method of random sampling was used for the study by Venkatesh et al. [14] which revealed that effort expectation, social influence, and facilitation conditions were significantly related to the AABB owners' intention to use and use social media marketing in the American small businesses context. In their study in Thailand, 196 data samples were collected via questionnaires of online, [3] results show performance expectations, effort expectation, social influence, and facilitation conditions that significantly influence behavior prospects for implementing social media for work purposes. While extended by Dahnil et al. [20] UTAUT is tested and believed to be effective in investigating technology acceptance for the individual; found that elements four positive to adopting social media. According to him, UTAUT is still valid and suitable to investigate technology adoption by a business organization such as SMEs. A survey of 310 firms by Chatterjee and Kar [11] emphasizes that effort expectancy, social Influence, and social influence have a significant while facilitating conditions have an insignificant impact on the use of social media marketing by SMEs. Al Mursalin [23], found the positive effect of four direct validations studied in the UTAUT model on the information technology adoption behavior of 255 Bangladeshi SMEs. Based on Previous Studies and the above arguments, the following hypotheses were formulated:

H1. Expected performance is positively related BI to in Jordanian SMEs.
H2. Expected effort is positively related to BI use SM in Jordanian SMEs.
H3. Social impact effects is positively related to BI use SM in Jordanian SMEs.
H4. Facilitating condition is positively related to BI use SM in Jordanian SMEs.

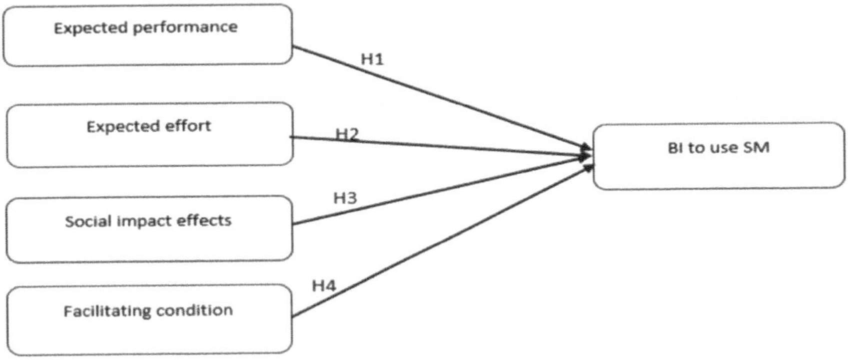

Fig. 1 Conceptual framework

As discussed in previous studies, a conceptual framework is shown in Fig. 1. It is shows the effect of EP, EE, SI, and FC on the managers' behavior, in relation to the Social Media adoption.

3 Methodology

A survey research design and quantitative research method were used for this study. It conducted in Amman capital city among all manufacturing SMEs [24]. And, data were collected once -cross-sectional- to answer the research questions of the study. Data is collected through email and personal surveys using a questionnaire. This study requires the respondents to possess certain knowledge of information technology, as well as general organizational infrastructure [25]. The respondents are made up of CEOs/owners and IT managers, supervisors, and those who have accounts and use social media to promote their products [18]. Of the total 202 questionnaires, it was found that 162 participants were SM users, and randomly sampling method was used [26]. While the remaining 40 respondents indicated not using SM in their daily operations. Table 1 explains the respondent's description. The survey instrument for the study was adapted Based on [13], the questionnaire includes five concepts of UTAUT, 4 item for Expected performance (EP), 4 item for Expected effort (EE), 3 item for social impact (SI),4 item for facilitating conditions (FC), and 3 item for Behavioral Intention to use SM (BI). Respondents answer questions that are arranged in the form of choices and scale questions using five Likert scale (1–5), ranging from 1 = "strongly disagree" to 5 = "strongly agree" to ensure stability between factors and avoid confusion among respondents. SPSS 21 was used to analyze the data [8, 14].

Table 1 Respondent's description

Demographic	Category	Frequency	%
Participants	CEOs/owners	130	80
	IT managers	32	20
Total		162	100
Experience (years)	1–4	16	10
	5–8	38	23
	9–12	66	41
	+ 12	42	26
Total		162	100
Age of firm (years)	<1	10	6
	2–6	38	23
	6–12	48	30
	>12	66	41
Total		162	100

Table 2 Reliability of variable

Variables	Number items	Reliability
BI	6	0.93
PE	4	0.88
EE	4	0.91
SI	4	0.89
FC	4	0.89

4 Finding

4.1 Reliability Analysis

Reliability is defined as the extent to which measurements of a given test can be repeated. The alpha coefficient for the different constructs was calculated using the SPSS reliability procedure and is presented in Table 2 [26] argued that a coefficient of 0.70 is desirable. However, the reliability of all fixtures falls within the acceptable range, which is above 0.70. Therefore, acceptable alpha of cronbach's for the scales used in this study is above the lower limit (alpha >70).

4.2 Correlation Analysis

Table 3 explains a summary of the findings of the association analysis. To understand the relationship between all study variables. Pearson correlations coefficients were

Table 3 Pearson correlations of study variables

	BI	PE	EE	SI	FC
BI	1.0				
PE	0.68[a]	1.0			
EE	0.66[a]	0.74[a]	1.0		
SI	0.55[a]	0.66[a]	0.65[a]	1.0	
FC	0.49[a]	0.54[a]	0.65[a]	0.75[a]	1.0

[a] Correlation is significant at the 0.01 level

calculated, and based on Table 3 the dimensions of the correlation coefficients (r) show the strength of the link between the dimensions. The overall correlation values of the variables show the correlation with positive dimensions above 0.49.

4.3 Regressions

4.3.1 Analysis of Multiple Regressions on Dimensions that Impact BI

To answer the research question about dimensions that indicate BI to use SM, a regression analysis was performed on predictive factors and BI to use SM. Table 4 shows that the relationship between PE, EE, SI, and FC and BI to use SM is significant ($F = 39.707$; Sig. $= 0.00$). The R2 obtained displays that the influencing dimensions account for 63% of the variation in BI to use SM. Of all the dimensions involved in the regression equation, four dimensions appeared as significant predictors of Behavioral Intention to use SM. Based on these findings, H1, H2, H3, and H4 hypotheses are supported. This leads to the conclusion that PE, EE, SI, and FC are support related to Behavioral Intention to use SM. To measure which dimensions have the most influence on BI to use SM, Table 5, shows beta values were used. Based on the beta values of the four variables are significant, the predictor dimensions that exercise the most influence Behavioral Intention to use SM is PE($\beta = 0.334$), EE($\beta = 0.382$), SI ($\beta = 0.293$), and FC ($\beta = 0.281$).

Table 4 Summary of regression analysis for dimensions BI

Model	B	S.E	Β	T	Sig	Tolerance	VIF
1 (constant	275	304		905	368		
Expected performance	0.150	0.090	0.33	2.682	0.005***	0.421	1.920
Expected effort	346	084	0.38	4.234	0.004***	0.307	4.252
Social impact	0.037	0.078	0.29	0.481	0.002***	0.471	3.694
Facilitating condition	179	0.072	0.28	1.717	0.004***	0.550	2.222

Note Significant levels: *** p<0.00; ** p<0.01; * p<0.05

Table 5 Hypotheses testing result summary

	Hypothesis	Reject/accept
H1	Expected performance positive effects on BI to use SM in Jordanian SMEs	Accept
H2	Expected effort positive effects on BI to use SM in Jordanian SMEs	Accept
H3	Social impact positive effects on BI to use SM in Jordanian SMEs	Accept
H4	Facilitating condition positive effects on BI to use SM in Jordanian SMEs	Accept

For the regression of the independent dimensions on the behavioral intention to use the SM, the tolerance and condition index values for all independent dimensions are examined to detect multilinearity. The variance should be close to 1.00 to indicate little or no polylinearity [26]. ACI [24], a cut-off value of 10.00 has been suggested as an acceptable VIF. From the tolerance and VIF values shown in Table 5, the output indicates that there is no multi-line effect between the independent dimensions on the dependent dimension. The hypothesis test results are set out in Table 5 as follows.

5 Discussion and Theoretical Contribution

This research expands the overall concept of UTAUT in predicting factors of technology adoption, particularly in the SM applications adoption among Jordanian SMEs context. In other words, this research obtained a positive and significant relationship between Expected performance and BI. In other meaning, Expected performance could predict continuous improvement on BI to use SM in Jordanian SMEs [27]. This finding is stable with a past study by Dwivedi et al. [19] who obtained that Expected performance significantly influences BI. In the current study, the study also obtained a positive and significant relationship between Expected effort and BI [6, 17, 28, 29]. In other words, the Expected effort could predict continuous improvement on BI to use SM in Jordanian SMEs. However, this discovery is reliable with previous research by Abdat [17], and [6] which found that Expected effort significantly influences BI. This research found also a positive and significant relationship between Social impact and BI to use SM in Jordanian SMEs. In other meaning, the social impact could predict continuous improvement on BI to use SM in Jordanian SMEs. This result is consistent with past study findings by Abdat [17] that discovered a positive impact of social impact on BI to use SM in Jordanian SMEs. Finally, for facilitating of conditions were found significant and positive on BI. In other meaning, facilitating conditions could predict continuous improvement on BI to use SM in Jordanian SMEs. This finding is consistent with privacy study findings by Abdat [17] that discovered facilitating conditions influence BI to use SM in Jordanian SMEs.

6 Conclusion

The study aims to assess the dimensions that influence the adoption of social media in Jordanian SMEs to enhance their marketing performance and business technology use using the approach of the UTAUT model. The results of the research revealed that the (UTAUT) model was able to explain 65% of the variance in behavioral intention. The results of the study found that all dimensions of the (UTAUT) model were a positive and significant effect on Behavioral Intention to use Social Media in Jordanian SMEs. Another meaning, the search contributes in a sense that it confirms the influence of the UTAUT factors in Jordanian SMEs for adopting SM. However, despite this research supplied realizing of the influencing features of SM adoption for purposes of a business, there are some limitations to our study. Firstly, Jordan was the data collection center; the results should be accurately generalized to countries similar to Jordan's environment. Upcoming research can discover SM adoption cultures across different. Second, while a method of a survey of nature quantitative was chosen for this study, interview in-depth should supply more insights into SM adoption. Furthermore, future studies might examine comparatively among other accepted models of technology at the organizational and individual levels such as the DOI, TAM, and TOE models, to inform understanding of SM adoption for purposes of business.

References

1. Trawnih, A., Yaseen, H., Al-Adwan, A.S., alsoud, a.R., jaber, o.A.: Factors influencing social media adoption among SMEs during Covid-19 crisis. J. Manag. Inf. Decis. Sci. **24**(6), (2021)
2. Yaseen, H., Al-Adwan, A.S., Al-Madadha, A.: Digital marketing adoption among SMEs in Jordan: a mixed-method approach. J. Theor. Appl. Inf. Technol. **97**(4), 1396–1407 (2019)
3. Puriwat, W., Tripopsakul, S.: Explaining social media adoption for a business purpose: an application of the UTAUT model. Sustainability **13**(4), 2082 (2021)
4. Razak, S.A., Latip, N.A.B.M.: Factors that influence the usage of social media in marketing. J. Res. Bus. Manag. **4**(2), 1–7 (2016)
5. Ministry of Digital Economy and Entrepreneurship (MDEE).: Studies and Reports. https://www.modee.gov.jo/EN/List/Studies_and_Reports. (2021)
6. Alzougool, B., Al-Amad, A., Al-Tamimi, D., Al-Hind, A.: The adoption of social media by small and medium enterprises in Jordan: a descriptive study. J. Ahmed Bin Mohammed Mil. Coll. Adm. Sci. Law **1**(2) (2019)
7. Al-Weshah, G.A., Al-Zubi, K.: E-business enablers and barriers: empirical study of SMEs in Jordanian communication sector. Glob. J. Bus. Res. **6**(3), 1–15 (2012)
8. Alshourah, S.: Total quality management practices and their effects on the quality performance of Jordanian private hospitals. Manag. Sci. Lett. **11**(1), 67–76 (2021)
9. Alsharji, A., Ahmad, S.Z., Bakar, A.R.A.: Understanding social media adoption in SMEs: Empirical evidence from the United Arab Emirates. J. Entrepreneurship Emerg. Econ. **10**(2) (2018)
10. Wamba, S.F., Carter, L.: Social media tools adoption and use by SMEs: an empirical study. In: Social media and networking: concepts, methodologies, tools, and applications, pp. 791–806. IGI Globa (2016)

11. Chatterjee, S., Kar, A.K.: Why do small and medium enterprises use social media marketing and what is the impact: empirical insights from India. Int. J. Inf. Manage. **53**, 102103 (2020)
12. Hudders, L., De Jans, S., De Veirman, M.: The commercialization of social media stars: a literature review and conceptual framework on the strategic use of social media influencers. Int. J. Advert. **40**(3), 327–375 (2021)
13. Venkatesh, V., Morris, M.G., Davis, G.B., Davis, F.D.: User acceptance of information technology: toward a unified view. MIS quarterly, 425–478 (2003)
14. Venkatesh, V., Thong, J.Y., Xu, X.: Consumer acceptance and use of information technology: extending the unified theory of acceptance and use of technology. MIS quarterly, 157–178 (2012)
15. Waehama, W., McGrath, M., Korthaus, A., Fong, M.: ICT adoption and the UTAUT model. In: Proceedings of the international conference on educational technology with information technology, vol. 17, pp. 24–30 (2014)
16. Ikumoro, A.O., Jawad, M.S.: Intention to use intelligent conversational agents in e-commerce among Malaysian SMEs: an integrated conceptual framework based on tri-theories including unified theory of acceptance, use of technology (UTAUT), and TOE. Int. J. Acad. Res. Bus. Soc. Sci. **9**(11), 205–235 (2019)
17. Abdat, F.A.: Using UTAUT model to predict social media adoption among Indonesian SMEs. Saudi J. Econ. Fin. Cot. **4**(10), 498–505 (2020)
18. Altawalbeh, M.A.: The impact of using social sites on academic performance for students at Jordanian universities. Turkish J. Comput. Math. Educ. (TURCOMAT) **12**(7), 3264–3270 (2021)
19. Dwivedi, Y.K., Rana, N.P., Jeyaraj, A., Clement, M., Williams, M.D.: Re-examining the unified theory of acceptance and use of technology (UTAUT): Towards a revised theoretical model. Inf. Syst. Front. **21**(3), 719–734 (2019)
20. Dahnil, M.I., Marzuki, K.M., Langgat, J., Fabeil, N.F.: Factors influencing SMEs adoption of social media marketing. Procedia Soc. Behav. Sci. **148**, 119–126 (2014)
21. Hussein, L.A., Baharudin, A.S., Kiumarsi, S., Hilmi, M.F.: Factors influencing the intention to continue using B2B e-commerce in manufacturing SMEs. Eng. Technol. Appl. Sci. Res. **10**(2), 5528–5533 (2020)
22. Mandal, D., McQueen, R.J.: Extending UTAUT to explain social media adoption by micro businesses. Int. J. Managing Inf. Technol. **4**(4), 1 (2012)
23. Al Mursalin, J.: Information system adoption and usage: validating UTAUT model for Bangladeshi SMEs. Int. J. Manag. IT Eng. **2**(10), 1–19 (2012)
24. Amman chamber of Industry (ACI(.: Retrieved from 2022. http://www.aci.org.jo/development/en/. (2020)
25. Crammond, R., Omeihe, K.O., Murray, A., Ledger, K.: Managing knowledge through social media. Balt. J. Manag. **13**(3), 303–328 (2018)
26. Hair Jr, J.F., Anderson, R.E., Tatham, R.L., Black, W.C.: Multivariate data analysis: With readings: Prentice-Hall, Inc, (2010)
27. Ainin, S., Parveen, F., Moghavvemi, S., Jaafar, N.I., Shuib, N.L.M.: Factors influencing the use of social media by SMEs and its performance outcomes. Ind. Manag. Data Syst. **115**(3), 570–588 (2015)
28. Alalwan, A.A., Rana, N.P., Dwivedi, Y.K., Algharabat, R.: Social media in marketing: A review and analysis of the existing literature. Telematics Inform. **34**(7), 1177–1190 (2017)
29. Al Tawara, A., Gide, E.: A comprehensive literature review on the adoption of social media marketing in SME retailors in Jordan. Int. J. Bus. Manag. Studies. ISSN, 2158-1479 (2017)

The Impact of Entrepreneurial Orientation on Innovative Performance in Jordanian SMEs During the Covid-19 Pandemic

Mohammad Abdalkarim Alzuod and Farah Niaz Dalain

Abstract The purpose of this research was to evaluate the impact of entrepreneurial orientation (EO) in the setting of Jordanian SMEs, and to empirically examine the relationship between entrepreneurial orientation (EO) and its dimensions, namely innovativeness, risk-taking, and proactiveness, and innovative performance. As a result, data were gathered through a questionnaire survey given to 361 managers/owners of SMEs in Jordan, with a specific emphasis on those based in Zarqa city. 221 questionnaire surveys were recovered and declared suitable for analysis out of the 361 disseminated. IBM SPSS Statistics 22 was used in analysis of the data. The findings indicate that entrepreneurial orientation does have a significantly positive link with innovative performance. However, the entrepreneurial orientation aspects of innovativeness, risk-taking, and proactiveness were shown to have a negligible influence on innovative performance. According to these findings, the researcher suggests that Jordanian SMEs prioritize the entrepreneurial orientation aspects explored in this study in order to strengthen their entrepreneurial culture and innovative performance.

Keywords Innovative performance · Entrepreneurial orientation · SMEs · Jordan

1 Introduction

Innovation is often regarded as one of the most important resources of sustained competitive advantage, since it outcomes in enhancements of up-to-date goods and services, to the side of adds value to performance [1]. Also, innovation can be considered as an approach in which an organization illustrates and defines challenges before

M. A. Alzuod (✉) · F. N. Dalain
Zarqa University, Zarqa, Jordan
e-mail: malzuod@zu.edu.jo

F. N. Dalain
e-mail: fdalain@zu.edu.jo

A. Hamdan et al. (eds.), *The Implementation of Smart Technologies for Business Success and Sustainability*, Studies in Systems, Decision and Control 216,
https://doi.org/10.1007/978-3-031-10212-7_14

attempting to acquire innovative knowledge to solve them. While innovative performance (IP) is critical for increasing and maintaining competitive advantage, most firms, particularly SMEs, have difficulties in remaining innovative performance.

Numerous studies have been made to examine the association between entrepreneurial orientation (EO) and innovative performance. Several researches; [2–4] reviewed a favorable and statistically significant association between EO and IP. It was discovered that creative small and medium firms can assist in the processing, creation, and introduction of unique products/technologies, as well as ensuring optimal performance.

Khalili et al. [2], made a similar point, emphasizing the need of examining EO and IP across sectors, organizational kinds, and sizes. Serrano-Bedia et al. [5] discovered that IP seemed as comparable across sectors in this aspect. Finally, [6] advocated for further empirical study in a variety of sectors to examine the impact of EO on organizational learning and IP.

When comparing the level of innovative performance among the developed and developing countries, there are evidence showing the declining in innovative performance among countries. On the basis of the Global Innovation Index Ranking, Jordan went from 64 position in 2014 to 81 position in 2021 during the Covid-19 [7].

Covid-19 virus expanded rapidly in China in 2019, plunging the majority of nations into a state of panic about how to cope with the pandemic threat. The World Health Organization (WHO) classified it as a worldwide pandemic due to its rapid spread to several nations. Numerous governments have implemented lockdown procedures to prevent the virus from spreading. This Covid-19-related lockout has a huge impact on market and social activity. This lockout is particularly detrimental to Small and Medium-Sized Enterprises (SMEs) [8]. SMEs account for 95% of Jordan's national economic components, contribute 40% of GDP, and employ around 70% of the labor force in the country. According to [9], critical success elements for SMEs in Jordan include technical processes and technologies, organizational structure, monetary system, marketing and productivity, and human resource structure. The technological factor was at the forefront of success, since it connected to the inventive capacity of SMEs.

According to the reviewed literature, empirical studies on the influence of an EO and IP are still few, which served as the motivation for this effort to investigate the significance of innovative performance which Jordanian SMEs might embrace. The research is intended to have consequences for academic and practitioner circles about ways for enhancing the entrepreneurial orientation of the country's population, particularly SMEs.

2 Literature Review

2.1 Innovative Performance (IP)

According to [2] that IP is the novelty of products/services, which includes new projects and goods that lead to an increase in quality and the compatibility of the organizational structure with the requirements of the competitive environment. Hence, an organization's IP can be quantified by the amount of new commodities and projects, the volume of innovative methods of work, the number of patentable innovations, the renewal of the organization's structure to adapt to conditions of dynamic environment, the marketing of new products ahead of competition, and the new products rate in the line of production. While, [10] concentrated in one of their study that IP is the realization of ideas, which may take the shape of drawings and models of unique processes, products, procedures, and gadgets. According to [11], IP is defined as an organization's new product success that results in other sorts of efficiency such as advertising, production, and finance.

Halim et al. [12], proposed that firms require IP to foster novelty of new ideas, experimentation, and creative processes that result in the developments of new products/services, or technological improvements and the pursuit of creative and novel solutions. Comlek et al. [13], defined IP in terms of the extent to which a corporation introduced new commodities, devices, or process systems to the market. IP was assessed using various standards, including the amount of new items required to bring the first product/service to market, the pace with which the product/service is introduced to the market, the amount of newly patented products or processes that must be recreated. Additionally, IP may be defined as the total performance of the firm's goods and processes that adds to its economic performance. It is an evaluation that is represented in the firm's research and development stages, the introduction of new items, the copyrights scales, the trademarks testimonies, and marketing.

The concept of IP can be classified to product innovation and process innovation [14]. Also, [15] considered two measures for innovative performance, product innovation, and process innovation. While product innovation refers to the introduction of a new or upgraded product, piece of equipment, or service to the market, process innovation refers to the introduction of a particular or improved industrial production process or a creative form of social service. This is not to say that the two are mutually incompatible; innovativeness can lead to product development and inversely [16].

The dependent variable in this research is IP. It is quantified as the synthesis of achievements of organization's entire as outcomes of its rejuvenation and improvements of efforts in several creative aspects of the business, notably operations, commodities, and organization [17]. Additionally, a one-dimensional notion of IP is used in this research. It is more appropriate for the research setting since it is on the basis of the prior research by Alpkan et al. [18] that integrates the overall outcomes of product and process innovation and radical/incremental innovation.

2.2 Entrepreneurial Orientation (EO)

The topic of entrepreneurial orientation is also extensively studied in the literature [19], for instance, characterized EO as the decision-making processes, techniques, and activities that support new entrance. It is also an EO that is recognized and approved for more qualities, which it's include (e.g. innovativeness, riskiness, proactiveness, aggressive competitiveness and autonomy). On other hand, [2] suggested in another article that EO includes a desire to innovate, take calculated risks, operate self-guided procedures, and become more aggressive and proactive than rivals when it comes to new market possibilities. Nevertheless, [20], and [21] agreed that characteristics of EO reflect management's behavior in assumption of strategic choices and operational philosophies of management that are embodied in two aspects: proactivity and risk-taking. As with [21] defined entrepreneurial orientation as a managerial mentality and categorize it into three broad dimensions: (e.g. radical innovation, competitive orientation, and aggressive or proactive risky decisions). While in terms of family entrepreneurial orientation considered [22] EO as the attitudes and mind-sets of families to engage in entrepreneurial activity.

EO seems to be the variable that is operationalized as a firm's readiness to engage in taking risks and innovative actions in order to develop new products or services, enter new markets, and proactively seize new market opportunities ahead of competition [23]. Lastly, the most comprehensive dimensions used to conceptualize EO were summarized; innovativeness, risk-taking and proactiveness, accounting for over 80% of overall quantitative articles. As according [23], most EO factors are often significantly connected.

2.3 Entrepreneurial Orientation and Innovative Performance

Sanal et al. [24], urged for more studies on EO and its incorporation into a paradigm consisting; "market orientation and ambidexterity to IP". Another conception was offered with [2] that emphasized the requirement of examining EO and IP across sectors, organizational kinds, and sizes. In the same context, [5] discovered that the IP firms seemed to be comparable across sectors. Finally, [6] recommended doing more empirical study in a variety of sectors to determine the influence of EO on IP and organizational learning.

Furthermore, from previous literature review, the moderating roles of EO studies have conducted within another context. Case in point, [25] used EO as moderator through intellectual capital and innovation. Also, [26] study discovered that EO as moderator between environmental turbulence and IP relationship. While, [27] found that EO should be affected among ambidextrous capabilities (explorative and exploitative capabilities) and IP passively. On the same line, [28] was examined the balance dimension of organizational ambidexterity and IP relationship amongst Chinese high/new technology firms. [29], also EO has been investigated on the relationship of resource orientation and market orientation with IP at Dutch manufacturing SMEs. Lastly, [30] found risk taking dimension of EO as intra-cluster ties, extra-cluster ties with IP through Indonesian SMEs in the Cibaduyut footwear-manufacturing firms.

On the basis of this analysis, this research suggests and tests the following hypotheses:

H1: There is no statistically positive significant impact of entrepreneurial orientation on innovative performance.
H1a: There is no statistically positive significant impact of innovativeness on innovative performance.
H1b: There is no statistically positive significant impact of risk-taking on innovative performance.
H1c: There is no statistically positive significant impact of pro-activeness on innovative performance.

2.4 Model of Study

Entrepreneurial orientation is critical for the development of innovative performance in SMEs, and as a result, the above mentioned argument based on [2, 5, 6] that lead to the suggested model of research (Fig. 1).

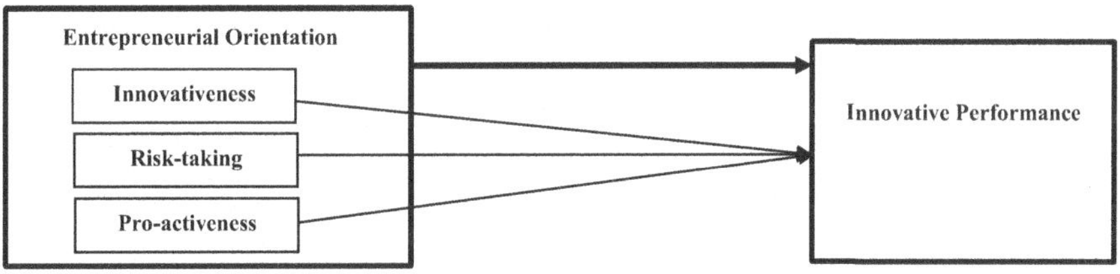

Fig. 1 Model of study [2, 5, 6]

3 Research Methodology

3.1 Population and Sampling of the Study

Considering practical reasons, 5950 enterprises in Zarqa city were chosen as the research sample [31]. The city of Zarqa was chosen since it is one of the top cities in Jordan, with the bulk of SMEs based on Ministry of Planning & International Cooperation [32]. The research used tables of Krejcie and Morgan [33] to choose the suitable size of sample, which are 361. The research used a systematic random sample procedure, followed by the administration of 361 questionnaire surveys. 221 questionnaires were collected and considered suitable for analysis from the administered surveys, showing a response rate of 61.2%. The respondents to the survey were owners/managers of SMEs based in Zarqa city, with the bulk of the firms being less than 10 years old (71.3%). According to the research, 44.6% of enterprises were classified as manufacturing, while 17% were classified as engineering and electrical. In terms of overall employment, the majority of enterprises employed between 20 and 99 people (43.7%).

3.2 *Measurements*

The items in this research were graded on a Likert scale ranging from (Disagree strongly to agree strongly). Seven elements from the IP Index are used to evaluate innovative performance [17]. EO is quantified using nine questions taken from [23] that assess; (innovativeness dimension, risk-taking dimension and pro-activeness dimension) respectively.

4 Results

4.1 *Data Analysis*

As previously stated, the descriptive statistic of data was conducted using IBM SPSS 22, and regression analysis was performed to evaluate the developed hypothesis.

4.2 *Reliability Test*

The reliability of the variables was specified using the minimal Cronbach's alpha level determined by [34]. The consistency of the variable's study is much more than the approved level, as suggested by Hair et al. [34]. As Table 1 the lowest acceptable level for (α) is 0.60. This demonstrates that the research tool is very reliable.

As a result, significance is classified into three categories: low, medium, and high, with the low category consisting of values between 1.33 and 2.33, the medium category consisting of values between 2.34 and 3.67, and the high category consisting of values between 3.68 and 5. Table 2 used offered a descriptive analysis of the relative relevance of the study's major variables based on the five Likert scales. The indicators

Table 1 Reliability test

Variable	Item no	Cronbach's alpha (α)
Innovative performance	7	0.83
Entrepreneurial orientation	9	0.87
Innovativeness	3	0.92
Risk-taking	3	0.89
Pro-activeness	3	0.81
Total	16	0.85

Table 2 Mean and std. deviation of the main variables

Items	Mean	Standard deviation	Level of importance
Entrepreneurial orientation	3.425	0.614	Medium
Innovative performance	3.517	0.566	Medium

in the table below used indicate the relative relevance of the study's major factors on a medium scale. This suggests that respondents had a favorable view about the prevalence of these characteristics in Jordanian SMEs.

4.3 Hypotheses Test

To evaluate the first hypothesis, the following methodology will be utilized, as the significance of the ANOVA table will be used to define the major hypothesis and sub hypotheses. Each of the given tables is analyzed in detail Tables 3, 4 and 5.

Table 3 Analysis of multiple regressions for the hypotheses

R	R-square	Adjusted R square
0.708	0.498	0.492
ANOVA table		
F-value	Sig	
88.71	0.000	

Table 4 Coefficients of hypotheses

Variables	Beta	*T*-calculated	Sig
Innovativeness	1.133	6.552	0.000
Risk-taking	0.123	3.251	0.010
Pro-activeness	0.016	0.346	0.414

Table 5 Testing hypotheses results

Main hypothesis and sub hypotheses	Result
H1: There is no statistically positive significant impact of entrepreneurial orientation on innovative performance	Rejected
H1a: There is no statistically positive significant impact of innovativeness on innovative performance	Rejected
H1b: There is no statistically positive significant impact of risk-taking on innovative performance	Rejected
H1c: There is no statistically positive significant impact of pro-activeness on innovative performance	Rejected

5 Discussion

The primary purpose of this study is to specify the impact of entrepreneurial orientation on innovative performance in Jordanian small and medium-sized enterprises (SMEs) during the covid-19 epidemic. According to the findings, EO as well as its aspects (innovativeness, risk-taking, and proactiveness) has a major impact on IP, a finding that is accurate with and reinforces the findings of prior studies by [6, 27, 29], as well as [30]. The strong effect of EO and IP may relate to entrepreneurial orientation's crucial role in increasing innovative performance in SMEs.

According to the study's statistical findings, the relevance of the study's primary factors, EO and IP, is medium. Such degrees have been shown to boost Jordanian SME innovation and entrepreneurship in their commodities and activities.

This finding may be attributed by the absence of an innovation culture inside the Jordanian SMEs analyzed, which is required for the investigation of their learning possibilities. Although the majority of respondent SMEs are situated in Zarqa (an economic region), creative activities are fairly limited, and innovation culture reflects current practices and a genuine dedication to product development.

A critical topic raised by this analysis is the Covid-19 epidemic that would impact the Jordanian market between 2019 and 2021, particularly affecting small businesses. These obstacles compelled the SMEs to work innovatively and entrepreneurially in order to overcome marketing obstacles and blockages.

Regardless of its theoretical and practical contributions, the current empirical study has some limitations that must be addressed in order to make way for future studies. Future research may expand the research framework to include big businesses, and the research mode may be expanded to a wider scale, involving a large number of participants, a large firm size, and a diverse range of sectors, in order to gain a more complete knowledge about the impacts of entrepreneurial orientation on innovativeness.

The research makes a contribution by emphasizing and describing how Jordanian SMEs might improve their chances of success. The study's findings are likely to have ramifications for both academic and practitioner circles about how to increase the degree of innovation in SMEs.

The researcher proposes that Jordanian SMEs concentrate their efforts on the EO factors evaluated in this study in order to foster an entrepreneurial culture and improve their innovation performance. Also, the Jordanian government should be to provide support all types to SMEs to advance them, especially what they have suffered during the period of the Covid—19 pandemic.

6 Conclusions

The purpose of this research was to investigate the entrepreneurial orientation dimensions of innovativeness, risk-taking and proactiveness, as well as their influence on innovative performance, from the perspective of owners/managers of SMEs in Jordan's Zarqa province. The data were acquired using survey questionnaires and processed using SPSS regression analysis. The findings indicate that the characteristics of EO have a considerable influence on IP, corroborated by earlier research indicating that entrepreneurial orientation has the greatest effect on innovative performance.

References

1. Delgado, M.: The role of intellectual capital assets on the radicalness of innovation: Direct and moderating effects Working papers published at UAM-Accenture. University of Madrid, Madrid (2011)
2. Khalili, H., Nejadhussein, S., Fazel, A.: The influence of entrepreneurial orientation on innovative performance: study of a petrochemical company in Iran. J. Knowl. Innov. China **5**(3), 262–278 (2013)
3. Madhoushi, M., Sadati, A., Delavari, H., Mehdivand, M., Mihandost, R.: Entrepreneurial orientation and innovation performance: the mediating role of knowledge management. Asian J. Bus. Manag **3**(4), 310–316 (2011)
4. Wiklund, J., Shepherd, D.: Knowledge-based resources, entrepreneurial orientation, and the performance of small and medium-sized businesses. Strateg. Manag. J. **24**(13), 1307–1314 (2003)
5. Serrano-Bedia, A.M., Concepción López-Fernández, M., García-Piqueres, G.: Complementarity between innovation activities and innovation performance: evidence from Spanish innovative firms. J. Manuf. Technol. Manag. **23**(5), 557–577 (2012)
6. Fernandez-Mesa, A., Alegre, J.: Entrepreneurial orientation and export intensity: examining the interplay of organizational learning and innovation. Int. Bus. Rev. **24**(1), 148–156 (2015)
7. Cornell University, INSEAD and WIPO. The global innovation index 2021: the human factor in innovation, second printing. Fontainebleau, Ithaca, and Geneva (2021)
8. Li, Z., Anaba, O.A., Ma, Z., Li, M.: Ghanaian SMEs amidst the covid-19 pandemic: evaluating the influence of entrepreneurial orientation. Sustainability **13**(3), 1131 (2021)
9. Al-Mahrouq, M.: Success factors of small and medium-sized enterprises (SMEs): the case of Jordan. Anadolu Univ. J. Soc. Sci. **10**(1), 1–16 (2010)
10. Hagedoorn, J., Cloodt, M.: Measuring innovative performance: is there an advantage in using multiple indicators? Res. Policy **32**(8), 1365–1379 (2003)

11. Hassan, M.U., Shaukat, S., Nawaz, M.S., Naz, S.: Effects of innovation types on firm performance: an empirical study on Pakistan's manufacturing sector. Pak. J. Commer. Soc. Sci. **7**(2), 243–262 (2013)
12. Halim, H.A., Ahmad, N.H., Ramayah, T., Hanifah, H.: The growth of innovative performance among SMEs: leveraging on organisational culture and innovative human capital. J. Small Bus. Entrepreneurship Dev. **2**(1), 107–125 (2014)
13. Comlek, O., Kitapçı, H., Çelik, V., Özşahin, M.: The effects of organizational learning capacity on firm innovative performance. Procedia Soc. Behav. Sci. **41**, 367–374 (2012)
14. Alzuod, M.A.K., Isa, M.F.M., Othman, S.Z.B.: Examining of innovative performance amongst SMEs in Jordan. J. Knowl. Manag. Appl. Pract. **2**(1), 53–69 (2019)
15. Carvache-Franco, O., Carvache-Franco, M., Carvache-Franco, W.: Barriers to innovations and innovative performance of companies: a study from Ecuador. Social Sciences **11**(2), 63 (2022)
16. Neely, A., Hii, J.: Innovation and business performance: a literature review. Working papers published at The Judge Institute of Management Studies, University of Cambridge, Cambridge, England (1998)
17. Gunday, G., Ulusoy, G., Kilic, K., Alpkan, L.: Effects of innovation types on firm performance. Int. J. Prod. Econ. **133**(2), 662–676 (2011)
18. Alpkan, L., Bulut, C., Gunday, G., Ulusoy, G., Kilic, K.: Organizational support for intrapreneurship and its interaction with human capital to enhance innovative performance. Manag. Decis. **48**(5), 732–755 (2010)
19. Lumpkin, G.T., Dess, G.G.: Clarifying the entrepreneurial orientation construct and linking it to performance. Acad. Manag. Rev. **21**(1), 135–172 (1996)
20. Antoncic, B., Hisrich, R.D.: Intrapreneurship: Construct refinement and cross-cultural validation. J. Bus. Ventur. **16**(5), 495–527 (2001)
21. Avlonitis, G.J., Salavou, H.E.: Entrepreneurial orientation of SMEs, product innovativeness, and performance. J. Bus. Res. **60**(5), 566–575 (2007)
22. Giner, B., Ruiz, A.: Family entrepreneurial orientation as a driver of longevity in family firms: a historic analysis of the ennobled Trenor family and Trenor y Cía. Bus. Hist. **64**(2), 327–358 (2022)
23. Soininen, J., Martikainen, M., Puumalainen, K., Kyläheiko, K.: Entrepreneurial orientation: growth and profitability of finnish small-and medium-sized enterprises. Int. J. Prod. Econ. **140**(2), 614–621 (2012)
24. Sanal, M., Alpkan, L., Aren, S., Sezen, B., Ayden, Y.: Linking market orientation and ambidexterity to financial returns with the mediation of innovative performance. J. Econ. Soc. Res. **15**(1), 31–54 (2013)
25. Wu, W.Y., Chang, M.L., Chen, C.W.: Promoting innovation through the accumulation of intellectual capital, social capital, and entrepreneurial orientation. R & D Manag. **38**(3), 265–277 (2008)
26. Al-Nuiami, M., Idris, W.M.S., Moh'd AL-Ferokh, F.A.: An empirical study of the moderator effect of entrepreneurial orientation on the relationship between environmental turbulence and innovation performance in five-star hotels in Jordan. Int. J. Bus. Adm **5**(2), 111–125 (2014)
27. Yang, D., Lang, C., Li, H.: Ambidextrous capabilities and innovation performance-The moderating effect of Entrepreneurial orientation. In: International Conference on E-Business and E-Government. pp. 1081–1083. IEEE (2010)
28. Bai, J., Ren, J.: Organizational ambidexterity and innovation performance: the moderating effect of entrepreneurial orientation. Front. Bus. Res. China **10**(4), 664 (2016)
29. Reulink, R.B.J.: Strategic orientation and innovation performance at Dutch manufacturing SME's: the overrated role of market orientation and entrepreneurial orientation. Master's thesis, University of Twente, Enschede, Netherlands (2012)
30. Gunawan, T., Jacob, J., Duysters, G.: Entrepreneurial orientation and network ties: innovative performance of SMEs in an emerging-economy manufacturing cluster (No. 2013/28). Maastricht School of Management (2013)
31. Zarqa chamber of industry: Annual report 2021. Zarqa, Jordan (2021)
32. Ministry of planning and international cooperation: Annual report 2020.Amman, Jordan (2020)

33. Krejcie, R.V., Morgan, D.W.: Determining sample size for research activities. Educ. Psychol. Measur. **30**(3), 607–610 (1970)
34. Hair, J.F., Black, W.C., Babin, B.J., Anderson, R.E.: Multivariate data analysis, 7th edn. Prentice Hall, Upper Saddle River, New Jersey (2010)

The Role of Information Technology to Fight the Covid-19 Pandemic (Jordan is a Model)

Majed Al Masadeh, Fatima O. Haimour, Siham O. Haimour, Maram Y. Al-Safarini, and Farah Al Masadeh

Abstract Jordan, like other countries, was affected by the global pandemic (COVID-19) which required the need for researchers' efforts to develop the latest electronic devices, modern technologies, and digital applications to limit the spread and mitigate the damage caused by it. Information systems and technology played a greater role in this global pandemic. In Jordan, researchers developed various technology innovations and applications to fight the COVID-19 pandemic. This paper studies the information technology investments to respond to this pandemic. It also presents a questioner about information technology investments from this research, the results conclude the strength and the ability of information technology have an important role to cope with damage from COVID-19.

Keywords Covid-19 · Evaluation of IT investment · Information technology

M. Al Masadeh
Economics and Administrative Sciences, Zarqa University, Zarqa, Jordan
e-mail: mmasadeh@zu.edu.jo

F. O. Haimour (✉) · S. O. Haimour · M. Y. Al-Safarini
Information Technology, Zarqa University, Zarqa, Jordan
e-mail: fhaimour@zu.edu.jo

S. O. Haimour
e-mail: shaimour@zu.edu.jo

M. Y. Al-Safarini
e-mail: maram@zu.edu.jo

F. Al Masadeh
Economics and Administrative Sciences, Jordan University, Amman, Jordan

A. Hamdan et al. (eds.), *The Implementation of Smart Technologies for Business Success and Sustainability*, Studies in Systems, Decision and Control 216,
https://doi.org/10.1007/978-3-031-10212-7_15

1 Introduction

The emergence of the Corona pandemic came at a time when the role of information technology and computer technology has escalated, and the ranks of companies, information institutions, projects, research, and products have risen to the top of the global pyramid in capabilities and influence [1].

Before the outbreak of the pandemic, information products and computers of all sizes, capabilities, and scope of work were racing against time to provide solutions and initiatives in the field of healthcare, so it was necessary to present what it had in the battle with the pandemic and its various effects and to employ its enormous potential to seize the historical moment [2].

The expansion in the use of data science technology, big data, artificial intelligence, communication infrastructure, and information technology was necessary and effective in the scientific achievements made during the past two years of In order to achieve the basic objectives of containing, stopping, and controlling the epidemic.

It was said that "something good came out of the bad", COVID-19 presents many opportunities for service and technology providers, it also provided a golden opportunity in a global competition for scientific and medical research, not only this but also invest in technology which is a reminder that digital technology brings a package of benefits and play an important role in reducing and managing the risks that are caused by the quarantine during and after the pandemic [1, 3].

It is well known that information technology and information systems have an important role in many sectors and fields such as: (government, education, industrial, medical, information technology, and Banking) [2, 4].

Many researchers are hard-working to help to fight the pandemic; they developed products, apps, and digital tools to predict and track the spreading of the virus to combat it. So that it has a huge impact to protect hospitals from this scary virus [5].

All workers in the information technology sector and future pandemics must contribute to fighting the pandemic by utilizing their previous knowledge and experience to limit the spread of this epidemic.

According to the great role that is played by information technology tools to help epidemiological investigation teams in speeding access to the infected and those in contact with them in Jordan, we highlighted in this paper the most important websites, digital tools, and applications that have been adopted in Jordan as a tool to limit the spread of the Corona virus and predict its presence to control it, as well as We created a questionnaire that surveys the opinions of the local community from several areas by asking a set of questions, then we analyzed them and discussed their results.

The remainder of the paper is organized as follows. In Sect. 2, related work is reviewed. In Sect. 3, the method and research design is described. In Sect. 4, the results of our study and the discussion is appeared in Sect. 5. Section 6 concludes the paper.

Table 1 Jordanian IT solutions during covid-19

No.	Name	Description
1	SanadJo	SANAD Mobile App allows access to Jordan Digital Services with single e-ID [6]
2	Amman Jordan covid-19 (AMAN)	AMAN App has been developed for Jordan's Ministry of Health (MOH) by COVID-19 JOTECH Community, a group of tech-savvy volunteers. AMAN is a privacy-conscious, exposure detection App. It helps keep the community and Jordan safe from COVID-19 [7]
3	CRadar	An electronic application (CRadar) to report injuries of all kinds, or suspected cases, with the ability to determine the exact location; [8]
4	Visit Jordan	A travel journey that is safe and secure [9]

2 Related Works

The world has developed many applications to take advantage of technology to handle the Corona pandemic. In the following, we briefly describe some of Jordanian IT solutions that are used during Quarantine, discuss how information systems and technology scholars could contribute knowledge and insights to help fight the pandemic.

2.1 Jordanian IT Solutions

Table 1 briefly describes some of the Jordanian IT solutions during covid-19 pandemic.

3 Methods

3.1 The Research Questions

The research questions formulation developed from the framework are:

- What is the role of information technology for the covid-19 pandemic?
- What is the role of information technology when facing a pandemic?

3.2 Study Design

This study used a self-reviewed online questionnaire. The sample consisted of the sample of the local community (100 individuals) in various fields, such as Government, Education, Industrial, Medical, Information Technology, Banks and others.

3.3 Data Collection

All members of the local community were contacted through their email addresses and invited to participate in the study. They were provided with the aim of the study and a link to the online survey, which was designed using the Google Forms online-survey platform. All responses were anonymous and confidential.

3.4 Data Analysis

The authors used SPSS version 23 to analyze the results of the study. Only completed questionnaires were analyzed. Descriptive statistics were employed for demographic characteristics such as gender, works. Moreover, acceptability and ethical consideration were analyzed using means and percentages.

4 Results

4.1 Demographic Variables

The total number who agreed to participate was 115 individuals. The number of females was 47 (40.87%) and the number of males was 68 (59.13%). The mean age = 29.5 (SD = 7.7). The participants were divided according to the nature of the work into seven categories (see Table 2); Government by a percentage of 15.65. Education, Industrial, and Medical sectors have the same percentage with (14.78), Information Technology with (17.39), Banks with (12.17), others sector under "other" categories with (10.43).

Table 2 Distribution of study participants according to their business sectors

Government	Education	Industrial	Medical	Information technology	Banks	Others
15.65%	14.78%	14.78%	14.78%	17.39%	12.17%	10.43%

4.2 *The Role of Information Technology to Combat Covid-19*

In the following, we will read the results that obtained from the questionnaire (see Table 3).

The vast majority of participants (94.78) think that the covid-19 is very dangerous, while (3.48) think the opposite and (1.74) neither agree nor disagree.

A percentage of 75.65 strongly wish to use government applications that were approved by the government during the Corona pandemic. Also a ratio (21.74) they want to use it with a total of (97.39). While a very small percentage (1.74) is not looking to use it and (0.87) neither agree nor disagree.

The percentage of participants who strongly consider technological applications useful during the quarantine (54.78) and the percentage (26.09) also consider it, while the percentage (10.34) does not consider and (8.70) neither agree nor disagree.

A (38.26) of participants strongly believe that IT applications helped to reduce the spread of the Corona virus, and (26.09) believe it also. (31.91) do not believe that. (21.74) neither agree nor disagree.

The vast majority of the participants in the study and their percentage (86.96) believe technological applications helped in social communication during the quarantine, while (7.83) do not agree with that. (5.22) neither agree nor disagree.

A percentage of (62.61) strongly considered as technology has helped in doing business and various activities. Also a ratio (27.83) they want to use it with a total of (90.43). While (5.22) is not considered that and (4.35) neither agree nor disagree.

The percentage of subscribers who consider that technology has helped spread health awareness is (59.13), also (23.48) considered that. While (10.43) don't consider that. (6.96) neither agree nor disagree.

The percentage of subscribers who consider that create job opportunities during the quarantine is (26.96), also (50.43) considered that. While (7.83) don't consider that. (14.78) neither agree nor disagree.

The ratio of participants who responded strongly to the Technology has helped spread the latest news about health developments is (55.65) as well as (35.65), while the ratio of participants who responded strongly against that is (3.48), while (5.22) neither agree nor disagree.

The ratio of participants who responded strongly to the Technology has helped spread rumors and false news about vaccinations is (55.65) as well as (32.17), while the ratio of participants who responded strongly against that is (10.43), while (1.74) neither agrees nor disagrees.

Percentage of subscribers who consider that Technology helped raise awareness of vaccinations is (50.43), also (50.43) considered that. While (7.83) don't consider that. (14.78) neither agree nor disagree.

Only a (39.13) of participants who believe Information technology has helped in reaching people infected with the covid-19, also (39.13) considered that. While (11.30) don't consider that. (10.43) neither agree nor disagree.

Table 3 The questionnaire's results

	Strongly disagree	Disagree	Neither agree nor disagree	Agree	Strongly agree
In your opinion, is the covidl9 dangerous?	1.74	1.74	1.74	42.61	52.17
Did you need to use one of the government applications designed to combat the covid-19?	0.87	0.87	0.87	21.74	75.65
Do you consider technological applications useful during the quarantine period?	1.74	8.70	8.70	26.09	54.78
Have IT applications helped reduce the spread of the Corona virus?	5.22	8.70	21.74	26.09	38.26
Technological applications helped in social communication during the quarantine	4.35	3.48	5.22	33.91	53.04
Technology has helped in doing business and various activities	3.48	1.74	4.35	27.83	62.61
Technology has helped spread health awareness	8.70	1.74	6.96	23.48	59.13
Technology has helped create job opportunities during the quarantine	3.48	4.35	14.78	50.43	26.96
Technology has helped spread the latest news about health developments	1.74	1.74	5.22	35.65	55.65

(continued)

Table 3 (continued)

	Strongly disagree	Disagree	Neither agree nor disagree	Agree	Strongly agree
Technology has helped spread rumors and false news about vaccinations	6.09	4.35	1.74	32.17	55.65
Technology helped raise awareness of vaccinations	2.61	4.35	13.91	28.70	50.43
Information technology has helped in reaching people infected with the covid-19	6.96	4.35	10.43	39.13	39.13
Information technology has helped reach quarantine breakers	2.61	2.61	15.65	20.00	59.13

A (59.13) of participants who believe Information technology has helped reach quarantine breakers, also (20.00) considered that. While (5.22) don't consider that. (15.65) neither agree nor disagree.

5 Discussions

Technology played an important and effective role during the Corona pandemic period in many fields of health, education, and various area. It helped to overcome many problems and try to control and mitigate the effects of global outbreaks. Below we discuss the most prominent things that resulted from our questionnaire.

Figure 1 shows the comparison between different components of the questionnaire.

The component of a questionnaire that got the highest percentage from the respondent's point of view uses one of the government applications designed to fight the covid-19, while the component "Technology has helped create job opportunities during the quarantine" got the least percentage with "strongly agree" opinion.

The majority believes that information technology has contributed to spreading awareness about the covid-19. It helped to reach quarantine breakers. It helped to spread health awareness. Information technology has had a significant impact in all areas and businesses, including health, education, and economics as well.

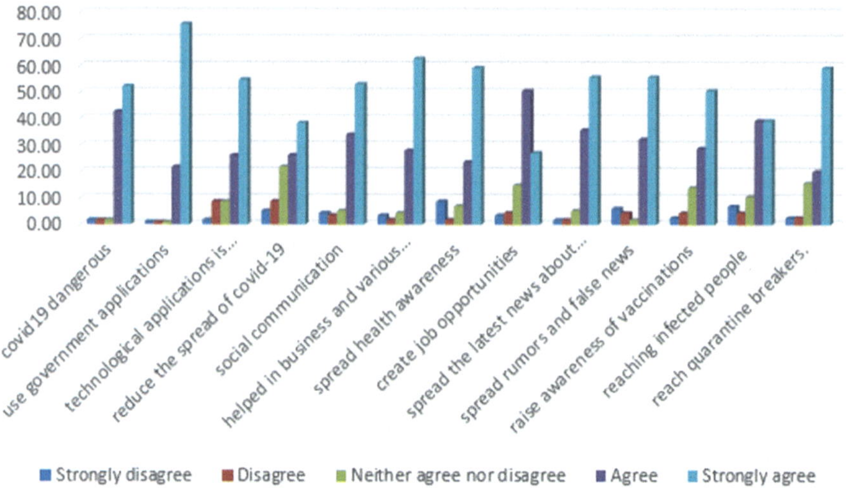

Fig. 1 Compare the different components of the questionnaire

Information technology also had an important role in communicating the latest developments in the epidemic, and on the other hand, it contributed in one way or another to spreading rumors about disease and vaccinations later.

The second lowest percentage of respondents to "IT applications helped in reduce the spread of the covid-19". And the third lowest percentage of respondents "IT helped in reaching people infected with the covid-19".

6 Conclusion

Until now, the epidemic is still around us and we still need every opportunity to face it in various ways. We never deny the prominent role of information technology in confronting the pandemic, and we see it necessary to unite all efforts between the technology sector and other sectors to try to connect to the highest levels of benefiting from modern technology, artificial intelligence and machine learning. And other sectors in the field of technology. Finally, we can say, after Corona, life is normal, but it is digital in everything. In the future, we aspire to explore the most prominent investments that information technology can provide to confront Covid 19.

From this research we can conclude that during the COVID-19 pandemic there was clear evidence that technology played an important role in the success of pandemic handling.

References

1. Sein, M.K.: The serendipitous impact of COVID-19 pandemic: a rare opportunity for research and practice. Int. J. Inf. Manage. **55**, 102164 (2020)
2. Ben-Assuli, O., Padman, R.: Trajectories of repeated readmissions of chronic disease patients: risk stratification, profiling, and prediction. MIS Q. **44**(1) (2020)
3. Thompson, S., Whitaker, J., Kohli, R., Jones, C.: Chronic disease management: how IT and analytics create healthcare value through the temporal displacement of care (2019)
4. He, W., Zhang, Z.J., Li, W.: Information technology solutions, challenges, and suggestions for tackling the COVID-19 pandemic. Int. J. Inf. Manage. **57**, 102287 (2021)
5. Mingis, K.: Tech pitches in to fight COVID-19 pandemic (2020). Retrieved from Computerworld: https://www.computerworld.com/article/3534478/tech-pitches-in-to-fightcovid-19-pandemic.html
6. SanadJo (2022). Available at (online): https://play.google.com/store/apps/details?id=com.modee.sanad&hl=en&gl=US. Accessed 17 Jan 2022
7. AMAN—Aman.jo JORDAN APP COVID-19 (2022). Available at (online): https://play.google.com/store/apps/details?id=jo.gov.moh.aman&hl=en&gl=US. Accessed 17 Jan 2022
8. eGovernment, J.: CRadar. App Store Preview (2022). Available at (online): https://apps.apple.com/jo/app/cradar/id1512613147. Accessed 17 Jan 2022
9. Available at (online): https://www.gateway2jordan.gov.jo/index.html. Accessed 17 Jan 2022

Capitals Investment Decision of Malaysian Family Business: From Traditional into Neoliberal Business Entity

Tahirah Abdullah, Siti Afiqah Zainuddin, Nur Farahiah Azmi, Nur Izzati Mohamad Anuar, Siti Rohana Mohamad, and Nadzirah Mohd Said

Abstract The purpose of this paper is to explore and understand how accounting is implicated in the neoliberalisation process of the traditional family entity into profit making entity. The researcher draws upon Bourdieu's notions of capital to frame the study and finally explained the ways of neoliberalism was infused in this transformation process. The study adopts an extended case study approach utilising multiple data collection methods. The data were analysed at both, a macro level where the government has steered the accountability of traditional family towards entrepreneur of the self by institutionalising them under Small and Medium Enterprises (SMEs) and their resulting accountability ramifications were discussed at a micro level. The results, though highly case-specific, indicate that accounting through accountability has become a powerful conduit for the exercise of the neoliberalism reforms by the government and implemented by managerial accountability control over neoliberalised family institutions. The paper also demonstrates that different form of capital plays an important role in transforming the traditional family into a business family. Despite the positive implications of neoliberalised family, the neoliberalism has created managerial accountability conflict in the neoliberalised family.

Keywords Neoliberal · Family business · Capital investment · Bourdieu · Accounting

1 Introduction

Neoliberalisation has been the most overarching global political agenda in the contemporary world since the 1970s. It is the 'most successful ideology in world history' [1]. Indeed, neoliberalism is everywhere. But what exactly is neoliberalism that has conquered the world. According to [2], neoliberalism "is in the first instance

T. Abdullah (✉) · S. A. Zainuddin · N. F. Azmi · N. I. M. Anuar · S. R. Mohamad · N. M. Said
Faculty of Entrepreneurship and Business, Universiti Malaysia Kelantan, Kota Bharu, Malaysia
e-mail: tahirah@umk.edu.my

A. Hamdan et al. (eds.), *The Implementation of Smart Technologies for Business Success and Sustainability*, Studies in Systems, Decision and Control 216,
https://doi.org/10.1007/978-3-031-10212-7_16

a theory of political economic practices proposing that human well-being can best be advanced by the maximization of entrepreneurial freedoms within an institutional framework characterised by private property rights, individual liberty, unencumbered markets and free trade. The role of the state is to create and preserve an institutional framework appropriate to such practices. The state has to be concerned, for example, with the quality and integrity of money. It must also set up military, defense, police, and juridical functions required to secure private property rights and to support freely functioning markets…It holds that the social good will be maximized by maximizing the reach and frequency of market transactions, and it seeks to bring all human action into the domain of the market". In this sense, neoliberalism changes the organisation and individuals to extract the income from the ownership resources in the form of entrepreneurial [3]. Therefore, neoliberalism addressed that all conduct is economic conduct [4]. However, how it processes and transform particular entity into economic sense is questionable.

The effect of neoliberalism is felt by many sectors including profit as well as non-profit organisations, public sector organisations, civil society organisations, non-government organisations, and others. This can be found in various disciplines of studies such as political-economic policies, financial system, business management, family business, education and accounting studies. Some accounting literatures dealing with this transformation for example [4–11]. The debate about the transformation has gained fresh prominence with many arguing that accounting is a part of neoliberal technologies in transforming the organisation and individual.

What more important, families have not been able to escape from this particular influence of transformation. The neoliberal changes in the domain of family relations have been discussed in some research, especially those related to accounting in Small and Medium Enterprises (SMEs). However, the discussion of the research only revolved around the issues from the perspectives of efficiency and effectiveness of using accounting techniques in SMEs rather than focusing on the method that causes the families to be neoliberalised by the development discourses which concentrate on the SMEs. Overall, this implies that the neoliberalisation of the family has not yet been taken seriously by the accounting academics.

This has opened a discussion that neoliberalism has never underestimated the role of the family in the neoliberal agenda, putting it as a crucial entity in restructuring the traditional family entity into manufacturing or business entity [12]. In doing so, the government, with the support of the International Monetary Fund and World Bank has created discourses to attract the society to transform from the family entity into performing or business entity. For instance, few banks and financial institutions aims to promote the business and entrepreneur activities, the formation of government agencies to offer the business seminars and entrepreneur training and to offer a capital for start-up the business and others. This has further explained that the neoliberalism project is "a political project to re-establish the conditions for capital accumulation and to restore the power of economic elites" [13]. Hence, this transformation aimed to restructure the economics of the family entity on the one hand and to exercise the power of the state in reinstating the capital accumulation on the other side. It offers a means that this transformation involves both a macro dimensions of structuring the political economic agenda and micro dimensions that is restructuring the traditional family entity into the business entity. Nevertheless, one thing that should be addressed here is this traditional family entity has mobilising the elements of cultural, traditional values and their religious accountability of which has become a part of the important element in discussing the neoliberalism of the family entity into business entity.

In doing so, the corporation or firm was established that is an entity into which capital is legalised, organised and differentiate from the firm or individual entity, in this case the individual entity is referred to a family entity. The business formation is one of the firms used to legalised and organised the activities of family business entity in the way of justifying their actions. For example, the logic of transforming family entity into a business entity in Malaysia is due to the fact that the composition of SMEs is mainly run and managed by family entities [14]. A considerable amount of past research on family business only investigated general topics which include succession, conflict, performance dimensions, and governance. Hence, greater effort is needed to address the complexity of family businesses in managing both the family and business entities while also mobilising a different form of capital. As highlighted by [15], it is important to understand how family business makes a specific decision such as capital investment decision. In other words, the way a family business entity makes a calculation practice is important to be understood because it directly favours the entities considering their function as a judgement system or social organisation that can assign "worth" [11, 16] further added that "this attribution of worth can have important social consequences and impact the way of society or firm work". Therefore, the current research does not deal with a normative perception that small business formed by family entity is only given limited access to financial information in the attempt of making a capital investment decision (e.g. [17–20]). Overall, the whole purpose of the current research is to understand how family is able to mobilise its capital through performing entity as a result of neoliberalism project or how social cultural-politics tends to shape the investment decision in a family business entity.

The paper is divided into four main parts. The first part highlights the cultural theorisations of capital which extend capital beyond the money. The second part provides a description of the method adopted by the study. The third part discusses the finding on the process of neoliberalising the traditional family field, types of capital and how accounting via accountability is implicated in this transformation. It provides a genealogy of political discourses and institutional reforms that aim to transform a traditional family field into a business field through neoliberalism. The implication, recommendation and conclusion are presented in the fourth part of the paper.

2 Cultural Theorisations of Capital

The issues of how entrepreneurial families operating in a developing economy context make their investment decisions amid multiple forms of capital is indeed a question that needs to answer with empirical data that reveals the 'substantive rationalities' that the social actors in their own local settings deploy in managing their economic enterprises. In other words, many of the answers to this fundamental question lie in the way that those 'entrepreneurs' conceive what capital is and the social mechanisms and forces that they employ to accumulate it. Hence, this study offers Bourdieu's theorisation of capital [21] that provide understandings of the notion of capital and its accumulation. The present study explores the implication and role of accounting as a technology of governance in a family setting under the neoliberalisation process of transforming the traditional family entity into a profit-making entity. Neoliberalism involves a continuous process, not one stake at one time. Similarly, the capitals (i.e. cultural, economic, symbolic and social capital) are going through the transformation process.

3 Extended Case Method

The present study adopted an extended case study as proposed by [22] to provide the case study with the capacity to deal with macro-micro connections. It is an appropriate methodological approach to understand both the historical politico-economic that contribute to the emergence of neoliberalism agenda and how the traditional family were reacted to align with the macro changes. It addresses different research aspects of the accounting phenomenon in its natural setting. Researching the processes of capital investment decision making in the context of family business requires an exploration through case study fieldwork, which allows the researcher to investigate "a contemporary phenomenon within its real-life context; especially when the boundaries between the phenomenon and context are not clearly evident" [23]. A request to conduct the study was sent to one family business which mainly operated in Kuala Lumpur, Malaysia.

Given its standing as a successful and leading family-owned business producing Hajj and Umrah equipment, it was felt that it would be worthwhile to conduct the study with this family business entity. This study gathered information relating to its history, structure and the nature of its operations such as products, business activities and others through internet resources such as the company's webpage, Instagram, Facebook and local newspapers. The data were collected through direct observations in the shop lots and factory during office hours, semi-structured and unstructured interviews with the owner, employees, apprentice and family members that involve in managing the business and some documents that were relevant to the study was also collected.

4 Data Analysis and Findings

4.1 Chronologies of Neoliberalising the Families into Business Entities

At the macro level, neoliberalising the families into business entities occurred at five phases as depicted in Fig. 1. These neoliberal state practices have reconstructed traditional families, social culture and the practices of the businesses. An expansion from being accountable in terms of social and religious rituals to being accountable to financial institutions and government agencies, and hence from simple accountability

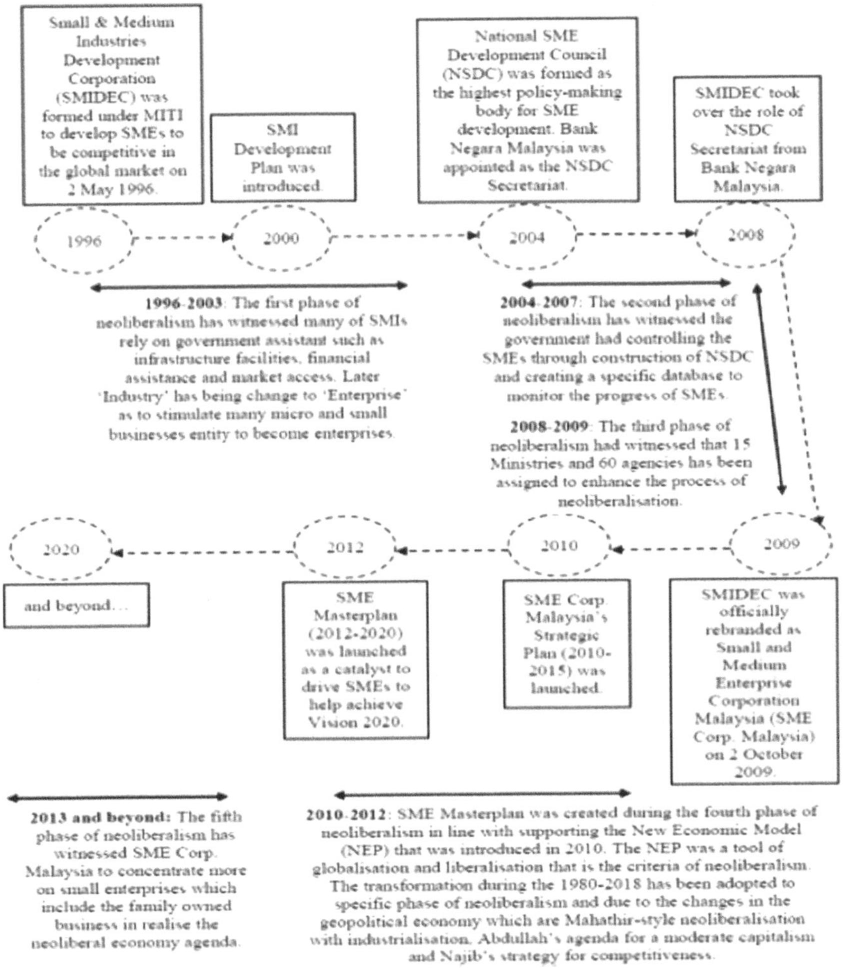

Fig. 1 A process of neoliberalising the families into business entities and neoliberal implications

to more complex accountability, has been a critical characteristic of this transformation. Under these neoliberal practices, families in the business field have been restructured and reorganised in accordance with the demands of this recent phase of governance. The SMEs (i.e. family businesses) sector, as a field, is a politico-economic phenomenon which has been structured, regulated and managed at the macro institutional level of the state. This transformation to a new economic model required a well-structured organisation to monitor and govern the activities and performance of SMEs. Thus, SME Corp. has played it more significant roles in implementing this transformation.

4.2 Collection of Capitals

At the micro level, this study found that a particular Malay woman and her husband, whom traditionally thinking and talking about a job at somewhere else. They work for a government institution and thus receive a monthly payment based on their work. The spouses are paid under a wages system whereby they work at a particular time and are given specific wages. However, a specific event changed them into entrepreneurs. As the conflict happened, neoliberalism appeared and offered an opportunity to resolve the problem. Under the context of entrepreneurial, they are not labour, but they are entrepreneurs. Thus, this transformation has changed the family's style, in particular, their accountability towards the family and business profession, and their way of living and spending their time. Somehow, this change has not altered their traditional values and accountability towards religion. Nevertheless, conflict occurs when the husband appears and sets a rule. This is where the result of neoliberalism has emerged in entrepreneurial activities.

Under the neoliberalisation process, the woman utilised her social capital including experience and networks to produce and accumulate capital. They were becoming an entrepreneur of the self-caused her relatives to give her negative feedback about her behaviour and ways of accumulating capital. These were due to their skeptical thinking, and traditional norm towards being an entrepreneur could not help in changing the economy of the family. The woman went through the neoliberalisation process for a certain period, and later she strategically transformed her children into entrepreneurs. Just as the notion of "strategies" in Bourdieu's framework is dependent on the types and volume of capital, so has this woman capitalised her children as symbolic capital in managing the business. The children need to go through many processes such as to have a business certificate from higher educational institution and to comply with the rules and regulations, which are both family and business requirements, as created by their mother, to enable them to become entrepreneurs.

4.3 *The Conflict of Neoliberalising Family Business*

Work Based on Gender

Despite these positive transformations, the study discovered conflict due to the neoliberalisation process of the family business, which happened in manufacturing and office management. The woman decided that her daughters should be assigned to manage administrative activities such as accounting, modelling and arrange meetings with the clients. Meanwhile, the sons were responsible for maintaining the manufacturing activities, including governing male workers from Bangladesh. Thus, under neoliberalism, the woman distributed the accountability of the children based on their gender, for instance, manufacturing is male work and administration at the office is female work. However, neoliberalisation was not implicated while the children performed their social accountability towards the family. For example, everyone was responsible for taking family members (brothers, sisters and grandfather) to the hospital if they were not in good condition.

The Rule of 6 O'clock

The interesting part of JR Hajj Niaga family business is how the husband makes use of his power as a husband to regulate the rule for the family business. The power is used to make a boundary between the family and the business. A possible explanation for this matter maybe because he is in a different world from his wife and children. For him, labouring is more important because he spends his work time in a separate world or place that is very bureaucratic and organisational and has a specific room called an office. He is a government worker and goes to his office to work from 8 am until 5 pm every weekday. He will then return home after office hours and by a common definition, home is referred to as the place where one lives permanently, especially as a member of a family or household. In other words, when the husband returns to his home, the home should contain his family which refers to his wife and children. Moreover, they should be doing something related to family or home activities that represent the family entity. However, for the wife and her children, business work or office work can be discussed anywhere and anytime and there is no difference between going home and work. As a result, they tend to bring business matters home and discuss family matters at the office. In response to this matter, the husband has created a rule that there should be no business discussions after 6 o'clock and the office and factory activities should be closed on the weekend and everybody had to be at home before night.

5 Implication and Conclusion

The study has made a significant academic and practical contribution that can be built on through future research. This study makes substantial to an academic contribution to the existing body of knowledge in the area of financial management particular capital investment decision making among family business. Furthermore, the findings of this study assist in updating the current literature in a related area such as form of capitals and family business management. From a practical perspective, this study contributed to various SMEs in planning their business strategies in managing capitals, managing accountability towards government, financial institution and family business.

First, future research could seek to have access to data collection at the family home of the case family-owned business. This is important because some capital investment decisions were discussed at home; as they were family business matters, they were not shared with non-family members. It is crucial to have an understanding of how the family discusses family business issues at home—a place where the neoliberalisation process has no implications. Thus, it might be interesting to explore the interactions of accounting from the house particularly during movement control order policies are implemented. Second, the political-economic focus has also changed as the Malaysia prime minister was also changed. The new government has put more emphasis on the role of women in improving the economy of the state. In other words, women have become another focus of the neoliberalisation process under the new governmentality. Moreover, women are perceived as 'worthy' of being capitalised under the new government. A closer understanding of how accounting is implicated in the neoliberalisation process of women would be of value and interesting.

References

1. Anderson, P.: Renewals. New Left Rev. (2000)
2. Harvey, D.: Neoliberalism as creative destruction. Ann. Am. Acad. Pol. Soc. Sci. **610**(1), 21–44 (2007)
3. Hisano, S., Akitsu, M., McGreevy, S.R.: Revitalising rurality under the neoliberal transformation of agriculture: experiences of re-agrarianisation in Japan. J. Rural. Stud. **61**, 290–301 (2018)
4. Cooper, C.: Accounting for the fictitious: a Marxist contribution to understanding accounting's roles in the financial crisis. Crit. Perspect. Acc. **30**, 63–82 (2015)
5. Ellwood, S., Newberry, S.: Public sector accrual accounting: institutionalising neo-liberal principles? Acc. Auditing Accountability J. **20**(4), 549–573 (2007)
6. Alawattage, C.: The calculative reproduction of social structures—the field of gem mining in Sri Lanka. Crit. Perspect. Acc. **22**(1), 1–19 (2010)
7. Andrew, J., Kaidonis, M.A., Andrew, B.: Carbon tax: Challenging neoliberal solutions to climate change. Crit. Perspect. Acc. **21**(7), 611–618 (2010)
8. Jayasinghe, K., Wickramasinghe, D.: Power over empowerment: encountering development accounting in a Sri Lankan fishing village. Crit. Perspect. Acc. **22**(4), 396–414 (2011)
9. Zhang, Y., Andrew, J., Rudkin, K.: Accounting as an instrument of neoliberalisation? Exploring the adoption of fair value accounting in China. Acc. Auditing Accountability J. **25**(8), 1266–1289 (2012)
10. Cooper, C., Graham, C., Himick, D.: Social impact bonds: the securitization of the homeless. Acc. Organ. Soc. **55**, 63–82 (2016)
11. Chiapello, E.: Critical accounting research and neoliberalism. Crit. Perspect. Acc. **43**, 47–64 (2016)

12. Bendell, B.L.: Environmental investment decisions of family firms—an analysis of competitor and government influence. Bus. Strategy Environ. (2021)
13. Harvey, D.: A Brief History of Neoliberalism. Oxford University Press, Oxford (2005)
14. Department of Statistics Malaysia (DOSM) (2016)
15. Pounder, P.: Family business insights: an overview of the literature. J. Fam. Bus. Manage. **5**(1), 116–127 (2015)
16. Boltanski, L., Thévenot, L.: On Justification: Economies of Worth. Princeton University Press, Princeton, Oxford (2006)
17. Sava, R., Mârza, B., Eşanu, N.: Financial reporting for SMEs–past and perspectives. Procedia Econ. Finance **6**, 713–718 (2013)
18. Albu, N., Albu, C.N.: Strategies for and implications of the IFRS for SMEs implementation in emerging economies. In: Comptabilités et Innovations (2012)
19. Sian, S., Roberts, C.: UK small owner-managed businesses: accounting and financial reporting needs. J. Small Bus. Enterp. Dev. **16**(2), 289–305 (2009)
20. Strouhal, J., Müllerová, L., Cardová, Z. Paseková, M.: National and international financial reporting rules: Trstiny the comparability of Czech reporting from the SMEs Perspective. WSEAS Trans. Bus. Econ. (2009)
21. Bourdieu, P.: The forms of capital. In: Richardson, J.G. (ed.) A Handbook of Theory of Research for the Sociology of Education, pp. 241–258. Greenwood Press, New York (1986)
22. Burawoy, M.: The extended case method. Sociol Theory **16**(1), 4–33 (1998)
23. Yin, R.K.: Case Study Research: Design and Methods. SAGE, London and Singapore (2009)

The Effect of Top Management Support and Support Supplier Development on Green Supply Chain Management in the Construction Jordanian

Sultan Alshourah, Maan Mansour, Khaled AlZeaideen, and Zakaria Azzam

Abstract There is a lack of systematic empirical evidence regarding the success factors of GSCM practices. To address these issues, this study examines the relationship between Top management support and Support supplier development practices of green supply chain management (GSCM) in the Jordanian construction sector. Based on a review of the previous studies, 27 items of the survey design were included. 136 manager's data was collected from a population study. The study hypotheses were tested by multiple regressions. The results of the study revealed that Top management support and Support supplier development were supported with practices of GSCM. The results of the study improve and expand the current literature on more efforts to adopt GSCM, as well as motivate researchers to develop this concept and expand studies on the factors that lead to the success GSCM in the world and Jordanian construction sector in especially. This research is one of the few studies that examine the success factors of GSCM practices in the construction sector in Jordan in particular and developing countries in general. Theoretical implications and managerial implications of these findings are discussed.

Keywords GSCM · Top management · Support supplier · Construction

1 Introduction

The construction industry contributes significantly and decisively to the economic and social development of all countries of the world. This industry uses a large amount of energy and natural resources. Hence, the active and effective participation of industry in the country's efforts to achieve sustainable development is essential [1]. Since 2020, the construction sector in Jordan has been facing grateful pressure due to the Coronavirus crisis. In the last decade, the construction sector has been economic growth, but it is in a downward trend after 2017. However, in 2018, the growth of the construction sector decreased from 40%. In 2020 amid the pandemic, the construction

S. Alshourah (✉) · M. Mansour · K. AlZeaideen · Z. Azzam
Zarka University, Zarqa, Jordan
e-mail: salshourah@zu.edu.jo

A. Hamdan et al. (eds.), *The Implementation of Smart Technologies for Business Success and Sustainability*, Studies in Systems, Decision and Control 216,
https://doi.org/10.1007/978-3-031-10212-7_17

sector in Jordan has seen a massive drop in activity in both the private and public sectors [2]. According to [3], the rate of decline in the construction sector in Jordan during and after the Corona crisis amounted to more than 50%, indicating that this is a very large number. He pointed out that the construction sector is the backbone of the national economy and is the main operator of a number of sectors estimated at 150 professions that support the sector and work with the contractor. An empirical study by [4], revealed that the accumulation of financial obligations and workflow, new policy issues, and a review of the impact of COVID-19 on the construction sector are the most major challenges facing the construction sector in Jordan. However, they confirm that the need is necessary to pave the method for the development of additional emergency plans and a new business strategy in anticipation of extending the duration of the COVID-19 pandemic over the coming years. Hardan and Al-Najjar [4], emphasized that the best solutions to meet the challenges and environmental changes are to change the strategies of the construction sector through such as the adoption of GSCM and the introduction of new strategies. Recommended the need for further surveys on factors affecting GSCM practices.

Recently, GSCM is acquisition more attention among researchers, operations academics, and managers of supply chain management [5]. The growing interest in GSCM is mainly due to the economic and environmental, and social benefits of implementing practices of green such as being environmentally friendly products, reducing material waste, and reducing energy consumption [1]. In recent years, GSCM has received application and adoption in Western environmental studies, mainly in Europe and North America [6]. However, studies of the relationships between factors affecting GSCM practices in developing countries have not been widespread [4, 7]. This study seeks to contribute to the GSCM literature. And that by presenting empirical data from a developing country especially Jordan, which is characterized by the weakness of literary studies in it, especially construction literature, this search also supplies insights to managers and decision-makers in Jordan, to enhance and increase their awareness of the GSCM strategy adopt to enhance their construction performance as well as the improving factors for implementing GSCM in Jordan.

Previous research highlights the core role of capabilities, skills, and resources in ensuring a company's competitiveness [7]. In time, based on the theory of dynamic ability, it is believed that the competitive position of any firm is built on utilizing its capabilities. However, previous studies have emphasized still not clear what kind of capabilities and resources can continue to create value in a rapidly changing environment [8]. Moreover, the integration of organizational efforts (internal and external) and GSCM development and implementation can improve organizational efficiency and increase the value of organizations, enhancing competitiveness [9]. In addition, this coordination has been shown to contribute to positive environmental and organizational performance outcomes, for example, reduced energy costs and waste disposal. Therefore, this study constricted, top management support, and support supplier development as resources and compatibilities enhance GSCM implementation then leading to greater organizational performance outcomes. Hence, this study addresses the following research question. How does Top management support, and Support supplier development influence the GSCM practices in the Jordanian construction sector?

2 Literature Review

2.1 Green Supply Chain Management (GSCM)

In the existing literature and past, there is no unanimous definition of GSCM, but there are several, approximately 22 [10]. The reasons behind the availability of different definitions about GSCM are; GSCM is an interdisciplinary field Academics professional, and researchers from various disciplines interpret the term GSCM differently and thus emphasize the different issues in these definitions [10].

Many definitions of green supply chain management have appeared in previous studies [11], these definitions are almost all agreed that it is a link between the purchase and the customer [12, 13]. According to [6] GSCM is "integrating environmental thinking into supply-chain management, including product design, material sourcing, and selection, manufacturing processes, delivery of the final product to the consumers as well as end-of-life management of the product after its useful life". Ninlawan et al. [13], defined it as "closing the loop that range from green purchasing to customer and reverse logistics". Al-Ma'aitah [5] GSCM defined is "viewed as an operational cycle (purchasing, manufacturing, distribution, etc.) with environmental concern in order to achieve environmental, economic, and operational performance for an organization".

In the previous research, various dimensions of GSCM practices such as investment recovery, environmental design, reverse logistics, customer cooperation, internal environmental management, green purchasing, waste management supplier, and cooperation were found as GSCM practices [11]. However, to measure GSCM practices, after a thorough analysis of the literature, this study chose internal management refers to "the extent to which management, employees and departments within the organization have embraced GSCM practices within the organization", Green purchasing "examines the extent to which organizations consider environmental and green concerns in procuring goods and services", and Customer Collaboration refers to "the extent to which organizations work closely with their customers to incorporate environmental considerations into the products and services of the organization, as well as to institute recycling and product return initiatives" [12, 14, 15].

2.2 Antecedent of GSCM

2.2.1 Top Management Support

Top management support for GSCM is the one critical factor in organization involvement that should be placed in the highest priority when GSCM is to be implemented [8, 11, 16]. Previous studies have emphasized that top management support is critical to ensuring awareness, understanding, and support to implementing the environmental vision and corporate policy throughout the organization. Financial support and an adequate supply of resources should be provided. Strategic planning must include environmental issues and the organization must participate in environmental projects and build a culture where knowledge is a vital organizational resource. Therefore, the top management support is essential to the success of GSCM programmers [17].

Systematic empirical evidence for [18] 141 reviewed, compiled, and screened papers on top management support for the adoption of GSCM practices. Their findings reveal that top management support and is the main driver for organizations to adopt GSCM practices. They suggest that managers need to considered and identify key sustainability risks in the past, present, and future to have a simple organizational design for innovative management approaches to address their sustainable supply chain practices. They said that top management should instill governance frameworks to help sustainable supply chains adapt in the organizations. According to [19], found that the relationship between top management support and GSCM was positively connected in the Manufacturing industry in Korea. The empirical study [20], discovered that the strong influence and driving force of the nature of relationships between support of top management and supply chain partners on the implementation of GSCM practices in the operate in the Gulf countries (Middle East region). Faunally, a survey of senior construction project managers in Indian construction companies has been published by [21], the results indicate that among the enabling factors, "support of senior management to green issues" is positively correlated with the implementation of GSCM focused on achieving the twin goals of the industry, and Development and environmental protection through innovation and technological development. Based on this and the discussion above the following supposition is formulated:

H1: Top Management Support is positively related to GSCM practices in the construction sector in Jordan.

2.2.2 Support Supplier Development

Previous studies have revealed that companies tend to be more concerned in enhancing their environmental efficiency within the organization by building a strong network of supplier relationships and enhancing the efficiency of GSCM practices, yet very few companies consciously cooperate with supply chain network partners [18].

A survey of senior construction project managers in Indian construction companies has been published by [21], indicated that "suppliers' efforts/support organization" has the highest strength within influencing adaptability in green SCM, and empowerment enabling factors such as support for construction project suppliers, the commitment of project owners, government support and incentives for green building practices can help companies adopt GSCM practices. Nkrumah et al. [12], found that the development of green supplier capabilities had positive and significant effects on adoption of GSCM practices in Ghana Manufacturing. In addition, an empirical study in the operate in the Gulf countries (Middle East region) by [20] revealed that influence of strong and driving power of the nature of the relationships between top management and supply chain partners on the implementation of GSCM practices. Furthermore, [17], suggests that Supplier management GSCM practices correlate to more effectively managing top management in the Brazilian automotive battery industry. Faunally, the study results by [8] show that supplier management is the most critical dimension in GSCM initiatives in the electrical and electronic industries. They also add establishing environmental requirements for purchasing items and implementing green purchasing are the two important factors in supplier management. Based on this and the argument above the following hypothesis is formulated:

H2: Support Supplier Development is positively related to GSCM practices in the construction sector in Jordan.

3 Theoretical Framework

The research framework has been developed based on a theoretical review of the literature on dynamic theory and competitive advantage that examines the relationship between senior Top management support, Support supplier development , and GSCM practices. Figure 1 shows the effect of between the study variables.

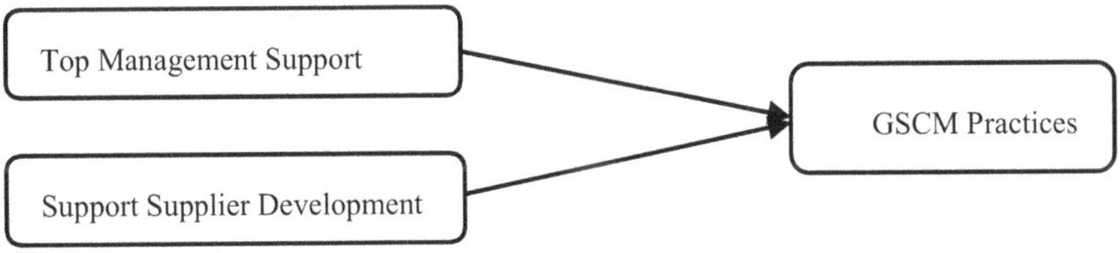

Fig. 1 Conceptual framework

Table 1 Items number, dimensions, previous studies and scale used study's

Dimensions	Items	Previous studies	Scale
GSCM dimensions	12	[1]	Five-point scale
Top management	7	[17]	Five-point scale
Support supplier development	6	[11]	Five-point scale

4 Research Methodology

This study relies on a quantitative and investigative design approach to achieve the research objectives. Based on Literature, The designed instrument of the research constructed of Top Management and Support Supplier Development Factors, and GSCM Dimensions [12, 13, 17, 20–22] and targeted General Manager, Project Manager, and Purchasing Manager, and those are the most documented ones to provide more accurate perceptions of GSCM practices and the factors affecting them [5]. According to the Jordanian Building Contractors Association (JBCA), the contractors in Jordan are classified as three stars [23]. The sampling used in this study is stratified random sampling based on the category of contractors, Data was collected via emailed. 330 were total questionnaires and was able to analyze 136 usable questionnaires resulting in a 45% response rate. However, the following Table 1 shows the Items' number and dimensions of the study and the previous studies that were relied upon to develop the study's questionnaire and measurement tool.

5 Findings

5.1 Descriptive Analysis

Table 2 shows that level contractor, it was found are operating for level third, comprising 55% of the total. Position it was found is operating for general managers comprising 49% of the total.

Table 2 Samples' characteristics

Level contractors					
Category	Freq.	%	Position	Freq.	%
First	30	22	General manager	66	49
Second	51	38	Project manager	40	29
Third	55	40	Purchasing manager	30	22
Total	136	100	Total	136	100

Table 3 Coefficients of reliability for the dimensions in the study

Factors	Cronbach's alpha	Mean	SD
GSCM dimensions	69	4.12	6.45
Top management	79	4.01	6.78
Support supplier development	90	3.89	6.39

5.2 Reliability

The alphas coefficient for the variables was calculated using the SPSS reliability procedure and is shown in Table 3 below. Hair et al. [24], argued that a factor of 0.70 is desirable. However, the results indicate that the reliability of all variables falls within the acceptable range, which is higher than 0.70.

5.3 Result Confirmatory Factor Analyses

The outcome of the study was analyzed using the Statistical Package for the Social Sciences (SPSS) version 25.0 for Windows. The analysis of factor was performed to determine the variables in combination with the principal components as a method for rotation and Varimax extraction. The consequences for each factor analysis carry out are potted below.

5.3.1 GSCM Dimensions

For the GSCM dimensions, factor analysis was performed based on the twelve questions about the GSCM dimensions. These elements represent three dimensions. According to Table 4, the factor analysis performed to explain GSCM dimensions

Table 4 Factor analysis on GSCM dimensions

Factors	Loading	Factors	Loading
Customer cooperation	0.89	Internal management	0.91
	0.88		0.86
	0.81		0.76
Green purchasing	0.85		0.71
	0.84	Reliability	0.88
	0.75	Eigenvalue	11.18
	0.74	Variance %	75.90
	0.70	KMO	0.75

Table 5 Factor analysis top management support and support supplier development

Factors	Loading	Factors	Loading
Top management support	0.89	Support supplier development	0.81
	0.88		0.85
	0.84		0.84
	0.80		0.79
	0.79		0.75
	0.77		0.74
	0.75		
Reliability	0.88		0.85
Percentage of variance	75.90	KMO	0.83

had a Kaiser-Meyer-Okin value of 0.75, exceeding the recommended value of 0.5 [24], and the Barlett test for sphericity was highly significant ($p = 0.00$), this supports the portability of the correlation matrix factors. Additionally, as shown Table 4, factor loadings range between 0.70 and 0.89. Reliability (Cronbach's alpha) for this factor is 0.89, which indicates high reliability. Thus supporting the inclusion of all scale items.

5.3.2 Top Management and Support Supplier Development Factors

The antecedents of GSCM were two factors: Top Management, and Support Supplier Development. However, factor analysis was performed based on the eighteen questions about the four factors. According to the Table 5 the factor analysis performed to explain antecedents of GSCM had a Kaiser-Meyer-Okin value of 0.75, exceeding the recommended value of 0.5 [23], and the Barlett test for sphericity was highly significant ($p = 0.00$), this supports the portability of the correlation matrix factors. Additionally, as shown Table 5 factor loadings range between 0.74 and 0.85. Reliability (Cronbach's alpha) for this factor is 0.88, which indicates high reliability. Thus supporting the inclusion of all scale items.

5.4 Correlation Analysis

According to Table 6 shows a swift of the outcomes of the correlation analysis. Pearson correlation coefficients (r) were calculated to explain an understanding of the association between all variables in the search. The values of (r) have shown in Table 6 Determine the strength of the relationship between dimensions. As revealed

Table 6 Pearson correlations of search dimensions

	GSCM	TM	SSD
OP			
GSCM dimensions	1.0		
Top management	0.58(**)	1.0	
Support supplier development	0.43(**)	0.60(**)	1.0

** Correlation is significant at the 0.01 level (2-tailed),
* Correlation is significant at the 0.05 level

Table 7 Multiple regression analysis for top management and support supplier development factors influencing GSCM dimensions

Model	Unst C		St C	T	Sig.	C S	
	B	Std. E	B			Tolerance	VIF
Top management	0.05	0.08	0.04	0.58	0.00	0.381	2.7
Support supplier	0.14	0.08	0.13	0.18	0.00	0.440	3.4

in Table 6, the overall dimensional correlation values show the (r) with the above positive values 0.40.

5.5 *Multiple Regressions on Top Management and Support Supplier Development Factors that Influence on GSCM Dimensions*

To answer the research question about the Top Management and Support Supplier Development factors influencing the GSCM dimensions, a regression analysis was performed on the predicted Top Management and Support Supplier Development factors and the GSCM dimensions. Table 7 shows that the relationship between Top Management and Support Supplier Development factors and GSCM dimensions is significant ($F = 41.107$; Sig. $= 0.00$). The R2 obtained indicates that the influencing Top management support and Support supplier development variables account for 66% of the variation in GSCM dimensions. Through the variables included in the regression equation, two dimensions emerged as significant predictors of GSCM dimensions. These are Top management and Support supplier development. Based on the output of the analysis, hypotheses H1 and H2 are supported. This leads to the conclusion that both factors are positively correlated with GSCM dimensions.

6 Discussions

The research question relates to Top management and Support supplier development affecting GSCM practices. The analysis showed that the foregoing factors can be categorized into two: top management support and commitment and Support Supplier Development were found to be highly relevant to GSCM practices. The following explains the discovery of each influencing factor.

The idea that Top management support is the most important factor influencing successful GSCM practices can be interpreted in light of the construction industry. First and foremost, the support of Top management are fundamental to successful GSCM practices as it enables to catalyze change through communication and reinforce the values embedded in the detailed vision of the building. The results of the study are in covenant with previous studies. For example, [20], reveals a strong influence and driving power of the nature of the relationships between top management and partners of the supply chain on the adaption of GSCM practices in the operate in the Gulf countries. This research also found a significant relationship between Support supplier development compatibilities and GSCM Dimensions in the Jordanian construction industry. The results of this study confirmed the fundamental role of Support supplier development compatibilities in determining GSCM success and adaption. The findings of the study are in agreement with previous studies. For example, [17], found that Supplier management practices of GSCM adaption correlate to more effectively managing top management in the Brazilian automotive battery industry.

7 Conclusion and Limitations

There is some limitation to research. First, this research was conducted in the Jordanian construction sector. Data were collected on a study sample of the target population by the general manager, project manager, and purchasing manager, which limit the generalizability of the study. Second, this study found effects top management and Support Supplier Development of GSCM adaption. Future research may focus on mediating and moderator in the top management and Support Supplier Development and relationship of the GSCM, as well as the practice of GSCM and sustainability performance more generally. Third, the collection of data of the research was limited to the Jordanian construction sector, and therefore the outcome of the research may not be generalized or applied to other countries or sectors. In light of this, we call for further research to examine whether the antecedent is a factor for the adoption of GSCM in another sector or country. Therefore, academics or researchers can reflect these factors to comprise other sectors or regions.

References

1. Habib, M.A., Bao, Y., Ilmudeen, A.: The impact of green entrepreneurial orientation, market orientation and green supply chain management practices on sustainable firm performance. Cogent Bus. Manage. **7**(1), 1743616 (2020)
2. JEA Jordan Engineers Association (2020). Report https://www.jea.org.jo/portal/.last. Accessed 21 Dec 2021
3. Yaqoub, A.: The decline in the size of projects and urban activity in Jordan is much more than what was announced. Jordanian Construction Contractors Association. PwC. Retrieved journey 20 (2022) from PwC website: http://www.jcca.org.jo/News.aspx?id=13515
4. Hardan, A., Al-Najjar, E.: Challenges and obligations floundering, in the Jordanian construction sector owing to Covid-19 pandemic. Int. Eng. J. Res. Dev. **6**(Icmrd21), 9–9 (2021)
5. Al-Ma'aitah, N.: Green supply chain management (GSCM) practices and their impact on performance: an insight from the Jordanian construction sector. Int. J. Constr. Supply Chain Manage. **8**(2), 87–104 (2018)
6. Srivastava, S.K.: Green supply-chain management: a state-of-the-art literature review. Int. J. Manag. Rev. **9**(1), 53–80 (2007)
7. Asif, M.S., Lau, H., Nakandala, D., Fan, Y., Hurriyet, H.: Adoption of green supply chain management practices through collaboration approach in developing countries—from literature review to conceptual framework. J. Cleaner Prod. 124191 (2020)
8. Wu, H.H., Chang, S.Y.A.: Case study of using DEMATEL method to identify critical factors in green management. Appl. Math. Supply Chain Comput. **256**, 394–403 (2015)
9. De Oliveira, U.R., Espindola, L.S., Da Silva, I.R., Da Silva, I.N., Rocha, H.M.: A systematic literature review on green supply chain management: research implications and future perspectives. J. Clean. Prod. **187**, 537–561 (2018)
10. Abdellatif, H., Graham, S.: Green supply chain management practices in developing countries—a case study from Jordan. Management **31**, 36 (2019)
11. Alshourah, S.: Role of strategic orientation dimensions of green supply chain management practices in Jordanian manufacturing companies. Zarqa J. Res. Humanit. Accepted Unpublished (2022)
12. Nkrumah, S.K., Asamoah, D., Annan, J., Agyei-Owusu, B.: Examining green capabilities as drivers of green supply chain management adoption. Manage. Res. Rev. (2020)
13. Ninlawan, C., Ninlawan, P., Tossapol, K., Pilada, W.: The implementation of green supply chain management practices in electronics industry. In: Proceedings of the International Multiconference of Engineers and Computer Scientists, vol. 3, pp. 17–19 (2010)
14. Zhu, Q., Sarkis, J., Lai, K.H.: Confirmation of a measurement model for green supply chain management practices implementation. Int. J. Prod. Econ. **111**(2), 261–273 (2008)
15. Rajaguru, R., Matanda, M.J.: Effects of inter-organizational compatibility on supply chain capabilities: exploring the mediating role of inter-organizational information systems (IOIS) integration. Ind. Mark. Manage. **42**(4), 620–632 (2013)
16. Braunscheidel, M.J., Suresh, N.C.: The organizational antecedents of a firm's supply chain agility for risk mitigation and response. J. Oper. Manag. **27**(2), 119–140 (2009)
17. Chiappetta Jabbour, C.J., Mauricio, A.L., Jabbour, A.B.L.D.S.: Critical success factors and green supply chain management proactivity: shedding light on the human aspects of this relationship based on cases from the Brazilian industry. Prod. Planning Control **28**(6–8), 671–683 (2017)
18. Moshood, T.D., Nawanir, G., Mahmud, F., Sorooshian, S., Adeleke, A.Q.: Green and low carbon matters: a systematic review of the past, today, and future on sustainability supply chain management practices among manufacturing industry. Cleaner Eng. Technol. 100144 (2021)
19. Chu, S.H., Yang, H., Lee, M., Park, S.: The impact of institutional pressures on green supply chain management and firm performance: top management roles and social capital. Sustainability **9**(5), 764 (2017)

20. Agi, M.A., Nishant, R.: Understanding influential factors on implementing green supply chain management practices: an interpretive structural modeling analysis. J. Environ. Manage. **188**, 351–363 (2017)
21. Mojumder, A., Singh, A.: An exploratory study of the adaptation of green supply chain management in construction industry: the case of Indian construction companies. J. Cleaner Prod. **295**, 126400 (2021)
22. Yu, Y., Zhang, M., Huo, B.: The impact of relational capital on green supply chain management and financial performance. Prod. Planning Control **32**(10), 861–874 (2021)
23. JCCA Jordan Construction Contractors Association (2018). http://www.jcca.org.jo/. Accessed 1 Dec 2021
24. Hair, JF., Risher, J.J., Sarstedt, M., Ringle, C.M.: When to use and how to report the results of PLS-SEM. Eur. Bus. Rev. (2010)

Evaluation of Using Computer Assisted Telephone Interview Mode of Data Collection in the Time of COVID–19 Pandemic—Palestine Case

Aya Amro

Abstract Despite the risk and the threats posed through COVID-19 pandemic in successfully conducting censuses and surveys through delays, interruptions, diversion of funding, the Palestinian Central Bureau of Statistics (PCBS) is determined to continue collecting data on a timely basis and of a quality that is fit for purpose. The contribution of this paper is twofold. Firstly, it introduces the adoption of CATI (Computer Assisted Telephone Interview) mode of data collection during the pandemic in the current surveys and the influence it may has on data quality. It also shed more light on the main differences between CATI and CAPI (Computer-assisted personal interviewing) modes in household surveys in particular. Secondly, it focuses on proposing a strategy regarding the samples frame of household surveys conducted using CATI, through adopting a new methodology as an additional data resource for carrying out surveys during this period. This methodology is based on the integrative role played by PCBS and other National Statistical System pillars (NSS) including a relevant private sector company; in order to establish a central geographical database (Geodatabase) of households with fixed line numbers at PCBS, where data can be used to develop a more comprehensive sample frame for CATI-based household surveys.

Keywords Data collection · Data quality · Household surveys · CATI · CAPI · COVID-19

1 Introduction

During the COVID-19 pandemic and the associated lockdowns, many constraints for producing reliable and timely data have tightened, as the interviewers could not be able to conduct fieldworks due to the risk associated with face-to-face interviews. In this sense, the need for an alternative method of data collection has arisen, and CATI (Computer Assisted Telephone Interview) approach seems to be the available

---

A. Amro (✉)
Senior Statistician, Palestinian Central Bureau of Statistics (PCBS), Ramallah, Palestine
e-mail: aamro@pcbs.gov.ps

A. Hamdan et al. (eds.), *The Implementation of Smart Technologies for Business Success and Sustainability*, Studies in Systems, Decision and Control 216,
https://doi.org/10.1007/978-3-031-10212-7_18

alternative at the time. One of the household surveys this paper is presenting and was conducted in PCBS through this approach is "the Effect of COVID-19 on Socio-Economic Conditions of the Palestinian Households of 2020" (COVID-19 survey).[1]

Many countries recently carried out surveys using CATI mode in data collection even before COVID-19 pandemic. A review of the literature on "Quality report of the European Union Labour Force Survey 2017" reveals that the percentage of interviews using CATI mode for this household survey reached 46.8% among the European Union countries.[2] However, this mode of data collection in PCBS is a recent use. The need for this kind of approach arises through the time of COVID-19 pandemic to be used as the main tool of data collection in many surveys. It mainly depends on using the phone to contact the respondent, where the interviewer can conduct the interview with the respondent in the form of a telephone conversation, and the interview is conducted through an application on the tablet device that reflects the survey form. The application is designed in a way that allow the researcher to move automatically through questions whenever needed. Data can be easily entered and submitted to PCBS headquarter server for the purpose of data checking and auditing.

Using CATI mode as the only tool of data collection for the first time will subject the PCBS to a critical issue, due to the lack of a comprehensive household sample frame for fixed line numbers or phone numbers, since the sample of household surveys carried out using CATI mode must be drawn from a unified general sample frame which is not available yet. In this regard, the paper presents a proposed method on how to develop a central Geodatabase of households with fixed line numbers through connecting to other data sources.

The main objective of this paper is to discuss the effect that CATI approach have on data quality through studying some measurements related to the accuracy dimension of COVID-19 survey. Also, this paper compares between COVID-19 survey (CATI-based) with Socio-Economic Conditions survey (SEFSec)[3] that was carried out in 2018 using Computer-assisted personal interviewing approach (CAPI) where both surveys have the same sampling units, in order to know how much the quality of data vary between CATI and CAPI adopted approaches in household surveys during the current pandemic.

2 Data Quality of CATI-Based Surveys

A deeper understanding of how the adopted data collection approach affects data quality is crucial. Data Quality has various Dimensions in which each of them opens

---

[1] Impact of COVID-19 Pandemic (Coronavirus) on the Socio-economic Conditions of Palestinian Households Survey (March–May), 2020: https://www.pcbs.gov.ps/Downloads/book2554.pdf.

[2] Quality report of the European Union Labour Force Survey 2017, 2019 edition—statistical reports, Eurostat.

[3] Socio-Economic Conditions Survey, 2018: https://www.pcbs.gov.ps/Downloads/book2431.pdf.

doors against new challenges. These dimensions are: Relevance, Accuracy and Reliability, Timeliness and punctuality, Coherence and Comparability, Accessibility and Clarity based on the European Statistics Code of Practice (ES CoP).[4] One of the quality dimensions this paper focuses on is "Accuracy"; as the scope of this paper is the quality of survey data, which can be measured by related indicators of accuracy dimension. A primary step for understanding data quality dimension can help us to improve it [9]. In this sense, the paper applies data measurement and data assessment on data of COVID-19 survey.

2.1 Data Assessment

As Part of examining the accuracy of any survey, it is recommended to conduct several comparisons against reference values. The comparison of survey results with independent and more accurate information about the population parameters is a well-known method to analyze sample quality.[5]

In order to assess the relevant data quality dimension in this paper, we set multiple comparisons by the most in common comparable social indicators including the average household size, the categories of household size, and the sex of household head of the following housing surveys: SEFSec 2018 and COVID-19 survey, considering census data of 2017 as a reference.

The Results of data assessments regarding indicators comparison are illustrated below:

As illustrated in the Table 1, regarding average household size and household size categories, one can notice that the overall percentage of households of COVID-19 survey in both indicators is much closer to SEFSec survey than the census 2017 with a slight difference. This may lead to many explanations, one of these explanations is that SEFSec survey is considered a panel survey; which means that each household is being visited more than once, so that every change to the household including its size is being monitored over time, considering the split households that can be tracked and included in the sample as well. This result give us a close percentage to the one of census 2017. On the other hand, the sample of COVID-19 survey represents the same completed households used in SEFSec survey of 2018. Therefore, Implementing COVID-19 survey based on this sample increases the percentage of having small households with a size category of (1–3) individual, especially of those split households. The reason for this is that the response of small households can be higher than the households of a greater size. In turn, it affects the overall average household size. However, regarding the sex ratio indicator, there is no evident difference between each survey when compared to the census 2017.

---

[4] European Statistics Code of Practice: https://ec.europa.eu/eurostat/web/products-catalogues/-/KS-02-18-142.

[5] https://www.europeansocialsurvey.org/methodology/ess_methodology/data_quality.html.

Table 1 Percentage of Households among Selected Indicators

Average household size						
Region	COVID-19 survey		SEFSec 2018 survey		Census 2017	
West bank	5.1		4.9		4.8	
Gaza strip	6.0		5.8		5.6	
Palestine	**5.5**		**5.3**		**5.1**	
Household size categories						
Categories						
1–3	21.0		25.5		27.8	
4–6	47.0		44.7		45.4	
7–9	26.0		25.5		22.9	
10 +	6.0		4.3		4.0	
Total	**100.0**		**100.0**		**100.0**	
Sex ratio of the head of household						
Region	Malc	Female	Male	Female	Male	Female
West bank	91.0	9.0	89.7	10.3	89.5	10.5
Gaza strip	91.2	8.8	90.7	9.3	90.6	9.4
Palestine	**91.1**	**8.9**	**90.0**	**10.0**	**90.0**	**10.0**

Source The Effect of COVID-19 on Socio-Economic Conditions of the Palestinian Households survey database [5], the Socio-Economic Conditions Survey database, 2018 [6], census 2017 database [7]. Palestinian Central Bureau of Statistics PCBS–Ramallah–Palestine

2.2 *Data Measurement*

Data quality measurement is associated to the calculation of non-sampling errors [2]. A review accomplished by Groves [8] on survey error and data quality reveals four identifiable sources of error: coverage, non-response, sampling and measurement or response error [3].

As we focus in this paper on the "Accuracy" dimension, we consider using the following indicators regarding this dimension based on the National Guidelines for Data Quality in Surveys[6] in order to measure the accuracy of COVID-19 survey compared to SEFSec survey:

- Achieved Coefficient of Variations (CVs) of key variables in domains of interest.
- The rate of over-coverage: The proportion of units accessible via the frame that do not belong to the target.
- Response and non-response rates.

The Results of data measurements regarding selected indicators of accuracy dimension are as follow:

[6] National Guidelines for Data Quality in Surveys: https://ndqf.in/wp-content/uploads/2021/07/National-Guidelines-for-DATA-QUALITY-in-Surveys.pdf.

Achieved CVs of Key Variables in Domains of Interest:

In order to compare the computed CV of COVID-19 survey with SEFSec 2018 survey, we first choose some of the key variables of both surveys as follow:

The key variables regarding Covid-19 survey:

1. Percentage of households in Palestine that receive assistance from one of the social protection programs.
2. Percentage of households that declaring a state of emergency because of COVID-19 pandemic is the main reason that made the main income earner to stop working during the lockdown period (March–May), 2020.
3. Percentage of households that the monthly household income decreased by the half or more during the lockdown period (March–May), 2020.
4. Percentage of households that no internet available at home is the main reason for children do not participate in educational activities.

The key variables regarding SEFSec 2018 survey:

1. Percentage of households that government wage and salary is the main source of income.
2. Percentage of households that private sector wage and salary is the main source of income.
3. Percentage of households that wages from Israeli labor sectors is the main source of income.
4. Percentage of households in Palestine that receive assistance.

The value of computed CV's for the key indicators of COVID-19 survey ranges between 2.1 and 5.4%, while the value of computed CV's for the key indicators of SEFSec 2018 survey ranges between 2.9 and 6.3%.

Through those results, one can notice that the value of the CV's for COVID-19 survey lies within an acceptable range (Survey Data Interpretation Guide),[7] which means that data is consistent. In addition, the values of similar indicators of both surveys are quiet close.

The Rate of Over-Coverage:

Coverage errors result from inadequate representation of the target population based on the units in the sampling frame. Over-coverage occurs due to the inclusion of units that do not belong to the target population. Factors contributing to over-coverage regarding COVID-19 survey are:

1. The phone number of the household is out of service.
2. The phone number of the household is incorrect.

The calculated over-coverage rate through those factors of the total sample is considered an estimate of the whole sample frame. The over-coverage rate for both

---

[7] Survey Data Interpretation Guide: https://www.toronto.ca/wp-content/uploads/2017/12/93c0-tph-survey-data-interpretation-guide-aoda.pdf.

Table 2 Response and non-response rates

	COVID-19 survey	SEFSec 2018 survey
Response rate	93.6	90.2
Non-response rate	6.4	9.8
Total	100.0	100.0

Covid-19 survey and SEFSec 2015 survey is 6.26%, 5.64% respectively. The over-coverage rate of COVID-19 survey indicates that the sampling frame is adequately representative of the target population.

Response and Non-Response Rates:

The term response usually refers to the level of participation in survey or interview research. Nonresponse error represents the gap between the sample and the respondents [1]. As measured by continually worsening response rates, the acceptance of general population surveys has significantly declined in the last few years and decades [4]. However, In the case of COVID-19 survey, the cases of completed or partially completed households are considered response cases, whereas, non-response cases were attributed to different factors, which are (Table 2):

1. The household refuses to cooperate.
2. The phone of the household is switched off.
3. No one of the households' member answered the call.
4. The respondent is an unqualified member to give answers.

The results are as follow:

The results show high values in response rate in favor of COVID-19 survey. This can be explained by the period of which this survey was conducted in. Due to the lockdowns across Palestine in that period, many households were forced to stay home. In this sense, we can come up with a theoretical conclusion that the household was more likely to be a respondent than a non-respondent. The response rate of COVID-19 indicates that the sample itself is representative and the possibility of a bias is very slight when comparing the non-response rate of COVID-19 survey with some of the European Union's countries such as Sweden (43.4), Denmark (45.0), and Netherlands (48.4).[8]

In general, we can conclude that data quality is not significantly affected by using CATI mode. On the contrary, results of both data assessment and data measurement of the accuracy dimension for survey output data has showed that this mode is of a good quality as CAPI mode. However, several constraints can affect data quality of CATI-based surveys other than the one we have discussed in this paper. One of the important indicators must be taken into account regarding data quality is the response burden indicator,[9] which is used to measure and compare the average length

---

[8] Quality report of the European Union Labour Force Survey 2017, 2019 edition, statistical reports–Eurostat.

[9] Statistics Code of Practice for the European Neighborhood South countries (based on the European Statistics Code of Practice), principle (9), Eurostat, April, 2016.

of completion of the questionnaire. This indicator is a crucial one in assessing CATI-based surveys as the length of the questionnaire plays a major role in the interview made by telephones; this we may discuss in other studies or papers.

3 Sample Frame of CATI-Based Household Surveys

Survey samples should reflect the underlying target population adequately. Samples of conducted household surveys by PCBS are selected either from a master sample of the total household's frame of census 2017, or from other large-scaled household survey sample frame.

In order to build an integrated central geodatabase of households with fixed line numbers, we need to get benefit of other data sources, including mutual cooperation between PCBS and other NSS's and private sectors associations that provide relative data of the fixed line or mobile phone networks systems across Palestine. NSOs need to build a broad coalition of all segments of society and make sure all producers and users of data are counted and benefit from the systematic implementation of open data principles across the NSS [10]. In this section, we discuss a proposed method of how to extend the sample frame of CATI-based household surveys in general as an initial step to create a central geodatabase that includes all the available fixed line numbers of Palestinian households along with spatial identification data of these households.

3.1 Methodology

The main idea behind this method is based on the mutual relations between the pillars of the national statistical system, and the telecommunications company in Palestine in particular. This company owns a database that includes the entire fixed line numbers available to Palestinian households provided with some identification data of each number including the name of relative customers on the national level. There are main distributors of the wired phone numbers set by the company in residential areas depending on the population density of each area. Those distributers are either small boxes containing up to 60 landlines in areas with low-density population per box for different housing units, or they are on a form of large boxes (Lockers) containing up to 800 fixed line numbers in areas with high density population, especially in city centers. In addition, each distributor is associated with spatial coordinates x and y based on the Palestinian coordinates system used in PCBS.

In order to achieve the proposed strategy, we need to work on a random sample of around 1000 households to be selected from the households' database of census 2017 for around one or two Enumeration Areas (EA). This strategy relied mainly on the process of linking the identification data of these households including the name of household head or any relative names, along with the spatial coordinates x and y

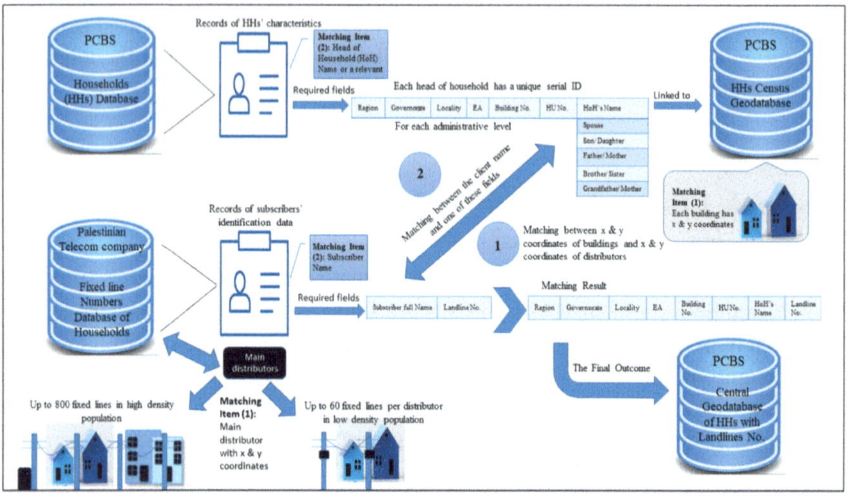

Fig. 1 Linking process. Source: Designed by the author

of each building in the selected areas. Data is disaggregated at each administrative level (Region, Governorate, Locality, EA) with the registered landline numbers of the households in the database of the telecommunication company.

The first step of the linking process is based on matching item (1), through locating the housing units connected to each distributor within its enumeration area by matching each building x and y coordinates in the selected enumeration areas from PCBS with x and y coordinates of the distributors. Then, each household is being identified from the located housing units or buildings. The second step is based on matching item (2), which represents the name of household head or a relevant name according to the relation of head of household data in the census database, and the subscriber name of which the fixed line number is registered by at the company database. Eventually, a list of households with their available fixed line numbers and relative x and y coordinates of the building at each administrative level is generated (see Fig. 1).

3.2 Associated Limitations

The proposed linking process is subject to many limitations that must be considered such as:

- The extent of the company's cooperation in the process of linking and providing data at the individual level.

- Defining the ownership and the reference of data between the two parties after the completion of the linking process, and establishing a geographic database of households and how it may influence data confidentiality.
- The subscriber may have more than one fixed line number registered by his/her name, which can affect the matching process between the subscriber name and household head name.
- The possibility that the utilization of the housing unit at PCBS database is not for habitation only but for habitation and work, and the fixed line number could be registered by the name of the work place.

4 Conclusion and Recommendations

Due to COVID-19 pandemic that invade the whole world in early 2020, many constraints faced the interviewers in fully achieving their field works. In this sense, PCBS showed a high level of risk management through switching the initially adopted mode of data collection in conducting many household surveys. The use of CATI mode in household's surveys during this pandemic was a successful experience; as the results of data quality showed a high accuracy in COVID-19 survey data. This may lead to the conclusion that this mode of data collection can be adopted not only when personal interviews are difficult to be carried out, but it also can be considered for the long term of conducting surveys in PCBS, either as a secondary tool or as the main tool of data collection, or by moving to mixed-mode data collection.

The current crisis is somehow serving to steer our thinking to keep looking for the alternatives, through seeking for a better change in the adopted strategies regarding data collection even after the age of COVID-19 pandemic. Some of the recommendations to take in consideration in the foresee future are:

- Consider adopting CATI mode of data collection in carrying out future surveys at PCBS even after COVID-19 pandemic.
- Taking in Consideration the proposed method when carrying out household surveys using CATI mode
- Taking advantage of surveys and censuses that will be implemented during the coming period, such as the establishments census 2022, by adding a question related to fixed line or mobile phone number in the relative data collection questionnaire.
- Reconsidering the strategic plans regarding the use of administrative records as an alternative resource of some surveys, as well as, moving toward a Register-Based Census.
- Establishing data sharing platforms between PCBS, civil society and private sectors associations in order to allow statisticians to share a large number of indicators derived from big data sources that can be linked as much as possible.

- Embracing open data principles and practices, in order for NSOs to raise their standing as the trusted institution that ensures all users have ready access to high-quality data and statistics that meet national and international demand for information, while protecting privacy and confidentiality in line with the Fundamental Principles of Official Statistics [10].

References

1. Biemer, P.P., Lyberg, L.E.: Introduction to Survey Quality (2003)
2. De Leeuw, E.D., Collins, M.: Data collection methods and survey quality: an overview (1997)
3. De Leeuw, E.D.: The effect of computer-assisted interviewing on data quality: a review of the evidence (2008)
4. Hubrich, S., Wittwer, R.: Effects of improvements to survey methods on data quality and precision—methodological insights into the 10th wave of the cross-sectional household survey "Mobility in Cities–SrV." Transp. Res. Procedia **25** (2017)
5. Palestinian Central Bureau of Statistics, Impact of COVID-19 Pandemic (Coronavirus) on the Socio-economic Conditions of Palestinian Households Survey (March–May) (2020)
6. Palestinian Central Bureau of Statistics, Socio-Economic Conditions Survey, 2018—Main Findings, Ramallah–Palestine (2019)
7. Palestinian Central Bureau of Statistics, Population Final Results—Detailed Report Palestine—Population, Housing and Establishments Census 2017, Ramallah–Palestine (2019)
8. Robert M. Groves, Floyd J. Fowler, Jr. Mick P. Couper, James M. Lepkowski, Eleanor Singer and Roger Tourangeau Survey Methodolgy (2004)
9. Sidi, F., Shariat Panahy, P.H., Affendey, L.S., Jabar, M.A., Ibrahim, H., Mustapha, A.: Data quality: a survey of data quality dimensions. In Proceedings—2012 International Conference on Information Retrieval and Knowledge Management, CAMP'12 (2012)
10. United Nations Statistics Division (UNSD) Open data practices in official statistics and their correspondence to the fundamental principles of official statistics (2019)

Driving Innovative, Sustainability, and Creative Practices

Impact of Innovation and Government Support on the Business Performance of SMEs in Oman

Ammani Ammal and Mohammed Al-Sakiti

Abstract This paper aims to draw attention to the impact of innovation and government support on the business performance of Small and Medium Enterprises (SMEs) in Oman. This conceptual paper was developed with a critical review of past studies, journal articles, conference proceedings and government websites that are relevant to the study and the targeted keywords. This conceptual paper aimed to analyze the constructs of innovation and government support that contribute to the business performance of SMEs. It was found based on the previous studies that innovation and government support have a significant effect on business performance and government support plays a key role in encouraging people to start SMEs. It was also found that government support mediates the relationship between innovation and the business performance of SMEs. This study concludes that the Government of Oman is supporting SMEs by providing financial support, technical support, marketing facilities and networking opportunities in addition to some initiatives to help SMEs to combat the effects of the Covid-19 Pandemic.

Keywords SMEs in Oman · Innovation · Government support · Business performance

1 Introduction

1.1 Background of the Study

SMEs are seen as the backbone of modern economies as they contribute to the global economy and sustainable development [24]. The number of SMEs registered in the Authority for SMEs Development in Oman till August 2021 exceeds 56,687 [2].

A. Ammal · M. Al-Sakiti (✉)
University of Technology and Applied Sciences, Ibri, Sultanate of Oman
e-mail: mohammed.al-sakiti@ibrict.edu.om

A. Ammal
e-mail: Kannivadi.Ammani@ibrict.edu.om

A. Hamdan et al. (eds.), *The Implementation of Smart Technologies for Business Success and Sustainability*, Studies in Systems, Decision and Control 216,
https://doi.org/10.1007/978-3-031-10212-7_19

It implies that SMEs play a key role in the development of the Omani economy by creating employment opportunities, utilizing natural resources efficiently, and improving the standards of living for Omani nationals.

As SMEs play a key role in the development of the economy, innovation has received considerable attention in the literature [27]. Innovation is considered a crucial driver for business competitiveness and performance, particularly for SMEs [4]. Not only this, but it also assists SMEs to grow into large-scale enterprises [19]. Innovation has a positive impact on the performance and productivity of SMEs [4, 10, 25].

1.2 Significance of the Study

By reviewing the relationship between innovation, business performance and government support in the context of SMEs, there will be a clear understanding of how the local authorities can support SMEs to sustain and grow into large-scale businesses. Specifically, this study will benefit policymakers and entrepreneurs in making effective strategies to promote innovation activities. Furthermore, the evaluation of the current measures taken by the Government to support SMEs may serve as a tool for making better improvements.

The paper is arranged into five sections. Following this introduction, the next section presents a literature review and hypothesis development with emphasis on three main aspects namely, innovation, business performance and government support. Section 3 focuses on the data collection and analysis of this paper. Section 4 explains the Government support provided to SMEs in Oman. The final section summarizes the key findings and conclusions drawn from the study including some recommendations for future research.

1.3 Objectives of the Study

The study was designed to advance knowledge of innovation, business performance and government support, with emphasis on SMEs in Oman. To this end, the paper specifically aims at achieving the following objectives:

- To identify the relationship between innovation, government support and business performance in the context of SMEs in Oman.
- To assess the current measures taken by the Government to support manufacturing SMEs in Oman.

It is hoped that the findings of this study will provide helpful insights to policymakers and stakeholders to enhance the performance and productivity of SMEs in Oman while encouraging further research in the discipline of innovation and entrepreneurship.

2 Literature Review and Hypothesis Development

2.1 Innovation and Business Performance

Many studies have been conducted to evaluate the effect of innovation on the business performance of SMEs. Radicic et al. [25] conducted a quantitative study to evaluate the impact of innovation support programs on output innovation among manufacturing SMEs. They found that publicly funded innovation support programs have positive effects on the performance of SMEs in the traditional manufacturing industries with a typical increase in the probability of innovation by around 15%. Another interesting study aimed to examine the impact of the product, process and organizational innovations on the financial and operational performance of Spanish SMEs [4]. This study revealed some important findings. First, innovations play a key role in creating value for SMEs in terms of better business performance. Second, technical and non-technical innovations lead to different impacts on the business performance of SMEs. Lastly, in terms of financial performance, product innovation has a positive impact on sales increase, while organizational innovation contributes to cost reduction. Jalil et al. [12] studied the impact of innovation capability on the performance of SMEs in Malaysia taking into account the mediating role of technology adoption. They found that there is a positive relationship between innovation capability and SMEs performance in Malaysia. Moreover, the results indicated that knowledge management, planning and leadership management are considered as significant factors of innovation capabilities.

Oduro and Nyarku [19] examined the incremental innovation practices among SMEs in Ghana with emphasis on the performance, types, propensity and management challenges faced by SMEs while adopting incremental innovation to improve their existing products and services. In terms of the impact of innovation on organizational performance, the results indicated that SMEs enjoy increased competitiveness, enhanced customer satisfaction, improved commercialization capacity and high global market reach. Hall et al. [10] studied innovation and productivity in SMEs in the context of Italy. The key findings of their study revealed that product innovation has a positive impact on labour productivity, however, process innovation has a larger impact on the associated investment. Rosenbusch et al. [27] conducted a meta-analysis study to examine the relationship between innovation and performance in SMEs. Overall, the findings show that innovation has a positive impact on the performance of SMEs, however, the strength of such a relationship depends on different contextual factors, including the age of the firm, the type of innovation and the cultural context. Hence it is hypothesized that:

H_{01}: Innovation has a positive impact on the business performance of SMEs.

2.2 Government Support and Business Performance

Recently, the connection between government support and business performance has become the central attention of many researchers and policymakers around the world [17]. The role of government in supporting SMEs is realized as a facilitator, regulator and catalyst [7]. It is significant to note that initiatives of various governments and business support programmes are responsible for different levels of success in various countries around the world [24]. Government support programmes are developed to assist and encourage the success of the business activities of SMEs [1, 28]. Government agencies provide technical support, commercialization, grants and incentives for business ventures to adopt new technologies. Government institutions help SMEs to obtain financial capital which helps them to grow and promote their businesses [6]. Government financial support improves the innovation success of SMEs, which helps them to possess a competitive position in the market [3]. Moreover, government support, in terms of interest-free loans, tax relaxation and other financial and non-financial incentives, increases the chances of SMEs' survival and growth [8].

Nguyen et al. [17] suggest that financially supporting policies of the government like loans with low-interest rates or tax exemptions could be effective since they may help firms to improve their growth and profitability. SMEs' growth depends mainly on government support and developmental policies, which enhances the potential of SMEs to grow and overcome crises [11]. Ntiamoah et al. [18] and Peter et al. [24] established the fact that SMEs' performance is significantly influenced by government support. They also pointed out that government should eliminate all unnecessary bureaucratic bottlenecks and favouritism, establish strategic and robust venture capital sector, crowdfunding and other forms of the financial institution to make funds available at an affordable rate to critical sectors such as agriculture and manufacturing. Anwar et al. (2020) indicated that government support and entrepreneurial finance have a significant impact on the success of new ventures. They also established that government support strengthens the relationship between entrepreneurial finance and new venture success. Hence, it is hypothesized that:

H_{02}: Government support has a positive effect on the business performance of SMEs.

2.3 Innovation, Government Support and Business Performance

Based on the previous research, innovation is another factor that can affect the performance of SMEs [13, 15, 20]. Compared to large companies, SMEs are considered more innovative, because of their flexibility and ability to incorporate their findings quickly and efficiently into their business activities [33]. According to Guo et al. [9], the ideas and creativity of entrepreneurs are embodied into business strategies so as for them to be able to operate and outperform competitors, which may increase business performance. Ngisau and Ibrahim [16] found that encouragement of the local government to continue supporting innovation activities and the manufacturing firms will use the advantages offered by the government to shape future national innovation growth. Rita et al. [26] found that peer-to-peer lending and financial bootstrapping has a positive effect on innovation and business performance. The availability of financial resources helps to stimulate the creation of opportunities for innovation and better business processes [5]. Previous studies pointed out that government support brings out innovation and both innovation and government support improve the business performance. Based on the reviews, it is hypothesized that:

H_{03}: Government support mediates the relationship between innovation and business performance of SMEs.

3 Research Methodology

The research design of this study is descriptive in nature as this conceptual paper was developed from the data collected from journal articles, magazines, newspapers, and government websites. Based on the review of previous studies, the researchers developed the research model and hypothesis for this study. This hypothesized model could be tested by collecting data from Manufacturing SMEs in Oman through structured questionnaires and analyzed through SPSS and Partial Least Square–Structural Equation Modelling (Fig. 1).

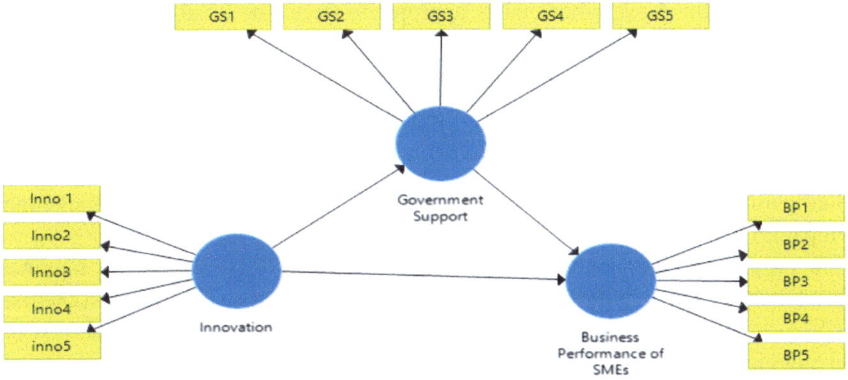

Fig. 1 Research Model

4 Government Support for SMEs in Oman

4.1 Financial Support

The Government of Oman provides financial support facilities to encourage the establishment and growth of SMEs in various industries. Oman Development Bank (ODB) provides short-term and long-term loans with appropriate grace periods of 3% and the financing program includes working capital and fixed assets financing with a maximum of 250,000 OMR [21]. Sharakah provides customized financial solutions that suit the needs of SMEs by offering loan schemes, bill discounting schemes and equity schemes. The owners of the SMEs should meet certain criteria to be eligible for financial support, including the availability of viable business plans, building competitive advantage and providing employment opportunities [30]. SME Development Fund (Inma) provides also different financial support facilities that suit the needs of SMEs in Oman, including asset financing, debt financing, bill discounting and projects financing [31].

4.2 Technical, Marketing and Legal Support

Sharakah provides some consultation services to SMEs to ensure the sustainability of these businesses. The consultation services include reviewing business plans, project financial projections, risk assessment, access to markets, networking opportunities, preparing marketing plans, monitoring performance and reviewing financial reports [30]. Legally, the Government takes appropriate decisions to support the growth and sustainability of SMEs in the country. The Ministry of Finance has issued a circular stating that all government contracts valued at 10,000 OMR or lower must be awarded to SMEs [22]. Madayn Innovation Academy plays an important role in enhancing and promoting innovation activities within the companies investing in industrial areas. It also provides facilities to SMEs and start-up businesses by coordinating with incubators, research and development companies and marketing companies [14].

In order to support and promote SMEs and innovators outside the national boundaries, Oman International Development and Investment Company (OMINVEST) signed a strategic partnership agreement with the General Commission of the Sultanate at Expo 2020. The agreement aims at supporting SMEs to present and showcase their innovative products and services through the Oman Pavilion [23]. The Ministry of Transport, Communications and Information Technology supports SMEs through some initiatives. One of the key initiatives is called SAS program. The focus of this program is mainly on fast-growing IT start-ups. Different facilities are provided to SMEs, including providing space for start-ups, consultancy on different business aspects, training programs, and building relationships and partnerships [29]. SME Development Fund provides different facilities to SMEs, including consultancy services, mentoring and monitoring the business performance. In addition to these facilities, it also supports SMEs to deal with legal concerns by coordinating with the key government bodies such as Tender Board, Free Zones, Industrial Areas and larger companies [31].

4.3 Supporting SMEs During Covid-19 Pandemic

The economy of Oman has been affected by the Covid-19 Pandemic like many countries across the world. The Ministry of Economy took some initiatives to accelerate the recovery of SMEs from the repercussions of Covid-19 in cooperation with the SME Development Authority [32]. The initiatives include:

- E-commerce Programme Initiative: It supports SMEs to access local and international digital platforms.
- Winter Nights Festival: It provides SMEs with an opportunity to display their products and services on special occasions and holidays.
- Creative Industries Incubator: It targets SMEs, entrepreneurs and business owners in the field of creative industries by providing technical support, consultations and guidance.
- Manufacturing Incubator: This initiative will be implemented by Suhar University which targets SMEs and owners of manufacturing industries. It aims at providing marketing services, consultations and technical support.
- Establishing Emerging Agricultural Manufacturing SMEs in Cooperation with Haya Water: It aims at supporting and developing SMEs in the field of agricultural wealth by finding suitable sites for SMEs near water treatment plants.
- Industrialization Initiative in Cooperation with Mazoon Dairy Company: It aims at providing business opportunities for SMEs in different areas, including manufacturing packaging materials, dairy boxes, raising and selling livestock, and package labelling.
- Entrepreneurship Education Platform: This initiative focuses on creating a digital training platform to develop the theoretical and practical skills of Omani entrepreneurs and craftsmen.
- Consultation to Support SMEs Affected by the Pandemic: This initiative is expected to benefit 100 SMEs by providing legal advice and consultations.

5 Discussion and Conclusion

SMEs are recognized all over the world as they play a key role in the development of economies and regions by creating employment opportunities, enhancing the standards of living, providing innovative products and services to people. Oman Vision 2040 aims for private sector partnership, economic diversification and enhancing the entrepreneurial capabilities of the young population. In line with the vision, the

Government is taking various steps through government institutions and policies to encourage and promote SMEs, especially in the non-oil sectors.

The Government of Oman provides financial support to SMEs through different institutions, such as Oman Development Bank, SME Development Fund and Sharakah. In terms of technical, marketing and legal support, the Government provides different facilities such as consultation, marketing and promotion opportunities, networking opportunities inside and outside Oman, incubation for start-ups and new ventures and legal facilities. Since Covid-19 Pandemic has affected businesses to a greater extent, especially SMEs, the Government took some actions and initiatives to support SMEs tackle the effects of the pandemic. To encourage innovation in SMEs in recent times, Government institutions are organizing various discussion forums, exhibitions, conferences and competitions. They are also introducing various policies to attract investors in industrial areas by providing all the required facilities, granting financial support and exemption of taxations.

It is evident, based on the previous studies, journal articles, government websites and magazines, that innovation and government support have a significant effect on the business performance of SMEs and government support mediates the relationships between innovation and business performance of SMEs. This study concludes that the Government of Oman is supporting innovation and both innovation and government support have a significant impact on the business performance of SMEs in Oman. This conceptual paper recommends conducting an empirical study in future on the impact of innovation and government support on business performance in the context of manufacturing SMEs in Oman.

References

1. Awojide, S.O.: We need more government support for SME's in Nigeria (2015)
2. Businessgateways (2022). Retrieved 1 Mar 2022, from https://businessgateways.com/news/2021/10/19/Over-56000-SMEs-registered-Oman-till-August-2021
3. Doh, S., Kim, B.: Government support for SME innovations in the regional industries: the case of government financial support program in South Korea. Res. Policy **43**(9), 1557–1569 (2014). https://doi.org/10.1016/j.respol.2014.05.001
4. Exposito, A., Sanchis-Llopis, J.A.: Innovation and business performance for Spanish SMEs: new evidence from a multi-dimensional approach. Int. Small Bus. J. **36**(8), 911–931 (2018)
5. Farkas, G.: The effects of strategic orientations and perceived environment on firm performance. J. Competitiveness **8**(1), 55–65 (2016)
6. Fonseka, M.M., Yang, X., Tian, G.L.: Does accessibility to different sources of financial capital affect competitive advantage and sustained competitive advantages? Evidence from a highly regulated Chinese market. J. Appl. Bus. Res. (JABR) **29**(4), 963–982 (2013)
7. Gede, D.: Mengembangkan UKM melalui pemberdayaan peran pemerintah daerah. Jakarta, ID: Bakrie School of Management (2009)
8. Guan, J., Yam, R.C.: Effects of government financial incentives on firms' innovation performance in China: evidences from Beijing in the 1990s. Res. Policy **44**(1), 273–282 (2015)
9. Guo, H., Tang, J., Su, Z., Katz, J.A.: Opportunity recognition and SME performance: the mediating effect of business model innovation. R&D Manage. **47**(3), 431–442 (2017)
10. Hall, B.H., Lotti, F., Mairesse, J.: Innovation and productivity in SMEs: empirical evidence for Italy. Small Bus. Econ. **33**(1), 13–33 (2009)

11. Ibrahim, M.I., Keat, O.Y., Abdul-Rani, S.H.B.: Government support policy as a potential moderator on the relationship between entrepreneurial orientation, contemporary marketing and SMEs performance in Nigeria a proposed framework. Kuwait Chapter Arab. J. Bus. Manage. Rev. **6**(10), 32–42 (2017)
12. Jalil, M.F., Ali, A., Kamarulzaman, R.: Does innovation capability improve SME performance in Malaysia? The mediating effect of technology adoption. Int. J. Entrepreneurship Innov. 14657503211048967 (2021)
13. Love, J.H., Roper, S.: SME innovation, exporting and growth: a review of existing evidence. Int. Small Bus. J. **33**(1), 28–48 (2015)
14. Madayn Academy. Madayn (2022). Retrieved 27 Feb 2022, from https://www.madayn.om/EN/Pages/Madayn-Academy.aspx
15. Mustikowati, R.I., Tysari, I.: Orientasi kewirausahaan, inovasi, dan strategi bisnis untuk meningkatkan kinerja perusahaan (studi pada UKM sentra Kabupaten Malang). Jurnal Ekonomi MODERNISASI **10**(1), 23–37 (2015). https://doi.org/10.21067/jem.v10i1.771
16. Ngisau, N., Ibrahim, N.: Technological innovation adoption in manufacturing sector: the moderator role of government support. J. Econ. Bus. Manage. **8**, 200–205 (2020)
17. Nguyen, H.T.T., Van, H.V., Bartolacci, F., Tran, T.Q.: The impact of government support on firm performance in Vietnam: new evidence from a dynamic approach. Asian Acad. Manage. J. **23**(2) (2018)
18. Ntiamoah, E.B., Li, D., Kwamega, M.: Impact of government and other institutions' support on performance of small and medium enterprises in the agribusiness sector in Ghana. Am. J. Ind. Bus. Manag. **6**(05), 558 (2016)
19. Oduro, S., Nyarku, K.M.: Incremental innovations in Ghanaian SMEs: propensity, types, performance and management challenges. Asia-Pac. J. Manage. Res. Innov. **14**(1–2), 10–21 (2018)
20. Otero-Neira, C., Lindman, M.T., Fernández, M.J.: Innovation and performance in SME furniture industries: an international comparative case study. Mark. Intell. Planning (2009)
21. Odb.om (2022). Retrieved 27 Feb 2022, from https://odb.om/
22. Oman Observer (2022). Government contracts valued at RO 10,000 reserved for SMEs. Oman Observer. Retrieved 27 Feb 2022, from https://www.omanobserver.om/article/728/Business/government-contracts-valued-at-ro-10000-reserved-for-smes
23. Ominvest.net (2022). Retrieved 27 Feb 2022, from https://www.ominvest.net/news-details190.html
24. Peter, F., Adegbuyi, O., Olokundun, M., Peter, A.O., Amaihian, A.B., Ibidunni, A.S.: Government financial support and financial performance of SMEs. Acad. Strateg. Manag. J. **17** (2018)
25. Radicic, D., Pugh, G., Hollanders, H., Wintjes, R., Fairburn, J.: The impact of innovation support programs on small and medium enterprises innovation in traditional manufacturing industries: an evaluation for seven European union regions. Eviron. Plann. C. Gov. Policy **34**(8), 1425–1452 (2016)
26. Rita, M.R., Kristanto, A.B., Nugrahanti, Y.W., Utomo, M.N.: Entrepreneurial orientation and emotional bias in MSMEs' financing and performance. Jurnal Ekonomi dan Bisnis **24**(2), 237–270 (2021)
27. Rosenbusch, N., Brinckmann, J., Bausch, A.: Is innovation always beneficial? A meta-analysis of the relationship between innovation and performance in SMEs. J. Bus. Ventur. **26**(4), 441–457 (2011)
28. Shamsuddoha, A.K., Ali, M.Y.: Mediated effects of export promotion programs on firm export performance. Asia Pac. J. Mark. Logist. **18**(2), 93–110 (2006)
29. Sas.om (2022). Retrieved 27 Feb 2022, from https://www.sas.om/SASEN/Pages/services.aspx
30. Sharakah (2022). Retrieved 27 Feb 2022, from https://www.sharakah.om/
31. SMEF (2022). Retrieved 27 Feb 2022, from https://www.smefoman.com/
32. Times of Oman. Eight new initiatives to support SME sector in Oman announced (2022). Retrieved 27 Feb 2022, from https://timesofoman.com/article/106039-eight-new-initiatives-to-support-sme-sector-in-oman-announced

33. Verhees, F.J., Meulenberg, M.T.: Market orientation, innovativeness, product innovation, and performance in small firms. J. Small Bus. Manage. **42**(2), 134–154 (2004)

Digitalization and Success of Peruvian Micro-Enterprises in the Retail 4.0 Sector

Hernán Ramírez-Asís, Rosa Vílchez-Vásquez, Antonio Huamán-Osorio, Tatiana Gonzales-Yanac, and Jorge Castillo-Picón

Abstract Changes in the retail market are urging businesses to adopt a modern manufacturing paradigm known as agile development. In a market that is extremely competitive with changing consumer demands and dramatic improvements in performance, it is seen as a good strategy. This paper focuses on modern technology and the link between two machines and contact between all supply chain elements, also known as digital production. The fourth technological revolution is this modern technology, regarded as the Internet of Things or retail 4.0. It raises multiple challenges and prospects for the industrial world. Business processes, approaches and collaboration with sellers and end consumers must be adapted. This thesis explores the requirements and the usage of technology in this transition. Furthermore, after a sample of over 550 respondents, 334 responses are obtained the inquiry would incorporate a quantitative approach. The test was evaluated using the SPSS statistical method, and regression analysis was used to verify the conclusions. The findings indicate that digitalization and the distribution chain in the retail industry have a direct connection.

Keywords Digitalization · Transformation · Micro-enterprises · Retail 4.0

H. Ramírez-Asís (✉)
Universidad Señor de Sipán, Chiclayo, Perú
e-mail: ramirezas@crece.uss.edu.pe

R. Vílchez-Vásquez · A. Huamán-Osorio · T. Gonzales-Yanac · J. Castillo-Picón
Universidad Nacional Santiago Antúnez de Mayolo, Huaraz, Perú
e-mail: rvilchezv@unasam.edu.pe

A. Huamán-Osorio
e-mail: ahuamano@unasam.edu.pe

T. Gonzales-Yanac
e-mail: dgonzalesy@unasam.edu.pe

J. Castillo-Picón
e-mail: jcastillop@unasam.edu.pe

A. Hamdan et al. (eds.), *The Implementation of Smart Technologies for Business Success and Sustainability*, Studies in Systems, Decision and Control 216,
https://doi.org/10.1007/978-3-031-10212-7_20

1 Introduction

The notion of digitalization has arisen over the years. Digitalization has provided the ability to change and adjust companies. Virtual industry transition is the way to concentrate on IT daily routine and promote connections through the supply chain to create incomparable organizational excellence.

Changes are pushing businesses to follow a modern development paradigm known as agile production. It is seen as the winning formula by the suppliers, who aspire to significantly boost output in an intensely challenging environment with fast-changing client demands, to become national and international leaders [1].

Research has found that product creation ventures can be more effective when focused on sophisticated consumer requirements than on emerging opportunities in technology [2]. Indeed, standardized goods and industrial manufacturing can no longer gain a significant market share as long as systems and practices remain scalable and agile. Development industries must predict a high-precision shift in demand whilst still reacting rapidly to unforeseen additional changes. The key restrictions of a supplier are the high prospect of quality and lower cost of staying successful on the market with its consumers. Companies that cope with the technology that maintains their market position are able to provide rapid product creation and flexible output [3].

AccOn the internet, the coming industrial revolution would cause contact between people as well as cyber-physical machinery (CPS) through large networks according to experts from industry and academia [4, 5]. The networks would attach goods and computers to employees and enable remote control and enhancement of activities. The commodity details are collected in the cloud storage and software systems are the link to track all indicators in real time. Manufacturing firms would be able to fulfill their needs and please their clients by utilizing these systems more flexibly and agilely. Significant improvements in processing and the flow of the company's operations would have to be made in order to incorporate this technology [6].

Because of the COVID 19 pandemic, the retail industry has been directly impacted because all the operations needed to access the goods have to be operational, forcing all staff and managers to engage physically.

When the COVID 19 pandemic happened, the retail market was significantly impacted and several retail businesses have been lowered and threatened with bankruptcy.

Although technology innovation raises challenges, digital transformation is a way to innovate and redefine industries. Digital innovation aims to develop scalable enterprises that in coming years will be critical to organization growth in Peru.

A pandemic of COVID 19 pressuring the Government to revisit its growth models by relying more on emerging infrastructure and transformation is leading many countries worldwide to face challenging economic conditions.

Environmental, economic and technological challenges are highly aggressive for retail sectors around the globe. Retail companies in Peru must manage their supply chain agilely to overcome this and the viability of retail companies in Peru [7].

Industrial businesses have big odds, but challenges [4, 8]. The benefits of Retail 4.0 contribute to economic development of manufacturing enterprises. First, we will define a company, the fundamental characteristics and the importance of this business category's organizational and operational durability, and then discuss the key obstacle to or delay new technology adoption.

Manufacture is the selling practice of products. The products, machinery and processing are produced. The last items can be offered to the final customer or any business. The manufacturing method includes operation that renders raw materials the finished product. The definition of the object begins and is normally connected to the configuration of the elements and the materials. Production processes make these materials the final component required [9].

Various studies examine that digitization is an important factor in the success of microenterprises [5, 6], the present study wishes to demonstrate that this reality is replicated in Peruvian microenterprises in the retail 4.0 sector, which is why the objective of the present study is to identify the relationship between digitization and the success of Peruvian microenterprises in the retail 4.0 sector.

1.1 Organizational and Operational Agility

Producers need to constantly update their products and services to develop a strategic advantage in an increasingly digital environment. They must be flexible to create value and separate themselves from the rivalry in businesses and business [10].

The organization's agility depends on its members' facts, experience and inventiveness [11]. Activity, speed, mobility and skill include organic agility. In comparison, organizational resilience is continuously innovative. Fresh opportunities are generated through technical development in the retail industry, including the outsourcing of additional non-value operations utilizing equipment and agile software. This enables organizations to adapt quickly to trends and environmental changes.

Expanding the idea of organizational agility implies agility will adapt to change accomplished by unrelenting ingenuity and the use of flexible and reconfigurable technologies [12]. Quick and precise printing of judgments and the development of a valid and real-time knowledge system are also synonymous with agility. Lean production is also the first step in improving retail and external process control, both in the detection and recycling of waste sources. Lean and effective technological advancement leads to organizational efficiency and improved resource output [13].

Organizational and organizational resilience are the key principles that enable the growth of Business 4.0 businesses worldwide. It also allows decision-making easier and more effective. Operational resilience lets us quickly and easily prepare for customer demands through machines and smartphones.

1.2 Customer Experience

Consumer engagement is the attempt made to give the customer more than just a commodity, through architecture, related support, and connectivity across the whole product cycle [14]. Using technology and digital technology, new market models can be built and new consumer value can be generated. Customers' active involvement in the phase of product creation often addresses their unique requirements and needs [10]. However, in the numerous fields of customer service, this may be feasible and we will discuss some of these areas in the following section:

Customer understanding: Companies must consider their consumer desires and expectations. Others utilize social media to see what keeps their customers comfortable or sad. They use their past encounters and learnings. Companies frequently market their goods on numerous digital channels.

Topline Growth: With salespeople using them for meetings and for sales pitches, the usage of multimedia platforms is growing. Tablets and web tools in particular, enable users to connect and communicate with them. Some people use CRM to connect local references to customers. These networks provide real-time sales and discounts over Ethernet.

Customer touchpoints: Digital platforms may be utilized to enhance after-sales and customer support. They can be used for a variety of internet consulting work that can be completed that can be physically done to save the client or company. These instruments help the consumer to save time and money for the business. Many businesses already provide consumer solutions to boost consumer contact points [15].

Digitally modified businesses: Besides the usage of modern technology technologies, businesses adjust their way of doing business to help their companies. Some use conventional goods for digital platforms.

New digital businesses: Besides the usage of digital businesses, some companies often offer new digital products to manage the product lifecycle. The retailer will keep these goods aware of its' own products. It would also support the consumer by estimating the wear period of their product to avoid loss. Digital Globalization: Convert multinationals into global corporations. These businesses benefit from multinational pooled finance, HR resources, and also key development and design skills. Services exchanged around the globe encourage productivity and reduce risk.

Leadership is a matter of digital market change. In order to accomplish digital transition and shift, good leadership is required. But it is important to develop a plan to know with which fields to start. The transition of everything cannot be done at once by a corporation. Reinventing corporate practices to optimize computer technologies and promote the collaboration of the supply chain involves analyzing any part of each work. All operations must be analyzed and re-engineered and each organization must start with a separate portion [16, 17].

Through widening the concept of organizational agility, agility takes the shape of adaptation to improvements and can be accomplished by constant creativity and the usage of scalable and reconfigurable technology. Agility is often a question of quick and successful decisions and of building up a relevant information system in real-time. Lean development also is the first phase in improving production and external management of operations and in the identification and removal of waste sources. Lean welding and effective technology use contribute to operational stability and increased resource performance. Organizational and organizational resilience is the basis for manufacturing firms to work within the Retail 4.0 climate. It also makes for easier and more effective decision-making. Operational resilience helps them to efficiently and effectively respond to the demands of the consumer by using computers and devices [18, 19].

1.3 The Potential of Retail 4.0

Retail 4.0 is a solution for the future by developments by businesses in channel coordination structures, cyber-physical automation, and robotics. Its aim is to adapt by requiring customized goods, to shifts in customer behavior. This causes the company to shift its paradigms and procedures to switch to customized mass manufacturing, often called mass personalization. Interoperability, virtualization, decentralization, real-time capacity, service orientation and modularity are main concepts for Retail 4.0. Interoperability enables a person or a computer to execute several operations. It also aims to establish a popular programming language that allows communicating between various computer systems such as computers and software simpler [2, 20].

Virtualization is characterized by means of process simulation or the process-intensive sensor devices as the capacity to generate a virtual picture of the floor and the production process. Decentralization is the decentralized policy of regulation. It is able to make decisions rapidly available for employees and machines (particularly thanks to ICT's) [21, 22].

The orientation service reflects the reorientation of its business model based on accessible evidence. It includes repenting the way to do business with the consumer by providing not just a commodity, but also a full and optimized solution via a combination of services and functionality now allowed by integrated sensors. Lastly, modularity implies that all can be separated into tiny sets. The agile development processes are a special principle. To help handle the complexities of a procedure, the product or method should. Smart and linked modern emerging technology is the gateway to a 4.0 world. Indeed, commodity insight is the capacity for data collection and review. Connectivity requires data to be moved from one item to another in order to enable data communication, decision-making, and review [8, 23].

1.4 Technologies Supporting Retail 4.0

A corporate policy must be enforced, be it human capital, finances, strategies, and content. This strategy must be applied. Resources are designated by a business that develops implements and implements policies to enhance its productivity and effectiveness as 'all properties, skills, organizational processes, qualities, intelligence, and knowledge' [24].

The ubiquity of the Internet and the exponential growth of all sorts of innovations imply that businesses will recognize and pick the ones most relevant to their truth. New technology classes relating to retail 4.0 may be illustrated in the literature review.

The final list of activated innovations has been concluded: advanced manufacturing; additive manufacturing; growing reality; simulation; cloud computing; industrial IoT; cyber protection [25, 26].

Advanced manufacturing reflects the technological improvements businesses can utilize for product and procedure creation. This involves the use of digital modeling, computer-aided design, and manufacture. Several firms provide various more or less common solutions in the form of mass customization [27, 28].

The method includes successively filing layers of material according to a blueprint designed on the device, enabling the creation of prototypes and componential pieces of special and uncommon shapes. Materials used for additive manufacturing can be metallic, ceramic, fiber or bi-fabricated. The process represents the technologies used for developing 3D structures by software.

It may also be used to put digital knowledge into existence to promote repair, assembly, control devices, product design etc. In order to imagine and communicate with phenomena, virtual reality simulates a fictional environment.

Simulation: combining the numerous IT methods and advanced systems applications to model and simulate the device performance. The simulation of configuration, performance, and functions can be designed to be accurate, thus improving the design efficiency of a product and enhancing its success rate for one-time inventions. Cloud storage: it enables connectivity between computer software and hardware over the Internet. It's characterized as storing and viewing data via the web rather than a local hard drive Cloud computing provides businesses and consumers with a high degree of scalable news, high flexibility, and high dependability [29, 30].

2 Methodology

Desk analysis has found a thorough view of the digital transformation in the manufacturing business in the most influenced processes of the digital transformation of the processing field. In addition, a link has been formed between the numerous company abilities and the digital retail industry transition.

Nevertheless, a methodological methodology has been implemented for the gathering of data from several businesses operating in the manufacturing field. For different companies depending on scale, a broad spectrum of the study was intended (small, medium, and large). The number of questionnaires answered was 334. Questionnaires were distributed in Google formats and the data stored were analyzed using the SPSS predictive process. The research has randomly chosen personnel who act as mechanics, supervisors, and quality officials in order to address the survey.

In addition, the quantitative method was addressed with a survey questionnaire which was entered into Google formats. This Google form has been granted a survey of 550 participants. The questionnaires were responded to by just 334 respondents. Data are maintained for mathematical analysis and validation hypothesis when downloading and uploading the excellent file to SPSS from Google.

3 Results

3.1 Regression Analysis

The SPSS statistical program was used to execute the regression analysis, and the hypotheses were tested with a margin of error of 5%. The outcomes are as follows (Tables 1 and 2):

Table 1 Model summary

Model	R	R square	Adjusted R square	Std. error of the estimate
1	0.399[a]	0.159	0.149	0.08172

[a] Predictors: (Constant), External Factors Average, Strategic Average, Organizational Average, Cultural Average, Managerial Average

Table 2 Coefficients[a]

Model		Unstandardized coefficients		Standardized coefficients	T	Sig.
		B	Std. error	beta		
1	(Constant)	0.369	0.015		24.240	0.000
	Cultural average	0.203	0.071	0.226	2.855	0.005
	Strategic average	−0.529	0.089	−0.516	−5.946	0.000
	Managerial average	0.218	0.074	0.133	2.945	0.013
	Organizational average	−0.235	0.079	−0.225	−2.975	0.003
	External factors average	0.166	0.054	0.156	3.097	0.002

[a] Dependent Variable: Performance

Table 3 Component matrix[a]

Component	Component	
	1	2
Cultural average	0.876	0.081
Strategic average	0.904	-0.108
Managerial average	0.895	0.011
Organizational average	0.877	0.006
External factors average	0.571	0.525
Extraction method: principal component analysis		

[a] 2 components extracted

- Performance and cultural issues are intimately linked (0.005).
- Strategic factors have a significant impact on performance (0.000).
- Performance and management issues are intimately linked (0.013).
- Organizational variables and performance have a significant association (0.003).
- There is a strong correlation between performance and external circumstances (0.002).

3.2 *Factor Analysis*

A Cronbach Alpha indicator was used in the Factor Analysis to determine the validity or validity of the data acquired (Table 3).

The validity and reliability of the analysis were examined by examining each variable separately.

When it comes to internal and external aspects of a company, the Cronbach Alpha values for each of these components ranged from 0.877 to 1.07, with the exception of external elements, which had an alpha of 0.57.

Validity and reliability analyses were found to be acceptable in all except one variable, the "External Factors" variable (Cronbach Alpha 0.571), which is regarded as acceptable but likely to include some bias.

3.3 *Interview Findings*

The goal of this study is to understand digital transformation care among operating retail companies.

This report offers insights into retail businesses. In order to identify how retail businesses, manage digital manufacturing transitions, both quantitative and qualitative methodologies were applied. This study seeks to examine in detail the problem of digital transitions and the relationship of the digital process with the dynamic structure of capabilities.

Secondary data are also collected from the internet and advisory sources to further emphasize how the transition takes place, in the context of the era of new technologies, in industrial businesses. The way new digital technologies are applied is to enormous adjustments to retail firms. The design method is turned into a completely interactive one with 3D applications built on paper-based technology. In addition, automatic machines support manual labor to improve productivity and accuracy by manual processing. Lastly, during the distribution time, warehouse procedures and coordination of the entire operation were converted into an automated approach to improve process accuracy and productivity. Research indicates that supermarket businesses are still experiencing transition.

Knowledge of robotics, the internet of things, improved realism and simulation remain the key method. Exhibitions and forums are some platforms for the development of information in the topic used by respondents. Rarely do organizations have an internal R&D mechanism that allows them to consider patterns more thoroughly.

However, R&D not only lets the companies realize the trend but also allows it simpler and faster to spot market opportunities and developments. R&D serves its clients not only. The sensing goal is primarily to collaborate with external personnel and to externalize R&D programs that enable micro and small businesses to identify opportunities [25, 27]. After sensing, the necessary processes are important for an entity to use certain opportunities. Retail companies in Peru are restructuring their business model to allow emerging technologies to fulfill demand. Small firms benefit from complimentary assistance from strategic partners to identify and deter assets that may occur during this transition process. Moreover, retail businesses can not only restructure their strategies but also rationalize decision-making protocols in the digital transition phase to remove errors and partialities [10, 30].

In addition, retail businesses have succeeded in seizing the value that new innovations can achieve by inspiring and including their staff in the digital phase.

Summarizing the capability mission, the reorganization of the business paradigm and creating a disruptive workforce plan is encouraged. Successful recognition and adaptation to technological opportunities lead to retail companies' development and profitability [31]. It is important for maintaining the competitive advantage to become or reconfigure assets and structures. Both interviewees demonstrated the importance of employee interest in decision-making.

Through doing so, companies could acquire greater responsibility for strategic acts to "a more thorough and expedient recognition of opportunities and threats" Awareness of the transition capability indicated by both respondents is another significant aspect. The enrichment and development of workers and current company information resources to extract benefits from modern technologies is of considerable importance.

In addition, incentives are designed to promote and reward transition employees. Finally, it is primarily by enhancing workers' skills for manufacturing digitalization and granting them incentives to engage in a transition that the role of retail firms is improved. Retail businesses can therefore participate in expanding their staff so that innovative technologies can be implemented. Furthermore, to overcome the problems of inadequate demand research and an easier and reliable approach for managing competition in the face of the significant difficulties in the garment industry, the finding that digital transformation is important. This report highlights the expected effect on the whole supply chain activities in the retail field of digital transformation. Digital transformation through Internet practice and artificial intelligence continues to show that digital transformation plays a significant role in the success of the supply chain.

The study also outlined the obstacles that retailers could encounter as digital transformation is being introduced in their supply chain activities. Companies must consider and join additional demand channels and more digital customers as regards the role of creative business models that add value to their product portfolio. It is also essential for supply chain management to understand how these new innovations may be utilized for their existing supply chain processes and how these technologies can be used.

4 Conclusion

The study shows that engaging in and implementing new technology would provide businesses with a permanent competitive edge by greater access to knowledge, cost savings, better quality of goods, sensitivity, and collaboration skills. Superb changes will arise in the digitalization of supply chain control. In this article, we will highlight some of the issues which underline the significance, challenges, and competitiveness of the management of the digital supply chains.

Combined digital supply chain management will greatly encourage supply chain convergence to reduce market fragmentation and have a competitive solution to digital technology. The theory of supply chain management could be significantly changed in a mid-range diffusion approach for the introduction and application of new technologies. The main limitation of the study is that since it deals with micro companies, it is necessary to extend the research in order to generalize to South American companies, as future research is proposed to expand to large companies in Latin America to verify the behavior of the variables in this group of companies.

References

1. Glas, A.H., Kleemann, F.C.: The impact of retail 4.0 on procurement and supply management: a conceptual and qualitative analysis. Int. J. Bus. Manage. Invention **5**(6), 2319–8028 (2016). https://doi.org/10.1016/j.procir.2016.03.162
2. Dijkkamp, J.: The Recruiter of the Future, a Qualitative Study in AI Supported Recruitment Process. (Master's thesis, University of Twente), Netherlands (2019)
3. Bauernhansl, T.: Industrie 4.0 in Produktion, Automatisierung und Logistik Anwendung · Technologien Migration. In: Ten Hompel, M., Vogel-Heuser, B. (eds.) Springer Vieweg, Wiesbaden (2014)

4. Brettel, M., Friederichsen, N., Keller, M., Rosenberg, M.: How virtualization, decentralization and network building change the retail landscape: an retail 4.0 perspective. Int. J. Mech. Ind. Sci. Eng. **8**(1), 37–44 (2014)
5. Ramírez, E.H., Huerta, R.M., Concepcion, R.J., Norabuena, C.H.: Community banks and their effect on the rural micro-enterprises from the Ancash area in Peru. In: Proceedings of ADVED 2019—5th International Conference on Advances in Education and Social Sciences 21–23 Oct 2019, pp. 141–147. Istanbul, Turkey (2019). https://www.ocerints.org/adved19_e-publication/abstracts/a161.html
6. Heavin, C., Power, D.J.: Challenges for digital transformation–towards a conceptual decision support guide for managers. J. Decis. Syst. **27**(Sup1), 38–45 (2018). https://doi.org/10.1080/12460125.2018.1468697
7. Fuster-Guillén, D., Ocaña-Fernández, Y., Salazar, D.E., Ramirez, E.H.: Human development and family integration: study from the comprehensive service of the elderly in Peru. Revista Venezolana de Gerencia **25**(90), 477–490 (2020). https://doi.org/10.37960/rvg.v25i90.32392
8. Zhang, J., Raza, M., Khalid, R., Parveen, R., Ramírez-Asís, E.H.: Impact of team knowledge management, problem solving competence, interpersonal conflicts, organizational trust on project performance, a mediating role of psychological capital. Ann. Oper. Res. 1–21 (2021). https://doi.org/10.1007/s10479-021-04334-3
9. Weinman, J.: Digital Disciplines: Attaining Market Leadership via the Cloud, Big Data, Social, Mobile, and the Internet of Things (2015)
10. Shaikh, A.A., Lakshmi, K.S., Tongkachok, K., Alanya-Beltran, J., Ramirez-Asis, E., Perez-Falcon, J.: Empirical analysis in analysing the major factors of machine learning in enhancing the e-business through structural equation modelling (SEM) approach. Int. J. Syst. Assur. Eng. Manag. (2022). https://doi.org/10.1007/s13198-021-01590-1
11. De Carolis, A., Macchi, M., Negri, E., Terzi, S.: A maturity model for assessing the digital readiness of retail companies. In: IFIP International Conference on Advances in Production Management Systems, pp. 13–20. Springer (2017)
12. Euchner, J.B.: Navigating digital transformation: an interview with Satya Nadella: Satya Nadella talks with Jim Euchner about the challenges and opportunities of digital transformation. **61**(4), 11+ (2018). http://link.galegroup.com/apps/doc/A546959807/AONE?u=ndul&sid=AONE&xid=064d7362
13. Gates, B.: Lean retail in the age of the industrial internet. Supply Demand Chain Executive **15**(5), 24–27 (2014). https://doi.org/10.1007/s13398-014-0173-7.2
14. Rathi, R.A.: Artificial intelligence and the future of HR practices. Int. J. Appl. Res. 113–116 (2018)
15. Hmoud, B., Laszlo, V.: Will artificial intelligence take over human resources recruitment and selection? Netw. Intell. Stud. 21–30 (2019)
16. Mathur, S.: Artificial Intelligence: Redesigning Human Resource Management, Functions and Practices. ResearchGate, Delhi (2019)
17. Sharma, N.: Recruitment strategies: a power of e-recruiting and social media. Int. J. Core Eng. Manage. (IJCEM) 15–35 (2014)
18. Manju.: The impact of e-recruitment in HRM effectiveness. Int. J. Res. 753–760 (2017)
19. Malik, M., Mujtaba, M.: Impact of e-recruitment on effectiveness of HR department in private sector of Pakistan. Int. J. Hum. Resour. Stud. 80–94 (2018)
20. Makridakis, S.: The forthcoming artificial intelligence (AI) revolution: its impact on society and firms. Futures 46–60 (2017)
21. Johansson, J., Herranen, S.: The Application of Artificial Intelligence (AI) in Human Resource Management: Current State of AI and its Impact on the Traditional Recruitment Process. Jonkoping University, Jonkoping (2019)
22. Ramírez, E.H., Maguiña, M.E., Huerta, R.M.: Attitude, satisfaction and loyalty of customers in Municipal Savings Banks of Peru. Retos **10**(20), 313–325 (2020). https://doi.org/10.17163/ret.n20.2020.08
23. Jia, Q., Guo, Y., Li, R., Li, Y., Chen, Y.: A conceptual artificial intelligence application framework in human resources management. In: Proceedings of the International Conference on Electronic Business, pp. 106–114. AIS Electronic Library, Guilin (2018)

24. Stoyanova, M.: Good practices and recommendations for success in construction digitalization. TEM J. **9**(1), 42–47 (2020). https://www.ceeol.com/search/article-detail?id=838205
25. Kerpedzhiev, G.D., et al.: An exploration into future business process management capabilities in view of digitalization. Bus. Inf. Syst. Eng. **63**(2), 83–96 (2021). https://doi.org/10.1007/s12599-020-00637-0
26. Salmen, A.: Does digitalization boost companies 'new product launch success? An empirical study. SCENTIA Int. Econ. Rev. **1**(1), 62–83 (2021). https://doi.org/10.52514/sier.v1i1.6
27. Kysh, L.M.: Improving the business processes management at enterprises in the context of digitalization. Восточно-европейский научный журнал **1**(65), 15–22 (2021)
28. Sanchez-Segura, M.I., Medina-Dominguez, F., de Amescua, A., Dugarte-Peña, G.L.: Knowledge governance maturity assessment can help software engineers during the design of business digitalization projects. J. Softw. Evol. Process **33**(4), e2326 (2021). https://doi.org/10.1002/smr.2326
29. Menchini, F., Russo, P.T., Slavov, T.N.B., Souza, R.P.: Strategic capabilities for business model digitalization. Revista de Gestão **29**(1), 2–16 (2022). https://doi.org/10.1108/REGE-10-2020-0086
30. Schuh G., Frank J., Holst L., Müller D., Leiting T., Bruhns L.: Digitalization as an enabler of subscription business models in the manufacturing industry. In: Voigt, K.I., M. Müller, J. (eds.) Digital Business Models in Industrial Ecosystems, pp. 49–70. Springer, Cham (2021). https://doi.org/10.1007/978-3-030-82003-9_4
31. Barykin, S.Y., Rasskazova, O., Evseeva, O., Evseeva, S., Ostapenko, G.: Staff training digitalization as value creation process in companies. Int. J. Entrepreneurship **25**(4), 1–17 (2021)

Exploring the Dimensionality of Psychological Contract in Islamic Perspective: Empirical Analysis of Indonesian Employees

Tien Suhartini

Abstract Research on psychological contract originating from the Western view has received immense attention by human resources management researchers. However, psychological contract from the religion or spiritual insight has not been much investigated. This study arranges to extend the understanding of the psychological contract from Islamic perspective. A total of 98 Muslim employees in profit and nonprofit organizations in Indonesia, Malaysia, United Kingdom, and Taiwan, participated as respondent in this study. The construct validity of Islamic psychological contract (IPC) via Statistical Product and Service Solutions (SPSS) bring out to the main finding; the IPC measurement model in the profit and non-profit organizational context is clarified by three dimensions; transactional, relational, and liturgical contract.

Keywords Liturgical contract · *Ghariza* · Islamic psychological contract

1 Introduction

The term psychological contract (PC) introduced originally by Argyris [1] who defined "Psychological Work Contract" to define the implicit relationship between workers and their foreman. The concept was expanded by organizational scholar Rousseau [2] in contemporary research as psychological contract. Rousseau [3] viewed PC as a term in the organization which defines the fulfilment and non-fulfilment about obligations, expectations, and promises of organizational relationships reciprocally. It is also stated as an unwritten and understanding obligations system between employers and workers which contain shared expectancy and responsibilities about how they should act [4]. The three critical main constructs are hope, trust, and positive emotions to create long-term relationships between leaders and followers [5]. Schein concluded that the way we define, manage, understand, apply PC principles in building relationships inside and outside of work, will show our

T. Suhartini (✉)
Alma Ata University, Yogyakarta 55283, Indonesia
e-mail: tien.ssmm@almaata.ac.id

A. Hamdan et al. (eds.), *The Implementation of Smart Technologies for Business Success and Sustainability*, Studies in Systems, Decision and Control 216,
https://doi.org/10.1007/978-3-031-10212-7_21

human side. Qualities of respect, compassion, trust, empathy, fairness, objectivity, characterize a person's overall outlook on life [6]. People's attitudes about the world around them are shaped by beliefs and values, and will determine how they behave in certain situations [7]. There are 2 main variants proposed by Rousseau and McLean [8] in defining a theoretical framework for psychological contracts, namely promissory and social and conclude that contracts as a representation of economic exchange.

For a Muslim, the world is only a temporary residence and the main task is to worship Allah, doing anything in order to get Allah's pleasure. This belief should be reflected in all areas of a Muslim's activity, including in scientific research, which the author does not encounter much in the Western PC literature. According to the Islamic point of view, however, consideration of the public interest (*maslaha*) is paramount and always must be upheld, because Islam encourages its followers to be careful in their actions, including in organizing [9].

In view of EER (employee employer relationship) in organization, perhaps what has not been sufficiently explored are: (1) PC in Islamic perspective; (2) PC from religion perspective conducted in Muslim majority country; and (3) PC from religion perspective conducted by Muslim (believers) employees. The current study will discuss all of these three gaps in prior researches, therefrom can broaden the understanding of PC, including PC from an Islamic point of view.

This study tries to explain PC from an Islamic perspective in the Indonesian context. It is expected to be one of the pioneering research works exploring PC from an Islamic perspective. The study conducted for Indonesian workers with the following reasons: (1) Indonesia is a country with the largest Muslim population in the Southeast Asia region [10]; (2) the limited number of studies in this field have been conducted both in profit and nonprofit organizations in Indonesia; (3) Muslim workers constitute the largest workforce in Indonesia [11]; and (4) Indonesia owning Pancasila as the principle of the state which is adopted, mostly, from Islamic point of view, therefore the nation maintains Islamic teachings, both values and culture [10, 11]. The research questions are: (1) after reviewing the western literature. Is it necessary to study PC from an Islamic (religion) perspective? (2) What are the constituents that clarify Islamic PC in the context of a profit and nonprofit organizations?

2 Literature Review

2.1 Psychological Contract

PC is a term in the organization as to obligations, expectations, and promises, reciprocally between employer and employee [3], In the worker's perspective, PC is a personal belief about the obligations that must carry out to the foreman, in order to gets certain rewards in return for the contribution to the company in reciprocal exchange by mutual agreement. The three main constructs that are critical in building long-term relationships between employers and their employees are expectancy, trust, and positive emotions. Followers will be more confident and willing to involve in risk-taking behavior when trust in leader's abilities, integrity, and benevolence [5].

Events at work place affecting employee's emotions at work and work attitudes [12]. Therefore, PC might be affected by economic market situations, similarly, and congruence to SET (social exchange theory) theorizing that social settings shapes exchange relations [13]. PC research in Indonesia is dominated by its correlation with Leadership, commitment, turnover intention, satisfaction, and OCB. (Based on Google scholar data in 2021). Meanwhile, PC research based on Herzing's Publish or Perish, is more varied, with additional correlation on job performance, trust, motivation, emotional exhaustion, favoritism, work engagement, leadership efficacy, and workplace incivility. What they have in common is that the breaches of contract studied were all based on breaches of transactional and relational contracts.

2.2 Islamic Psychological Contract (IPC)

The IPC is an individual (Muslim) belief system regarding obligations with exchange partners based on faith and belief that carrying out work properly and seriously is a form of practicing the verses of the Qur'an, therefore, that apart from partners, they are also bound by Allah SWT [14, 15] and practicing religion commands, which will be bound to partners and Allah SWT as well [16]. First dimension is Transactional contract which views exchange as something that is temporary, narrow, tangible, to fulfil external (material) welfare, such as adequate pay, suitable work, and opportunities for skills development. The conclusions that can be drawn regarding TPC from an Islamic perspective [14] are, Workers: (1) get a decent wage, to meet their basic needs; (2) get assignments fit to their skill and abilities; (3) are required to seek worldly happiness and enjoyment; (4) have the same rights and opportunities; (5) are entitled to the promised compensation after completing their obligations as in the agreement; and (6) have the right to receive a reward for their achievements and capacities.

The second is Relational contract, that is a contract for a relatively long period of time to fulfil mental welfare. Work is seen as means to stay in touch (*silaturahim*), getting appreciation, and increasing creativity. The conclusions regarding RPC from an Islamic perspective [16] are, Workers: (1) who fulfil promises (agreed contracts) must be confident and believe that it is the best thing in the employer-employee relationship; and not to feel sad and worried/fearful, whatever the result; realize if break the promise, the loss or consequence will return to themselves; (3) try to make improvements by honing creativity and innovation; (4) will always keep the mandate and promises; believe that work with sustainable value will be better and can be expected; and (5) Leaders and followers love each other in goodness.

Third is Liturgical contract, that is a contract to achieve happiness in the hereafter or eternal welfare, by collecting sustain charity (*amal jariyah*), reward, and Allah's blessing. The conclusions regarding LPC from an Islamic perspective [16] are, Workers: (1) who fulfil promises (agreed contracts) will get heaven; (2) realize that all their actions are witnessed by other believers and Allah SWT; (3) don't care about people's judgments, for the truth is only in the sight of Allah SWT; (4) do the job as best they can, since their deeds will be witnessed by Allah and His Messenger; (5) believe that work is worshiping to Allah SWT, doing the best work due to get eternal rewards and happiness in this world and the hereafter; and (6) Leaders and those followers pray for each other for the common good.

3 The Islamic Worldview

3.1 The Principle of Ghariza *(Basic Instinct)*

Ghariza, or instinct, is the natural potential in humans to maintain and preserve their survival, protect their species, and receive instructions regarding the existence of *Al Khaliq* (the Creator). Instinct cannot be sensed with the senses directly, but the human mind is able to sense its existence through its appearances [17]. Three principles of *ghariza* i.e., *Al Baqa, Al Nau'*, and *Al Tadayyun* [18]. *Ghariza al-Baqa* (Self-defense instinct) is seen when humans defend themselves, their homeland, desire to lead, and dominate others. Human being has the desire to own, feel afraid, be brave, like joining in groups, which are carried out in the context of self-defense. It has explained in QS 16:80 that Allah has made houses, household utensils and jewelry for human being until a certain time and in QS 3:14 that human (men) love women, children, wealth, gold, silver, horses, livestock and farms as the pleasure of living in this world.

Ghariza al-nau (Instinct to preserve offspring) is instinct which is can be fulfilled by man and woman relationship. Allah declared in QS 49:13 that He created humankind from a man and a woman and made them into nations and ethnic groups to know one another. In QS 4:1 Allah said that He created human being from Adam, then created his wife from him and from them Allah brought forth many males and females.

Ghariza al-tadayyun (religious instinct) is instinct to worship that awakens it by thinking about the verses and signs of Allah SWT, the Day of Judgment, the beauty of God's creation, or something that is correlated with it. The manifestation of this instinct is to purify the Creator who governs everything, then becomes worship. In QS 30:30 Allah asked human to turn face straight to the religion of Allah (Islam); (fit to) the nature (*fitrah*) of Allah who has created human being in tune to that nature (*fitrah*). Which is no change. (That is) the straight religion; but most people don't know.

3.2 *The* Maqasid *(Objectives) of Sharia*

Maqasid Syaria (the purposes of establishing Islamic law) is the ultimate goal of life (world and the hereafter) in bringing benefit and avoiding harm, by maintaining faith (*al-din*), soul (*al-nafs*), intellect (*al-'aql*), posterity (*al-nasl*), property (*al-mal*), and dignity (*al 'ird*) [19]. Everybody has a right to embrace and believe in a religion, freely and without interference to protect the religion. The way to protect the soul is states that a human being should not be hurt, injured, or murdered. Protecting mind or intellect includes everything that causes loss of mind, by consuming anything that can damage the mind (drugs and liquor), as well as freedom to express opinion. Protecting property is by guaranteeing everyone to own property wealth by buying, selling and working. Protecting offspring is by keeping human descendants protected from biological, psychological, economic, social, lineage, inheritance laws or other negative impacts, i.e., by carrying out the law of marriage. Protecting dignity is by guaranteeing the honor of everyone i.e., by freeing people to choose the profession they want. The benefit that is designed by humans, must be attached to the benefit referred to the text (postal) if there is a discrepancy between the two [20].

4 Research Methodology

The objective of this study is to define and clarify the hypothesized measurement model of IPC in organizational context. The unit of analysis is Indonesian Muslim employee who works at profit and nonprofit organization. The first step in this research is by examining and determining the dimensions of psychological contract (PC), which are characterized as Transactional and Relational PC by Rousseau [20], and the last dimension is liturgical PC which is in line with the Islamic teaching. The new PC framework according to results of this study is termed the Islamic Psychological Contract (IPC).

4.1 *Sampling Procedure*

This study takes company (profit and nonprofit organization) employees as respondents. The unit of analysis is Indonesian Muslim employees. Sampling used a disproportionately stratified random sampling procedure to obtain a more representative sample which is better than other sampling methods. Random sampling method allows all individuals in the population to have an equal and independent opportunity to be selected as respondents for the research [21, 22].

The researcher used the stratification technique to limit the sample to a group of respondents who were suitable for the study (Muslim employees). The stratification technique was sufficient for this study, in accordance with its objective, to answer research questions addressed to Muslim employees in the general population of employees. Redistribution of numbers within strata in disproportionate sampling may be considered more appropriate and representative for studies suspecting variability within respondent strata than in proportional sampling designs [22, 23]. The questionnaire used was a self-developed questionnaire by the author which was distributed via private delivery to a randomly selected What's Apps Group. A total of 98 questionnaire forms were successfully collected, and all of them were usable and valid for computer input. Thus, the returned and usable questionnaire is 100%.

4.2 Sample Characteristics

Judging from the demographic analysis, the data describes that the gender of the survey respondents was almost balance between female (56%) and male (44%). Respondents aged 30–40 were the largest (31%) followed by aged 20–30 (28%), aged 40–50 comprised 20%, aged 50–60 constituted 18%, and the last was aged 60 and above made up 3%. The last education showed that majority of the respondents (45%) were magisters, followed by under graduated (33%), below under graduated (13%), and PhD (9%).

5 The Dimensionality of Islamic Psychological Contract (IPC)

This IPC study is a new breakthrough in PC research. Therefore, the measures employed in this study were developed independently, based on the verses of the Qur'an and Al-Hadith. Initially, the hypothesized IPC measure consisted of 44 items. After the pilot survey was conducted and analyzed through SPSS, there were 3 invalid items, but all of them were reliable. In view of that, the author ensures that 41 items are valid and reliable. This study sought to identify the construct validity of the IPC based on data collected from 98 respondents ($n = 98$) who are Muslim employees of companies in Indonesia, Malaysia, Taiwan, and UK. The dimensionality of IPC was searched through Pearson Correlation and Sig. (2-tailed) to confirm validity, and Cronbach's Alpha to obtain reliability.

This section presents the results of SPSS analysis to support the construct validity of IPC. The hypothesized measurement model of IPC, contains the three latent variables, i.e., Liturgical, Relational, and Transactional, loaded on 41 items. The value of significance for validity of the three latent factors were positive between 0.000 and 0.010. While for the Pearson correlation were more than 0.202, i.e., between 0.259 and 0.859.

The reliability of 3 latent variable of IPC (Liturgical, Relational, and Transactional) were checked using the reliability analysis section of the IBM SPSS Statistics

23 text output, which showed that the coefficient was statistically reliable. The results showed, through the coefficient of Cronbach's alpha, that all values were more than 0.600, i.e., 0.826, 0.894, and 0.911, respectively. This is the justification for the researcher to confirm that all of indicators were valid and reliable.

6 Discussion

This research has hypothesized that IPC is defined by 3 dimensions i.e., Liturgical, Relational, and Transactional. Using SPSS Statistics 23, validity and reliability results have confirmed the dimensionality of the IPC components. The PC concept originally came from the Western concept. In this study, it is slightly different by including the concept of Islamic teachings. It should be noted that the findings of this study partially replicate the prior researches of Rousseau and Schein, who found that PC consists of two general dimensions, Relational and Transactional. In IPC, Relational (to seek mental wellbeing) and Transactional (to seek material wellbeing) contract must be in line with the tenets of sharia.

The effort to support high moral standards is a basic principle of Islam that is captured as a psychological contract in this study. Employers strive to develop employees' moral and ethical behavior, like Argyris said as the aim of psychological work contract. This is in line with sharia, as noted in QS 28:77 which explains that humans are recommended to prioritize achieving eternal happiness in the place of the hereafter (liturgical) by not neglecting happiness in the world (transactional) and to establish good relationships (doing good) with others (relational) without destroying the balance of nature and ecosystems in the world. This verse leads us to believe that calling for behavior appropriate to high moral standards is the behavior in Islamic teachings, which can lead to an increase in positive organizational commitment and outcomes as well.

The study findings are the IPC dimensions in this study based on Islamic literatures, i.e., Alquran and Al hadith. Muslims consider work as part of worshiping Allah [23] and always encouraged to do their best, as doing so can bring them closer to Allah [9] IPC not only leads to higher positive interpersonal bonds and organizational or individual income, which can be found in this world, as found in the Western PC literatures, but also reinforce the Muslim bond with Allah, The Creator, because the work itself is seen as an act of worship. Thus, in implementing IPC, employees also strive to purify themselves, which can bring better results in the world by living in peace and prosperity, and in the future, entering the heaven. Through this spiritual/religious motivation, workers believe that work is worship, to seek Allah's Grace, the desire to fulfill IPC can be higher and in turn, will produce better organizational and individual outcome in a way that is pleasing to Allah.

7 Conclusion and Recommendation

After studying the PC literature of Western value systems, it is necessary to investigate PC in religion perspective. It can be concluded that this study has contributed new knowledge to the existing psychological contract (PC) literature. Firstly, by studying the construction of PC in Western concept, and then explore it based on Islamic teachings, which are sourced from the two main heritages, Al-Quran and Al-Hadith. This research, found a new dimension (liturgical) of the existing PC and proposed a new construction called Islamic Psychological Contract (IPC). Furthermore, this study also produces a scale to measure employee IPC in the context of profit and non-profit organizations. The scale for measuring employee IPC offered in this study is in 41 items, with details for liturgical 17 items, relational 8 items, and transactional 16 items. The IPC construct measurement model in the context of profit and non-profit organizations is explained by three components, namely: Liturgical, Relational, and Transactional.

From this research that resulted in the psychological contract construct from an Islamic perspective, organizational researchers and practitioners are encouraged to utilize the findings of this study to develop this pioneering research in the area of the psychological contract, to enrich the understanding of the IPC construct. The author urges other researchers to carry out further studies in a wider geographical context to increase the scale of the IPC yielded by this study.

References

1. Argyris, C.: Understanding Organizational Behavior. APA Psyc.Net, Dorsey (1960)
2. Rousseau, D.: Psychological and implied contracts in organizations. Empl. Rights Responsib. J. **2**, 121–139 (1989)
3. Rousseau, D.: New hire perceptions of their own and their employer's obligations: a study of psychological contracts. J. Organ. Behav. **11**, 389–400 (1990)
4. Rousseau, D.: Psychological Contracts in Organizations: Understanding Written and Unwritten Agreements. SAGE Publications, Inc. (1995)
5. Avolio, B., Gardner, W., Walumbwa, F., Luthans, F., May, D.: Unlocking the mask: a look at the process by which authentic leaders impact follower attitudes and behaviors. Leadersh. Q. **15**(6), 801–823 (2004)
6. Schein, E.: Organizational Psychology. Prentice-Hall, Englewood Cliffs, N.J (1980)
7. Rokeach, M.: The Nature of Human Values. The Free Press, New York (1973)
8. Rousseau, D., McLean P.: The contracts of individuals and organizations. In: Cummings, L.L., Staw B.M. (eds.) Research in Organizational Behavior, pp. 1–47. JAI Press, Greenwich, CT (1993)
9. Kamil, N., Sulaiman, M., Ghani, A., Ahmad, K.: Investigating the dimensionality of organisational citizenship behaviour from Islamic perspective (OCBIP): empirical analysis of business organisations in southeast Asia. Asian Acad. Manage. J. **19**(1), 17–46 (2014)
10. Kettani, H.: World Muslim population. In: Proceedings of the 8th Hawaii International Conference on Arts and Humanities 2010, pp. 1–61. Honolulu, Hawaii (2010)
11. Junaidah, H.: Islamic revival in human resource management practices among selected Islamic organizations in Malaysia. Int. J. Islam. Middle East. Financ. Manag. **2**(3), 251–267 (2009)

12. Weiss, H., Cropanzano, R.: Affective events theory. Res. Organ. Behav. **18**(1), 1–74 (1996)
13. Cropanzano, R., Mitchell, M.: Social exchange theory: an interdisciplinary re-view. **31**(6) (2005)
14. Suhartini, T.: The role of Islamic psychological contract on authentic leadership and organizational commitment. In: International Conference on Business and Technology (ICBT) 2021. Springer (2021)
15. Suhartini, T.: Implementasi kepemimpinan otentik dan person-organisation fit serta hubungannya dengan kontrak psikologis dan komitmen organisasional Islami. Jurnal Ekonomi Syariah Indonesia **10**(2), 132–149 (2020)
16. Anwar, M., Maharani, D.: Konsep kepribadian islam menurut taqiyuddin an nabhani. Jurnal Islamika **2**(2), 132–143 (2019)
17. Al-Nabhani, T.: Nizham al-islam. peraturan hidup dalam islam. Penerjemah, Abu Amin, dkk. HTI-Press, Jakarta (2007)
18. Moneim, A.: Towards Islamic Maqasid education philosophy for sustainable development: quranic perspective with special attention to Indonesia. Millah: Jurnal Studi Agama **17**(2), 221–266 (2018)
19. Khatib, S.: Konsep Maqashid Al- Syariah: Perbandingan antara Pemikiran Al Ghazali dan Al Syathibi. MIZANI: Wacana Hukum, Ekonomidan Keagamaan. **5**(1) 47–62 (2018)
20. Rousseau, D.: Psychological contract inventory technical report. Version 3. Re-search Gate, Pennsylvania (2000)
21. Gay, L., Airasian, P.: Educational Research: Competencies for Analysis and Application, 6th edn. Prentice Hall, Upper Saddle River, NJ (2000)
22. Sekaran, U., Bougie, R.: Research Methods for Business: A Skill-Building Approach, 5th edn. Wiley, Haddington (2010)
23. Suhartini, T.: The meaning of work for preschool teachers. In: The 1st International Conference on Governance, Public Administration and Social Science 2018, ICOGPASS, vol. 1, p. 49. STIA LAN, Jakarta (2018)

Introducing e-Disclosure (via XBRL) in Jordan, the Benefits, Cost and Level of Actual Compliance

Mansour I. Saaydah

Abstract The objective of this paper is to acquaint readers with the basics of e-Disclosure, using Extensible Business Reporting Language (XBRL) newly mandated in Jordan, and to find out the level of compliance with it, and the impact of it on companies' annual reports (release) timing in 2021, as the first year of application. The reports timing and languages used in reporting by all listed companies were obtained from ASE website. This included a comprehensive listing of 171 companies from all economic sectors. The study used SPSS to analyze data, and the simple t-test to evaluate the hypotheses. The results showed that the majority of listed companies (89.5%) have complied, and reported in XBRL format in 2021. However, there was a significant delay in reporting reached 29.01 days in average (spanning from just few days to up to 242 days). This first year experience in XBRL reporting gives good indication that Jordanian companies did not face major obstacles, or have inadequate technology or qualified financial statement preparers to cope with the new reporting requirement. Furthermore, considering Coronavirus Pandemic impact, an initial result, indicates no significant difference in the mean delay in XBRL reporting between the government-declared most hurt economic sectors (Specifically the Hotels and Tourism and Transportation sectors) and one of the economic sectors that was not declared as such (banking sector). Further studies, of course, are needed to dig deeper in the issue and employ explanatory variables to understand corporate behavior in dealing with this new requirement.

Keywords e-Disclosure · XBRL · Listed companies · Amman stock exchange · 2021

1 Introduction

Accounting information disclosed in the company's annual report must be of good quality (relevant, faithful, timely, understandable, verifiable, and comparable) to

---

M. I. Saaydah (✉)
The University of Jordan, Amman 11942, Jordan
e-mail: msaaydah@ju.edu.jo

A. Hamdan et al. (eds.), *The Implementation of Smart Technologies for Business Success and Sustainability*, Studies in Systems, Decision and Control 216,
https://doi.org/10.1007/978-3-031-10212-7_22

serve the objectives of variety of users, and help them in their analysis and economic decision-making [14]. Quality financial reports require skilled preparers and efficient use of todays', widely available and relatively cheap, information technology for further processing and analysis of those reports.

The extensive use of internet in today's business activities and the rapid e-commerce growth are behind the revolution of data interchange between all kinds of parties including financial reporting to stakeholders by listed companies. A recent development in this direction, is the use of Extensible Business Reporting Language (XBRL), which is becoming a global standard for electronic financial reporting to provides standardized reports, consistent and accurate data, in a machine-readable format that can be compared across the world [26]. Many financial markets around the world, including Amman Stock Exchange (ASE), are now requiring the use of XBRL for e-Disclosure.

The provision of timely information in emerging financial market, such as ASE, has great importance because other sources of information beside annual reports are relatively rare compared to the situation in advanced countries. For example, business news conferences, economic press releases and financial analysts' forecasts are almost nonexistent. Therefore, firms' annual reports are considered one major source of data for financial market transactors, and this source should be as timely as possible for the benefit of all stakeholders.

The objective of this paper is to acquaint readers with the basics of XBRL, its benefits and cost, and the actual level of compliance, and timing of reporting in its first year of application in Jordan (2021).

The organization of this paper is follows: the problem statement is given next followed by a paragraph on the nature and evolution of XBRL. Then the benefits and cost of applying XBRL and ASE adoption of it is briefly reviewed. After that a short theoretical background on the timing of annual reports filing and hypotheses development of the study is given followed by past studies section. Research method and results of the study are then given followed by a conclusion.

2 Problem Statement and Research Questions

E-Disclosure is imposed on listed companies for the first time in Jordan starting annual reports of 2021. Amman stock exchange has claimed that XBRL reporting, among other benefits, will Increase the speed of sending and receiving data and disclosure reports among various stakeholders. This development in the method of reporting has taken place while Jordan economy is still under the pressure of Coronavirus (Covid 19) pandemic, which impacted negatively on the activity and financial conditions of many companies, and called the government to declare some economic sectors as most economically hurt sectors. Therefore, as a new development in financial reporting in Jordan, while Covid 19 burden still on for the third year, it worth investigating how smooth and speedy the transition from typical Pdf reporting to E-Disclosure via XBRL has taken place. The research questions are as follows:

1. What is the level of compliance with XBRL reporting by Jordanian listed companies in 2021?
2. How quickly has been the XBRL reporting by Jordanian listed companies in 2021?
3. Is there a timing difference in the 2021 XBRL reporting between the sectors declared as most economically hurt and the other sectors?

3 Nature and Evolution of XBRL

Extensible Business reporting Language (XBRL) is a language for the electronic transmission of financial and non-financial data which is advancing companies reporting around the world. It is based on extensible Markup Language (XML). The impact of this development in reporting and data communication is being compared to the impact of barcodes introduction in merchandising entities. XBRL is a system for coding and decoding information that provides an identifying tag for each item of the data. The use of XBRL tags enables automated processing of business information by computer software. This increases the speed of handling data, reduces the possibility of human errors and allows auto-checking of information [9].

The US Securities and Exchange Commission (SEC) mandated the reporting of public company financials in XBRL format starting in 2009, in order to make financial information easier for investors to analyze, and to assist in automating regulatory filings and business information processing. Many US companies (especially small firms) have outsourced their XBRL creation and filing process [1].

Today there is a not-for profit consortium by the name of "XBRL International". Its goal is to improve reporting in the public interest by replacing older, paper-based reports with more useful, more effective and more accurate digital versions. XBRL is used around the world, in more than 50 countries, including Japan, UK, Netherlands, Sweden, South Frica, UAE, Saudi Arabia, Jordan and others (XBRL Website).

4 Benefits and Cost of XBRL

Amman Stock Exchange [4] issued a brochure entitled "E-disclosure using XBRL" in which it listed the expected benefits from using XBRL reporting. The following is a brief summary of these benefits:

1. **Reliability:** Data entered once for all possible uses. Entered data has to be reviewed and validated by the auditor before submission.
2. **Flexibility:** XBRL gives enough flexibility to customize multiple business forms commensurate with the needs of all companies.
3. **Efficiency**: Encoding and entering data only once will require minimum resources, and make information retrieval and dissemination save time, efforts and cost.

4. **Data Integrity**: Assembling information electronically provide sophisticated validation mechanisms to ensure its integrity and validity which will improve and accelerate the analysis and other uses of financial information.
5. **Providing** continuous reporting, and keeping them electronically usable in case of need.
6. **Serves** the needs of Investors, Analysts and Others better, through preparing various financial reports, annual and periodic reports and providing them in an electronic format.
7. **Reduce the cost** of exchanging information and financial statements due to adopting the Internet as a method for this end.
8. **Increase the speed** of sending and receiving data and disclosure reports among various stakeholders, which reinforce transparency principle while disseminating the information.
9. **Contribute** to increase the disclosure level by adding new disclosure requirements, for instance, risk management, governance, prospectus, listing and membership applications, material facts disclosure, corporate actions and any other requirements by ASE and JSC.
10. **The possibility** of publishing information in different languages like Arabic, English, French and others.

5 Cost of Applying XBRL Reporting

There are no local studies about the cost of adopting and applying XBRL by Jordanian companies, probably due to the fact that 2021 is the first year of application, however, taking a look at other countries experience, especially the US, should be useful.

The CFA Institute, in the US, 2017 [6], surveyed large and small companies about the cost of XBRL reporting and found that large companies have paid annually an average of $21,000 in 2013 and the majority of small companies spent $10,000 or less each annually on outside service for preparing and reviewing their XBRL reporting.

Similarly, the AICPA and XBRL/US have conducted a pricing study in 2014 and regenerated it in 2018 (based on 2017 reports) to assess the true cost of XBRL formatting by small companies and found that the average cost was $5476 annually, which represent a decline of about 45% from 2014 average price of $10,000. They further discussed with the providers of service, and found that companies who paid higher than average annual fees have complexities in their financial statements and incurred rush charges due to last-minute changes to the filings.

6 Amman Stock Exchange Adoption of XBRL

On its homepage, ASE explains the justification for demanding that listed companies apply E-Disclosure through XBRL starting 2021 reports as follows "*A number of financial markets started using the XBRL as an international standard to view the financial reports and non-financial information, which enable investors with different languages to read and use the information easily and fast. Using a unified language for the e-Disclosure that is universally acceptable like the XBRL will cause significant development on the quality of disclosure in financial markets and thereby increase transparency in these markets. In light of the desire of the Capital Market Institutions in Jordan to apply the latest international practices in the field of financial markets, the Amman Stock Exchange (ASE) has initiated, in cooperation with the Jordan Securities Commission (JSC) during 2016 to move forward to apply the e-Disclosure System using XBRL in the Jordanian Capital Market, which would reflect positively on it and all interested parties".*

7 Theoretical Background on Timing of Annual Reports

As stated in the introduction to the paper, timing of annual reports release is one important characteristic of the quality of financial reporting to meet the need of different users including economic decision-makers. The Jordanian laws and regulation (especially companies law No. 22 and its amendments and Securities law number 18) require that listed companies disclose their annual reports within 3-months of the end of the fiscal year. Therefore, any delay would not be tolerated. In fact, the bourse penalizes the company for disclosure delay by, first stopping its share trading, and second delist the share if the company doesn't meet this requirement within the grace period [4].

Several theories were developed to deal with some aspects of annual reports release timing. The **agency theory** developed by Jensen and Meckling [16] worn from information asymmetry between agents (managers) and principles (shareholders), where agents are always accused of holding private information for their own benefit in the first place, rather than for the interest of shareholders. To fight the risk of information asymmetry governments around the world obligate listed companies to publish their audited financial statements within certain time limit of company year-end. Spence [21] developed the **signaling theory** based on asymmetric information between companies and outsiders due to the fact that management realize company's prospect and future chances of success more than outsiders. Therefore, management may want to convey some good news (superior performance) to the market to distinguish itself from other companies with just normal performance. Still the **compliance theory** is a basic idea that encourage companies to provide their annual report on time to avoid any penalty by regulators [24].

Many researchers have attempted to identify the reasons for financial reporting timing, using explanatory variables, such as firm size, profitability, type of auditor report (whether standard or qualified), the presence of extraordinary items, quality of internal control, financial risk or stress situation and other similar variables. For example, Abdulla [2], using Bahraini data, found a significant negative relationship between companies timing of reporting and their profitability, size and dividend distributed. Similarly, Owusu-Ansah [17] confirmed that company's size, profitability and age important factors in explaining annual report timing in companies listed on Zimbabwe stock exchange. Rober [18] on the other hand, thought of earnings management as main reason for reporting delay. He empirically found positive relation between earnings management and the timing of US companies annual reports filing. Dogan et al. [11] confirmed that firm's profitability, in the form of good news early bad news late, by Turkish listed companies on Istanbul stock market determines the timing of annual reports release. He also found that company size and increased risk are factors in the timing of companies annual reporting. Al-Ajmi [3] also confirmed that company size, profitability and leverage are important determinants of annual report timeliness for companies listed on Bahraini stock market.

However, Wu et al. [25] disagree with Dogan et al. [11] that annual report release has systematic relation with earnings news. They conclude that the board of directors has the ultimate responsibility for the implementation of corporate governance code and therefore board characteristics are important determinants for the timeliness of annual reports. Their empirical study on Taiwanese companies showed that board's independence has significant influence on reporting lag.

Still Conover et al. [8] compared the impact of firm's performance and capital market scrutiny on reporting lag between Common law and Code law countries. They found that filing on time is less frequent in Code law countries, while poor performance and longer reporting lag are more strongly linked in Common law countries. They also found that greater capital market scrutiny and more timely filing are related.

Finally, Dewi et al. [10] disagree that firm's characteristics has any impact on annual report timing because there are regulations that govern the timing of publishing financial statements. They arrived at this conclusion based on not finding any effect of audit tenure or financial destress on the delay in annual reports timing by Indonesian companies.

Given these contradictions in finding about the determinants of annual reports timing, this study is confined to just showing the pattern of annual reports timing distribution by Jordanian companies, via a new medium in 2021, as the first year in which firms are required to apply e-Disclosures using XBRL, and will not use explanatory variables reasoning due to the limited available data so far (one year).

8 Hypotheses Development

Due to compliance theory, and probably because of the preparation steps taken by ASE to familiarize companies with the importance and necessity to use XBRL in reporting starting 2021, the first hypothesis in the study is stated as follows:

HO1: The majority of Jordanian listed companies are able to submit their 2021 annual reports using XBRL format.

Due to the claim that XBRL reporting saves time and efforts and reduce human errors the second hypothesis is expressed as follows:

HO2: The majority of Jordanian listed companies are able to submit their 2021 annual report in XBRL format on time.

Covid 19 pandemic is an emerging factor that may influence the timing of 2021 corporate annual report release. Because of the negative impact of the pandemic on most economic sectors, the Jordan Securities Commission (JSC) has extended the deadlines for 2020 annual report submission by 40 days, however, such extension was not offered for 2021 reports (JSC news 16/06/2020). Nevertheless, it is still possible that some companies in the government-declared most hurt economic sectors, due to Covid 19 pandemic were unable to prepare and publish their reports on time, therefore, the third hypothesis is expressed as follows:

HO3: Companies in the government-declared most hurt economic sectors issued their 2021 annual reports in XBRL format late in comparison to non-declared sectors.

9 Past Studies

The following paragraph includes a number of past studies on the impact of XBRL reporting, mainly, on the quality of financial reports and stock market efficiency, most of them are done abroad, and two of them are done in Jordan.

Cong et al. [7] examined the impact of XBRL reporting on market efficiency in the US. Based on the disclosure theories about imperfect markets, they hypothesized that XBRL reporting facilitates the generation and infusion of idiosyncratic information into the market and thus improves market efficiency. They employed the event approach to measure the impact of XBRL reporting on market liquidity using 9992 firm-year observations, and found a synchronous increase of information asymmetry and trading volume as posited in the disclosure theory.

Faboyede et al. [12], stated that the purpose of XBRL is not to redefine accounting terms or replace existing accounting principles but to provide universal definitions for existing terms so that business information on the Internet can be read and understood by different reporting and analytical software without the need for human intervention, and without regard to which system originates the information. They added that XBRL, plays a central role in developing and redefining standards for communication globally, to help businesses who today face more requests to provide a variety of information, yet most businesses are not well equipped to cope with. XBRL is an XML based information format that places self-describing tags around discrete pieces of information. As tags are assigned, it is possible to extract only desired information, rather than having to view or print the entire document. XBRL works on any current operating system, and interface with almost any software. The flexibility of XBRL allows a financial report created in one software to be transferred to someone even using an entirely different software. XBRL allows the recipients to import the data immediately into their own software application.

Ashoka and Abhishek [5] focused on the impact of XBRL reporting on the quality of financial reporting system in India and on the disclosure of both financial and non-financial information in single annual report. They used a questionnaire and obtained 24 respondents (from academicians, chartered accountants, company secretaries, financial analysts and research scholars). They analyzed data by one-sample t-test and concludes that XBRL Reporting positively impacts the quality of financial reporting system in India and it also helps to communicate both financial and non-financial information under single financial report.

JUNUS and IRWANTO [15] examined the reaction of stock prices on the implementation of XBRL in companies listed on the Indonesia Stock Exchange (IDX). They employed the event study method to calculate abnormal returns of 462 companies for 2015 financial statements. Their findings, based on one-sample t-test, showed that market reacted to both companies that have implemented or not implemented XBRL; however, based on the independent t-test there were differences between companies that have not applied XBRL and those who have implemented XBRL.

Gatea et al. [13] examined the contribution of XBRL language in providing information that help users in decisions-making to enhances transparency and comparability. They considered that XBRL helps in providing an integrated services model for management, supervision of operations, and improving the quality of financial reports. They assumed that the use of XBRL will provide stakeholders with credible financial information about the companies of interest. XBRL can help create data comparison more quickly and less costly. Overall, they conclude that XBRL facilitates reporting for internal users as well as external users, and thus it is able to meet the expectations of users, regardless of the type of data they wish to collect for investment purposes.

Sassi et al. [19], examined the impact of the adoption of the XBRL on the development of stock markets using a large international sample. The analysis is based on panel data from 18 countries for the period 2000 to 2019. The study found evidence that the adoption of XBRL has negative impact on the development of stock markets. One reason for that could be the cost of applying XBRL may differentiate between the market participants in terms of their IT resources and analytic capacities. This situation could lead to more information asymmetry between the users and reduce liquidity, as well as market efficiency. Another reason may be that the value of XBRL adoption is obscured by the lack of knowledge on the benefits of XBRL reporting.

Tohang and Lan [23] examined the impact of the adopting XBRL in reducing the information risk of financial reporting disclosures. They applied the Event return volatility (ERV) to represent the occurrence of information risk, using a sample of listed companies in the Scandinavian region of the European Union. They found that XBRL adoption had lowered the incidence of information risk, which support prior research stating that a decrease in the degree of ERV, as a proxy of lower information risk, may impact the information quality, efficiency and transparency.

Slehat [20] investigated the impact of environmental, organizational, and technological factors on the ability of Jordanian companies to implement XBRL. He used a questionnaire survey to collect data from 102 participants from the finance departments in 20 industrial companies selected randomly. The findings of the study revealed that there are significant differences caused by environmental factors (such as external agencies and competitive environment pressures), organizational factors (such as supportive management, interpersonal skills, sufficient knowledge, qualifications and experience of staff), and technological factors (such as the availability of expertise programmers and trainers, and ease of understanding and use) are influencing the XBRL adoption in Jordan.

Tawfik et al. [22], tried to identify the role of XBRL in developing electronic reporting information in Jordan. They relied on a questionnaire that was distributed to Heads of finance departments, chief accountants, Accountants, and internal auditors in 20 randomly selected industrial companies. Based on 90 responses, they conclude that the Laws and regulations, organizational factors, individuals working, technical environment are supportive to the application of the language that expanded business reporting.

10 Research Methodology

This study is exploratory in nature, to observe the level of compliance with, and to measure the impact on the timing of annual report release, as a result of implementing XBRL format by listed Jordanian companies in 2021, as the first year of e-Disclosure mandated by ASE. The paper investigated, and classified all companies reporting data on the following bases; whether the company filed its annual report in any mean or not, whether the company used XBRL in reporting or not, and whether the company reported on, before or after the latest proper date of reporting (31/3/2021). The finding, based on actual filing data on ASE website, is detailed in the following Table 1.

It appears from the table that 18 (or 10.5%) of listed firms did not apply XBRL in their 2021 reports, most of them however, reported using pdf format. Similarly, 10.5% of companies reported earlier than the deadline, and 9.4% reported on time, while the majority of companies (42.1%) reported late 1–30 days. Overall, 89.5% of listed companies were able to report using XBRL format and 27.5% of them were late in reporting (15.2% + 9.4% + 2.9%). Some of the late reporters required more than 30 days to complete and file and some even required more than 90 days.

Future studies, using explanatory variables should be helpful in understanding the characteristics of late reports, and why they were unable to report using XBRL although it is required by Jordanian financial authorities (JSC and ASE).

Table 1 Timing of 2021 annual reports filing by sectors

Economic sector	Not reported In XBRL format	Reported earlier than Deadline (31/3/21)	Reported on time (On 31/3/2021)	Reported late 1–30 days	Reported late 31–60 days	Reported late 61–90 days	Reported late more than 90 days	Grand total
Banks	1	1	2	8	4			16
Insurance	2	2	3	7	4		2	20
Diversified financial service	4	5	1	12	6	1	1	30
Real estate	5	2	3	13	5	3		31
Health care				1	1	1		3
Education		1	1	2		1		5
Hotels and tourism	2	2		3		1	1	9
Transportation			1	6	1			8
Technology and communication				2				2
Utilities and energy				3		1	1	5
Commerce	1	1	1	4	1	1		9
Manufacturing	3	4	4	11	4	7		33
Grand total # %	18 10.5	18 10.5	16 9.4	72 42.1	26 15.2	16 9.4	5 2.9	171 100%

To summarize the delay, by economic sector and average number of days, Table 2 is prepared to show the average delay, the standard deviation, and the minimum and maximum.

It appears from Table 2 that the sector with the highest average delay in XBRL reporting is Utilities and Energy sector (47.7 days) followed by the Hotels and Tourism sector (with 43.67 days). It may be understood that Hotels and Tourism sector is late in reporting because it was government-declared as one of the most hurt economic sectors during 2020 because of COVID-19 pandemic, but the utilities and Energy sector was not declared as such. The lowest average delay was by the Commercial service sector (with 23.11 days), probably because the need for the commercial services was not significantly affected during the pandemic and therefore, there was no excuse for any delay in reporting.

The highest standard deviation was associated with Hotels and Tourism sector which is almost twice as much as the average delay, while most other standard deviation numbers are close to their related averages. The minimum delay is 0 and the maximum is 242 days in two cases.

Table 2 Summary of XBRL disclosure timing delay by sector

Economic sector	Number of companies	Average delay of XBRL reporting (Days)	Standard deviation of delay (Days)	Minimum delay (Days)	Maximum delay (Days)
Banks	15	26.6	16.74	0	60
Insurance	20	32.3	26.93	0	242
Diversified financial service	30	28.67	37.91	0	200
Real estate	32	23.16	22.92	0	74
Health care	3	42.67	20.6	27	66
Education	5	36.5	35.51	0	87
Hotels and tourism	9	43.67	78.22	0	242
Transportation	8	27.75	13.27	0	49
Technology and communication	2	28.5	0.71	28	29
Utilities and energy	5	47.4	36.26	14	96
Commercial services	9	23.11	27.34	0	84
Industrial companies	33	29.39	27.02	0	83
Overall statistics	171	29.01	35.619	0	242

11 Results

Looking at the numbers and percentages at the bottom of Table 1, it is clear that the majority of Jordanian listed companies (89.5% = 100% − 10.5%), regardless of the delay that has taken place, were able to comply and submit their 2021 annual reports in XBRL format. This, without the need to run the t-test, confirms hypothesis number 1, and leads to accepting it, and conclude the majority of listed Jordanian companies have complied with ASE mandate to apply e-Disclosure using XBRL format starting 2021.

The results above may indicate the relative ease of achieving this requirement by companies although it is the first year of application, despite the significant delay in many cases. Other possible explanations that also need future evidences could be, for example, the reasonable cost incurred in XBRL reporting and the less difficulty faced companies in doing so due to the adequate IT infrastructure available and trained employees or even because of the proper in advance preparation steps carried by ASE to help companies cope with this new development in annual reporting.

Also, Table 1 shows that only 19.9% of listed firms were able to file their 2021 annual reports in XBRL format either on time (9.4%) or even earlier (10.5%), while the remaining firms (80.1%) were late in filling from few days to up to 242 days. Examining the significance of this delay using the one sample t-test as shown in Table 3 indicate that the mean delay in XBRL filing is 29.01 (the standard deviation = 35.619) which is significant at 0.000 level.

This result leads to the rejection of hypothesis number 2, and conclude that e-Disclosure via XBRL did not help the majority of listed companies to file their 2021 annual reports on time. Unfortunately, the time constraint did not allow the researcher to contact number of companies to learn how they managed to deal with e-Disclosure for the first time and how helpful ASE actions were to file on time, and what causes the delay, which should be the subject of future work.

Because of COVID-19 negative impact on every aspect of life, including the economic activity of businesses, the governments in Jordan, as well as in other countries, arranged emergency support or relief programs directed to the sectors declared officially during 2020 as most hurt sectors, which make them illegible for the relief programs, hoping this will protect companies from failing or having to discharge their workforces.

Among the most hurt Jordanian economic sectors in 2020 declared by the minister of labor are the "Hotels and Tourism" sector and the "Transportation" sector, while, for example the banking sector was not declared as such. To find out if there is

Table 3 One sample T-test for listed companies reporting delay

Number of observations	Mean difference (From 0)	Standard deviation	Standard error	DF	T value	Sig. (2-tailed)
171	29.01	35.619	2.732	170	10.62	0.000

Table 4 Independent sample T-test for the mean difference in reporting delay

Panel A: Group statistics				
Sector	Number of observations	Mean delay	Standard deviation	Standard error
Banks	15	26.6	16.741	4.322
Hotels and tourism $ trans.	17	36.16	56.574	13.721

Panel B: Independent samples T-test				
Mean difference	Standard error	DF	T value	Sig. (2-tailed)
−9.576 Days	15.186	30	−0.631	0.533

* Equal variance assumed, F = 2.325, Sig. at 0.138

a difference in the amount of delay in e-Disclosure via XBRL between these two sectors and another selected sector not identified as most hurt, the banking sector, as a pilot study to get initial result. Table 4 Panel A shows the mean difference in reporting, although this difference seems to be big, approximately 10 days (36.18 – 26.6) it is not significant, as shown in Panel B of the same table (T. value = −0.631, sig. 0.533), equal variance assumed*, leading to the rejection of hypothesis 3 and concluding there is no excessive delay in XBRL format filing experienced by companies operating in the government-declared most hurt economic sectors in 2020. This result probably points to the relative ease of meeting this first-time requitement in2021 and/or the availability of adequate IT infrastructure and trained financial statement preparers.

12 Conclusion

The objective of this paper was to review the advancement in introducing e-Disclosure, via the Extensible Business Reporting Language (XBRL), in Jordan, and to find out the level of compliance with, and the impact of it on the timing of annual reports release in 2021, as a first year of application. The type of language used in reporting, and the dates of reporting, by all listed companies were obtained from ASE website. This includes all 171 listed companies from all economic sectors. The results showed that the majority of companies (89.5%) have, complied with, and reported in XBRL format in 2021, as mandated by Amman Stock Exchange. This first year experience in reporting via XBRL, however, gives good indication that Jordanian listed companies did not face major obstacles, or have inadequate IT resources or qualified financial statement preparers to cope with new reporting advancement. However, there was a significant delay in reporting averaged 29.01 days (spanning from just few days to up to 242 days in two cases). Furthermore, as initial result, there was no significant difference in the mean delay in reporting, due to Coronavirus

Pandemic, between the government-declared most hurt economic sectors (specifically, the Hotels and Tourism and Transportation sectors) and one of the economic sectors that was not declared as such (banking sector).

Future research will have more than one year of data to be able to dig deeper in the current issue, and resort to employing explanatory variables in order to explain corporate behavior in dealing with the new reporting environment.

References

1. AICPA and XBRL US: XBRL cost for small reporting companies have declined 45% since 2014, 1–2 (2018)
2. Abdulla, J.Y.A.: The timeliness of Bahraini annual reports. Adv. Int. Acc. **9**, 73–88 (1996)
3. Ai-Ajmi, J. : Audit and reporting delays: Evidence from an emerging market. Adv. Account. **24**(2), 217–226 (2008). https://doi.org/10.1016/j.adiac.2008.08.002
4. Amman Stock Exchange: e-Disclosure using extensible business reporting language (2018)
5. Ashoka, M., Abhishek, N.: The impact of XBRL reporting on the quality of financial reporting system in India. **8**(1) (2018). Available online at www.lbp.world
6. CFA Institute: The cost of structured data: myth vs. reality. 1–12 (2017). www.cfainstitute.org
7. Cong, Y., Hao, J., Zou, L.: The impact of XBRL reporting on market efficiency. J. Inf. Syst. **28**(2) (2014). https://doi.org/10.2308/isys-50794
8. Conover, M.C., Miller, R.E., Szakmary, A.: The timeliness of accounting disclosure in international security markets. Int. Rev. Financ. Anal. **17**(5), 849–869 (2008)
9. Deloitte: Extensible business reporting language, moving to a global standard for electronic business reporting (2009)
10. Dewi, K.I.K., Subekti, I., Saraswati, E.: The determinants of delay in publication of financial statement. Int. J. Multicultural Multireligious Underst. (2019). https://doi.org/10.18415/ijmmu.v6i1.470
11. Dogan, M., Coskun, E., Celik, O.: Is timing of financial reporting relating to firm performance?—An examination on ISE listed companies. Int. Res. J. Financ. Econ. **12**, 220–233 (2007)
12. Faboyede, S., Nwobu, O., Akande, O., Oladipo, O.: Extensible business reporting language (XBRL): a tool for accounting education in the 21st century. J. Acc. Financ. Auditing Stud. **2**(3), 85–97 (2016)
13. Gatea, A., Alnawas, A., Gali, Z.: The effect of XBRL financial reporting on enhancing the transparency of information in the financial statements. Turk. J. Comput. Math. Educ. **12**(11), 4945–4953 (2021)
14. IASB: conceptual framework (2018)
15. JUNUS, O., IRWANTO, A.: Stock reaction to the implementation of extensible business reporting language. J. Asian Financ. Econ. Bus. **8**(1), 675–685 (2021). Online ISSN: 2288-4645. https://doi.org/10.13106/jafeb
16. Jensen, M., Meckling, W.: Theory of the firm: managerial behavior, agency costs and ownership structure. SSRN # 94043
17. Owusu-Ansah, S.: Timeliness of corporate financial reporting in emerging capital markets: empirical evidence from the Zimbabwe stock exchange. Acc. Bus. Res. **30**(3), 241–254 (2000)
18. Rober, W.R.: SEC regulation of corporate 10K filing dates: the effect of earnings management and market recognition. A Dissertation for PhD Degree at Virginia Commonwealth University (2005)
19. Sassi, W., Ben Othman, H., Hussainey, K.: The impact of adoption of the XBRL on the development of stock markets. Asian-Pac. Manage. Acc. J. **16**(2), 95–127 (2021)

20. Slehat, Y.: Affecting factors on extensible business reporting language (XBRL) adoption among public listed companies in Amman stock exchange. Aust. J. Basic Appl. Sci. **12**(6), 36–40 (2018). https://doi.org/10.22587/ajbas
21. Spence, M.: Job market signaling. Q. J. Econ. **87**(3), 355–374. Published By: Oxford University Press (1973). https://doi.org/10.2307/1882010
22. Tawfik, O., Hayek, A., Al-Smady, A.: The extend of applying the extensible business reporting language at Jordanian industrial companies. Int. Rev. Manag. Mark. **7**(1), 66–75 (2017)
23. Tohang, V., Lan, M.: The impact of adoption of XBRL on information risk in representative countries of Scandinavian region. Jurnal Keuangan dan Perbankan **21**(4), 515–526 (2017)
24. Tyler, T.: Why People Obey the Law? Yale University Press, Newhaven (1990)
25. Wu, C.-H., Wu, C.-S., Liu, V.W.: The Release timing of annual reports and board characteristics. Int. J. Bus. Financ. Res. **2**(1), 103–108 (2008). Available at SSRN: https://ssrn.com/abstract=1543851
26. XBRL international: the business reporting standard: https://www.xbrl.org/the-standard/what/ixbrl/. Accessed 10 Feb 2022

Digital Audit During Covid-19 in Jordanian Audit Firms a Study of the Reality and Outlook the Future

Omar Fareed Shaqqour, Ayman Saleh Mustafa Harb, Osama Mohammad Khaleel Ballout, and Raed Jameel Jaber

Abstract This study aims to investigate the effect of the Covid-19 on audit procedures, through studying the availability of digital audit requirements of modern technologies and artificial intelligence techniques in Jordanian audit firms during the pandemic, as well as outlook the future impact of applying digital audit on the auditing profession, through studying the impact of the application of digital audit on the auditor's personal qualities. The study population comprises Jordanian audit firms, and the study sample consists of 60 firms, randomly selected. A questionnaire was distributed to an auditor in each of the study sample firms, where (60) questionnaires were distributed, (52) restored, representing a percentage of (87%). The results of the study showed the availability of digital audit requirements of modern technologies and artificial intelligence techniques in Jordanian audit firms. Also, applying digital audit affects the auditor's personal qualities.

Keywords Digital audit · Auditing profession · Covid-19 pandemic

1 General Framework

1.1 Introduction

With the rapid and modern developments and the acceleration of technology and modern technology, as well as the impact Covid-19 pandemic had on the world at the end of 2019, leading to a state of spacing between people in all aspects of life. To prevent the spread of this virus through human populations, many commercial, industrial and service companies have resorted to working remotely by employing technology. Education has followed suit as well as audit firms, to alleviate the spread

O. F. Shaqqour (✉) · A. S. M. Harb · O. M. K. Ballout
Zarqa University, Zarqa, Jordan
e-mail: oshaqqour@zu.edu.jo

R. J. Jaber
Arab Open University, Aman, Jordan

A. Hamdan et al. (eds.), *The Implementation of Smart Technologies for Business Success and Sustainability*, Studies in Systems, Decision and Control 216,
https://doi.org/10.1007/978-3-031-10212-7_23

of the disease and keep a balance between health and economy. Mandatory lockdown measures imposed by governments to control the spread of the virus have led to individuals having to work from home, whenever possible. Auditors could no longer visit the company under audit nor their own offices, so audit procedures were done remotely, and auditors could no longer perform audit and verification processes and procedures as they did before the pandemic, thus things became quite challenging for them as they moved towards digital auditing. Consequently, companies investing in digital power have become more capable of coping with the new situation. Digital auditing, which uses modern technology and artificial intelligence, emerged as an alternative to traditional auditing, being done without the need for rapprochement between humans, and thus it was possible to continue to provide audit services during Covid-19.

This study aims to investigate the effect of Covid-19 on audit procedures, through studying the availability of digital audit requirements of modern technologies and artificial intelligence techniques in Jordanian audit firms in Jordan during the pandemic, as well as outlook the future impact of the application of digital audit on the auditing profession, through studying the impact of the application of digital audit on the auditor's personal qualities.

1.2 Research Problem

The repercussions of Covid-19 have led to auditors having to work remotely, as they can no longer go to the company under audit, nor to their own offices, and audit procedures are done remotely, and they could no longer do the audit and verification processes and procedures as before the pandemic, thus digital audit emerged as a reliable alternative. The research problem lies in studying the reality of digital audit in Jordan, by analyzing the availability of digital audit requirements of modern technologies and artificial intelligence techniques in Jordanian audit firms during Covid-19, as well as outlook the future impact of the application of digital audit on the auditing profession, through studying the impact of the application of digital audit on the auditor's personal qualities and audit procedures.

Elements of the Research Problem:

The study will try to address the following main question:

What effect does digital audit application have on the auditing profession during Covid-19?

The following sub-questions branch from the main one:

1. How available are the digital audit requirements of modern technologies and artificial intelligence in Jordanian audit firms during Covid-19?
2. What effect does the application of digital audit have on the auditor's personal qualities?

2 Theoretical Framework

The world has been affected since the end of 2019 by a serious pandemic, Covid-19, which had an impact on health and economy, leading to closures that affected the economy. Jordan was affected in 2020 by the pandemic, resulting in closures in March 2020 to protect the health of citizens. This has affected Jordan's economy and its citizens. Many economic sectors were closed, and a large part of these sectors shifted to the use of technology and remote work. In a similar way, the auditing profession was affected in relation to how to apply digital auditing to the auditing profession.

The auditing profession has developed in all countries after the separation of ownership from management [6]. The auditing profession defined as a method of control, which includes a set of procedures taken by the independent, competent person in order to examine records, financial and accounting data and information to verify the fairness of its preparation in accordance with the generally accepted accounting principles [5]. Mahmod and Al-Kabbiji [10] believe that the impact of the quality of external audits on the performance of industrial companies should be investigated.

The importance of auditing is due to the fact that it is a means rather than an end, and aims to serve the parties that use the audited financial statements to be used in the decision-making and policy-making process [9].

The audit sector is considered to be a challenging sector [13], as the auditing profession is one of the many areas that have been influenced by developments that have challenged the whole world, and due to the technological and information revolution, paper auditing has become a thing of the past although there are audit firms that still apply it, since paper auditing has been completely replaced by electronic auditing, in which electronic records are recorded. Today we are witnessing the emergence of a new type of audit, digital audit.

The decision to opt for the application of an electronic audit instead of a paper audit depends on the nature of the corporate records [8]. Many benefits can be driven from implementing an electronic audit, since it is possible that electronic audits can reduce the activities of the client and the audit department to complete the audit.

Digital auditing defined [3] as the use of automated audit software in the audit process to examine, verify and report the audit process.

There are three main digital trends affecting auditing: data analysis, artificial intelligence, robotics and block chain [11].

The availability of mass data, and the use of advanced systems in financial reporting such as the emergence of cloud accounting, made the auditor able to modify the audit procedures towards more centralized application [15], and the Applying the XBRL in financial market [2].

The International Federation of Accountants (IFAC) has organized a number of steps to plan and manage the Covid-19 crisis in professional review and audit procedures as follows [7]:

1. Creating a crisis management plan through the formation of a knowledgeable diverse team.
2. Creating business plans that work on virtual communication through creating a work environment.
3. Creating a communication strategy to exchange information during crises.
4. Equipping members with resources so that members of the auditing profession are qualified to provide support to those affected by the crisis.
5. Making plans for future contingencies.

3 Literature Review

The study [1] aimed to explore the impact of digital transformation on audit quality. Results showed that Egyptian auditors expect some important changes in the audit quality determinants as a result of the spread of digital transformation, especially in relation to the qualifications of auditors, some applied audit procedures, and the timing of the audit report issuance. Auditors agree that digital transformation will not greatly affect the audit quality determinants relating to the no-confidence of users to the audit process,

The research [3] aimed to define the role of artificial intelligence technologies using digital auditing in achieving audit quality and supporting the audit strategy used in audit companies in the Kingdom of Bahrain. The research found that the use of artificial intelligence technologies contribute to achieving quality in the audit process as well as supporting the implementation of audit strategy in the Kingdom of Bahrain.

This study [4] aimed to provide and apply an integrated electronic audit program by using electronic computer as well as identify the main challenges and difficulties facing this application and any pitfalls, while indicating the areas of artificial intelligence and computer techniques in the completion of the audit process. The study recommends applying other artificial intelligence techniques in the audit process, and using computers in the implementation of logical operations to work mathematical analysis directly from electronic records.

The study [12] aimed to measure the role of digital audit in achieving competitive advantages in Jordanian audit companies. The study found that the use of digital audit contributes to achieving competitive advantages such as a reduction in cost, quality, flexibility and market share. The study also found that there are obstacles that limit the use of electronic audit, including the cost of specialized auditing software and the high prices of public programs and their unsuitability for all business establishments, in addition to the need for scientific and practical qualification of auditors.

The study [14] aimed to identify the possibility of using artificial intelligence techniques in controlling the quality of internal audit in Jordanian public shareholding companies. The study concluded that there is an impact of using artificial intelligence techniques on controlling the quality of internal audit.

4 Research Methodology

The study used the descriptive-analytical approach to achieve the research objectives in answering the following questions:

1. How available are the digital audit requirements of modern technologies and artificial intelligence in Jordanian audit companies during Covid-19?
2. What effect does the application of digital audit have on the auditor's personal qualities?

4.1 Research Hypotheses

This study aims to determine the availability of digital audit requirements of modern technologies and artificial intelligence techniques in Jordanian audit companies during Covid-19, and the future impact of digital audit on the auditor's personal qualities, audit procedures, and audit outputs. Based on the above, the following main hypothesis was formulated:

HO1: Digital audit requirements of modern technologies and artificial intelligence are not available in Jordanian audit firms during Covid-19.

HO2: There is no statistically significant impact of digital auditing on the auditor's personal qualities.

4.2 Data Collection Resources

The researchers used the field survey method for audit companies in Jordan, to identify the availability of digital audit requirements of modern technologies and artificial intelligence techniques in audit firms in Jordan during the corona pandemic, and the future impact of digital audit on the auditor's personal qualities, audit procedures, and audit outputs.

4.3 Research Population and Sample

The study population comprised Jordanian audit firms. The study sample consisted of 60 firms, which were selected by the random sample method. A questionnaire was distributed to an auditor in each of the study sample firms, where (60) questionnaires were distributed, and (52)questionnaires were restored, representing (87%).

Table 1 Descriptive statistics of the availability of digital audit requirements

Item	N	Mean	SD	Ranking
1. Modern technological and computer devices and equipment are used to develop audit procedures	52	4.50	0.50	5
2.Modern technological software is used in the audit process	52	4.92	0.27	1
3. Auditors in the the audit office are constantly trained to keep abreast of technological developments in the field of audit	52	4.65	0.56	3
4. Advanced artificial intelligence technologies are used to support audit procedures	52	4.79	0.41	2
5. Smart programs are applied to develop the audit process	52	4.54	0.73	4
Availability of digital audit requirements in Jordanian audit firms during Covid-19	52	4.68	0.24	

5 Analysis of the Research Questionnaire's Answers

The study questionnaire consisted of 25 questions divided into two sections:

Section I: The availability of digital audit requirements of modern technologies and artificial intelligence techniques in audit firms in Jordan during Covid-19, measured by questions (1–5).

Section II: Impact of digital auditing on the auditor's personal qualities, measured by questions (6–13).

5.1 The Availability of Digital Audit Requirements During Covid-19

Table 1 shows the arithmetic mean values and the standard deviation of the items of the axis "the availability of digital audit requirements in Jordanian audit firms during Covid-19". By analyzing the answers to the questions of the axis of the availability of digital audit requirements in Jordanian audit firms during Covid-19, it can be concluded that Jordanian audit firms meet the requirements of digital audit, where the average answers to all items of the axis was more than 4.5, or more than 90%, and the average answers to this axis was 4.68 and 93.6%.

5.2 The Effect of Applying Digital Audit on the Auditor's Personal Qualities

Table 2 shows the values of the arithmetic mean and the standard deviation of each item of the axis "the impact of digital audit on the auditor's personal qualities", and through the analysis of the answers to the questions of the axis "the impact of digital

Table 2 Descriptive statistics of the impact of digital audit on auditor's qualities

Item	N	Mean	SD	Rank
1. The application of digital audit requires the auditor to have the appropriate knowledge and education in the field of Information Technology	52	4.63	0.45	1
2. The application of digital audit requires the auditor to have the appropriate experience in the field of Information Technology	52	4.51	0.46	2
3. The application of digital audit requires the auditor to have the appropriate training in the field of Information Technology	52	4.40	0.57	3
4. The application of digital audit requires the auditor to have less apparent independence when conducting the audit	52	3.98	0.73	6
5. The application of digital audit leads to a number of auditors leaving the profession as a result of the requirements of Information Technology	52	1.25	0.44	7
6. The application of digital audit requires the auditor to exert less professional care when carrying out the task	52	4.25	0.65	4
7. The application of digital audit requires reliance on automation and reduces reliance on humans when conducting audit procedures	52	4.09	0.80	5
8. The application of digital audit leads to a decrease in the demand for auditors and the loss of a number of auditors of their jobs	52	1.24	0.48	8
The effect of digital audit on the auditor's personal qualities	52	3.55	0.21	

audit on the auditor's personal qualities, it can be concluded that the application of digital audit requires IT knowledge, education, and experience from the part of the auditor. The arithmetic mean for these items showed more than 4, indicating a percentage of 80%. Answers also showed that the application of digital audit requires less professional care on the part of the auditor when performing the task and that the application of digital audit requires reliance on automation and reduces reliance on humans when conducting audit procedures, and that the application of digital audit requires the auditor to have less apparent independence when conducting the audit. However, the answers showed that the digital audit does not lead to a number of auditors leaving the auditing profession due to the requirements of Information Technology and does not lead to a decrease in the demand for auditors and the loss of a number of auditors for their jobs, where the average answers to these items was low and less than 25%.

5.3 *Testing the Validity and Reliability of the Intrument*

To verify the internal consistency and reliability of the resolution questionnaire's questions, the Cronbach-Alpha test was used, since the calculation of the reliability coefficient based on this method indicates stability, reliability and dependence to a large extent.

Table 3 Results of applying the alpha coefficient to the study questions

Variable	Questions	Alpha coeffi.
All questions	1–25	0.9268
Availability of digital audit requirements	1–5	0.8343
Impact of digital audit on auditor's personal qualities	6–13	0.8124

When the Cronbach-Alpha test was applied to questions measuring the study variables of 25 questions, the alpha value was 0.9268, which showed that the questions are highly correlated and that there is a high degree of consistency for all questions. Table 3 shows the results of applying the alpha coefficient to the study questions.

6 Hypotheses Test Results

The researchers used EXCEL and SPSS for Social Sciences to analyze the data and identify the descriptive statistics of the study variables and the results of the hypotheses test.

HO1: Digital audit requirements of modern technologies and artificial intelligence are not available in Jordanian audit firms during Covid-19.

The availability of digital audit requirements of modern technologies and artificial intelligence techniques in Jordanian audit companies during Covid-19 was measured by questions (1–5) in the study questionnaire.

The Paried Sample T-test was used to test the first hypothesis. Table 4 showed the results of testing the availability of digital audit requirements of modern technologies and artificial intelligence techniques in Jordanian audit firms during Covid-19, where *P*-value sig. = 0.000, which is a value below the significance level $\alpha = 0.05$, therefore the null hypothesis is rejected and the alternative one accepted, i.e. digital audit requirements of modern technologies and artificial intelligence techniques are available in Jordanian audit firms during Covid-19.

HO2: There is no statistically significant impact of digital auditing on the auditor's personal qualities.

The impact of digital audit on the auditor's personal qualities was measured by questions (6–13) in the study questionnaire.

Table 4 Study hypotheses test results

Hypothesis	Value of t	Stat. Sig.
HO1	65.048	0.000
HO2	35.168	0.000

The Paried Sample T-test was used to test the second hypothesis. Table 4 showed the results of testing the impact of digital audit on the auditor's personal qualities, where the *P*-value sig = 0.000, which is a value below the significance level $\alpha = 0.05$, therefore the null hypothesis is rejected and the alternative one accepted, i.e. digital audit affects the auditor's personal qualities.

7 Conclusions and Recommendations

7.1 Conclusions

1. Jordanian audit firms meet the requirements of digital audit, where the average answers to all the axis' items amounted to more than 4.5 or more than 90%, and the average answers to this axis reached 4.68 representing 93.6%.
2. The application of digital audit requires the auditor to have the knowledge, education, experience and appropriate training in the field of Information Technology, where the average answers to these items was more than 4, i.e. more than 80%.
3. Answers showed that the application of digital audit requires the auditor to exercise less professional care when carrying out the task, that the application of digital audit requires the reliance on automation and reduces the reliance on humans when conducting the audit procedures, and that digital audit requires that the auditor have less apparent independence when conducting the audit.
4. Answers showed that digital audit does not lead to a number of auditors leaving the auditing profession as a result of the requirements of Information Technology, and does not lead to a decrease in the demand for auditors and the loss of a number of auditors for their jobs, where the average answers to these items was low by less than 25%.

7.2 Recommendations

The researcher suggests the following recommendations:

1. Encouraging audit firms to provide appropriate IT training to their auditors.
2. Encouraging audit companies to increase their interest in digital auditing.
3. Further studies on digital auditing.

References

1. Almaleeh, N.: The impact of digital transformation on audit quality: exploratory findings from a Delphi study. Sci. J. Bus. Res. **42**(3), 9–36 (2021)

2. Al-Ramahi, N., Abuhussein, A., Shahwan, Y.: The impact of applying the XBRL on the quality of annual reports of the firms listed in Amman stock exchange. In: 2021 22nd International Arab Conference on Information Technology (ACIT), pp. 1–5. IEEE (2021)
3. Alsammarraee, A.E., Al-Shareeda, N.A.: The role of artificial intelligence techniques using digital auditing in achieving audit quality and supporting the audit strategy from the point of view of auditors (A field study in audit firms in the Kingdom of Bahrain). Glob. J. Econ. Bus. **8**(1), 15–31 (2021)
4. Anbar, O., Sami, J., Ahmed, H.: Quality of auditing by adopting artificial intelligence, applied research in a sample of supervisory bodies working in the federal office of financial supervision. J. Acc. Financ. Stud. University of Baghdad, Iraq, **11**(34), 27–76 (2011)
5. Arens, A.A., Elder, R.J., Beasley, M.S., Hogan, C.E.: Auditing and assurance services. Auditing Assur. Serv. (2016)
6. Harb, A.S.M.: The Effect of internal audit on accounting information technology in the public joint stock pharmaceutical industries in Jordan. Acad. Acc. Financ. Stud. J. **24**(1), 1–8 (2020)
7. IFAC (2020). Seven essential crisis management planning steps for PAOs. Retrieved (2 Nov 2020), from http://www.ifac.org/knowledge-gateway. Mercia and Podcast;. (6 June 2020). COVID-19 and audit report implications
8. Jaber, R.J., Abu Wadi, R.M.: Auditors' usage of computer-assisted audit techniques (caats): challenges and opportunities. In: Conference on e-Business, e-Services and e-Society, pp. 365–375. Springer, Cham (2018)
9. Kumar, R., Sharma, V.: Auditing: Principles and Practice. PHI Learning Pvt. Ltd. (2015)
10. Mahmod, A.L., ALkabbji, R.F.: Impact of external auditing quality and audit committees on accounting conservatism and the performance of industrial firms listed at the Amman stock exchange. Int. J. Financ. Res. **11**(4), 556–562 (2020)
11. Meuldijk, M.: Impact of digitalization on the auditing profession. Audit Comitee News 33–35 (2017)
12. Okab, R.: Electronic audit role in achieving competitive advantages and support the strategy of the external audit in auditing offices in the Hashemite Kingdom of Jordan. Int. Bus. Res. **6**(6), 181 (2013)
13. Olasanmi, O.O.: Computer aided audit techniques and fraud detection. Res. J. Financ. Acc. **4**(5), 67–80 (2013)
14. Othman, O., Hussein, A., Adel, J.: The possibility of using artificial intelligence techniques in controlling the quality of internal auditing (field study in Jordanian public shareholding companies). In: The Eleventh Annual Scientific Conference on Business Intelligence and Knowledge Economy. Zaytouna University, Faculty of Economics and Administrative Sciences , Aman Jordan (2012)
15. Shaqqour, O.F., Alkabbji, R.F., Masoud, A.A.: The impact of appling cloud accounting on the financial fraud at Jordanian companies. In: 2021 22nd International Arab Conference on Information Technology (ACIT), pp. 1–6. IEEE (2021)

The Role of Global Reporting Initiative (GRI) for Achieving Sustainability Reporting

Talal M. A. Aljajawy, Yousef Dolab, and Eman J. A. Alkhfajy

Abstract Sustainability deals with companies as a social unit that influences society and is affected by it. Therefore, this society needs to know the social, environmental and economic impacts of the activities carried out by companies and report them. The objective of this research is to define the importance of both sustainability and reporting it through the sustainability accounting system through preparing sustainability reports. This research also aims at identifying the role of GRI and ESG indicators in supporting and enhancing reporting on sustainability and the extent to which they contribute to company's sustainability. It was concluded that GRI provides objective criteria and indicators that can be adopted by companies and various sectors to achieve objective measurement and reporting the level. It was found that the level of sustainability reporting conducted by the companies listed in the Iraqi market for securities is weak by 19.76% due to lack awareness of sustainability importance.

1 Introduction

Since the importance of sustainability grows and is reported, attention has been paid to it by business organizations, organizations accounting and professional associations and academics, especially after accounting collapses suffered by major international companies at the beginning of the twenty-first century. This leads to an increase for the demand for validation of these reports ensuring corporate transparency in their environmental, social and governance (ESG). One of the most important global

T. M. A. Aljajawy (✉) · Y. Dolab
Al-Esraa University College, Baghdad, Iraq
e-mail: tjajawy@esraa.edu.iq

Y. Dolab
e-mail: Dr.yousef@esraa.edu.iq

E. J. A. Alkhfajy
University of Warith Al-Anbiyaa, Karbala, Iraq
e-mail: iman.ja@g.uowa.edu.iq

A. Hamdan et al. (eds.), *The Implementation of Smart Technologies for Business Success and Sustainability*, Studies in Systems, Decision and Control 216,
https://doi.org/10.1007/978-3-031-10212-7_24

sustainability reporting initiatives is the Global Reporting Initiative (GRI), a non-profit organization that promotes economic, environmental and social sustainability. It provides a comprehensive report (and reporting) which provides a framework for all worldwide companies and institutions established under the auspices of the Global Sustainability Standards Board (GSSB) which develops and endorses standards for the implementation of a globally consistent basis, interest to inform and understand the impact of companies on the economy, environment and society increasing the quality and ensuring the transparency of information to stakeholders. Thus, it provides stakeholders with greater accountability to identify and compare the impact of different reporting companies. It identifies the theoretical issues as concrete and specific ones helping to understand and manage the impacts of sustainability developments in the company's activities and strategy. In addition, it assess the risks and opportunities facing companies to ensure that they continue to meet the needs of the current generation without any chance of compromising the ability of future generations to achieve their needs and thus achieve a sustainable global economy.

2 Previous Research

In his research, "Accounting Response to Environmental Conservation and Support for Sustainable Development", Al-Rekaby [11] aimed to know the response of the accounting branches to the preservation of the environment and support for sustainable development. He concluded that there was no standard for environmental accounting, which led to the availability of accounting data, in which it was possible to provide measures related to the environment. Onaiza and Ali [12] discussed the impact of the disclosure of social responsibility in the financial statements in the decisions of the users of these lists—an operational and exploratory study aiming at shedding light on the importance of corporate social responsibility. The users of the financial statements are interested in disclosing social responsibility and this disclosure affects their decisions. Al-Armoty (2013) discussed the impact of sustainability accounting for the financial reports of the Jordanian public joint stock companies listed on Amman Stock Exchange. The aim of the study was to identify the impact of sustainability accounting for the financial reports of the Jordanian public joint stock companies listed on Amman Stock Exchange Statistical significance of the elements of sustainability accounting for environmental, social and economic elements for the financial reporting reports of the Jordanian public joint stock companies from the point of view of financial managers, accountants, auditors and employees of Accounting Departments For accounting. In their research "The Assurance Process of GRI Sustainability Reports—Influence on Accountability and Transparency" Johannson and Lundberg (2012) introduced an analysis of the proven data for sustainability of the Swedish companies to assess how the sustainability process of "The Assurance Process of GRI Sustainability Reports—Influence on Accountability and Transparency" such companies affect transparency and accountability. They also found that the use of reporting standards and the guarantee of sustainability greatly reduce the risk of administrative capture to ensure process and thus be an open and standardized process that reviews the quality of information in sustainability reports increasing transparency and enhancing the credibility of those reports. Bidari [27] stated his objective to highlight the relationship between factors that influence the size, age and profitability of banks (and), the structure of ownership and levels of disclosure of CSR. He concluded that all the banks in their annual reports disclose information about their social responsibility in order to obtain the community's recognition of the suitability of their social behavior, however, most of such information is descriptive and does not disclose the environmental performance.

3 Practical Framework

Examine the level of companies' contribution to sustainable development using the GRI and ESG criteria:

How to calculate the reporting ratios for any ESG axis of sustainability:

$$\text{Reported requirements} \times \frac{100}{27} = (\%)\ \text{Ratio of reporting of economic}$$

Information

$$14 \times \frac{100}{27} = 51.851\%\ \text{Ratio of reporting of economic}$$

3.1 *First: The Economic Axis*

1. Al-Mansour Company for Pharmaceutical Industries and Medical Supplies (Table 1):
2. Al-Khatem Communications Company (Table 2):
3. The Modern Company for Animal and Agricultural Production (Table 3)
4. Gulf Commercial Bank (Table 4):
5. Al Mamoura Real Estate Investment Company (Tables 5 and 6):

3.2 *Second: The Environmental Axis*

1. Al-Mansour Company for Pharmaceutical Industries and Medical Supplies (Table 7):

Table 1 Reporting information on the economic axis 2014–2015

The sides	2014 (%)	2015 (%)
1. Economic performance	18.518	18.518
2. Market presence	11.111	11.111
3. Indirect economic impacts	11.111	11.111
4. Procurement practices	3.704	3.704
5. Anti-corruption	0	0
6. Anti-competitive behavior	7.407	7.407
Total	51.851	51.851

Table 2 Reporting information on the economic axis

The sides	2014 (%)	2015 (%)
1. Economic performance	14.815	14.815
2. Market presence	3.704	7.407
3. Indirect economic impacts	11.111	11.111
4. Procurement practices	3.703	3.704
5. Anti-corruption	0	0
6. Anti-competitive behavior	0	3.704
Total	33.333	40.741

Table 3 Reporting information on the economic axis

The Sides	2014 (%)	2015 (%)
1. Economic performance	22.222	14.815
2. Market presence	7.407	11.111
3. Indirect economic impacts	7.407	7.407
4. Procurement practices	3.704	0
5. Anti-corruption	0	0
6. Anti-competitive behavior	3.704	3.704
Total	44.444	37.037

Table 4 Reporting information on the economic axis

The sides	2014 (%)	2015 (%)
1. Economic performance	25.926	25.926
2. Market presence	7.407	7.407
3. Indirect economic impacts	11.111	11.111
4. Procurement practices	0	0
5. Anti-corruption	0	0
6. Anti-competitive behavior	3.704	3.704
Total	48.148	48.148

Table 5 Reporting information on the economic axis

The sides	2014 (%)	2015 (%)
1. Economic performance	18.519	18.519
2. Market presence	7.404	11.111
3. Indirect economic impacts	7.404	7.404
4. Procurement practices	0	3.704
5. Anti-corruption	0	0
6. Anti-competitive behavior	0	0
Total	33.333	40.741

Table 6 Percentage of reporting information on the economic axis of the research sample

Company	2014 (%)	2015 (%)	Medium (%)
Al-Mansour Company for Pharmaceutical Industries and Medical Supplies	51.85	51.85	51.85
Al-Khatem Communications Company	33.33	40.74	37.04
The Modern Company for Animal and Agricultural Production	44.44	37.04	40.74
Gulf Commercial Bank	48.15	48.15	48.15
Al Mamoura Real Estate Investment Company	33.33	40.74	37.04
Medium	37.407	42.22	43.70

2. Al-Khatem Communications Company (Table 8):
3. The Modern Company for Animal and Agricultural Production (Table 9):
4. Gulf Commercial Bank (Table 10):
5. Al Mamoura Real Estate Investment Company (Tables 11 and 12):

Table 7 Reporting information on the environmental axis

The sides	2014 (%)	2015 (%)
1. Materials	1.389	1.389
2. Energy	4.167	8.333
3. Water	0	0
4. Biodiversity	0	0
5. Emission	0	0
6. Effluents and waste	4.166	4.166
7. Environmental compliance	0	1.389
8. Supplier environmental assessment	0	0
Total	9.722	15.277

Table 8 Reporting information on the environmental axis

The sides	2014 (%)	2015 (%)
1. Materials	1.389	1.389
2. Energy	6.944	6.944
3. Water	0	0
4. Biodiversity	0	0
5. Emission	0	0
6. Effluents and waste	0	0
7. Environmental compliance	0	1.389
8. Supplier environmental assessment	0	0
Total	8.333	9.722

Table 9 Reporting information on the environmental axis

The sides	2014 (%)	2015 (%)
1. Materials	1.389	1.389
2. Energy	4.166	6.944
3. Water	0	0
4. Biodiversity	0	0
5. Emission	0	0
6. Effluents and waste	4.167	4.166
7. Environmental compliance	1.389	1.389
8. Supplier environmental assessment	0	0
Total	11.111	13.888

Table 10 Reporting information on the environmental axis

The sides	2014 (%)	2015 (%)
1. Materials	0	0
2. Energy	6.944	6.944
3. Water	0	0
4. Biodiversity	0	0
5. Emission	0	0
6. Effluents and waste	0	0
7. Environmental compliance	1.389	1.389
8. Supplier environmental assessment	0	0
Total	8.333	8.333

Table 11 Reporting information on the environmental axis

The sides	2014 (%)	2015 (%)
1. Materials	0	0
2. Energy	0	0
3. Water	0	0
4. Biodiversity	0	0
5. Emission	0	0
6. Effluents and waste	0	0
7. Environmental compliance	0	0
8. Supplier environmental assessment	0	0
Total	0	0

Table 12 Reporting information on the environmental axis of the research sample

Company	2014 (%)	2015 (%)	Medium (%)
Al-Mansour Company for Pharmaceutical Industries and Medical Supplies	9.72	15.28	12.50
Al-Khatem Communications Company	8.33	9.72	9.03
The Modern Company for Animal and Agricultural Production	11.11	13.89	12.50
Gulf Commercial Bank	8.33	8.33	8.33
Al Mamoura Real Estate Investment Company	0.00	0.00	0.00
Medium	7.638	7.50	9.44

3.3 *Third: Social Focus*

1. Al-Mansour Company for Pharmaceutical Industries and Medical Supplies (Table 13):
2. Al-Khatem Communications Company (Table 14):
3. The Modern Company for Animal and Agricultural Production (Table 15):
4. Gulf Commercial Bank (Table 16):
5. Al Mamoura Real Estate Investment Company (Tables 17 and 18):

3.4 *Fourth: General Disclosures*

1. Governance (Table 19)
2. Other general disclosures:
3. Al-Mansour Company for Pharmaceutical Industries and Medical Supplies (Table 20):

Table 13 Reporting information on the social axis

The sides	2014 (%)	2015 (%)
1. Employment	4.348	4.348
2. Labor/Management relations	0	0
3. Occupational health and safety	0	0
4. Training and education	4.348	4.348
5. Diversity and equal opportunity	2.898	2.898
6. Non-discrimination	0	0
7. Freedom of association and collective bargaining	0	0
8. Child labor	0	0
9. Forced or compulsory	0	0
10. Security practices	0	0
11. Rights of indigenous peoples	0	0
12. Human rights assessment	0	0
13. Local communities	0	0
14. Supplier social assessment	0	0
15. Public policy	1.449	1.449
16. Customer health safety	0	0
17. Marketing and labeling	2.898	2.898
18. Customer health safety	0	0
19. Socioeconomic compliance	0	1.449
Total	15.941	17.390

4. Al-Khatem Telecom Company (Table 21):
5. The Modern Company for Animal and Agricultural Production (Table 22):
6. Commercial Bank of Qatar (Table 23):
7. Al Mamoura Real Estate Investment Company (Table 24):

From the above table, it is possible to extract the medium to the degree of reporting information on other public disclosures of the companies in the research sample according to schedule Table 25.

Table 26 summarize the level of reporting of ESG indicators for each of the research sample companies according to GRI standards:

4 Conclusions

1. The level of sustainability reporting for the listed Iraqi companies was 19.76%. Although Al-Mansour Pharmaceutical Industries recorded the highest percentage of sustainability reporting at 24.11, it is still weak. This is a sign of poor reporting of sustainability by Iraqi companies listed on the Iraq Stock Exchange.

Table 14 Reporting information on the social axis

The Sides	2014 (%)	2015 (%)
1. Employment	4.34	2.89
2. Labor/Management relations	0	0
3. Occupational health and safety	0	0
4. Training and education	0	2.89
5. Diversity and equal opportunity	2.89	2.89
6. Non-discrimination	0	0
7. Freedom of association and collective bargaining	0	0
8. Child labor	0	0
9. Forced or compulsory	0	0
10. Security practices	0	0
11. Rights of indigenous peoples	0	0
12. Human rights assessment	0	0
13. Local communities	0	0
14. Supplier social assessment	0	0
15. Public policy	0	0
16. Customer health safety	0	0
17. Marketing and labeling	1.44	1.44
18. Customer health safety	0	0
19. Socioeconomic compliance	0	1.44
Total	8.69	11.59

2. Sustainable development requires companies to adopt the concept of social and environmental performance as well as financial performance for the social and environmental efficiency of the company. These companies need to carefully carry out social and environmental activities and the efficient assessment of compliance. Based on this, the company's progress towards sustainability can be determined.
3. The issue of sustainability accounting and reporting in developing countries, including Iraq is still in its early stages, thus, it needs further research.
4. Sustainability reporting is one of the most important stages of accounting development due to an increase in size and capacity of companies in different sectors.
5. The possibility of adopting GRI standards by companies listed in the Iraq Stock Exchange as objective indicators of sustainability reporting.

Table 15 Reporting information on the social axis

The sides	2014 (%)	2015 (%)
1. Employment	2.89	2.89
2. Labor/Management relations	0	0
3. Occupational health and safety	0	0
4. Training and education	0	0
5. Diversity and equal opportunity	0	0
6. Non-discrimination	0	0
7. Freedom of association and collective bargaining	0	0
8. Child labor	0	0
9. Forced or compulsory	0	0
10. Security practices	0	0
11. Rights of indigenous peoples	0	0
12. Human rights assessment	0	0
13. Local communities	0	0
14. Supplier social assessment	0	0
15. Public policy	0	0
16. Customer health safety	0	0
17. Marketing and labeling	1.44	1.44
18. Customer health safety	0	0
19. Socioeconomic compliance	1.44	1.44
Total	5.79	5.79

Table 16 Reporting information on the social axis

The sides	2014 (%)	2015 (%)
1. Employment	4.348	4.348
2. Labor/Management relations	0	0
3. Occupational health and safety	0	0
4. Training and education	2.899	2.899
5. Diversity and equal opportunity	1.449	1.449
6. Non-discrimination	0	0
7. Freedom of association and collective bargaining	0	0
8. Child labor	0	0
9. Forced or compulsory	0	0
10. Security practices	0	0
11. Rights of indigenous peoples	0	0
12. Human rights assessment	0	0
13. Local communities	0	0

(continued)

Table 16 (continued)

The sides	2014 (%)	2015 (%)
14. Supplier social assessment	0	0
15. Public policy	1.449	1.449
16. Customer health safety	0	0
17. Marketing and labeling	1.449	1.449
18. Customer health safety	0	0
19. Socioeconomic compliance	1.449	0
Total of Percentage	13.043	11.594

Table 17 Reporting information on the social axis

The sides	2014 (%)	2015 (%)
1. Employment	4.348	4.348
2. Labor/Management relations	0	0
3. Occupational health and safety	0	0
4. Training and education	0	0
5. Diversity and equal opportunity	1.449	1.449
6. Non-discrimination	0	0
7. Freedom of association and collective bargaining	0	0
8. Child labor	0	0
9. Forced or compulsory	0	0
10. Security practices	0	0
11. Rights of indigenous peoples	0	0
12. Human rights assessment	0	0
13. Local communities	0	0
14. Supplier social assessment	0	0
15. Public policy	0	0
16. Customer health safety	0	0
17. Marketing and Labeling	0	0
18. Customer health safety	0	0
19. Socioeconomic compliance	1.449	1.449
Total	7.246	7.246

Table 18 Reporting information on the social axis

Company	2014 (%)	2015 (%)	Medium (%)
Al-Mansour Company for Pharmaceutical Industries and Medical Supplies	15.94	17.39	16.67
Al-Khatem Communications Company	8.70	11.59	10.14
The Modern Company for Animal and Agricultural Production	5.80	5.80	5.80
Gulf Commercial Bank	13.04	11.59	12.32
Al Mamoura Real Estate Investment Company	7.25	7.25	7.25
Medium	10.14	10.72	10.43

Table 19 Average reporting information on corporate governance performance

Company	2014 (%)	2015 (%)	Medium (%)
Al-Mansour Company for Pharmaceutical Industries and Medical Supplies	0	0	0
Al-Khatem Communications Company	0	0	0
The Modern Company for Animal and Agricultural Production	0	0	0
Gulf Commercial Bank	0	0	0
Al Mamoura Real Estate Investment Company	0	0	0
Medium	0	0	0

Table 20 For reporting on private information the other general disclosures

Data	2014 (%)	2015 (%)
1. Organization profile	37.209	37.209
2. Strategy	0	0
3. Ethics and integrity	2.325	2.325
4. Stakeholder engagement	0	0
5. Reporting practice	0	0
Total	39.534	39.534

Table 21 Reporting on private information and other general disclosures

Data	2014 (%)	2015 (%)
1. Organization profile	37.209	37.209
2. Strategy	0	0
3. Ethics and integrity	2.325	2.325
4. Stakeholder engagement	0	0
5. Reporting practice	0	0
Total	39.534	39.534

Table 22 Reporting on private information and other general disclosures

Data	2014 (%)	2015 (%)
1. Organization profile	30.233	30.233
2. Strategy	0	0
3. Ethics and integrity	2.325	2.325
4. Stakeholder engagement	0	0
5. Reporting practice	0	0
Total	32.558	32.558

Table 23 Private information reporting rates the other general disclosures

Data	2014 (%)	2015 (%)
1. Organization profile	37.209	37.209
2. Strategy	0	0
3. Ethics and integrity	2.325	2.325
4. Stakeholder engagement	0	0
5. Reporting practice	0	0
Total	39.534	39.534

Table 24 Private information reporting rates the other general disclosures

Data	2014 (%)	2015 (%)
1. Organization profile	30.233	30.233
2. Strategy	0	0
3. Ethics and integrity	2.325	2.325
4. Stakeholder engagement	0	0
5. Reporting practice	0	0
Total	32.558	32.558

Table 25 Reporting of information on other public disclosures

Company	2014 (%)	2015 (%)	Medium (%)
Al-Mansour Company for Pharmaceutical Industries and Medical Supplies	39.53	39.53	39.53
Al-Khatem Communications Company	39.53	39.53	39.53
The Modern Company for Animal and Agricultural Production	32.56	32.56	32.56
Gulf Commercial Bank	39.53	39.53	39.53
Al Mamoura Real Estate Investment Company	32.56	32.56	32.56
Medium	36.74	36.74	36.74

Table 26 Sustainability reporting level (ESG indicators) for each of the companies

Company	Economic axis (%)	Environmental axis (%)	Social axis (%)	Corporate governance performance (%)	Other general disclosures (%)	Sustainability reporting level (ESG indicators) (%)
Al-Mansour Company for Pharmaceutical Industries and Medical Supplies	51.85	12.50	16.67	0	39.53	24.11
Al-Khatem Communications Company	37.04	9.03	10.14	0	39.53	19.15
The Modern Company for Animal and Agricultural Production	40.74	12.50	5.80	0	32.56	18.32
Gulf Commercial Bank	48.15	8.33	13.32	0	39.53	21.87
Al Mamoura Real Estate Investment Company	37.04	0.00	7.25	0	32.56	15.37
Medium	42.96	8.47	10.63	0	36.74	19.76

References

First: Reports and Official Documents

1. Reports of the board of directors of companies for the period (2015–2014)
2. Reports of the auditors of companies for the period (2015–2014)
3. The annual report of the Iraqi market for securities for the years (2015–2014)
4. Guideline for notifying the restricted companies about the performance of sustainability, the draft of the preliminary version issued in (2016)
5. Corporate financial statements sample search for the period (2015–2014)
6. Sustainability reporting guidelines G4—implementation guide (2013)
7. Guidelines for the preparation of sustainability reports G4—principles for the preparation of the report and the normative clarifications issued in (2013)
8. United Nations Conference on Trade and Development (2013)

Second: Books

9. Omar bin Lakhdar, K.: Sustainable Development of Organizations Quality, Environment, Health and Safety, 1st edn. Dar Al-Ayyam Publishing and Distribution, Amman, Jordan (2016)

Third: Research and Periodicals

10. Abu-Zer, A.I., Al-Atoum, R.A.: An introduction to the review of accounting sustainability using analytical procedures—an empirical study on industrial companies in the Arab Gulf states. Jordanian J. Bus. Adm. University of Jordan **12**(1), 2016 (2016)
11. Al-Rekaby, A., Khalaf, S.: Accountability response to sustainability and support for sustainable development. J. Baghdad Coll. Econ. Sci. University of Science 431–454 (2014)
12. Onaiza, H.H., Ali, M.N.: The effect of disclosing social responsibility in financial statements in the resolutions of the users of these lists, an applied and exploratory study. Al-Ghary J. Econ. Adm. Sci. **9**(26) (2011)
13. Al-Mashhadany, Tawfiq, O.I., Khalaf, A., Al-Rikaby.: The role of accounting in environmental conservation and sustainable development support. In: Proceedings of the Second International Forum on the Outstanding Performance of Organizations and Governments, Second Edition: The Growth of Institutions and Economies between Achieving Financial Performance and Performance Challenges Held at the University of Ouargla on 22 and 23 Nov (2011)
14. Al-Mashharawi, Zaher Husni.: The role of accounting for rationalization in support of the sustainability strategy from an accounting perspective. Al-Azhar Univ. J. Gaza Hum. Sci. Ser. **17**(1) 145–176 (2015)

Fourth: Thesis's and Dissertations

15. Rahim, A., Mohamed, G.K.: Measuring the Effect of Implementing the Egyptian Index of Corporate Responsibility for Sustainable Development in Controlling Financial Performance with a Field Study on Egyptian Companies Listed in the Stock Exchange. Suez Canal University, Egypt, Ph.D (2015)
16. Al-Armoty, Ahmed, A.: The effect of sustainability accounting on the financial reporting reports of the Jordanian joint stock companies listed on the Amman stock exchange. Master Thesis in Unpublished Accounting, presented to the Board of the Business School, Middle East University (2013)
17. Aweida, Najwa.: The role of governance in enhancing the performance of a sustainable institution: a case study at the Sidal complex—Constantine factory. Master of Economic Sciences, Introduction to the Council of the Faculty of Economic, Commercial and Management Sciences, Republic of Algeria (2013)
18. Ghalab, F.: The role of the audit function in the field of corporate governance to achieve the principles and criteria of sustainable development, a study of some industrial institutions. Master Thesis Presented to the Council of the Faculty of Economic Sciences, Commercial Sciences and Management Sciences, University of Farhat Abbas Setif University, Algeria (2011)

Foreign References—First: Books

19. Brockett, A., Rezaee, Z.: Corporate Sustainability: Integrating Performance and Reporting. Wiley, Hoboken, New Jersey (2012)
20. Schaltegger, S., Martin, B., Roger, B.: Sustainability Accounting and Reporting. Springer, the Netherlands (2006)

Second: Periodicals

21. Alazzani, A., Wan Hussin, N.: Global reporting environmental reporting: a study of oil and gas. Ecol. Ind. **32**, 19–24 (2013)
22. Kuzey, C., Uyar, A.: Determinants of sustainability reporting and its impact on firm value: evidence from the emerging market of Turkey. J. Cleaner Prod. (2016)
23. Christofi, A., Christofi, P., Sisaya, S.: Corporate sustainability: historical development and reporting practices. Manage. Res. Rev. **35**(2), 157–172 (2012)
24. Hyršlová, J., Becková, H., Kubáňková, M.: Sustainability accounting: brief history and perspective. In: The Ninth International Days of Statistics and Economics, Prague, Sep 10–12 (2015)
25. Utile, B.J.: Effect of sustainability reporting on firm's performance: a review of literature. Int. J. Bus. Manage. **4**(7), 203–208 (2016)

Third: Thesis's and Dissertations

26. Bhat, S.H.: Disclosures and performance in the semiconductor manufacturing industry (2010–2014). Master Thesis, the Faculty of the Department of Environmental Studies, San José State University (2016)
27. Bidari, G.: Factors affecting CSR disclosure in Nepalese banks: a global reporting initiative perspective. Doctor Thesis, Edith Cowan University (2016)
28. Ntourou, C., Nayak, S.: G4 Guidelines regarding social aspects—a study of listed companies at Nasdaq OMX Stock Holm On social aspects. Master Thesis, School Of Business, Economics and Law, University Of Gothenburg (2015)
29. Fagerström, P.: Analyzing sustainability reporting by best performing companies in global sustainability indices—describing the contents and appearance of the reports. Master Thesis, Department Of Geosciences and Geography, University Of Helsinki (2016)
30. Johansson, D., Lundberg, T.: The assurance process of GRI sustainability reports—influence on accountability and transparency. Master Thesis, Jönköping International Business School, Jönköping University (2012)
31. Lisene, M.: Sustainability practices and reporting by the South African banking sector. Master Thesis, School Of Accounting Sciences, North West University (2015)

An Empirical Study of Governance Challenges: Lesson Learnt from Australian Islamic Charitable Institution

Nur Farahiah Azmi, Hazriah Hasan, and Mohd Nor Hakimin Yusoff

Abstract This study aims to examine the governance challenges faced by Islamic charitable institutions. A case study approach has been adopted in this study where the data were collected via semi-structured interviews and documents. The interviews were conducted among board members and executive management of the selected Islamic charity. The data were then analysed using ATLAS.TI program. The findings of the study show that there are three main themes developed such as unclear job description, absence of member requirement and scarcity of human resources. Finally, the results of this study can be a guideline to the Islamic charitable institutions to improve in terms of governance practices in the future.

Keywords Challenges · Contingency theory · Governance · Islamic charity

1 Introduction

The third sector, often known as the nonprofit sector, has gotten more attention throughout the world since the preceding decades. This heightened worry stems from several issues, including the third or nonprofit sector's major contribution to the country's economic and social well-being [11, 12, 14]. As providers of many services, such as health, training, culture and recreation, safety and civic participation, the third sector has played an essential role. Furthermore, religious charities are one of the organisations that fall under the third or nonprofit sectors, which have primarily become a centre point in various nations.

Islamic charitable institution has been referred to as Islamic endowment, Islamic non-governmental organisation (NGOs), and Islamic non-profit organisations (NPOs) in various studies. Most Islamic philanthropic organisations around the world arose as a result of the founding of Islam. In particular, the Islamic charity organisations' major contribution is poverty reduction, which contributes to the development of education and the improvement of economic conditions for

N. F. Azmi (✉) · H. Hasan · M. N. H. Yusoff
Faculty of Entrepreneurship and Business, Universiti Malaysia Kelantan, Kota Bharu, Malaysia
e-mail: nurfarahiah@umk.edu.my

A. Hamdan et al. (eds.), *The Implementation of Smart Technologies for Business Success and Sustainability*, Studies in Systems, Decision and Control 216,
https://doi.org/10.1007/978-3-031-10212-7_25

the less fortunate [6, 25]. Sulaiman and Zakari [27], also emphasised the societal contributions of these organisations, such as the advantages received by less fortunate persons from donations made by Islamic charities, which can help to close social gaps and inequities.

Despite the aforementioned contributions of Islamic charitable institutions, governance issues remain to be present. Islamic charities in Australia, for example, have been linked to money laundering and terrorism financing. According to the Australian Institute of Criminology, Islamic organisations have been scrutinised owing to uncontrolled contribution procedures and low transparency and accountability standards [7]. Several cases related to Islamic charity in Australia were reported, including misconduct and mismanagement of the organisation, lack of accountability and transparency, money laundering, misappropriation of funds, and inefficient and ineffective management of Islamic charities [4]. Therefore, the study attempts to address the governance challenges faced by the Islamic charitable institutions through examining the lens of the contingency approach.

The remainder of this paper is organised as follows. The second section provides a brief review of past studies on governance issues and contingency theory. The third section presents the research methodology. The next section offers a discussion of the findings from the collected data. Last but not least, Sect 5 summarises the study and presents some recommendations for future research.

2 Literature Review

2.1 Governance Challenges in Australian Islamic Charitable Institutions

Islamic charities have a long history of issues and challenges in the effective management of the charity. Several cases related to Islamic charity in Australia were reported, including misconduct and mismanagement of the organisation, lack of accountability and transparency, money laundering, misappropriation of funds, and inefficient and ineffective management of Islamic charities [4]. These issues linger AFIC for a long time and there is a need for close monitoring of its top management which perhaps lessens the governance issues that occurred. Furthermore, it also reported that the Islamic Development Organisation has been involved in money laundering and terror financing [26]. Thus, the ACNC has revoked the Islamic charity registration status.

2.2 Contingency Theory in Corporate Governance

Contingency is defined as the alignment or fitting of a firm's features to circumstances that represent the firm's condition [13]. Any variable that modifies the impact of organisational features on performance is referred to as a contingency. Environmental [11–17] and strategic contingencies are examples of external and internal aspects of a company [17]. The determinants of a firm's effectiveness are explained using contingency theory. However, a firm's efficacy can be defined in a variety of ways, including profitability [30], customer satisfaction [18], or a combination of non-financial and financial indicators [18, 19]. Effectiveness is a general term that refers to how well something works [13]. In management control system (MCS) research, the contingency method is commonly employed [21]. Contingency theory in MCS is concerned with observing the consequences of interdependence between organisational structure and contingency factors (environment and strategy) in the creation of perverse outcomes.

The contingency theory argues that corporate governance success would depend on organisational variables, such as strategy, size, environment and other organisational resources and capabilities [2]. Therefore, the form and efficiency of company governance would depend on elements of urgency in response to internal and external variables. Companies can organise their corporate governance in a way that takes into account environmental and organisational variables such as competition and strategy to guarantee effective control via corporate governance and implementation of the plan. They also stated that corporate governance's function might be varied in different settings, which implies that corporate governance's efficacy would rely on the size, age, stage of operations and nature of industries and innovation.

As for the contingency approach of corporate governance in NPO, previous study Chelliah et al. [9] explored what some NPO leaders believe are crucial governance challenges and their effects on the theory and practice of NPO governance. The authors surveyed leaders of NPOs involved in governance to demonstrate the distinct internal and external contingencies faced in determining the most effective governance systems. Findings of the study highlighted that there is a need for an improvement in funding of the NPOs which may greatly affect the governance model of a NPO since the NPOs are greatly dependant on the government grant on its funding [9]. Therefore, this paper answers a call to conduct a thorough examination of the Australian Islamic charitable institution. This study aimed to define some of the main issues and challenges that influenced the performance of the Islamic charitable institutions by utilising the contingency approach. In this respect, this paper focused on the internal factor that influences organisational performance in terms of board roles and board composition in the Islamic charitable context.

3 Methodology and Research Design

The study was qualitative and interpretive on a case study [28]. In addition, the qualitative method has shown an increasing trend for this area of study [5], thus this approach is the most appropriate. The Islamic charity ABC was chosen as a case study is because of the size of the charity which is denoted as a large charity. The data were collected such as documentation and interview. According to Yin [31], the use of multiple data sources was identified as triangulation, which improved the case study's construct validity and relevance. It was also elaborated by Merriam [23] that the use of triangulation is a method to "compare and cross-check the data collected through observations at different times or in different places or interview data collected from people with different perspectives or follow-up interviews with the same people" (2009, p. 216). In this study, the use of proof from diverse sources was important to enhance the rationality of the data.

Interviews were conducted in one of the Australian Islamic councils, which is denoted as Islamic charity ABC, to gain a better understanding of the response to the governance issues experienced by this Islamic charity ABC. A semi-structured interview was used in this study because it allowed the researcher to be more flexible in asking questions while still performing the interview efficiently. The interview was also the primary method used by earlier researchers on governance studies to acquire data. [20, 22]. The total number of research participants in the study was six, which included members of the board of trustees and management. Table 1 presents detailed information on the research participants' profiles.

The next data collection is through the document. The documents used in this study were governance policy, annual reports, minutes of board meetings, articles, and other related documents. These documents were then reviewed to collect information regarding the background of the organisation, policies, and regulations of the organisations.

As for data analysis, this study conducted a thematic analysis based on six phases of thematic analysis recommended by Clarke et al. [10] such as (1) data familiarisation; (2) initial codes generations; (3) themes searching (4) themes reviewing (5) defining and naming themes (6) producing the report. Furthermore, the study also utilised ATLAS.TI software for data management. This software is used to analyse text data from interviews, focus groups, documents, notes and open-ended surveys

Table 1 The research participants' profiles

Pseudonym	Gender	Venue of interview session	Designation (during the interview conducted)
Research participant A	Male	Office	Treasurer/Portfolio leader
Research participant B	Male	Office	Service manager
Research participant C	Male	Meeting room	Executive member/Portfolio leader
Research participant D	Female	Meeting room	Secretary/Portfolio leader
Research participant E	Female	Coffee shop	Executive member/Portfolio leader
Research participant F	Female	Meeting room	Executive member/Portfolio leader

[29], assisted researchers to better understand the findings of the study [24] and able to improve data quality and reliability [3]. Transcribed data will then be coded and used to understand the governance challenges faced by the Islamic charity.

4 Findings and Discussion

4.1 Theme 1: Unclear Job Description

To guarantee that members of the Islamic charity are capable of completing their obligations, the roles and obligations of the members must be simple and unambiguous. As a result of the interview with members of Islamic charity ABC about the difficulties and problems of governance procedures, noteworthy conclusions were obtained. One of the research participants specifically mentioned "blurred responsibilities," which happened when the board of trustees and management members were busy with operational tasks.

The following sentence exemplifies the situation:

> So I guess I've learnt a lot since then. I would say that during that time, I didn't know the separation difference between a good governance strategy and operations. I think I took too many operational roles…Sometimes I get confused between the roles of active operations or the driver of the organisation… I guess understanding my role was one of the challenges in the past. (Research Participant E)

Furthermore, research participant A mentioned the dangers of being unduly preoccupied with portfolio tasks, citing the impact on the board of trustees and management members' functions. The following is the participant's statement:

> For example, we have Sister A conducting a leadership program. Although we certainly have to perform some tasks and programmes, we need to be careful so that it would not overwhelm us and we would not be too involved in the portfolios. We also play the role of the decision-maker, which is another job to be fulfilled… So I think that could be the issue. (Research Participant A)

The research participant B also commented that the board trustees' work ranges were too large, affecting their performance. The following statement illustrates this situation:

> The major challenge of the organisation ABC is the significantly large scope. When you state your wish to represent the interest of the Muslim community, the staff should be present. Therefore, I think the challenge is the ability to reduce the scope so that our staff executive can work effectively. We are not able to solve all problems and challenges. Sometimes, it is difficult to say no to something, I say no, but we often help you around. The challenges are the reduction of the scope, what should be focused on, and what we should say no to. (Research Participant B)

In addition, the study also conducted a cross-check with Islamic charity ABC document. An examination of the number of duties allotted to Islamic charity ABC members was undertaken using the document titled "Standing Committee and Operational Guidelines" of Islamic charity ABC. As a consequence, the highest number of portfolios controlled by board trustees was three, while the lowest number of portfolios was one.

4.2 *Theme 2: The Absence of Member Requirement*

The next interview question presented by the researcher was "is there any specific skill, qualification or expertise required?" Most participants said that the members of Islamic charity had no special criteria, as evidenced in the following statement:

> There is no set of specific skills, qualifications, and expertise. The main checklist is that you are a part of the society.... (Research Participant F)

Another research participant stated that the requirements for a person to be selected as a board trustee were established by the organization's constitution, as demonstrated in the following statement:

> I think the constitution might outline some of the general aspects of it, about what's required for such board.... (Research Participant E)

Following that, the organization's constitution was evaluated in this study to cross-check the interview results. The following are the specifics of the organization's constitution:

i. Be a financial society member of the Islamic charity ABC for a minimum of six months.
ii. Not included in the receipt of any direct emolument from the Islamic charity ABC or a society member
iii. Has a record of active service to the Muslim community and possesses good character
iv. Be a citizen or permanent resident of Australia instead of an employee of a foreign government, including diplomatic mission
v. Be a Muslim
vi. Has the capacity and willingness to make a significant commitment of time and energy
vii. Does not hold office in any other organisation, in which the objectives conflict with the objectives of Islamic charity ABC

(Source: Islamic charity ABC Constitution)

The interview results and document analysis showed that there was no clear guideline for the precise abilities and knowledge necessary for the nomination of a board trustee. The study also revealed that there was no specific regulation specified in the organization's constitution addressing the competence and competence necessary for a board trustee. Furthermore, as demonstrated in the following remark, one of the participants suggested an attempt to assemble the skilled and competent board and management members:

> We need to restructure the organisation constitution to ensure that the services originate from individuals who are specialised in that area. This element encompasses the board or the way to the board. The officers, staff members, and advocacy should be again reflected by the individuals who have the skills and expertise in those particular areas as well.
>
> (Research Participant E)

4.3 *Theme 3: Insufficient Human Resources*

Insufficient human resources were highlighted by the research participants during the interview, as indicated in the following statements:

> I think that might be some of the staff members. (Research Participant B)

> We also need to compensate for the lack of members to perform the operation… compared to another office, they have more staff. (Research Participant F)

> The core issue is that the staff… The project needs someone to refer to… (Research Participant E)

5 Conclusion and Recommendation

Figure 1 presents the analysis of the study regards on governance challenges faced by Islamic charity ABC. The study's findings highlighted the issue of blurred roles among Islamic charity members. Islamic charity ABC may consult AICD [1] and Brown et al. [8] to encounter this issue. This study then recommends Islamic charity to provide comprehensive board training to board trustees and management members. As a result, the board will be better able to understand its roles and responsibilities. Managing human resources is also important for Islamic charity ABC to ensure that the charity's mission and vision are met. As previously discussed, there was a lack of human resources in Islamic charity ABC because volunteers are one of the driving forces behind the programme and activities. Therefore, the study suggests that the Islamic charity develop a volunteer policy, as suggested by AICD [1], in order to encourage and increase volunteer interest in their programme and activities. In sum, the study also recommends comparative case studies to be investigated in the future.

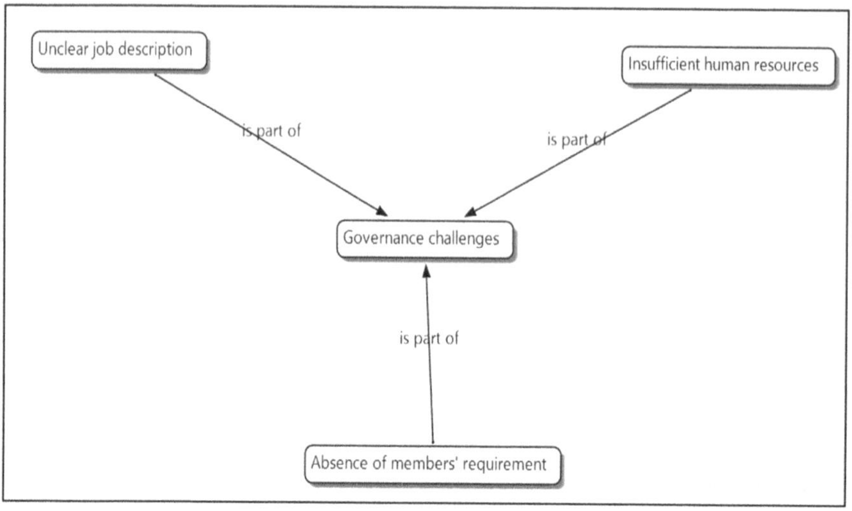

Fig. 1 Network view governance challenges

References

1. AICD: Good governance for not-for-profit organisations. New South Wales (2013)
2. Aguilera, R.V., Filatotchev, I., Jackson, G.: An organizational approach comparative governance: costs, contingencies, and complementaries. Organ. Sci. **19**(3), 475–492 (2008)
3. Ang, C.K., Embi, M.A., Md Yunus, M.: Enhancing the quality of the findings of a longitudinal case study: reviewing trustworthiness via ATLAS.ti. Qual. Rep. **21**(10), 1855–1867 (2016)
4. Azmi, N.F., Hasan, H., Yusoff, M.N.H.: Islamic charities in Australia and the governance predicaments. Int. J. Law Gov. Commun. **5**(18), 19–25 (2020)
5. Azmi, N.F., Hasan, H., Yusoff, M.N.H., Mohamad, S.R.: A review on literature of Islamic charities between 2008–2018. In: International Seminar on Entrepreneurship and Business (ISEB 2018) (2018)
6. Bremer, J.: Islamic Philanthropy: reviving traditional forms for building social justice: a tool for sustainable community development. In: Centre for the Study of Islam and Democracy 5th Annual Conference, pp. 1–26 (2004)
7. Bricknell, S., Mccusker, R., Chadwick, H., Rees, D.: Money Laundering and Terrorism Financing Risks to Australian Non-Profit Organisations. Research and Public Policy Series, xvii, AIC Reports (2011)
8. Brown, W.A., Hillman, A.J., Okun, M.A.: Factors that influence monitoring and resource provision among nonprofit board members. Nonprofit Voluntary Sect. Q. **41**(1), 145–156 (2012)
9. Chelliah, J., Boersma, M., Klettner, A.: The role of a national regulator in developing solutions to governance challenges in the not-for-profit sector. J. Developing Areas **49**(6), 449–458 (2015)
10. Clarke, V., Braun, V., Hayfield, N.: Thematic analysis. In: Qualitative psychology: a practical guide to research methods, pp. 222–248 (2015)
11. Cornforth, C.: Nonprofit governance research: limitations of the focus on boards and suggestions for new directions. Nonprofit Volunt. Sect. Q. **41**(6), 1116–1135 (2012)
12. Dalton, B., Butcher, J.: The rise of big charity in Australia. In: Association for Research on Nonprofit Organizations and Voluntary Action. Denver (2015)
13. Donaldson, L.: The Contingency Theory of Organizations. Sage, Thousand Oaks, CA (2001)

14. Elliott, S., Haigh, D.: Advocacy in the New Zealand Not-for-profit Sector: "Nothing stands by itself". Third Sector Review, 157–178 (2013). Retrieved from http://search.ebscohost.com/login.aspx?direct=true&profile=ehost&scope=site&authtype=crawler&jrnl=13239163&AN=94342523&h=vCqgtClyzBYNWVuoTpDPLGTrD/AE83Gaiit46ifbPwinx3KV1gLtfyOHgNmHrw/0DORHJG6lT9fHtzhFM
15. Geiger, S.W., Ritchie, W.J., Marlin, D.: Strategy/structure fit and firm performance. Organ. Dev. J. **24**(2), 10–22 (2006)
16. Hambrick, D.C.: Some tests of the effectiveness and functional attributes of Miles and Snow's strategic types. Acad. Manag. J. **26**(1), 5–26 (1983)
17. Hoque, Z.: A contingency model of the association between strategy, environmental uncertainty and performance measurement: impact on organizational performance. Int. Bus. Rev. **13**, 485–502 (2004)
18. Ittner, D.C., Larcker, D.F.: Are nonfinancial measures leading indicators of financial performance? An analysis of customer satisfaction. J. Acc. Res. **36**, 1–35 (1998)
19. Kaplan, R.S., Norton, D.P.: The balanced scorecard-measures that drive performance. Harvard Bus. Rev. **70**, 71–79 (1992)
20. Kumar, P., Zattoni, A.: Farewell editorial: exiting editors' perspective on current and future challenges in corporate governance research. Corp. Gov.: Int. Rev. **27**(1), 2–11
21. Langfield-Smith, K.: A review of quantitative research in management control systems and strategy. In: Chapman, C., Hopwood, A., Shields, M. (eds.) Handbook In Management Accounting, vol. 2, pp. 753–783. Oxford, Elsevier (2007)
22. McNulty, T., Zattoni, A., Douglas, T.: Developing corporate governance research through qualitative methods: a review of previous studies. Corp. Gov. Int. Rev. **21**(2), 183–198
23. Merriam, S.B.: Qualitative research a guide to design and implementation revised. Jossey-Bass, San Francisco (2009)
24. Paulus, T., Woods, M., Atkins, D.P., Macklin., R.: The discourse of QDAS: reporting practices of ATLAS.ti and NVivo users with implementations for best practices. Int. J. Soc. Res. Methodol. 1–13 (2015)
25. Razak, S.H.A.: Zakat and waqf as instrument of Islamic wealth in poverty alleviation and redistribution: case of Malaysia. Int. J. Sociol. Soc. Policy **40**(3–4), 249–266 (2020)
26. Shanahan, R.: Typology of terror: the backgrounds of Australian jihadis. Australas. Policing **12**(1), 32 (2018)
27. Sulaiman, M., Zakari, M.A.: Efficiency and effectiveness of waqf institutions in Malaysia: towards financial sustainability. In: Access to Finance and Human Development—Essays on Zakah, Awqaf and Microfinance (2015)
28. Walsham, G.: Doing interpretive research. Eur. J. Inf. Syst. **15**(3), 320–330 (2006)
29. Woods, M., Paulus, T., Atkins, D.P., Macklin, R.: Advancing qualitative research using Qualitative Data Analysis Software (QDAS)? Reviewing potential versus practice in published studies using ATLAS.ti and NVivo, 1994–2013. Soc. Sci. Comput. Rev. 1–21 (2015)
30. Yeung, M.C.H., Ennew, C.T.: From customer satisfaction to profitability. J. Strateg. Mark. **8**, 313–326 (2000)
31. Yin, R.K.: Case Study Research and Applications: Design and Method, 5th edn. Sage Publications, Thousand Oaks (2018)

Islamic Sukuk and Its Role in Economic Recovery and Growth Post-Covid

Iyad A. A. Khanfar and Nehad A. A. Khanfar

Abstract Islamic Sukuk witnessed consistent growth over the past years and is expected to follow the same trend. Sukuk issuance continues by states and corporates alike. Sukuk serves two purposes. That is to say, the issuers of Sukuk aim at diversifying the sources of funds, especially under difficult economic situations. Meanwhile, investor demand remains intact due to the shortage of new Sukuk supply and the global quest for profit. While Sukuk usually enjoys healthy annual yields, also it is unlikely to default. Compared to overall issuances by the end of 2021, the defaulted Sukuk stands at a meagre rate of 0.27%, making it an attractive and secure financial instrument. Therefore, Sukuk issuance becomes part of the value proposition of world financial centres such as the U.K., Luxembourg, Hong Kong, and others. This paper analysis Sukuk's impact on the country's economic development by focusing on specific countries with different economic-financial experiences. Furthermore, it will investigate the feasibility of Sukuk in supporting diverse sources of funds and liquidity. In addition, this paper examines the role of Sukuk in achieving economic growth and development post-Covid pandemic. It also attempts to ascertain the critical phases of economic growth and links them to Sukuk in order to assess the effectiveness and significance of Sukuk in being a potential financial instrument that could support the economies in the post-Covid period.

Keywords Covid 19 · Economic development · Islamic finance · Islamic bonds · Sukuk · Securitization

I. A. A. Khanfar (✉)
Zarqa University, Zarqa, Jordan
e-mail: i.khanfar@zu.edu.jo

N. A. A. Khanfar
Islamic College, London, UK
e-mail: n.khanfar@islamic-college.ac.uk

A. Hamdan et al. (eds.), *The Implementation of Smart Technologies for Business Success and Sustainability*, Studies in Systems, Decision and Control 216,
https://doi.org/10.1007/978-3-031-10212-7_26

1 Introduction

Islamic bonds, known as Sukuk, are rapidly becoming one of the main attractive instruments of Islamic finance in Muslim and non-Muslim countries. Sukuk witnessed continuous innovations that brought the attention of the capital markets by corporates and sovereign financial players globally. Despite a humble start of Sukuk issuance in 2000 with a total of USD336 million, according to Al Madani et al. [1], the Sukuk issuance increased massively to a total of USD252.3 billion in 2021. "Central banks, governments, and multilateral institutions" dominated the issuances (Fitch Ratings, ibid). There is a growing interest in issuing Sukuk; the issuers vary between sovereign bodies, multinational corporations, and financial institutions. They use Sukuk as an alternative to conventional debt-based bonds. Therefore, Sukuk is becoming an essential vehicle of the international financial system. In addition, many policymakers treat Sukuk as one of the long-term financing sources that are the only Sharia-compliant instrument for financing. Thus, Sukuk would be attractive for specific investors who are not interested, for religious reasons, in investing in any conventional types of bonds. Therefore, the practical experience shows that Sukuk would be vital to recruit financial resources to fund particular projects aiming to aid society in general and poor people in particular, contributing to the economic cycle. The idea of Sukuk is built on investing the excess of institutional and personal funds to benefit society as a whole. It is crucial to understand that Sukuk's objective is to attain Maqasid Al-Shari'ah (the objectives of Islamic law), such as enhancing people's life quality. This includes investing and improving the infrastructure of all kinds and levels like energy, transportation, agricultural projects, which encourage economic development that generates job opportunities and creates broader economic activities [2]. This paper outlines specific countries' experiences on the relation between Sukuk and economic growth. It also analytically reviews the elements of success stories that can be benefited from by other countries, especially in times of economic crises like the one of the Covid pandemic. Furthermore, through practical experiences, it explores the possible role of Sukuk in the economic development and recovery post-Covid.

2 Hypothesis

Developing and issuing Sukuk can address financial and economic needs in targeted countries that lead to economic growth. Therefore, it will encourage more investors to be part of the economic cycle that meets their investing requirements in Sharia-compliant financing instruments.

3 Method

A qualitative methodology is employed in this paper to analyse different Sukuk issuances programs based on empirical studies that focused on different cases in different countries. Nonetheless, the data were collected through secondary sources, which were obtained employing a content analysis method of diverse publications of research papers and annual reports on Sukuk programs of projects and activities.

4 Objectives

This paper's main objective is:

- To empirically examine to what extent Sukuk financing can boost economic growth.
- To examine whether the Sukuk financing can be used in the post-Covid to aid economic recovery and growth.

5 Obtaining Data

The data were obtained mainly from secondary sources; thus, data comes from information published and presented by specific institutions or research papers and reports. Some data analyses were relied on and referred to some empirical and case studies carried out by experts and specialized economic/financial institutions focusing on the role of Sukuk recovery in the real economy represented by SMEs post-covid.

5.1 Literature Review

Abdur Rahman et al. emphasised that Sukuk's growth was remarkable in many countries worldwide, Muslim and non-Muslim, in the past decades. Expanding on the same trend of discussion [3] suggested that Sukuk, at large, can be instrumental in supporting the economy and alleviating poverty. Therefore, according to Ledhem [4], Sukuk financing contributes to enhancing economic growth, which places a significant role of Sukuk in financing investments and empowers the country's economy. In addition, Jallow and Joof [5] concluded that, in the long term, there is a clear connection between "Sukuk return and economic growth" that leads to an increase in GDP growth. According to Karimzadeh [6], The Sukuk market can play a significant role in providing sovereign institutions and private corporations access to the vast pool of Islamic financing as an alternative instrument for liquidity. Thus, Sukuk as a financing instrument consistently proved to have a positive impact on economic activities and, as a result, on economic growth [7]. Some studies suggested that Sukuk can be in a favorable position to lead economic growth and increase the capacity of positive impact on the GDP, as in the case of Turkey [2].

6 Overview on Sukuk Development

Sukuk is widening its financing ground steadily on an annual basis. For many investors, the central attractive aspect of Sukuk is a financing instrument that is Shariah-compliant structured. Therefore, Sovereign bodies, multinational corporations, and financial institutions rely on Sukuk as one of the most effective structures for raising funds from international capital markets [8]. Sukuk is also attractive as an Islamic-ethical-based alternative to syndicated financing widely known and offered in the conventional financial markets [9]. It is believed that developing Sukuk would improve the exposure to financial services, generate Sharia-compliant options for "small and/or risk-averse investors," and strengthen capital markets [6].

Being a long-term Islamic financing instrument, Sukuk is distinguished from conventional bonds where the main difference lies in the fact that Sukuk investment focuses on the real economy's transactions while it is not necessarily the case with bonds [10]. In addition, Sukuk plays a significant role in advancing the economy on the country and world level [11]. This role becomes more feasible knowing that Sukuk is "the second-largest" product of the Islamic finance industry after Islamic banking [12]. Furthermore, it holds the lion's share of the Islamic capital market, giving options for a more comprehensive set of investors regardless of their faith, culture, or religion [13, 14]. The banks will be even better when offering well-functioning Sharia Compliant Bonds to lessen information irregularity, so encouraging the effective distribution of resources would be critical for economic growth [15].

7 Sukuk Issuances, Different Preferences

The Sukuk issuances included many Muslim and non-Muslim countries with a leading position to Malaysia as it issued almost half of Sukuk on the global level. Municipal issuance of Sukuk is a crucial method to finance various activities. Different countries adopted this kind of issuances, such as Turkey, Iran, Germany, and Malaysia. However, every country is at different phases of capacity and preparation in creating its own municipal Sukuk. The countries also have a variation of Sukuk preferences where Malaysia and Indonesia, for example, often prefer long-term sovereign Sukuk. However, some Gulf countries adopt different models of issuances [16]. It is worth mentioning that other countries that have actively issued Sukuk in recent years, like Qatar, Kuwait, UAE, Oman, and Bahrain, also prefer sovereign Sukuk [17].

Countries and other entities also issue different models of Sukuk. That is to say that Sukuk preferences would vary to include Sukuk al-Ijara, sukuk al Murabaha, sukuk al Mudarabah, Sukuk al istisna, and others. Every issuer would consider its Sukuk specification to address the needs of targeted projects and economic activities. One fact that can be confirmed is that Sukuk can cater and address different levels of financing needs, especially when having well-developed Sukuk. As presented later in the paper, Sukuk served a wide range of projects and activities with reasonable yield or return. It is worth mentioning that Sukuk, with its diverse set of preferences, showed a good level of resilience in different economic environments. It did work in developed, developing, and Less-Developed Countries (LDCs), where it offered a vital source of financing as there is limited capital availability [18].

8 An Analysis to Correlative Conjunctions of Sukuk and Economic Growth

Theoretically, looking at the correlation between the capital market and economic growth goes back to Schumpeter's discussion on the theory of economic development [19]. He suggested that establishing a robust financial system can enable innovations that positively impact economic growth by providing investors with financial resources and services. This argument was adopted and developed later by Kusuma et al. [20] and Reuters [21]. In supporting their hypothesis, it is argued that developing countries have to give great attention to creating an efficient financial system to assist capital growth, leading to overall economic growth [22].

The endogenous growth models supported and enhanced the previous hypothesis and gained formal ground and popularity through the work of Dewi et al. [23]. His work overtly denotes the modeling that bridging between "the financial intermediation role of capital markets and growth indicators."

The modeling established that the capital market is a vehicle that positively impacts the economic growth of emerging economies. The same variables also clarify the economic growth in the most developed countries [22]. This link between capital market and economic growth is evident in the case of Islamic bonds (Sukuk); evidently, studies suggested that there is a close correlation "between Sukuk markets and Islamic banks and their role in the economic development process." The relation proves the importance of Sukuk to economic development as it represents an active availability of financing resources to motivate and fund economic development. Thus, encouraging economic development through Sukuk assist the process of providing viable solutions to unemployment and poverty. As a result, issuing Sukuk can achieve the national and governmental developmental aims and vision of sustainability through sustainable projects to improve viable economic growth and development [24].

Namely, offering project-based Sukuk would help expand and diversify financing resources of infrastructure with a better possibility of motivating productive partnerships between the public and private sectors. Issuing Sukuk for infrastructure projects also encourages private investors to grasp projects with high economic value, which would promptly make them part of the economic growth process [25]. Many Sukuk issuers appeared to agree on the fact that Sukuk would increase productivity, "empowerment, and an overall increase in economic activities" [2].

The relation between Sukuk issuance and economic growth is evident based on report data analysis. It is believed that Sukuk can impose positive change on the world economy, not just for investors; it can also impact the lives of ordinary people as they would enjoy the benefit of the growth driven by Sukuk [10]. As a reflection on that, a study suggested that, empirically and econometrically, the Malaysian Sukuk issuance is an obvious reflection of the positive correlation between Sukuk and the country's economic growth over the years [7]. The same can be concluded regarding Indonesia and its Sukuk issuances. Studies suggest that Sukuk issuances positively impacted Indonesian GDP growth in the long term. The association between Sukuk returns and economic growth in Indonesia provides further evidence that Sukuk issuance can significantly benefit the economic development that can directly increase the income of households and corporations [4].

Consequently, one of the early positive findings is to confirm that the experience of Sukuk issuance in Indonesia plays a crucial role in motivating economic growth. Namely, if Sukuk issuance increases, Indonesia will witness its economic growth. The opposite will happen if the development of Sukuk issuance declines. It is essential to mention that the positive impact of Sukuk issuance on economic growth would occur in the long run, not in the short term [18]. Many countries around the world showed great interest in Sukuk. Among those countries that achieved pronounced success in utilising Sukuk and integrating them into their financial systems, enhancing their economic growth. Below, the paper will present some of these countries.

9 Sukuk, Practical Cases and Role Models of Success

Malaysia is one of the first and prominent pioneers of Islamic finance and Sukuk issuance. About 60% of total outstanding global Sukuk securities are believed to be issued in the country, mainly to finance infrastructure development. Through DanaInfra Nasional Berhad, a state-owned entity specialized in infrastructure financing, Malaysia issued 61 billion R.M. worth Sukuk to develop a three-line mass rapid transit system (MRT) in Kuala Lumpur. The same issuances were also utilized to finance several privately-held expressway enterprises [26].

Due to its success story of Islamic finance in general, and Sukuk issuance in particular, Malaysia is positioned as the world's largest market for Sukuk issuance in the global capital market. What makes the Malaysian experience worth focusing on is that Sukuk as a financing instrument prompted its economic growth through directing internal financial assets for infrastructure development. In addition, Sukuk became one of the main constituents of the GoM's budget management. Hence, the country's efforts succeeded in putting it the leading Islamic Capital Market player [27]. However, it is worth mentioning that Malaysia accumulated a significant and remarkable experience in Islamic financial practices from the 90s and on, which allowed for more creativity and pioneering abilities in most Islamic financial products, not to mention Sukuk and Islamic insurance (Takaful). Sukuk should not be treated as an exception or strange to Malaysian economic achievements as an economic success story in Malaysia. That is to say, Malaysia laid down a significantly developed infrastructure for Islamic finance that became an intrinsic element and prominent component of its legal system and economy. Therefore, the national Sukuk market in Malaysia maintains a solid position to perform as a powerful and profitable instrument for state and private corporates to enhance long-term funds for varied economic activities and infrastructure growth needs [28].

Indonesian case follows the footsteps of Malaysia; since 2009, when its Sukuk first appearance in the international market, the country developed its experience in that area to become the largest issuer of $USD Sukuk on the global level. This journey continued consistently and firmly; between 2015 and 2019, the country executed massive infrastructure projects as part of its Long-Term Development plans. These projects were successfully financed through Sukuk. The projects included but were not limited to "energy, telecommunications, transportation, agriculture, manufacturing, housing, and others." Remarkably, Indonesia's issuances of Sukuk to finance major infrastructure projects provided critical assistance to the country to avoid using other debt instruments. Indonesian Sukuk experience showed that this financial instrument is relevant to infrastructure financing methods.

Furthermore, Sukuk also provided more substantial evidence that it can be an alternative to conventional bonds; thus, it would help developing economies "to avoid the debt trap in financing development infrastructure," which is crucial to boost the economic growth, especially in the long run [29]. However, the Indonesian Government is adamant about motivating internal and external Islamic financial institutions and funds to subscribe to infrastructure Sukuk issuances. It is fascinating to see that the issuances were oversubscribed, whether regular Sukuk of ten years tenor or green Sukuk of 5 and half year tenor. Proving the viability of Sukuk, the Government encouraged and gained the trust of community funds like Zakat, waqf, and Hajj to participate in the issuances and be an integral part of infrastructure development [26].

As a result, this mechanism of attracting community funds for Sukuk will make financial and economic players contribute to overall economic activities that motivate a healthy economic cycle. The positive influence of Sukuk in Indonesia can be evident by the rise of the per capita income of ordinary people in Indonesia, which positively affected the country's overall economic development. The consideration of the rise of per capita income was apparent in the middle-income class [30]. In order to prove that, the data analysis between 2014 and 2019 supported the assumption that Sukuk has positively affected economic growth while conventional bonds did not affect economic growth in Indonesia [5].

This enormous Sukuk development raised Indonesia to position itself highly in Islamic capital markets. As a result, Indonesia is classified as the fifth country on the global level [31].

Bangladesh Sukuk issuance is rather exciting and promising to encourage further Sukuk issuances in countries with weak economies or modest economic structures. Despite that, the country is no stranger to Islamic financial products that include different financial models to provide finance "through a Micro Enterprise Investment Scheme (MEIS)" to different industries that vary between agricultural projects, rural infrastructure, and housing [32]. However, it has never been exposed to Islamic Sukuk before 2021 [33]. Nonetheless, the country is making progress on that financial front. Short-term Sukuk issuance has become regularly issued by the central bank to help Islamic banks meet liquidity needs. As part of this progress, the Government towards the end of 2021 to finance water supply projects issued the first sovereign Sukuk worth (US $926.29 million). Notably, all issuances were "oversubscribed by almost four times," indicating a notable success. Therefore, the enthusiasm for participation made Sukuk a prospective game-changer for the country's economy [33]. This statement aligns with the Central Bank's prediction, which considers Sukuk an essential instrument to generate needed capital for private and governmental funds to finance major infrastructure projects. In addition to the governmental Sukuk issuance 2020, further sovereign Sukuk are expected to be issued in the next few years to advance the market's development [1].

Bangladesh government presented Sukuk regulations that took place in 2021. The regulations indicate that the first issuance positively impacted and inspired the Government to enhance Sukuk's position as a reliable financial instrument to boost the economy and fill the gap of infrastructure funds needed in the country, estimated to be about $US 300 billion by 2031 [33].

10 Sukuk Post-Covid, Direct and Indirect Possible Impact on Economic Recovery and Growth

Hauser, [3] in his Speech at U.K. Islamic Finance Week, December 2, 2020, he predicted that Islamic finance models of risk-sharing contracts would increase and become more prevalent in response to Covid economic crisis. According to him, they "pose materially lower medium-term risks to stability." He suggested that the Bank of England encouraged risk-sharing arrangements for a long time "which could be packaged in Sukuk form." The enthusiasm in the U.K. towards Sukuk to be part of the capital instrument to help in economic recovery post-Covid is very clear and does show a long-standing position of the country regarding Islamic finance with the apparent belief that Sukuk can be the way forward. In support of the previous view, it is believed that risk-sharing Sukuk can play a crucial role post-Covid in financing the infrastructure of the education and health sectors [34]. Therefore, Sukuk would be a reliable mechanism for a profit-generating instrument to investors, granting them to reduce their financial liabilities. Thus, Sukuk is becoming a means of financial relief inspired to support micro-entrepreneurs or (SMEs) lacking the capital to make them macro businesses. Consequently, many countries consider Sukuk issuance to restrain the financial liabilities that arose from the Covid pandemic [35]. One of the recent prominent cases is the Malaysian announcement of issuing RM500 million of 'Sukuk Prihatin' to help enhance the economy post-Covid [36].

With its long and accumulative experience in Islamic finance and being one of the most pioneering issues of Sukuk, Malaysia seems to conclude that Sukuk can be the way forward for enhancing the economic recovery post-Covid. For that purpose, they issued 'People's Sukuk.' The issuance aimed at funding micro enterprises (SME). In addition, it will finance school development in rural regions and provide grants for infectious disease research. The Prime Minister introduced this Sukuk as one of the assisting substitutes to support the governmental measures post-Covid for the economy [37]. It is believed that Sukuk would be a vital financial source to spur the reconstruction of the economy of the country post-Covid [38]. The post-pandemic era created new needs priorities that focus on social demands. Social Sukuk worth USD1.5 Billion already issued by Islamic Development Bank as "Debut Sustainability Sukuk in Response to COVID-19" [39].

Furthermore, a new trend of Sukuk issuance is coming into the scene, driving Sukuk to be mainly utilized to rebuild economic endeavors, particularly supporting the recovery (SMEs) that were affected negatively by the Covid "either in the form of capital assistance or in business development through business programs and assistance." This kind of Sukuk issuance would achieve the main objective of Islamic finance, which is to support the development of the real economy [40]. One of the remarkable experiences is the Islamic Development Bank (The IsDB) that issued the first Covid-19-related Sukūk in the market. The IsDB Sukūk aims to impact the Covid effect on economic activities to create resilience for any future shocks. In assuming this kind of Sukuk, the IsDB may open the way for other financial institutions, public or private, to follow suit and help economic growth and recovery. This Sukuk of IsDB comes under its Sustainable Finance Framework, which became the "first-ever AAA-rated Sustainability Sukūk in addition to being the first Covid-19-related Sukūk in the global capital markets".

This issuance strongly proves the bank's deep belief that Sukuk can be the way forward in achieving economic recovery and growth, which directly contribute to sustainability. In response to this successful experience, the Algerian Government is at an advanced phase of IsDB's approval to facilitate the Government's issuance of Sukūk to assist raise funds and sustain recovery endeavors post-Covid-19 [41].

11 Conclusion and Findings

In conclusion, Islamic finance plays a significant role in enhancing economies generally. It can be noticed that issuing Sukuk is developing and evolving in Muslim and non-Muslim countries. Sukuk proved to be one of the reliable financial methods in the capital markets in addition to conventional bonds. In addition, this paper established that the Islamic Sukuk broadened its financial attractiveness for further and potential increase as an alternative to conventional bonds where Sukuk would be widely used to empower economic growth.

By this means, Sukuk needs more pioneering collaborations with other Islamic financial elements that focus on utilizing Islamic social instruments to help economic recovery and growth post-Covid, as mentioned in the discussion earlier. Notwithstanding, governments of Muslim countries should encourage new products of Sukuk to accumulate on existing experiences of Sukuk issuances that focus on improving people's livelihoods through economic development post-Covid. This plan would provide investment to eliminate poverty through economic empowerment.

Research, empirical, and case studies indicated a positive impact of Sukuk on economic growth and recovery. Therefore, Sukuk financing proved to be one of the financial instruments that can boost the economic growth in the issuing countries. Hence, it can validate the hypothesis that Sukuk financing can positively impact economic growth. This paper shows that achieving economic sustainability and alleviating poverty can be attained through different models of Sukuk as a low-cost fund to finance economic development in general. Thus, examining this possibility to be utilized post-Covid time. As shown in the discussion above, the successful cases can be a motivation for more issuances of Sukuk to increase its contribution in the economic activities post-Covid that empower the societies to overcome the economic difficulties through offering more job opportunities.

Nevertheless, further rigorous determination is required to be done. Countries with weak economies and high poverty rates should be encouraged to profit from Sukuk as an essential financial instrument for economic empowerment. Many underdeveloped countries worldwide would be in serious need of Sukuk to resolve the problems facing the financing abilities of its infrastructural projects, poverty alleviation, and other economic challenges. This paper confirms that Sukuk can be ideal for supporting the economy post-Covid without being exposed to interest rate or inflation risk; this is since Sukuk represent tangible assets where more investments can be attracted. Therefore, this will encourage more people to inject their money into the economic cycle, which positively impacts economic growth.

References

1. Al Madani, H., Alotaibi, K., Alhammadi, S.: The role of Sukuk in achieving sustainable development: evidence from the Islamic Development Bank. Banks Bank Syst. **15**(4), 36–48 (2020)
2. Alkhawaja, M.I.: Sukuk: measuring the role of Sukuk as an economic growth enhancer in Turkey. Aust. J. Basic Appl. Sci. **13**(2), 95–99 (2019). https://doi.org/10.22587/ajbas.2019.13.2.12
3. Hafsa, Y., Oumaima, B.: An overview of Sukuk issuance and their contribution in economic empowerment. Int. J. Zakat Soc. Finance **2**(1), 33–48 (2021). https://media.neliti.com/media/publications/340667-an-overview-of-sukuk-issuance-and-their-31946fa0.pdf. https://doi.org/10.51377/azjaf.vol2no1.42

4. Islamic Finance News: Bangladesh: could corporate Sukuk take off? (2021). https://www.bluemountcapital.com/wp-content/uploads/2021/04/IFN-newsletter.pdf. Accessed on 15 Feb 2022
5. Jallow, O., Joof, F.: Investigating the Asymmetric Effect of Sukuk Returns on Economic Growth—Evidence from Indonesia, a NARDL Perspective (2020). https://mpra.ub.unimuenchen.de/102939/1/MPRA_paper_102939.pdf. Accessed 2 Feb 2022
6. Karimzadeh, M.: Role of Sukuk in Islamic capital market: experience of Iran (1994–2011). Arab. J. Bus. Manage. Rev. **1**(7), 105 (2012)
7. Ayaz, M., Shah, H.S., Hussain, T., Iqbal, M.: Islamic capital markets and economic growth. Islamic Banking and Finance Rev. **6**(1), 39–58 (2019). https://doi.org/10.32350/ibfr.2019.06.03
8. Ahmad, R., Maiyaki, A.A., Yusoff, M.: Effect of leverage and selected control variables on performance of firms listed on Bursa Malaysia. Paper presented at the third annual conference of economic forum of entrepreneurship and international business, Oxford, UK (2012)
9. Aloui, C., Hammoudeh, S., Hamida, H.: Global factors driving structural changes in the co-movement between Sharia stocks and sukuk in the Gulf Cooperation Council countries. North Am. J. Econ. Finance **31**, 311–329 (2015). https://doi.org/10.1016/j.najef.2014.12.002
10. Ibrahim, A.A., Elatrash, R.J., Farooq, M.O.: Hoarding versus circulation of wealth from the perspective of maqasid al-Shari'ah. Int. J. Islam. Middle East. Financ. Manag. **7**(1), 6–21 (2014). https://doi.org/10.1108/IMEFM-06-2012-0053
11. Khan, N.: The role of Islamic finance in economic development. Lingnan J. Banking, Finance Econ. (6), 79–83 (2017). Retrieved from http://commons.ln.edu.hk/ljbfe/vol6/iss1/6. Accessed on 03 Feb 2022
12. Sari, N., Syamsurijal, A.K., Widiyanti, M.: The impact of Islamic capital market development on economic growth: the case of Indonesia. J. Smart Econ. Growth **3**(2), 37–141 (2018). www.jseg.ro. Accessed on 12 Feb 2022
13. Altaleb, G.S., Alkhatib, A.Y.: Sovereign Islamic Sukuk and the economic development: a case study of Jordan. Eur. J. Bus. Manage. **8**(32), 115–118 (2016)
14. Hasan, S.B.M.: Sukuk: Taking Indonesia by storm. Ethis Crowd. Blog (2018)
15. RM500mil Prihatin Sukuk for Malaysians to help rebuild economy, NST Business (2020). Retrieved from https://www.nst.com.my/business/2020/06/598686/rm500mil-prihatin-Sukuk-malaysians-help-rebuildeconomy. Accessed 25 Dec 2021
16. Al Natoor, B.: Bangladesh policy steps may spur Islamic finance development, Fitch ratings (2021). https://www.fitchratings.com/research/islamic-finance/bangladesh-policy-steps-may-spur-islamic-finance-development-10-11-2021
17. Jalil, A.: Islamic finance to lessen Covid-19 impact (2020). https://themalaysianreserve.com/2020/10/07/islamicfinance-to-lessen-covid-19-impact Accessed on 25 Dec 2021
18. Parker, M.: The emergence of COVID-19 mitigation Sukuk—Malaysia set to join IsDB (2020). Retrieved from https://www.bernama.com/en/thoughts/news.php?id=1854830. Accessed on 26 Dec 2021
19. Response of The Islamic Development Bank Group Discussion Draft: The Covid-19 crisis and Islamic finance (2020). Retrieved from https://www.isdb.org/sites/default/files/media/documents/2020-10/1.%20IsDB%20Group%20Report%20on%20Covid-19%20and%20Islamic%20Finance__FINAL.pdf. Accessed on 28 Dec 2021
20. Kusuma, K.A., Caputo Silva, A.C.: Sukuk markets: a proposed approach for development. Policy research working paper no.7133. World Bank (2014). Retrieved from https://openknowledge.worldbank.org/handle/10986/20702. License: CC BY 3.0 IGO. Accessed on 11 Jan 2022
21. Reuters, T.: State of the Global Islamic Economy Report 2019/20. Thomson Reuters, DinarStandard (2020). Available at: www.salaamgateway.com/SGIE19-20. Retrieved 3 Feb 2022
22. Smaoui, H., Khawaja, M.: The determinants of sharia compliant bonds market development. Emerg. Mark. Financ. Trade **53**(7), 1501–1518 (2017). https://doi.org/10.1080/1540496X.2016.1224175

23. Dewi, S., Majid, A., Shabri, M.A., Salina, K.: Dynamics of financial development, economic growth, and poverty alleviation: the Indonesian experience. South East Eur. J. Econ. Bus. **13**(1), 17–30 (2018). https://doi.org/10.2478/jeb-2018-0002
24. Al-Sayed., O.: Sukuk risk: analysis and management. Eur. J. Appl. Soc. Sci. Res. **1**(3), 67–76 (2013). Retrieved from www.ejassr.org. Accessed on 8 Feb 2022
25. Ben Salem, M., Fakhfekh, M., Hachicha, N.: Sukuk Issuance and economic growth: the Malaysian case. J. Islamic Econ. Banking Finance (2016). Gale Academic OneFile, link.gale.com/apps/doc/A598310339/AONE?u=anon~ed89e782&sid=googleScholar&xid=912546fd. Accessed 13 Feb 2022
26. Comcec Coordination Office: The role of Sukuk in Islamic capital markets (2018). https://sbb.gov.tr/wpcontent/uploads/2018/11/The_Role_of_Sukuk_in_Islamic_Capital_Markets.pdf.p73
27. Basyariah, N., Kusuma, H., Qizam, I.: Determinants of Sukuk market development: macroeconomic stability and institutional approach. J. Asian Finance, Econ. Bus. **8**(2), 201–211 (2021). https://doi.org/10.13106/JAFEB.2021.VOL8.NO2.0201
28. Hauser, A.: Why Islamic Finance has an Important Role to Play in Supporting the Recovery from Covid—And How the Bank of England's New Alternative Liquidity Facility Can Help. Markets, U.K. Islamic Finance Week (2020). Retrieved from https://www.bankofengland.co.uk/-/media/boe/files/speech/2020/why-islamic-finance-has-an-important-role-to-play-in-supporting-the-recovery-from-covid.pdf?la=en&hash=58B28008CE026D33B0590F7DB5CDA84DAE723CAD. Accessed on 12 Feb 2022
29. Greenwood, J., Jovanovic, B.: Financial development, growth, and the distribution of income. J. Polit. Econ. **98**(5), 1076–1107 (1990)
30. Shaw, E.S.: Financial Deepening in Economic Development. Oxford University Press, New York (1973)
31. Nila, D., Mulyati, Y., Tantular, Y., Abdulhadi, D., Islahuzzaman.: Towards sustainable development: Indonesia project-based Sukuk for infrastructure. Int. J. Eng. Technol. **7**(4.34), 217. https://doi.org/10.14419/ijet.v7i4.34.23893
32. Kamiya, M., Zhang, L.-Y., Yin.: Finance for City Leaders (2016). https://www.mypsup.org/library_files/downloads/Finance%20for%20City%20Leaders%20Handbook%20-2nd%20Edition.pdf. First published in Nairobi in 2016 by UN-Habitat. United Nations Human Settlements Programme
33. IIFM Sukuk Report 2019: International Islamic Financial Market, 8th edn. (2020). Available at: www.iifm.net/sukuk-reports/. Retrieved 14 Feb 2022
34. Reuters, T.: Islamic finance development in 2017: towards sustainability. J. Islamic Finance Accountancy **2**(1), 58–60 (2018). Retrieved from https://forms.thomsonreuters.com/ifgbusiness. Accessed on 10 Feb 2022
35. Khoutem, J.D.B.: Islamic banks-Sukuk markets relationships and economic development. J. Islamic Acc. Bus. Res. **5**(1), 47–60 (2014)
36. Pagano, M.: Financial markets and growth an overview. Eur. Econ. Rev. **37**, 613–622 (1993)
37. Mat Ali, S.A., Arshad, A., Ibrahim, I.: Relief measures of Islamic finance combating adverse impacts of Covid 19 in Malaysia: an analysis. Int. J. Law Gov. Commun. **6**(24), 116–134 (2021). https://doi.org/10.35631/IJLGC.624008
38. Ishak, K., Bengkalis, S.S., Syariah, D.P.: The role of Islamic finance in economic development, pp 131–143 (2018). https://ejournal.stiesyariahbengkalis.ac.id/index.php/iqtishaduna/article/download/135/135. Accessed on 18 Feb 2022
39. Khiyar, K., Al Galfy, A.: The role of Sukuk (Islamic bonds) in economic development. J. Financ. Actuarial Math. Manage. 31–35 (2014). http://www.psp-ltd.com/JFAMM_10_2_2014.pdf. Accessed on 15 Feb 2022
40. Schumpeter, J.A.: The Theory of Economic Development. Harvard University Press, Cambridge, MA (1911)
41. McKinnon, R.I.: Money and Capital in Economic Development. The Brookings Institution, Washington, D.C. (1973)

Sukuk Investment in Islamic Banks: Risk and Solutions

Mohammed Abdullah Al-Mekhlafi

Abstract The study aimed to identify the risks and sources of Sukuk investment in Islamic banks and Identify mechanisms for managing the risks of Islamic banks' investment in Sukuk. The researcher used an inductive and deductive technique to reach the goal of the study. The study findings: Risk is one of the components of the Islamic investment method, but this does not preclude the development of a scientific, administrative approach to dealing with them, through a series of steps in which the risks of Islamic banks' Sukuk investments. The researcher recommended Islamic financial institutions (including Islamic banks) that issue Sukuk should follow Shariah standards for issuing and trading that Sukuk, since these controls serve as a safety valve for managing the risks associated with Islamic financial institutions investing in Sukuk.

Keywords Risk · Islamic bank · Investment · Sukuk

1 Introduction

Islamic banking has been in constant development for the past four decades, resulting in its global expansion and a 15% yearly growth rate. Due to their flexibility and attractiveness in terms of return and risk due to the variety of methods and structures of issuance, Islamic sukuks have emerged as one of the most important financial instruments, with growth rates in the volume of their issuances exceeding 40% annually (Mohammed 2012).

The Islamic financial industry's most well-known product is the Islamic Sukuk. It has grown at a breakneck pace in recent years, becoming the fastest-growing

M. A. Al-Mekhlafi (✉)
Accounting Department, College of Business Administration, University of Business and Technology, Jeddah, Saudi Arabia
e-mail: m.mekhlafi@ubt.edu.sa

A. Hamdan et al. (eds.), *The Implementation of Smart Technologies for Business Success and Sustainability*, Studies in Systems, Decision and Control 216,
https://doi.org/10.1007/978-3-031-10212-7_27

component of the Islamic financial market. At the end of 2021, global issuances of Islamic Sukuk totaled $147 billion (www.alghad.com 2021).

The risk of Islamic banks investing in Sukuk and how to handle them will be identified in this study.

1.1 Research Problem

Can be managed the risks of Islamic banks investing in Sukuk?

1.2 Research Objective

The research aims to:

1. Identifying the risks and sources of Islamic banks investing in Sukuk.
2. Identifying mechanisms for managing the risks of Islamic banks' investment in Sukuk.

1.3 Research Importance

The significance of the research can be seen in its exposure to the risks of Islamic banks investing in Islamic Sukuk, as well as the risks that it may be exposed to, as well as the development of administrative mechanisms to deal with those risks in a way that combines economic efficiency and legitimacy credibility.

1.4 Research Design

The following two methodologies were used in the study:

1. Inductive approach: by extrapolating the risks to which Islamic banks' investments in Sukuk are exposed and determining their sources and types.
2. The deductive approach: by devising mechanisms to manage the risks of Islamic banks' investment in sukuk.

2 Literature Review and Theoretical Framework

2.1 Literature Review

Dwaba's [10] study aimed to seek to identify Islamic Sukuk in terms of its concept, characteristics, types, risks, types, and sources of those risks and to develop mechanisms to address those risks based on Islamic financial engineering tools. The study findings: there are many mechanisms of Islamic financial engineering to manage the risks of Islamic Sukuk, and among these mechanisms are the feasibility study, the efficiency of management and control over them, and the guarantees that they contain a guarantee, mortgage, and third-party guarantee, and among these mechanisms are also cooperative insurance, reserves, and hedging.

Al Sayed's [6] study aimed to address the concept and meaning of Sukuk risks, as well as the sources and causes of Sukuk risks, as well as the mechanisms for dealing with and regulating Sukuk risks, with a focus on the Islamic perspective and Al Shariah laws for dealing with such risks. The study findings: Islamic finance is built on the ideals of justice and equal distribution of financial resources in society, a function that the traditional financial system has failed to fulfill. Practices for asset risk management should be implemented. The regulator must take steps to encourage product innovation and documentation. It is necessary to develop Islamic credit rating organizations that will rate assets in accordance with Shariah. Islamic financial experts demand that the Sukuk be closely monitored and audited. It will result in complete Shariah compliance.

Bin Zaid and Nasser's [9] study aimed to investigate the problem of pricing Islamic Sukuk using the global interest rate LIBOR, as well as the significant risks that this entails for the safety of investment in this Sukuk, and thus investment operations in the Islamic financial industry as a whole, as we attempted to demonstrate through this research the inefficiency of relying on interest rates in general, and the LIBOR rate in particular, due to the scandal of interest rate manipulations used in its calculation. According to the results of the study, sukuks that do not rely on the LIBOR index to determine their returns, such as the Sudanese government Sukuk (Shahada) and (Sarh), are deemed to be less hazardous than those that do, such as those included in the Dow Jones Citigroup Islamic Sukuk Index. Numbers and statistical indicators are used to do this.

Alswaidan et al. [7] study aimed to reviews Sukuk risk classification schemes are based on extending and customizing traditional finance risk classification schemes. The study findings: Debt-based, equity-based, assets-based, agency-based, and hybrid structures are the most common Sukuk structures. They all stem from Shariah and the fundamental ethical principles of Islamic banking and commerce. In addition, Sukuk risk profiles are linked to Sukuk structures. As a result, the risks associated with the Sukuk structure are fundamentally Shariah hazards. Based on an assessment of Sukuk structure hazards, the study proposes a Sukuk risk classification matrix.

2.2 *Theoretical Framework*

2.2.1 Risk

Definition of Risk

Risk refers to the uncertainty and severity of an activity's events and repercussions (*or outcomes*) in relation to anything that human's value [8, pp 1–11]. The impact of uncertainty on objectives is known as risk [14].

Types of Risk

There are two type of risk [13, p 256].

General Risks (Market Risk)

Excessive market risks are systemic risks that cannot be avoided through diversification since they affect the returns of all assets, independent of the organization producing those securities.

Special Risks

It depicts the sporadic hazards that can be avoided or mitigated by diversification, as this type of risk is caused by the facility's or industry's conditions.

2.2.2 Definition of Investment

Assigning an individual's liquidity at a specific time and for a specific length of time, which may be long or short, and tying it to one or more assets that he keeps for that period of time in order to get future cash flows [17, p 13].

2.2.3 Islamic Bank

Definition of Islamic Bank

Islamic bank is a non-interest based financial organization that pursues creative and progressive financial engineering initiatives to provide efficient and competitive banking, investment, trade finance, commercial, and real estate financing services in accordance with Islamic regulations [16, p 662].

Characteristics of Islamic Bank

Islamic banks are based on [1]:

1. First and foremost, the money must come from a halal source and be used in a halal manner.
2. Second, money investments must be free of any suspicion of usury.
3. Third, the returns are distributed between the money's owners and those who manage and employ it.
4. Fourth, the poor have a title to the assets of the wealthy because of the zakat commitment.
5. Fifth, Sharia supervision is the foundation for Islamic banks' evaluation and oversight.
6. Sixth, there is no distinction between the material and spiritual and moral dimensions of life.

2.2.4 Sukuk

Definition of Sukuk

Investment Sukuk are certificates of equal value that represent undivided shares in ownership of tangible assets, usufruct, and services, or (in the ownership of) the assets of specific projects or special investment activity, but this is only true after receipt of the Sukuk's value, the closing of subscriptions, and the use of funds received for the purpose for which the Sukuk were issued [4, p 468].

Characteristics of Islamic Sukuk

The most important characteristics of Islamic sukuk are [4, p 471 and 472]:

1. Investment Sukuk are certificates of equal value issued in the owner's or bearer's name in order to prove the owner's claim to the financial rights and liabilities represented by the certificate.
2. Investment Sukuk represent a common share of ownership in the assets available for investment, whether they be nonmonetary assets, usufructs, services, or a combination of all of these, as well as intangible rights, debts, and monetary assets. The holder of these Sukuk does not owe the issuer any money.
3. Investment Sukuk are issued under the terms of a Shari'ah-nominated contract and are subject to Shari'ah requirements for issuance and trade.
4. The terms that regulate the trading of Investment Sukuk are the same terms that govern the trading of the rights they represent.
5. The owners of these certificates share the profit as specified in the subscription prospectus, and they bear the losses in proportion to the number of certificates they own (hold).

3 Risks of Islamic Banks Investing in Sukuk and Its Sources

Because Islamic sukuk are founded on Islamic financing formulas, they are vulnerable to the risks that Islamic investment projects are exposed to, hence we discover that Islamic sukuk are exposed to many total risks [10]:

1. **Credit Risk**

   These risks arise as a result of the customer's inefficiency, bad reputation, and unwillingness or inability to pay, and all of this falls primarily on the shoulders of the institution issuing the instruments and granting credit to the customer, negatively impacting its returns, so these risks are considered own risks.
2. **Market risk**

   The quality of these risks is determined by the upward and downward trends that occur in markets as a result of economic, social, or political factors, regardless of whether they are real asset markets, financial markets, which include money markets, and capital markets, which have two markets: the issuance market and the trading market, and thus fall under general risks.

3.1 *Real Asset Market Risk*

The nature of Islamic sukuk in terms of being a common share in the ownership of assets is related to the quality of these risks, and because real assets of goods and services are sold in markets, they may be subject to a decrease in value due to supply and demand factors, government economic policies, and other market factors.

3.2 *Money Market Risk*

It can be divided into the following categories.

3.2.1 Exchange Rate Risk

Currency exchange rate changes in forward transactions create the quality of this risk in the money market. When buying goods in a foreign currency and the price of that currency falls, this results in losses proportional to the drop in the price of that currency against other currencies. Exchange rate risks also arise when issuing sukuk in a specific currency and investing the proceeds in other currencies, or if the institution issuing the sukuk maintains open positions towards some foreign currencies or payment obligations, particularly in Murabaha.

3.2.2 Interest Rate Risk

Changes in the level of interest rates in the market in general affect the quality of these risks, which affect all investments regardless of the kind and conditions of the investment. As a general rule, the higher market interest rates are, the lower the market value of traded assets is, and vice versa, which impacts the rate of return on investment [13, p 258].

Because interest rates have no place in dealing with Islamic sukuks, or in their operations and investment fields, they may be influenced by the interest rate if it is used as a reference price in Murabaha financing [11, p 17].

3.2.3 Securities Price Risk

The quality of these risks occurs as a result of changes in the prices of securities in the capital markets, whether caused by actual or artificial and immoral elements such rumors, monopoly, gambling, shame, buying and selling, and the like, which affect the market value of Islamic sukuk.

3.2.4 Inflation Risk

The quality of this risk emerges as a result of a decline in money's purchasing power coupled with an increase in prices, implying that the money invested will lose its real worth. As a variable-return instrument with components of money, debts, notables, and benefits, the impact of Islamic sukuk on inflation is exactly proportional to the increase in its components of money and debts.

3. **Operational Risks**

   These risks arise as a result of human or technical errors or accidents, and they fall under general risks if they are caused by external factors such as natural disasters, such as what disasters or accidents cause in the destruction of crops in Muzara'a bonds investments or the destruction of leased assets in Ijarah bond investments, and so on. Internal factors such as insufficient equipment, technical means, qualified and trained human resources, account corruption, or a lack of administrative capacity (i.e. *administrative competence*).
4. **Risks of Sharia Violations**

   These risks arise from the use of sukuk funds for banned purposes, whether prohibited for themselves, such as prohibited products and services, or prohibited to earn, such as usury, gambling, fraud, deception, treachery, and so on, and they are subject to particular hazards.

4 Mechanisms for Managing the Risks of Islamic Banks' Investment in Sukuk

The concept of risk management for Islamic banks' investment in sukuk: It is a scientific strategy to dealing with pure risk that involves anticipating probable incidental losses and devising and implementing strategies to reduce the likelihood of a loss or the financial effect of a loss [12, p 51].

4.1 Steps to Manage the Risks of Islamic Banks Investing in Sukuk

4.1.1 Risks Identification

The Islamic bank must assess the types of risks to which Islamic sukuks are exposed, whether public or private hazards, as well as the source and causes of those risks. The process of identifying hazards must be ongoing, with a thorough awareness of the risks at the level of each investment firm and the whole enterprise.

4.1.2 Risk Evaluation

This step is one of the steps in Islamic bank investment risk management that provides a comprehensive picture of the risks, as it measures and evaluates those risks to determine the likelihood of loss, as well as classifying them as high, medium, or low. It also determines whether the risks are high, medium, or low [15, p 57].

4.1.3 Examine and Select Appropriate Risk Management Options

In this step, Islamic banks investigate the many options for dealing with each form of sukuk risk, as well as make the necessary choice to choose the best option, whether it is to avoid, distribute, or accept the risk. The benefits of using the proper approach for dealing with hazards should outweigh the expenses associated with the sukuk's risks.

4.1.4 Putting Mechanisms into Place to Implement the Appropriate Alternative to Dealing with Risks

4.2 Mechanisms for Managing the Risks of Islamic Banks' Investment in Sukuk

There are many and varied mechanisms for managing the risks of Islamic banks' investment in sukuk, and these mechanisms can be called financial engineering mechanisms, including [11, pp 18–26].

4.2.1 A Sharia Economic Feasibility Study

The sharia economic feasibility study of projects is a great importance in avoiding the risks to which the sukuk proceeds are directed, as the evaluation of this study is governed by the so-called "Islamic profitability", by directing financing towards Islamic projects in terms of halal activity, as well as economic efficiency. To achieve the optimal allocation of resources through the preservation and development of money, and taking into account the Islamic priorities in investment in accordance with the necessities, needs and improvements, In addition to social efficiency and the social responsibility it includes for financing and achieving net social benefits that bring benefit to society.

4.2.2 Administrative Control and Efficiency

The secret to sukuk's profitability and growth is their administrative efficiency, and the more effective this management is, the more these sukuk combine the dilemmas of profitability, liquidity, and security, as well as ethical considerations and social duties. Keeping in mind that in speculative sukuks, sukuk holders are not allowed to interfere in management, even though they have the right to follow up on the speculator's actions to ensure that he complies with the terms of speculation, and risks can be avoided. Reduced credit and operational risks, as well as sharia violations.

4.2.3 The Guarantees

In Islamic economic philosophy, there are many different forms of guarantees that can be used to cope with the risks of Islamic banks investing in sukuk, the following are the most essential of these guarantees.

The Guaranty

It is the obligation of one party to bear the burden of another as a contribution, and guaranteeing a person of good character is a successful technique to limit the risks of using the proceeds of the sukuk, particularly credit risks, risks of legitimate violations, and asset and return risks.

The Mortgage

Because mortgages are one of the mechanisms for mitigating risks, particularly credit risks, the creditor may compel the debtor to authorize him to sell the mortgage when the obligation is due in order to recover the loan's price without resorting to the courts Even if the mortgage is approved, the property must be able to be owned and sold, must be defined by reference, designation, or description, and must be capable of delivery [5, pp 968–970].

4.2.4 Cooperative Insurance

This insurance is one of the risk management techniques used by Islamic banks to defend against the risks associated with sukuk, particularly credit and asset risks, by insuring investment assets against all hazards associated with such assets.

4.2.5 Reserves

Because the establishment of these reserves is believed to create profits, this technique is suited for dealing with all of the hazards associated with Islamic sukuk. The profit rate reserve can be created by deducting a portion of the profits of the Sukuk holders after deducting the Mudarib share in order to protect against future losses that the Sukuk may be exposed to, and the investment risk reserve can be created by deducting a portion of the profits of the Sukuk holders after deducting the Mudarib share in order to protect against future losses that the Sukuk may be exposed to Shari'ah Standard No [3, p 297].

4.2.6 Hedging

Hedging against exchange rate risks is accomplished by as closely as possible unifying the employment currency with the sukuk issuance currency, selecting stable currencies for dealing [2, pp 52–58].

4.2.7 Diversification

Diversification in Islamic banks' sukuk assets in order to reduce risk while maintaining a positive return is referred to as efficient diversification. Diversification can be based on a variety of factors, the most important of which are:

Diversification of Issuers

There are two approaches to diversifying the issuer.

Simple Diversification

This technique is based on diversifying issuers while taking care not to overdo it due to its numerous disadvantages, including the complexity of managing assets, the high cost of searching for new investments, the risk of making poor investment judgments, and high average purchase costs.

Diversification of Markowitz

This strategy is founded on the assumption that risks are determined not just by investment risks, but also by the relationship between investments returns [13, pp 283 and 284]. This means that the investments to which the revenues of the sukuk are directed must be carefully chosen in order to account for the nature of the correlation between the returns generated by them. Because they have an inverse connection, this strategy of diversification may be able to eliminate private hazards while also reducing general risks.

Variation of Due Dates

Inflation is more likely to affect long-term sukuks than short-term sukuks. Due to the nature of their components and participation in profit and loss, Islamic sukuks are less vulnerable to inflation risks than standard sukuks.

Sectoral Diversification

Expanding the issuance of Islamic sukuk to include all sectors of the economy diversifies the risks and protects against their shocks.

International Diversification

It is possible to issue Islamic sukuk in more than one Islamic countries at the same time in the same currency, which helps to hedge against exchange rate risks.

5 Conclusions and Recommendations of the Study

5.1 Conclusions of the study

The findings are as follow:

1. Islamic sukuk are assets that contain a mix of cash, debts, assets, and benefits, or some of these components separately, and are subject to the same risks as other Islamic investment projects.
2. Credit risk, operational risk, market risk, etc.... Be consider in Islamic sharia.
3. As one of Islamic Banks bases is participation in the profit and loss, risk is one of the components of the Islamic investment method.
4. There are few mechanisms for managing the risks of Islamic banks' investment in sukuk (these mechanisms are known as financial engineering mechanisms).

5.2 Recommendations of the study

Based on the results, the researcher can determine a number of recommendations as follows:

1. Islamic financial institutions (including Islamic banks) that issue sukuk should follow Shariah standards for issuing and trading those sukuk, since these controls serve as a safety valve for managing the risks associated with Islamic financial institutions investing in sukuk.
2. It is feasible to learn from others' experiences in developing acceptable financial tools to manage the risks of sukuk investing, as long as they do not make what is forbidden or forbid what is permissible, and truly combine legitimacy and economic efficiency.
3. An invitation to pay attention to Islamic financial institutions' scientific research at the Shariah and economic levels in order to devise financial tools that improve the management of these institutions' investment risks, particularly the risks of Islamic financial institutions investing in sukuk.

References

1. Abu Obaid, A.: Islamic banks and their role in deceiving the banking sector. www.Kantakji.org
2. Accounting and auditing organization for Islamic financial institutions, Shari'ah Standard No. (1): Trading in Currencies, Bahrain (2017)
3. Accounting and auditing organization for Islamic financial institutions, Shari'ah Standard No. (11): Istisna'a and Parallel Istisna'a, Bahrain (2017)
4. Accounting and auditing organization for Islamic financial institutions, Shari'ah Standard No. (17): Investment Sukuk, Bahrain (2017)
5. Accounting and auditing organization for Islamic financial institutions, Shari'ah Standard No. (39): Mortgage and Its Contemporary Applications, Bahrain (2017)
6. Al-Sayed O.: Sukuk risk: analysis and management. Eur. J. Appl. Soc. Sci. Res. (EJASSR) **1**(3) (2013)
7. Alswaidan, M.W., Daynes, A., Pasgas, P.: Understanding and evaluation of risk in Sukuk structures. J. Islamic Acc. Bus. Res. **8**(4) (2017)
8. Aven, T., Renn, O.: On risk defined as an event where the outcome is uncertain. J. Risk Res. **12** (2009)
9. Bin Zaid, R., Nasser, S.: The relationship of Islamic Sukuk pricing with LIBOR rate as a threatening factor for the safety of investment in it—an applied study on a sample of Islamic Sukuk. In: The Ninth World Conference on Islamic Economy—Growth, Justice and Stability from an Islamic Perspective, Istanbul, Turkey (2013)
10. Dwaba, A.M.: Islamic Sukuk risk management. In: Yemeni Islamic Banks Conference, Reality and Future Challenges, Yemeni Businessmen Club, Sana'a, Republic of Yemen (2010)
11. Dwaba, A.M.: Banks' Benefits: Justifications and Questions. Dar Al-Salaam Publishing, Cairo (2008)
12. Hammad, T.A.A.: Risk Management (Individuals—Departments—Companies—Banks). University House, Alexandria (2003)
13. Hindi, M.: Modern Thought in the Field of Investment. Mansha'at Al Maaref, Alexandria (1999)
14. ISO: Risk Management—Principles and Guidelines, ISO 31000:2009 (2009)
15. Karasneh, I.: Basic and Contemporary Frameworks in Banking Supervision and Risk Management. Arab Monetary Fund, Institute of Economic Policies, Abu Dhabi (2006)
16. Naveed, A.K., Kashif, U.R.: Customer satisfaction and awareness of Islamic Banking system in Pakistan. Afr. J. Bus. Manage. **4**(5) (2010)
17. Ramadan, Z.: Principles of Investment (Real and Financial), vol. 1. Dar Wael, Amman (1998)

Pro-environmental Behaviour Impacts on the Willingness to Pay for Bio-based Sustainable Food Packaging

Nur Hafizah Muhammad, Abdullah Muhamed Yusoff, Nor Maizana Mat Nawi, Nurul Fardila Abd Razak, and Derweanna Bah Simpong

Abstract Food packaging waste become an important issue which it has had adverse environmental effects. Manufacturers ought to take the initiative to produce sustainable food packaging and acquire positive feedback from the grocer to use the packaging. With legislative responsibilities, food manufacturers are considering the best option to pack their products with minimal materials. This study aims to determine consumers' view of new bio-based plastic material as food packaging; while to recognise and identify consumers who are willing to pay for this new material and aware of the environmental benefits. Results were analysed through a Tobit regression model which recommend that a compelling marketing plan for this packaging and an imperative basis of Malaysian consumer segmentation are essential. Additionally, raising awareness of the benefits of saving the environment would appeal to several people to choose this packaging, hence improving the environment.

Keywords Pro-environmental behaviour · Sustainable packaging · Willingness to pay · Malaysia

1 Introduction

Food packaging waste has become very important concerns for the Malaysian government and the public. In 2016, the amount of waste generated was 38,200 tonnes per day generated in 2016. Among reported accumulation of the waste, food packaging and food waste are the major components dumped to the landfills [1]. Consequently, the environmental quality worsens due to the poor waste management process which resulted to increase of pollution around the country. For example, during incineration process of untreated waste at the landfill, the by-product of the process is polluting the soil as the organic compound penetrates into the ground and eventually will enter the water resources [2]. Moreover, the air and odour pollution of the dumpsite and it

N. H. Muhammad (✉) · A. M. Yusoff · N. M. M. Nawi · N. F. A. Razak · D. B. Simpong
Faculty of Hospitality, Tourism and Wellness, Universiti Malaysia Kelantan, 16100 Kota Bharu, Malaysia
e-mail: hafizahm@umk.edu.my

A. Hamdan et al. (eds.), *The Implementation of Smart Technologies for Business Success and Sustainability*, Studies in Systems, Decision and Control 216,
https://doi.org/10.1007/978-3-031-10212-7_28

processes are also significantly increase if waste situation remains unchanged over the years in Malaysia, particularly. The Environmental Quality Act and National Policy on the Environment were established and constantly amended, altered and improved to pursuit the increasing complexity of environmental issues [3]. Moreover, retailers and food producers initiatively are improving the products with more environmentally friendly option of food packaging to shown supportive act towards sustainable campaign in Malaysia. Even more, 10 cents have been charged on the certain days to consumer who wish to have plastic bag during the grocery shopping at the malls such as Tesco, AEON and local supermarkets.

Sustainable packaging is a packaging concept been defined as packaging that adds real value to society by effectively containing and protecting food products during movement across the supply chain and designed to use materials and energy efficiently and made up of materials that are cycled continuously and does not jeopardize to human health or ecosystems [4]. It has lately been mushrooming in Southeast Asia countries where Thailand has become the hub of the new substance called bio-based packaging material also known as bioplastic [5]. Meanwhile in Malaysia this bioplastic industry is still at infancy stage which has relevance potential to advance in recent years due to the increasing waste accumulation and emission of carbon dioxide [6]. Bio-based material is a substance derived from plant-based resources or renewable biomass by-product such as sugarcane husk and corn starch, crop leaves, rice straw or tapioca roots. It is designed with a similar property as fossil fuel plastic which is strong, durable and anti-bacteria. However, this bio-based plastic is believed to be safer to the environment and produce less carbon dioxide [5, 7]. Fossil-fuel plastic is undeniably one of the greatest inventions in the last century and food is almost impossible to pack without it. However, the environment issue is often associated with plastic materials used in food packaging. For example, litters on the streetways are plastic food packaging and same goes in the ocean, where a media reported that plastic island was appeared on North Pacific Ocean, recently [8]. This has gained much attention to the consumers, and they are becoming more environmental conscious with the groceries available in the market [9]. Therefore, the specific objectives are to identify the impact of environmental awareness on the willingness to pay for this new material and to recognize the characteristic of consumers who would willingness to spend money. This paper is organised as followed: (2) Methodology; (3) Results and discussion; and (4) Conclusion.

2 Methodology

2.1 Research Instrument

A questionnaire was constructed with three sections with the consent statement at the beginning of the instrument. The statement stated the purpose of this study and explanation of the bio-based packaging material as it is new to the consumers. Then, four environmental variables were placed to capture the psychological constructs of study. There were New Environmental Paradigm (NEP) perspective, awareness of consequences of waste accumulation, ascription of responsibility as a consumer and personal norms on how to control one's own behaviour. These variables are adopted from the Value-Belief-Norm model by Stern et al. [10]. The model provides a framework for investigating psychology factors that promote pro-environmental behaviour towards buying bio-based packaging. The model end with the pro-environmental behaviour where a person will or will not react upon the environmental values such as buy a sustainable food packaging to save the environment. Those 4 psychological variables were measured using 5-points Likert scale ranging from strongly disagree to strongly agree.

The questionnaire continued with the contingent valuation method. This is a popular stated preference method that has been applied in marketing and economic studies. This method used to test the economic value of a good without a market or new product where consumers were asked to state the maximum amount of money from their weekly groceries budget to give up for this new material [11]. This section aimed to elicit consumers' willingness to pay as the actual pro-environmental behaviour at the market with expression in monetary term. Lastly, the questionnaire end with socio-demographic section. In the previous years, marketers strived to understand the emerging sustainable product market and how consumer makes purchasing decision [12]. Even though socio-demographic variables are either very weak predictors or are conflicting in nature, but some authors believe these variables able to yield profiling consumer environmental locus of control [13–15].

2.2 Data Collection

The study was conducted in early December 2020 until March 2021 via online platform. This quantitative research yielded 152 valid questionnaires through Google Form platform. Table 1 shows the structure of the sample based on socio-demographic profile.

2.3 Tobit Regression Model Specification

To access the willingness to pay (WTP) from the contingent valuation stated preference method, Tobit model chosen as the equation model to determine the ideal price for the final product packed with the bio-based packaging material. Tobit model is a censored regression model that is designed to estimate linear relationships between variables when there is left (below) or right (top) censoring in the dependent variable [16, 17]. Suppose in this study respondents express their WTP (y^*) with a monetary term that zero is the minimum threshold and value below than zero is a censored observation. Then, zero expenditure can be interpreted as a left-censored variable that

Table 1 Tobit analysis results of WTP between sociodemographic and psychological variables

Variables	WTP*	SD	WTP = y > 0	WTP = y*
Number of children				
1 child	**21.78****	**8.53**	**13.04**	**17.64**
2 children	6.20	9.39	3.26	4.59
3 and more children	**23.92****	**8.84**	**14.56**	**19.56**
Age groups				
25–34 years old	**32.28****	**10.81**	**15.04**	**21.05**
35–44 years old	**36.41****	**11.86**	**17.62**	**24.51**
45 years old and older	**40.84****	**13.59**	**20.57**	**28.36**
Income level				
Middle	**−22.42****	**11.57**	**−13.73**	**−18.42**
High	**−22.18***	**13.29**	**−13.61**	**−18.25**
Psychological constructs				
NEP	**11.10***	**6.22**	**6.30**	**8.71**
Constants	−22.81	34.90		

Notes * and ** denote 0.10 and 0.05 significance levels, respectively
y > 0 = truncated WTP; y* = uncensored WTP

equal to zero when $y^* \leq 0$; while the model only observes y* when y* > zero expenditure. This model chosen due to the nature price of a product in the market, which is impossible to have below zero value, also to observe the robustness of the results and to achieve the maximization model by considering the truncated expenditure. Below is the Tobit model specified as an unobserved latent WTP* (y*):

$$(\boldsymbol{WTP*}) = \boldsymbol{X}_0'\beta_0 + \boldsymbol{X}_1'\beta(\text{psychological}) \\ + \boldsymbol{X}_2'\beta(\text{sociodemographic profiles}) + \varepsilon$$

where;

$$WTP* = \begin{cases} y* \; if \; WTP > 0 \\ 0 \quad Otherwise \end{cases}$$

Diverse sets of variables have been used as basis to determine the pro-environmental behaviour in the consumer studies [18–21]. It is found that the used of socio-demographic variables to predict pro-environmental behaviour had produced either very weak indicators or caused conflict in nature [22–24]. Thus, some studies implied psychological variables such as environmental theory which believed to explain better the pro-environmental behaviour than socio-demographic [25–27]. On the other hand, using only psychological variables in the prediction model also shown inconsistent and many contradictory results [14, 15, 28, 29]. Therefore, it is suggested that using a combination of psychological and socio-demographic variables significantly will generate robustness and better insight for the consumers' WTP of the new material as shown in the model specification above. Using STATA MP 16, the proposed model specification estimation, two post-estimations analysed to predict how marginal changes in explanatory variables will affect the WTP when being uncensored and truncated. The derivations given as follows [17]:

1. The changes in unconditional of WTP being uncensored (at 0): $(\partial E(WTP*|\partial x))$
2. The changes in conditional of WTP being truncated (at 0): $(\partial E(WTP|x, WTP > 0)/\partial x)$.

3 Results and Discussion

The regression model in Table 1 demonstrated statistically significant at 0.05 significant level with log-likelihood of—723.51 and estimated standard error (sigma) of 990.51. Overall, the model presented those participants of this present study were willing to spend approximately RM24.78 on average with minimum WTP of—RM27.32 and maximum WTP of RM70.21 in Malaysian Ringgit for the bio-based food packaging.

3.1 Number of Children

Results indicated that households that has a child are willing to spend RM21.78 compared to the household with no children. Two marginal effects coefficient estimates in the last two columns of Table 1 showed the probability of marginal impact on the expected value of WTP given by the one-unit changes in the variable when the WTP is being censored (y > 0) and when WTP is being uncensored (y*). A unit change of the 'a child' construct will result in a positive WTP of RM13.04 (y > 0) and RM17.64 (y*) increase in comparison with 'no child' household. Kernel density graph in Fig. 1 proved that on average household with a child presence would willing to spend RM37.43 for that new packaging which is RM18.81 extra from the household with no children.

Similarly, the results also showed that the more children a household have, the more willing the parents would spend the money to buy this environmentally friendly packaging. A significant result of 3 and more children construct indicated that participants willing to pay RM23.92 if compared to no children household. Moreover, the marginal coefficients demonstrated RM14.56 (y > 0) and RM19.56 (y*) increase when one unit change of the construct. The distinct comparison seen in Fig. 1 where the more children presence in the household, the WTP is increasing on average of RM40.12 if compared to parents with no children presence at home.

In Asia, many households are built up by marriage institutions where when someone married, he or she advocates the active role in displaying environmental

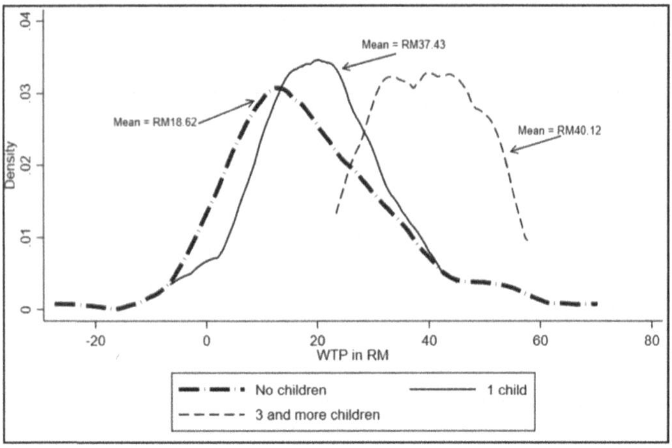

Fig. 1 Differences of WTP between household composition: number of children

behaviour, especially female of the family [30, 31]. This probably they want to demonstrate a good role model for the children and as part of environmental education at home to shape the children to protect the environment [15].

3.2 Age

All three age groups are shown substantially increasing as age increased among the participants. Young age group, between 25 and 34 years old, is willing to give up RM32.28 for the packaging if compared to 18–24 years old participants with marginal effects of RM15.04 ($y > 0$) and RM21.05 (y^*) of probability one unit change. Similarly, two other age groups, middle-aged participants (35–44 years old) and the elder participants (45 years old and above) also showed increasing trend of WTP at RM36.41 and RM40.84, respectively. The results also predicted that if one unit change, they also willing to spend extra for the bio-based packaging material at RM17.62 and RM20.57 of censored WTP and RM24.51 and RM28.36 for uncensored WTP as shown in Table 1.

Figure 2 demonstrated the average of WTP among four age groups. It clearly indicates the increasing WTP from young participants to older participants. Young participants were willing to spend RM13.69, and participants who aged 25–34 years old agreed on RM22.74 of their weekly grocery budget. Meanwhile, much older participants were set to give up RM30.03 for middle-aged group and RM32.90 for elderly group on that sustainable packaging. A study in Vietnam stated that young consumers perceived eco-friendly product as high in price, insufficient availability in Asian market, poor labelling and required extra time to pay attention to it which probably make them reluctant to spend more penny to purchase even though they

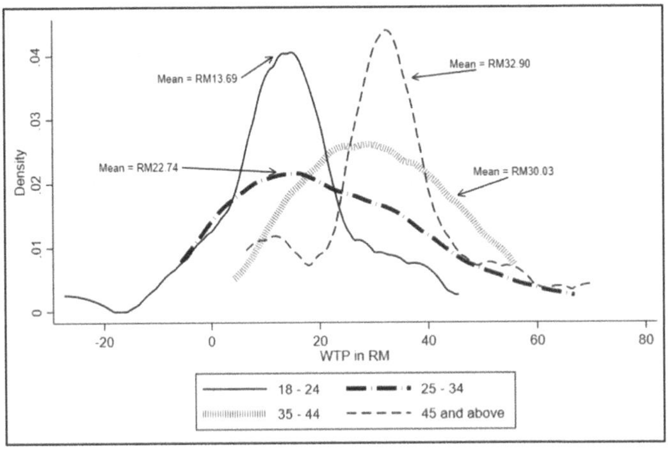

Fig. 2 Differences of WTP between age groups

are concern on the environment health [32]. On the other hand, much experiential necessities are needed to obtain prior make decision [33], where the older participants presumably are emotionally satisfied when making choice to buy the new bio-based food packaging during the contingent valuation exercise.

3.3 Income Level

Both income groups are shown a significant difference of WTP in comparison with low-income participants as based group. The results are showing an interesting outcome where both middle- and high-income groups indicated a negative sign of WTP in Table 1. Despite of their better financial condition, middle income participants are unwilling to spend—RM22.42 on this new bio-based packaging, the WTP of one unit change is also at—RM13.73 of positive WTP (y > 0) and declined by RM4.69 for uncensored WTP (y*). Equally, the high-income group also are unwilling to pay the new packaging at WTP—RM22.18 if compared to the low-income group. If one-unit changes in that income group, the WTP will decrease at WTP—RM13.61 (y > 0) and at—RM18.25 (y*).

These interesting finding indicated that having a high financial position will not guarantee someone will portray their environmental concern behaviour by purchasing this packaging. In contrary, study on a food organic product in Germany showed that highest expenditure power of a person has, the more pro-environmental attitudes and preferences for sustainable product [34]. Two similar trends were found in Asian studies recently where that the higher the income, the more involvement someone has towards saving the environment and higher the frequency of buying eco-friendly food products [35, 36]. However, studies in New Zealand and the US were supported

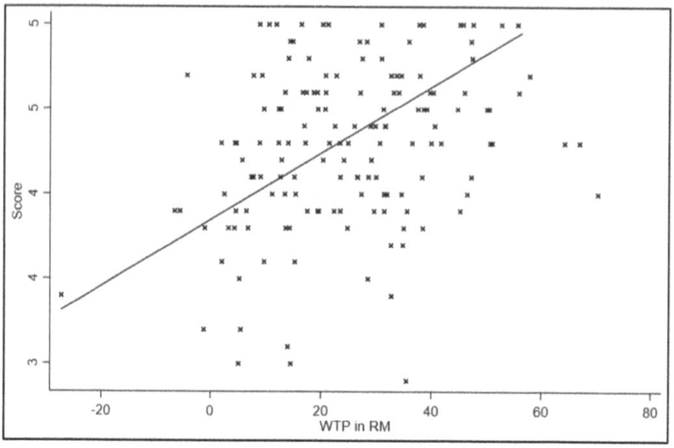

Fig. 3 WTP of NEP construct

present study outcome which low-income consumers were more likely to buy sustainable food product [37]. As this study focus on the new sustainable food packaging material, no such study found in the literature that solely emphasis on the packaging material and the finding suggests that income level does significantly impact the purchasing decision not be influenced by the higher the income, the more spending power on a sustainable product.

3.4 Psychological Constructs: NEP

In Table 1, the significant result of WTP of NEP construct seems to be consistent with the hypothesis of the model, where the WTP is increased at RM11.10 and if a unit change of NEP score, the WTP will change by RM6.30 for truncated WTP and RM8.71 when the data being uncensored. Particularly, Fig. 3 revealed that the WTP is positively inclining to the right when the NEP score increased. This result demonstrated that consumers enact their profoundly held environmental ideological convictions through their purchasing decisions and that affection plays an important role impacting their choices [33]. Participants with high environmentalism feelings tend to behave in such a way that matches their inner values.

4 Conclusion

The main purpose of this study was to examine the impact of socio-demographic and psychology aspects on a consumer's pro-environmental behaviour of purchasing state-of-the-art bio-based food packaging material as the alternative of fossil fuel plastic. The findings recommend that essential for formulating a compelling marketing plan for this packaging and an imperative basis of segmentation of Malaysian consumers. Heterogeneity in participants' WTP based on their household composition, age groups, income and environmental awareness are the segments that marketing strategies in need to promote that new sustainable packaging and refusal to accept the packaging can be expected in the future. Moreover, even though the results show a positive acceptance of the participants, but it appears that additional knowledge is likely to be embedded into the packaging to enhance the understanding of the consumers on the new food packaging. The information on the benefits of the packaging and how it will save the environment would well-informed the consumer and attract more of them to choose food products packed with the bio-based material. At the same time, the introduction of the packaging would be gradually implemented into the market to avert the drastically change on consumer's weekly budget.

References

1. Zainu, Z.A., Songip, A.R.: Policies, challenges and strategies for municipal waste management in Malaysia. J. Sci. Technol. Innovation Policy **3**(1), 18–22 (2017)
2. Ibrahim, N.R., Mat Noordin, N.N.: Understanding the issue of plastic waste pollution in Malaysia: a case for human security. J. Media Inf. Warfare **13**(1), 105–140 (2020)
3. Zailani, S., Shaharudin, M.R., Govindasamy, V., Ismail, M., Shaikh Mahdzar, S.: The eco-efficiency practices of the sustainable packaging and its effect towards sustainable supply chain performance. In: International Symposium on Technology Management and Emerging Technologies (ISTMET). Langkawi, Kedah, Malaysia (2015)
4. James, K., Fitzpatrick, L., Lewis, H., Sonneveld, K.: Sustainable packaging system development. In: Filho, W.L. (ed.) Handbook of Sustainability Research, pp. 371–407. Peter Lang Scientific Publishing, Frankfurt (2005)
5. Rattana, S., Gheewala, S.H.: Environment impacts assessment of petroleum plastic and bioplastic carrier bags in Thailand. J. Sustain. Energy Environ. **10**, 9–17 (2019)
6. Abdul Latif, N.S., Ong, M.Y., Nomanbhay, S., Salman, B., Show, P.L.: Estimation of carbon dioxide (CO_2) reduction by utilisation of algal biomass bioplastic in Malaysia using Carbon Emission Pinch Analysis (CEPA). Bioengineered **11**(1), 154–164 (2020)
7. Chaisu, K.: Bioplastic industry from agricultural waste in Thailand. J. Adv. Agric. Technol. **3**(4), 310–313 (2016)
8. Farrelly, T.A., Borrelle, S.B., Fuller, S.: The strengths and weaknesses of Pacific Islands plastic pollution policy frameworks. Sustainability **13**(1252), 1–36 (2021)
9. Orzan, G., Cruceru, A.F., Balaceanu, C.T., Chivu, R.G.: Consumers' behaviour concerning sustainable packaging: an exploratory study on Romanian consumers. Sustainability **10**(1787), 1–11 (2018)
10. Stern, P., Dietz, T., Abel, T., Guagnano, G.A., Kalof, L.: A value-belief-norm theory of support for social movements: the case of environmentalism. Hum. Ecol. Rev. **6**(2), 81–97 (1999)
11. Bateman, I.J., Carson, R., Day, B., Hanemann, M., Hanley, N., Hett, T., Lee, M.J., Loomes, G., Mourato, S., Ozdemiroglu, E., Pearce, D.W.: Economic Valuation with Stated Preference Techniques: A Manual. Edward Elgar, Cornwall (2002)
12. Albayrak, T., Caber, M., Moutinho, L., Herstein, R.: The influence of scepticism on green purchase behaviour. Int. J. Bus. Soc. Sci. **2**(13), 189–197 (2011)
13. Cleveland, M., Kalamas, M., Laroche, M.: Shades of green: linking environmental locus of control and pro-environmental behaviours. J. Consum. Mark. **22**(4), 198–212 (2005)

14. Paco, A., Raposo, M.: Green' segmentation: an application to the Portuguese consumer market. Mark. Intell. Plann. **27**(3), 364–379 (2009)
15. Patel, J., Modi, A., Paul, J.: Pro-environmental behaviour and socio-demographic factors in an emerging market. Asian J. Bus. Ethics **6**, 189–214 (2017)
16. Tobin, J.: Estimation of relationships for limited dependent variables. Econometrica **26**(1), 26–36 (1958)
17. Cameron, C.A., Trivedi, P.K.: Microeconometrics Using Stata, 2nd edn. Stata Press, Texas (2010)
18. Martin, V., Weiler, B., Reis, A., Dimmock, K., Scherrer, P.: 'Doing the right thing': how social science can foster pro-environmental behaviour change in marine protected areas. Mar. Policy **81**, 236–246 (2017)
19. Wi, A., Chang, C.H.: Promoting pro-environmental behaviour in a community in Singapore—from raising awareness to behavioural change. Environ. Educ. Res. **25**(7), 1019–1037 (2019)
20. Amoah, A., Addoah, T.: Does environmental knowledge drive pro-environmental behaviour in developing countries? Evidence from households in Ghana. Environ. Dev. Sustain. **23**, 2719–2738 (2021)
21. Wang, X., Van der Werff, E., Bouman, T., Harder, M., Steg, L.: I am versus we are: how biospheric values and environmental identity of individuals and groups can influence pro-environmental behaviour. Front. Psychol. **12**(618956), 1–11 (2021)
22. Shrum, L.J., Lowrey, T.M., McCarty, J.A.: Recycling as marketing problem: a framework for strategy development. Psychol. Mark. **11**(4), 393–416 (1994)
23. Klineberg, L.S., McKeever, M., Rothenbach, B.: Demographic predictors of environmental concern: it does make a difference how it's measured. Soc. Sci. Q. **79**(4), 734–753 (1998)
24. Schultz, P.W., Zelezny, L.: Values as predictors of environmental attitudes: evidence for consistency across 14 countries. J. Environ. Psychol. **19**(3), 255–265 (1999)
25. Wagner, S.A.: Understanding Green Consumer Behaviour: A Qualitative Cognitive Approach. Routledge, London and New York (1997)
26. Straughan, R.D., Roberts, J.A.: Environmental segmentation alternatives: a look at green consumer behaviour in the new millennium. J. Consum. Mark. **16**, 558–575 (1999)
27. Paul, J., Modi, A., Patel, J.: Predicting green product consumption using theory of planned behaviour and reasoned action. J. Retail. Consum. Serv. **33**, 198–202 (2016)
28. Getzner, M., Grabner-Krauter, S.: Consumer preferences and marketing strategies for 'green shares. Int. J. Bank Mark. **22**(4), 260–278 (2004)
29. Steg, L., Vlek, C.: Encouraging pro-environmental behaviour: an integrative review and research agenda. J. Environ. Psychol. **29**(3), 309–317 (2009)
30. Sakellari, M., Skanavis, C.: Environmental behaviour and gender: an emerging area of concern for environmental education research. Appl. Environ. Educ. Commun. **12**(2), 77–87 (2013)
31. Gronhoj, A., Thogersen, J.: Why young people do things for the environment: the role of parenting for adolescents' motivation to engage in pro-environmental behaviour. J. Environ. Psychol. **54**, 11–19 (2017)
32. Pham, T.H., Nguyen, T.N., Phan, T.T.H., Nguyen, N.T.: Evaluating the purchase behaviour of organic food by young consumers in an emerging market economy. J. Strateg. Mark. **27**(6), 540–556 (2018)
33. Joshi, Y., Rahman, Z.: Consumers' sustainable purchase behaviour: modelling the impact of psychological factors. Ecol. Econ. **159**, 235–243 (2019)
34. Schaufele, I., Hamm, U.: Organic wine purchase behaviour in Germany: exploring the attitude-behaviour-gap with data from a household panel. Food Qual. Prefer. **63**, 1–11 (2018)
35. Dangi, N., Gupta, S.K., Narula, S.: Consumer buying behaviour and purchase intention of organic food: a conceptual framework. Manage. Environ. Qual.: Int. J. **6**(31), 1515–1530 (2020)
36. Nam, S.J.: Moderating effects of consumer empowerment on the relationship between involvement in eco-friendly food and eco-friendly food behaviour. Int. J. Consum. Stud. **44**, 297–305 (2020)
37. Jinghan, L., Zepeda, L., Gould, B.W.: The demand for organic food in the U.S.: an empirical assessment. J. Food Distrib. Res. **38**(3), 54–69 (2007)

Environmental, Social and Governance (ESG) Activity and Corporate Controversies in South Africa: The Interacting Role of a Skillful Board

Ahmad Al-Hiyari, Abdulsalam Mas'ud, and Mohamed Chakib Kolsi

Abstract Corporations have been involving in controversies through violation, unhealthy competitions, negligent behaviors, and scandals which result in huge financial and non-financial losses. Therefore, this paper investigates the nexus between ESG performance and corporate controversies and the moderating effects of the skillful board on this nexus with respect to public firms in South Africa. Using a sample of 958 firm-years observations with available ESG data from the Thomson Reuters Eikon database as listed in the Johannesburg Stock Exchange (JSE) during the period 2010–2019, we find that ESG performance practices are positively and significantly related to corporate controversies, indicating the likelihood of using ESG practices as 'greenwashing mechanism' by managers which eventually fuels corporate controversies instead of mitigating them. Much more interesting, the result revealed that the existence of board members with specific skills relating to ESG in the corporate boards would likely weaken attempts of managers in the opportunistic implementation of ESG practices, thereby reducing the corporate controversies. Our results have interesting implications for stakeholders, regulators, and scholars who may consider ESG as a shield used opportunistically by managers to cover up irresponsible ESG activities. In addition, the empirical findings highlight the importance of the skillful board members in moderating the ESG—corporate controversies nexus.

Keywords ESG practices · Corporate controversies · Board specific skills

---

A. Al-Hiyari (✉) · M. C. Kolsi
Department of Accounting and Finance, Liwa College of Technology, Abu Dhabi, UAE
e-mail: Ahmed.alhiyari@ect.ac.ae

A. Mas'ud
Department of Taxation, Federal University Dutse, Dutse, Jigawa State, Nigeria

A. Hamdan et al. (eds.), *The Implementation of Smart Technologies for Business Success and Sustainability*, Studies in Systems, Decision and Control 216,
https://doi.org/10.1007/978-3-031-10212-7_29

1 Introduction

In recent years, there have been growing concerns over corporate controversies through inactions or incidents involving firms. The environmentally related corporate controversy involving the Deepwater Horizon oil platform explosion in the Gulf of Mexico in 2010 is an example. This was an incident that caused a massive oil spill, and reparations costing more than USD 40 billion [25]. There are also growing ethically related controversies such as impositions of fines due to socially irresponsible investing, malpractices and violations. For instance, there is an imposition of a fine of more than €8bn on Google in relation to anti-competitive practices in the EU, and also a $5bn fine that could possibly be imposed on Facebook in relation to privacy violations [8]. In a broader governance perspective, an analysis involving 24 companies performed by Bank of America revealed more than $500bn was wiped off the value of US large companies in the last five years due to certain controversies [8]. These controversies revolve around accounting scandals, breaches of data policies, sexual harassments and other ESG issues, which eventually resulted in market value losses of $534bn to the affected companies compared to their counterparts in S&P 500. As a result, it is unarguable that corporate controversies jeopardize firm's reputation and could negatively affects its performance [34]. In fact, these series of corporate controversies are gradually becoming important concerns for institutional investors and asset managers due to mounting fears of climate change, conspicuous scams and injurious corporate governance failures [8]. It is arguable that these events could be mitigated through sound ESG practices. In fact, ethical theories postulate that managers can come up with morally sound approaches that work in the best interest of the business, thereby preventing it from unforeseen corporate controversies [4, 6, 19].

In line with the above background, we argue that an important link has been neglected in the literature—that is, the ESG performance practices and corporate controversies relationship. Corporate controversies are firms' unethical activities that have negative environmental, social and governance effects on stakeholders. These controversies arise through the involvement of a firm in an incidents or actions in relation to environmental, social, and corporate behavior [23]. They could be news stories relating to suspicious social behavior and environmental scandals which put a firm under the media spotlight that will eventually attract the attention investors [2]. Eventually, these controversies may jeopardize a company's reputation and by extension impact negatively on its performance [2, 12]. In relation to this, there are two links to be studied here-first, the impact of ESG practices on these corporate controversies, and then that corporate controversies and firms' performance. While the later has been studied [26], the former remains unexamined, which highlights the relevance of this study.

The ESG performance and corporate controversies relationship can be explained by the ethical and opportunistic theories. On the one hand, ethical theories demonstrates that managers can come up with morally sound approaches that work in the best interest of the business, in this, ESG practices could mitigate corporate controversies [21]. Under this scenario, managers pursue ESG practices in such a way to prevent a company from social, environmental, and corporate scandals or controversies. The prior research has revealed that firms with better ESG performance provide more transparent and reliable accounting information and exhibit a decline in information asymmetry [1]. On the other hand, the agency theory [16] demonstrates that managers may use ESG activities as a greenwashing policy to hide corporate misbehaviour or to legitimise their social image and reputation in the eyes of stakeholders. Therefore, whether ESG practices lead to reduced corporate controversies is still an open empirical question.

Prior research provides evidence that corporate governance plays a role in mitigating managers' irresponsible ESG activities [11]. In this study, we hypothesize that board members with specific skills can moderate the ESG—corporate controversies nexus. Skillful board members are those with specific skills, knowledge and experience that make a company more effective with potential to stimulate the ESG Performance-Corporate controversies relationship. In this context, Dass et al. [5] argue that directors with specific skills can bring industry knowledge, expertise, and connections and therefore improve the quality information to the boardroom. Consequently, these directors provide better strategic advice as well as monitor top executives more effectively. Consistent with this argument, Dass et al. [5] find clear evidence that directors with specific skills have favorable effects on firm value, especially when the information environment is of low quality. They also highlight the importance of skillful board members in enhancing the ability of firms to effectively manage industry shocks and minimizing their cash gap in the cash conversion cycle. Based on resource dependency theory [28], firms are likely to benefit from the presence of skillful board members by seeking their advice to mitigate corporate controversies and accessing to relevant information about social impacts, good governance and environmental practices, new eco-friendly technologies, etc. (Homroy and Slechten 14). However, despite a sizable body of research conducted on the relationship between board characteristics and corporate sustainability practices, relatively limited attention has been devoted to examining the effects of board specific skills on corporate ESG practices [10]. Therefore, we wonder whether the presence of skillful board members can moderate the ESG—corporate controversies relation.

Our paper contributes to the literature in several ways. First, the study was conducted using data from South Africa, an emerging market. This bridge an important gap in the literature. The fact is that studies on ESG are much more prevalent from developed countries. In relation to this, Friede et al. [9] argued that ESG is not only popular in developed economies, ESG has strong implications to the emerging markets as investment in ESG creates positive outcomes through returns in emerging markets. While the aforesaid can serve as a justification for the selection of South Africa, it can further be justified from the three perspectives of ESG controversies. In terms of environment, the 2019 Greenpeace report ranked South Africa among the highest in the world for emissions of two dangerous pollutants, sulfur dioxide and nitrogen dioxide which is 10 times higher than those proposed by the World Health Organization's (WHO) guidelines as safe for human health [35]. From the social perspective also, the environmental problems have cause hundreds of early deaths every year and eventually fueling corporate controversies including lawsuit against the companies concerned [35]. From the governance perspective, South Africa has experience series of financial scandals among its private sector companies through inflation of their profits and assets, which amounted to over R250 billion, thus, demonstrating a clear lack of ethics, and was rated among the biggest business wrongdoing in African continent [3]. Second, we understand the existence of paucity of evidences that link ESG practices and corporate controversies despite theoretical insights. Towards this end, we study the relationship between ESG performance practices and corporate controversies from array of different theoretical frameworks. By considering different insights of these theories for the aforesaid relationship, we gain wider and objective view in which ESG practices can either mitigate or amplify corporate controversies. Finally, due to divergent views—either ESG practice to mitigate corporate controversies as postulated by ethical theories [21]—or amplify them as highlighted by agency theory [16], we understand the need for a variable that will exact a stimulating role in the relationship—either to strengthen or weaken the initial direct relationship. In this, we are motivated to examine moderating role of skillful board on the proposed direct relationship.

2 Hypotheses Development

Shakil [30] had earlier posited that a good disclosure of ESG can improve the company's reputation, which invariably can save firms from controversies and scandals. Interestingly, recent evidences from practitioners' perspective revealed that ESG expertise is now becoming a criterion for nomination into board of directors [22]. A survey showed that directors focus much on ESG in their boards' decisions—indicating that 80% of the directors reported that their boards are focusing much on ESG [15]. This survey further documented that 78% of a selected S&P 100 proxy statements revealed that companies appoint at least one board committee to be responsible for ESG matters. It further revealed that 42% of companies have at least one director with ESG expertise. These are made in such a way to mitigate risks and take advantages of opportunities [22]. Therefore, the emerging evidences highlighted above indicate the possibility for ESG performance practices to mitigate corporate controversies.

Moreover, emerging empirical evidences further highlight the likelihood of ESG practices to mitigate corporate controversies. For instance, Utz [32] assessed the reliability of ESG assessments in the case of corporate scandals using three different scenarios; before, during and after scandals. The study established that positive inverse relationship between retrospective indicators of ESG and corporate scandals. Other empirical evidences deployed corporate controversy as an independent variable that explains either firm's performance [26] or firm value [2]. Another study explored the moderating effect of corporate controversies in the relationship between board gender diversity and firms' performance, though the effect was insignificant

[31]. In a more recent study, Shakil [30] examined the moderating effect of ESG corporate controversies in the relationship between on ESG performance and financial risk and found a significant moderating effect of ESG controversy on the aforesaid relationship. It is evident that all these studies examined corporate controversy either as an independent variable or as a moderator variable, while none explored the variable as a dependent variable. Therefore, to the best of the researchers' knowledge only one study considered corporate controversy as dependent variable, event in that, only single dimension of the controversy was examined—environmental controversy [24]. Specifically, the study of Marsat et al. [24] examined the effect of environmental performance practices on environmental controversies, however, it focused only on a single dimension of ESG, neglecting the other two important aspects of social and governance. Thus, implying the need to examine the composite effect of ESG performance practices on corporate controversies.

In addition to practical and empirical support on the effect of ESG performance practices on corporate controversies, there is also theoretical support from number of theories which highlight that ESP performance practices are related with corporate controversies, however, in divergent ways. In one hand, the 'ethical theories' signal a possible negative relationship between ESP performance practices and corporate controversies. On the other hand, the agency theory [16] highlights contrasting position, which signals a positive relationship. This means that increasing ESG activities could fuel corporate controversies due to opportunistic behaviors of managers. Therefore, based on the two possible directions of ESG practices and corporate controversy relationship, a non-directional hypothesis is formulated as follows:

Hypothesis 1: There is relationship between ESG performance practices and corporate controversies.

Johnson et al. [18] posited that board specific skills refer to the human and social capital that stems from the education, knowledge, and experience acquired outside the firms by certain members of the board directors. Hillman et al. [13] further clarified that board of directors bring important resources to the firms including expertise, access to business stakeholders and important information. On the basis of their various roles, Hillman et al. [13] provide four classifications of board of directors. Firstly, there are inside directors which provides firms with information relating to company finance and regulation, formulate strategies and lead the firm in decision making. Secondly, there are also directors with business expertise, who are either present, or former top managers in larger profit-oriented firms or top analyst of these firms. This category provides firm with guidelines on strategy making, problem solving and professional advice. Thirdly, there are also directors with specific skills who support the firms based on their respective specific fields such as bankers, lawyers, actuaries, environmentalists, and others with expertise in public relations among others. Finally, there are also directors drawn from influential members of community such as faculty members from universities, religious, political or social leaders. In this study, skillful board refer to the third category which means members with specific skills in the board relative to ESG practices, which now becomes an important criterion for nomination into board of directors [22].

The relevance of skillful board in resolving corporate controversies has been highlighted by resource dependency theory [17]. The theory highlights independent directors with specific skills on board can provide firm with more assistance in gaining the desirable resources. They highlight that an outside director with low specific skill having being practice in a law firm could offer legal advises during board meetings with executive directors which can save firm from controversies. Specifically, resource dependence theory argues that existence of directors with specifics skills in the boards could increase firms' accessibility to resources, which essence will strengthen the operation of the company and save it from cooperate controversies.

Drawing from the earlier theoretical insights which postulate the relationship between ESP performance practices and corporate controversies, we can recall that divergent postulations were found from the theories. While 'ethical theories' signal a possible negative relationship, the agency theory signals a positive relationship. On the premise of this divergent possibilities, we proposed the examination of the moderating effect of skillful board. We predict that when ESG practices mitigate corporate controversies, the existence of skillful board members will support such mitigation effort, thereby by strengthening the ESG Performance Practice-Corporate Controversies relationship. On the other hand, when ESG performance practices further amplify corporate controversies due to opportunistic behaviors of managers as highlighted by agency theory [16], the presence of skillful board members will mitigate such adverse effect, thereby weakening the adverse effect of ESG practices (via opportunistic behavior) on the corporate controversies. This means that the relationship could be weaker where managers deployed ESG practices opportunistically as "greenwashing" mechanism which eventually can further amplify corporate controversies, or that the relationship could stronger where managers come up with morally sound approaches to pursue ESG practices to mitigate corporate controversies. These arguments lead us to develop the following hypothesis.

Hypothesis 2: The relationship between ESG performance practices and corporate controversies can be moderated by skillful board.

3 Research Design

3.1 Data and Sample Construction

The final sample consists of 958 observations from 2010 to 2019. Of the 958 observations, 204 (21.29% of the sample) have ESG issues while 754 (78.71% of the sample) do not have ESG issues. Financial, ESG, and board characteristics data have been obtained from Refinitiv.

3.2 Variable Design and Measurement

Dependent and Independent Variables
Consistent with previous studies, we obtained data about ESG issues from the Thomson Reuters Asset4 database. Thomson Reuters ASSET4 classifies ESG controversies into 23 different topics, which covers environmental, social and governance issues. In order to have comparable data on ESG controversies, we follow Aouadi and Marsat [2] and assign a binary variable (ESGCON) that equals to one if a firm experience at least one controversy, and zero otherwise. For independent variable, we also use the ASSET4 ESG database to obtain data on ESG scores. The score is ranged between 0 and 100, with a higher score indicating excellent CSR practices.

3.3 Regression Model and Specification

We employed fixed-effects panel-data estimator to rigorously test the hypotheses developed in this study. The Hausman (12) test confirms that the fixed effects estimation is best fitted to our panel dataset. Thus, we examine the hypothesized relationship between the independent variables and dependent variable by estimating a fixed-effects model that is outlined as follows:

$$\begin{aligned} \mathrm{ESGCON_{it}} &= \beta_0 + \beta_1 \mathrm{ESGSCORE_{it}} \\ &\quad + \beta_3 \sum \mathrm{Control} + \mathrm{Fixed\ effects} + \varepsilon_{it} \end{aligned} \tag{1}$$

Table 1 displays the variable definition. The above equation is estimated using probit model because our dependent variable is constructed as dichotomous variable that represents whether the firm faces at least one controversy or not, as in Aouadi and Marsat [2]. To test our second hypothesis on the moderating role of board skills on the nexus between ESG performance and corporate controversies, we introduce the interaction term between ESGSCORE and BRDSKILL in the regression model as follows:

$$\begin{aligned} \mathrm{ESGCON_{it}} &= \beta_0 + \beta_1 \mathrm{ESGSCORE_{it}} + \beta_2 \mathrm{BRDSKILL_{it}} \\ &\quad + \beta_3 \mathrm{ESGSCORE} \times \mathrm{BRDSKILL_{it}} \\ &\quad + \beta_4 \sum \mathrm{Control} + \mathrm{Fixed\ effects} + \varepsilon_{it} \end{aligned} \tag{2}$$

Table 1 Variable's description and measurements

Variable	Measure
CONT	A dichotomous variable equal to 1 if firm faces at least one controversy during the year t; and 0 otherwise
ESGSCORE	Thomson Reuters score for environmental, social, and governance performance. The score has a minimum value of 0 and a maximum of 100
BRDSKILL	Percentage of board members who have either an industry specific background or a strong financial background
BRDIND	The number of independent non-executive directors divided by the total number of board members
SEPCHAIR	A dichotomous variable equal to 1 if the roles of chairman and CEO are separate, and 0 otherwise
BRDSIZE	The natural logarithm of total number of directors
BRDMEETING	The natural logarithm of the number of meetings during the financial year
SIZE	The natural logarithm of total assets
LEV	Total liabilities/total assets
GROWTH	The percentage change in total sales over the previous year
ROA	Net income before extraordinary items/total assets
DY	Cash dividend per share/market price per share * 100
R&D	R&D expense/total sales

4 Empirical Results

4.1 Multivariate Analyses

Table 2 contains the results of estimating models (1) and (2) using the fixed effects probit regression, with standard errors clustered at the firm level. The models are significant and the adjusted R-square ranges from 24.55 to 25.08%. This implies that the model displayed a satisfactory fit with the data utilized. In model (1), we regress ESG performance practices on corporate controversies without considering the moderating variable. In line with the predication of the opportunistic theory, the results show that the estimated coefficient of ESGSCR is positive and statistically significant at conventional levels ($\beta = 0.011$, $p = 0.031$), indicating that firms are more likely to engage in ESG practices to mask their unethical behavior like toxic waste spills (environmental), human right violations (social) accounting frauds (governance). Therefore, ESG can be employed as a 'greenwashing mechanism' to mitigate investors' activism, thereby mitigating the undesirable effects of these unethical ESG activities. The result supports the studies of Ferrero and Sánchez [7], Prior et al. [29] and Velte (33), who demonstrated how the firm's managers can take advantages of ESG practices for their own private objectives at the expense of shareholders. Overall, consistent with H1, there is a positive relationship between sustainability practices and corporate controversies.

Table 2 Regression analysis

Variable	Model (1)		Model (2)	
	Coeff	*p*-value	Coeff	*p*-value
Intercept	−6.201	0.000***	−7.127	0.000***
ESGSCR	0.011	0.031**		
BRDSKILL			−0.037	0.083*
ESGSCORE × BRDSKILL			−0.008	0.001***
CEODUL	0.362	0.225	0.315	0.285
BRDIND	−0.013	0.076*	−0.011	0.046**
BRDMEETING	1.055	0.000***	1.090	0.000***
BRDSIZE	0.562	0.186	0.642	0.123
LEV	−0.064	0.010**	−0.061	0.001***
FIRMSIZE	0.282	0.003***	0.282	0.000***
GROWTH	−0.026	0.540	−0.026	0.551
ROA	−0.942	0.031**	−0.944	0.001***
DY	−0.006	0.850	−0.001	0.959
R&D	0.095	0.883	0.139	0.824
IMR	0.085	0.138	0.082	0.170
Year effect	Included		Included	
Industry effect	Included		Included	
Adjusted R^2 (%)	24.55		25.08	
Sample period	2010–2019		2010–2019	
Controversy firm-years	204		204	
No-controversy firm-years	754		754	

Notes ***; **; * denote significance at the 0.01, 0.05, and 0.10 level, respectively
See Table 1 for variable definitions

In model (2), we present the results with the interaction effect of skillful board on the ESG performance–corporate controversies nexus. The results in Table 2 show that the coefficient of interactive variable ESGSCORE × BRDSKILL is overwhelmingly significantly negative ($\beta = -0.008108$, $p = 0.001$) which is consistent with H2. This result suggests that directors with specific skills weaken the positive association between sustainability practices and corporate controversies. The finding is consistent with resource dependency theory [17] that argued that independent directors with specific skills on board can provide firm with more assistance in gaining the desirable resources which can save firm from controversies. Specifically, the theory postulates that when firms appointed directors with specifics skills into their corporate boards, such increases their access to more resources including skills and expertise, and eventually strengthen its operation and save it from cooperate controversies. Thus, drawing from the first hypothesis, which predicts that

ESG practices mitigate corporate controversies based on the postulation of stakeholder theory [20] and resource dependency theory [28, 27], the existence of board members with specific skills will support such mitigation effort, thereby by strengthening the negative relationship (becoming more negative). Alternatively, when ESG performance practices further amplify corporate controversies due to opportunistic behaviors of managers based on the postulation of agency theory [16], the presence of board members with specific skill will weaken the initial direct positive relationship (becoming negative) following the intervention of the skillful board. Interestingly, the result supports the second postulation. It revealed that the existence of skillful board dampens the direct positive relationship between ESG performance practices and corporate controversies. Specifically, the result revealed that the existence of board members with specific skills in the board will reduce attempts by managers to deploy ESG practice as 'greenwashing mechanism' or to opportunistically deploy such practices to drive undue advantage at the expense of the corporations. This means that even in instances where managers deploy ESG practice as 'greenwashing mechanism' which could lead to corporate controversies, existence of skillful board will mitigate such controversies.

5 Conclusion and Implications

The paper deployed variety of corporate governance theories to examine the effects of ESG performance on corporate controversies from the perspective of an emerging market. It further explores how skillful board play a vital role within this framework by moderating the aforesaid relationship. Our empirical analysis is based on a sample of non-financial firms listed on the Johannesburg Stock Exchange (JSE) from 2010 to 2019. We use fixed-effects panel-data estimator to answer our research questions. We document a positive relationship between ESG performance practices and corporate controversies, which is consistent with opportunistic theory [16] that emphasized that managers use ESG practices as a greenwashing mechanism to extract private benefits. We also document that skillful board members negatively and significantly moderates the ESG–corporate controversies relationship. The results support resource dependency theory and explain that the presence of skillful board members enhances capabilities to deal with ESG issues. These finding possesses important implications to theory, practice and future research.

The results of this study have significant implications to stakeholders. For instance, capital providers, policy makers and regulators have serious concerns regarding corporate social and environmental issues, as more ESG issues can jeopardize firm's reputation, which ultimately harm its performance. This study provides evidence that ESG performance practices in South Africa are opportunistically used to hide corporate irresponsible ESG activities. Furthermore, the study supports the recent practical evidence by larger firms which revealed that ESG expertise is now becoming a criterion for nomination into board of directors. This further portray the relevance of board members with specific skills in corporate boards. For instance, a recent survey reported that 78% of a selected S&P 100 proxy statements revealed that companies appoint at least one board committee to be responsible for ESG matters [22]. This means that companies are now focusing on appointing members with specific skills relating to ESG into their boards. The survey further revealed that 42% of companies have at least one director with ESG expertise [22]. This development is made in such a way to mitigate risks, including corporate controversies. Consequently, the finding from this study imply to the firms that they are in the right track by appointing directors with specific skills relating to ESG into their corporate board as such will assist them to mitigate ESG related risk and controversies and deter managers from using ESG practices as 'greenwashing mechanism' or for opportunistic purposes.

References

1. Al-Hiyari, A., Kolsi, M.C.: How do stock market participants value ESG performance? Evidence from Middle Eastern and North African Countries. Glob. Bus. Rev. (2021)
2. Aouadi, A., Marsat, S.: Do ESG controversies matter for firm value? Evidence from international data. J. Bus. Ethics **151**(4), 1027–1047 (2018)
3. Business Insider: The biggest South African business scandals over the past decade (2020). https://www.businessinsider.co.za/the-top-south-african-business-scandals-the-past-decade-2020-1
4. Carroll, A.B.: A three-dimensional conceptual model of corporate performance. Acad. Manag. Rev. **4**(4), 497–505 (1979)
5. Dass, N., Kini, O., Nanda, V., Onal, B., Wang, J.: Board expertise: do directors from related industries help bridge the information gap? Rev. Financ. Stud. **27**(5), 1533–1592 (2014)
6. Donaldson, T., Preston, L.E.: The stakeholder theory of the corporation: concepts, evidence, and implications. Acad. Manag. Rev. **20**(1), 65–91 (1995)
7. Ferrero, J.M., Sánchez, I.M.G.: Is corporate social responsibility an entrenchment strategy? Evidence in stakeholder protection environments. RMS **9**(1), 89–114 (2014)
8. Flood, C.: ESG controversies wipe $500bn off value of US companies. Financ. Times (2019). https://www.ft.com/content/3f1d44d9-094f-4700-989f-616e27c89599
9. Friede, G., Busch, T., Bassen, A.: ESG and financial performance: aggregated evidence from more than 2000 empirical studies. J. Sustain. Finance Investment **5**(4), 210–233 (2015)
10. Gallego-Álvarez, I., Pucheta-Martínez, M.C.: Environmental strategy in the global banking industry within the varieties of capitalism approach: the moderating role of gender diversity and board members with specific skills. Bus. Strateg. Environ. **29**(2), 347–360 (2020)
11. Gerged, A.M., Albitar, K., Al-Haddad, L.: Corporate environmental disclosure and earnings management—the moderating role of corporate governance structures. Int. J. Fin. Econ. 1–22 (2021)
12. Hausman, J.: Specification tests in econometrics. Econometrica, **46**, 1251–1271 (1978)
13. Hillman, A.J., Keim, G.D., Luce, R.A.: Board composition and stakeholder performance: do stakeholder directors make a difference? Bus. Soc. **40**(3), 295–314 (2001). https://doi.org/10.1177/000765030104000304
14. Homroy, S., Slechten, A.: Do board expertise and networked boards affect environmental performance? J. Bus. Ethics. Springer Netherlands, **158**(1), 269–292 (2019)
15. Huennekens, K.: ESG Disclosure in 2020 Proxy Statements. Nasdaq (2020). https://www.nasdaq.com/articles/esg-disclosure-in-2020-proxy-statements-2020-05-13
16. Jensen, M.C., Meckling, W.H.: Theory of the firm: managerial behavior, agency costs and ownership structure. J. Financ. Econ. **3**(4), 305–360 (1976)

17. Johnson, J.L., Daily, C.M., Ellstrand, A.E.: Boards of directors in SMEs: a review and research agenda. J. Manag. **22**(3), 409–438 (1996)
18. Johnson, S.G., Schnatterly, K., Hill, A.D.: Board composition beyond independence: social capital, human capital, and demographics. J. Manag. **39**(1), 232–262 (2013). https://doi.org/10.1177/0149206312463938
19. Jones, T.M.: Instrumental stakeholder theory: a synthesis of ethics and economics. Acad. Manag. Rev. **20**(2), 404–437 (1995)
20. Jones, T.M., Wicks, A.C.: Convergent stakeholder theory. Acad. Manag. Rev. **24**(2), 206–221 (1999)
21. Kim, Y., Park, M.S., Wier, B.: Is earnings quality associated with corporate social responsibility? Account. Rev. **87**(3), 761–796 (2012). https://doi.org/10.2308/accr-10209
22. Lee, A.H.: Climate, ESG, and the board of directors: "You cannot direct the wind, but you can adjust your sails." (2021) https://www.sec.gov/news/speech/lee-climate-esg-board-of-directors
23. Li, J., Haider, Z.A., Jin, X., Yuan, W.: Corporate controversy, social responsibility and market performance: international evidence. J. Int. Finan. Markets. Inst. Money **60**, 1–18 (2019)
24. Marsat, S., Pijourlet, G., Ullah, M.: Does environmental performance help firms to be more resilient against environmental controversies? Int. Evid. Finan. Res. Lett. (2021)
25. Merry, M.K.: Framing Environmental Disaster: Environmental Advocacy and the Deepwater Horizon Oil Spill. Routledge (2013)
26. Nirino, N., Santoro, G., Miglietta, N., Quaglia, R.: Corporate controversies and company's financial performance: exploring the moderating role of ESG practices. Technol. Forecast. Soc. Change **162**(June 2020), 120341 (2021)
27. Pfeffer, J.: Size, composition, and function of hospital boards of directors: a study of organization-environment linkage. Adm. Sci. Q. **18**(3), 349–364 (1973). https://doi.org/10.4324/9780203337394-50
28. Pfeffer, J., Salancik, G.R.: The External Control of Organizations: A Resource Dependence Perspective. Harper & Row, New York, NY (1978)
29. Prior, D., Surroca, J., Tribó, J.A.: Are socially responsible managers really ethical? Exploring the relationship between earnings management and corporate social responsibility. Corp. Gov.: Int. Rev. **16**(3), 160–177 (2008)
30. Shakil, M.H.: Environmental, social and governance performance and financial risk: moderating role of ESG controversies and board gender diversity. Resour. Policy **72**(May), 102144 (2021). https://doi.org/10.1016/j.resourpol.2021.102144
31. Shakil, M.H., Tasnia, M., Mostafiz, M.I.: Board gender diversity and environmental, social and governance performance of US banks: moderating role of environmental, social and corporate governance controversies. Int. J. Bank Mark. (2020). https://doi.org/10.1108/IJBM-04-2020-0210
32. Utz, S.: Corporate scandals and the reliability of ESG assessments: evidence from an international sample. RMS **13**(2), 483–511 (2019)
33. Velte, P.: Environmental performance, carbon performance and earnings management: Empirical evidence for the European capital market, Corp. Soc. Responsib. Environ. Manag. **28** (1), 42–53 (2021)
34. Walsh, G., Mitchell, V.W., Jackson, P.R., Beatty, S.E.: Examining the antecedents and consequences of corporate reputation: a customer perspective. Br. J. Manag. **20**(2), 187–203 (2009)
35. Williams, C.: As South Africa Clings to Coal, a Struggle for the Right to Breathe. Opera News (2020). https://e360.yale.edu/features/as-south-africa-clings-to-coal-a-struggle-for-the-right-to-breathe?ct=t(RSS_EMAIL_CAMPAIGN

Modern Social Media, Public Relations, and Its Effects on Society and Business

The Influence of Visual Risk Communication on Community During the COVID-19 Pandemic: An Investigation of Twitter Platform

Haneen Mohammad Shoaib

Abstract Major events and unconventional circumstances impose a series of challenges on leaders around the world. The COVID-19 pandemic crisis which is characterized by constant change and improbability, forced the countries' leadership to ramp up risk communication efforts. Under this condition, social media platforms played a critical role for quick dissemination of information. Consequently, the current study aims to understand the role of the Saudi Monarchy King Salman bin Abdul-Aziz Al Saud on visual health risk communication during the COVID-19 pandemic through social media platforms like Twitter. It explores how virtual space builds an emotional connection to deal with the COVID-19 situation with the local and international community. The research incorporates a multidisciplinary perspective to understand the role of virtual space in constructing an emotional connection with the local and international community in the critical period of the COVID-19 pandemic. The study uses Lasswell's communication model and Goffman's dramaturgical framework to review and analyze the official Twitter account "@KingSalman" of King Salman bin Abdul-Aziz Al Saud. It explores how social media interaction supports the Saudi Monarchy's leadership role, comparing front and backstage scripts. The findings indicate that Saudi Monarchy leadership's influence on virtual space plays an essential part in coping with the COVID-19 crisis and building an emotional connection with the community. It supports enhancing the social capital through bonding, bridging, and linking across Saudi Arabia regions to fight against the Covid-19 pandemic. This study contributes to understanding leadership influence on virtual space and supports dealing with the COVID-19 crisis.

Keywords Leadership · COVID-19 · Social media · Monarchy-online · Visual risk communication · King Salman · Saudi Arabia

H. M. Shoaib (✉)
College of Business Administration, University of Business and Technology, Jeddah, Saudi Arabia
e-mail: haneen@ubt.edu.sa

A. Hamdan et al. (eds.), *The Implementation of Smart Technologies for Business Success and Sustainability*, Studies in Systems, Decision and Control 216,
https://doi.org/10.1007/978-3-031-10212-7_30

1 Introduction

The first case of Coronavirus Disease (COVID-19) was identified in December 2019 in Wuhan City, China [9, 30], and it spread around China and then around the world, encountering of several human lives, as well as the imposition of enormous social and economic interruption [38]. Considering the spread of the deadly virus on March 11th, 2020, "The World Health Organization (WHO) declared a Public Health Emergency of International Concern and pandemic" [50]. Since 2022, different variants of the virus, i.e., "Delta, Alpha, Beta, and Omicron," have appeared and become dominant in several regions; as of February 24th, 2022, Worldwide, a total of 431,647,262 confirmed cases, 5,946,699 deaths, and 360,658,193 recovered reported from the virus [49].

This global public health crisis has since cascaded into a series of challenges for leaders around the world, threatening both the health and economy of populations. As a result of pandemic, the countries' leaders and healthcare institutions are forced to design dynamic strategies and standard operating procedures (SOPs) to sustain social, economic, educational, and cultural activities [16]. During such unconventional circumstances, the overall response of general public is usually a greater search for information. Therefore, the use of social media as a tool for professional communication and education in healthcare has been increasing. Social media have become popularly use to seek for medical information and have fascinated the general public to collect information regarding corona virus pandemics in various perspectives. During these days, people are forced to stay at home and the social media have connected and supported awareness and pandemic updates. Never before in human history has it been possible to communicate so quickly during a pandemic, social media platforms have been a key piece for the dissemination of information.

In this regard, social media broadly supports leaders and institutions to disseminate health prevent movement control orders at regional, national, and international levels [39]. As a result, it becomes a straightforward approach for global leaders to interact with people by posting online content. Thus, social media plays an intermediary role between leaders and the public to interact and address the health issues mainly, COVID-19 preventions [36, 39]. The way content is used, shared, and interpreted via social media, from just distribution to real impact and change, has been transformed by different social networking sites, i.e., Facebook, YouTube and Twitter [1, 42, 43]. Therefore, with the rising number of social media users, Twitter has become the most popular social media channel of all social media sites. It confirmed that it influences political changes in the regions [45]. The most vigorous social media spaces in which leaders connect with the public are Twitter and Facebook [13].

Additionally, visual risk communication through social media channels like Twitter offered a broad platform to healthcare professionals and country leadership to educate people on maintaining social distancing, following COVID-19 prevents, and staying at home [29]. As a result, Twitter has played a vast role by allowing leaders to easily develop visual interaction with individuals. They educate and motivate them to follow SOPs to avoid the maximum spread of the virus [42, 43]. Mengu et al. [32] emphasized that leadership's visual risk communication through Twitter enables and motivates people to make their own health decisions. Drylie-Carey et al. [17] concluded that the leaders' visual messages significantly engage and affect individuals' emotions and persuasiveness. In this regard, emotional attachment makes individuals follow SOPs and health recommendations more consciously to reduce the spread of the virus.

According to Sleigh et al. [42, 43], "as one of the most used and well-established social media platforms, with a history of use during public emergencies, natural disasters and epidemics, Twitter holds great potential for strategic and cost-effective visual health risk communication" (p. 2). Twitter allows healthcare institutions and leadership of a country to reach and communicate millions of people, but it also enables research, evaluation, and monitoring of health risk communication campaigns.

Uysal et al. [46] suggested that government institutions and leaders use social media to build visual risk communication with local and international audiences. Moreover, Hussain [25] and Tsoy et al. [45] mentioned that Twitter offers a dynamic opportunity for the public and leaders to interact better on subject matters. As a result, leaders tried to create their virtual identity in line with real-world leadership influence [20]. In the era of social media, leaders have changed their workstyle via posting content on social media and getting instant feedback [13]. Through this means, virtual space provides extended support to leaders across the nations to mitigate the risk of the COVID-19 pandemic. Accordingly, Saudi Arabian healthcare institutions and Monarchy also develop visual risk communication through social media channels like Twitter to motivate and educate people to prevent the spread of the virus.

Considering the case of Saudi Arabia, citizens are immensely engaged in social media [30]. Massive engagement of the audience via social media is noteworthy in Saudi Arabia. As per Global Web Index [19] and We Are Social [47] reports, there are 27.8 million users on social media; internet penetration rate up to 93%; time spent per day is around 8 h; streaming TV content via internet 61%; average time spent on social media is around 3 h. Users in Saudi Arabia prefer to choose YouTube (89.5%), WhatsApp (80.5%), Instagram (76.4%), Facebook (73.9%), Twitter (71.4%), Facebook messenger (54.1%) and Snapchat (53.4%) [19, 47]. Saudi citizens have the highest social media penetration worldwide regardless of strong cultural and traditional roots. Hence, social media provides an opportunity to showcase leadership identity, especially during the COVID-19 pandemic [50].

Given the prominent role of social media in performing leadership roles, there is a scarcity of research using dramaturgical perspectives and communication models to understand the influence of frontstage and backstage Monarchy leadership. Empirical studies covered social media interaction and mapping issues in different time spectrums [22]. However, the Saudi Monarchy @KingSalman on virtual space during the COVID-19 pandemic in Saudi Arabia is understudied. The current study explores the virtual space identity of the Saudi Monarchy "King Salman bin Abdul-Aziz Al Saud" in Saudi Arabia. It evaluates its presence on frontstage and backstage via Tweets.

Further, this study identifies the significance of social capital as an outcome of the Monarchy's leadership role during the COVID-19 pandemic. This study structure is as follow: first, it discusses the state of COVID-19 in Saudi Arabia and the part of Monarchy online identity; second, it elaborates on the dramaturgical perspectives and Lasswell's communication model using the Twitter data of @KingSalman as a contributor of social capital; third, it analyzes the Monarchy communication being part of backstage and frontstage interactions; fourth, it tackles the implications of the shades of dark emotions in leadership expression. Finally, it concludes with the implications of building leadership identity online and calls for future international research.

2 Covid-19 Visual Risk Communication in Saudi Arabia

The practical and timely decision from the leadership and government agencies on the first wave of the COVID-19 pandemic (2nd March 2020 to 31 July 2020) controlled the maximum spread of the virus. Dynamic visual communication of the leadership potentially controlled the COVID-19 crisis around the country. The leadership of Saudi Arabia educated and delivered the essential prevention of COVID-19 through social media, especially Twitter. Furthermore, preventive and control measures by the Saudi Arabia Leadership are presented in Table 1.

Saudi Arabia's government has taken quick response and preventive measures to overcome the spread of COVID-19. The first case was confirmed on the 2nd of March 2020. World Health Organization declared COVID-19 a global pandemic on the 11th of March 2020 when the total number of patients was 45 in Saudi Arabia. Saudi Arabia implemented strict policies to deal with the pandemic to ensure everyone stay-at-home, keep social distancing, quality care at the hospital, and availability of protective equipment. Moreover, Saudi Arabia donated US $10 million for COVID-19 treatment US $10 million to Yemen for the arrangement of equipment and supplies [6].

3 Covid-19 Vaccination Process in Saudi Arabia

Saudi Arabia was one of the first countries in the Gulf Cooperation Council (GCC) to initiate the COVID-19 vaccination programs by approving three vaccines, and such are; "Moderna (mRNA-1273), Pfizer/BioNTech (BNT162b2), and Oxford/AstraZeneca (AZD1222)." As a result, the COVID-19 vaccination campaign in Saudi Arabia began on 17 December 2020 [2]. Therefore, The Ministry of Health, Saudi Arabia, quickly opened mass vaccination centers to vaccinate the maximum number of individuals.

The Monarchy of Saudi Arabia pledges to donate USD500 Million to fund the development and availability of COVID-19 vaccinations for everyone around the country [12]. Additionally, through the motivation and instruction from the Monarchy of Saudi Arabia, several research centers and Universities were instructed to start experiments and trials of Saudi-based COVID-19 vaccine with the collaboration of

Table 1 Preventive and control measures by the Saudi Arabia government

Date (2020)	Measures
2 Feb	Suspension of direct flights to/from China; evacuation of Saudi students from Wuhan
6 Feb	Suspension and ban entry travel to China for 14 days
27 Feb	Suspension of tourist and Umrah pilgrims; flights from affected countries
2 Mar	First COVID-19 case reported
4 Mar	Closure of Umrah
8 Mar	Closure of educational institutes; started remote education; Qatif lockdown
14-Mar	Closure of mosques; suspension of events and activities
15-Mar	Closure of shopping malls and restaurants; suspension of domestic and international travel
21-Mar	Nationwide curfew from 7 pm to 6 am
24-Mar	Curfew and lockdown from 3 pm to 6 am in Makkah, Madinah, Riyadh
28-Mar	Lockdown in Madinah Suburbs; extended suspension of domestic and international flights
29-Mar	Lockdown in Jeddah
30-Mar	Provided free healthcare to COVID-19 patients; lockdown in Makkah suburbs
31-Mar	Saudi government asked Muslims to put Hajj plans on hold until further notice
6-Apr	Curfew in most cities, 24 h
12-Apr	Saudi Arabia joined the World Health Organization solidarity trail to find COVID-19 treatment
15-Apr	Massive Covid-19 testing in the community

Source Composed by Author (2021)

CanSino Biologics China. Thus, The Ministry of Health Saudi Arabia conducted a phase-III clinical trial of CanSino. Furthermore, they recruited around 5000 male and female volunteers to participate in clinical trials of the COVID-19 vaccine. Moreover, according to Alobaidi et al., "Russian Direct Investment Fund (RDIF) spokesman, the Gam-COVID-Vac-Lyo vaccine being developed in Moscow, Russia, maybe conducting their phase-III clinical trial in Saudi Arabia among other countries as well" (p. 1746).

After the final approval of the COVID-19 vaccination, the Ministry of health installed 587 vaccination centers around the country and demonstrated the first vaccine in December 2020. Subsequently, King Salman Bin Abdul Aziz and Crown Prince Mohammed Bin Salman received their vaccines early; the event was broadcasted on several television channels and disseminated via social media to help enhance vaccine uptake by the locals and immigrants.

In early January 2021, Saudi Arabia had secured the second-highest number of vaccines of all countries in GCC, after the UAE. Additionally, Saudi Arabia is one of the countries that controlled the maximum spread of the COVID-19 virus and vaccinated the vast majority of local and international people around the country [15].

4 Dramaturgical Perspectives

This study uses Goffman's dramaturgical framework theoretically and methodologically to understand better how social interactions comprise actors, scripts, and staging [21]. The critical elements of dramaturgy are based on identity formation and embodiment. When considering culturally sensitive issues, dramaturgy and frame analysis are appropriate approaches to understanding incidents better. It focuses more on collective information that people draw upon in understanding social performances. Some researchers have focused on front-stage interactions in the literature, and others have drawn attention towards the backstage, while some emphasize both. It contributes to the creation of identity through face-to-face communication. It helps individuals to consider social performance based on mutual information systems. The frame analysis of Goffman [21] supports exploring the microanalysis of virtual social experiences. This research extends this theory's use to social interactions in the online virtual space and identity creation through social media posts. This research explores Twitter, the backstage scripts of the Saudi Monarchy-online identity, to understand the social option of virtual space contact between the Monarchy, the Saudi people, and the global audience. The concept of this study builds on a dramaturgical approach that analyses tweets from King Salman during the COVID-19 pandemic. The emphasis will be on the popularity of the scripts used by the Twitter account @KingSalman on the backstage of social media communication over the front stage scripts restricted to communication through conventional media [40]. This analysis scrutinized the elements of COVID-19 in Saudi Arabia. Scripts may be any words or phrases used by the Saudi Monarchy in its virtual space to establish contact with people to ensure a collective response to the COVID-19 pandemic. Understanding the value of such scripts in a virtual backstage platform like Twitter is imperative. The main focus will be on the scripts channeled by the Saudi Monarchy-online identity via King Salman's official Twitter account to understand better the value of such scripts in a virtual backstage platform.

5 Lasswell's Communication Model

Lasswell's communication model elaborates the process of communication with a more significant number of people. It elevates effective communication using different media channels. It reviews content posted online about the news or facts and their effects on the community. The primary goal of this model is to break down the communication components. It is organized as the "scientific study of the process of communication"; 'who' refers to a person or organization who formulates the message, 'what' indicates the content of the message, 'channel' shows the medium of transmission, 'whom' describes individual recipient, 'effect' deals with the outcome of the message. This model supports reviewing the feedback in social relations [28]. The step-by-step procedure supports reviewing and analyzing the communication process through media channels using questions "Who," "Says what," "In which channel," "To whom," and "With what effect." Analysis techniques comprise control (individual), content (messages), media, audience, and effects. This model supports developing the case for King Salman's Twitter feed from March to November 2020, considering the role during the COVID-19 pandemic.

6 Social Capital and Covid-19 Pandemic

Mapping Lyda Hanifan introduced the concept of social capital in 1916, defining "those tangible assets [that] count for most in the daily lives of people: namely goodwill, fellowship, sympathy, and social intercourse among the individuals and families who make up a social unit." It relates to the activities performed by a group of people to achieve common goals and objectives. OECD defined social capital as "networks together with shared norms, values, and understandings that facilitate co-operation within or among groups." Social relationships bring collective benefits for society. Mainly, there are three types of social capital, i.e., bonding, bridging, and linking. Bonding social capital describes connections within a community and group with shared attributes and similarities in demographics, attitude, and resources. Bridging social capital describes relationships among people across a cleavage that divides society based on race, class, or religion. Linking social capital describes groups in different social hierarchies and distinguishes them based on power dynamics, social status, and wealth. It is a subset of bridging social capital [14, 48].

Social capital has been widely recognized in recent years through social media platforms. Facebook and Twitter have created an excellent example for social capital where people get connected and collectively respond to a social cause. For instance, it has been noticed that during the COVID-19 pandemic, leaders get associated with communities through social media and ensured safety measures to control the widespread of Coronavirus in the country. It creates a strong social network among people from different regions and collectively responds to the pandemic.

In Saudi Arabia, Monarchy King Salman used virtual space to establish bonds, bridges, and links with social capital people. King Salman used the Twitter platform to communicate with people and disseminate messages in response to the COVID-19 pandemic. It indicates a strong connection and social relationship with the local and international community. It ensures the collective efforts from leaders to people in the community to collaborate for a single cause. Through tweets, it also depicts the emotional bonding to the people and delivers a message that leaders are already aware of the pandemic and are responsible for ensuring people's safety. Considering the concept of social capital, Saudi Monarchy has created a great example of being in touch with people through virtual space.

7 Saudi Monarchy Presence in Virtual Space

Salman bin Abdul-Aziz Al Saud has been the seventh king of Saudi Arabia since 2015. Hence, it is difficult for the King and the people to interact face-to-face and deal with the challenges and opportunities. Thus, it became essential to opt for other channels like Twitter to communicate with people directly. After coming into power, King Salman owned the official Twitter account "@KingSalman" and recognized his social media presence worldwide. Moreover, Saudi citizens endorsed his virtual identity by retweeting his tweets. It confirms Saudi Monarchy's online identification through Tweets on certain occasions, such as national disasters, especially in the pandemic [7]. This ensures accessibility and serves as face-to-face contact by developing online scripts that are individually channeled to people to create a sense of cyber interaction. The Twitter space Monarchy has built an online identity and presence, which is a significant part of virtual leadership and can be seen in the use of media objects defined by symbols such as the King's profile picture, content, and images. Tweets on @KingSalman indicate backstage script linked with front stage script via media channels, live streams, and physical presence. Baker and Rowe [10] stated that social media networks offer power and voice to individuals normally bared from conventional media involvement. Twitter provides a forum to become influential for those who do not influence. However, it is a place that can often be insecure.

Still, the Monarchy's online persona is created with the maximum power, and it was willing to take risks associated with social media to communicate with its people. The Monarchy appreciated its connectivity and openness by taking the initiative and directly communicating with people through an official Twitter account [25]. It represents King Salman's role is evolving with time in the virtual space, comfortably engaging with a larger audience nationally and globally. Uysal et al. [46] mentioned that governments need to use social media to enhance tactical contact and establish relationships with their national and foreign audiences. Concerning establishing an online Monarchy identity in Saudi Arabia, it is imperative to understand the significant real-time power of Twitter and its implications specifically in this study. To illustrate the theoretical and practical consequences of using social media's influence on virtual space in crucial periods such as the COVID-19 pandemic, the sequence of events, and the Monarchy's response to a pandemic are studied. King Salman's tweets are analyzed to bond, bridge, and link social capital to the COVID-19 pandemic preparedness.

7.1 Content Analysis of King Salman Tweets

Worldwide, Twitter holds a broad potential for strategic visual health risk communication. It allows the leaders and other institutions to interact with millions of people in limited time space. Thereby, like other social media channels, along with this great potential, Twitter faces corresponding ethical challenges [31, 37]. Accordingly, the current study was based on a dramaturgical perspective that supports reviewing the content and establishing a united effort to deal with a pandemic. The focus is to differentiate the expression in backstage scripts on the official Twitter account and frontstage scripts on other media channels [44]. Goffman stated that it is not always the case, as both stages can be unified [11]. This study extended the domain to capture insights usually missed in qualitative research. This study examines the Saudi Monarchy's dark emotional expression through backstage scripts on virtual space Twitter from March 26 to November 21, 2020. It comprises seven tweets posted by King Salman during the COVID-19 pandemic.

The analysis indicates the bonding, bridging, and linking with social capital via social media platforms to express emotions. It affects building connection, readiness, and preparedness to deal with the COVID-19 crisis. The backstage scripts concentrate on the words or phrases used by the Saudi Monarchy in its virtual space to establish contact with its nation and the world to convey shades of dark emotions as appeared due to the COVID-19 pandemic.

The first tweet was related to the COVID-19 pandemic (backstage), and the message conveyed through the G20 summit (frontstage) deals with economic and healthcare challenges through uniting efforts. It raised a call for the collaboration of leaders and communities worldwide to cope with the COVID-19 crisis. At that time, Coronavirus was spreading quickly from one region to another. In Saudi Arabia, preventive and control measures started on the 2nd of February after the suspension of direct flights to and from China and Saudi students' evacuation from Wuhan city in China. To stop the spread of Coronavirus, the suspension of flights for tourists and Umrah pilgrims started on the 27th of February, especially from the most affected countries. After reviewing the daily case meter in March, the Saudi government announced lockdown and curfew in several cities. It was critical for leaders, Saudi Monarchy, and the community to be united and follow the standard operating protocols in the new normal. Hence, social media platforms support a better way to unite social capital and encourage everyone to respond to COVID-19 responsibly. On 12th April, Saudi Arabia joined the World Health Organization solidarity trail to find COVID-19 treatment and released funds for this cause. Considering King Salman's backstage script, it clearly states the unified efforts to cope with the COVID-19 crisis. At the front stage, the execution of measures was in place at the national and international levels. King Salman showed concerns and responsible behavior via three tweets on religious events, "Islamic month Ramadhan, Eid Al-Fitr and Eid Al-Adha." He prayed for strength, goodness, and blessing to lift people from the world's all affliction. From King Salman's Twitter account, 5 out of 7 tweets mentioned the COVID-19 pandemic either directly or through the word (affliction) and reflected emotional bonding and connection to foster social capital worldwide for the same cause.

Moreover, King Salman also addressed the COVID-19 pandemic with the meeting of the leaders of G20 countries through a tweet on 21st November. It depicts the responsible leadership response to mitigate the effects of the COVID-19 pandemic on the world. It recalls global leaders' roles from the start of 2020 to the end of the year to remember the leadership role and encourage them to seek a prosperous future.

The way leaders exchange certain information with people gives a clearer understanding of who they are and what they care about. It also enables leaders to remain connected with people and address their matters wisely. The most crucial influence of a leader's social media tweet or post concentrates on other individuals or circumstances. Leaders' display of dark feelings can be negotiated differently on the front and backstage of media outlets. Typically, voicing negative emotions supports maintaining social capital via social media. Social media platforms are considered safe outlets for people to exchange thoughts that they would not otherwise convey if they were in the direct presence of others. It provides an alternate platform for those hesitant or unable to address problems in person. It supports building the connection with leaders and people in their countries.

7.2 Leadership Exposed via Shades of Dark Emotions

The shades of dark emotion are revealed to connect with followers through social media platforms. Leaders express their concerns and support to the vulnerable community during difficult times, such as the COVID-19 pandemic. The emotional, social capital bond is built through various dark emotional speech shades on front and backstage media. It formulates the choice of dark emotions using multiple media platforms. The content analysis of Tweets posted by Saudi King Salman indicated shades of dark emotion backstage. It addressed the global pandemic challenges worldwide and expressed a joint response to the COVID-19 crisis. The first tweet expressed emotion favoring humanity and stated, "May God spare humanity from all harm." The expression of this emotion reflects a clear message to social capital across the globe that together, we will cope with the C COVID-19 crisis. Reinforcing the shades of dark emotion, the second indicates a clear message from backstage that God Almighty accepts our prayers and lifts us from the world's all affliction. This expression of emotion provided excellent support in forming social capital via bonding, bridging, and linking global leaders and people who have faced a tough time due to the COVID-19 pandemic. King Salman's tweets establish a bond with a group of people who have common attributes. It also creates a bridge with communities divided by race, class, or religion.

Further, it supports establishing a link with different social strata based on power dynamics [14, 48]. Dark emotion shades in three types of social capital, including sorrow, grief, and compassion. They were all a common theme in all the dark feelings depicted on King Salman's Twitter account. The shadow emotions also appeared based on the sad event the COVID-19 pandemic was going on worldwide and their mass media coverage. Hence, the expression of shades of dark emotion appeared systematically. In the third tweet, King Salman state that "we see hope in our coming days, showing resolve and positivity to overcome all affliction," and our wishes come true "when we (world) are at our best of states." Thus, it clearly expresses emotional support to people in different regions.

On 31st July 2020, one of the crucial tweets clearly expressed the bonding, bridging, linking social capital, stating, "May God take away from our country (Saudi Arabia) and the world this affliction of Corona pandemic by his mercy and grace." The peak time when the COVID-19 cases increased drastically in Saudi Arabia and worldwide. His tweet showed the shades of dark emotions that all global leaders were putting joint efforts into to cope with the COVID-19 crisis. The establishment of virtual space leadership can be seen as a mode of control as it focuses on exploiting impressions through defined scripts and expressions to serve specific objectives. In this view, the power to preserve social order and comply with broader cultural frameworks of social interactions occurs within social interactions. Year-end tweet on 21st November 2002, King Salman acknowledged the efforts to deal with the COVID-19 pandemic with the meeting of the leaders of the G20 countries. It delivers a great message to the people across the globe that "our responsibility was—and will remain—to move towards a better, healthy and prosperous future for all." It indicates a social capital bridge to mitigate the effects of the COVID-19 pandemic on the world. In the case of the dark emotional expression of the Saudi Monarchy, the King's emotional intelligence and emotional decisions had to be much more aware and careful in recognizing, knowing, and handling his audience's emotions in positive ways. This was to connect effectively with them and represent their online presence and identity. This is the Monarchy's only social network forum in which it has participated [41].

8 The Outcome

The Monarchy and the Ministry of Health of Saudi Arabia have dedicated every effort to control the maximum spread of the COVID-19 virus, ordering the health and safety measurements to the residents and emigrants. However, the economic cost to the country ran into billions of dollars and affected almost all the public and private sectors. In this concern, the Monarchy has been making a serious and strong decision in the meantime. Some of the significant strategies measures are listed below:

(a) COVID-19 infection rate is low, with 474 individuals infected per million compared to 620 in the United States of America (Worldometer 2020).
(b) According to the Ministry of Health in Saudi Arabia, the recovery rate from the infection has been rising dramatically, up to 96%, because of the early implementation of safety measurements [34].

(c) Overall, the "case fatality rate" is widely lower than the international average (1.6% vs. 2.2%), which reflects the healthcare availability of mechanical ventilators and ICU beds [5].
(d) Educating the people on basic safety measurements and awareness on reducing the maximum spread of the COVID-19 virus and its complications, particularly on children and older people, helped enforce effective precaution measures [4].
(e) With the effective and early decision on the COVID-19 infection, the Ministry of Health added an extra 30% ICU bed, and the maximum infection rate reached around 18% [8].
(f) By 21 August 2021, the COVID-19 vaccination rate was 96.2 doses per hundred people (~34.81 million). At this rate, Saudi Arabia could have 70% of people vaccinated (2 doses) in 58 days (or by Oct 18, 2021) [15].
(g) According to a recent report by the World Economic Forum, after China, Saudi Arabia ranked second for the most optimistic country to recover economically within a year of the COVID-19 Pandemic [18].

9 Discussion

Social media, mainly Twitter, is considered one of the world's leading platforms for communication, particularly in a health crisis. World leaders and other institutions widely used social media to spread maximum knowledge on how to reduce the extreme spread of the virus and sustain social life during pandemic crises [35]. Considering the dynamic role of Twitter, they introduced a "warning label" in March 2020. That label highlights the tweets containing potentially misleading or harmful information about COVID-19. Thus, Twitter has played a fundamental role in visual health risk communication during the COVID-19 pandemic [31]. In addition, various world leaders and institutions utilized the platform to instruct, boost, prompt, and morale political discussion [37]. Accordingly, the Saudi Arabian leadership interacts with the public through Twittersphere.

Therefore, the findings indicate that King Salman bin Abdul-Aziz Al Saud's role in visual communication was followed through the principle of social capital in terms of bonding, bridging, and linking global leadership towards the COVID-19 crisis and encouraged them to perform their strategic role for the sake of humanity and well-being. It expresses the emotional connection with the local and international community. Nonetheless, King Salman's leadership reflected that global leadership is concerned about the pandemic and ready to protect people by implementing effective measures and controls. It depicts the front leadership role in the tweets. Woolcock [48] claimed that the relationship between emotional intelligence and leadership is built on self-awareness and empathy, apparent in King's tweets concerning the people.

In the Monarchy's deliberate choice, such empathy is evident in deciding what to tweet and what not to base on the emotional choice that empathizes with the Monarchy's national and foreign audiences to conquer challenges, defuse situations and conflicts, and unite people. Many researchers have shown that being in contact with the emotions of those around us provides an advantage in social capital [33]. For exemplary leadership, therefore, having emotional intelligence is essential. A leader can have powerful means to focus and transform emotional experiences through high emotional intelligence [24]. This is evident in the dark emotional expression of the Saudi Monarchy on Twitter, especially in its tweets' emotional choices and the kind of dark emotions expressed on that particular platform.

Emotional intelligence also requires the proper control of emotions and relationships with others [3]. Emotional management abilities are interdependent as they rely on the recognition, comprehension, and management of emotions and their implications [26]. Kilduff et al. [27] indicate that determining when negative emotions such as fear and anger may impede or facilitate individual goals is related to emotional intelligence because the expression of negative emotions is full of danger due to the precarious state of such emotions [23]. Feelings are at the center of people's decisions in daily life activities, especially the COVID-19 pandemic. Emotions can affect and form pathways of thought, incredibly dark emotions.

This type of emotional intelligence was observed in King Salman's tweets since the start of the COVID-19 pandemic by focusing on a global issue, connecting leaders, and ensuring joint efforts to deal with the crisis. As the world experienced significant dark events in the COVID-19 pandemic, King Salman's Twitter account plays an essential role in expressing emotional feelings and standing at the front line as a global leader to tackle challenges wisely. This demonstrated repercussions for representing the dark emotions of leadership on virtual space (Twitter) and the use of emotional intelligence to control emotions and improve relationships. It also gave rise to the value of emotional choice in describing the dark emotional expression of the Saudi Monarchy online.

There has been comprehensive research on social media use; however, the shades of dark emotions remain less explored. Therefore, investigating and comparing the impact of dark emotions on social media and connecting it with social capital is the main contribution of this study.

10 Conclusion

The current study is the first to interpret the trends and characteristics of visual risk communication through social media channels during the COVID-19 pandemic. This study's results show that the influence of leadership through visual communication plays an essential role in coping with the COVID-19 crisis and builds an emotional connection with the community. This study aimed to understand the Saudi Arabian Monarchy @KingSalman's role in the COVID-19 pandemic through visual risk communication. As discussed earlier, visual risk communication is a novel term in several countries. Thus, limited scholars addressed the important role of visual communication during the COVID-19 pandemic. Therefore, the current study contributed to describing the key indicators of visual communication and its dynamic role in interacting the country's leadership with the people, likely during the COVID-19 pandemic. Additionally, It complemented research to build an emotional connection with the local and international community through visual communication by the country's leadership to cope with the COVID-19 crisis. It highlighted an edge to connect with social capital through bonding, bridging, and linking to mitigate the risk of COVID-19 spread in Saudi Arabia. Saudi Monarchy-online identity is manifested by virtual space on Twitter; the various facets of the Saudi Arabian leadership were reflected. The Saudi Monarchy's unparalleled social media interaction with the public enhances the general public's ability to have easy access to information. It helps create the identity of virtual and face-to-face leadership that affects the public in terms of having more faith and trust in decisions and future activities. Thus, there is an immense need to have international research on online leadership identity to support how governments can foster social capital through social media. Therefore, it is vital to investigate how leaders use various media outlets to express dark emotions and their impact on the public, nationally and internationally. Hence, future studies are suggested to examine the role of other government agencies, i.e., the Ministry of Health Saudi Arabia, in developing visual risk communication during the COVID-19 Pandemic in Saudi Arabia.

References

1. Abbas, J., Wang, D., Su, Z., Ziapour, A.: The role of social media in the advent of COVID-19 pandemic: crisis management, mental health challenges and implications. Risk Manage. Healthc. Policy **14**, 1917 (2021)
2. Al-Mansour, K., Alyahya, S., AbuGazalah, F., Alabdulkareem, K.: Factors affecting COVID-19 vaccination among the general population in Saudi Arabia. In: Healthcare, vol. 9, No. 9, p. 1218. Multidisciplinary Digital Publishing Institute (2021)
3. Al-Saggaf, Y., Simmons, P.: Social media in Saudi Arabia: exploring its use during two natural disasters. Technol. Forecast. Soc. Chang. **95**, 3–15 (2015)
4. Alahdal, H., Basingab, F., Alotaibi, R.: An analytical study on the awareness, attitude, and practice during the COVID-19 pandemic in Riyadh, Saudi Arabia. J. Infect. Public Health **13**(10), 1446–1452 (2020)
5. AlFattani, A., AlMeharish, A., Nasim, M., AlQahtani, K., AlMudraa, S.: Ten public health strategies to control the Covid-19 pandemic: the Saudi experience. IJID Reg. **1**, 12–19 (2021)
6. Algaissi, A.A., Alharbi, N.K., Hassanain, M., Hashem, A.M.: Preparedness and response to COVID-19 in Saudi Arabia: building on MERS experience. J. Infect. Public Health (2020)
7. Alobaidi, S.: Predictors of intent to receive the COVID-19 vaccination among the population in the Kingdom of Saudi Arabia: a survey study. J. Multidiscip. Healthc. **14**, 1119 (2021)
8. Alqahtani, F., Khan, A., Alowais, J., Alaama, T., Jokhdar, H.: Bed surge capacity in Saudi Hospitals during the COVID-19 pandemic. Disaster Med. Public Health Prep. 1–7 (2021)
9. Andrews, M.A., Areekal, B., Rajesh, K.R., Krishnan, J., Suryakala, R., Krishnan, B., Muraly, C.P., Santhosh, P. V.: First confirmed case of COVID-19 infection in India: a case report. Indian J. Med. Res. **151**(5), 490 (2020)
10. Baker, S.A., Rowe, D.: The power of popular publicity: new social media and the affective dynamics of the sport racism scandal. J. Polit. Power **6**(3), 441–460 (2013). https://doi.org/10.1080/2158379X.2013.846553
11. Baum, M.A., Potter, P.B.K.: Media, public opinion, and foreign policy in the age of social media. J. Polit. **81**(2), 747–756 (2019)

12. Brazil-Arab News Agency: Saudi Arabia donating USD 500 million to fight COVID (2020). Available on: https://anba.com.br/en/saudi-arabia-donating-usd-500-million-to-fight-covid/. Retrieved on: 12 Oct 2021
13. Carlsen, H.B., Toubøl, J., Brincker, B.: On solidarity and volunteering during the COVID-19 crisis in Denmark: the impact of social networks and social media groups on the distribution of support. Eur. Soc. **23**(sup1), S122–S140 (2021)
14. Claridge, T.: Functions of social capital–bonding, bridging, linking. Soc. Capital Res. **20**, 1–7 (2018)
15. Coronavirus, E.: 13,968 cases and 223 deaths (2020). https://www.worldometers.info/coronavirus/country/ethiopia. Accessed on, 27. covidvax.live. Saudi Arabia; 2021. Available from: https://covidvax.live/location/sau. Accessed 10 Dec 2021
16. Dhahri, A.A., Iqbal, M.R., Khan, A.F.A.: A cross-sectional survey on availability of facilities to healthcare workers in Pakistan during the COVID-19 pandemic. Ann. Med. Surg. **59**, 127–130 (2020)
17. Drylie-Carey, L., Sánchez-Castillo, S., Galán-Cubillo, E.: European leaders unmasked: Covid-19 communication strategy through Twitter. Profesional Información, **29**(5) (2020)
18. Forum, W.E.: These countries are the most optimistic about economic recovery from the pandemic. Available from: https://www.weforum.org/agenda/2021/08/ipsos-post-pandemic-economic-recovery/. Accessed 10 Dec 2021 (2021)
19. Global Web Index: (2020). Accessed on 25 Jan 2021. https://www.globalwebindex.com/reports/social
20. Goel, A., Gupta, L.: Social media in the times of COVID-19. J. Clin. Rheumatol. (2020)
21. Goffman, E.: Frame Analysis. Penguin, London (1974)
22. Hancock, J.T., Gee, K., Ciaccio, K., Lin, J.M.-H.: I'm sad you're sad: emotional contagion in CMC. In: Proceedings of the 2008 ACM Conference on Computer Supported Cooperative Work, pp. 295–298 (2008)
23. Hansen, P., Hansen, R.: Finding the face in the crowd: an anger superiority effect. J. Pers. Soc. Psychol. **54**, 917–924 (1988)
24. Harms, P.D., Credé, M.: Emotional intelligence and transformational and transactional leadership: a meta-analysis. J. Leadersh. Organ. Stud. **17**(1), 5–17 (2010)
25. Hussain, W.: Role of social media in COVID-19 pandemic. Int. J. Front. Sci. **4**(2), 59–60 (2020)
26. Joseph, D.L., Newman, D.A.: Emotional intelligence: an integrative meta-analysis and cascading model. J. Appl. Psychol. **95**(1), 54–78 (2010)
27. Kilduff, M., Chiaburu, D.S., Menges, J.I.: Strategic use of emotional intelligence in organizational settings: exploring the dark side. Res. Organ. Behav. **30**, 129–152 (2010)
28. Lasswell, H.: The structure and function of communication in society. In: Bryson, L. (ed.) The Communication of Ideas, p. 117. Institute for Religious and Social Studies, New York (1948)
29. Latiff, D.I.A., Kamal, S.: Information of Covid-19 on Ott Services Influences the Intention to Follow "The Stay at Home" Orders (2021)
30. Ma, Y., Diao, B., Lv, X., Zhu, J., Liang, W., Liu, L., Zhang, S., Shen, B., Wang, H.: COVID-19 in hemodialysis (HD) patients: report from one HD center in Wuhan, China. MedRxiv (2020)
31. Malik, A., Khan, M.L., Quan-Haase, A.: Public health agencies outreach through Instagram during the COVID-19 pandemic: crisis and emergency risk communication perspective. Int. J. Disaster Risk Reduction **61**, 102346 (2021)
32. Mengu, S., Mengu, M., Gunay, K.: Value-based communication during COVID-19 pandemic: a study on the twitter messages of Turkish Ministry of Health. Athens J. Mass Media and Commun. **7**(1), 23–44 (2021)
33. Mikolajczak, M., Menil, C., Luminet, O.: Explaining the protective effect of trait emotional intelligence regarding occupational stress: exploration of emotional labour processes. J. Res. Pers. **41**(5), 1107–1117 (2007)
34. MoH, M.O.H.: COVID 19 Dashboard: Saudi Arabia (2020)
35. Mohamed, S.: An analysis of risk communication by the Finnish and Scottish government on twitter during the Covid-19 pandemic (2020)

36. Naeem, M.: The role of social media to generate social proof as engaged society for stockpiling behaviour of customers OECD.: What is social capital? (2020). Retrieved from https://www.oecd.org/insights/37966934.pdf
37. Rahmanti, A.R., Ningrum, D.N.A., Lazuardi, L., Yang, H.C., Li, Y.C.J.: Social media data analytics for outbreak risk communication: public attention on the "New Normal" during the COVID-19 pandemic in Indonesia. Comput. Methods Programs Biomed. **205**, 106083 (2021)
38. Saleem, M., Kamarudin, S., Shoaib, H.M., Nasar, A.: Retail consumers' behavioral intention to use augmented reality mobile apps in Pakistan. J. Internet Commer. 1–29 (2021)
39. Shoaei, M.D., Dastani, M.: The role of social media during the COVID-19 crisis: a narrative review. Health Technol. Assess. Action **4**(1) (2020)
40. Shoaib, H.M.: The Saudi Monarchy@ Kingsalman: emotional choice on virtual space. J. Theor. Appl. Inf. Technol. **98**(21) (2020a)
41. Shoaib, H.M.: The Saudi Monarchy@ Kingsalman leadership on virtual space. J. Theor. Appl. Inf. Technol. **98**(12) (2020b)
42. Sleigh, J., Amann, J., Schneider, M., Vayena, E.: Covid-19 on Twitter: an analysis of risk communication with visuals (2021)
43. Sleigh, J., Amann, J., Schneider, M., Vayena, E.: Qualitative analysis of visual risk communication on twitter during the Covid-19 pandemic. BMC Public Health **21**(1), 1–12 (2021)
44. Tsao, S.F., Chen, H., Tisseverasinghe, T., Yang, Y., Li, L., Butt, Z.A.: What social media told us in the time of COVID-19: a scoping review. Lancet Digital Health (2021)
45. Tsoy, D., Tirasawasdichai, T., Kurpayanidi, K.I.: Role of social media in shaping public risk perception during COVID-19 pandemic: a theoretical review. Int. J. Manage. Sci. Bus. Adm. **7**(2), 35–41 (2021)
46. Uysal, N., Schroeder, J., Taylor, M.: Social media and soft power: positioning Turkey's image on Twitter. Middle East J. Cult. Commun. **5**, 338–359 (2012)
47. We are Social: Accessed on 25 Jan 2021. https://wearesocial.com/digital-(2020)
48. Woolcock, M.: Microenterprise and social capital: a framework for theory, research, and policy. J. Socio-Econ. **30**(2), 193–193 (2001)
49. Worldometer: Corona virus. Available on: https://www.worldometers.info. Retrieved on: 25 Feb 2022
50. Yang, H., Bin, P., He, A.J.: Opinions from the epicenter: an online survey of university students in Wuhan amidst the COVID-19 outbreak1. J. Chin. Gov. **5**(2), 234–248 (2020)

Management of Arab Media Organizations in Artificial Intelligence Era

Suzan Elkalliny and Yasir Aboualgasim

Abstract Arab media are facing challenges represented by the rapid and successive technological developments after the passage of the media from the traditional era to the era of electronic digitization, and the necessary changes in management modes, content, and the development of electronic platforms, so that the name of media have changed in the current era to become e-journalism. Arab media is not far from being impacted by this technology considering the rapid digital transformation. Therefore, this prospective study comes to uncover the administrative preparations of the Arab media system from the perspective of media experts in the era of artificial intelligence and the application of Metaverse. The study was conducted among 400 media experts from Arab countries through an electronic questionnaire to find out their vision of the preparations for the management of Arab media in the era of artificial intelligence that will come soon. This study reached a set of results, the most important of which are experts' expectations regarding the effects on media professionals were positive, and the percentage of experts' expectations of the negative effects on the Arab media has increased and including the change in media coverage patterns for all media content and form. Experts believe that Arab media institutions have not yet taken appropriate measures. Therefore, the main study recommendation was Arab media institutions must adopt new administrative patterns and models based on the use of artificial intelligence technology.

Keywords Arab media · Media management · Artificial intelligence era

1 Introduction

The world's media is witnessing rapid and successive technological developments that have greatly affected the performance of media institutions worldwide. The Arab media has not been far from these technological developments since the launch of satellites until the era of artificial intelligence, where Arab media institutions

S. Elkalliny (✉) · Y. Aboualgasim
Liwa College of Technology, Abu Dhabi, UAE
e-mail: suzan.elkalliny@ect.ac.ae

A. Hamdan et al. (eds.), *The Implementation of Smart Technologies for Business Success and Sustainability*, Studies in Systems, Decision and Control 216,
https://doi.org/10.1007/978-3-031-10212-7_31

are racing to use technology and train media professionals on it to keep pace with development at the level of performance, content, broadcast, and media distribution. Considering these developments, the media department is one of the factors that ensure the success of the Arab media and its various institutions in achieving its goals, as the development of plans, strategies, organizational structures, follow-up, and coordination are among the most important elements that chart the way to keep pace with development and make optimal use of modern technology.

2 Research Method

2.1 The Research Problem

The research problem lies in the following central question:

To what extent are Arab media organizations administratively ready to receive artificial intelligence technologies? especially after Mark Zuckerberg, founder of Face-book, announced on the twenty-eighth of October 2021 that it had changed its name to become (META) and at the same time referred to the company's work in developing meta verse technology [1], which combines virtual reality, augmented reality and the Internet.

2.2 The Research Objectives

This research seeks to achieve several goals, including the following:

1. A survey of experts' opinion on the extent of Arab media institutions' readiness to receive artificial techniques
2. Predicting the future of Arab media institutions management in the age of artificial intelligence.

2.3 The Research Questions

1. What is the evaluation of media experts in the Arab countries for artificial intelligence techniques in the media?
2. What is the impact of artificial intelligence techniques on media work in Arab countries from the point of view of experts?
3. What is the influence of artificial intelligence technologies on Arab media professionals?
4. What are the expected effects of artificial intelligence technologies on Arab media organizations?

5. To what extent do Arab media professionals benefit from these technologies from the point of view of experts?
6. What are the expected effects of artificial intelligence technologies on media legislation and laws in Arab countries from the point of view of experts?

2.4 The Research Type

This research is considered one of the prospective and exploratory studies of a future phenomenon and its impact on Arab media institutions.

2.5 Methodology

The study is based on the survey approach by conducting a survey of media experts in the Arab world, who are a qualitative audience specialized in using appropriate research tools to deduce their vision of the expected effects of artificial intelligence technologies on Arab media institutions.

2.6 Data Collection Procedures

The study relies on an electronic questionnaire to collect data that includes six main themes related to the topic of the study, to ask media experts about.

2.7 Research Sample

This research includes a community of experts and specialists in the Arab media, whether they are academics or practitioners. These were targeted in various Arab countries.

The research relied on the elite deliberate sample of media experts in the Arab countries depending on their backgrounds that they are the most capable of evaluating the uses of artificial intelligence techniques, and their impact on Arab media institutions according to their specialization and experience. 400 electronic forms were distributed via e-mail and WhatsApp to experts from academics and professionals in 12 Arab countries.

The framework of the sample includes 12 Arab countries (Table 1).

This study included academic and professional experts with more than ten years of experience, and the questionnaire included 400 experts from 12 Arab countries (Table 1).

Table 1 Distribution of the study sample

Demographic characteristics		Count	Percentage
Kind	Male	279	69.75
	Female	121	30.25
	Total	**400**	**100**
Nationality	Egypt	83	20.75
	Sudan	78	19.5
	United Arab Emirates	46	11.5
	Iraq	39	9.75
	Kuwait	32	8
	Tunisia	28	7
	Syria	22	5.5
	Oman	16	4
	Bahrain	16	4
	Yemen	15	3.75
	Morocco	13	3.25
	Jordan	12	3
	Total	**400**	**100**
Experience	Academic	182	45.5
	Professional	118	29.5
	Total	**400**	**100**
Years of experience	10–less than 15	71	17.75
	15–less than 20	132	33
	20 and more	197	49.25
	Total	**400**	**100**

2.8 The Research Terminology

- Arab Media Institutions: In this study, it means all traditional and digital Arab media institutions.
- Media experts: specialists in the field of media, including academics and practitioners with long experience (ten years or more).

3 Literature Review

There have been numerous research and studies on the management of media institutions. Some researchers were interested in studying the management of media content in media organizations, while others have been interested in studying the management of media institutions during digital transformation, as [2] study which concluded that the technology of the virtual world in the media will pose huge challenges to the Media institutions by the speed of media coverage, publishing and radio in addition to the part of the personalization of the media, so that the media will become personal media that requires the provision of new and distinct services that are not currently recognized, in addition to the integration of media institutions with the Blockchain system that facilitates the process of recording data (Based on a network of protected data blocks connected together) which makes it easier for the user to activate all media services through the virtual assistant and by voice.

This study concluded that media organizations in the era of artificial intelligence need complete transparency and uniqueness in the services they provide to the public, and that the institutions must find a way to train their employees of the media to integrate their capabilities with new technology to preserve its continuity and its audience.

Moreover, according to [3] which is concerned with studying the uses of one of the artificial intelligence techniques within media organizations to produce short video clips, which is Algorithm technology. The study concluded that this technology can help media organizations produce short and impressive video clips to be broadcast through smart platforms and social media, and it also allows users to easily obtain all the visual data through huge databases and reproduce it in a different and attractive way to viewers, but it may carry fabricated information resulting from deletion and addition by the user himself and what he adds to the videos of his thoughts, beliefs and personal opinions, which is reflected in the form of Informational chaos that cannot be controlled.

In the same frame, [4] study focused on the effects of using virtual reality technology and artificial intelligence on the local cultures of different countries. Localization in the culture of globalization and the decline in national production, which created many internal cultural problems and complexities for many countries of the world and negatively affected the cultural identity of countries, and many countries of the world began to lose their cultural uniqueness.

In addition to [5] study (applied on the US elections) concluded that the media management through smart platforms produced much false, misleading, and incomplete information that would distort the news and mislead public opinion, and this was strongly demonstrated in the official American discourse that lacks transparency.

The American interest in using artificial intelligence technology in media production institutions will be exploited in affecting the integrity of the elections, and the moral and democratic issues in the official American discourse will not be real and will be deliberately falsified to mislead public opinion. Therefore, among the proposals of this study was the necessity of researchers' interest in analyzing the content of official discourses based on artificial intelligence techniques to study their negative effects on societies.

On the other side, [6] conducted a study on traditional and modern media in Malaysia concluded that traditional media have benefited from modern means of communication, as both pushed the other to technological development and the development of content and media forms, and the integration between media helped the Malaysian public to read news The paper press on social media and the follow-up of radio and television programs on modern applications and new platforms, which attracted a large number of audience to follow the contents of traditional means through modern means and fulfilled the desires and needs of the Malaysian public from the media.

Furthermore, some studies revealed the obvious transformation in managing the media as [7] that concluded the sharp changes in management patterns of media organizations at every stage of technological development through content industry, community media management and the presence of the same institutions on electronic platforms.

Besides In the study of [8], the researcher analyzed 30 performance reports of 17 international media organizations to analyze their current administrative status with technological development and concluded that the transformation of institutions into the era of digital media led to the merging of traditional means with modern means, which created radical changes in the types of jobs and tasks assigned to media professionals, he also showed administrative and technical challenges within media institutions that required abandoning the traditional models recognized in management.

The same idea was confirmed by Mills et al. [9] which indicated that media organizations that do not operate according to a strategy for employing digital media so as to integrate with new electronic applications and platforms will make these institutions unable to compete and market themselves in the media, for instance, the Department of Media at the University of South Africa failed to market the university's brand among its competitors because there was no media strategy based on modern technology.

While [10] indicated in their study of the technological development that imposes on media organizations changes in the form and content of media materials after he performed a second-level analysis of media studies conducted on administrative performance within media organizations in the era of digital transformation and the new media the study revealed that the integration of new technology with social media had profound effects on the administrative, technical and professional levels within the media institutions.

While the study of [11] concluded that the uses of modern technologies in media institutions allowed newer and faster means of communication using multimedia—social media—on-line games, and the production itself has become faster and more accurate. And all of this did not negatively affect societies. On the contrary, different societies took advantage of this new technology and benefited from it in trade and education, and it was employed optimally in many countries for the welfare of their people and their economy.

Finally, it is clear from the review of previous studies that as the technology develops more, the more media organizations need to develop management patterns and use administrative models that are appropriate to the nature of the age of technology, as well as developing content and production methods and creating new patterns of media jobs that fit the requirements of the age of artificial intelligence.

Thus, the era of artificial intelligence needs to develop and manage the media system and employ modern technology in a manner commensurate with the requirements of Arab societies. This directed the researchers to survey the opinion of experts with knowledge, experience, and competence to benefit from their vision regarding the requirements of Arab media institutions in the age of artificial intelligence.

4 Results

4.1 Expert Evaluation of Artificial Intelligence Techniques

Experts differed in evaluating artificial intelligence techniques and their effects on media organizations. It is noted that the percentage of those who expect their negative effects on Arab media organizations has increased (84.25%), although (82.5%) of experts expect that new technologies will provide smart services to Arab media organizations.

(80.75%) expect that new artificial intelligence techniques may eliminate traditional media in favor of smart media, and (74%) confirm that smart technical means are a new colonial scheme to control the media in developing societies and countries.

4.2 The Effects that Experts Expect on the Arab Media (Table 2—Effects of Artificial Intelligence)

The experts expected several effects on the Arab media institutions, and they are arranged according to their importance to the experts as follows:

- Lead to the rise in the financial profit value at the expense of the media values (91%)
- Lead to an increase in the entertainment content at the expense of the serious content (78.5%)
- Lead to dominating the Arab media with virtual reality technology (77%)
- Lead to changing media coverage patterns for all media content and forms (76.75%)
- Transformation of Arab media into smart platforms (76.75%)
- Lead to a decline in values, norms, customs and traditions in the Arab media (60.5%).

Table 2 Experts' insights

	Agree		Neutral		Disagree	
	C	%	C	%	C	%
First: evaluating artificial intelligence techniques and their purposes						
Provide smart electronic services to media organizations	330	82.5	46	11.5	24	6
Negatively affect Arab media institutions	337	84.25	46	11.5	117	29.25
Reduce the number of media workers and rely on smart technologies	260	65	14	3.5	94	23.5
Eliminate traditional media in favor of modern media	323	80.75	23	5.75	44	11
Surpass the other social media	284	71	90	22.5	26	42.25
Support freedom of media and freedom of expression in Arab countries	56	14	71	17.75	273	68.25
A neo-colonialist scheme to control the media in developing societies and countries	296	74	22	5.5	82	20.5
Engaging the Arab media in the struggle of unequal competition	239	59.75	53	13.25	108	27
Second: the effects of artificial intelligence on media practices						
Allow excellence of media practices from the professional point of view	205	51.25	121	30.25	74	18.5
Undermine the Arab media	112	28	106	26.5	28	7
The development of the Arab media in terms of coping up with events and achieving a media scoop	244	61	114	28.5	42	10.5
Eliminate traditional Arab media	112	28	136	34	142	35.5
Transformation of Arab media into smart platforms	307	76.75	51	12.75	42	10.5
Providing more comprehensive and in-depth informational contents	187	46.75	123	30.75	90	22.5
Diversifying and developing media content	267	66.75	99	24.75	34	8.5

(continued)

Table 2 (continued)

	Agree		Neutral		Disagree	
	C	%	C	%	C	%
Changing media processing patterns for all types of media content and media forms	307	76.75	49	12.25	44	11
Greatly facilitate the Arab media to carry out its work	269	67.25	96	24	26	6.5
Dominance of the advertising and marketing aspect of the Arab media	209	52.25	67	16.75	44	11
Rising of the entertainment content at the expense of serious content	314	78.5	74	18.5	12	3
Deterioration of values, norms, customs and traditions in the Arab media	242	60.5	90	22.5	68	17
Decline in the cultural role of the Arab media	225	56.25	91	22.75	84	21
Dissemination of misleading and incorrect information	228	57	146	36.5	26	6.5
Rise of the financial profit values at the expense of media values	364	91	26	6.5	10	2.5
Controlling the Arab media by virtual reality technology	308	77	58	14.5	34	21.25
Third: the effects of artificial intelligence techniques on media professionals in Arab media institutions						
Fast and easy access to sources and information	317	79.25	79	19.75	64	16
Facility of conducting interviews with senior figures and officials whose meeting is difficult in reality	340	85	72	18	48	12
Reducing media censorship procedures	267	66.75	121	30.25	72	18

(continued)

Table 2 (continued)

	Agree		Neutral		Disagree	
	C	%	C	%	C	%
Improving the level of performance of media professionals in Arab countries	228	57	152	38	80	20
Increasing competition between Arab media professionals and their international counterparts	144	36	96	24	220	55
Shrinking of media job opportunities	202	50.5	176	44	82	20.5
Introducing new functional patterns	372	93	80	20	8	2
Opening new training and education opportunities for media professionals	384	96	10	2.5	6	1.5
Changing media policies for media training in the Arab world	314	78.5	72	18	14	3.5
Modernization of education programs in the field of media in the Arab world	305	76.25	73	18.25	22	5.5
Fourth: the effects of artificial intelligence techniques on Arab media institutions						
Decline in state and government censorship of the media	299	74.75	59	14.75	42	10.5
Imbalance of the administrative structures of the Arab media	243	60.75	114	28.5	43	10.75
Decline in the economic situation of the Arab media	201	50.25	138	34.5	61	15.25
The rapid professional and technical development of the Arab media	276	69	90	22.5	34	8.5

(continued)

Table 2 (continued)

	Agree		Neutral		Disagree	
	C	%	C	%	C	%
The Arab media's adoption of international administrative and technological methods	275	68.75	107	26.75	18	4.5
Establishing media globalization in the Arab media	288	72	90	22.5	22	5.5
The growth of Massive media entities in the Arab world	169	42.25	113	28.25	18	4.5
Expanding the margins of media pluralism and media diversity in the Arab world	252	63	114	28.5	34	8.5
The independence of the Arab media from Arab governments	155	38.75	107	26.75	138	34.5
Achieving the principle of privacy in managing media work in Arab countries	155	38.75	179	44.75	66	16.5
Increasing the spread of private institutions at the expense of government institutions	268	67	98	24.5	34	8.5
The control of massive economic institutions on the Arab media	356	89	34	8.5	10	2.5
Fifth: preparations of Arab media institutions for artificial intelligence techniques						
Lack of technicians and experts in this field	340	85	42	10.5	18	4.5
Lack of awareness of media departments in this field and its importance	347	86.75	35	8.75	18	4.5
No urgency to use this technology	135	33.75	132	33	133	33.25

(continued)

Table 2 (continued)

	Agree		Neutral		Disagree	
	C	%	C	%	C	%
Absence of training and qualification institutions specialized in this technology	348	87	18	4.5	34	8.5
Lack of qualified Arab media organizations to receive the age of artificial intelligence	368	92	–	–	32	8
Sixth: legislation and laws for media institutions in Arab countries in the age of artificial intelligence						
Renewal of legislation and laws for media institutions	227	56.75	123	30.75	50	12.5
Protect media professionals and preserves their rights	114	28.5	195	48.75	91	22.75
Cause legislative and legal problems for the Arab media	244	61	146	36.5	10	2.5
Help for better organization for the media work	128	32	149	37.75	123	30.75
Reducing the control role of governments on the Arab media	204	51	131	32.75	65	16.25
Changing the laws of production, broadcasting and distribution of media content	360	90	28	7	12	3
Contribute to preventing the unqualified from practicing the profession	123	30.75	–	–	287	71.75
Maximizing the role of the private sector in the media	304	76	90	20.5	26	6.5
Governments lose control of smart media	235	58.75	83	20.75	82	20.5
Abolition of sanctions against media professionals	81	20.25	116	29	203	50.75
Make smart media a tool in the hands of non-specialists	291	72.75	91	22.75	18	4.5

(continued)

Table 2 (continued)

	Agree		Neutral		Disagree	
	C	%	C	%	C	%
Provide greater privacy and the right to keep confidential information sources for media professionals	382	95.5	84	21	34	8.5
Allow freedom of opinion and expression for the media professionals	211	52.75	115	28.75	74	18.5
Recommendations for achieving the artificial intelligence readiness:						
It is necessary to develop appropriate legislation and laws for smart digital media	360	90	16	4	24	6
Media professionals must be trained and qualified to deal with new technologies	352	88	24	6	24	6
The strategies, plans and policies of Arab institutions in the Arab world must be updated	360	90	32	8	8	2

It is noted from the above a rise in the percentage of experts' expectations of negative influences. On the other hand, there were expectations of many positive influences, especially with regard to professional development and media performance.

Experts (67.25%) believe that it leads to great facilities for the Arab media to practice their work and leads to the development and diversification of media content (66.25%), and (61%) believes that it will develop in the Arab media in terms of keeping in line with the development of events and achieving a media scoop. However, there is a fear among experts of the spread of misleading and fabricated information (57%), and there is a fear of deterioration of the Arab media cultural and developmental role accompanied by the growth of the entertainment and superficial role of the Arab media (56.25%).

4.3 The Effects that Experts Expect on Media Professionals in Arab Media Institutions

In this context, it is noted that all expectations regarding the effects on media professionals are positive and no negative expectation appears, as the positives included opening new training and education opportunities for media professionals (96%)—introducing new job patterns (93%)—helping media professionals to facilitate interviews with VIPs and officials (85%)—It also helps media professionals to quickly access information sources (79.25%). Experts also expect the speedy change of media training policies as well as the modernization of media education programs in Arab countries. Some scholars expected (66.75%) that it will reduce the censorship procedures on media professionals.

4.4 The Expected Effects of Artificial Intelligence Techniques on Arab Media Institutions

Experts expect a decline in Arab countries' control over media institutions (74.75%), and this will be reflected in the control of large economic institutions and entities over media institutions (89%) and the increase in the spread of private media institutions over governmental or semi-governmental institutions (67%), as well as the imbalance in the administrative structures of the existing media outlets (60.75%), as well as expanding the margin of media pluralism and diversity in media institutions in form and content (63%).

But one of the most important advantages that Arab media institutions will witness is the rapid professional and technical development (69%), as well as the institutions' adoption of advanced administrative and technological methods (68.75%), which are indispensable for survival and competition.

Experts (72%) also expect media globalization to be consolidated by all its institutions, objectives, patterns, and content in this challenging era.

4.5 *Arab Media Institutions Readiness for the Age of Artificial Intelligence from the Point of View of Experts*

Experts (92%) believe that Arab media institutions are not yet ready to receive new technology from several aspects, perhaps the most important of which is that Arab media institutions are still not aware of the importance of this field (86.75%), and they have not yet taken the appropriate measures. They still lack qualified technicians and experts. To work in this field (85%) and it still lacks thc prcsence of specialized training and rehabilitation institutions in this field (87%),

Sixth: The expected effects on media legislation and laws from the point of view of experts.

Experts expect positive effects on legislation and laws that allow greater privacy and the right to keep information sources confidential for media professionals (95.5%), as well as the laws of production, publishing, broadcasting, and radio will witness changes commensurate with the nature of this age (90%), but new legislation may maximize the role of the private sector (76%), making smart media a tool in the hands of non-specialists (72.75%).

It is noted that the decline of approval percentage on some statements, especially those related to the freedom of journalists and the practice of the profession. Slightly more than half of the experts (52.75%) expect that the new media legislation will guarantee freedom of opinion, while the percentage of experts' disapproval has increased (71.75%) that the new legislation will prevent non-specialists from practicing the media profession, and researchers believe that this part is dangerous for the future of media in an era that needs qualified and trained specialized cadres from graduates of media colleges.

5 Recommendations

The proposals of this study came through the suggestions of experts, as the last main point of the survey form was about the experts' propositions to be well-prepared for the age of artificial intelligence. The proposals included three main insights that included the following:

1. Developing legislation and laws regulating institutions that are compatible with smart media.

   The researchers believe that this can be done through:

- Introducing legislation and laws to manage media institutions with setting general rules and policies for the work of private institutions and the so-called citizen journalism, and rules for merging traditional means with modern means, as well as penalties in case of falsification of news and information and media misrepresentation of public opinion
- Introducing legislation for regulating the ownership of artificial intelligence media and the rules for operating, broadcasting, and publishing
- Introducing laws with new job titles for media professionals and their work patterns, with job descriptions for different professions, and confining work in artificial intelligence only for qualified and trained people who can take on this responsibility.

2. Updating the strategies, plans and policies of the media institutions with the adoption of non-traditional administrative models. The researchers believe that this can be done through:

- Arab media institutions adopt new administrative patterns and models based on the use of artificial intelligence technology
- Studying global administrative models in the major international media institutions, and then settling the appropriate models for Arab societies and adapting them to suit the characteristics of Arab societies.
- Adopting a policy of administrative and technological integration between the Arab countries and working on a joint Arab production to fit the customs, traditions, and the qualitative nature of Arab societies.

3. Training and qualifying media professionals to deal with new technologies.
The researchers believe that this can be done through:

- Arab educational institutions adopt new programs in media studies that are compatible with the requirements of the age of artificial intelligence and qualify graduates for the labor market and new jobs.
- Establishing centers for continuous training and media qualification, some of which are Arab national, to qualify Arab media professionals in any Arab country, and some are national within each country in proportion to the needs of its labor market.

Finally, the two researchers stress on the necessity of applying what is called media literacy at all academic stages in schools and universities, as well as for the Arab family, to educate the Arab public about the selection of media materials and safe dealing with the media of the age of artificial intelligence.

References

1. Al Ain News: https://al-ain.com/article/metaverse-zuckerberg-anticipates-end-facebook. Last accessed 2021/11/15
2. Newman, N.: Digital news report 2021. Reuters Institute. https://reutersinstitute.politics.ox.ac.uk/digital. Accessed 2021/9/21
3. Lu, Z., Nam, I.: Research on the influence of new media technology on internet short video content production under artificial intelligence background. Complexity **2021**. https://www.hindawi.com. Accessed 2021/12/16
4. Chen, G.-M.: The impact of new media on intercultural communication in global context (2021). https://www.mepiforum.org/essay/essay-analysis-of-guo-ming-chens-the-impact-of-new-media-on-intercultural-communication-in-global-context-2578. Accessed 2022/1/2
5. Landon, M.: Disinformation in contemporary US foreign policy: impacts and ethics in an era of fake news, social media, and artificial intelligence. Public Integrity **24**(3), 512–522 (2022). https://www.tandfonline.com/journals/mpin20. Accessed 2021/12/5
6. Salman, A., et al.: The impact of new media on traditional mainstream mass media. Innov. J. Public Sect. Innov. J. **16**(3), 1–11 (2011). https://www.innovation.cc/scholarly-style/2011_16_3_7_ali_saman_media-impact.pdf. Accessed 2021/12/3
7. Goessel, B.M., et al.: Trends in technology, and the future of media management (research)? pp. 2–5. Ilmedia (2018). ilm1(2019200041). https://d-nb.info/118479734X/34. Accessed 2021/12/5
8. Sylvie, G., Weiss, A.: Putting the management into innovation and media manage-ment studies: a meta-analysis. Int. J. Media Manag. **14**(3), 183–206 (2012). https://www.econbiz.de/Record/putting-the-management-into-innovation-media-management-studies-a-meta-analysis-sylvie-george/10009625412. 2021/12/7
9. Mills, A., et al.: Managing the new media: tools for brand management in social media. https://www.researchgate.net/publication/312809487_Managing_the_New_Media_Tools_for_Brand_Management_in_Social_Media. Accessed 2021/12/5
10. Hollifield, C., et al.: Media management: a casebook approach (2015). https://scholar.google.com/citations?view_op=view_citation&hl=en&user=CoUKIOcAAAAJ&citation_for_view=CoUKIOcAAAAJ:2osOgNQ5qMEC. Accessed 2021/12/5
11. Levinson, P.: Digital media in an international context (2013). http://www.semesteratsea.org/wp-content/uploads/2012/05/A-Lewis_SEMS2500_DigitMedia.pdf. Accessed 2021/12/6

The Role of the Voice of the Kingdom (Sawt Al-Mamlaka) Program in Spreading Awareness of the Corona Virus According to Students in the Faculty of Media at Zarqa University

Othman Mansour Tahat, Ahmed Ananbeh, and Ashraf Faleh Al-Zoubi

Abstract This study attempted to reveal the role of The Voice of the Kingdom (Sawt Al-Mamlaka) program in spreading awareness of the Corona virus according to students in the Faculty of Media at Zarqa University. The Voice of the Kingdom (Sawt Al-Mamlaka) program broadcast by the Kingdom (Al-Mamlaka) Channel in raising the level of awareness of the Corona pandemic, and determining the level of respondents' dependence on The Voice of the Kingdom (Sawt Al-Mamlaka) program in spreading awareness of the Corona virus according to the variables of gender and department. In this study, the researchers used the descriptive analytical approach through a field questionnaire method that was distributed to an intentional sample of (115) male and female students from the Faculty of Media at Zarqa University who watched The Voice of the Kingdom (Sawt Al-Mamlaka) program broadcast on The Kingdom (Al-Mamlaka) Channel. The study revealed no significant differences in the role of The Voice of the Kingdom (Sawt Al-Mamlaka) program in spreading awareness of the Corona virus from according to students in the Faculty of Media at Zarqa University due to the variables: gender and department, where the values of (F) reached (2.674, 1.057, 0.059), respectively, gender and department, and the interaction between gender and department, which are non-significant values at the significance level (0.05). This indicates that students of different genders and departments in the College of Media are aware of the positive role of The Voice of the Kingdom (Sawt Al-Mamlaka) program in spreading awareness of the Corona virus.

Keywords Kingdom (Al-Mamlaka) · Awareness · Corona · Zarqa university

O. M. Tahat (✉) · A. Ananbeh · A. F. Al-Zoubi
Faculty of Media, Zarqa University, Zarqa, Jordan
e-mail: otahat@zu.edu.jo

A. Ananbeh
e-mail: aananbeh@zu.edu.jo

A. F. Al-Zoubi
e-mail: azoubi@zu.edu.jo

A. Hamdan et al. (eds.), *The Implementation of Smart Technologies for Business Success and Sustainability*, Studies in Systems, Decision and Control 216,
https://doi.org/10.1007/978-3-031-10212-7_32

1 Introduction

Individuals connect to multiple mass media means to see what information, opinions and ideas these media provide to the receiving audience, regardless of their social, scientific, and cultural tendencies and levels, with the aim of raising awareness and achieving intended goals that make the citizen a valid element in society [1]. The media has become the largest means as a provider of news of all kinds, whether cultural, political, artistic or educational for individuals and peoples. It fills their time, enhances their values, and affects their behavior more than the family, school and university [2].

In recent years, programs directed at young people have spread through the Kingdom (Al-Mamlaka) satellite channel, and they have become attracting young people who spend hours watching them, looking forward to developing and developing their life skills. These young people's interests also forced those in charge of television programs to express those interests and aspirations in order to meet their gratifications through qualitative media content.

Kingdom (Al-Mamlaka) TV is one of the Jordanian media institutions that is unique in broadcasting enhanced audio and video content and presenting it using multimedia, and a sophisticated and attractive modern production, which contributes to spreading awareness among generations. It also broadcasts talk shows based on dialogue between a presenter and one or more guests, including The Voice of the Kingdom (Sawt Al-Mamlaka) program, which is the focus of this study, which was titled "The Role of The Voice of the Kingdom (Sawt Al-Mamlaka) Program in Spreading Awareness of the Coronavirus according to students in the Faculty of Media at Zarqa University."

The study aims to identify the role played by The Voice of the Kingdom (Sawt Al-Mamlaka) program in spreading awareness among students in the Faculty of Media at Zarqa University who are of The Kingdom (Al-Mamlaka) channel viewers.

2 Previous Studies

After reviewing many previous studies related to the role of the media in spreading awareness of the Corona virus and its contribution to raising awareness of communicable and non-communicable diseases and epidemics, it is clear that there is research interest in this field.

A study [3] was applied to a sample of 486 Jordanian pharmacists, found a positive effect of frequent follow-up to the media, including television programs and other sources of information; to realize the risks and the roles of these means in health awareness of the virus among the respondents.

Another study [4] aimed to monitor how the talk shows on the Egyptian public and private satellite channels dealt with the Corona pandemic through delivering awareness messages to people through implying precautious procedures for the safety of citizens.

Likewise, a study [5] proved that pharmacists' awareness of the coronavirus that causes the epidemic is much higher than the levels of awareness among students. In a similar fashion, a study [6]; which was applied to a sample of (384) citizens from the Saudi Al-Jawf region; there are differences in the level of knowledge of the Coronavirus according to the variable of age and educational level; although (93.2%) of the respondents were aware that the cause of the Middle East Respiratory Syndrome was a viral disease.

The study of Ali [7] also concluded that the most prominent roles played by the Facebook network in health awareness among the public were represented in "increasing health awareness" and "deepening the level of communication with health authorities", while the most prominent gratifications achieved by the respondents were represented in "increasing health awareness" and "deepening the level of communication with health authorities." "health culture", and "increasing attention to personal hygiene".

In the same vein, Choi et al. [8]; indicated that television and newspapers did not provide sufficient information about the Corona virus, and this result is consistent with the study of Abu Sunna [9]; which concluded that there is a direct relationship between the rate of respondents' exposure to paper and electronic newspapers and their level of knowledge of Corona disease.

3 Theory Used in the Study

This study is based on the theory of media dependence developed by Melvin, DeFleur and Sandra Paul Rokich in 1974 [10]. The theory is based on two main pillars represented in goals, where individuals seek to achieve personal and social goals as a result of obtaining information from various media and non-media sources, and the second pillar is information sources where media systems control the sources that achieve the goals of individuals and disseminate the information provided by the sources through many portals and stages, such as: preparing, arranging, coordinating, and publishing in different forms and images that express the original image you submitted [11].

The theory of media dependence is based on a set of main assumptions that have been identified based on the relationship of members of the public to the media and their levels of exposure to it. These assumptions are as follows [12]:

- The relationship of individuals' dependence on the media is determined by the strength of their motives to obtain information that is expected to be useful to achieve their personal goals.
- The degree of individuals' dependence on the media is linked to a direct positive relationship to the level of exposure to media means, so that the higher the degree

of dependence on the media, the higher the levels of individuals' exposure to them.

- The difference in the goals of understanding, guidance, relaxation and expression among members of the public affects the degree of their dependence on the media, as these four goals work as a single component of dependence and related to the achievement of certain goals among the public.

The public's dependence on the media results in three types of effects, as follows:

1. Cognitive effects: These effects are represented in the arrangement of the public's concerns regarding the topics and ideas published by the media, which become more important by increasing the focus of those means on them. And the correctness of these events, and the problem of ambiguity can be solved with the interpretation and supplementation provided by the media, and then the formation of trends through the unlimited dissemination of opinions, topics and personalities of interest to the public, especially when crises, turmoil or wars occur, and the provision of information that share divergent values, not only as news coverage, but also to highlight the basic conflict between the department that shares it with the audience [13].
2. Affective or emotional effects: it means the feelings and emotions that a person has towards his surroundings of events, and this effect appears when certain information is presented through the media that affects the feelings of individuals and their responses, and thus the direction targeted by the media messages, for example that individuals are presented to topics of violence in the media leads to emotional apathy and unwillingness to help others, and the moral support resulting from the media playing important communicative roles, such as expressing the community and its culture, and raising the morale of the public as a result of their feeling of social integration and participation in community activities with the rest of the individuals [14].
3. Behavioral influences: include the elements of activation and inactivity, and the element of activation is related to the individual doing something, as a result of exposure to the media, such as taking a supportive behavioral stance such as: demanding equality and justice or opposition such as engaging in violence and other activities, and for the element of inactivity, it means avoiding doing what leads to negativity, such as: not participating in political and social activities within society, such as elections and events that serve the interest of society [15].

Based on the foregoing, the benefit of this theory is represented in its application and use in the current study, especially in defining the study's objectives and questions in order to know what is the role of The Voice of the Kingdom (Sawt Al-Mamlaka) Program in raising awareness of the Corona virus according to students in the Faculty of Media at Zarqa University.

4 Study Problem and Questions

Jordan is witnessing special circumstances that imposed on young people the necessity of their participation in the issues of their society and their environment, to increase their awareness of what is going on around them and to achieve themselves and their independence in making sound decisions that enhance their contributions and participation in facilitating and developing their lives for the better.

The democratic life in the Kingdom has promoted the development of awareness among Jordanian youth—especially university students—about the various issues of society, and universities are among the national institutions that enhance knowledge that would help students understand life, respect the opinion and the other opinion, and understand their rights and duties. And thus participate in decision-making at various levels of life. The greater the awareness of university students, the greater their pride in their identity and the lessening of their feelings of alienation, thus promoting the values of freedom, belonging, justice and equality.

From here lies the problem of the study, in the following main question: What is the role of The Voice of the Kingdom (Sawt Al-Mamlaka) program in spreading awareness of the Corona virus according to students in the Faculty of Media at Zarqa University?

The following sub-questions arise from this question:

1. Are students in the Faculty of Media at Zarqa University exposed to The Voice of the Kingdom (Sawt Al-Mamlaka) program?
2. What is the role of The Voice of the Kingdom (Sawt Al-Mamlaka) program in spreading awareness of the Corona virus according to the gender variable?
3. What is the role of The Voice of the Kingdom (Sawt Al-Mamlaka) program in spreading awareness of the Corona virus according to the variable of the department?
4. What is the role of The Voice of the Kingdom (Sawt Al-Mamlaka) program in spreading awareness of the Corona virus?

5 Study Objectives

The main objective of the study is to know the role of The Voice of the Kingdom (Sawt Al-Mamlaka) program in spreading awareness of the Corona virus according to students in the Faculty of Media at Zarqa University, and other sub-goals branched from this goal, which are:

1. Determining the degree of respondents' dependence on The Voice of the Kingdom (Sawt Al-Mamlaka) program in spreading awareness of the Corona virus.
2. Determining the degree of respondents' dependence on The Voice of the Kingdom (Sawt Al-Mamlaka) program in spreading awareness of the Corona virus according to the gender variable.

3. Determining the degree of respondents' dependence on The Voice of the Kingdom (Sawt Al-Mamlaka) program in spreading awareness of the Corona virus, according to the variable of the department.
4. Determining the extent to which the students in the Faculty of Media at Zarqa University are exposed to The Voice of the Kingdom (Sawt Al-Mamlaka) program.

6 Procedural Definitions

The Voice of the Kingdom (Sawt Al-Mamlaka): A program broadcast on Al-Mamlaka TV. It is a news talk show that deals with events and real-time news by commenting and describing the situation by hosting the stakeholders, with a focus on political and social news and people's problems and concerns.

Corona virus: It is the severe acute respiratory syndrome virus Corona, and the disease caused by the Corona virus is called "Covid 19" and is the cause of the outbreak of a disease that appeared in China in 2019, and was classified by the World Health Organization in 2020 as a global epidemic due to its spread in various countries of the world.

Students in the Faculty of Media at Zarqa University: A group of students registered in the Faculty of Media at Zarqa University for the undergraduate stage in the first semester of the academic year 2021–2022, compared to the numbers of the Admission and Registration Department at Zarqa University.

7 Study Methodology

This part described the procedures undertaken by the researchers to achieve the objectives of this study, which included a description of the study population from which the sample was drawn and the method in which it was chosen, as well as a description of the study tool and procedures that were followed to ensure its validity and reliability, and how to apply them to the sample members, and a description of the method of data collection and correction method, as well as a reference to the statistical methods that were used.

7.1 Quality of the Method Used in the Study

The study is one of the descriptive studies that aims at an accurate description of the trends and opinions of a group of individuals "they are linked by common personal characteristics and characteristics" towards a phenomenon in a way that allows to know its impact and the impact of its components and characteristics on that group of individuals "the study sample" and to determine the relationship between their characteristics and personal characteristics and the variables of the phenomenon studied [16].

The study adopted the descriptive analytical approach as it is appropriate to achieve the objectives of the study, answering its questions and reaching its results by providing the main data from the respondents about their reliance on The Voice of the Kingdom (Sawt Al-Mamlaka) program in spreading awareness of the Corona virus.

7.2 Study Population

The sample was some students in the Faculty of Media at Zarqa University numbered (230) male and female students. An intentional sample was chosen by random method at a rate of (50%) from the study community with (115) male and female students, and the study tool was distributed to them and the recovery rate was (100%) of the distributed questionnaires, due to the accuracy of the method of applying and following up the questionnaires by researchers, and the following Table 1 shows the demographic distribution of the study sample members.

Study tool

The tool (scale) was built for the role of the Voice of Jordan program in spreading awareness of the Corona virus according to students in the Faculty of Media at Zarqa University, according to the theoretical literature and previous studies. The study scale has two parts:

First: demographic information, consisting of: gender, department, and extent of follow-up to The Voice of the Kingdom (Sawt Al-Mamlaka) program.

Second: study questions consisting of (10) statements, all relate to the role of The Voice of the Kingdom (Sawt Al-Mamlaka) program in spreading awareness of the Corona virus according to students in the Faculty of Media at Zarqa University.

Table 1 Study population by demographic variables

Variable	Category	Frequency	Percentage (%)
Gender	Male	50	43.5
	Female	65	56.5
Department	Journalism and digital media	64	55.7
	Radio and television	51	44.3
How often you watch (Sawt Al-Mamlaka)	High	18	15.7
	Intermediate	63	54.8
	Low	34	29.5
Total (n) = 115			

7.3 *Validation Procedures*

- **Content validation**

The scale was presented after preparing the initial image to (11 arbitrators) of media professionals, specialists and faculty members at Zarqa University, to express opinions on the validity of the content and the affiliation of the phrases to the scale and their suitability for measuring what was set to measure, and the degree of clarity, and then, modifications were suggested, and a standard (80%) was adopted to indicate the validity of the statement, based on the opinions of the arbitrators, some statements were modified in terms of wording to increase their clarity. Some statements were deleted due to their similarity and proximity with other statements, and some statements were deleted because they were not suitable for the purposes of the study and some of them were not suitable for the dimension to which they belong. As a result, the scale became composed of (10) statements; this was an indication of the validation of the content of the study tool.

- **Consistency of study tool**

To calculate the stability of the study tool, the researchers extracted Cronbach's alpha coefficient for the fields of study and the total score of the scale, where it was found that the value of the alpha Cronbach's coefficient for scale items as a whole was (0.902), which is an acceptable value [17].

7.4 *Study Variables*

First: The independent variables including:

- (Gender and Department).
- The Voice of the Kingdom (Sawt Al-Mamlaka) Program.

Second: The dependent variable, Spreading awareness of the Corona virus according to students in the Faculty of Media at Zarqa University.

Statistical methods used:

(SPSS v.21) was used to enter and analyze the study data, with the help of the necessary statistical methods, to achieve the objectives. These methods were as follows:

- (Frequencies and Percentages): for demographic data description purpose for the sample members.
- Mean: to identify how high or low were the responses of the sample members on the statements and fields of the study.
- The Standard Deviation: to identify the deviation in the responses of the sample members on each statement from its mean, the closer the value from Zero the more focused responses and the lowest the distraction.

- (Cronbach's Alpha Scale): to measure the stability of the data.
- (One Sample T-Test): to check the statistical differences in the mean of the responses of the sample members from the complete mean.
- Analyzing two independent samples test: to check the statistical differences of the responses of the sample members due to the variables of gender and department.

8 Study Results

This part of the study includes answering the questions of the study, where the means and standard deviations were extracted to identify the responses of the study sample members to "the role of the Voice of Jordan program in spreading awareness of the Corona virus according to students in the Faculty of Media at Zarqa University", and the following are the answers to the study questions:

Results related to the first question: Is there a role for the Voice of Jordan program in spreading awareness of the Corona virus according to students in the Faculty of Media at Zarqa University?

To answer the first question, the means and standard deviations were extracted to identify the responses of the study sample members to the role of The Voice of the Kingdom (Sawt Al-Mamlaka) program in spreading awareness of the Corona virus according to students in the Faculty of Media at Zarqa University, and the following are the results:

It is clear from Table 2 the means of (the role of the Voice of Jordan program in spreading awareness of the Corona virus according to students in the Faculty of Media at Zarqa University), ranged between (4.14 and 3.75), where the role obtained a total mean of (3.99), which is from the high level, and statement No. (6) had the highest mean, which reached (4.14), and with a standard deviation (0.90), which is from the high level, and the statement stated (I find The Voice of the Kingdom (Sawt Al-Mamlaka) program contributes to spreading awareness among the public to adhere to social distancing), and statement No. (2) came in second place with an mean (4.11) and a standard deviation (0.75), which is of the high level, and the statement stipulated that (I find that The Voice of the Kingdom (Sawt Al-Mamlaka) program contributes to spreading awareness of Corona through the experts and specialists hosted by the program). Statement No. (5) came in second place with an mean (4.09) and a standard deviation (0.67), which is from the high level, and Statement No. (5) states (I find The Voice of the Kingdom (Sawt Al-Mamlaka) program that contributes to persuading the public to receive anti-Corona vaccines).

In the last place, statement No. (3) came with a mean of (3.75) and a standard deviation of (0.72), which is from the high level, as the statement states (I find that The Voice of the Kingdom (Sawt Al-Mamlaka) program contributes to spreading awareness of Corona through the topics discussed in the program).

The results of this study agree to some extent with the study [3], whose results indicated that there is a positive effect of frequent follow-up of the media, including television programs and other sources of information on the awareness of risks and

Table 2 The role of the Voice of the Kingdom program in spreading awareness of the Corona virus according to students in the Faculty of Media at Zarqa University

No	Statement	Mean	Standard deviation	Order	Role
6	I find that the Voice of the Kingdom program contributes to spreading awareness among the public to adhere to social distancing	4.14	0.90	1	High
2	I find that the Voice of the Kingdom program contributes to spreading awareness of Corona through the experts and specialists hosted by the program	4.11	0.75	2	High
5	I find that the Voice of the Kingdom program contributes to persuading the public to receive the anti-Corona vaccines	4.09	0.67	3	High
4	I find that the Voice of the Kingdom program contributes to spreading awareness of Corona by following the developments of the epidemic in the Kingdom's hospitals	4.05	0.78	4	High
1	I find that the Voice of the Kingdom program contributes to spreading awareness of the Corona virus	4.04	0.69	5	High
8	I find that the Voice of the Kingdom program contributes to spreading awareness among the public by the obligation to wear masks to prevent corona	4.04	0.94	5	High
9	I find that the Voice of the Kingdom program contributes to spreading awareness among the public by conducting tests related to Corona when a person feels his symptoms	3.95	0.71	7	High
7	I find that the Voice of the Kingdom program contributes to spreading awareness among the public by implementing defense orders related to the issue of addressing Corona	3.90	0.91	8	High
10	I find that the Voice of the Kingdom program contributes to spreading awareness among the public to adhere to the guidelines related to Corona until the Jordanian economy recovers	3.85	0.76	9	High
3	I find that the Voice of the Kingdom program contributes to spreading awareness of Corona through the topics discussed in the program	3.75	0.71	10	High

(continued)

Table 2 (continued)

No	Statement	Mean	Standard deviation	Order	Role
Overall		3.99	0.57		High

the roles of these media in health awareness of the virus among the respondents, it also agrees to some extent with the study [9], which concluded that there is a direct relationship between the rate of respondents' exposure to paper and electronic newspapers and their level of knowledge of Corona disease, while it differs somewhat with the study [8], which concluded that television and newspapers did not provide enough information about the Corona virus.

There are different subjects that were addressed in the "Sawt Almamlaka" program in the fight against the virus such as the awareness campaigns regarding the importance of vaccination and the importance of following all health precautions from hand sanitizations and wearing face masks. The results of this study indicated that the sample has highly followed the different subjects addressed by the "Sawt Almamlaka" program. The most prominent subjects of which were the discussions and interviews that were presented by the presenter of the program with government officials as well as health officials and policymakers in the health sector. In addition to all interviews carried out with health experts talking about the virus and how to combat it. All these interviews were targeted to increase the audiences' awareness about the negative effects of the pandemic on all social, health, and economic levels.

In order to identify the statistical significance of the role of The Voice of the Kingdom (Sawt Al-Mamlaka) program in spreading awareness of the Corona virus according to students in the Faculty of Media at Zarqa University, the One Sample T-test was used, the results of which appear in the following Table 3.

In Table 3 the mean of **the role of the Voice of Jordan program in spreading awareness of the Corona virus according to students in the Faculty of Media at Zarqa University** was 3.99, with a standard deviation of (0.57), and the calculated T-values amounted to (18,544), which is a higher value of its tabulated value (±1.96), the results show that there are statistically significant differences at the level (0.05) between the mean of the scale; and the hypothetical mean (3.00), and accordingly, it is clear that there is a positive, high-level role for the Kingdom's voice in spreading awareness of the Corona virus according to students in the Faculty of Media at Zarqa University.

Table 3 One sample T-test identifies the role of the Voice of the Kingdom program in spreading awareness of the Corona virus according to students in the Faculty of Media at Zarqa University

Mean	Standard deviation	Calculated (T) value	Tabled (T) value	Freedom level	Statistical significance
3.99	0.57	18.544	1.96	115	0.000[a]

[a] Function at the level (0.05) the value of T = 3.00

Table 4 Two way ANOVA identifies the significance of the differences the role of the Voice of Jordan program in spreading awareness of the Corona virus according to students in the Faculty of Media at Zarqa University attributed to gender and department

Contrast source	Total squares	Freedom levels	Square average	(F) value	Statistical significance
Gender	0.870	1	0.870	2.674	0.105
Department	0.344	1	0.344	1.057	0.306
Gender[a] department	0.019	1	0.019	0.059	0.809

[a] Statistical function at the level (0.05)

Results of the second question: Are there statistically significant differences at the level (0.05) of the role of the Voice of Jordan program in spreading awareness of the Corona virus according to students in the Faculty of Media at Zarqa Universitydue to gender, the department, and the interaction between them.

To answer the second question, the Two Way ANOVA test was used to identify the significance of the differences in the role of The Voice of the Kingdom (Sawt Al-Mamlaka) program in spreading awareness of the Corona virus from the point of view of the students of the Faculty of Media at Zarqa University due to gender and department, and Table 4 illustrates this.

It is clear from Table 4 that there are no statistically significant differences in the role of The Voice of the Kingdom (Sawt Al-Mamlaka) program in spreading awareness of the Corona virus according to students in the Faculty of Media at Zarqa Universitydue to gender and department, where the (F) values reached (2.674, 1.057, 0.059), respectively for gender and department and the interaction between gender and department, which are non-significant values at the significance level (0.05).

This indicates that students of different genders and departments in the Faculty of Media are aware of the positive role of The Voice of the Kingdom (Sawt Al-Mamlaka) program in spreading awareness of the Corona virus.

9 Recommendations

Based on the findings of the study, the researchers recommend the following:

1. The necessity for Jordanian stations and channels to prepare programs that focus on disseminating information related to Corona.
2. Increasing the awareness of media and journalism faculties students about the positives of following up on various media outlets, especially television programs, with the aim of raising their awareness of events and their developments at the local and international levels.
3. Enhancing cooperation between media colleges in Jordanian universities and media institutions to develop and employ programs to serve community awareness and the educational and learning process.

4. The challenges of the Corona pandemic require the media to develop their serious and honest programs and messages directed to the receiving public to raise their awareness of the Corona pandemic and to avoid the irresponsible publication that is sometimes used through social media.

References

1. Tahat, O.: The role of Jordanian television in political participation from the point of view of students of the College of Media at the University of the Middle East. Dirasat Hum. Soc. Sci. J. **44**(2) (2017)
2. Tahat, O.: Electronic press and its role in fighting against the perspective of students of the Faculty of Arts at University of Jordan. Opción Rev. de Cien. Humanas y Sociales **36**(91) (2020)
3. Karasneh, R., Al-Azzam, S., Muflih, S., Soudah, O., Hawamdeh, S., Khader, Y.: Media's effect on shaping knowledge, awareness risk perceptions and communication practices of pandemic COVID-19 among pharmacists. Res. Soc. Adm. Pharm. **17**(1), 1897–1902 (2021). Available: http://doi.org/10.1016/j.sapharm.2020.04.027
4. Abdel-Dayem, M.: Treating the talk shows on the Egyptian public and private satellite channels for the crisis of the Corona virus (Covid-19): an analytical study. J. Media Res. **4**(54) (2020)
5. Basheti et al.: Pharmacists' readiness to deal with the coronavirus pandemic: assessing awareness and perception of roles. Res. Soc. Adm. Pharm. **17**(3), 514–522 (2021). Available: http://doi.org/10.1016/j.sapharm.2020.04.020
6. Nooh, H., Alshammary, R., Alenezy, J., Alrowaili, N., Alsharari, A., Alenzi, N., Sabaa, H.: Public awareness of coronavirus in Al-Jouf region, Saudi Arabia. J. Public Health **29**(5), 1107–1114 (2021)
7. Ali, M.: The role of Facebook network in promoting health awareness among the public. Unpublished master's thesis. Middle East University, Faculty of Media, Jordan (2017)
8. Choi, D.H., Yoo, W., Noh, G.Y., Park, K.: The impact of social media on risk perceptions during the MERS outbreak in South Korea. Comput. Hum. Behav. **72** (2017)
9. Abu Sunna, N.: The relationship of exposure to Saudi and online newspapers to the level of knowledge of corona disease. J. Media Res. (1) (2015)
10. Ball-Rokeach, S., DeFleur, M.: Dependency model of mass media effects. Commun. Res. **3**(1), 3–21 (1976)
11. Devler, M., Rokich, P.: Theories of Communication. International Publishing House, Cairo (2002)
12. Ball-Rokeach, S., Rokeach, M., Grube, J.: Great American Values Test: Influencing Behavior and Belief Television. The Free Press, New York (1984)
13. Abdel Hamid, M.: Media Theories and Influence Trends. The World of Books for Distribution, Egypt
14. Hijab, M.: Communication Theories. Dar Al-Fajr for Publishing and Distribution, Egypt (2010)
15. Makkawi, H., El-Sayed, L.: Communication and Its Contemporary Theories. Egyptian Lebanese House, Cairo (1998)
16. Hijab, M.: The Basics of Media and Social Research. Dar Al-Fajr for Publishing and Distribution, Egypt (2003)
17. Hair, F., Black, C., Babin, J., Anderson, E., Tatham, L.: Multivariate Data Analysis, 7th ed. New York (2010)

Artificial Intelligence and the Challenges of New Journalistic Formats

Abdulkrim Ziani, Ismail Noori, and Mohammed Fyadh

Abstract Digital revolution imposed its effects on the reality of knowledge production in various fields and disciplines. The media and the press gained a large share of this transformation, as the response was through the emergence of new forms based on innovation and renewal. The origin of these journalistic forms remains in need of study. And analysis to stand on its impact and track its characteristics and expected future changes. This is due to the calculation that innovation and innovation grew significantly in 2019, especially in digital journalism based on the Internet and interactive texts that invest written text, images, video, and news games. Journalism practice, an information professional, specializing in collecting and retrieving data and information, is responsible for information service. The reality of innovation and renewal will impose its direct effects on the public, institutions working in the field of news, and the way to deal with available resources, in addition to the content.

Keywords New journalistic formats · Immersive journalism · News media · Structured journalism

1 Introduction

The news media is fueling the march towards benefiting from artificial intelligence applications, especially in the news industry. Benefiting from the sub-fields of artificial intelligence represented in [1]; machine learning [2]; computer vision [CV] [3]; speech recognition [4]; natural language processing [NLP] [5]; planning, scheduling,

A. Ziani (✉) · M. Fyadh
Mass Communication College, Umm Al Quwain University, Umm Al Quwain, UAE
e-mail: drabdul.karim@uaqu.ac.ae

M. Fyadh
e-mail: drmohammed.f@uaqu.ac.ae

I. Noori
Arts and Sciences College, Ahlia University, Manama, Kingdom of Bahrain

A. Hamdan et al. (eds.), *The Implementation of Smart Technologies for Business Success and Sustainability*, Studies in Systems, Decision and Control 216,
https://doi.org/10.1007/978-3-031-10212-7_33

and optimization [6]; expert systems and [7] robotics [1]. The successive developments in the field of artificial intelligence made those in charge of the news media reconsider more technologies that constituted a qualitative leap in journalistic practice at the time of their emergence. For example, virtual reality technology has been of exceptional importance. Still, the growth and interaction of immersive technology is the most prominent result that had a practical impact in stimulating journalistic practice towards broader horizons [2]. This is with the importance of acknowledging that artificial intelligence has had a significant effect on the complete change of journalism and at various levels in the world; collection, production, and distribution. The matter today is based on the importance of focusing on the aspects of innovation and renewal and the constant endeavor towards improving and developing the journalist's skills. This is because of the mechanism and ready-made templates provided by artificial intelligence, at various levels, if it is in terms of ready-made designs, ready-made texts, and editing ready-made news and articles until the journalist has no choice but to watch the mass-produced mechanisms of journalistic products. In particular, the most critical challenge arises in human presence and working to make these technologies aids [3]. Use virtual 3D graphics, work on the use of the screen and sensors installed in the head helmet, and work on monitoring interactions based on [1]; immersion and three-dimensionality [2], subjective presence [3], sensory illusion, and [4] interaction with motion tracking and its effect on the way narrative content is received. I have the future [4]. Effective communication is what journalistic practice is meant to look like. Digital technology has its feasibility, value, and even necessity. It is the controlling and systematic way of working. But the human touch remains the practice, as the possibility that makes the journalist the link between media coverage and the subject. It is the ability to represent an incident of social empathy and to make the audience sympathetic to the news story, based on the way it shot it. The journalist here can draw the features of (the victim or the executioner) by employing language, style, and context [5].

2 Good and Compelling News Story

From the reality of data and information produced by the digital age, data-driven journalism [DDJ] has emerged, based on the analysis and filtering of data [6] and the trend towards building a good and compelling news story. The importance of this process stems from the massive increase in the number of news content producers, the entry of more mechanisms and dynamics that are based on investing digital technology in design and editing, and the adoption of an interdisciplinary approach at the level of computer science, statistics, and human sciences. This process contributes to the openness of the news media to the enormous transparency produced by the digital revolution, and thus the ability to deal with reality flexibly and make room for innovation, renewal, and effective communication. By taking advantage of the open sources provided by the Internet environment regarding publishing [7], data, and software. This approach adopts the traditional journalistic method of follow-up and monitoring and uses leaked materials and reports to build a well-structured news story. To create a stronger relationship with the audience of the recipients, and work to consolidate the bonds of the relationship by presenting a professionally built news story that has a direct impact in helping the public to absorb more situations, based on reports provided by trained and qualified journalists and Professionalism, away from the confusion and confusion caused by the statements made by more intruders and inexperienced amateurs.

The idea of structured journalism is about satisfying an audience's need for information on a topic [8], not a desire to know the latest happenings, in the sense of striving towards structuring data in an organized manner and making it a treasury of information that is invested in supporting and sustaining journalism. It is how a journalist creates content and makes it usable to the maximum extent possible, given its usability in creating new news stories.

Immersive Journalism is a type of journalistic production that seeks to employ all available digital technologies, the main objective of which is to make the audience participate [9] amid the news story through the use of image, three-dimensional games, and sounds, to create an atmosphere that deepens the recipient's sense of presence and emotional interaction with the event. From the reality of virtual reality, this form of journalism allows the audience to immerse themselves in the experience by entering the virtual version and making it a participant and witness.

360° video reports are based on an omnidirectional spherical camera and can use several cameras to cover the target area. This technology has become popular today, having become necessary to download on the smart phone [10].

3 Conclusions

The fields of emotional computing expanded to include various fields and specializations; for example, its active and vital role emerges at the level of the interactive learning environment, where the orientation towards identifying the student's feelings, while receiving the lesson, through the direction towards monitoring the expressions issued by the face, in addition to speech Issued by the student, which reflects his condition. With the importance of taking into consideration the method of directing the lesson to the students, based on the experience that the teacher possesses, the behaviors issued by him, and the talents that he enjoys, in addition to the persuasive method through which students are attracted to the lesson, and interact positively with it. The main objective here is to improve students' motivation to learn.

From this reality emerges the importance of awareness that emotional computing in education interacts in Intelligent Tutoring Systems, ITS, based on comprehensive knowledge content while relying on accurate educational goals and keenness on high discipline. An educational system with the task of providing the main notes and precise explanations to the students through a computer program. A video movie, Word, or PowerPoint program, and this system contribute to facilitating the student's homework. And provide more explanation for the topics that are difficult for some students [11]. And suppose the effectiveness is based on activating the computer path and focusing on automated action. In that case, the matter, in the end, is an expression of [the interaction of human feelings], where the endeavor is towards the experience of education as a path based on creativity, innovation, development, development and qualification, and setup. Talking about observing and observing students' reactions in terms of facial expressions, manner of speaking, and emotional responses all interact with a relationship between one teacher and a broad group of students. While analyzing information related to the classroom remains, it represents a shared experience between the teacher and the students, which needs more data analysis, actions, and behaviors. Considering that the main objective is to increase teaching effectiveness through the development of data related to the classroom and work to improve performance and development of curricula and create an atmosphere of interaction between the teacher and the student, profound and practical [12]. It is the aspiration towards investing this technology in monitoring the response emanating from feelings, by processing data and information, by focusing on sensors that monitor different cases of behavioral and physiological reactions, and working to employ them to serve the way of dealing with digital technology in a way Practical, easier and more useful [13]. The exciting thing is that the primary trend is to invest feelings as information and data, which can be integrated into the communication system and directly contribute to the decision-making method. However, the ethical aspect remains present, given the importance of standing on the correct and accurate way in monitoring feelings, expressing them in line with reality, and telling them honestly and truthfully [14].

References

1. de-Lima-Santos, M.F., Ceron, W.: Artificial intelligence in news media: current perceptions and future outlook. Journalism Media **3**(1), 13–26 (2022). http://doi.org/10.3390/journalmedia3010002
2. Godulla, A., Planer, R., Wolf, C., Lück, A., Vaaßen, F.: An immersive journey through flawed technology: users' perceptions of VR in journalism. Journalism Media **2**(3), 454–468 (2021). https://doi.org/10.3390/journalmedia2030027
3. Tejedor, S., Vila, P.: Exo journalism: a conceptual approach to a hybrid formula between journalism and artificial intelligence. Journalism Media **2**(4), 830–840 (2021). https://doi.org/10.3390/journalmedia2040048
4. Mills, K.A., Brown, A.: Immersive virtual reality (VR) for digital media making: transmediation is key. Learn. Media Technol. (2021). http://doi.org/10.1080/17439884.2021.1952428
5. Roberts, J.: Empathy cultivation through (pro)social media: a counter to compassion fatigue. Journalism Media **2**(4), 819–829 (2021). https://doi.org/10.3390/journalmedia2040047
6. Hammond, P.: From computer-assisted to data-driven: journalism and big data (2015). http://jou.sagepub.com/content/early/2015/12/15/1464884915620205.a
7. Lopezosa, C., Codina, L., Fernández-Planells, A., Freixa, P.: Journalistic innovation: how new formats of digital journalism are perceived in the academic literature. Journalism (2021). https://doi.org/10.1177/14648849211033434
8. Caswell, D.: Structured journalism and the semantic units of news. Digit. J. **7**(8), 1134–1156 (2019). https://doi.org/10.1080/21670811.2019.1651665

9. Soler-Adillon, J., Sora, C.: Immersive journalism and virtual reality. In: Pérez-Montoro, M. (eds.) Interaction in Digital News Media. Palgrave Macmillan, Cham (2018). http://doi.org/10.1007/978-3-319-96253-5_4
10. https://pie.en.softonic.com/iphone
11. Keles, A., Keles, A.: Intelligent tutoring systems: best practices. In: Stankov, S., Glavinic, V., Rosic, M. (eds.) Intelligent Tutoring Systems in E-Learning Environments: Design, Implementation and Evaluation, pp. 1–26. IGI Global (2011). http://doi.org/10.4018/978-1-61692-008-1.ch001
12. Su, S., Lin, H.K., Wang, C., Huang, Z.: Multi-modal affective computing technology design the interaction between computers and human of intelligent tutoring systems. Int. J. Online Pedagogy Course Des. (IJOPCD) **6**(1), 13–28 (2016). http://doi.org/10.4018/IJOPCD.2016010102
13. Tara, J.: Brigham merging technology and emotions: introduction to affective computing. Med. Ref. Serv. Q. **36**(4), 399–407 (2017). https://doi.org/10.1080/02763869.2017.1369289
14. Richardson, S.: Affective computing in the modern workplace. Bus. Inf. Rev. **37**(2), 78–85 (2020). http://doi.org/10.1177/0266382120930866

Levels of Psychological Stress and Social Isolation Among Palestinian Media Professionals Working in Palestinian Media Institutions During Covid-19 Pandemic

Oqab Jabali, Abed Alkarim Ayyoub, and Mohammad Suliaman

Abstract This study aimed to identify the level of psychological stress and social isolation among a sample of Palestinian media professionals working in Palestinian media institutions. Participants of the study incorporated (200) media professionals, aged between (60–25) years; they were selected using purposeful sampling. Two scales of psychological stress and social isolation were developed by the researchers. The results of the study showed that the level of psychological stress among the media professionals was high while the level of social isolation was medium due mainly to Israeli occupation and spread of Covid-19 pandemic. The study results also showed that there were no statistically significant differences at the level ($\alpha \leq 0.05$) in the level of feeling psychological pressure and social isolation attributed to a number of demographic variables including age, marital status and years of experience.

Keywords Covid-19 pandemic · Psychological stress · Social isolation · Palestinian media professionals

1 Introduction

Psychological stress is related to many professions and fields especially those that entail coming into contact with individuals and following up on events and realities. The media profession is important as it is the fourth authority in society; it reveals facts, follows up on events and conveys them to public opinion; journalists should stand before their professional responsibilities while even if they are in a state of

O. Jabali (✉)
Faculty of Humanities, Language Center, An-Najah National University, Nablus, Palestine
e-mail: oqab.jabali@najah.edu

A. A. Ayyoub
Faculty of Economics and Social Sciences, An-Najah National University, Nablus, Palestine
e-mail: ayyoub@najah.edu

M. Suliaman
Jordan University, Morogoro, Tanzania

A. Hamdan et al. (eds.), *The Implementation of Smart Technologies for Business Success and Sustainability*, Studies in Systems, Decision and Control 216,
https://doi.org/10.1007/978-3-031-10212-7_34

tension and anxiety, which, by turn, may produce psychological pressures that accumulate over time. One of the reasons that result in existing pressures is the inability to separate the media professional life from the personal and social life especially during crises and epidemics or pandemics such as the current Covid-19. Conflict in the instructions issued to the media organizations may be another cause of more stressful situations that workers face in the media field [7].

The Palestinian media operates in a complex environment which affect its performance. The Israeli occupation constitutes one of the main obstacles that stand in the way of the Palestinian media cadres and prevent them from exercising their work freely. The occupation does not suffice with abusing journalists, it rather insists on violating the media freedoms and limiting their movement and communication with the outside world. This has impacted the media professionals and their performance negatively and made them unable to practice their work freely [2].

Another major hindrance that impacts media cadres negatively is the spread of Covid-19; it limited the mobility of most workers and employees and decreases their access to their job to the minimum [22]; it also caused many journalists to lose their jobs as in the case of An-Najah Broadcasting Channel (Nbc) and Ajyal Radio Network in Ramallah [16]. They felt more pressures and were pushed to live deep isolation beside the isolation caused by quarantine.

The nature of the work done by the media professionals dictates that they are often in the middle of the event and in the focus of the conflict, which makes them vulnerable to threats and dangers; it requires them to be on a high degree of alertness and they should be ready to work around the clock [10].

Psychological stress has different meanings for individuals, groups and professionals depending on their specializations and areas of interest. For psychologists, it is the psychological state that pushes the individual to a state of anxiety or distress and a feeling of discomfort and instability [4]. Sarafino [26] defines it as an event or group of events that the individual perceives as threatening and harmful to the extent that they make him/ her feel in a state of tension and instability; while Smither (2009) states that psychological stress as a physiological or psychological response that a person feels when being exposed to unpleasant events and situations. For [18] psychological stress is a disturbing psychological state for the individual that may make him/her feel tense and depressed and cause him/her a state of anxiety and disharmony.

The researchers believe that stress is a state of inequity especially when pressures exceed the human's ability to bear them. There are many sources that lead individuals to suffer from psychological stress. Duck [19] indicates that there are external sources that cause stress (e.g., family conflicts, financial and economic pressures, occupational such as working long hours, fluctuations and political events that afflict the Arab region and Palestine in particular). The internal sources may result from individual's suffering from psychological and health problems, the way of thinking and the way of interpreting events and situations [13]. Many theories explain psychological stress. Lazarus' theory considers that cognitive processes and the way in which the individual perceives situations and the way he interprets them is what causes stress, while the psychoanalytic theory argues that the feeling of psychological pressure arises as a result of repressed instincts, motives, needs and desires. In Selye's theory, stress is a general response to multiple environmental events [14].

A lot of research was conducted the various types and levels of psychological stress among various sectors of professionals and workers including journalist (e.g., [1–3, 20, 21] other researchers investigated the level of burn-out and motivation among professionals [7, 11, 24]. All these studies found that pressure levels and burnout were somehow high and that burn-out is manifested in stress, dullness and lack of achievement. According to these studies, females experience more pressure and stress than males.

Psychological pressures may lead to an individual's feeling of social isolation which is a stressful experience related to lack of satisfaction of the need to belong to others and lack of social integration. Al-Rawajfa [8] defines social isolation as the extent to which the individual feels lonely and isolated or distant from them; it leads to the weakness of the social network to which the person belongs. Schaefer and Millman [27] argue that social isolation is an extreme pattern of disturbances in relations with the group, as the individual prefers to separate from others and remain alone most of the time due to circumstances that may be beyond his/her control. Maroldo [23] defines social isolation as the failure to participate in social situations in an appropriate and effective manner.

Social isolation greatly affects the personality of the individual and his/her relationship with social environment; it indicates a decrease in the individual's ability to effective social interaction. People who suffer from social isolation have lack of learning, social awareness, and cognitive growth; they tend to distrust others. Social isolation affects the individual at the personal and social levels. Many theories have explained the problem of social isolation. The psychoanalytic theory considers that the feeling of isolation and loneliness represents a state of repression of the surrounding experiences in the unconscious that was acquired during early childhood; the individual resorts to isolation in case s/he fails to obtain warmth and intimate relationships with others [17]. For the behavioral theorists social isolation is a learned habit from the surrounding environment; social learning theory sees that the individual's isolation behavior is caused by the failure to learn the most appropriate ways of dealing with people or the individual's lack of appropriate social skills [5].

The researchers believe that psychological stress and social isolation are linked to each other; both affect the human personality, especially the personality of the Palestinian media person. Normal people can adapt and coexist with others in natural relationships that enhance the level of psychological health of the individual. Avoiding effective communication leads the individual to engage in defensive and isolationist behavior. The loss of a sense of belonging and psychological security and the lack of clarity in the image of the future in front of individuals in society, especially the Palestinian media may result in psychological disorders.

Social isolation and seclusion was also investigated by researchers (e.g., [5, 6, 9, 15]. Most of these studies tackled different types of professionals rather than the media profession, they found that social isolation levels were medium among females but high among males and that the unemployed feel more isolation than those with jobs. The current study is the first to investigate isolation among media cadres.

The state of alienation that Palestinian media professionals live in is attributed to the absence of a unified media body that brings them together under its umbrella and reflects the true picture of the current Palestinian situation due to the state of division and dispersion. The Palestinian territories have been subjected to many political, social and economic events that imposed many psychological pressures on media cadres as they do not receive sufficient attention and investigation. Palestinian media professionals are exposed to violations committed by the Israeli occupation which still practices its crimes within an organized and systematic policy to conceal the truth and silence mouths [12].

2 Statement of the Problem and Questions

The media profession is a profession full of psychological troubles especially in the Palestinian environment which abounds with risky and harsh events represented by incidents of martyrdom, Israeli atrocity and settles attacks. Most of these events impose new pressures on media cadres and institutions. The study's problem lies in identifying the level of psychological stress and its relationship to social isolation among a sample of media professionals working in media institutions operating in the Palestinian territories using a number of variables. Put simply, the current study aims to reveal the levels of psychological stress experienced by media professionals in the various media outlets operating in the Palestinian territories, and the extent of the impact of the variables (e.g., marital status, age, experience). The researchers meant to answer the following questions:

1. What are the levels of psychological stress and social isolation among media professionals working in media institutions operating in the Palestinian territories?
2. Are there statistically significant differences at the significance level ($\alpha \leq 0.05$) in the level of psychological stress and social isolation among workers in media institutions attributed to the three variables?

3 Method and Procedure

The current study used the descriptive/ correlative approach. The study population consisted of all the (600) Palestinian media professionals; the study sample which included (200) media professionals whose ages ranged between (25–60) years was chosen purposefully due to the fact that it is cost-effective in time and cost and it helps choose the participants who are mostly more knowledgeable of the study phenomenon (Look at Table 1). Data was normally distributed and was analyzed

Table 1 Distribution of study sample based on the three variables

Variable		N
Age	25–40	118
	41–60	82
Total	200	
Marital status	Married	102
	Single	98
Total	200	
Experience	1–10 years	114
	11 years or more	86
Total	200	

using descriptive statistics; t-test was used to calculate the mean differences between the scores using SPSS version 26.

4 Study Results and Discussion

The results related to the first question.

"What are the levels of psychological stress and social isolation among media professionals working in media institutions operating in the Palestinian territories?" In order to answer this question, means and standard deviations of the sample members were calculated on the scale with its total degree. The mean was (3.88) and the standard deviation was (0.651). This means that the level of feeling psychological pressure among Palestinian journalists was high.

This can be attributed to the fact that Palestinian journalists are exposed daily to great psychological pressure and to a state of severe psychological and physical stress as a result of the nature of the work they do. A large number of media professionals is subjected to continuous, organized and systematic targeting by the Israeli occupation; it takes some of them checkpoints and interrogate them for long hours; it also prevents a large part of them from traveling in order to confuse them and deprive them of a sense of stability and the ability to perform their work completely freely. There is no sanctity or immunity for any journalist, place or person before the occupation and its security services. This is evident through the martyrdom of many journalists and the wounding of dozens of them.

Another important cause of psychological stress among journalists is the disruption of social life in the media's life; many media professionals suffer from continuous change in the social environment and geographical location. In addition, there is the pressure caused by the financial situation, economic instability, accumulation of needs, security threats, overwork, poor professional relations in some media institutions, and the absence of an official incubator, in the true sense of the word, to protect and defend media professionals and give them their rights and provide them with their needs. The political division between the West Bank and Gaza has confused many media institutions and professionals; it also decreases their motivation. Such results are consistent with [2] and [7] which state that media cadres are exposed to stress and their motivation to do a better job decreases proportionally.

In terms of social isolation, the means (3.62) and standard deviations (0.189) of the sample members were calculated. This means that the level of social isolation of Palestinian media people was moderate; Palestinian media people watch many harsh and painful events on different television channels or through their direct contact with incidents of martyrdom, the funerals of the martyrs, and the continuous incursions into the various Palestinian villages, cities and camps. This creates a general feeling of fear and threat, accompanied by anxiety and anticipation of danger.

Some Palestinian journalists suffer from a feeling of alienation and confusion about their personal, social, academic and professional future, which ultimately leads to a state of helplessness and despair in life, dissatisfaction with life and psychological and social maladaptation that ultimately leads to a state of social isolation. This result is consistent with the results of the study [25], which indicated that individuals who were exposed to traumatic experiences in their lives had expectations of the future bearing a pessimistic view and an expectation of weak social relations in the future.

Social isolation among Palestinian media professionals can be explained by the life difficulties; it affects their existence, lives, affiliation and sense of the meaning of life. Many media professionals receive relatively little income and this leads them to working in other additional jobs in order to secure the harsh requirements of life especially if they are married and have children. They may resort to seclusion or to compensatory methods for direct social life such as social communication to compensate or enhance their social isolation. Such results are consistent with Crowe (2016) in that social support is one of the factors that promote mental health.

The results related to the second question:

"Are there statistically significant differences at the significance level ($\alpha \leq 0.05$) in the sense of psychological stress and social isolation among media professionals, attributed to the three variables?" To answer this question, the means and standard deviations of the study members' estimates of the total score of the psychological stress and social isolation scale were calculated and a t-test was conducted. Table 2 shows the results:

The study results showed no statistically significant differences at the significance level ($\alpha \leq 0.05$) in the sense of psychological stress and social isolation among media professionals, attributed to the three demographic variables.

The researchers attribute this result to the fact that the Palestinian media person, regardless of their age group, marital status, and experience, is subjected to the same conditions and the same policy of violation and prosecution. A large part of the media cadres is subject to targeting and arrest by the Israeli occupation; this makes the Palestinian media person to live in a state of psychological pressure which, by turn, prevents them from defining their future goals and achieving their dreams and ambitions. Suffering constant psychological stress helps isolate them from the social life.

Table 2 T-test of the average estimates of the study members in psychological stress and social isolation according to the three variables

Domain	Age	Mean	SD	t	p
Psychological stress	40–25	3.91	0.541	0.892	0.11
	60–41	3.85	0.492		
Social isolation	40–25	3.64	0.671	0.764	0.62
	60–41	3.59	0.645		
Domain	**Marital status**	**Mean**	**SD**	**t**	**p**
Psychological stress	Married	3.89	0.732	0.742	0.54
	Not-married	3.86	0.682		
Social isolation	Married	3.63	0.671	0.683	0.18
	Not married	3.61	0.547		
Domain	**Marital status**	**Mean**	**SD**	**t**	**p**
Psychological stress	Married	3.89	0.732	0.742	0.54
	Not-married	3.86	0.682		
Social isolation	Married	3.63	0.671	0.683	0.18
	Not married	3.61	0.547		

Moreover, the suffering experienced by the Palestinian journalist is one; it begins since the journalist joins the journalistic work and continues with until advanced years of his life. The researchers believe that the years of experience of the media person are not an indication of his enjoyment of psychological strength, but rather it depends on psychological construction, the ability to withstand difficulties and the extent of his awareness and belief in the justice of the issue.

5 Conclusion

The current study aimed at investigating levels of psychological stress and social isolation among Palestinian media professionals working in Palestinian media institutions based on a number of demographic variables. The main study findings showed that Palestinian media cadres feel too much stress which results in moderate levels of social isolation and seclusion. The researchers also found that the demographic variables were not influential even during crises and tough situations like the spread of covid-19; in other words, cadres' ages, marital status and experience did not yield statistically significant differences on the level of stress and social isolation among these cadres. Each person involved suffers the same type of stress which is mostly attributed to the Israeli occupation as well as the various life hardships that prevail in Palestine. It is axiomatic that when people get stressful, they tend to isolate themselves socially so that they feel better or reduce the negative impacts of these feelings i.e., stress and isolation, on their families and surroundings.

The interested authorities and agencies should spare no effort in handling this problem and find solutions so that these cadres carry out their mission the right way. Counselling and guidance are supposed to be present by media institutions to their cadres; financial support is also needed to be given to those who may be victims of Israeli occupation and its practices.

6 Recommendations

The researchers call for the need to:

1. Work hard to provide qualified psychologists and social workers for early detection of stressors and psychological problems that media professionals may encounter especially during pandemics such as Corona.
2. Work on building counseling and training programs specialized in the psychological field and concerned with rehabilitating journalists on how to deal with the psychological stress or pressures that they may face.
3. Work to promote the concept of social integration through socializing and networking with relevant institutions in order to secure the economic situation and achieve stability and social welfare for them.

References

Arabic References

1. Abu Al-Hossein, M.: Psychological stress among nurses working for the government and its relationship to self-efficacy. Master's thesis, College of Education, Islamic University, Gaza, Palestine (2010)
2. Abu Own, D.: Psychological pressures and its relationship with motivation and self-esteem among a sample of journalists after Gaza war. Master's thesis, Gaza University, Palestine (2014)
3. Ahmed, H.: The Impact of the israeli siege on the media in the Gaza strip. Palestinian Center for Development and Media Freedoms, Ramallah, Palestine (2010)
4. Al-Farmawi, H., Abdullah, R.: Psychological Stress in Work and Life. Daar Al-Safa'a for Publishing and Distribution, Amman (2009)
5. Al-Fatalawi, A.: Social isolation among Iraqi refugees. J. Arts Baghdad Univ. **91**, 383–413 (2009)
6. Al-Harbi, M., Abdullah, H.: Efficiency of a counselling program based social isolation among primary school pupils. Education Study Institute, Jeddah, Saudi Arabia (2017)
7. Al-Qarni, S.: Media and self burn-out: a study of levels of occupational stress in Saudi media institutions. J. King Saudi Univ. **3**(2) (2004)
8. Al-Rawajfa, A.: The effect of a group counseling program on reducing the feeling of social isolation of a sample of students from the first secondary stage in Jordan. J. Educ. Sci. **12**, 15–43 (2004)

9. Al-Zain, M., Al-Shar'a, H.: The level of meaninglessness and social isolation among employed and unemployed persons in al-Jisa district: comparative study. Jordanian Educ. J. **2**(1), 213–240 (2017)
10. Al-Zeyoudi, M.: Sources of social isolation and burn-out among special education instructors in Karak Governorate based on some variables. J. Damascus Univ. **23**(2), 189–219 (2007)
11. Barakat, F.: Self bur-out among a sample of journalist working in Damascus in light of some variables. J. Arab Univ. Union Educ. Psychol. **12**(2) (2014)
12. Hammad, A.: Impact of Israeli Occupation on Mass Media in Gaza. Palestine Center for Media Development and Freedom, Ramallah (2010)
13. Ibrahim, A.: Psychology: its foundations and features. Dar Al Uloom: Riyadh (2003)
14. Jabareen, R., Banat, S.: Violence against women and its relationship to genderic thought and psychological stress in the Lower Galilee region. Master's thesis, Amman Arab University, College of Educational and Psychological Sciences, Jordan (2016)
15. Jbara, K.: The relationship between excessive use of internet and social isolation among students of education at Ha'el University. J. Irbid Commun. Univ. **19**(3) (2018)

Foreign References

16. Anadolu Agency: Corona ravages the Palestinian media (2021). https://www.aa.com.tr/ar (2012)
17. Capuzzi, D., Gross, D.R.: Introduction to the counseling profession. Routledge: New York (2012). https://doi.org/10.4324/9780203120064
18. Daivs, S.: The relationship between optimism, loneliness and death anxiety. Bull. Psychon. Soc. **30**(135–136) (2000)
19. Duck, S.: Human Relationships. Stage Publications, London (1999)
20. Feinstein, A.: The emotional health of war journalists. In: The Annual Conference of Journalists in Danger, Toronto (2001)
21. Hayes, C.: Walking mean streets: journalists and adaptation. In: The Annual Conference of Journalists in Danger, Toronto (2001)
22. International Labor Organization impact of the Covid-19 pandemic on the labour market in the occupied Palestinian territory: a forecasting model assessment. https://www.ilo.org/wcmsp5/groups/public/---arabstates/---ro-beirut/documents/publication/wcms_774731. Last accessed 2022/1/20
23. Maroldo, G.: Shyness and loneliness among college men and women. Psychol. Rep. **48**(3), 885–886 (1981). https://doi.org/10.2466/pr0.1981.48.3.885
24. Reinardy, S.: Women newsrooms show signs of in newspaper burn out from. http://www.allacademic.com/meta/p_mla_apa_research_citation/603030602/p633623_index. Last accessed 2022/1/13
25. Rialon, R.A.: A comparative analysis of the children's future orientation scale ratings of traumatized urban youth with and without posttraumatic stress disorder. Doctoral dissertation, Columbia University, New York, NY (2011)
26. Sarafino, E.P.: Health psychology: biopsychosocial interactions (4th Ed.). New York, NY, USA: John Wiley & Sons, Inc (2002)
27. Schaefer, C., Milman, H.: How to help children with common problems. Jason Aronson, Inc: England (1999)

Managing the Sale of Agricultural Land in Ukraine: Opportunities and Challenges

Nadiia P. Reznik, Victoria Nikolaevska, and Yuliia Havryliuk

Abstract The article examines the current state, conditions and main tasks of the land market and directions of state policy in the field of regulating land relations, highlights the main problems of its development, analyzes the experience of purchase and sale as well as renting agricultural land in European countries. It is proved that investment and innovation activity, as a function of land resources management, is a crucial section in the land policy of the state, as well as a conceptual approach to the formation of regulatory and legal regulation of land relations.

Keywords Land · Land relations · Chernozems · Sustainable development · Agrosphere · Management · Sales · Land market

The problem of introducing free circulation of agricultural land in Ukraine has become one of the priorities for both agricultural producers and agricultural economists, politicians and other specialists directly or indirectly connected with the agricultural sector. This is explained by the fact that land is not only the main means of production in agriculture, but also an important component of production capital, one of the most important economic resources, which occupies a central place in the world policy and strategy of economic development of any state. It should be noted that, having such a powerful natural resource potential, the agricultural sector of Ukraine significantly lags behind the leading countries of the world in terms of labor productivity. There is also a structural imbalance caused by an irrational production structure, excessive depletion of natural potential. The problems of inefficient organization of rural development, their depressiveness, insufficient diversification of activities, unemployment and impoverishment of the rural population, increasing

N. P. Reznik (✉)
Department of Management of Organization named after Y. S. Zavadskiy, National University of Life and Environmental Sciences of Ukraine, Kyiv, Ukraine
e-mail: nadya-reznik@ukr.net

V. Nikolaevska · Y. Havryliuk
Marketing and International Trade Department, National University of Life and Environmental Sciences of Ukraine, Kyiv, Ukraine

A. Hamdan et al. (eds.), *The Implementation of Smart Technologies for Business Success and Sustainability*, Studies in Systems, Decision and Control 216,
https://doi.org/10.1007/978-3-031-10212-7_35

anthropogenic burden on the environment, inefficient use of land resources are added to this [6].

However, it is obvious that ensuring resource-efficient ecological socio-oriented growth of the domestic agricultural sector requires urgent settlement of land relations. Today, 6.9 million owners of land shares (1.6 million of whom are over 70 years old) do not have the opportunity to independently cultivate land, rent it out profitably, sell it or pledge their own share. At the same time, non-transparent schemes of alienation of agricultural land are used under the veiled protection of the interests of peasants and the preservation of national wealth. That is, in fact, the shadow land market associated with the sale of land plots already exists today. And in the context of growing demand for agricultural products in the world, a significant reduction in natural resources, the existing format of land relations destructively affects the socio-economic processes in rural areas and hinders the economic growth of the agricultural sector.

With the entry into force of the law on the land market (Law of Ukraine No. 552-IX of 31.03.2020), which marked the beginning of the long-awaited land reform, Ukrainian citizens have the opportunity to freely alienate and purchase agricultural land plots (shares) with a total area of up to 100 ha, so they started to be increasingly interested in agricultural land as an object for investment [15].

Although this law is primarily important for owners of such land plots and their tenants (sole farmers, farms and agricultural holdings), whose long-term lease relations can be transformed into purchase and sale relations. At the same time, it is no less interesting for people who are not involved in rental relations or the agricultural sector in general. These are people who consider the relevant land plots as a potential investment object that can provide significant capital gains in the future.

The first task of land reform concerns deep production processes, the entire spectrum of human interaction with the natural environment. For the practical implementation of all this, a set of mutually agreed legal, economic, environmental, technical and organizational measures is needed, which together will ensure a painless, effective transition for the economy to a new land structure, new criteria for assessing the level of management on land. The latter should take into account: financial results (the ratio of income and expenses), applied capital (including land), deferred expenses (or capital losses), the number of jobs [1, 4].

The second task of the land market is to form rational land use and is closely related to the first, as it is an integral part of the overall problem of land turnover in order to create prerequisites for more efficient management. Here it is advisable to highlight the problems of rationality of land use sizes connected to the redistribution of land, the withdrawal of excess land plots from some users and their transfer to others. The organizational and legal form of land use of large-scale private agricultural production is unpredictable in Ukrainian legislation. The economic consequences of excessive concentration are reduced to the problem of farm manageability, which has different solutions in relation to different organizational and legal forms. The negative consequences of excessive concentration of land in one owner are most clearly manifested if the land owner is a natural person. This is where the center of speculation, rentier creation, and so on is formed. But large land holdings are created on the basis of combining a significant number of small sole land owners. If the collective of a large farm owns more land than it can effectively cultivate, then such a collective can turn into a group rentier, and, as a result, start speculating on land. From the point of view of social consequences, the task of rationalizing collective land use should also not be ignored, although the social consequences of turning collectives into land rentiers are less acute. The rational size of land owned (other than state-owned) should be determined from the ability to effectively cultivate the land. Hence two conclusions about the criteria for the rationality of land ownership areas:

- the size of the land which are granted for ownership must be consistent with the number of permanent able-bodied employees in the farms organized by them;
- the size of land use should take into account the specialization of farms.

The third task of the land market is to rationalize the forms of land use. The first stage of forming new land relations is almost over. Together with its completion, issues were identified that were taken into account when forming the mechanism for implementing land relations, the principles of which are laid down in legislative documents. In particular, one of them is related to the emergence of land shares owned by persons who both work and do not work in agriculture. Land shares in the amount of, as a rule, the average district norm were transferred on a free basis to almost all rural residents, including pensioners and social workers. Thus, there was a contradiction between the equalization principle adopted for the population by land shares and the requirements for effective management of land, rationalization of its use [6, 14].

A special problem has arisen in connection with the transfer of land to citizens who do not work in agriculture. This issue is not solved within the framework of the first task of the market, because most pensioners and social workers do not process their land plots (with the exception of family members of peasant farms). For the most part, they are transferred either to a common collective, or common joint ownership, or to lease [1, 11]. This leads to two possible options for forming the land share market.

The first one is relatively free. It provides for state non-interference in the process within the established restrictions (intended use, maximum size of land use, priority of local farm property, etc.), regardless of the forms of land ownership: rental, individual, private general or partial. In this case, the principle of freedom of choice of the share owner, priority of his interests is obtained. The lack of convincing statistical data on the advantages of a particular form of land use contributes to the development of various forms of land use [2, 14].

The second option is stimulating. Encouraging owners of land shares to exchange operations. Under all conditions, the land share should not even be temporarily out of production turnover.

The fourth task of the land market is to service the turnover of land that is individually owned by citizens, in cases of donation, inheritance, etc. Should any state regulation apply to them, or can they pass without any restrictions? In our opinion, they should also contain restrictions, including: targeted use of land, compliance with land use limits, requirements for the age and professional qualification of a possible new owner. If a citizen has received an agricultural land plot as a gift or by will, which he is not able to use himself, then the land must be sold at auction or by tender.

The analysis of the four main tasks related to land market regulation shows that they can be further developed, specified and detailed. Having first touched on one selected task, we immediately make sure that there are others, sometimes no less significant. Such work is necessary to form a holistic mechanism of regulators that contribute to the implementation of both individual and general tasks. Tasks related to land turnover are closely intertwined with others that appear in the process of land use, including those when the turnover of land plots does not occur [5, 9].

According to P. T. Sabluk and V. Ya. Mesel-Veselyak, the land market will become an effective mechanism for selecting the strongest owners on the land. Fears that at the same time speculation on land plots and the purchase of them by mafia structures will become widespread will disappear. Provided that the purchase and sale of land is carried out under the control of the state—through the land bank. It is also necessary to establish the procedure under which land can be sold to those who directly process it and are able to use it effectively [3, 5]. Thus, the development of a civilized agricultural land market on the proper legal basis in Ukraine [9]:

- will contribute to the creation of large, competitive enterprises;
- will increase the efficiency of production by differentiated use of land of different quality, by attracting investment to invest primarily in land that can give a high profit;
- will make any land purchase and sale transactions "transparent", remove them from the shadow sector of the economy, and help fill local budgets by expanding the tax base and correctly determining the amount of taxes;
- will give investors confidence in the absence of investment risk.

The functioning of the land market in Ukraine is closely linked to the development of mortgage relations. Since land is one of the largest real estate objects, mortgage lending is perhaps the most proven and reliable way to attract private investment in the real estate market in the world practice.

The development of mortgage relations in the land market in Ukraine requires the creation of economic, legal and institutional conditions for borrowers and lenders to realize their rights and opportunities to use real estate for mortgage operations.

The economic prerequisites for the introduction of a mortgage loan include: price stability; stability of the national currency; overall economic growth.

Legal prerequisites for the introduction of mortgage lending are: legal guarantees of private ownership on real estate; legal unity of the land plot and buildings, structures and perennial plantings located on it; legal security of mortgages; legal regulation of the organization and activities of mortgage banks. Among the institutional prerequisites for the development of mortgage relations are: the formation of specialized mortgage institutions; the existence of a wide network of cadastral services and the organization of state registration of rights on real estate; the development of notary services, valuation and insurance activities; availability of real estate markets and the exchange-traded securities market.

It is assumed that the mortgage lending system in its perfect form will be a functional improving set of rules and measures that will provide mortgage participants with the opportunity to perform all the necessary operations for lending secured by real estate. The introduction of this system will stimulate the development of medium and long-term lending to production (primarily in the agricultural sector) and create conditions for its increase and overall growth of the economy of Ukraine [1, 13].

Within the framework of this direction, in order to use the potential of land collateral, it is assumed [2]:

- the development of a mechanism that will guarantee the mortgage mortgagee, in case of non-fulfillment of the obligation secured by the mortgage, to receive satisfaction from the value of the pledged land plot;
- the development of measures to protect the rights of mortgagors;
- the use of mortgages to secure obligations under land purchase and sale agreements;
- the improvement of the regulatory framework for land mortgages;
- training and retraining of mortgage lending specialists.

Regulatory consolidation of collateral for land plots and measures for the development of mortgage relations is also covered in the new Land Code of Ukraine and Article 133 of the code defines [2]:

1. Land plots belonging to citizens and legal entities on the right of ownership may be transferred as collateral.
2. A land plot that is jointly owned may be pledged with the consent of all co-owners.
3. The transfer of a part of a land plot as collateral is carried out after its allocation in kind.
4. The mortgagee of a land plot can only be banks that meet the requirements established by the Law of Ukraine.
5. The procedure for pledge of land plots is determined by law. For the development of mortgage relations, the final provisions of the Land Code provide for measures for the further development of land legislation in this direction: the creation of a s State land (mortgage) bank with appropriate infrastructure, the introduction of state registration of land rights, the development of regulatory legal acts provided for in the code, including "On the State land (mortgage) Bank".

The Law of Ukraine "On mortgage" No. 898-IV of 05.06.2003, as amended in 2021, defines priority areas of activity for the creation and development of the mortgage market in Ukraine, namely:

- formation of a regulatory and legal framework on mortgage issues in order to effectively use the possibilities of mortgages as an additional source of raising funds to solve socio-economic problems;
- to create a system of guarantees and transparent conditions for the activities of mortgage market participants, primarily through the introduction of an effective system of state registration of rights on real estate, improvement of the real estate

valuation system, standardization of activities in the mortgage market, as well as through the definition of mortgage loan insurance mechanisms;
- establishing the types of mortgage-backed securities, determining the procedure for their start-up and circulation, ensuring proper control over their turnover;
- creation of specialized mortgage banks and other institutions that will operate in the mortgage market;
- expansion of scientific research on mortgage and mortgage market problems, training of mortgage specialists [14].

An important place in the land management system is occupied by land (mortgage) banks. Although land banks have not yet been established in Ukraine, they should be considered as financial organizations that will conduct loans, carry out operations related to the turnover of land plots, and so on. They will play an important role in organizing state control over the use and protection of land.

The creation of the state land bank is of strategic importance for Ukraine in the context of a free land market. It is correct to create it in cooperation with international financial institutions. Today, Ukraine's agriculture needs long-term financial resources. Practically no Ukrainian bank can provide such resources, because we are talking about land loans for a period of 5–10 years, and in a free market, this is the optimal time to return funds for the purchase of land. Currently, in Ukraine, only 5% of farmers now have the opportunity to get loans from foreign banks, so in Ukraine such loans could give a specialized land bank. The transfer of 5 million hectares of state-owned land as collateral to such a bank would bring $10 billion to the country at 1% per annum.

Taking into account the tasks that the land bank should perform, its activities should cover the following three main areas: commercial, economic and legal and regulatory. Each of these areas focuses on achieving the goal set by the land bank. Thus, the bank commercial activities should be primarily related to the turnover of land plots: purchase and sale, donation, exchange, pledge, etc. The implementation of these functions will ensure that the land bank receives funds in the form of bank interest, depending on the price of land, supply and demand for it.

The economic and legal direction should cover the following functions, including:

- mobilization of monetary resources for the implementation of state programs for the organization of rational use and protection of land;
- centralization of free money capital;
- providing loans to business entities on land; issuing securities; protecting the interests of business entities on land when resolving land disputes in court;
- control over the receipt and expenditure of funds by land owners or users [9].

The controlling function of the land bank is combined with the activities of state control services over the use and protection of land. In case of detection of violations, the land bank should apply sanctions to violators of financial discipline through such levers as: termination of financing, withdrawal of loans, etc.

A special place in the land management system belongs to land auctions, which are held in accordance with the Regulations on holding a land auction for the purpose

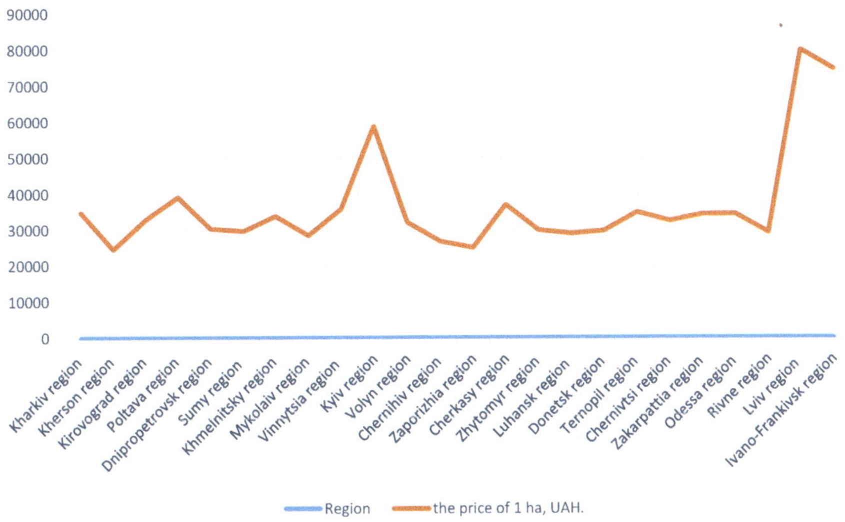

Fig. 1 The price of land plots per 1 ha, UAH (as of 21.02.2022)

of selling a land plot on the terms of a tender on the principle of "who will pay more". The auction is held in order to find such a land owner who will ensure the efficient use of the land fund and attract additional funds into economic circulation [13, 15].

The current land price is the market price, which was formed by the demand from buyers and the supply of share owners. In addition, the price is affected by the region of the site location, land quality and general political risks of Ukraine.

Most likely, there is every reason to expect that when the reform moves to the second stage—after 01.01.2024, there will be a new push to increase the price of agricultural land. This process is due to the fact that legal entities will be allowed to enter the market, and the total land bank, which can be acquired by one person, will increase to 10 thousand hectares.

It is noted that according to forecasts of the Ministry of Agrarian Policy, after the opening of the market, the price of 1 ha of agricultural land will grow annually by at least 7–10%. It is expected that the average market value of 1 ha of land will fluctuate between $1–2 thousand [12].

For six months of activity of the land market, the main results became known: the weighted average price was 40 thousand UAH per ha (for conducting commercial agricultural production), and the total volume from the sale of land plots amounted to 5.3 billion UAH (see Fig. 1) [8].

The land market continues to grow, setting new records. Compared to last year, the volume of the land market increased by 30% both in terms of the number of transactions and the area of land in circulation. The total area of land sold in December 2021 exceeded 49 thousand hectares, and the number of purchase and sale transactions reached almost 19.5 thousand (see Table 1) [7].

Table 1 Key indicators of the agricultural land market in Ukraine

Registered all land in the state land cadastre	69.5% (43.8 mln ha)
Registered agricultural land in the state land cadastre	75.8% (32.4 mln ha)
Average standard monetary valuation of arable land	27,520 UAH/ha
Average amount of rent for agricultural land plots	
State and municipal property (at land auctions)[a]	4039 UAH/ha
Private property[b]	2117 UAH/ha
Average purchase and sale price of an agricultural land plot[c]	38,557 UAH/ha
Credited on the pledge of agricultural land with an area of[d]	810 ha
Average number of purchase and sale transactions per day	470
Average size of a land plot in a purchase and sale transaction	2.42 ha
Number of completed purchase and sale transactions, total	64,262
Area of registered purchase and sale transactions, total	155,523 ha

[a] Since the beginning of 2021
[b] Average price as of the beginning of 2021
[c] Weighted average area price for the period from 01.07.2021 to 01.01.2022, larger plots get more weight when calculating the average price per hectare
[d] As of November 1, 2021

Most of the land, 65.8% of the total area of land plots in circulation, is land for commercial agricultural production. Land for personal farming accounts for 30.98% of the total land area in circulation. A small part is made up of farm land, 2.82% of the total area. All the other types of land reach only 0.35% of the land in circulation.

In terms of the number of transactions, we see the opposite trend. 50.8% of transactions were made with land for personal peasant farm, and only 43.9%—with commercial land. By the type of land—the vast majority of transactions were concluded with arable land (by area—86.2% of all land in circulation), another 10.6%—with pastures and hayfields. The remaining agreements concerned either a mixture of different types of land or other types of land.

According to statistics, more than a third of all agricultural land was sold in the first half of the year. The most active regions are the Kharkiv, Kherson and Kirovohrad regions. The outsiders are the Rivne, Lviv, and Ivano-Frankivsk regions (see Fig. 2) [7].

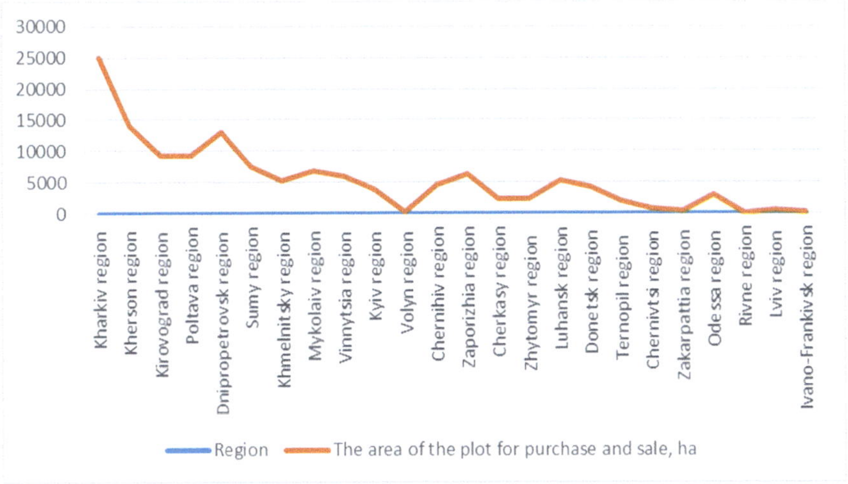

Fig. 2 The area of the plot for purchase and sale plots by regions of Ukraine as of 21.02.2022, ha

According to calculations, according to the Agropolit portal, the area of land during the purchase and sale operation was 134,074 ha. At the same time, the average size of the plot that was the object of purchase and sale was 3.60 ha, and the maximum share size was 67.05 ha.

During the period of operation of the agricultural land market, we have full information about the number and total area of completed purchase and sale transactions, but we are not able to measure the volume of the market in monetary terms. Because not all purchase and sale transactions have the specified price. According to the results of the first six months of the land market, the transaction price is available only for 55% of completed transactions.

We modeled the volume of the land market in monetary terms, using available data on the value of transactions and the characteristics of land sold. We built a regression model of the value of a land plot, taking into account its area, purpose, type of land and its location (at the district level). Due to the failure to take into account other factors affecting the value of the land plot (such as proximity to large cities, potential for alternative use of the land plot)—our model systematically underestimates the value of land plots (the calculated discrepancy for the available data is about 25%) (see Table 2) [8].

Therefore—the calculated volume of the land market in monetary terms should be taken as a conservative estimate. The total volume of land plots sold amounted to UAH 5.2 billion. The largest part of this market is former submoratory lands, the so-called commodity land, the total market volume of which amounted to UAH 3.2 billion. Land for personal agriculture occupies about a third of the agricultural land market and collectively brought landowners UAH 1.7 billion. In the third place in monetary terms is land for individual gardening, 2.4% of the market with a total value of UAH 127 million.

Table 2 The model of the volume of the agricultural land market in monetary form

	Number of transactions	Total area, ha	Total price, mln UAH[a]	% of the total value
All purchase and sale transactions	64,262	155,523	5272.4	
Distribution by purpose				
For conducting commercial rural production	28,181	102,401	3219.6	61.1
For farming purposes	543	4389	106.7	2.0
For personal farming	32,663	48,182	1765.0	33.5
For subsidiary farming	138	123	41.4	0.8
For individual gardening	2571	254	127.1	2.4
Distribution by land type				
Arable land	49,855	134,098	4566.7	86.6
Hayfields and pastures	11,466	16,433	475.9	9.0
Other[b]	2941	4992	229.8	4.4

[a] Estimated total price, for transactions without specifying the price—the transaction cost is modeled
[b] Other types of land include a mixture of different types of land

As of December 31, 2021, 77,880 land market transactions were registered in Ukraine. A total of 198.4 thousand hectares were sold [10].

Conclusion. It should be noted that investment and innovation activity, as a function of land management, is currently a crucial section in the entire economic, including land, policy of the state. But now, we can say with full confidence that neither economic theory nor the practical economic mechanism were prepared for the development and implementation of an effective innovation and investment policy. As a result, the effectiveness of economic reforms is sharply reduced, the reproduction process is slowed down, there is a low efficiency of market transformations, and so on.

Thus, the main tasks of the land market are defined, including: building a complex of mutually agreed legal, economic, environmental, technical and organizational measures that will ensure an effective transition to a new land structure; the formation of rational land use; rationalization of land use forms; servicing the turnover of land that is individually owned by citizens, in cases of donation, inheritance, etc.; as well as highlighting the main directions of state policy in the field of land relations regulation, which, in turn, will contribute to the creation of large competitive enterprises. It will also increase production efficiency by differentiating the use of land of different quality, by attracting investment to invest primarily in land that can generate high profits. It will also create conditions for ensuring transparency of land purchase and sale agreements, removing them from the shadow sector of the economy, help fill local budgets and give confidence to investors.

References

1. Fedorov, M.M.: Zemelna reforma i rozvytok rynku zemelnykh vidnosyn. Ekon. APK (Econ. AIC) (7), 55–60 (2011) [in Ukrainian]
2. Kulynych, P.F.: Rehuliuvannia rynku silskohospodarskykh zemel: idei ta zakonoproekty. Instytut derzhavy i prava NAN Ukrainy (Institute of State and Law of the National Academy of Sciences of Ukraine). http://www.agroconf.org/uk/node [in Ukrainian]
3. Laktionova, G.P.: Zemlja kak faktor proizvodstva v rynochnoj ekonomike. Visnyk KhDAU (Bull. KhSAU) (2), 33–36 (2015) [in Russian]
4. Lupenko, Yu.O., Sabluk, P.T., Mesel-Veseliak, V.Ya., Fedorov, M.M.: Rezultaty i problemy reformuvannia silskoho hospodarstva Ukrainy. Ekon. APK (Econ. AIC) (7), 26–38 (2014) [in Ukrainian]
5. Martyniuk, M.P.: Ekonomichni peredumovy ta modeli zapusku rynku zemel v Ukraini. Ekon. APK (Econ. AIC) (12), 16–21 (2016) [in Ukrainian]
6. Mesel-Veseliak, V.Ya., Fedorov, M.M.: Stratehichni napriamky rozvytku ahrarnoho sektoru ekonomiky Ukrainy. Ekon. APK (Econ. AIC) (6), 37–49 (2016) [in Ukrainian]
7. Ohliad stanu zemelnykh vidnosyn v Ukraini. https://kse.ua/wp-content/uploads/2022/01/Oglya-stanu-zemelnih-vidnosin_gruden.pdf [in Ukrainian]
8. Reznik, N.P., Popov, V.M., Podplietnii, V.V., Popova, S.P.: Financial support for the development of joint territorial communities. Test Eng. Manag. (2020). Available at: http://www.scopus.com/inward/record.url?eid=2-s2.0-85082999646&partnerID=MN8TOARS
9. Reznik, N.P.: Principles of formation of investment potential of agro-industrial complex of the state. Innov. Econ. (10), 14–16 (2012) [in Ukrainian]
10. Reznik, N.P., Dolynskyi, S.V., Voloshchuk, N.Y.: Retrospective analysis of basic risk as a part futures trading in Ukraine. Int. J. Sci. Technol. Res. (2020). Available at: http://www.scopus.com/inward/record.url?eid=2-s2.0-85078765334&partnerID=MN8TOARS
11. Rynok zemli v Ukraini. https://agropolit.com/spetsproekty/892-rinok-zemli-v-ukrayini-mapa-zemelnih-ugod-i-tsin-na-zemlyu-z-1-lipnya-2021-roku?utm_source=latifundist&utm_medium=widget [in Ukrainian]
12. Rynok zemli, znovu? Cherez pivroku pislia zapusku rozbyraiemos: kupuvaty chy ni. https://latifundist.com/blog/read/2857-rinok-zemli-znovu-cherez-pivroku-pislya-zapusku-rozbirayemos-kupuvati-chi-ni [in Ukrainian]
13. Shaporenko, O.I.: Upravlinnia stanom i vykorystanniam silskohospodarskykh zemel u rehioni. Nord-Press, Donetsk (2004) [in Ukrainian]
14. Shchorichnyk «Monitorynh zemelnykh vidnosyn v Ukraini. 2021–2022». http://land.gov.ua/wp-content/ [in Ukrainian]
15. Stupen, M.H., Bochko, O.I.: Ekonomiko-instytutsionalni zasady vykorystannia zemelnykh resursiv. Ekonomist (The Economist) (3), 15–16 (2015). http://nbuv.gov.ua [in Ukrainian]

The Effectiveness of the Public Relations Department of the Palestinian Police in Raising Public Awareness of Cyber Blackmailing During the Covid-19 Pandemic

Kareem Sarhan and Mais Hamadneh

Abstract The study aimed at identifying the effectiveness of the public relations department of the Palestinian police in raising public awareness of cyber blackmailing during the Covid-19 pandemic. The study adopted a descriptive research method, and a two-form questionnaire was developed as the research tool to obtain relevant data from two study samples of the target population. The first study sample included a public audience. The researchers used simple random sampling and collected (469) valid questionnaires that achieved conformity to the specified requirements. The second study sample included the Palestinian police personnel. Using simple random sampling, the researcher collected (50) valid questionnaires that achieved conformity to the specified requirements.

Keywords Palestinian police · Effectiveness · Public relations · Public awareness · Cyber blackmailing · COVID-19

1 Introduction

Public relations in the police force are of immense importance; given the importance and sensitivity of the security apparatus, as it is responsible for the security and stability of society. Public relations in the police apparatus takes on the task of building and strengthening relations with citizens to accept decisions related to controlling behavior and compliance with laws in terms of application and practice. Public relations are considered a link between the higher administrative levels in the police and employees on the one hand, and between the police and citizens on the other hand, which means that they have a great responsibility in the communication processes, facing crises, and phenomena facing society, such as: the phenomenon of cybercrime in general, and extortion Email. The crime of electronic extortion is one of the forms of electronic crime, a phenomenon that penetrates society and threatens its foundations, and strikes in the hindering of the most important goals

K. Sarhan (✉) · M. Hamadneh
An-Najah National University, Nablus, Palestine
e-mail: Kareem.sarhan@najah.edu

A. Hamdan et al. (eds.), *The Implementation of Smart Technologies for Business Success and Sustainability*, Studies in Systems, Decision and Control 216,
https://doi.org/10.1007/978-3-031-10212-7_36

of any civilized society in achieving security for its members. The reason behind criminalizing electronic extortion is the threat, extortion, and pressure exerted on the victim by threatening to divulge a secret as disgrace and shame. Additionally, forcing the victim to submit to the offender's desire, and to fulfill their legitimate or illegitimate demands and coercion in fear of scandal [1].

The problem of the study: A unit has been established in the Palestinian Police Service to investigate cybercrime and collect its evidence, consisting of a limited number of cadres, devices, and programs. However, the work of this department faces many difficulties, the most important of which are: the legislative shortcomings, and the inadequacy of the provisions of the Palestinian Penal Procedures Law No. 3 of 2001 to investigate and collect evidence of this type of crime, as the aforementioned law does not include texts that clarify how to deal with electronic evidence (digital/intangible), especially with regard to the collection of digital evidence, or its value in evidence. Based on the foregoing problem of the study is represented in the following main question: **How effective is the public relations of the Palestinian police apparatus in educating the public about the crimes of electronic extortion?**

Study Hypotheses: The study seeks to test the following null hypotheses:

1. There are no statistically significant differences at the level ($\alpha \leq 0.05$) in the average responses of the study sample members towards the effectiveness of the public relations of the police apparatus in educating the public about cyber-extortion crimes, according to the gender variable.
2. There are no statistically significant differences at the level ($\alpha \leq 0.05$) in the average responses of the study sample members towards the extent of the effectiveness of the public relations of the police apparatus in educating the public about electronic blackmail crimes, according to the educational qualification variable.
3. There are no statistically significant differences at the level ($\alpha \leq 0.05$) in the average responses of the study sample members towards the effectiveness of the public relations of the police apparatus in educating the public about electronic blackmail crimes, according to the work variable.
4. There are no statistically significant differences at the level ($\alpha \leq 0.05$) in the average responses of the study sample members towards the extent of the effectiveness of the public relations of the police apparatus in educating the public about electronic extortion crimes, according to the monthly income variable.
5. There are no statistically significant differences at the level ($\alpha \leq 0.05$) in the average responses of the study sample members towards the effectiveness of the public relations of the police apparatus in educating the public about electronic blackmail crimes, according to the marital status variable.

2 Study Tool

The researchers used the questionnaire tool.

3 Theoretical Framework

The theory of dependence on the media: Since the media has become an essential part of the lives of citizens in various societies, including the Palestinian society, as it provides them with satisfaction for some of their needs, to become a part of citizens' lives and the basic structures of society. The importance of the media is enhanced by its ability to influence public opinion, when the media perform their functions and roles effectively, so citizens depend on them as social systems, and dependent relationships arise among them on the means of communication, which provide and direct those goals, which in turn guide individuals (Devler and Rokic 1999).

3.1 Cognitive Effects

The cognitive effects of the media, according to the theory of media dependence, include the following: Ambiguity, Attitude Formation, Agenda Setting and Enlargement Ideas, by influencing individuals' belief systems by increasing them, either by increasing the categories that join these beliefs and organize them in a specific framework, or by increasing the beliefs in each category.

3.2 Affective Effects

Which are the effects related to feelings and sensations, such as love, hate, and others, where the media plays a significant role in emotional and emotional influence through these effects: Desensitization, Fear and Anxiety, Morale support and alienation (Morale and Alienation).

3.3 Behavioral Effects

Among the most important behavioral effects of the media are: the act, or the loss of desire for this act, and these two effects are a direct result of the cognitive and emotional effects, and the behavioral effects are divided according to the dependence model into: First: Activation, which means that the individual does something as a result of exposure to the means The media, which is the final product to link the cognitive and emotional effects, and activation may be a socially useful act as in quitting smoking, political participation, or socially harmful, as in simulating bad behaviors such as violence and crime. Second: Deactivation, and inactivity is the avoidance of action, and this effect may occur because of exaggerated media coverage, which is reflected on the personal behavior of the recipient, such as: reluctance to participate in politics, or stay away from community activities.

The theory of media dependence is based on two main pillars: goals and sources; For to achieve the personal and social goals of individuals, groups, and organizations, they must rely on resources controlled by other persons, groups, or organizations, and vice versa. Devler and Rokich believe that the goals that individuals seek to achieve are understanding (Understand social world), guidance and counseling (Act Meaning Fully and Effectively, Fantasy and Escape), and providing information.

The Concept of Electronic Extortion

Electronic extortion is defined as: obtaining money or benefits from a person by force and coercion by threatening to reveal his secrets that he possesses, where extortion involves the use of threats of physical or psychological harm, or damage to reputation and social standing by fabricating scandals, accusing, and publishing secrets, forcing the blackmailer to Payment is forced to those who extort them. And extortion is the behavior of pressure and coercion to do a forbidden act, either because of the victim's weakness or need [2, p. 428].

It is a process of threatening to publish photos, videos, or personal and sensitive information if the victim does not submit to the extortionist's requests, as defined by Mada Company, a Palestinian company specialized in providing communications, (internet) and added services, as it considers that most of the requests are related to the crime of extortion The email is as follows:

1. Pay money.
2. Doing illegal activities.
3. Carrying out immoral acts.
4. Disclosure of confidential institutional or political information.
5. Working with the enemy [3].

The Role of the Palestinian Police in Confronting Corona

The Palestinian Police Authority declared a state of emergency in the face of the Corona pandemic in the Palestinian territories, as it did the following:

1. Follow-up procedures for banning shifts, and not moving between regions.
2. Disseminate awareness and guidance messages or any message that the police want to convey to the public, such as: awareness of preventive measures to confront the Coronavirus.
3. Conducting the necessary sterilizations for vital streets in the center of Palestinian cities.
4. Sterilization of buildings, mosques, and churches in cities.

5. A special corner for the locations of the police stations on the map linked with the (GPS) of the phone, making it easier for users to reach the nearest police station when needed.

It is worth noting that the Palestinian Police receives and deals with more than one million emergency calls annually, i.e., an average of 3500 calls per day, and handles more than 30,000 real emergency events annually (the website of the Palestinian Police).

In recent years, through the establishment of units and sections specialized in law enforcement and investigation in search of evidence. The digital blackmail associated with cybercrime, the most important of which is electronic extortion, as the investigation stage in cybercrime goes through two phases, as explained [4: 286], which are:

The first stage: includes the procedures related to the crime scene, such as: closing the crime scene to avoid losing evidence.

The second stage: includes a set of tasks performed by specialists and others from the crime scene team.

The Anti-Cybercrime Prosecution was established for the first time in Palestine, in (2016); To follow up on issues that fall within the framework of the (internet), such as: extortion, information theft, hacking, defamation, spreading viruses, and violations that take place through the (internet), and later, specifically in (2018), a crime law was introduced. The electronic system, where it allocates a specific penalty for each crime between threats and extortion (Scholi 2020).

According to the statistics issued by the Palestinian General Investigation Department in the year (2019), there is a noticeable increase in this crime, which was monitored by the police for years, where the year (2015) recorded the occurrence of (502) cases, and in (2016) 1327 cases were recorded. It also increased in (2017) to (2025) cases, but rose in (2018) to (2568) cases, an increase of (26.6%) over the year (2017) [5]. Hence, the Public Relations Department of the police has a great responsibility; As one of the most widely used communication patterns, which has the ability to communicate its communication messages to the target audience, through the implementation of a set of communication activities, programs and events, which are consistent with the rapid global developments, especially with regard to technology and digital communication tools, and this is reflected in the increasing use The Internet and social networking sites educate the public about the crime of electronic extortion, and the mechanism for dealing with it and confronting it [6].

Palestinian Police Service

Police work is governed by a set of laws, including: Criminal Procedures Law No. (3) of (2001), Palestinian Correctional and Rehabilitation Centers Law No. (6) of (1998), Traffic Law No. (5) of (2000), and General Assembly Law No. (12) of the year (1998), the Firearms and Ammunition Law No. 2 of the year 1998, the Palestinian Police Draft Law, in addition to the regulations, instructions and regulations issued by the Minister of Interior and the Director General of Police.

The objectives and tasks of the Palestinian Police Service are as follows:

1. Maintaining public order (internal security).
2. Protection of the public and its property.
3. Work to spread a sense of security and reassurance.
4. Adopt all measures to prevent crime in accordance with the law.
5. Protection of public facilities, infrastructure, and places of special importance.
6. Confronting acts of violence and terrorism and preventing incitement to commit them.
7. Carrying out regular police work [7].

Public Relations Department in the Palestinian Police Service:

The Public Relations Department is one of the most important specialized departments in the Palestinian Police Service, which works on:

1. Linking the police institution to its internal and external audience.
2. Enhancing the image of the police institution in society.
3. Enhancing the affiliation and loyalty of its members to the institution.
4. Strengthening cooperation between civil society institutions and the police to provide the best service to citizens.
5. Promote a culture of rule of law.

First: Study Methodology:

To achieve the objectives of the study, the researcher used the descriptive-analytical approach, which attempts to assess the effectiveness of the public relations of the police apparatus in educating the public about cyber-extortion crimes considering the Corona pandemic (COVID-19), and the descriptive-analytical approach tries to compare, interpret, and evaluate in the hope of reaching meaningful generalizations. It increases the balance of knowledge about the subject, and data has been collected from secondary and primary sources.

Second: Population and sample of the study:

The study community is represented by the people living in the West Bank, using the simple random sampling method, and the number of valid questionnaires for analysis reached (384).

Third: The study tool:

According to the nature of the research, the researchers used the questionnaire as a tool in their study due to its relevance to the study's objectives, methodology, and society.

The characteristics of the study sample:

Frequencies and percentages were calculated for the study sample members according to gender, educational qualification, work, and marital status.

Distribution of the study sample by gender:

It was found through the results that (23.4%) of the study sample members are males, while females constituted (76.6%) of the study sample.

1. Distribution of study sample members according to academic qualification:

The results show that (20.8%) of the study sample have a scientific qualification (high school), and (13.8%) have a diploma (diploma), and (51.3%) have a bachelor's degree, while (13.3%) of the individuals the study sample had a scientific qualification (Master's), while (0.8%) of the study sample had a scientific qualification (Ph.D.).

2. Distribution of the study sample according to work:

It was found through the results that (42.7%) of the study sample members work, while (57.3%) do not work.

3. Distribution of the study sample members according to marital status:

It was found through the results that (36.5%) of the study sample have their marital status (single/single), and (57.8%) of the study sample have their marital status (married/married), and that (3.4%) have their marital status (separated/separated), while (2.3%) of the study sample have their marital status (widower).

Analysis of the study axes:

The two researchers used the five-point Likert scale; To obtain the study vocabulary responses, according to the following degrees of agreement: (strongly agree—agree—neutral—disagree—strongly disagree). And then expressing this scale quantitatively by giving each of the previous statements a score, according to the following: Strongly agree (5) degrees, agree (4) degrees, neutral (3) degrees, disagree (2) two points, strongly disagree (1) one degree.

The researchers conducted an analysis of the study axes, to know the reality of these axes in the study community, and the results of the analysis of the study axes can be clarified through the following:

4 Result Analysis and Discussion

1. The results of the analysis of the first axis (the programs and activities that the citizen receives from the police to raise awareness of the crimes of electronic extortion):

To identify the responses of the study sample members about the programs and activities that the citizen receives from the police to raise awareness of the crimes of electronic extortion, the arithmetic mean, the standard deviation, the relative weight of each of the paragraphs of the axis, and the total score of the axis were calculated. And the standard deviation is equal to (0.820), and the relative weight is equal to (53.60%), and this indicates the approval of the members of the study sample on the programs and activities that the citizen receives from the police with the aim of raising awareness of cyber-extortion crimes to a moderate degree.

It is clear from the results that the fourth paragraph (I follow guidance and awareness programs on the Internet in relation to cyber-extortion crimes), is the most important paragraph of this axis, as it obtained the highest relative weight (67.20%) and that the lowest paragraph is the seventh [I receive tips and awareness instructions using the messaging method Direct (SMS)] with a relative weight (44.40%), and based on the relative weight of the axis as a whole, which reached (53.60%), it turns out that the members of the examined sample agree to a moderate degree on the programs and activities that the citizen receives from the police in order to raise awareness of the crimes of electronic extortion.

The researchers attribute the result of the highest paragraph to the citizens' interest in following up guidance and awareness programs via (internet) networks; To protect themselves and their families from the danger of cyber-extortion crimes, and thus their feelings of complacency and heading towards a better life, free of any loopholes.

The researchers attribute the result of the lowest paragraph to the fact that the broad base of subscribers to direct messaging services (SMS) cannot be reached all, and therefore it is difficult to warn and provide instructions through it, which naturally reduces its use, in addition to the fact that some warnings may be lengthy, and therefore cannot be Send it via SMS.

2. Results of the second axis analysis (public satisfaction with the effectiveness of public relations in raising awareness of cyber-extortion crimes):

To identify the responses of the study sample members about the public's satisfaction with the effectiveness of public relations in raising awareness of cyber-extortion crimes, the arithmetic mean, the standard deviation, the relative weight of each of the paragraphs of the axis and the total score of the axis were calculated, where it was found that the arithmetic mean of all paragraphs equals (3.14), The standard deviation is equal to (0.689), and the relative weight is equal to (62.80%), and this indicates the approval of the members of the study sample on the public's satisfaction about the effectiveness of public relations in raising awareness of cyber-extortion crimes to a moderate degree.

It is clear from the results that the fifth paragraph (the method of providing awareness of cyber-extortion crimes suggests trust and seriousness) is the most important paragraphs of this axis, as it obtained the highest relative weight (67.60%) and that the lowest paragraph is the third (there is awareness among citizens of safe methods when using the computer And the smartphone) with a relative weight (58%), and based on the relative weight of the axis as a whole, which reached (62.80%), it turns out that the members of the examined sample agree to a medium degree with the public's satisfaction about the effectiveness of public relations in raising awareness of cyber-extortion crimes.

The researchers attribute the result of the highest paragraph to the interest of the staff working in the police department, which is public relations, to highlight their interest in seeking to provide the necessary awareness, which shows the extent of their seriousness and interest, as well as complete confidence in providing awareness, which comes through public satisfaction with the police apparatus and the Public Relations Department This would indicate that the work is progressing in full swing, according to what is planned and what achieves the acceptance and conviction of the public.

The researchers attribute the result of a lower paragraph to the fact that in recent times it has become clear that all segments of society have used computers and smartphones in the processes of communicating with others, and accessing (internet) sites of all kinds, as they are browsing more, which makes them vulnerable to hacking and piracy, especially If there is sensitive information for them, and thus the increase in the public's demand for those sites, it warns of a lack of awareness about safe methods, in addition, their recklessness and blind trust in the sites that are browsed.

The second question: What is the extent of the public's satisfaction with the effectiveness of public relations in raising awareness of cyber-extortion crimes?

1. The results of the analysis of the second axis (the means used by the public relations department to educate the public about the crimes of electronic extortion):

To identify the responses of the study sample members about the means employed by the Public Relations Department to educate the public about crimes of electronic extortion, the arithmetic mean, the standard deviation, thc relative weight of each of the paragraphs of the axis, and the total degree of the axis were calculated, where it was found that the arithmetic mean of all the paragraphs is (3.32), the standard deviation is equal to (0.773), and the relative weight is equal to (66.40%), and this indicates the approval of the study sample members on the means employed by the Public Relations Department to educate the public about crimes of electronic extortion to a moderate degree.

It is clear from the results that the tenth paragraph (there is an independent unit or department that follows up on crimes of electronic extortion) is the most important paragraphs of this axis, as it obtained the highest relative weight (77.20%), and that the lowest paragraph is the eighth (the necessary financial resources are available in the police force to Confronting the crimes of electronic extortion), with a relative weight (58%), and based on the relative weight of the axis as a whole, which reached (66.40%), it turns out that the members of the examined sample agree to a moderate degree on the means employed by the Public Relations Department in educating the public about the crimes of electronic extortion.

The researchers attribute the result of the highest paragraph to the fact that security is a necessity to maintain the confidentiality of information, as well as to combat any electronic extortion, as a special and independent department has been allocated, its first goal being to follow up on cyber-extortion crimes, as the developments that the world has witnessed in recent decades, due to the communications and transportation revolutions And information technology has reflected on all human activities with its positives and negatives, and this is what has established in the police apparatus the importance of dedicating a full circle to the task of following up on any electronic extortion crimes.

The researchers attribute the lowest paragraph result to the reluctance of the department responsible for allocating the financial resource necessary for the electronic extortion department to follow up on its work to the fullest, as some of the work entrusted to them requires the presence of a financial resource that contributes to obtaining the necessary programs, through which the owners of cybercrime can be reached and followed up and reach them as quickly as possible.

2. The results of the analysis of the third axis (programs and activities implemented by the public relations department of the police to educate the public about extortion crimes):

In order to identify the responses of the study sample members about the programs and activities implemented by the Public Relations Department of the Police Service in educating the public about extortion crimes, the arithmetic mean, the standard deviation, the relative weight of each of the axis's paragraphs, and the total score of the axis were calculated, where it was found that the arithmetic mean of all the paragraphs (2.94), the standard deviation equals (0.976), and the relative weight equals (58.80%), and this indicates the approval of the study sample members on the programs and activities implemented by the Public Relations Department of the police in raising public awareness of extortion crimes to a moderate degree.

It is clear from the results that the fourth paragraph (Guidance and Awareness via the Internet) is the most important paragraphs of this axis, as it obtained the highest relative weight (66.40%), and that the lowest paragraph is the eighth [Awareness using the direct message method (SMS)] with relative weight. (45.60%), and based on the relative weight of the axis, which reached (58.80%), it turns out that the members of the examined sample agree to a moderate degree on the programs and activities implemented by the Public Relations Department of the police in educating the public about extortion crimes.

The researchers attribute the result of the highest paragraph to the fact that in the last decade, all segments of society have increased their demand for the use of the Internet and its websites, which make it easier for them to access and deal with them. (Internet), and providing them with the correct ways to evade any extortion, and thus the (Internet) network is considered easy to reach for the largest mass base that launches for them tips and advice through its various sites.

The researcher attributes the result of the lowest paragraph to the fact that it is difficult to reach the largest possible number of audiences through direct messaging services (SMS); Due to the vast number of users of the websites, in addition to the excessive cost of using these messages periodically to reach the masses, and this limits their use.

3. The results of the analysis of the third axis (the corona pandemic and the crimes of electronic extortion):

To identify the responses of the study sample members about the Corona pandemic and cyber-extortion crimes, the arithmetic mean, standard deviation, the relative weight of each of the paragraphs of the axis, and the total score of the axis were calculated, as it was found that the arithmetic mean of all paragraphs equals (3.52), and the standard deviation equals (0.686), and the relative weight is equal to (70.40%), and this indicates the approval of the study sample members on the Corona pandemic and cyber-extortion crimes to a high degree.

It is clear from the results that the third paragraph (the Corona pandemic has caused an imbalance in the educational sector in Palestine and the world) is the most important paragraphs of this axis, as it obtained the highest relative weight (87.20%), and that the lowest paragraph is the fourteenth paragraph (there is an independent unit or department to confront Corona), with a relative weight (53.60%), and based on the relative weight of the axis as a whole, which reached (70.40%), it turns out that the members of the examined sample highly agree with the Corona pandemic and cyber-extortion crimes.

The researchers attribute the result of the highest paragraph to the fact that the Corona pandemic ravaged all educational levels, because of its danger and ease of transmission between people, which may lead to their lives to death, and which caused cracks and gaps at the educational level, as well as at the level of cognitive achievement, and accordingly the demand of students to use websites increased. The website, which requires them to be aware of the dangers of using these sites and falling prey to others.

The researchers attribute the result of the lowest paragraph to the lack of prior planning for the police apparatus to develop an emergency plan to work with in the presence of an unprecedented and planned situation, so that it can continue to work and continue its work without obstacles in front of it.

5 Hypothesis Testing

- First: Testing the public questionnaire hypotheses

The third question: Are there statistically significant differences between the views of the study sample members about the effectiveness of the public relations of the Palestinian police in educating the public about cyber-extortion crimes according to demographic variables (sex, educational qualification, work, marital status, and place of residence)?

1. Differences according to the sex variable:

To find out whether there are statistically significant differences in the responses of the study members according to the gender variable, an (Independent Sample T-test) test was conducted; To clarify the significance of the differences in the responses of the study members according to the gender variable, it is clear from the results that there are no statistically significant differences at the level (0.05) or less in the study members' attitudes about (public satisfaction with the effectiveness of public relations in raising awareness of cyber-extortion crimes) according to a variable Gender, while it is clear that there are statistically significant differences at the level (0.05) and less in the study members' attitudes about (programs and activities that the citizen receives from the police in order to raise awareness of cyber-extortion crimes) according to the gender variable in favor of males.

2. The differences according to the educational qualification variable:

To identify whether there are statistically significant differences in the responses of the study members according to the difference in the educational qualification variable, "One Way ANOVA" was used; To clarify the significance of the differences in the responses of the study members according to the difference in the educational qualification variable, it is clear from the results that there are statistically significant differences in the attitudes of the study members about (programs and activities that the citizen receives from the police system with the aim of raising awareness of crimes of electronic extortion, public satisfaction about the effectiveness of public relations in awareness with electronic extortion crimes) according to the educational qualification variable.

In order to determine the validity of the differences between the categories of academic qualifications, the (LSD) test was used, as it becomes clear from the results that there are statistically significant differences at the level of (0.05) and less between the study members whose academic qualifications are (Bachelor) and those whose educational qualifications are (high school, diploma, master's, and doctorate) on (programs and activities that the citizen receives from the police with the aim of raising awareness of cyber-extortion crimes), for the benefit of the study members whose scientific qualification is (Ph.D.), and the study members who have their scientific qualification (Ph.D.), and the study members whose qualification is (Ph.D.). High school, diploma, bachelor's, and master's degrees), for the benefit of study members who have a scientific qualification (Ph.D.).

3. The differences according to the work variable:

To find out whether there are statistically significant differences in the responses of the study members according to the difference in the work variable, the (T: Independent Sample T-test) was tested to clarify the significance of the differences in the responses of the study members according to the difference in the work variable, it is clear from the results that there are no differences Statistically significant at the level (0.05) or less in the study members' attitudes about (programs and activities that the citizen receives from the police in order to raise awareness of cyber-extortion crimes, and public satisfaction about the effectiveness of public relations in raising awareness of cyber-extortion crimes) according to the work variable.

4. The differences according to the marital status variable:

To identify whether there are statistically significant differences in the responses of the study members according to the difference in the marital status variable, the "One Way ANOVA" was used; To clarify the significance of the differences in the responses of the study members according to the difference in the marital status variable, it is clear from the results that there are no statistically significant differences in the attitudes of the study members about (programs and activities received by the students).

6 Results

1. The approval of the study sample on the programs and activities that the citizen receives from the police to raise awareness of the crimes of electronic extortion to a moderate degree, as the arithmetic mean of all the paragraphs of the axis was (2.68), while the relative weight of all the paragraphs of the axis was (53.60%).
2. The approval of the study sample members of the public's satisfaction about the effectiveness of public relations in raising awareness of cyber-extortion crimes to a medium degree, as the arithmetic mean of all the items of the axis was (3.14), while the relative weight of all items of the axis was (62.80%).
3. There are no statistically significant differences at the level (0.05) or less in the study members' attitudes about (public satisfaction with the effectiveness of public relations in raising awareness of cyber-extortion crimes) according to the gender variable.
4. There are statistically significant differences at the level of (0.05) or less in the study members' attitudes about (programs and activities that the citizen receives from the police to raise awareness of cyber-extortion crimes) according to the gender variable in favor of males.
5. There are statistically significant differences at the level of (0.05) and less between the study members whose academic qualification is (Bachelor) and the study members who have their academic qualification (high school, diploma, master's, and doctorate) regarding (programs and activities that the citizen receives from the police force with the aim of Awareness of cyber-extortion crimes), for the benefit of study members who have a scientific qualification (Ph.D.), and between study members who have a scientific qualification (Ph.D.) and study members who have a scientific qualification (high school, diploma, bachelor's, and master's), for the benefit of study members who have a scientific qualification (Ph.D.).
6. There are statistically significant differences at the level (0.05) and less between the study members who have a scientific qualification (Ph.D.), and the study members who have a scientific qualification (diploma, bachelor's, and master's) regarding (public satisfaction with the effectiveness of public relations in raising awareness of cyber-extortion crimes), for the benefit of the study members whose academic qualification is (Ph.D.), and between the study members who have their scientific qualification (high school) and the study members whose academic qualification is (diploma, bachelor), for the benefit of the study members who have their scientific qualification (high school).
7. There are no statistically significant differences at the level (0.05) or less in the study members' attitudes about (programs and activities that the citizen receives from the police to raise awareness of cyber-extortion crimes, and public satisfaction about the effectiveness of public relations in raising awareness of cyber-extortion crimes) with a variable difference the work.
8. There are no statistically significant differences in the attitudes of the study members about (the programs and activities that the citizen receives from the

police to raise awareness of cyber-extortion crimes, and public satisfaction about the effectiveness of public relations in raising awareness of cyber-extortion crimes) according to the variable of social status.

7 Recommendations

Based on the above-mentioned findings, the following recommendations can be given:

1. Strengthening the Police Department's Public Relations Department for its programs and activities to serve the public and strengthen relations with them.
2. The importance of spreading awareness among citizens about safe methods when using computers and smartphones considering the era of rapid technological development.
3. The importance of letting the employees in the Public Relations Department of the Police Service convey the opinions and attitudes of the masses about the nature and their satisfaction with the work of the police.
4. Allocating the necessary financial resources in the police apparatus to confront electronic extortion crimes.
5. Allocate a unit or department to operate under any emergency, so that the work of the police does not stop in following up on any acts of extortion that occur among citizens.

The two researchers suggest conducting future studies related to the following: the role of public relations in limiting the spread of the phenomenon of electronic extortion, the effectiveness of public relations in improving the image of the Palestinian police apparatus in front of the public, and the extent of the public's satisfaction with the nature of public relations in warning them against electronic extortion, social responsibility and national affiliation among Workers in the Palestinian Police Service.

References

1. Abdulaziz.: Criminal responsibility for the crime of electronic extortion in the Saudi system, a comparative study. J. Gener. In-depth Legal Res. (25), pp. 27 (2018)
2. Shater, W.: The legal framework for the crime of electronic extortion in electronic games—a comparative study according to the Saudi system and Kuwaiti law. Arab J. Sci. Publishing (16) (2020)
3. mada.ps, 2020: Mada Internet, 7.11.2020 (2020)
4. Abdel-Baqi, M.: Investigation and proof of cybercrime in Palestine: a comparative study. Dirasat J. High Sharia Law **45**(4), Supplement (2) (2018)
5. Dunia Al-Watan 2019: dunia al-watan, 7.11.2020 (2019)
6. Al-Anazi, N.: The effectiveness of public relations in addressing the phenomenon of electronic extortion (a field study of the activities of the Iraqi Ministry of Interior and University Youth for the period 1/4/2019–1/5/2020). Al-Farabi University College, Iraq (2020)

7. Al-Jazeera.Net 2016: Al jazeera, 7.11.2020 (2016)
8. Palestine... the triple threat electronic Extortion, 7.11.2020 (2020)
9. Theories of mass media

Facebook Posts for Academic and Non-academic Events: Higher Education Institutions in Abu Dhabi

Imen Gharbi, Mohammad Hani Al-Kilani, Ajayeb Salama AbuDaabes, and Walaa Saber Ismail

Abstract The article aims to explore social media strategies used by Higher Education Institutions through their Facebook platforms. The study analyzed the post content related to academic and non-academic events, format and language that might influence the level of engagement in terms of reactions. The authors used content analysis of 893 posts to examine Facebook posts from 16 HEIs in Abu Dhabi. Results show that non-academic events generate more reactions than academic events. Besides, the format and language used in the posts are other factors that contributes to engagement. The results also show that reactions in private HEIs are higher compared to public HEIs. This study provides guidance on how HEIs can use social media strategically to improve users' engagement.

Keywords HEIs · Engagement · Social media reaction · Facebook · Post formats

1 Introduction

To sustain a highly competitive marketplace, Higher Education Institutes (HEIs) put a great effort in their marketing strategies to engage with prospective students by building identity and connection [18]. HEIs need to understand the needs and perceptions of students to achieve academic success and social activities, connections and experiences [27].

During the past two decades, social networking platforms have become important tools for HEIs to communicate and interact with their customers. HEIs in UAE had also to respond and adapt to rapid technological development and embrace the change by using different social media platforms along with the traditional branding tools. The new digital era and the intensification of competition have pushed universities like the rest of the companies to modify their communication and marketing strategy and change their engagement tactics by planning their social media content to attract more students [1]. HEIs need to identify the engagement needs of their users

I. Gharbi (✉) · M. H. Al-Kilani · A. S. AbuDaabes · W. S. Ismail
Liwa College of Technology, Abu Dhabi, UAE
e-mail: imen.gharbi@ect.ac.ae

A. Hamdan et al. (eds.), *The Implementation of Smart Technologies for Business Success and Sustainability*, Studies in Systems, Decision and Control 216,
https://doi.org/10.1007/978-3-031-10212-7_37

on social media platforms and then tailor their content and align their marketing communication strategy to facilitate interaction and engagement. Some studies had demonstrated that HEIs are using social media platforms to disseminate valuable information [22], to share resources with students and to allow them to interact with their instructors [20], to promote the work of faculty [2]. Other studies found that HEIs use social media for recruitment and admissions [16]. Nevertheless, there are very limited studies that investigated the impact of social media platforms for engaging with the target audience in HEIs and specially the evaluation of social media messages and their effectiveness in terms of interactions. In fact, few studies had explored how HEIs use social media for non-pedagogical purposes [23]. HEIs offering a lot of events on social media would attract students looking for a great university experience [4] such as high-level sporting events [3].

The main aim of this study is to identify the importance of Facebook content type and format to foster reaction within the target audience in HEIs in Abu Dhabi. Facebook is one of the most popular social media platforms in the UAE in 2021 after YouTube and WhatsApp. The number of Facebook users in UAE is 79% with a total number of 7.89 M in 2021 [11]. The study will mainly analyze the impact of posting academic and non-academic events on Facebook and the language used on the level of reaction. The level of Facebook post engagement will be quantified by the number of reactions (comments, emotes and shares) to analyze the efficiency of the post content type. Our research questions are the following:

- Does the type of post content (academic/non-academic events) affect the level of users' engagement?
- Does the type of post format (image, video, link, text) affect the level of users' engagement.
- Is there a difference between Public and private HEIs in terms of reaction to posts?

The rest of this paper is organized as follows. Section 2 provides an overview of the literature on related works. The study's methodology is described in Sect. 3. Section 4 provides the results of the analysis. Finally, we conclude the article and provide the implications for Abu Dhabi HEIs and directions for future research.

2 Literature Review

2.1 Social Media Content in Higher Education

With the rapid change in marketing communication environment and the shift from traditional promoting channels to more advanced information technology, HEIs had to move to digital format and adopt social media to communicate with their students. The new generation is less convinced by traditional promotional tools and printed materials [15].

In recent years, the use of social media platforms in higher education have increased. Many factors influence the use of Social media by HEIs. Kietzmann et al. [13] state that social media platforms enable users to search for people, engage in conversations, interact with groups, and share content. Universities use social media to enhance their brand image. Social media will facilitate the quick dissemination of the information to a large audience [5]. Many students will use Facebook to decide which is the best college to choose for their studies and what types of activities, events and social activities are offered [6]. Beyond the use of social media to socialize, social media platforms enhance organizational performance and reduce costs according to Parveen et al. [17]. The Facebook content was categorized it into entertainment, information and remuneration. The impact of each category on engagement was analyzed [19].

Previous studies had shown that different strategies might be used by HEIs to boost the engagement. Social media engagement is defined as the way people interact with the organization and with each other through social media platforms [22]. The content disseminated by HEIs is crucial to influence the level of engagement [21]. Some researchers focused on how HEIs use their websites enhance their brand image. It is also important for universities to recognize the importance of non-academic aspects to enhance the engagement and the involvement of students [14]. Few studies focused on the Facebook content disseminated by HEIs [23]. The author analyzed universities' Facebook posts in US and Thailand and categorized the posts in 12 types research, faculty, curriculum, campus, students, alumni, industry, events, products, image and reputation, announcements, and others. The Events are mainly academic, conferences and seminars but also non-academic held by or at the university such as sport, art and commencement. Chapleo et al. [7] classified HEIs in UK based on their way of communication through their websites. The authors identified mainly a set of variables that should be included in their website that are: 'teaching', 'research and management excellence', 'international projection', 'universities environment', and 'innovation and corporate social responsibility'. Bélanger et al. [4] identified the following categories of Facebook posts in Canadian HEIs: campus news, campus events, faculty and staff related posts, student engagement, student inquiries and promotional posts. Fähnrich et al. [10] categorized HEIs' Facebook content themes into themes: research, teaching, self-promotion, awards and prizes, addressing fans, and events. They found that 13% of the posts are related to events.

2.2 *Social Media Format and Reactions in Higher Education*

Prior studies explored the different strategies and tactics used by HEIs to enhance users' engagement; they indicate that each format type generate different level of engagement. Social media engagement can be measured in terms of activities such as liking, commenting, and sharing [19]. Some researchers found a negative association between longer Facebook texts and users' engagement [8, 21]. Fähnrich et al. [10] found that HEIs mostly use images and links while posts with video and status updates are less. They also found that 13% of the posts are related to events. Posts with images are more effective in terms of engagement among students compared to older users as they are digital natives [9]. On the other hand, images generate more likes and comments compared to other post formats. In addition, posts with images generate more likes and Shares while post with videos tend to be more shared compared to post with plain text as videos require more time to click on the video and to watch it compare to other formats [10]. Facebook users may feel more attracted to more straightforward and common messages that contain photos and text, rather than those that require clicking on a link or involve watching a video [12]. Valerio et al. [25, 26] analyzed the case of Mexican universities and their Facebook posts, they found that the most and least frequently used format by universities is links and images respectively. Nevertheless, their results show that image is the most conducive format in terms of engagement and links is the least efficient format to boost users' reactions.

3 Methodology

This study uses content analysis to categorize HEIs Facebook posts [23]. The paper addresses the 16 active HEIs operating in Abu Dhabi (5 public and 11 private). A total number of 4148 Facebook posts between 1/9/2019 to 31/8/2020 had been considered in this study, which represents a period of pre- and mid-Covid time. The extraction of Facebook posts was performed through scraping method using Python [24]. After extraction, the 4148 Posts had been categorized into different themes based on the content of the post of which 893 posts were categorized as academic events (397 posts) and non-academic event (496 posts). The academic events are mainly seminars, conferences, webinars, learning programs or trainings etc. As per non-academic events they are mainly related to sport, art, competitions, graduation ceremony, celebrations etc.

Post format is including: plain text, link, image or video. HEIs types are public and private. The language used in the post is: English, Arabic, English, and Arabic. Facebook post engagement/reaction was measured based on the study of [25, 26] and [1] who measured engagement in terms of likes, comments, and shares. As per our study we will also use Comments, Shares and Emotes (sum of like, love, wow, care, sad, angry, haha for each post).

Facebook, as one of the most popular social media platforms in the UAE, represents an interesting platform to analyze. This study is part of a larger project, for which the HEIs' post content analysis is performed for different social media platforms with a combination of qualitative and quantitative methods.

4 Results

To have a better understanding of how users are reacting to HEIs' Facebook posts, Fig. 1 will present the reactions in terms of emotes, comments and shares.

The results show that overall, non-academic events foster more reactions compared to academic events. The highest percentage of posts' reactions for non-academic events fall into the category of comments (11.02%), followed by Emotes (8.07%) and then Shares (6.94%). Results show that academic events received lower reactions in terms of Emotes (3.42%), shares (3.32%) and comments (3.05%).

Figure 2 shows that overall reactions for non-academic events are higher in Private HEIs compared to public with a percentage of 76.91 and 23.09 for emotes, 78.98 and 21.02 for comments and 73.12 and 26.88 for shares, respectively. Academic events received also more reactions in Private HEIs compared to public with a percentage of 72.79 and 27.21 for emotes, 73.33 and 26.67 for comments and 70.16 and 29.84 for shares, respectively. This result could be due to the higher number of private HEIs compared to the public. To address this issue, we conducted the same results for the average mean of reaction per post and we found similar results.

Figure 3 shows that HEIs receive more reactions when the post contains an image. For non-academic events, comment reactions to posts with image have the highest percentage in comparison with academic events (7.05–2.77%), followed by emotes (5.53–3.01%) then shares (3.1–2.82%). For posts related to academic events, all reactions to posts containing videos and links occur less (than 1% and 2% respectively) and no reactions for plain text. For posts related to non-academic events, the format that foster the highest reactions is image followed by video, link and text.

As per the reactions to posts based on the language used, the results from Fig. 4 show that the highest percentages of reaction are to the posts written in English for non-academic events (emotes 5.14%, comments 6.5% shares 3.64%), followed by almost equal reactions to the posts written in English and Arabic with 2.96%, 2.9% and 2.65% for comments, shares, and emotes respectively. The reactions to posts in Arabic are the lowest and range between 0.66 and 2.01%. For academic events, Posts in English and Arabic foster the highest reactions (range between 1.56 and 2.27%). Reactions to posts in Arabic only are very low (less than 1%).

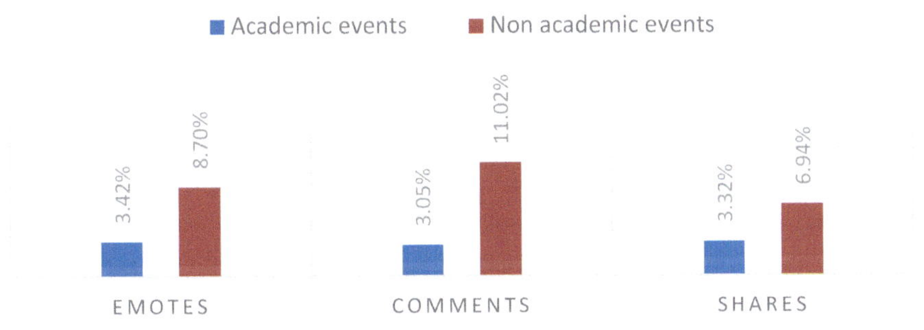

Fig. 1 Reactions per post content

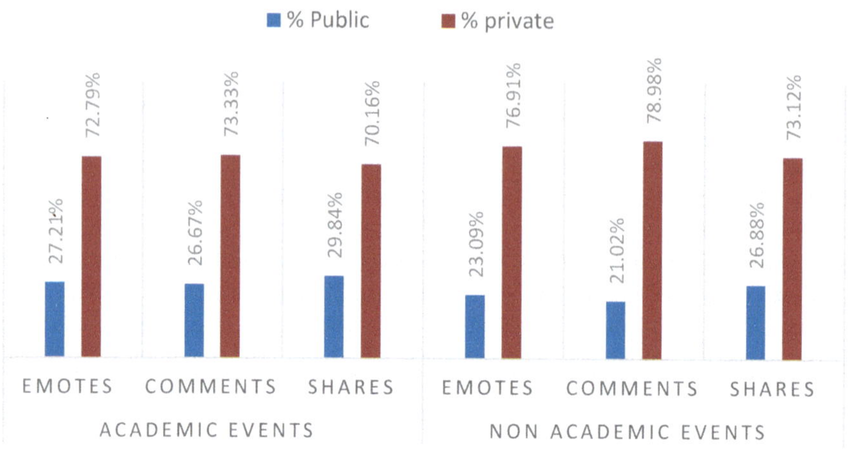

Fig. 2 Reactions per post content/public–private

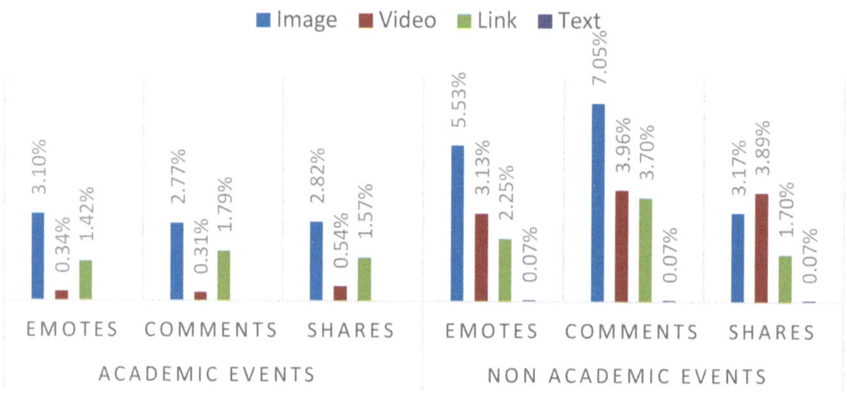

Fig. 3 Reaction per post content/per format

5 Conclusion

This study contributes to previous research by exploring how Higher education institutions can best engage their audience by using the best tactics and strategies in terms of content, format, and language to manage their Facebook platforms. In summary, the results of the study suggest that non-academic events foster more reactions than academic events. Besides, private HEIs seems to generate more reactions than Public HEIs. Image is the most conducive format to increase engagement while plain text is less efficient to generate reactions. Images are processed faster than text, are easier to recall, and create stronger emotional reactions. The language used in the post is also an indicator of the users' engagement.

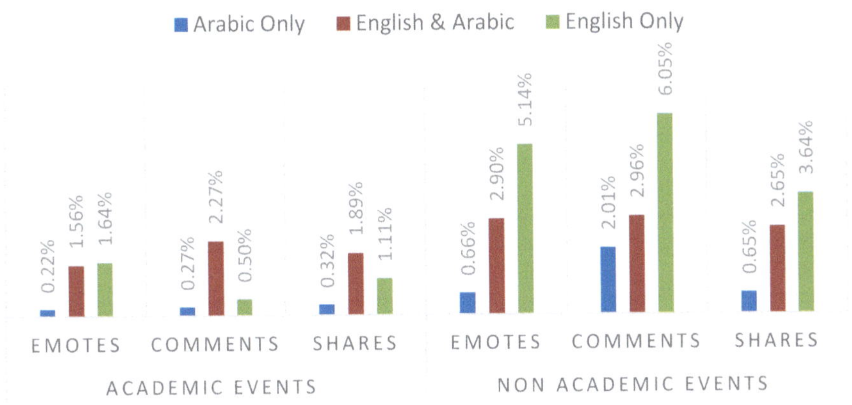

Fig. 4 Reaction per post content per language

The findings of the study can provide guidance for HEIs in Abu Dhabi to develop a plan for their social media content specially related to academic and non-academic events as part of their communication strategy to improve their digital content and foster the audience engagement composed mainly of a millennial generation that is deeply involved with technology. This study also has practical implications for HEIs to guide them to assess their online strengths and weaknesses in social media communication and exploit the advantages of the virtual marketplace to promote their positions over their rivals by selecting the right content, format, and language for posts, HEIs can achieve higher visibility and engagement and generate more likes, comments, and shares through their Facebook content.

This working paper focused mainly on academic and non-academic activities as this study represents a part of a main major project for internal grant, other articles are planned to cover other themes and other social media platforms such as Twitter, YouTube, and Instagram. Future research will also use diversified statistical analysis techniques.

References

1. Alsufyan, N.K., Aloud, M.: The state of social media engagement in Saudi universities. J. Appl. Res. High. Educ. **9**(2), 267–303 (2017). https://doi.org/10.1108/JARHE-01-2016-0001
2. Alsurehi, H.A., Youbi, A.A.A.: Towards applying social networking in higher education: case study of Saudi universities. Int. J. Acad. Res. **6**(5), 221–229 (2014). http://doi.org/10.7813/2075-4124.2014/6-5/B.33
3. Arnett, D.B., German, S.D., Hunt, S.D.: The identity salience model of relationship marketing success: the case of nonprofit marketing. J. Mark. **67**, 89–105 (2003)
4. Bélanger, C.H., Bali, S., Longden, B.: How Canadian universities use social media to brand themselves. Tert. Educ. Manag. **20**(1), 14–29 (2014). https://doi.org/10.1080/13583883.2013.852237

5. Caers, R., De Feyter, T., De Couck, M., Stough, T., Vigna, C., Du Bois, C.: Facebook: a literature review. New Media Soc. **15**(6), 982–1002 (2013). https://doi.org/10.1177/1461444813488061
6. Cain, J., Scott, D.R., Akers, P.: Pharmacy students' Facebook activity and opinions regarding accountability and E-professionalism. Am. J. Pharm. Educ. **73**(6) (2009)
7. Chapleo, C., Durán, M.V.C., Díaz, A.C.: Do UK universities communicate their brands effectively through their websites? J. Mark. High. Educ. **21**(1), 25–46 (2011). https://doi.org/10.1080/08841241.2011.569589
8. De Vries, L., Gensler, S., Leeflang, P.S.H.: Popularity of brand posts on brand fan pages: an investigation of the effects of social media marketing. J. Interact. Mark. **26**(2), 83–91 (2012). https://doi.org/10.1016/j.intmar.2012.01.003
9. Delello, J.A., McWhorter, R.R.: Reducing the digital divide: connecting older adults to iPad technology. J. Appl. Gerontol. **36**(1), 3–28 (2017). https://doi.org/10.1177/0733464815589985
10. Fähnrich, B., Vogelgesang, J., Scharkow, M.: Evaluating universities' strategic online communication: how do Shanghai ranking's top 50 universities grow stakeholder engagement with Facebook posts? J. Commun. Manag. **24**(3), 265–283 (2020). https://doi.org/10.1108/JCOM-06-2019-0090
11. Globalmediainsight 2020 (n.d.). Retrieved March 29, 2021, from https://www.globalmediainsight.com/blog/uae-facebook-users/
12. Joo, S., Choi, N., Baek, T.H.: Library marketing via social media: The relationships between Facebook content and user engagement in public libraries. Online. Info. Rev. **42**(6), 940–955 (2018). https://doi.org/10.1108/OIR-10-2017-0288
13. Kietzmann, J.H., Hermkens, K., McCarthy, I.P., Silvestre, B.S.: Social media? Get serious! Understanding the functional building blocks of social media. Bus. Horiz. **54**(3), 241–251 (2011). https://doi.org/10.1016/j.bushor.2011.01.005
14. Libbey, H.: Measuring student relationships to school: Attachment, bonding, connectedness, and engagement. J. Sch. Health. **74**(7), 274–283 (2004) https://doi.org/10.1111/j.1746-1561.2004.tb08284.x
15. Maringe, F.: University and course choice: implications for positioning, recruitment and marketing. Int. J. Educ. Manag. **20**(6), 466–479 (2006). https://doi.org/10.1108/09513540610683711
16. Nyangau, J., Bado, N.: Social media and marketing of higher education: a review of the literature. J. Res. Cent. Educ. Technol. **8**(1), 38–51. http://rcetj.org/index.php/rcetj/article/viewArticle/180
17. Parveen, F., Jaafar, N.I., Ainin, S.: Social media usage and organizational performance: reflections of Malaysian social media managers. Telematics Inform. **32**(1), 67–78 (2015). https://doi.org/10.1016/j.tele.2014.03.001
18. Peruta, A., Shields, A.B.: Marketing your university on social media: a content analysis of Facebook post types and formats. J. Mark. High. Educ. **28**(2), 175–191 (2018). https://doi.org/10.1080/08841241.2018.1442896
19. Pletikosa Cvijikj, I., Michahelles, F.: Online engagement factors on Facebook brand pages. Soc. Netw. Anal. Min. **3**(4), 843–861 (2013). https://doi.org/10.1007/s13278-013-0098-8
20. Rutter, R., Roper, S., Lettice, F.: Social media interaction, the university brand and recruitment performance. J. Bus. Res. **69**(8), 3096–3104 (2016). https://doi.org/10.1016/j.jbusres.2016.01.025
21. Sabate, F., Berbegal-Mirabent, J., Cañabate, A., Lebherz, P.R.: Factors influencing popularity of branded content in Facebook fan pages. Eur. Manag. J. **32**(6), 1001–1011 (2014). https://doi.org/10.1016/j.emj.2014.05.001
22. Smith, T., Lambert, R.: A systematic review investigating the use of twitter and facebook in university-based healthcare education. Health Educ. **114**(5), 347–366 (2014). https://doi.org/10.1108/HE-07-2013-0030
23. Taecharungroj, V.: Higher education social media marketing: 12 content types universities post on Facebook. Int. J. Manag. Educ. **11**(2), 111–127 (2017). https://doi.org/10.1504/IJMIE.2017.10002819

24. Thomas, D.M., Mathur, S.: Data analysis by web scraping using Python. In: Proceedings of the 3rd International Conference on Electronics and Communication and Aerospace Technology, ICECA 2019, pp. 450–454 (2019). http://doi.org/10.1109/ICECA.2019.8822022
25. Valerio, G., Herrera-murillo, D.J., Villanueva-puente, F., Herrera-murillo, N., Rodríguez-martínez, M.C.: The relationship between post formats and digital engagement: a study of the Facebook pages of Mexican universities. Int. J. Educ. Technol. High. Educ. **12**(1) (2015)
26. Valerio, G., Herrera-Murillo, D.J., Villanueva-Puente, F., Herrera-Murillo, N., del Rodríguez-Martínez, M.C.: Relación entre los formatos de publicación y el engagement digital: Estudio de las páginas de Facebook de las universidades Mexicanas. RUSC Univ. Knowl. Soc. J. **12**(1), 50–63 (2015). https://doi.org/10.7238/rusc.v12i1.1887
27. York, T.T., Gibson, C., Rankin, S.: Defining and measuring academic success. Pract. Assess. Res. Eval. **20**(5), 1–20 (2015)

The Effect of Employer Branding Through Social Media on Employee Engagement and Employer Attractiveness: A Case Study Research

Menatalla Kaoud and Mostafa ElBolok

Abstract Nowadays companies successfully use social media to promote the employer brand along with traditional recruitment tools. Developing an appealing employer brand can increase employee satisfaction, engagement, and retention. The promotion of brand exterior to the organization will also make it attractive to the prospective employees. A central question is addressed in this paper: how employer branding via social media can affect employee engagement and employer attractiveness? Using a case study methodology in an insurance brokerage firm, this paper presents an advanced understanding of employer branding through social media and in-depth analysis of the effect of good employer brands on employee engagement and employer attractiveness. This research will help companies in the implementation of effective employer branding strategies via internal and external communications allowing them to improve their brand image and employee engagement.

Keywords Employer branding · Employee engagement · Employer attractiveness

1 Introduction

Nowadays companies successfully use social media to promote the employer brand along with traditional recruitment tools [1]. Employer branding campaigns are mainly motivated by an aspiration to promote employee engagement which can be defined as 'the ability to capture the heads and hearts of your employees to instill an inherent desire and passion for excellence' [2]. Developing an appealing employer brand can offer a number of advantages for businesses, including increased employee satisfaction, engagement, and retention [3]. Promotion of brand inside the organization will

M. Kaoud (✉) · M. ElBolok
School of Business Administration, Nile University, Giza, Egypt
e-mail: mkaoud@nu.edu.eg

M. ElBolok
e-mail: melbolok@nu.edu.eg

A. Hamdan et al. (eds.), *The Implementation of Smart Technologies for Business Success and Sustainability*, Studies in Systems, Decision and Control 216,
https://doi.org/10.1007/978-3-031-10212-7_38

increase employee loyalty, although the promotion of brand exterior to the organization will make it attractive to the prospective workforces [4] 'Employer attractiveness is the intended gains that a prospective employee recognizes in working for a certain organization' [5].

Existing literature indicates a positive relationship between employee engagement and employer branding [6–8] which primarily considers good employer branding strategies as drivers of engagement. The digitalization of HRM practices enables the use of more internet-based recruitment tools, including the potential of social media, to accomplish a targeted recruitment strategy [9]. However, there are limited academic studies connecting employer branding, the attitude of employees and engagement [3, 10, 11]. Organizations devoted to employees' expectations, according to [6], use employer branding to increase employee engagement, even though they acknowledge that it is still unclear why employees respond with varied levels of engagement. Most of previous studies reveal the importance of employer branding by creating employer attractiveness, employee engagement, and employee retention, in addition to focusing on employee retention as a major element for gaining a competitive advantage [12]. Hence, it is crucial to further understand how they affect each other [3, 13–15]. Studies clearly demonstrate that social media has a future in employer branding, yet it still needs further investigation [16–18]. Therefore, this paper aims to answer the following question: how employer branding through social media can affect employee engagement and employer attractiveness? Using a case study methodology in an insurance brokerage firm, this paper presents an advanced understanding of employer branding through social media and in-depth analysis of the effect of good employer brands on employee engagement and employer attractiveness. On a managerial level, it is intended to help organizations to effectively retain employees' engagement and attract potential employees by enhancing employer branding practices. This paper is structured into three main parts. At first, a review of the literature presents relevant elements concerning employer branding and social media, employee engagement, and employer attractiveness. Subsequent sections present the research methodology and the research findings of the case study. The paper concludes with a summary, the research contributions, a critical reflection and potential opportunities for future research.

2 Literature Review

This review examines previous studies of employer branding and social media. It also presents the link between employer branding, employee engagement and employer attractiveness.

2.1 Employer Branding

In marketing literature, a brand is a sign that embodies all of the associations that have been created with a name [19]. Branding can be applied to a wide range of items [20]. The primary objective of branding cattle was to indicate ownership or, to put it another way, to differentiate. Differentiation in marketing has clear benefits and is viewed as a competitive advantage, but the concept is also gaining traction in creating relationships with employees, who like to be a part of distinctive organizations [21]. The concept of employer branding lies in the idea of brand as 'a mixture of attributes, tangible and intangible, represented in a trademark, which if managed appropriately, adds value and creates influence' [22]. Employer branding is the 'targeted long-term strategy to raise employee awareness and perceptions of a company among current and future employees, as well as other stakeholders. It is the method of creating recognizable and unique employer identity, as well as brand that differentiates it from its rivals' [23]. One objective of employer branding is to make the employer stand out in the minds of the employees [21].

2.2 Employer Branding and Social Media

Creating a good brand is established through the external and internal communications of a company [24]. Employer branding is often defined as the sum of organizational efforts to communicate that it is a desirable place to work [25]. Values and ethics are considered to be essential parts of employer branding [26] and public relations experts make strategic communication plans and disseminate messages to communicate to future employees about values and ethics. Internal communications play a key role in employer branding as it launches a critical value for the function in establishing the first primary driver of emotional engagement, meaningfulness, through consistent communication about an organization's vision and goals in the future [27]. Employer brand consists of two parts, according to [28], internal employer brand and external employer brand. To effectively align employer brand perception among current and potential employees and customers, these two aspects must be reconciled. This is where marketing communication comes in, with its primary goal to harmonize the perception of the employer brand among internal and external audiences [29].

Social media provides a unique communication channel which permits the message to reach as many audiences and as quickly as possible, the way employer brands modify their style of communication may be vital to relationships they seek to form with their audiences, e.g., employees and candidates [30, 31]. The effects and benefits of information dissemination via social media may be of particular relevance to businesses seeking to build a good employer branding reputation and attractiveness [17, 32] which in turn is positively linked to intentions to apply for a job [16, 18]. The study of [32] demonstrates that the employees' exposure to a company's profile on social media (Facebook or LinkedIn) had an impact on their perceptions of company's brand and its attractiveness. That means that the usage of social media can enhance the employer brand image and attractiveness.

2.3 Employee Engagement

Engagement involves functional as well as physical existence in performing an organizational role [33] and allows organizations innovate and compete in the market [34]. It is a term used to label management's efforts to cooperate with and encompass employees [35, 36]. Engagement is something employees convey to the workplace and is consequently linked to a higher level of customer satisfaction, loyalty, productivity, motivation, social responsibility behavior, and profit [37, 38]. Employee engagement can be defined as the individual's involvement, satisfaction and eagerness for work [38]. It is also the ability to capture the heads and hearts of your employees to instill an inherent desire and passion for excellence [2]. Employee engagement can be described as emerging an environment, which contributes to motivating and connecting employees with their job [39].

Employer branding aims to improving employee satisfaction. Satisfied employees create better relationships with customers [21]. Employees become emotionally attached to the company as a result of a strong employer branding strategy. The employer branding ought to endorse a positive response from the employee [21]. Communication plays a key role in increasing employee engagement [40, 41]. Relationships between engagement and organizational reputation [42], job satisfaction [43, 44] have been established, demonstrating the importance of engagement. Involved personnel have been described as a critical force in achieving business competitiveness, especially in the service sector [45]. The case study investigates the effect of employer branding through internal and external communication on employee engagement in service sector, in particular, insurance sector.

2.4 Employer Attractiveness

The perception of employer attractiveness has been used in various studies [18, 46, 47] because an attractive employer gives a decent yield on investment by attracting and retaining talented job candidates [48]. The attractiveness of an employer is related to an individual or job seekers perception of the organization as a good place for employment [49, 50]. The job experience and perks gained during the course of employment are also termed as employer attractiveness [5]. The employer value propositions that an organization bids to the potential employees play a big role in its image as an attractive employer [23].

Employer branding offers a competitive edge to gain, foster and retain employees [4]. Employer image considerably affects the attractiveness of the organization [12,

51, 52]. Also, firms that seek to preserve their status of attractive employer among existing and potential employees should deliberate who will represent the brand and how the response will be framed [30]. External employer branding activities highly influence employer attractiveness, while internal activities directed at employees have a major impact on both employer attractiveness and employer image [53]. Social media platforms have boomed in the recent years by which employers should encourage employees to provide information of the organizational cultures [54].

To win over and keep the best talent, firms are increasingly brought to focus branding themselves as desirable employers [48, 55–57]. The case study investigates the role of employer branding especially through the use of social media in enhancing employer attractiveness.

3 Research Methodology

The research methodology is based on a qualitative approach. The authors adopted the case study research method [58]. The choice of the case method is justified by the fact that the study of employer branding through social media represents a contemporary phenomenon [16, 30, 32] in a context of real life [58]. In addition, there is a clear need to study in-depth the link between employer branding, employee engagement, and employer attractiveness [3, 12–15]. This method provides a more in-depth investigation of the phenomenon than a large-scale survey or an experimental research study [59].

3.1 Presentation of the Case Study

Deraya Insurance Brokerage is an insurance brokerage firm, established in Egypt, specializing in medical insurance. The company works with international and local medical insurance providers. The company is a medium-sized enterprise that has 130 employees.

Deraya Insurance Brokerage was awarded the Top Worldwide Distributor of Bupa Global medical insurance for nine years in a row. The company is developing its employer branding practices engaging and retaining its employees and attracting qualified candidates in order to maintain its leadership position in the market and sustain its competitive advantage. Recently, the company adopted employer branding campaigns through social media channels (Facebook, Instagram and LinkedIn) with the participation of its employees. Therefore, this case is interesting for the research objective.

3.2 Data Collection

Twenty semi-structured interviews were carried out with employees participating in the social media campaigns (who got recognized and/or posted their testimonials about the workplace on social media pages: Facebook, Instagram and LinkedIn) from different departments (Sales, Accounting, Operations, Training and Quality) and with all hierarchal levels. In addition, four interviews were also conducted with Human Resources and Marketing teams in order to understand the objectives and the consequences of using the social media in employer branding.

A one-month observational study (from 1st December to 29th December 2021) was also conducted to observe the internal communication forms inside the company.

Besides, the authors consulted secondary data (e.g., archival documents, performance reports, website and social media pages) to extend the empirical analysis.

3.3 Data Analysis

Most interviews were recorded (a total of 12 h). They were translated and transcribed to accomplish the coding and the analysis of the data. The data in this study, the verbatim texts of the participant interview responses, were managed and analyzed manually. The authors investigated the effect of employer branding through social media on employee engagement and employer attractiveness.

4 Research Findings

This section presents summaries of key findings related to employer branding and social media. Table 1 presents some illustrative quotes of participant responses for each major area.

4.1 Brand Awareness Through Social Media

The case study shows the role of social media in the management of brand awareness and the perceptions of employees as well as the external audience. Deraya Insurance Brokerage uses social media channels (Facebook and Instagram) to promote its insurance services in a simple way. One of the main objectives of their social media campaigns is to increase awareness about the insurance services in a simple way as the insurance sector is always linked to complicated statistical techniques.

The second objective is to recognize its employees and to promote its work environment. They publish regular posts on social media channels about the

Table 1 Key findings and illustrative quotes from interviews with employees participating in the social media campaign

Brand awareness through social media	"Deraya focuses on the employees…" "We always share knowledge and experiences…" "Sharing useful information about our services and healthy lifestyle on Facebook and Instagram…This is becoming more important after the pandemic"
Impact of employer branding on employee engagement – Intentions to work	"Due to the learning culture, I became a manager now…" "I have knowledge and experience about my current position, and I also know about the work of my colleagues in other departments…this helps a lot at my work" "Best distributor in middle east" "We are close to our supervisors and managers, and this motivates us…" "Feeling appreciated on social media improved my productivity…" "The contest motivates us to achieve more…"
– Emotional attachment	"Loyal more when recognized…" "I like people with whom I work" "I feel more satisfied by appearing on social media"
Reconciliation between internal and external employer brand	"Both ways of recognition are important…" "I feel happy when I see my photo on the wall at the workplace…" "Internal and external communication are important…yet I prefer to be recognized on social media…My parents, my family and my friends are aware about my superior performance…"
Employer branding and employer attractiveness	"I feel proud being part of Deraya family" "This kind of appreciation and the supportive environment make me more attached and committed towards Deraya"

employees who demonstrate great performance in their work that they are called 'Star employees'. They also present the fun and supportive work environment and the team spirit through their social media channels. Moreover, they post employee testimonials on their website and social media pages (Facebook, Instagram and LinkedIn) using hashtags #GetToKnowUs and #Derayians. The employees participate in the development of communication itself.

The last objective is to share useful information and knowledge about healthy lifestyles on Facebook and Instagram pages using hashtags like #healthylifestyle, #healthfirst, and #covid19. In this way, they brand their culture image of knowledge sharing and they engage their audience. In addition, they present their achievements and awards through their digital channels (social media pages and website). Social media helped the company communicate and disseminate messages about its values (customer-oriented, employee satisfaction, healthy work environment, etc.) to its current and potential employees and customers. This shows how social media channels can contribute to the brand perception of employees as well as the external audience.

4.2 Impact of Employer Branding on Employee Engagement

Intentions to work. The case study shows that the branding efforts of the company to manage the awareness and perceptions of its employees and to convey that it is a desirable place to work increased the motivation and intentions of the employees to work as well as their passion for excellence. Extrinsic motivation in the workplace focuses on external rewards such as salaries, perks, benefits, workplace characteristics, and relationships with colleagues and supervisors, whereas intrinsic motivation underlines rewards derived from the work itself, as well as appreciation and accomplishment.

Deraya Insurance Brokerage promotes a learning culture in the workplace through different training programs to help its employees develop knowledge and competence, improve their work performance, and achieve their career goals. During the interviews, some employees also stated that one of the advantages they like in their organization is the non-existing barriers between different departments and management as well; by which they find management and colleagues paying attention to their ideas. Furthermore, they point out that the vision statement of their company affects them positively making them more passionate about their work. In addition, the HR team organizes, regularly, sales contest to motivate the sales teams achieve their targets in which monetary recompenses are offered based on their sales. The superior performance of the employees is well communicated via social media pages of the company and this recognition also improves their engagement and their intentions to work.

Emotional attachment. The employees stressed, during the interviews, that they have developed an emotional attachment to their company becoming more loyal in response to the various ways the company motivates and encourages them. The friendly and supportive work environment improves employees' perception towards the company's image. Also, being recognized on social media and at the workplace on their work achievements makes them feel satisfied with their jobs as well as being a part of their organization. This demonstrates how strong employer branding can promote a positive emotional response from the employees.

4.3 *Reconciliation Between Internal and External Employer Brand*

Employer brand consists of two parts, according to [28]: internal employer brand and external employer brand. These two aspects must be reconciled to effectively align employer brand perception among current and potential employees and customers.

The Human Resources department in collaboration with the Marketing department organizes internal reward and recognition initiatives such as 'Star Employees' to acknowledge the outstanding work performance of its employees. This employee recognition in the workplace is also promoted via social media channels (Facebook and Instagram) of the company. Most of the interviewed employees agreed on the importance of both ways of recognition and how it can increase their engagement. The internal communication is valued by the employees as they feel appreciated in the workplace from their coworkers, supervisors and managers, and the external communication via social media is more valued by them as they feel appreciated and acknowledged among their families and friends as well.

Furthermore, the Human Resources department and the Quality department are responsible for regular internal communications about the organization's vision, mission, values and future purposes to employees in all departments. This is to ensure the delivery of a consistent message about the company and its services to customers, prospects and future employees.

4.4 *Employer Branding and Employer Attractiveness*

Deraya Insurance Brokerage pays particular attention to attracting qualified employees in order to maintain its competitive advantage. The use of social media channels (Facebook and Instagram) to promote the fun and supportive work environment and the teamwork spirit and to recognize the superior performance of the employees increased the number of qualified candidates. The company also emphasized its values that reflect its focus on employees (healthy work environment, employees' satisfaction and well-being, work-life balance) in its recruitment campaigns through social media channels (Facebook and Instagram). This demonstrated how the employer branding via social media can positively affect the employer attractiveness. In addition, the case study revealed that the employees' exposure to the company's profile on social media recognizing their superior performance increased their retention.

5 Discussion and Conclusion

The case study shows an illustrative example of creating a good employer brand through external and internal communications. It also emphasizes the role of Human Resources and Marketing departments to harmonize the perception of the employer brand among internal and external audiences (existing and future employees). It stresses the importance of using social media channels (Facebook, Instagram and LinkedIn) in employer branding along with the internal efforts of communication. Setting clear objectives of communication for social media campaigns and ensuring the consistency of information disseminated through the different digital channels (social media pages and website) are also crucial for employer branding strategies. It also reveals that employee recognition with exposure to the company's profile on social media can have more effect than their recognition in the workplace on their engagement towards the company. Furthermore, the case study underlines that a strong employer brand via social media can have a positive impact on increasing the pool of qualified candidates and the retention of the existing employees.

In line with previous studies, the use of social media can improve the employer brand image and attractiveness [16, 32] Internal employer brand and external employer brand must be reconciled to effectively align employer brand perception among current and potential employees and customers [28] improving employer attractiveness [60]. Major factor that inspires engagement is the work environment [61]. Vision inspiration also refers to a strong positive belief in and emotional attachment to an organization's vision [62]. Employee engagement can be derived from employees who are attached to the values and mission of the company [63]. Moreover, with introducing a boundary less behavior more employees across all functions can be coordinated with the company's goals [64].

This research has contributed to the need of further investigation about the potential of social media in employer branding [16–18]. The case study provides deeper insights about the connection between employer branding, employee engagement and employer attractiveness and how they affect each other [3, 10, 13, 14]. From a managerial perspective, the case study in a multi-award-winning company allowed to share certain best practices in the fields of Human Resources Management and Marketing.

There are some limitations to the study, many of which point to future research opportunities. The study is based on a single case study with the goal of examining in depth the impact of employer branding via social media on employee engagement and attractiveness. The research examines an Egyptian brokerage firm specialized in medical insurance, so the findings may not be completely generalized to other sectors or countries. As a result, international studies would be useful in testing the generalizability of findings from other countries and sectors. Large-scale quantitative studies might also be interesting.

References

1. Groß, J., Koch, B.: Employer Branding in klein-und mittelständischen Unternehmen (2018)
2. Fleming, J.H., Asplund, J.: Where employee engagement happens. Gallup Manag. J. **3**(1) (2007)
3. Špoljarić, A., Verčič, A.T.: Internal communication satisfaction and employee engagement as determinants of the employer brand. J. Commun. Manag. **26**(1), 130–148 (2022)
4. Minchington, B., Thorne, K.: Measuring the effectiveness of your employer brand. Hum. Resour. Mag. **12**(4), 14–16 (2007)
5. Berthon, P., Ewing, M., Hah, L.L.: Captivating company: dimensions of attractiveness in employer branding. Int. J. Advert. **24**(2), 151–172 (2005)
6. Burawat, P.: The relationships among perceived employer branding, employee engagement, and employee expectation in service industry. Abs. Econ. Financ. Manag. Outlook **2**, 1–1 (2014)
7. Heilmann, P., Saarenketo, S., Liikkanen, K.: Employer branding in power industry. Int. J. Energy Sect. Manag. **7**(2), 283–302 (2013)
8. Kunerth, B., Mosley, R.: Applying employer brand management to employee engagement. Strateg. HR Rev. **10**(3), 19–26 (2011)
9. Tanwar, K., Kumar, A.: Employer brand, person-organisation fit and employer of choice: investigating the moderating effect of social media. Pers. Rev. **48**(3), 799–823 (2019)
10. Priyadarshi, P.: Employer brand image as predictor of employee satisfaction, affective commitment and turnover. Indian J. Ind. Relat. 510–522 (2011)
11. Lievens, F.: Employer branding in the Belgian Army: the importance of instrumental and symbolic beliefs for potential applicants, actual applicants, and military employees. Hum. Resour. Manage. **46**(1), 51–69 (2007)
12. Easa, N.F., Bazzi, A.M.: The influence of employer branding on employer attractiveness and employee engagement and retention: ten years of literature. Int. J. Customer Relat. Mark. Manag. (IJCRMM) **11**(4), 48–69 (2020)
13. Hadi, N.U., Ahmed, S.: Role of employer branding dimensions on employee retention: evidence from educational sector. Adm. Sci. **8**(3), 44 (2018)
14. Sahu, S., Pathardikar, A., Kumar, A.: Transformational leadership and turnover: mediating effects of employee engagement, employer branding, and psychological attachment. Leadersh. Organ. Dev. J. **39**(1), 82–99 (2017)
15. Piyachat, B., Chanongkorn, K., Panisa, M.: The mediate effect of employee engagement on the relationship between perceived employer branding and discretionary effort. DLSU Bus. Econ. Rev. **24**(1), 59–72 (2014)
16. Kucherov, D., Zhiltsova, E.: Social media in employer branding in FMCG in Russia: millennials' perspective. J. East-West Bus. **27**(2), 160–183 (2021)
17. Bondarouk, T., Ruël, H., Axinia, E., Arama, R.: What is the future of employer branding through social media? Results of the Delphi study into the perceptions of HR professionals and academics. In: Social Media in Human Resources Management. Emerald Group Publishing Limited, UK (2014)
18. Sivertzen, A.M., Nilsen, E.R., Olafsen, A.H.: Employer branding: employer attractiveness and the use of social media. J. Prod. Brand Manag. **22**(7), 473–483 (2013)
19. Gardner, B.B., Levy, S.J.: The product and the brand. Harv. Bus. Rev. **33**(2), 33–39 (1955)
20. Levitt, T.: Marketing success through differentiation-of anything, pp. 83–91. Graduate School of Business Administration, Harvard University (1980)
21. Davies, G.: Employer branding and its influence on managers. Eur. J. Mark. **42**(5/6), 667–681 (2008)
22. Swystun, J.: The brand glossary. Interbrand, Palgrave Macmillan, New York, NY (2007)
23. Backhaus, K., Tikoo, S.: Conceptualizing and researching employer branding. Career Dev. Int. **9**(5), 501–517 (2004)
24. Ruchika, Prasad, A.: Untapped relationship between employer branding, anticipatory psychological contract and intent to join. Glob. Bus. Rev. **20**(1), 194–213 (2019)

25. Lloyd, S.: Branding from the inside out. BRW **24**(10), 64–66 (2002)
26. Neill, M.S.: The influence of employer branding in internal communication. Res. J. Inst. Public Relat. **3**(1), 1–25 (2016)
27. Ruck, K., Trainor, S.: Developing internal communication practice that supports employee engagement. Diunduh dari. http://www.pracademy.co.uk/wp-content/uploads/2012/07/Developing-Internal-Communication-Practice-That-Supports-Employee-Engagement (2012)
28. Knox, S., Freeman, C.: Measuring and managing employer brand image in the service industry. J. Mark. Manag. **22**(7–8), 695–716 (2006)
29. Moroko, L., Uncles, M.D.: Characteristics of successful employer brands. J. Brand Manag. **16**(3), 160–175 (2008)
30. Kapuściński, G., Zhang, N., Zeng, L., Cao, A.: Effects of crisis response tone and spokesperson's gender on employer attractiveness. Int. J. Hosp. Manag. **94**, 102884 (2021)
31. Coon, M.: Social media marketing: successful case studies of businesses using Facebook and Youtube with an in-depth look into the business use of Twitter. Unpublished term project, Stanford University (2010)
32. Carpentier, M., Van Hoye, G., Weijters, B.: Attracting applicants through the organization's social media page: signaling employer brand personality. J. Vocat. Behav. **115**, 103326 (2019)
33. Kahn, W.A.: Psychological conditions of personal engagement and disengagement at work. Acad. Manag. J. **33**(4), 692–724 (1990)
34. Welch, M.: The evolution of the employee engagement concept: communication implications. Corp. Commun. Int J. **16**(4), 328–346 (2011)
35. Verčič, A.T., Vokić, N.P.: Engaging employees through internal communication. Pub. Relat. Rev. **43**(5), 885–893 (2017)
36. Verčič, A.T.: The impact of employee engagement, organisational support and employer branding on internal communication satisfaction. Pub. Relat. Rev. **47**(1), 102009 (2021)
37. Pandita, D., Bedarkar, M.: Factors affecting employee performance: a conceptual study on the drivers of employee engagement. Prabandhan Indian J. Manag. **8**(7), 29–40 (2015)
38. Harter, J.K., Schmidt, F.L., Hayes, T.L.: Business-unit-level relationship between employee satisfaction, employee engagement, and business outcomes: a meta-analysis. J. Appl. Psychol. **87**(2), 268 (2002)
39. Bridger, E.: Employee Engagement, vol. 10. Kogan Page Publishers, UK (2014)
40. Pugh, S.D., Dietz, J.: Employee engagement at the organizational level of analysis. Ind. Organ. Psychol. **1**(1), 44–47 (2008)
41. Wiley, J.W., Kowske, B.J., Herman, A.E.: Developing and validating a global model of employee engagement. In: Handbook of Employee Engagement. Edward Elgar Publishing, UK (2010)
42. Men, L.R.: CEO credibility, perceived organizational reputation, and employee engagement. Pub. Relat. Rev. **38**(1), 171–173 (2012)
43. Kaliannan, M., Adjovu, S.N.: Effective employee engagement and organizational success: a case study. Procedia Soc. Behav. Sci. **172**, 161–168 (2015)
44. Meng, J., Berger, B.K.: The impact of organizational culture and leadership performance on PR professionals' job satisfaction: testing the joint mediating effects of engagement and trust. Pub. Relat. Rev. **45**(1), 64–75 (2019)
45. Bhasin, J., Mushtaq, S., Gupta, S.: Engaging employees through employer brand: an empirical evidence. Manag. Labour Stud. **44**(4), 417–432 (2019)
46. Mesmer-Magnus, J., Guidice, R., Andrews, M., Oechslin, R.: The effects of supervisor humour on employee attitudes. J. Manag. Dev. **37**(9/10), 697–710 (2018)
47. Elving, W.J., Westhoff, J.J., Meeusen, K., Schoonderbeek, J.W.: The war for talent? The relevance of employer branding in job advertisements for becoming an employer of choice. J. Brand Manag. **20**(5), 355–373 (2013)
48. Lievens, F., Slaughter, J.E.: Employer image and employer branding: what we know and what we need to know. Annu. Rev. Organ. Psych. Organ. Behav. **3**, 407–440 (2016)
49. Ehrhart, K.H., Ziegert, J.C.: Why are individuals attracted to organizations? J. Manag. **31**(6), 901–919 (2005)

50. Highhouse, S., Lievens, F., Sinar, E.F.: Measuring attraction to organizations. Educ. Psychol. Measur. **63**(6), 986–1001 (2003)
51. Turban, D.B., Greening, D.W.: Corporate social performance and organizational attractiveness to prospective employees. Acad. Manag. J. **40**(3), 658–672 (1997)
52. Gatewood, R.D., Gowan, M.A., Lautenschlager, G.J.: Corporate image, recruitment image and initial job choice decisions. Acad. Manag. J. **36**(2), 414–427 (1993)
53. Kalińska-Kula, M., Staniec, I.: Employer branding and organizational attractiveness: current employees' perspective. Eur. Res. Stud. **24**(1), 583–603 (2021)
54. Pernkopf, K., Latzke, M., Mayrhofer, W.: Effects of mixed signals on employer attractiveness: a mixed-method study based on signalling and convention theory. Hum. Resour. Manag. J. **31**(2), 392–413 (2021)
55. Saini, G.K., Rai, P., Chaudhary, M.K.: What do best employer surveys reveal about employer branding and intention to apply? J. Brand Manag. **21**(2), 95–111 (2014)
56. Banerjee, P., Saini, G.K., Kalyanaram, G.: The role of brands in recruitment: mediating role of employer brand equity. Asia Pac. J. Hum. Resour. **58**(2), 173–196 (2018)
57. Theurer, C.P., Tumasjan, A., Welpe, I.M., Lievens, F.: Employer branding: a brand equity-based literature review and research agenda. Int. J. Manag. Rev. **20**(1), 155–179 (2018)
58. Yin, R.K.: Designing case studies. Qual. Res. Methods **5**, 359–386 (2003)
59. Dufour, S., Fortin, D., Hamel, J.: L'enquête de terrain en sciences sociales: l'approche monographie et les méthodes qualitatives. Les Éditions Saint Martin, Montréal, Canada (1991)
60. Kalinska-Kula, M., Staniec, I.: Employer branding and organizational attractiveness: current employees' perspective. Eur. Res. Stud. J. **24**(1), 583–603 (2021)
61. Waqas, A., Bashir, U., Sattar, M.F., Abdullah, H.M., Hussain, I., Anjum, W., Arshad, R.: Factors influencing job satisfaction and its impact on job loyalty. Int. J. Learn. Dev. **4**(2), 141–161 (2014)
62. O'Connell, D., Hickerson, K., Pillutla, A.: Organizational visioning: an integrative review. Group Org. Manag. **36**(1), 103–125 (2011)
63. Ewing, M., Men, L.R., O'Neil, J.: Using social media to engage employees: insights from internal communication managers. Int. J. Strateg. Commun. **13**(2), 110–132 (2019)
64. Patel, J.: How can leaders cultivate a culture of organizational trust and implement effective strategies to drive high-performance? (2021)

The Persuasive Effects of a Beauty Influencer of an Iconic Local Skincare Brand in Indonesia

Indah Fatmawati and Nabilah Permatasari

Abstract This study investigates the impact of influencers' credibility and influencer/product congruence on purchase decision becomes the focus of this study by considering the mediating effects of attitude toward the advertisement and the brand. Our results have proven that influencers' credibility positively affects attitude towards the ad but has no significant impact on attitude towards the brand and purchase decision. Meanwhile, influencer/product congruence positively affects the advertisements' attitude but not the brand and the purchase decision. Attitude toward the ad has a significant positive impact on attitude toward the brand. Attitude toward the brand has significant effects on purchase decisions. Attitude toward the advertisement mediates the influencers' credibility on attitude toward the brand. Furthermore, attitude toward the brand mediates the attitude toward the ad on the purchase decision.

Keywords Influencers' credibility · Influencer/product congruence · Attitude toward the advertisement · Attitude toward the brand · Purchase decision

1 Introduction

Marketing communication is a company's strategy used to convey, convince, and remember clients about the items or services they sell [1]. There are eight platforms in the marketing communication mix: advertising, personal selling, mobile marketing, direct and database marketing, internet and social media marketing, public relations and publicity, event and experiences, and sales promotion [2]. In Indonesia, the number of people who use the internet and actively participate in social media has expanded. A survey from We are a Social in partnership with Hootsuite discovered

I. Fatmawati (✉)
Doctoral of Management, Postgraduate Program, Universitas Muhammadiyah Yogyakarta, Yogyakarta, Indonesia
e-mail: indahfatmawati@umy.ac.id

N. Permatasari
Department of Management, Faculty of Economics and Business, Universitas Muhammadiyah Yogyakarta, Yogyakarta, Indonesia

A. Hamdan et al. (eds.), *The Implementation of Smart Technologies for Business Success and Sustainability*, Studies in Systems, Decision and Control 216,
https://doi.org/10.1007/978-3-031-10212-7_39

that the growth rate for social media active users was 13% in 2018–2019 [3]. One of the growing trends in marketing is using social media influencers. The use of influencer marketing is increasing, according to a United States-based survey [4].

Influencers may entertain, enlighten, and perhaps change their followers' mindsets, behavior, and habits [5]. In Indonesia, using influencers increases awareness, educates target consumers, increases sales, increases followers, and search engine optimization [6]. Socia Buzz survey [7] stated that to be a credible and successful influencer, they must have a strong reputation and competency to impact their followers positively. Aside from the influencers' credibility, advertising must notice the proper fit between the influencer and the product they want to market to improve the ads' effectiveness and persuasiveness [8]. Several previous researchers have learned about "match-up," which refers to congruence [9–13]. The term congruity relates to the attributes connected with the promoted product/brand [14] and [12]. Gaied and Ben Rached [12] stated that appropriate and reliable influencers would result in more favorable views toward advertising, brands, and buying decisions.

The trend of local skincare products is increasing, particularly in Indonesia. Tania [15] stated that from female respondents who acknowledge using skincare products, 74.7% of respondents claimed it was essential to get healthy skin by maintaining it as early as possible. Inspired by several previous studies [16] and [17], our study learned about the persuasive effect of beauty influencers on local skincare products in Indonesia.

2 Literature Review

2.1 Persuasion

Persuasion is an attitude change when a person receives written or verbal information exposure from another party by adapting their beliefs, values, or attitudes [18–20]. According to Eagly and Chaiken [21] in Fatmawati et al. [20], there are three types of persuasive approaches in communication studies like the cognitive response approach (Elaboration Likelihood Models, [22]), the heuristic processing mode (Six Principles of Persuasion, [23]) and the attributional reasoning approach (Prospect Theory, [24]). Persuasive approaches that matched this study are Elaboration Likelihood Models (ELM). According to Petty and Cacioppo [22] and cited by Larson [25], ELM is one of the persuasive communication theories explaining how a message can influence or change a persons' attitude and behavior.

2.2 Meaning Transfer

Our study refers to the Meaning Transfer theory. The meaning transfer process is the transition of the symbolic meaning of a message from the company to the customer. There are three stages in this series of transfer meaning. In the first stage, influencers have to develop some values of the symbolic meaning, such as status, class, gender, age, lifestyle, film roles, and other achievements. The symbolic meaning will properly tie a product/brand during the endorsement process in the second stage. The meaning will move from product to consumer in the third stage when purchasing the endorsed product/brand. The effectiveness of the influencers typically depends on how they communicate the meanings within the endorsement process [26].

2.3 Influencers' Credibility

Influencer credibility does matter in advertising. It can influence beliefs, opinions, attitudes, and behavior [9, 27–31]. Messages conveyed by someone who has confidence will be more readily accepted and more easily lead to attitude change, meaning a positive relationship between influencers and attitudes towards the brand [28, 32, 33]. Therefore, an influencer directly influences purchasing decisions [34].

H1a: Influencers' Credibility positively and directly influences attitude toward the advertisement.

H1b: Influencers' Credibility positively and directly influences attitude toward the advertisement.

H1c: Influencers' Credibility positively and directly influences purchase decisions.

2.4 Influencer/Product Congruence

Previous research showed some potential effects are caused by influencer/product congruences, in the form of rand satisfaction, preference, emotional attachment, choice, and purchase intention [35]. Kamins [11] and Misra and Beatty [36] in Morimoto [37] stated that the better the influencer and the advertised product, the greater the advertisement and the results for the brand, which means it has a positive and significant relationship with purchasing decisions [38–40].

H2a: Influencer/product congruence positively and directly influences attitude toward the advertisement.

H2b: Influencer/product congruence positively and directly influences attitude toward the brand.

H2c: Influencer/product congruence positively and directly influences purchase decisions.

2.5 The Relationship of Attitude Toward the Advertisement and Attitude Toward the Brand

McCracken [26] states that consumers must express their meanings and judgments in response to their perceptions. When consumers like an advertisement, this will affect the brand, which means that there is a positive and direct relationship between the two variables [17, 27, 28, 41, 42].

H3a: Attitude toward the advertisement positively and directly influences attitude toward the brand.

2.6 Attitude Toward the Brand and Purchase Decision

Attitude toward the brand represents the response that occurs in the minds of consumers to the brand by evaluating it internally [31]. Chaudhuri [43] states that consumers will carry out a comprehensive brand evaluation. In conclusion, attitude towards a brand is an evaluation process regarding a thorough assessment of a brand, and customers will take action to make purchasing decisions. Previous research has found a positive relationship between attitudes and purchasing decisions [44].

H3b: Attitude toward the brand positively and directly influences purchase decisions.

2.7 Mediation Effect of Attitude Toward the Advertisement on Influencers' Credibility and Attitude Toward the Brand

The information provided by the credible influencer will get a favorable response from the customer [42]. The high value of source credibility will ease message acceptance and typically create a more significant attitude change [37]. The intended response represents the attitude toward the advertisement. In addition to increasing attitudes toward the ad, influencers' credibility also leads to improved attitudes toward the brand [27, 28].

H4a: Attitude toward the advertisement mediates the relationship between an influencers' credibility and attitude toward the brand.

2.8 Mediation Effect of Attitude Toward the Advertisement on Influencer/Product Congruence and Attitude Toward the Brand

The influencer/product congruence concept strengthens the advertising result [16]. The stronger congruity between Influencers and endorsed products, the higher the value of attitudes toward the advertisement and the brand [12]. Moreover, Choi and Rifon [17] explained that indirect effects could occur through attitude toward the correlated advertising.

H4b: Attitude toward the advertisement mediates the relationship between influencer/product congruence and attitude toward the brand.

2.9 The Mediating Effect of Attitude Toward the Brand on the Attitude Toward the Advertisement and Purchase Decision Relationship

Consumers' attitude is one of the critical factors influencing customers to purchase a product [45]. If the customer has a favorable attitude toward the advertisement, their clear behavioral intention could be affected [16, 46]. The previous explanation shows a direct relationship between attitude toward the ad and the brand [27, 42]. The results of previous researchers show that there is a significant positive relationship between the three variables [12, 16, 17, 35, 42].

H4c: Attitude toward the brand mediates the relationship between attitude toward the advertisement and purchase decision.

3 Research Model

The research model is shown in Fig. 1.

4 Research Method

We employed a quantitative approach with the causal survey design [47]. Our research setting is a local viral skincare brand: Scarlett Whitening. We investigate a viral female beauty influencer, Rachel Vennya (RV), by ensuring whether the respondents saw her content about the recommended product on social media before purchasing. The sample selection is using purposive sampling with the following criteria: (1) young adult female (18–35 years old), (2) has known Rachel Vennya as the influencers for Scarlet Whitening, (3) seeing RV content and review before purchasing, and (4)

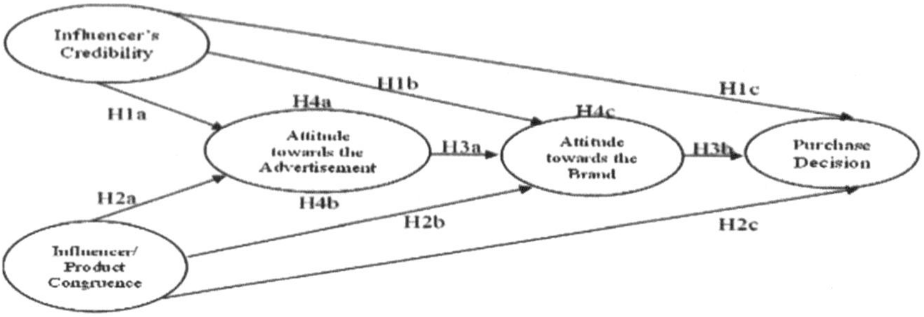

Fig. 1 Research model

has already purchased Scarlett Whitening products. The sample size is 5–10 times the total number of indicators, while for factor analysis, the sample must be greater than 100 [48]. Using 34 indicators multiplied by five, the minimum targeted sample is 170 respondents. We used the Structural Equation Model (SEM) for data analysis using AMOS 24 program.

5 Result and Discussion

One hundred eighty-nine people filled out the questionnaire in total. The potential respondents polled were all female. We gathered primary data by administering online questionnaires through Google Forms. According to the findings, 95.8% of respondents were between 18 and 23. Most respondents are university students (142 people or 85.53%). According to the results, most respondents have a monthly expenditure of Rp2,000,000 (128 respondents), and 59 out of them live with expenses ranging above Rp2,000,001 to Rp4,000,000.

Based on the validity test result, the value of standardized loading estimates above 0.5 and, ideally, 0.7 or higher. Hence, all indicators are valid. C.R. values between 0.6 and 0.7 are still acceptable and reliable [48]. Test results of the reliability test showed the following score of C.R.: Influencers' Credibility (0.964), Influencer/Product Congruence (0.940), Attitude toward the Advertisement (0.934), Attitude toward the Brand (0.930), and Purchase Decision (0.951).

Evaluation of multivariate outliers shows the value of Mahalanobis Distance. The criteria used are at the level of $p < 0.001$. In this study, the indicators used were 34, so the CHIINV value obtained was 65.24722. The result showed that three outliers exceed 65.24722, namely data 22, 144, 21, where each Mahalanobis d-squared value is 70,963, 65,868, and 70,661, then we delete those three data. The normality test is analyzed using the Z value of the skewness and kurtosis values. The critical value is in the range of ±2.58. If a value exceeds the value of the required ratio, then the data is said to be ordinarily undistributed. The normality tests results showed that our data were normal. Meanwhile, multivariate data is abnormal because it has a C.R. Value of 18,118, far from the range of ±2.58. According to Ghozali [49], one proper way to overcome multivariate non-normal data is bootstrapping. Bootstrapping results show a Bollen-Stine Bootstrap *p*-value of 0.053 or above 0.05 which means the data is standard.

The result of Goodness of fit is five indicators of the Influencers' Credibility variable from Modification Indices used to modify the model. The criteria of cut off value for CMIN/DF is 1.817 $\leq$ 2 (Fit), AGFI 0.736 $\geq$ 0.80 (Marginal fit), TLI 0.895 $\geq$ 0.90 (Fit), CFI 0.904 $\geq$ 0.90 (Marginal fit), RMSEA 0.066 $\leq$ 0.08 (Fit). The modified model is "fit" with existing data. Our model is fit based on three fit indices. Following Hair et al. [48], if a model has three or four fit indices can provide evidence that the model is fit. Therefore, researchers can proceed to the next stage, namely hypothesis testing.

In this study, the hypothesis is supported if the probability value is less than 0.05 and the value of c.r is more than 2000. The following are the results of the regression weight test.

Only H1a was supported based on the data analysis for our first three hypotheses. Hypothesis 1a, which stated that influencers' credibility significantly affects the attitude toward the advertisement, showed a probability value of 0.05 and the value of C.R. of 3.833 > 2.000. As a result, hypothesis 1a of this study was supported. It means the higher level of credibility, the better the consumer attitude towards the advertisement. For Hypothesis 1b, the results of the variable relationship test between influencers' credibility and attitude toward the brand showed a probability value of 0.324 ($p > 0.05$) and a C.R. value of 0.986 (C.R. > 2000). Therefore, the two variables possess positive and no significant effect. Hence, hypothesis 1b of this study was not supported. The result of this study contradicts the outcomes of research conducted by Refs. [12, 28]. In comparison, for Hypothesis 1c, according to the regression weight test, influencers' credibility positively affected purchase decisions. The result of probability was 0.438 > 0.05, and the value of C.R. was 0.776 < 2.000. This hypothesis, however, was insignificant due to the data result, which could not follow the requirement of probability and C.R. value. Then, hypothesis 1c of this study was not supported. So, no matter how high the influencer's credibility value is, it will not affect the purchase decision variable.

The following are the results of our second three hypotheses. For Hypothesis 2a, the result of regression weight showed that influencer/product congruence had a positive and significant effect on attitude towards the advertisement with a probability value of 0.013 < 0.05 and the value of C.R. of 2.491 > 2.000. Therefore, hypothesis 2a of this study was supported [17]. Customers' attitudes towards the advertisement are better when the customer feels there is a match between the product advertised and the influencer. For Hypothesis 2b, the number *p* is 0.173, which shows the number *p* above 0.05. Then, the value of C.R. is 1362, which means lower than the C.R. value. Therefore, H2b is not supported. Influencer/product congruence has a positive and insignificant effect on attitude toward the brand. No matter how high the value of influencer/product congruence is, it will only slightly change the value of attitude towards the brand. For Hypothesis 2c, the parameter estimation of regression weight coefficient values obtained a probability value of s 0.511 > 0.05 and the C.R. value of

−0.658 < 2000, which means the relationship between variables is the negative and not significant effect [17]. Insignificant results of this finding showed the average results of the respondents' answers, which stated that customers believe that the influencer matches with the products.

The next are the results of our third two hypotheses. Hypothesis 3a shows that attitude towards the advertisement positively and significantly affects attitude towards the brand. The obtained probability value was 0.003 < 0.05, and the value of C.R. was 2.977 > 2.000. Hypothesis 3a of this study was supported. The findings show that the better attitude toward the advertisement, the better customer attitude toward the brand. For Hypothesis 3b, the result showed a significant influence attitude towards the brand on purchase decisions with a probability value of 0.000 < 0.05 and the value of C.R. of 6.292 > 2.000. These results indicate that attitude towards the brand has a positive and significant effect on purchase decisions [17, 42]. Customers who have purchased the products stated that the company's advertising had succeeded in convincing them that the product has many benefits.

The mediation effect test compares the value of standardized direct effects with the indirect standardized impact, known as path analysis and Sobel test, to find out the significance value. A variable can have an indirect effect (mediation) if it has a sig <0.05. These results support hypothesis 4a [29]. Increasing the influencer's credibility will affect attitude towards the advertisement. The comparison result of the indirect effects of influencer/product congruence on attitude towards the brand through attitude towards the ad shows the value is 0.125. This value is smaller than the direct effect, with a value of 0.208 and a p-value of 0.056 (not significant). Therefore, these results do not support hypothesis 4b [16]. It cannot influence customers in responding to the product, although it has been helped by increasing the value of customer attitude toward the advertisement. Hypothesis 4c states that the advertisement's attitude did not directly relate to the purchase decision, based on the value of the direct effect is 0.000. It turns out that after a comparison, there is an indirect effect value of 0.278, and the p-value was 0.007 (significant). Therefore, the results support hypothesis 4c [17, 28]. These results indicate the better the value of attitude toward the advertisement, the bigger the positive effect on attitude toward the brand.

References

1. Kotler, P., Keller, K.L., Armstrong, G.: Marketing Management. Prentice-Hall, New Jersey (2012)
2. Kotler, P., Keller, K.L.: A Framework for Marketing Management, Sixth Edition Global Edition. Pearson, Boston (2016)
3. Kemp, S.: Digital 2019: Indonesia. DataReportal (2019). https://datareportal.com/reports/digital-2019-indonesia. Accessed April 13, 2019
4. Mediakix: Influencer marketing 2019 industry benchmarks (2019). https://mediakix.com/influencer-marketing-resources/influencer-marketing-industry-statistics-survey-benchmarks/. Accessed July 15, 2019

5. Dhanesh, G.S., Duthler, G.: Relationship management through social media influencers: effects of followers' awareness of paid endorsement. Pub. Relat. Rev. **45**(3), 101765 (2019). https://doi.org/10.1016/j.pubrev.2019.03.002
6. Yusra, Y.: Melakukan Kegiatan Pemasaran Memanfaatkan 'Influencer'. Dailysocial.id (2018). https://dailysocial.id/post/melakukan-kegiatan-pemasaran-memanfaatkan-influencer. Accessed May 23, 2019
7. SociaBuzz: The state of influencer marketing 2018 in Indonesia—Kupas Tuntas Tren Pemasaran Endorse. Slideshare (2018). https://www.slideshare.net/sociabuzz/the-state-of-influencer-marketing-2018-in-indonesia-kupas-tuntas-tren-pemasaran-endorse. Accessed May 23, 2019
8. Yang, W.: Star power: the evolution of celebrity endorsement research. Int. J. Contemp. Hosp. Manag. **30**(1), 389–415 (2018). https://doi.org/10.1108/IJCHM-09-2016-0543
9. Erdogan, B.Z.: Celebrity endorsement: a literature review. J. Mark. Manag. **15**(4), 291–314 (1999). https://doi.org/10.1362/026725799784870379
10. Lynch, J., Schuler, D.: The match-up effect of spokesperson and product congruency: a schema theory interpretation. Psychol. Mark. **11**(5), 417–445 (1994). https://doi.org/10.1002/mar.4220110502
11. Kamins, M.A.: An investigation into the 'match-up' hypothesis in celebrity advertising: when beauty may be only skin deep. J. Advert. **19**(1), 4–13 (1990). https://doi.org/10.1080/00913367.1990.10673175
12. Gaied, A.M., Ben Rached, K.S.: The congruence effect between celebrity and the endorsed product in advertising. J. Mark. Manag. **5**(1), 27–44 (2017). http://doi.org/10.15640/jmm.v5n1a4
13. Santos, A.L., Barros, F., Azevedo, A.: Matching-up celebrities' brands with products and social causes. J. Prod. Brand Manag. **28**(2), 242–255 (2019). https://doi.org/10.1108/JPBM-03-2017-1439
14. Kirmani, A., Shiv, B.: Effects of source congruity on brand attitudes and beliefs: the moderating role of issue-relevant elaboration. J. Consum. Psychol. **7**(1), 25–47 (1998). https://doi.org/10.1207/s15327663jcp0701_02
15. Tania, S.: Skin care product: the trend and usage experience. Jakpat (2018). https://blog.jakpat.net/skin-care-product-the-trend-and-usage-experience/
16. Lim, X.J., Mohd Radzol, A.R., (Jacky) Cheah, J.-H., Wong, M.W.: The impact of social media influencers on purchase intention and the mediation effect of customer attitude. Asian J. Bus. Res. **7**(2), 19–36 (2017). http://doi.org/10.14707/ajbr.170035
17. Choi, S.M., Rifon, N.J.: It is a match: the impact of congruence between celebrity image and consumer ideal self on endorsement effectiveness. Psychol. Mark. **29**(9), 639–650 (2012). https://doi.org/10.1002/mar.20550
18. Simons, H.W.: Persuasion: Understanding, Practice, and Analysis. Addison-Wesley, Boston (1976)
19. Wood, W.: Attitude change: persuasion and social influence. Annu. Rev. Psychol. **51**, 539–570 (2000). https://doi.org/10.1146/annurev.psych.51.1.539
20. Fatmawati, I., Dharmmesta, B.M., Purwanto, B.S., Nugroho, S.S.: Promoting young adults to perform energy-saving behavior through message framing: a lesson learned from Indonesia. Acad. Strateg. Manag. J. **17**(5), 1–20 (2018)
21. Eagly, A.H., Chaiken, S.: Cognitive theories of persuasion. Adv. Exp. Soc. Psychol. **17**, 267–359 (1984). https://doi.org/10.1016/S0065-2601(08)60122-7
22. Petty, R.E., Cacioppo, J.T.: The elaboration likelihood model of persuasion. Adv. Exp. Soc. Psychol. **19**, 123–205 (1986). https://doi.org/10.1016/S0065-2601(08)60214-2
23. Cialdini, R., Rhoads, K.: Human behavior and the marketplace. Mark. Res. **13**(3), 8–13 (2001)
24. Kahneman, D., Tversky, A.: Prospect theory: an analysis of decision under risk. Econometrica **47**(2), 263 (1979). https://doi.org/10.2307/1914185
25. Larson, C.U.: Persuasion-Reception and Responsibility, 12th ed. Wadsworth Publishing (2009)
26. McCracken, G.: Who is the celebrity endorser? Cultural foundations of the endorsement process. J. Consum. Res. **16**(3), 310 (1989). https://doi.org/10.1086/209217

27. Lafferty, B.A., Goldsmith, R.E., Newell, S.J.: The dual credibility model: the influence of corporate and endorser credibility on attitudes and purchase intentions. J. Mark. Theor. Pract. **10**(3), 1–11 (2002). https://doi.org/10.1080/10696679.2002.11501916
28. Muda, M., Musa, R., Naina, R., Borhan, H.: Celebrity entrepreneur endorsement and advertising effectiveness. Procedia Soc. Behav. Sci. **130**, 11–20 (2014). https://doi.org/10.1016/j.sbspro.2014.04.002
29. Sallam, M.A.A., Wahid, N.A.: Endorser credibility effects on Yemeni male consumer' s attitudes towards advertising, brand attitude, and purchase intention: the mediating role of attitude toward brand. Int. Bus. Res. **5**(4), 55–66 (2012). https://doi.org/10.5539/ibr.v5n4p55
30. Mackenzie, S.B., Lutz, R.J., Belch, G.E.: The role of attitude toward the ad as a mediator of advertising effectiveness: a test. J. Mark. Res. **23**(2), 130–143 (1986). https://doi.org/10.1177/002224378602300205
31. Mitchell, A.A., Olson, J.C.: Are product attribute beliefs the only mediator of advertising effects on brand attitude? J. Mark. Res. **18**(3), 318–332 (1981). https://doi.org/10.1177/002224378101800306
32. Lou, C., Yuan, S.: Influencer marketing: how message value and credibility affect consumer trust of branded content on social media. J. Interact. Advert. **19**(1), 58 (2019)
33. Shimp, T.: Advertising Promotion Additional Aspects of Integrated Marketing Communication. Erlangga, Jakarta (2000)
34. Kautsar, A.P., Widianto, S., Abdulah, R., Ph, D., Amalia, H.: Relationship of consumer involvement, credibility of the source of information and consumer satisfaction on purchase decision of non-prescription drugs. Procedia Soc. Behav. Sci. **65**, 449–454 (2012). https://doi.org/10.1016/j.sbspro.2012.11.147
35. Pradhan, D., Duraipandian, I., Sethi, D.: Celebrity endorsement: how celebrity–brand–user personality congruence affects brand attitude and purchase intention. J. Mark. Commun. **22**(3), 456–473 (2014). https://doi.org/10.1080/13527266.2014.914561
36. Misra, S., Beatty, S.E.: Celebrity spokesperson congruence: an assessment and brand of recall and affect. J. Bus. Res. **21**(2), 159–173 (1990). https://doi.org/10.1016/0148-2963(90)90050-N
37. Morimoto, M.: Congruence and celebrity endorser credibility in Japanese OTC drug advertising. Int. J. Pharm. Healthc. Mark. **12**(3), 234–250 (2018). https://doi.org/10.1108/IJPHM-09-2017-0049
38. Premeaux, S.R.: The attitudes of middle-class male and female consumers regarding the effectiveness of celebrity endorser. J. Promot. Manag. **11**(4), 33–48 (2005). https://doi.org/10.1300/J057v11n04
39. Afifah, O.N.: Pengaruh Celebrity Endorser (Raisa Adriana) terhdap Keputusan Pembelian produk Kosmetik Pond's White beauty (Studi Kasus pada Mahasiswi Universitas Islam Negeri Maulana Malik Ibrahim Malang). Universitas Islam Negeri Maulana Malik Ibrahim Malang (2016)
40. Yunior, S.T.: Analysis of the impact of attractiveness, credibility, and commercial star congruence on purchase decision of Yamaha motorcycle product (2018)
41. Hoyer, W.D., Maclnnis, D.J., Pieters, R.: Consumer Behavior, 6th ed. South-Western Cengage Learning, Mason (2012)
42. Seiler, R., Kucza, G.: Source credibility model, source attractiveness model, and match-up-hypothesis: an integrated model. J. Int. Sci. Publ. Econ. Bus. **11**, 1–15 (2017). http://doi.org/10.21256/zhaw-4720
43. Chaudhuri, A.: Brand loyalty mediates brand equity outcomes? J. Mark. Theor. Pract. **7**(2), 136–146 (1999)
44. Kidane, T.T., Sharma, R.R.K.: Factors affecting consumers' purchasing decision through ECommerce. In: Proceedings of the 2016 International Conference on Industrial Engineering and Operations Management Kuala Lumpur, Malaysia, vol. 8, no. 10, pp. 159–165 (2016)
45. Vahdati, H., Mousavi, N., Tajik, Z.M.: The study of consumer perception on corporate social responsibility towards consumers attitude and purchase behavior. Asian Econ. Financ. Rev. **5**(5), 831–845 (2015)

46. Chen, M.F.: Consumer attitudes and purchase intentions in relation to organic foods in Taiwan: moderating effects of food-related personality traits. Food Qual. Prefer. **18**(7), 1008–1021 (2007)
47. Sekaran, U., Bougie, R.: Research Methods for Business: A Skill-Building Approach. Wiley, Hoboken (2016)
48. Hair, J.F., Black, W.C., Babin, B.J., Anderson, R.E.: Multivariate Data Analysis, 7th ed. Prentice-Hall (2014)
49. Ghozali, I.: Model Persamaan Struktural. Konsep dan Aplikasi Dengan Program AMOS 22.0. In: Model Persamaan Struktural. Konsep dan Aplikasi Dengan Program AMOS 22. Update Bayesian SEM (2014)

Electronic Word of Mouth (E_WOM) Adoption via Social Media and Its Impact on Online Shoppers' Purchasing Intention During Corona Pandemic. A Case of Jordan

Zakaria Azzam and Sondos Al-Shaer

Abstract In digitalization era the role of e-word of mouth is very useful to be considered especially during the conditions of Corona pandemic outbreak. This study aims to identify the impact of Electronic Word-of-Mouth adoption via social media on online shoppers' purchasing intention during Corona Pandemic in Jordan, to analyze the impact of e-word of mouth elements on the purchasing intention through the independent variables which are represented in Quality, Quantity, Credibility, Source Expertise, and Source Trust. The researchers designed a questionnaire based on previous studies which have been given out to 400 respondents consisting of the sample of the study and only 384 respondents were valid for analytical descriptive study. SPSS used to analyze the data. The main results of this study are that e-word of mouth has a significant impact on consumers' purchasing intention and that the quantity of e-word of mouth has the most contribution to the impact on consumers' purchasing intention. However, the credibility of e-word of mouth has the least contribution to the impact of consumers' purchasing intention. Corona Pandemic level has statistically significant role at ($a \leq 0.05$) on the relationship between e-word of mouth and purchasing Intention Based on the findings of the study, it recommended that companies should develop a better e-word of mouth marketing strategies which aims to create a professional review system that encourages users to write detailed information. They should also create a special department to manage the way of presenting products or services and to figure out what attracts consumers' attention. And lastly, Review systems could provide guidance to assist shoppers by providing ratings of the products and describing their experiences in online shopping.

Keywords Electronic word of mouth · Purchasing intention · Social media · Corona pandemic · Jordan

---

Z. Azzam (✉) · S. Al-Shaer
Marketing Department, Zarqa University, Zarqa, Jordan
e-mail: zazzam@zu.edu.jo

S. Al-Shaer
e-mail: 20209036@zu.edu.jo

A. Hamdan et al. (eds.), *The Implementation of Smart Technologies for Business Success and Sustainability*, Studies in Systems, Decision and Control 216,
https://doi.org/10.1007/978-3-031-10212-7_40

1 Introduction

It is undeniable that Corona Pandemic in addition to the rapid changes in technological developments have left their mark on every aspect of our lives and affecting everyone on a global level. With that change, people began showing interest and heading towards adopting these technologies due to the various benefits that they offer.

The noticeable enhancements in technology nowadays have opened up various internet-based communication channels which are represented in social media platforms such as Facebook, Instagram, Twitter, and Snap chat. These platforms gave people the opportunity to share their opinions through the availability of chatting services and the ability to comment on posts shared on these platforms specifically during Corona Pandemic.

Due to the existence of opinions of consumers and potential customers on the different social media platforms mentioned above arose the concept of Electronic Word-of-Mouth, which can be defined as all informal communications directed at consumers through Internet-based technology related to the usage or characteristics of particular goods and services, or their sellers [1] and [2]. The usage of social media platforms as a way of expressing opinions has been adopted by people in Jordan, especially with 61.5% of the population owning social media accounts, which results in 6.3 million social media users in Jordan based on data analysis done in January 2020 until January 2021 [2].

2 Research Problem and Questions

Based on literature review portrayed in relation to the role of e-word of mouth on consumer buying decisions and purchase intention, the study emphasizes the existence and possible importance of e-word of mouth in the forms of reviews and recommendations on consumers in the process of researching information on social media while shopping online during Corona Pandemic [3]. According to the model illustrated by Farzin and Fattahi [4] consumers might draw their attention to the existence of information spread electronically which is demonstrated by e-word of mouth while recognizing the existence of information usefulness in the minds of consumers. The main problem in relation to e-word of mouth lies in the distrust amongst consumers which could potentially be negative e-word of mouth or positive e-word of mouth whether it is towards products, services, or brands. From this standpoint marketers should prioritize and pay attention to this topic by understanding whether the impact is negative or positive and how to handle and react to this impact. In addition, the real problem in this current study is the existence of a gap between consumers' opinions and thoughts towards e-word of mouth which makes it difficult for marketers to predict consumers' thoughts especially in relation to the elements of the independent variable which could possibly have an impact on the consumers' intentions. This study further explores Electronic Word of Mouth and its impact by raising the main question: Is there an impact of e-word of mouth adoption via Social Media on the Purchasing Intention of Online Shoppers during Corona Pandemic? [5–7] and [8].

3 Research Hypotheses

H0: There is no statistical significant impact at the significance level ($\alpha \leq 0.05$) of e-word of mouth adoption (Quality, Quantity, Credibility, Sources Expertise and Source Trustworthy) via Social Media on online Purchasing Intention in Jordan.

H.02: There is no statistically significant role at ($a \leq 0.05$) of Corona Pandemic as a moderating variable on the relationship between e-word of mouth adoption via Social Media on online Purchasing Intention in Jordan.

3.1 Research Model

Based on the research problem and objectives, the researcher designed a research model represented in Fig. 1 which shows the independent variable e-word of mouth with its elements which are (e-word of mouth Quality, e-word of mouth Quantity,

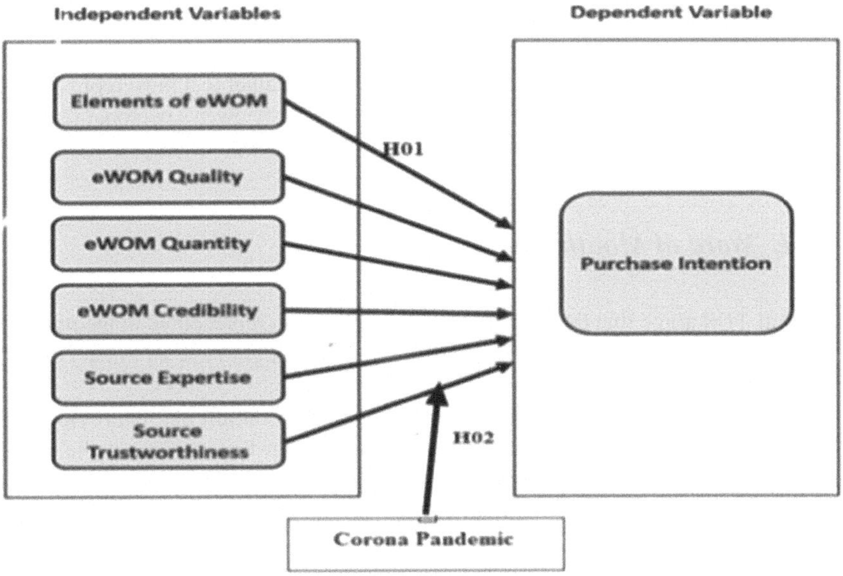

Fig. 1 Research model

e-word of mouth Credibility, Source Expertise, Source trustworthiness) based on the previous studies and the dependent variable which is represented by purchase intention [4, 9–15] and [16].

4 Literature Review

E-word of mouth Quality is defined as "the characteristics of online reviews which influence the perceived usefulness of the reviews". These specific characteristics were represented as five elements, which are: [12] and [17]

- Whether the review has been based on the attributes or the experience itself.
- Whether the review is positive or negative.
- The valence of the review.
- The timeliness factor of the review.
- How elaborate the review is.

The quality of e-word of mouth can also be defined as the quality of the review's contents in relation to the perspective of information characteristics, which have identified as: [12] and [17]

- **Relevancy**: The review must be relevant and related to the product or company.
- **Understandability**: It's essential for the review to be easily understood by different kinds of customers, this can be done by using simple terms targeted to all customers.
- **Sufficiency**: The review must have enough amount of information in order to be considered sufficient.
- **Objectivity**: In order for the review to be considered valuable it must represent an objective point of view which isn't influenced by the reviewer's personal feelings and taste.

4.1 E_Word of Mouth Quantity

Zhao et al. [18] argue that the volume of e-word of mouth represented in the number of ratings and comments on websites and social media might have an impact on the acceptance of products by the customers.

Reference [3] suggests that the quantity of e-word of mouth that is provided by a social network user is most likely influenced by another factor which is the usage intensity. It is also supported that people who heavily rely on social interactions and have a great need for using social networks tend to engage more in brand-related e-word of mouth.

According to [19] the consumer buying intentions are not only influenced by the quality of online reviews but they are specifically influenced by the quantity of them as well, especially when it's provided by the consumers who have the product [20].

States that most of the quantity of negative e-word of mouth online comes from customers who are unsatisfied and they have stopped purchasing from the store or they continue to purchase from it simply because it would be difficult for them to switch to a different brand, which then creates heavy quantity of e-word of mouth for the store/brand [17].

4.2 Credibility

According to literature review credibility is related to e-word of mouth refers to not only the online message itself but also the source of the message, which is called "the communicator" of the message. That being said, if an individual considers the information of the message credible, they will more likely be more confident in adopting the e-word of mouth message and later on use them in their purchasing decisions. Due to the fact that e-word of mouth communication is based on anonymous interactions, the readers of the messages have to carefully evaluate the information source's credibility before proceeding with a purchase decision. Pham [21] indicates that there is a positive association between the purchase intention of online customers and the credibility as well as the usefulness they perceive when reading online reviews [21–23], and [24].

The existence of connectivity allows the customers to express their opinions that other customers would listen to, which ultimately changes their mindset. It is easier for customers to take recommendations of strangers and consider them more credible than recommendations taken from sponsored celebrities [24].

4.3 Source Expertise

Reviewers should be encouraged to write more reviews in order to upgrade their reviewer status, and companies can achieve obtaining more reviews by giving the reviewers an incentive which is the Badges System. Some reviewing sites give reviewers a badge next to their names to indicate their level of experience, and their number of reviews. Giving reviewers an incentive such as the badges system would increase the number of reviews as well as, the engagements between customers [22] and [17].

4.4 Source Trustworthiness

According to literature review source trustworthiness can be referred to as the extent to which a certain individual's statement to be taken as genuine [25]. Ismagilova et al. [1] states that in order to understand the messages and sources of WOM in the most effective way, you must first understand the way that individuals judge the sources as well as their trustworthiness online.

Based on the Integrative model of word-of-mouth as a recruitment source represented by the source of information depends on various elements which are the personality, The expertise, Motives and Tie strength [9, 17] and [25].

4.5 E_Word of Mouth and Online Shopping

People during Corona Pandemic depending more and more on online shopping which requires people to be more aware of every aspect of online shopping, online shoppers must be aware of payment services, scams and frauds, as well as, how to stay safe when using online shopping services When shopping online a person must take advantage of the benefits of online shopping which include: Price comparison sites, Review sites, Coupons and offers, and various alternatives. Social networking websites are widely used as a mean of spreading the experience of products among consumers, customers are very interested in reading both negative and positive reviews of other customers' experience. Based on previous studies it was found that trust is positively associated with purchase intention. Social media users tend to seek information about product or services before they proceed with making a purchase decision [15, 26] and [18].

Based on previous studies related to, e-word of mouth and Online Shopping the following are the main observations: found that the positive, e-word of mouth which is generated by users on the Social Networking Site Facebook have a significant influence on Brand Attitude and it also has a significant impact on the Purchase Intention of consumers [23, 27] and [28]:

- WOM whether it's in an online or offline form, it has an economic impact that is represented in contrasting occurrences. Social media is certainly a big deal because it gives people the empowerment to share ideas with each other.

  That people nowadays believe that social media has replaced word-of-mouth and that it represents the essence of WOM, however, they argued that social media is merely a component of the entirety of WOM, whether it is online or offline.
- The essence of Social media which includes online platforms as well as tools that people use in order to share their experiences and opinions. Social media can be represented in various forms not just in texting form, but it also includes images, videos, and audio.
- People have adopted social media as a futuristic mean of engagement in its different sources including Facebook, Twitter, YouTube and other communication channels.
- Millions of people use social media sites on daily basis, the existence of these technologies caused information to be shared rapidly and the process of engaging with others easier.

5 Methodology

This study is based on the descriptive analytical method to achieve the objectives of this study to identify the Electronic word of mouth via Social Media and its impact on online shoppers' purchasing intention a field study on online shoppers in the city Amman. The data were collected by reviewing the theoretical literature and previous studies related to the subject of the study. The quantitative survey questionnaire was used to obtain data from the sample of the study.

This method is based on a scientific, accurate and integrated description of the existing situation or problem using a descriptive analysis. It is also based on the facts associated with it and unlimited to describing the phenomenon. It includes analyzing data, measuring, and interpreting it, arriving at an accurate analysis and results using statistical and inferential analysis and evaluating solutions and proposals to address them.

5.1 Research Population

The study community consists of online shoppers in Amman city, the reason behind choosing the selected sample is that they deal with the main concepts related to the study variables. The research sample consisted of (384) based on [29] in order to reach the maximum population size being 1,000,000.

5.2 Research Sample

It consisted of (384) participants, and they were randomly selected from the population of the study, the questionnaire was distributed electronically to ten regions in Amman, Tables 1 and 2 illustrate the sample distribution according to the personal variables.

The researcher used five Likert scale to include five degrees, were given Strongly agree (5), Agree (4), Neutral (3), Disagree (2), Strongly Disagree (1), and by setting signal (√) in front of the answer, which reflects the degree of consent. Cronbach's Alpha factor was used for internal consistency in order to verify the reliability of

Table 1 Frequency and percentage for the participants according to "Did Corona pandemic motivate you to search for information before making an online purchase?" variable (n = 384)

Category	Frequency	Percentage
Male	195	50.9
Female	189	49.1
Total	**384**	**100.0**

Table 2 Percentage for "making an online purchase" highest reached (97.90) for (yes), but the lowest percentage was (3.0%) for (no)

Category	Frequency	Percentage
Yes	372	97.0
No	12	3.0
Total	**384**	**100.0**

Table 3 Results of the Cronbach-alpha stability coefficients for the study sections and questionnaire (n = 384)

Section	Cronbach's alpha	Number of items
eWOM quality	6	0.80
eWOM quantity	4	0.82
eWOM credibility	5	0.84
Source expertise	5	0.88
Source trustworthiness	6	0.83
Purchase intention	6	0.86
Total	**32**	**0.94**

the questionnaire as a tool to collect data necessary for the current study. The reliability equation of the questionnaire (Cronbach-alpha) was applied to all sections of questionnaire and questionnaire as a whole, Table 3 illustrates that.

6 Results and Discussion

Multiple linear regression is applied to test the first hypothesis which is shown in Table 4.

Table 4 shows that there is statistical significant positive impact of, e-word of mouth on purchase intention, at the significance level ($\alpha \leq 0.05$), where (t) value was (11.005), by Sig. (0.000), by impact ratio (65.3%). The researcher rejects the null hypothesis and accepts the alternative one which states that: There is statistically significant effect ($\alpha \leq 0.05$) of, e-word of mouth on Purchase Intention. Table 4 shows the study dependent variable (purchasing intention) and the set of the independent variables factor (e-word of mouth Quality, e-word of mouth Quantity, e-word of mouth Credibility, and Source Trustworthiness and Source Expertise), are significant because F significant is (0.000) which is less than (0.05), and the calculated F value is (121.111), which is more than F value table (2.22). Therefore, we rejected the null hypothesis and accepted the alternative hypothesis, which states that: There is statistically significant relationship at ($a \leq 0.05$) between, e-word of mouth (e-word of mouth Quality, e-word of mouth Quantity, e-word of mouth Credibility, and Source Trustworthiness and Source Expertise) and consumer purchasing intention.

Table 4 The results of (multiple linear regression) analysis to detect the impact of eWOM on purchase intention table: results of multiple regressions of the main hypothesis

Dependent variable	R	R2	F	DF	Sig.	Independent variable	B	T	Sig.
Purchase intention	0.653	0.426	121.111	4	0.000	eWOM quality	0.472	6.837	0.00
				435		eWOM quantity	0.455	6.532	0.00
						eWOM credibility	0.588	9.293	0.00
				439		Source trustworthiness	0.506	7.464	0.00
						Source expertise	0.587	9.268	0.000

The relationship between the dependent and independent variables is strong and positive. It is more than (0.5) (Cohen 1988), R = 0.653. Also, the $R^2 = 0.426$, which means that the independent variables contribute to the dependent variable in a percentage of 42.6%. Since the values of the calculated t values of the variables (e-word of mouth Quality 6.837, e-word of mouth Quantity 6.532, and, e-word of mouth Credibility 9.293, Source Trustworthiness is 7.464 and Source Expertise is 9.268), are more than the t value table (1.986). Also, the t significant is less than (0.05). It means that there is a statistically significant relationship for all EWOM variables on the dependent variable (purchasing Intention) H02.1: There is no statistically significant role at ($a \leq 0.05$) of Corona Pandemic as a moderating variable on the relationship between EWOM and purchasing Intention. One-Way ANOVA analysis was used to test first-second main hypothesis. Table 5 shows the results.

Table 5 shows that, calculated "F" value for (e-word of mouth 8.79) are more than "F" tabled value (3.00), and the significant value of F are less than (0.05). While calculated "F" value for Consumer purchasing intention (2.695) is less than "F" tabled value (3.00). Also the significant value of F is more than (0.05). Based on the result showed in Table 5, Corona Pandemic level has statistically significant role at ($a \leq 0.05$) on the relationship between, e-word of mouth and purchasing Intention. The study used (LSD) (least significant difference) test, it used in the context of the

Table 5 One way ANOVA results

Variables		Sum of square	DF	F	Sig.
eWOM	Between group	4.255	2	8.79	0.000
	Within group	95.567	382		
	Total	99.822	**384**		
Purchasing intention	Between group	2.116	2	2.695	0.001
	Within group	171.606	382		
	Total	173.722	**384**		

analysis of variance, when the F-ratio suggests rejection of the null hypothesis H0, that is, when the difference between the population means is significant.

7 Recommendations

1. The study proved that there is a significant Impact of, e-word of mouth via Social Media on consumers' Purchase Intention, therefore companies that sell products online should monitor comments and reviews spread by the consumers on Social Media Platforms.
2. Companies should develop a better, e-word of mouth marketing strategies aimed to create a professional review system that encourages users and expertise to write detailed information about products or services and to express both positive and negative feedback to improve and maintain credibility.
3. Review systems could provide guidance to assist shoppers to produce the quality of products by rating the quality, adding pictures of the products, and describing their experience in online shopping.
4. Companies should create a special department to manage the way of presenting products or services and to figure out what attracts consumers' attention.

References

1. Ismagilova, E., Dwivedi, Y.K., Slade, E., Williams, M.D.: Electronic word-of-mouth (eWOM). In: Electronic Word of Mouth (eWOM) in the Marketing Context. Springer Briefs in Business. Springer, Cham (2017)
2. www.dos.jo2021
3. What is electronic word-of-mouth (ewom). IGI Global (2021). Retrieved from https://www.igi-global.com/dictionary/electronic-word-of-mouth-ewom/9572
4. Farzin, M., Fattahi, M.: eWOM through social networking sites and impact on purchase intention and brand image in Iran. J. Adv. Manag. Res. **15**(2) (2018)
5. Fu, X., Zhang, B., Xie, Q., Xiao, L., Che, Y.: Impact of quantity and timeliness of EWOM information on consumer's online purchase intention under C2C environment. Asian J. Bus. Res. **1**(2), 37–52 (2011)
6. An, J., Do, D.K.X., Ngo, L.V., Quan, T.H.M.: Turning brand credibility into positive word-of-mouth: integrating the signaling and social identity perspectives. J. Brand Manag. **26**(2), 157–175 (2019)
7. Liu, W., Ji, R.: Examining the role of online reviews in Chinese online group buying context: the moderating effect of promotional marketing. Soc. Sci. **7**, 141 (2018)
8. Yu, M., Liu, F., Lee, J.: Consumers' responses to negative publicity: the influence of culture on information search and negative word-of-mouth. J. Brand Manag. **26** (2019)
9. Abd-Elaziz, M.E., Aziz, W., Khalifa, G.S., Abdel-Aleem, M.: Determinants of electronic word of mouth (EWOM) influence on hotel customers' purchasing decision. Int. J. Heritage Tourism Hospitality **9** (2015)
10. Ahmad, N., Vveinhardt, J., Ahmed, R.: Impact of word of mouth on consumer buying decision. Eur. J. Bus. Manag. **6**(31) (2014)

11. Bhat, N.: The influence of electronic word of mouth (eWOM) on consumers purchase intention: a review and analysis of the existing literature. IOSR J. Eng. **10**(6) (2020)
12. Chowdhury, N.: Influence of electronic word of mouth on consumer purchase intention. EconStor Preprints (2016)
13. Gohar, F., Mohammadian, M., Seyyedi, P.: The impact of eWOM on purchase intention: the role of subjective norms, brand attitude and pester power. J. Crit. Rev. **7**(17) (2020)
14. Nee, I.: Managing Negative Word-of-Mouth on Social Media Platforms: The Effect of Hotel Management Responses on Observers' Purchase Intention. Innovatives Markenmanagement, 1st ed. Springer Gabler, Germany (2016)
15. Sulthana, A.N., Vasantha, S.: Influence of electronic word of mouth eWOM on purchase intention. Int. J. Sci. Technol. Res. **8**(10), 1–5 (2019)
16. Zulkiffli, W., et al.: The Effectiveness of Electronic Word-Of-Mouth (Ewom) on Consumer Purchase Intention Among Generation-Y, vol. 2, no. 6 (2017)
17. Huete-Alcocer, N.: A literature review of word of mouth and electronic word of mouth: implications for consumer behavior. Front. Psychol. **8** (2017)
18. Zhao, Y., Wang, L., Tang, H., Zhang, Y.: Electronic word-of-mouth and consumer purchase intentions in social E-commerce. Electron. Commer. Res. Appl. **41** (2020)
19. Hennig-Thurau, T., Gwinner, K.P., Walsh, G., Gremler, D.: Electronic word-of-mouth via consumer-opinion platforms: what motivates consumers to articulate themselves on the internet? J. Interact. Mark. **18**(1), 38–52 (2004)
20. Anastasiei, B., Dospinescu, N.: Electronic word-of-mouth for online retailers: predictors of volume and valence. Sustainability **11**(3) (2019)
21. Pham, T.D.: Determinants of electronic word of mouth perceived credibility a study of cosmetic purchasing behaviour. Unpublished Master Thesis, University of Twente, Netherlands (2016)
22. Altaher, A.: The impact of online shopping services quality on customers' attitudes towards online shopping. Unpublished Master Thesis, Middle East University, Jordan (2017)
23. Kudeshia, C., Kumar, A.: Social eWOM: does it affect the brand attitude and purchase intention of brands? Manag. Res. Rev. **40**, 310–330 (2017)
24. Reichelt et al.: How credibility affects eWOM reading: the influences of expertise, trustworthiness, and similarity on utilitarian and social functions. J. Mark. Commun. **20**(1–2), 65–81 (2014)
25. Rashedi, H., Seyed Siahi, S.F.: Role of source trustworthiness, experience sources and source expertise in the formation of WOM and purchase intention (Case study of Tehran mobile market customers). SSRN Electron. J. (2020)
26. Saleem, A., Ellahi, A.: Influence of electronic word of mouth on purchase intention of fashion products on social networking websites. Pak. J. Commer. Soc. Sci. 597–622 (2017)
27. Zaraket, S.: Consumer behaviour and social network sites: the impact of negative word of mouth. Routledge, Taylor & Francis Group (2021)
28. Word of mouth marketing. Oxford Reference (2021). Retrieved from https://www.oxfordreference.com/view/10.1093/acref/9780191844386.001.0001/acref-9780191844386-e-5243?rskey=Uqogch&result=2
29. Sekaran, U., Bougie, R.: Research Methods for Business: A Skill-Building Approach, 6th ed. Wiley, Chichester (2013)

The Degree of Dependence of Jordanian University Youth on Social Media During the Covid (19) Corona Pandemic: A Field Study

Kamel Kurshid Murad, Maha Al-Sarawi, Ahmed Oreqat, and Mazen Al-Fedawi

Abstract The study aimed to reveal the dependence of Jordanian University Students on social platform networks as a source of information on the Corona Pandemic (Covid-19). The survey method was used, as the study population consists of Jordanian University Students who are regular in the academic year 2020/2021. A representative and available random sample was taken from those who could be reached electronically by distributing an electronic survey questionnaire, and the final number of the sample was (421). Social networks can play a positive role in disseminating and circulating information related to the Corona Pandemic, as it showed that the respondents supported increasing their knowledge about the Pandemic through social networks. Motivation based on want of knowledge tops the results with (high) level. As for the emotional effects, they were generally rated (moderate). The results showed that the behavioral influences were also (moderate).As for the follow-up topics, the study revealed that the search for the production of antivirus-vaccines was the forerunner of all the follow-up issues, as was the methods of prevention and side effects of the virus, while the level of follow-up of government instructions and the spreading status of the pandemic in the world decreased to lower levels. With regard to the demographic variables relating to the other variables, it appeared that there were no significant statistical differences, while these differences only appeared in relation to the master's degree students.

K. K. Murad (✉) · A. Oreqat · M. Al-Fedawi
Middle East University, Amman, Jordan
e-mail: kmurad@meu.edu.jo

A. Oreqat
e-mail: aoreqat@meu.edu.jo

M. Al-Fedawi
e-mail: malfedawi@meu.edu.jo

M. Al-Sarawi
Jordan Television, Amman, Jordan
e-mail: maha.alsarawi@gmail.com

A. Hamdan et al. (eds.), *The Implementation of Smart Technologies for Business Success and Sustainability*, Studies in Systems, Decision and Control 216,
https://doi.org/10.1007/978-3-031-10212-7_41

Keyword Dependence theory · Social platform networks · Source of information · Corona pandemic (Covid 19)

1 Introduction

Social networks actually play an important role in creating a relationship between the public and what is happening in the world, especially in areas where individuals do not have direct knowledge or experience. In this research, the Degree of Dependence of Jordanian University Youth on these networks will be addressed during the Corona-Virus (Covid-19) pandemic. The role of this association is to build trends and public attitudes to form health awareness about the outbreak of the Corona Virus, which creates a global health crisis. This pandemic has a negative impact on the whole world, but the safety measures are already applied to address the spread of the pandemic which requires new human actions. And this neglected use normally affects the prevailing patterns and habits such as physical distancing, adherence to quarantine, and wearing masks and gloves.

2 Research Problem

The study problem is identifying the degree of dependence of Jordanian university youth on social networks during the pandemic. Despite the efforts being made globally to combat this corona-virus and limit its spread, there is a huge amount of information, behaviors and feelings spread on social networks. These behaviors and different information come from multiple and contradictory sources. This has led to the spread of a kind of psychological disorder among individuals due to a process of mixing news together. The research problem is to identify the degree of reliance of university youth on social networks during the Corona pandemic.

3 Research Questions

This study aims at answering the main question: What is the degree of Jordanian university youth's dependence on social networks during the Corona pandemic? This question leads to other questions such as:

1. What are the cognitive motives resulting from the reliance of Jordanian university youth on social networks during the Corona pandemic?
2. What are the emotional motives resulting from the reliance of Jordanian university youth on social networks during the Corona pandemic?
3. What are the behavioral motives resulting from the reliance of Jordanian university youth on social networks during the Corona pandemic?

4. What are the topics that Jordanian university youth follow on social networks regarding the pandemic?

4 Theoretical Framework of the Study

The basic idea of the dependence theory is summarized as follows: The ability of communication means to achieve greater cognitive, emotional, and behavioral motivation that will be increased when these means perform the functions of transmitting information in a distinct and intense manners. This theory assumes a number of assumptions that come at the forefront [1]

1. The degree of stability and balance of the social system vary as a result of the continuous changes and according to this difference the need for information and news increases or decreases. In case of social instability, the need for information increases, and also increases a public's exposure to the media in society. According to this hypothesis, the study will be applied to the crisis of the Jerusalem Intifada.
2. The media system is important to society, and the degree of public dependence also increases if it fulfills its needs. For this reason, mobile phone journalism was required to be studied during this corona pandemic.
3. The public issue varies in the degree of its dependence on the media as a result of some differences in the field of individual goals and needs. The situation includes young people who are considered their sources for obtaining information via quick telegrams and via adapting modern (mobile) technology, which is not available to all people.

The motives that make individuals depend on the media can be summarized as follows:

Cognitive aspects could be represented in arranging the public's priorities by increasing or decreasing the importance of an event according to the degree of its focus or neglect. These cognitive aspects could overcome the problem of ambiguity and form general trends for the public [2].

There are certain behavioral aspects of the media that are represented by the revitalization. This is the fruit of linking the cognitive and emotional motives, and the behavioral activity may be positive or negative [3].

5 Previous Studies

A number of previous studies were conducted about the topic of the degree of university youth's dependence on the media to obtain information about the Corona-Covid-19 pandemic. These studies aimed at raising health awareness about this dangerous disease. Including the Karasneh's et al. [4] study entitled "The Media's Motivation in Forming Knowledge, Perception and Awareness of the Risks of the Corona Covid 19 Epidemic among Pharmacists". The study aimed at investigating the role of the media in shaping the awareness of pharmacists in Jordan and their levels of awareness of the dangers of the emerging corona virus (Covid-19). The study adopted the survey method. The sample was 486 pharmacists. The study showed that all the participants in the study had good knowledge about the virus and its symptoms. The most prominent sources of their information were through social networks, radio and television.

In the same context, Irsheidat's [5] study was about "The Role of Jordanian Satellite Channels in Awareness of the Jordanian Society on the Corona Pandemic Survey Study". His study aimed at defining educational media, and activities in Jordanian satellite channels that aim at creating a state of awareness among individuals about Corona pandemic. The study concluded that the largest proportion of the participants is 89%. These participants relied on Jordanian satellite channels to follow the news of Corona pandemic. The study also confirmed that providing information about the ways the virus spreads among individuals aimed at spreading the awareness in the field of health.

Nooh's et al. [6] study was about the public's awareness of Corona virus (Covid-19) in Saudi Al-Jawf region. The researcher followed the survey method, and it appeared that there was an average level of knowledge about the virus among the respondents category. It was proved that there are differences in the level of knowledge of the respondents according to the variables of age and education level. Mejia et al. [7] conducted a study about "The media and its role in confronting the Corona virus". It verified the awareness of fear and the size of the problem. It aimed at revealing the role of the media in combating the Corona pandemic. The study used the field survey method on samples of epidemiological, health and public experts in the state of Peru. The study concluded that health experts exaggerate information about the Corona virus, and the media also exaggerates the consequences of the emerging virus. On the other hand, Ali's [8] study dealt with "the role of the Facebook network in promoting health awareness among the public". It is a survey study from the worker's points of view in Jordanian Ministry of Health. The study aimed at investigating the worker's opinions in Jordanian Ministry of Health regarding the health information published by the Facebook network. It was about the most attractive topics. These topics are to monitor the gratifications achieved as a result. The study adopted the descriptive approach using the questionnaire as a tool for the study. The sample of the study was nearly 273 participants. The most important results of the study indicated the network information that are characterized by diversity. The most attractive health topics on Facebook are: weight loss, diet, fitness, the benefits of food, aesthetic medicine, healthy culture, modern medical discoveries and Arab medicine. The topics for promoting health awareness are concerned with raising health awareness.

Mohamed [9] conducted a study about "The degree of Egyptian university youth's dependence on social media as a source of news". It is a social media. It dealt with a source of news and nature of youth's use of social networking sites. This study was carried out through a field study of a deliberate sample of male and female youth of public and private universities. The study sample consisted of 400 subjects at Cairo University and Sinai University. There was another sample from specialized news pages. They were pages belonging to press organizations that have a hard copy, as well as pages belonging to press organizations that do not have paper copies, and news pages were affiliated with television channels. Some pages belonged to foreign institutions, as they had pages issued in Arabic. Their source was from social networking sites such as Facebook and Twitter. The study concluded included that the political content was the most prevalent type among the various news pages with a rate of 44.5%. The seventh day's page was the most preferred among young people and the most widespread, as indicated by the interaction elements like comments and participation. The study revealed that most news pages neglected to clarify their source in the title of the news or the accompanying picture. It depends on the reader who should follow the link to its official website in order to read the full details of the news. The study also referred to the results of the study topic, as 28.7% of university youth clicked on the electronic link for more details.

6 Research Findings

After analyzing the participant's responses, the following results were as follows:

First: As shown in Table 1: The cognitive motives come from resulting the degree of reliance of university youth on social networks during the Corona pandemic. It scores a total arithmetic mean of (2.2094) with a standard deviation of (0.40278), and it comes at the average level. Then this parameter ranked first, with an arithmetic mean (2.4133) with a standard deviation (0.55653). It means that one of the most prominent cognitive motives for relying on social networks during the Corona pandemic is the increase in the researcher's information about the pandemic, which is within the high level. This result was consistent with the Issa [10] study, which confirmed that the participants relied on mobile journalism as a form of social networking with a rate of up to 78%. This was followed by an arithmetic mean (2.4014) and a standard deviation (0.58789) which indicated that one of the most prominent cognitive motives come from the degree of university youth reliance on social networks in raising awareness of the Corona pandemic. It contributed to raise the level of public knowledge of the pandemic, which is also within the high level. This result is consistent with Karasneh's [4] study, which showed that all participants had a good knowledge of the virus as a result of their exposure to social networks first, then, radio and television. As for the penultimate rank, an arithmetic mean scored (2.0024) with a standard deviation (0.66008) which showed that one of the cognitive motives is to see the hidden things, and it is within the intermediate level. At last, an arithmetic mean scored (1.8361) with a standard deviation (0.74593), which means that one of the cognitive motives is considered its disclosure of information that was secret or hidden. It is actually in the middle level. It follows from the results of this parameter that one of the most prominent cognitive motives was the increase in information about the pandemic, as

Table 1 The cognitive motives come from resulting the degree of reliance of university youth on social networks during the Corona pandemic

No.	Item	Standard mean	Standard deviation	Rank	Estimation
1	My information regarding Corona pandemic is increased	2.4133	0.55653	1	High
5	It contributed to increase my knowledge about the Corona pandemic	2.4014	0.58789	2	High
7	It supplied me of what I needed	2.3705	0.56525	3	High
2	My protective abilities are improved	2.3135	0.57448	4	Middle
6	It added new values for me	2.1283	0.65297	5	Middle
3	It lets me have a clear idea about this dangerous disease	2.0024	0.66008	6	Middle
4	It discovered hidden and secret information for me	1.8361	0.74593	7	Middle
The total arithmetic mean of the parameter		2.2094	0.40278	–	Middle

well as raising the level of public knowledge about it. This finding is consistent with the findings of the Karasneh's [4] study.

As well as Irsheidat's [5] study revealed that the majority of the participants in the study had enough information about the pandemic due to the degree of their dependence on the specific media in that study, which is the Jordanian satellite channels. But this result differs from what came out of the Al-Khawaldeh's [11] study, which revealed that the largest percentage of participants is rarely followed up on health programs on Jordanian television. Even though, they are concerned with health awareness and forming the right habits. Social networking sites during the Corona pandemic appeared that these motives reached a high level for three paragraphs, and they related to increase the information about the pandemic. It informed the participants of the information they need. While the results showed that these sites improved the preventive capabilities that added new values to the participating group. It also revealed to them information that was hidden. These results are attributed to the importance of these networks in the knowledge additions to the media audience and the emergence of their role during the Corona pandemic among the studied group. In additions, this result is consistent with the theory of the degree of university youth's dependence on the media in terms of its inclusion in the field of uncovering ambiguity, forming trends, arranging priorities for public attention, and expanding beliefs and values [3].

Second: As shown in Table 2: The emotional motives came from the degree of university youth reliance on social media during the Corona pandemic. It appears that the parameter of emotional motives got the degree to which university youth depend on social networks during the Corona pandemic. It has a total mean of (1.9979), with a standard deviation of (0.35726), and it reaches the average level. Within this parameter, an arithmetic mean (2.4347) and a standard deviation (0.68912) indicates that one of the most prominent emotional motives is the degree of university youth's adoption of social networking sites during the Corona pandemic (the occurrence of nightmares and disturbing dreams about the pandemic). It is within the high level. This was followed an arithmetic mean (2.3017) and a standard deviation (0.65226), which concludes that one of the most prominent emotional motives is the degree of inclusive youth dependence on social networks during the Corona pandemic (the pain a lot for the victims and the deceased). It is within the intermediate level. As for the penultimate rank, an arithmetic mean (1.5701) and a standard deviation (0.65317) concludes that one of the emotional motives is the degree to which university youth adopt social networking sites during the Corona pandemic. It enhanced rapprochement with the family. It is within the average level. The last rank came with an arithmetic mean (1.4489) with a standard deviation (0.68339) which indicates that among the emotional motives, the degree to which university youth adopt social networks during the Corona pandemic, is the feeling of social solidarity, which came within the level low

Third: As shown in Table 3: Behavioral motives came from the degree of reliance of university youth on social networks in raising awareness of the Corona pandemic. The arithmetic averages and standard deviations of the sample responses show that the behavioral motives for university youth's dependence on social networks during the Corona pandemic had a total mean of (2.0232) with a standard deviation of (0.41858). It came at the average level. Within this parameter, an arithmetic mean (2.3302) with a standard deviation (0.57963) showed that one of the most prominent behavioral motives got the degree to which university youth adopt social networks during the Corona pandemic. It promoted the culture of volunteering to help others, which is in the middle level. This was followed by a mean (2.2993) and a standard deviation (0.56965)which showed that one of the most prominent behavioral motives got the degree to which university youth rely on social media sites during the Corona pandemic (motivation to new healthy behaviors), that is within the intermediate level. As for the penultimate rank, an arithmetic mean (1.7648) with a standard deviation (0.73634) indicated that one of the behavioral motives got the degree to which university youth adopt social networking sites during the Corona pandemic. This contributed to the respondent's comprehension of physical and physical distancing measures, which is within the average level. An arithmetic mean (1.6105) with a standard deviation (0.73048) came at last to show that one of the behavioral motives got the degree to which university youth relies on social networks during the Corona pandemic. It prompted the respondents to participate in publications, which is at

Table 2 The emotional motives came from the degree of university youth reliance on social media during the Corona pandemic

No.	Item	Standard mean	Standard deviation	Rank	Estimation
11	I had nightmares and bad dreams about the pandemic	2.4347	0.68912	1	High
10	I have suffered greatly for the victims and the dead	2.3017	0.65226	2	Middle
15	The precautionary measures put me in social isolation	2.1876	0.70403	3	Middle
8	Social media increased my anxiety	2.1259	0.75329	4	Middle
13	These networks made me psychologically stable	2.0499	0.73347	5	Middle
9	It made me sympathize with the injured	1.8646	0.74528	6	Middle
14	It strengthened my closeness with my family	1.5701	0.65317	7	Middle
12	It made me feel social	1.4489	0.68339	8	Low
The total arithmetic mean of the parameter		1.9979	0.35726	–	Middle

the low level. These results are consistent with both Ali's [8] study and Nooh's [6] study regarding to the behavioral motives resulting from the degree of university youth's dependence on social networks during the Corona pandemic among Jordanian university youth. Thus, the participants confirmed that the content of the networks they visited had strengthened their culture of volunteering to help others. They motivated them to adopt new healthy behaviors, make them aware of the importance of health measures, and motivate them to express their opinion and comments on what is published about the pandemic. As for their urge to share publications, it appeared with a low rating among behavioral motives.

Fourth: As shown in Table 4: The reasons for the degree of reliance of university youth on social networks during the Corona pandemic showed that the reason has got a total mean of (2.4494) with a standard deviation of (0.41591), and it came at the highest level. An arithmetic mean (2.5273) with a standard deviation (0.53660) means that one of the most prominent reasons for the degree of adoption of social networks during the Corona pandemic achieved its aim in which the respondent found what he needs. This result is consistent with Irsheidat's [5] study Karasneh's [4] study. Then, an arithmetic mean (2.5249) with a standard deviation (0.58348) showed that one of the most prominent reasons for the degree of adoption of social

Table 3 Behavioral motives came from the degree of reliance of university youth on social networks in raising awareness of the Corona pandemic

No.	Item	Standard mean	Standard deviation	Rank	Estimation
21	I have strengthened the culture of volunteering to help others	2.3302	0.57963	1	High
20	It motivated me to new healthy behaviors	2.2993	0.56965	2	Middle
23	I feel responsible for adhering to health measures	2.2067	0.67086	3	Middle
22	I trusted my family relationships	2.1639	0.66849	4	Middle
18	It made me comment on some posts	1.9976	0.73517	5	Middle
17	It motivated me to express an opinion on what is published about the pandemic	1.8124	0.77793	6	Middle
19	It contributed to my understanding of physical and physical distancing measures	1.7648	0.73634	7	Middle
16	It prompted me to participate in posts	1.6105	0.73048	8	Low
The total arithmetic mean of the parameter		2.0232	0.41858	–	Middle

networks during the Corona pandemic got the characteristic of interaction with others, which is also within the high level. As for the penultimate rank, an arithmetic mean (2.3967) with a standard deviation (0.57062) showed that one of the reasons for the degree of adoption of social networks during the Corona pandemic is presenting the latest developments, which is also within the high level. An arithmetic mean (2.3777) with a standard deviation (0.61130) indicated that one of the reasons for the degree of adoption of social networks during the Corona pandemic is followed up on government decisions first, and it is within the high level as well.

Fifth: As shown in Table 5: The follow-up topics in social networks regarding the pandemic showed that the focus of the topics is followed by the social networks regarding the pandemic. It has a total arithmetic average of (2.3235) with a standard deviation (0.43265), and it came at the average level. An arithmetic mean (2.6580) with a standard deviation (0.53172) showed that one of the most prominent topics followed in social networks regarding the pandemic got the production of vaccines against the virus, which is within the high level. This was followed by an arithmetic mean (2.4632) with a standard deviation (0.58692) which concluded that one

Table 4 The reasons for the degree of reliance of university youth on social networks during the Corona pandemic

No.	Item	Standard mean	Standard deviation	Rank	Estimation
33	I find it what I need	2.5273	0.53660	1	High
28	It has the ability to interact with others	2.5249	0.58348	2	High
25	Multiple sources	2.5249	0.58348	3	High
32	Easy access	2.4869	0.57565	4	High
30	Covering all parts of the kingdom	2.4228	0.56640	5	High
26	Presenting a variety of topics	2.4204	0.56185	6	High
29	It has the feature to see the experiences of countries and others	2.4133	0.58976	7	High
27	It is fast and instant	2.3990	0.56265	8	High
24	Presenting the latest developments	2.3967	0.57062	9	High
31	Follow up on government decisions	2.3777	0.61130	10	High
The total arithmetic mean of the parameter		2.4494	0.41591	–	High

of the most prominent topics followed in social networks regarding the pandemic got preventive methods. These results are consistent with the Karasneh's [4] study, Irsheidat's [5] study, and Nooh's [6] study, but they differ from Mejia's [7] study. As for the penultimate rank, an arithmetic mean (2.1401) with a standard deviation (0.74439) showed that one of these topics follows social networks regarding the pandemic. It relates to the government instructions regarding defense orders, which is within the average level. Then, an arithmetic mean (2.1330) with a standard deviation (0.71139) indicated that one of the topics followed in social networks regarding the pandemic has the epidemiological situation in the Arab world, which is within the average level.

Sixth: The results of the Independent Samples T-Test. It clarifies the statistical differences according to the participants' points of view about the degree of dependence of the awareness of media students in Jordanian universities. This T-Test also clarifies the results of the role of social networks during the Corona pandemic. The results show the presence of significant statistical differences. The significance value was (0.004), which is less than (0.05). This means that it functions. The differences are attributed to the master's class, as the arithmetic mean was (2.2694), while the arithmetic mean for the bachelor's class was (2.1939), and this rejected the null hypothesis.

Table 5 The follow-up topics in social networks regarding the pandemic

No.	Item	Standard mean	Standard deviation	Rank	Estimation
37	Production of vaccines against the virus	2.6580	0.53172	1	High
40	Preventive methods on Jordanian society	2.4632	0.58692	2	High
39	The motives of the pandemic	2.4466	0.64003	3	High
43	Pandemic stats	2.3967	0.62632	4	High
47	The fate of the pandemic	2.3325	0.62763	5	Middle
41	Therapeutic methods	2.3302	0.58779	6	Middle
34	Epidemiological situation in the world	2.3254	0.64442	7	Middle
44	Medical health bulletins	2.3112	0.64409	8	Middle
46	Pandemic updates	2.2874	0.59806	9	Middle
42	Protection of self and family	2.2850	0.62067	10	Middle
36	The epidemiological situation in Jordan	2.2138	0.65240	11	Middle
45	Instructions from the relevant authorities	2.2067	0.72542	12	Middle
38	Government instructions regarding defense orders	2.1401	0.74439	13	Middle
35	The epidemiological situation in the Arab world	2.1330	0.71139	14	Middle
The total arithmetic mean of the parameter		2.3235	0.43265	–	Middle

The above results show that there are no statistically significant differences regarding the participants' responses according to the variables: age, school year, and university name. The test: One-Way Analysis of Variance (ANOVA) showed that the statistical significance values of the variables (age, school year, and university name) were all higher than the value of the statistical significance (0.05). This means that the scientific hypothesis is accepted.

7 The Most Important Results

1. Regarding the cognitive motives that result from the degree of university youth's dependence on social networks during the Corona pandemic, these motives got a high level for three paragraphs: They relate to increasing information about the

pandemic and supply the respondents with the information they need. While the results showed that these networks improved the protective capabilities and added new values to the studying group. That also revealed that these networks shed light on some hidden information. Besides, these paragraphs got a high average level, and thus attributed to the importance of these networks in the knowledge additions to the media audience and the emergence of their role during the Corona pandemic among the studying group. This result is compatible with the theory of dependence on the media in terms of its inclusion in uncovering ambiguity, forming trends, arranging priorities for public attention, and expanding beliefs and values.

2. Regarding the emotional motives for the contents of social networks on the respondents, the paragraph that indicates: "I had nightmares and disturbing dreams about the pandemic" appeared in a high degree, while the following paragraphs occupied a medium rating, namely: sympathy with the victims, social isolation, growing anxiety, psychological stability, and closeness with family. As for the paragraph that received a low rating in this regard was the paragraph: "it made me feel social together". These results are consistent with the theory of the degree of university youth dependence that links emotional processes (affective effects) with some terms related to feelings or emotions towards a phenomenon, and "Devler and Rokich" define these effects with feelings of emotional apathy, indifference, fear, anxiety, moral support and alienation.
3. Regarding the behavioral motives resulting from the degree of university youth's dependence on social networking sites during the Corona pandemic among Jordanian University Youth, the results showed that all the paragraphs of this parameter appeared moderate except one paragraph that appeared low, where the participants confirmed that the content of the networks they visited has strengthened their culture of volunteering to help others, motivate them to adopt new healthy behaviors, make them aware of the importance of health measures, and motivate them to express their points of view and comment on what is published about the pandemic. As for their urge to share publications, it appeared with a low rating among behavioral motives.
4. The results showed a high appreciation for the parameters of the reasons for the degree of university youth's dependence on social networks during the Corona pandemic, where the paragraph: "one of whom I need" ranked first, and the paragraph: "following up on government decisions" ranked last in this parameter. But the parameter as a whole was high among the respondents.
5. Regarding the topics that university youth follow on social networks that refer to Corona pandemic, the paragraphs have been occupied: the production of new vaccines, methods of prevention, the side motives of the pandemic on Jordanian society, and pandemic statistics. While these paragraphs appeared in high ranks in the field of parameter of topics. The other paragraphs appeared with an average rating, including: the fate of the pandemic, treatment methods, the epidemiological situation in Jordan, health guidelines, government instructions, and defense orders.

6. The study showed that there were no significant statistical differences due to the variables of gender, age, academic year, and the name of the university. This led to the acceptance of null hypothesis. As for the study program variable, the results showed significant statistical differences due to this variable in favor of the master's program.

8 Recommendations

According to the results of this study, the following recommendations could be included:

1. The research proposes expanding the investigation of the reasons for the public's lack of dependence on social networks during the Corona Virus pandemic. Despite the appearance of this result in a low percentage in the research, the power and dominance of these social networks make it important to research this topic scientifically.
2. The research recommends conducting more specialized media studies to fill the knowledge gaps in this research such as polling the opinion of the general public on social networks, or conducting certain studies about other specialized sectors of society in order to study the multiple effects of websites about these societies such as specialists in education, health, politics, economics, and social issues.
3. The necessity of elevating the proficiency degree of government agencies that depend on social networks in communicating various media messages, as there is a strong presence of these sites among all segments of society.
4. Conducting more research about Jordanian University Youth, and comparing the results of this study with other future studies dealing with other media such as satellite channels, the press, radio, and television.

References

1. Al-Qalini, S: The extent of the Egyptian elite's dependence on television in times of crisis: a case study on the Luxor incident. Egypt. J. Media Res. 4th issue, 35 (1998)
2. Abdel, H.: Books for Publishing and Distribution, p. 237 (2009)
3. Makkawi, H., Al-Sayed, L.: Communication and its Contemporary Theories, 13th edn, p. 329. The Egyptian Lebanese House, Cairo (2018)
4. Karasneh, R., et al.: Medicals effect on shaping knowledge, awareness rich perceptions and communication practices of pandemic Covid-19 among pharmacists. Res. Social Adm. Pharm. (2020)
5. Irsheidat, A.: The role of Jordanian satellite channels in raising the Jordanian society's awareness of the corona pandemic, survey study. Unpublished Master's thesis, Yarmouk University (2020)
6. Nooh, H., et al.: Public awarness of coronavirus in Al-Joufr region, Saudi Arabia. J. Public Health. https://doi.org/10.1007/s10389-020-01209-y

7. Mejia, C.R., et al.: The media and their informative role in the face of the coronavirus disease (COVID-19). Electron J. Gen. Med. **17**(6) (2020)
8. Ali, M.: The role of the Facebook network in promoting health awareness among the public. Unpublished Master's thesis, College of Mass Communication, Middle East University, Amman (2017)
9. Mohamed, M.: The Egyptian university youth's dependence on social media as a source of news, an applied study on Facebook and Twitter. Unpublished Master's thesis, Cairo University, Faculty of Mass Communication (2016)
10. Issa, T.: The Palestinian youth's dependence on mobile phone journalism as a source for news of the Jerusalem Intifada, a field study. Gaza J. Islamic. Univ. Humanitarian. Res. **27**(4), 347–375 (2019)
11. Khawaldeh, R.: The role of Jordanian television in increasing the health knowledge of viewers. Unpublished Master's thesis, Middle East University, Amman (2019)

Conceptual Fundamentals of Providing Long-Term Economic Growth as a Factor of Economic Security of Ukraine

Yaroslava Bazyliuk, Tetiana Heiko, Serhii Davydenko, Antonina Deshko, and Dmytro Pokryshka

Abstract The article provides an overview of factors ensuring long-term economic growth in Ukraine basing on the interpretation of its compliance with economic security. Upon the different theories of economic growth, an understanding of causes why Ukraine has not achieved high long-term economic growth, widening social and economic gaps with other countries and, consequently, not providing an acceptable level of economic security since the 1991. The authors conclude that Ukraine hasn't seized a window of opportunity provided by market transformation of the economy to increase productivity, which significantly reduced national economic security. Prolonged restrictive monetary and non-stimulating economic policy has led to restraint of money supply, scarce funding and non-strategic structural change, and has prevented effective capacity development and timely expansion of the internal market. It keeps small size of the national economy and led to the outflow of labour and capital resources from the country. The small size of the national economy and absence of significant results in the scalability of the economy are two sides of Ukraine's central economic problem, which leads to a critical compression of development resources and economic reproduction in the long run. Thus, economic policy of Ukraine should be modernized in terms for economic growth and economic security, focusing on measures to radically expand demand (expanding investments, rising household's income, increasing lending to the economy) while removing barriers to development from the supply side (in such spheres as entrepreneurship, industry, services and agriculture). The authors prove the prospects of optimizing budget expenditures to increase productivity in Ukraine basing on factors and criteria for ensuring national economic security. Output, create employment opportunities and improve global competitiveness.

Keywords Economic security of the state · Labour productivity · Economic growth · Government expenditures · Banking lending · Investments

Y. Bazyliuk (✉) · T. Heiko · A. Deshko · D. Pokryshka
National Institute for Strategic Studies (NISS), Pyrohova, 7a, Kyiv 01054, Ukraine
e-mail: arosava11@gmail.com; yaroslava@niss.gov.ua

S. Davydenko
Expert, Kyiv, Ukraine

A. Hamdan et al. (eds.), *The Implementation of Smart Technologies for Business Success and Sustainability*, Studies in Systems, Decision and Control 216,
https://doi.org/10.1007/978-3-031-10212-7_42

1 Introduction

The core problem of Ukraine's economic development for decades is the combination of critically small, especially in relation to European and other developed countries, size of the economy and low volatile average annual growth rate, which lags far behind the world average. In the long run, maintaining such a low-scale economy does not significantly improve the competitiveness of the national economy and living standards, which determine the prospects for ensuring an acceptable level of economic security.

Ukraine's economic "underdevelopment", characterized by an average annual GDP growth rate of −0.8% in 1991–2020, highlights the task of reviewing the causes of this situation and mechanisms for providing long-term economic growth in the context ensuring the economic security of the state. Since the methods of achieving economic growth may be a problem for the economic security in the medium and long term, their choice must correspond to the criteria for ensuring acceptable level of economic security. To do this, we compared some macroeconomic results of the national economy with the cost of its achieving, and compared with other countries. Based on the revealed results, we have substantiated the priorities of increasing long-term economic growth, which do not hinder the task of strengthening economic security.

We decided that the interpretation of the reasons for the almost twice lower rates of economic growth in Ukraine against the world on the basis of neoclassical theories of economic growth does not seem complete. We believe that intensification of geopolitical competition between countries, as well as structural digital and technological changes, play expanding role in the model of economic growth. Thus, we identified national priorities, which need to be more widely considered and implemented to the model of growth of Ukraine. To this end, we justified the feasibility of implementing a new structural policy focused on innovation, and targeted allocation of financial resources to create new digital and technological competitive advantages for the economy.

2 Literature Review

Escalation of economic competition and geopolitical opposition between different states predetermine an increasing importance of implementation of national interests and priorities, and further—economic security issues, in the strategy of economic development and growth. Leonard and others highly recommended to integrate economic and security considerations within the EU [1, 2]. The interpretation of economic security varied—from the interpretation of the International Labour Organization [3], which refers to economic security issues of financial and physical security of labour access to basic social infrastructure and decent work, to the objective state of national interests in ensuring sustainable development of the national economy and development of equal mutually beneficial economic cooperation with other states [4].

In our view, economic security is a complex characteristic that demonstrates the extent to which the results of economic growth and development ensure the security and well-being of nation, and in general—the economic sovereignty of the state. The maintenance of such sovereignty as the basis of national development is ensured by sufficient international competitiveness. In this context, the economic security of the state can be achieved through the consistency of physical and economic sufficiency of resources, efficiency of their usage and externally competitive scale and rate of economic growth. Accordingly, comparing the macroeconomic results of the national economy with previous periods and with other countries, will allow a more reasonable demonstration of real and potential protection of economic security of the state as a whole.

The aim of the article was to interpret the impact of factors ensuring long-term economic growth on the economic security of the state. To determine this, we selected key characteristics that link growth and security, namely: efficiency (assessed by economic dynamics and productivity); scale (estimated by nominal and parity GDP); financial balance (assessed by the main parameters of investment and public finance).

We assume that the basis for ensuring the economic security of the state is the state economic policy aimed at counteracting internal and external threats through the achievement of sustainable long-term economic growth. At the same time, it envisages the introduction of economic incentives that are consistent with the tasks of increasing the level of economic security of the state, which do not lead to the accumulation of macroeconomic imbalances and gaps with other countries, and creates promising structural changes and balances cash flows.

The common ideas about the reasons for achieving long-term economic growth are represented by the branching of neoclassical views based on factors of availability and expansion of supply (factors productivity) and demand (including measures to expand consumption, investment, government spending, lending, etc.).

In the theories that supported the factors of supply in solving problems of economic growth (Laffer, Mandel, Feldstein, etc.), there is a general ineffective state intervention in economic processes. Therefore, the direction of public spending is justified only to perform the most necessary functions of defence and social security. Similarly, Barro noted the deformation of market levers (high interest rates and tax rates) in increasing government consumption and maintaining the budget deficit, explaining this by the rational behaviour of the household not to save in anticipation of further tax growth [5]. Representatives of the Austrian school also focus on the need to minimize government intervention in the economy. At the same time, the idea of a positive attitude to state stimulation of the economy can be traced in the current work of Krugman, who supports the growth of public spending on infrastructure and support for the unemployed, as such policy has a stimulating effects [6].

Prospects for stimulating aggregate demand as causes of economic growth are identified in the works of Keynes as well as neo-Keynesian researchers Hicks, Modigliani, Samuelson. According to them, economic growth is achieved through increasing aggregate demand in conditions of full employment. At the same time, in other conditions, economic growth is constrained by excessive demand. Thus, the centralization of resources (in particular, through taxes) and their use to finance public and other investments under certain conditions will unlock economic growth. In these theories, a high share of national income redistributed through the budget is acceptable. At the same time, in the conditions of inefficiency of public expenditures, the prospects for the accumulation of macroeconomic imbalances increase significantly, and, accordingly, the gaps in national results increase [7–9].

3 Research Methodology

Countries are increasingly integrating into the global economy, and this process expands both opportunities and risks for their development and growth. So it's necessary to implement improved approaches to assessing the interdependence of economic trends and economic security of the country.

According to the approach used in Ukraine, the level of economic security of state is estimated on the basis of an integrated index, which includes 9 weighted average sub-indices. The calculation of these sub-indices involves normalization and weighing of 126 indicators [10].

Thus, macro indicators constitute the basis for determining the level of economic security and are used to interpret it in concordance with the established threshold values.

Applying such a method for this research would require calculating the level of economic security for other countries and comparing it with the level of economic security for Ukraine.

Nevertheless, there are significant methodological and information and statistical difficulties in applying this approach. They could negatively affect the accuracy of the results and their analysis. Therefore, at this stage of the research, we will limit ourselves to assessing macrotrends in Ukraine and the world on the basis of key characteristics, combining growth and security issues.

Thus, the comparison of efficiency will be carried out according to the indicators of: economic dynamics (real GDP growth rate, GDP per capita growth rate); productivity (GDP per capita and per employee); values (GDP in current international dollars and GDP at purchasing power parity (PPP)); proportionality (share of investments (gross fixed capital formation) in GDP, the level of budget expenditures and bank loans to GDP).

For the purpose of comparative assessments, we used different calculation methods for dynamic (growth rates) and structural macro-indicators. Dynamic macro-indicators were evaluated and compared according to the intensity of their cumulative change and on average over a certain period. The obtained results were used to rank countries.

For dynamic indicators, the intensity of the cumulative change is calculated by Formula (1):

$$CG_{i,n} = \left(IG_{i,n} - 1\right) * 100\% \quad (1)$$

where $CG_{i,n}$ the level of intensity of change of the *i-th* indicator for *n-years* relative to the base year; $IG_{i,n}$ chain growth index of the *i-th* indicator for the period of *n-years.*

Chain index of growth is calculated by Formula (2):

$$IG_{i,n} = R_{i,1} * R_{i,2} * .. * R_{i,n} \quad (2)$$

where R_i indices of growth rates of the *i-th* indicator for each year in the period of *n-years.*

The estimate of the average rate of change is calculated by Formula (3):

$$\overline{R}_i = \left(\sqrt[n]{CG_{i,n}} - 1\right) * 100\% \quad (3)$$

where $\overline{R}_i$ index of the average annual growth rate of the *i-th* indicator in the period of *n-years.*

Accordingly, countries with higher index values ov a certain period have better development efficiency and competitiveness, which contributes to strengthening their economic security.

We ranked different countries upon these indices. And the higher the rank of the country, the more noticeable the positive impact of the assessed parameter on its economic security. Ranking is performed in Excel using the function "RANK.EQ".

Structural indicators were compared according to the level of average annual changes for a certain period. The obtained results were also ranked and analysed.

We compared Ukraine's economic indicators both with the general sample, which includes all countries of the world, and with individual groups of countries, classified according to the relevant criteria. At this stage of the study, the World Bank's classification, which divides countries into four groups according to the level of income per capita, were used. This classification distinguishes 4 countries groups: high-income economies; upper-middle-income economies; lower-middle-income economies; low-income economies.

For each country groups we calculated the ranges from minimum to maximum values and determined the average values for relevant macro indicators. The comparison of the Ukraine's performance with corresponding indicators of each country group allows us to assess the level of its competitiveness, as well as compliance of Ukraine's structural characteristics with target levels of productivity, economic growth and therefore economic security in general.

Data sources for calculations are International Monetary Fund, the World Bank Group, Conference Board, the State Statistics Service of Ukraine, the National Bank of Ukraine, the Ministry of Finance of Ukraine.

4 Results

The economic security strategy of Ukraine for the period up to 2025, approved by the Decree of the President of Ukraine from 11.08.2021 No 347/2021 [11], defined the state of economic security of Ukraine during 2010–2019 as unsatisfactory. In 2019 the level of economic security was 43%, and in the first half of 2020–41%. Indicators of all main components of economic security were low. This means that risks of large-scale destabilization of the economy in the long run are high.

The study of factors of economic growth, which negatively affect the level of economic security of Ukraine, allowed us to make some important generalizations:

(1) *The instability of the growth rates and the weaknesses of recovery from the recession underlined the inefficiency of the process of realization of country's potential during the years of independence.*

Over the past 30 years the vast majority of the world's countries have multiplied the effectiveness and scale of their economies, whereas Ukraine is still approaching its potential at the beginning of the independence. The average annual GDP per capita growth rates in Ukraine during 1991–2020 reached negative values (−0.80%). This resulted in reduced productivity of each person in 2020 by 21.4% compared to 1990 (see Fig. 1).

The average growth rate for the high income countries in 1991–2020 was 1.3%, cumulatively increasing the GDP per capita by almost half. Upper-middle-income economies have doubled their productivity level over the same period, reaching almost 4.1% of average annual growth. And even low income countries increased productivity by 1.1% on average, increasing GDP by 37.2% overall.

Ukraine has not yet reached positive average annual growth rates for the entire period of independence. This is primarily due to the deep recession during the first 8 years of independence, in 1991–1998 GDP fell by more than 10% annually. And even almost 3% of the annual GDP growth in 2000–2020 was not enough to compensate for the economic "Armageddon" at the beginning of Ukraine's independence. The inconsistency of economic policy further limited the realization of the country's potential and its economic security.

(2) *Labour productivity reduction and retention at an unsatisfactorily low level.*

The significant threat to the economic security of Ukraine is the excessive volatility of labour productivity (calculated as the ratio of GDP to the number of employed persons), which has evolved on the basis of a high level of economic vulnerability to the crises of different genesis and intensity. Depressive dynamics of labour productivity in 2012–2015, caused by structural and technological parameters of economic development, as well as external aggression, only slightly changed in the next few years. At the same time labour productivity had a descending trend. The growth rate of productivity decreased from 3.5% in 2016 to 1.9% in 2019. In 2020, due to the negative impact of the corona-crisis, labour productivity in the Ukraine fell by 2%. Meanwhile, in 2016–2020 the average labour productivity growth in Ukraine (1.8%)

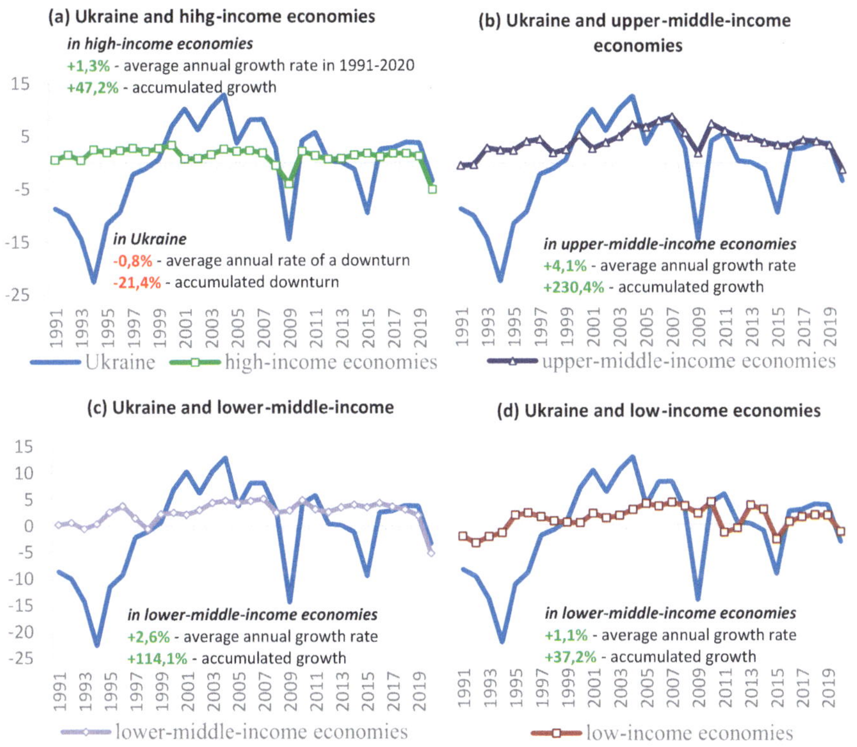

Fig. 1 GDP per capita growth rates in Ukraine and countries with different income levels in 1991–2020, % [12]

was one-third lower than in Poland (2.5%), was half as high as in Romania (3.4%). Labour productivity dynamics in Ukraine during this period was in line with a range of EU countries such as Czech Republic, Estonia, Latvia and Lithuania. Against the lower absolute value of labour productivity this indicates the lack of Ukraine's economy capability to leverage the factors that would allow to use capacity for catching-up growth and reduce the development gap with the EU countries.

As a result, in the long run the economy of Ukraine increases the gap with neighbouring countries in terms of labour productivity dynamics (see Fig. 2).

In 2020 the level of labour productivity in Ukraine was 98% of 1990 levels. During this period, Romania increased this indicator more than three times, Poland—by 1.7 times, Slovakia—by 1.3 times, Hungary and the Czech Republic—by more than 70%, Bulgaria—by 65%, Russia—by 29%.

Due to the low absolute indicators of labour productivity such growth rates do not create conditions and economic capacity to overcome threats to economic security, that Ukraine face during the formation of a new paradigm of innovative and technological development in the world.

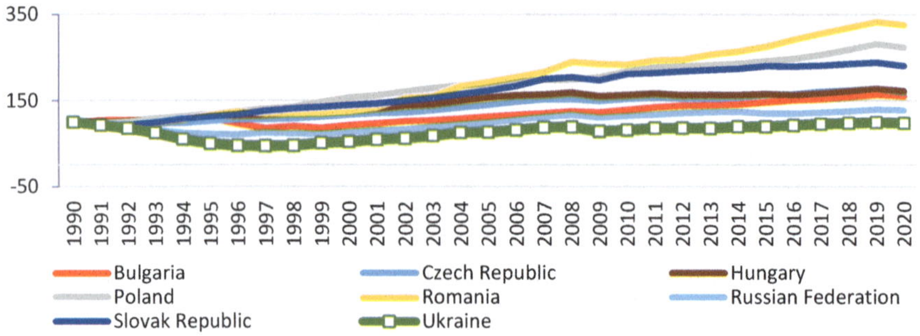

Fig. 2 Labour productivity growth, 1991–2020, % (1990 = 100) [13]

(3) *Weak potential for economic growth in Ukraine's is determined by low level of its capitalization and the curtailment of investment support, which disallowed the creation of stable foundation for the economic security of the state.*

The share of the investment component of Ukraine's GDP is generally on par with low-income countries. It emphasizes the critical dependence of economic development and competitiveness on investment (see Fig. 3).

Small differentiation in investment rates in high-income countries, on the other hand, demonstrates the importance of maintaining an appropriate level of investment over the long term in order to continually upgrade the economy.

Ukraine's economic security is facing significant threats due to the weakening investment capacity of the economy. These threats become more urgent in the context of the formation of a new development paradigm based on the fourth industrial revolution, which requires significant investments to ensure digital and technological transformations. In 2020 the level of gross fixed capital formation (GFCF) in Ukraine decreased to 13% of GDP. However, even in 2014–2019, its level did not correspond to the urgent investment needs and was in the range of 13.5–17.6%, significantly inferior to the period of export-oriented growth in 2003–2008 (over 20%). The low investment capability of Ukraine's economy becomes one of the key factors of structural inflexibility and, consequently, high vulnerability to threats, in particular slow economic growth over the long term (see Fig. 4).

In the context of economic security, the continuation of this situation is fraught with the obstruction of processes of technological and structural upgrading of the economy, introduction of more sophisticated production processes and the creation and development of new competitive activities.

(4) *Insufficient saturation of the economy with money and credit.*

Against the underdeveloped market of financial services bank lending remains the main source of financing for current and investment activities of economic entities in Ukraine. At the same time, despite the availability of sufficient financial resources, their transfer to the real sector remains problematic.

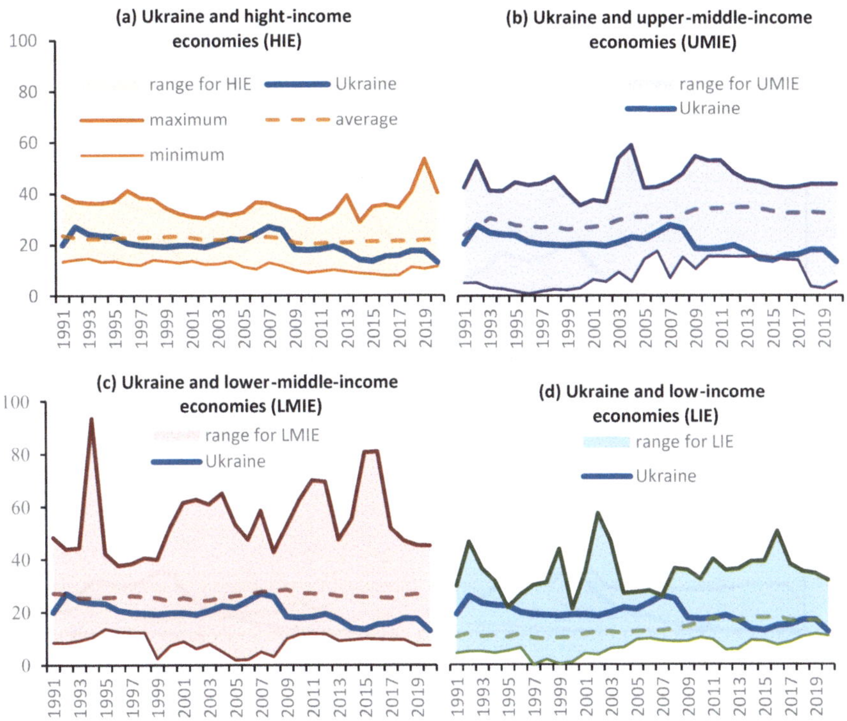

Fig. 3 The level of gross fixed capital formation in Ukraine and countries with different income levels in 1991–2020, % of GDP [12]

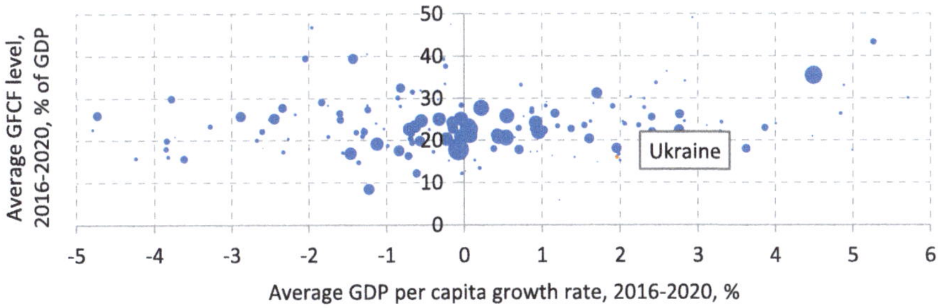

Fig. 4 Level of gross fixed capital formation and average GDP growth rates in Ukraine, 2016–2020

Ukraine remains an outsider in terms of lending to the economy compared to high- and middle-income countries, and even has a weak position among low-income countries (see Fig. 5).

This is backed by a lack of assets that can act as collateral for loans and a lack of transparent financial reporting.

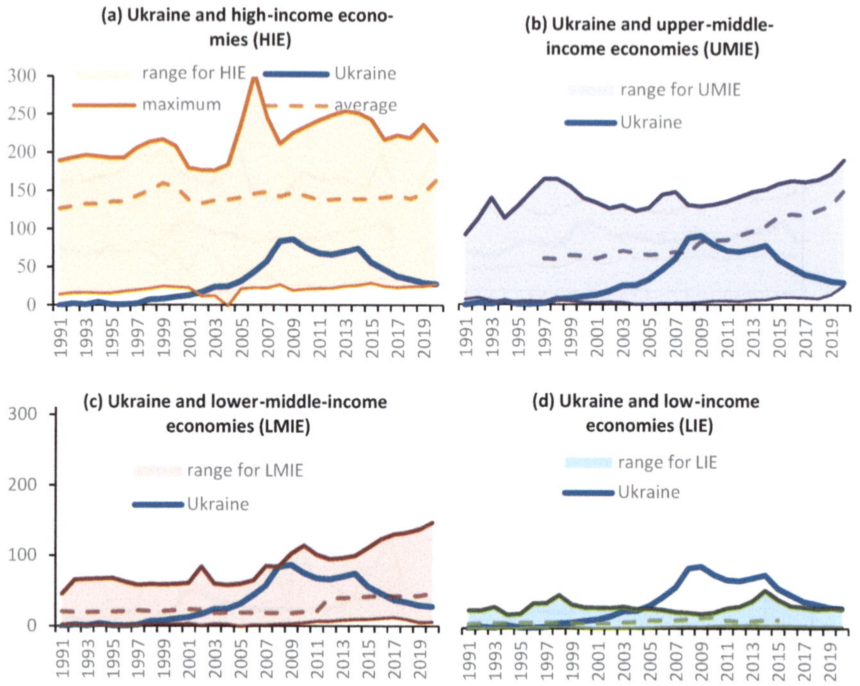

Fig. 5 Bank loans to the private sector in Ukraine and countries with different incomes in 1991–2020, % of GDP [12]

In general, lending to the Ukrainian economy in 2012–2020 remained at a consistently low level, declining by almost a third from 49% of GDP in 2014 to 17% of GDP in 2020.

The financial and foremost the banking sector is overly autonomous from the areas of economic governance and insufficiently motivated to lend to the real sector. Banks have taken into account the increased risks and this has led to an even greater deterioration of the difficult terms of the loan to entrepreneurs. Combined with increased regulatory requirements for banks, this has made it much more difficult to provide loans to the real economy and stimulated the placement of banking resources in less risky financial instruments.

The monetary aggregate M3 (money supply) has been constantly growing since 2012, while lending to the real economy sector has been on a downward trend. Since 2012 the trend of directing excess liquidity by banks to simple and risk-free transactions for the purchase of certificates of deposit of the National Bank of Ukraine and domestic government bonds (internal government loan bonds–IGLBs) has intensified (see Fig. 6).

It is due to the following features:

- **Complicated access of business entities to credit and high cost of it.** Specifically, local currency loans to the real economy sector were provided in 2014–2019

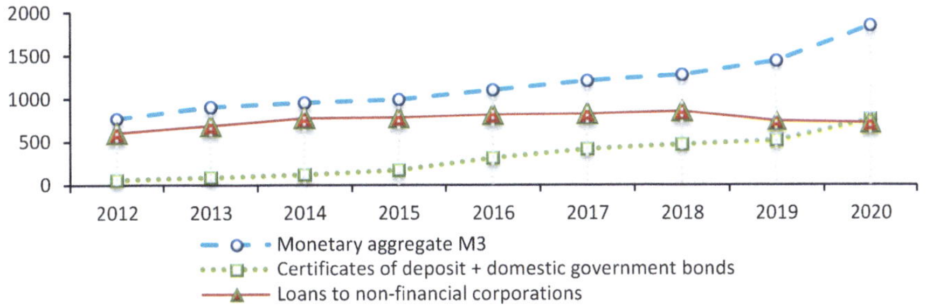

Fig. 6 Reversals tend in lending to the real economy sector towards the purchase of certificates of deposit of the National Bank of Ukraine and domestic government bonds, 2012–2020, UAH billion [14]

at average interest rates ranging from 14 to 20.8%. Average interest rates decreased only in 2020 to 10.3% (see Fig. 7).

Against the low profitability of operating activities of enterprises, the possibility of using profits as a source of payment for liabilities to the creditors has been narrowed. This has led to an increase in the deficit of working capital, the heightening of the risk of insolvency, the inability to develop and expand business, reduction in activity and therefore reduce jobs.

- **Significant influence of financial and economic crises and slow economic recovery, especially after the banking crisis of 2014–2017.** Substantial devaluation of the hryvnia against the background of hybrid aggression of the Russian Federation has worsened the servicing of foreign currency loans. The reform of the banking system due to the crisis stimulated a more than twice reduction in the number of banking institutions (from 175 in 2014 to 74 in 2021) and their structural units by 66.5% (from 21089 to 7073). All of this has caused the loss of banks clients (legal entities) that placed funds in banks that were withdrawn from

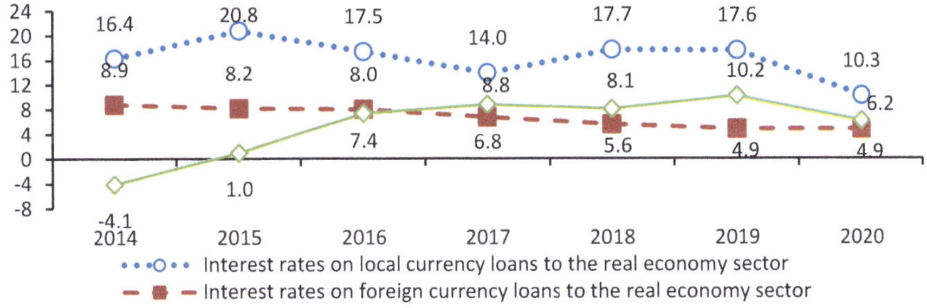

Fig. 7 Profitability (loss) level of operating activities of enterprises and interest on loans, 2014–2020, % [14]

the market, which received an extremely low probability of their return, due to being in the seventh queue waiting for reimbursement.

It indicates a low reflectivity of public policy to meet demand. Thus, since 2014, a restrictive policy regarding the saturation of the economy with money has been implemented, and its shift did not unblock business financing. Measures taken in 2014–2017 to slow down inflation significantly decreased business activity, reduced employment and production output. Banking sector reform and restrictive policies have also limited the ability of businesses to obtain bank loans, while they faced an increasing need for resources to recapitalize. Current containment policies, as well as other constraints, do not create the conditions for accelerating economic recovery and growth.

The issue of low economic growth can be addressed through a balanced expansionary policy—increasing the money supply and reducing money value. A possible mechanism to stimulate lending activity in the banking sector is to provide banks with subordinated loans from the National Bank of Ukraine. Their cost is within the refinancing rate. Such loan could be granted against liabilities of banks to direct these funds to targeted lending to real sector enterprises. In addition, the continuation of the existing Affordable Loans 5−7−9% program, business loan guarantee programs and revitalization of the work of the Export Credit Agency of Ukraine (ECA) could have a significant impact on economic growth.

(5) *Low efficiency of budget policy support for economic growth.*

Limited potential budget expenditures to support demand and low fiscal stimulus do not support economic growth. Although the level of redistribution of GDP through budget expenditures in Ukraine is at the level of average values in high-income countries and above average in upper-middle-income as well as lower-middle-income countries (see Fig. 8).

Thus, in Ukraine in 2000 this figure was 35.5% of GDP, in 2010 it has already reached 49.2% of GDP, decreased slightly in 2016–up to 40.6% of GDP and rose again to 47.1% of GDP in 2020. In 2000–2004 its level was explained by both the general growth of the economy and the increase in budget expenditures to stimulate it. Subsequently, the dynamics of this indicator was influenced by the need to support the economy through the financial (2008–2009) and security (2014–2018) crises, as well as corona crises (2019–2020) under the need to meet social obligations of the state under the conditions of declining GDP.

Note that the level of redistribution of GDP through the budget in Ukraine is more typical of countries with high-income and upper-middle-income countries (see Fig. 9).

Such a situation, takes place due to the existence of other imbalances, in particular, high fiscal burden on the economy, a significant size of shadow economy (31% of official GDP [18], disordered system of exemptions and social benefits, and weakness of state institutions.

According to the IMF, in 2020–2021 the trend of increasing the redistribution of GDP through budget expenditures was observed in other countries (primarily

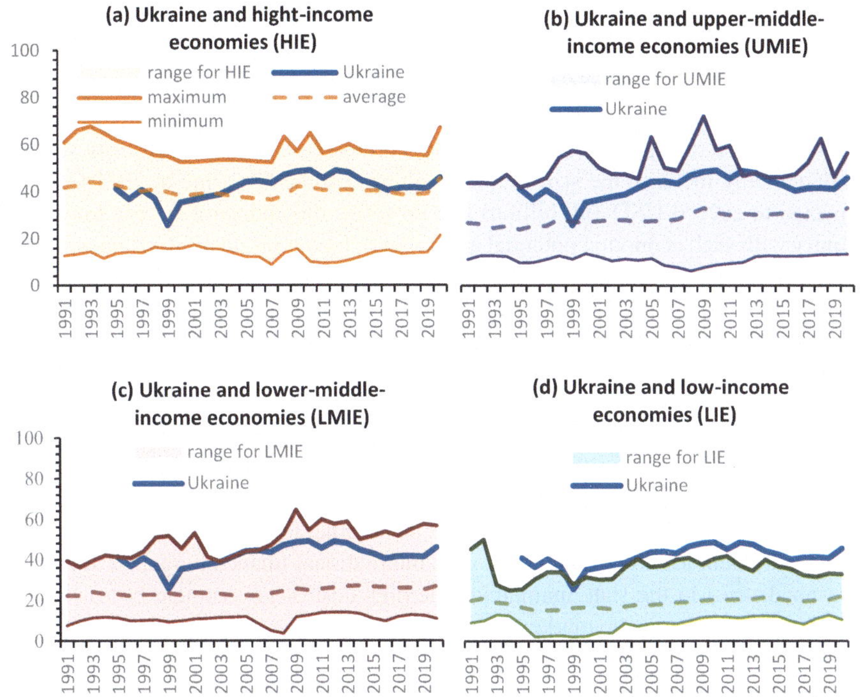

Fig. 8 Dynamics of budget expenditures in Ukraine and countries with different incomes, 1991–2020, percentage of GDP [15]

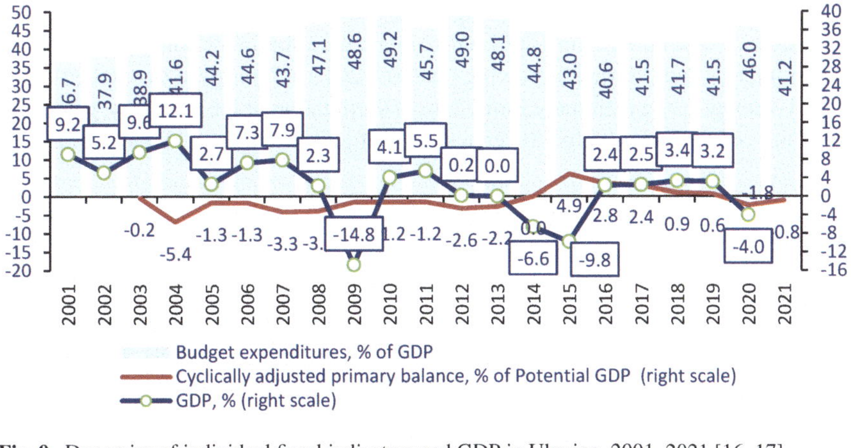

Fig. 9 Dynamics of individual fiscal indicators and GDP in Ukraine, 2001–2021 [16, 17]

developed) and was caused by crises in the world economy and unprecedented levels of state support for the economy (for example, in 2020 in Belgium it rose to 60.8%, and in France—to 62.4% of GDP) [17].

At the same time, the logic of optimizing the level of budget expenditures in Ukraine's GDP to support economic growth and increase economic security should fundamentally increase the size of the economy (for the first time in 2021 GDP is expected to exceed USD 190 billion) [4]; however, this indicator is very low for a country with such economic potential as Ukraine), labour productivity, domestic and foreign investments within an environment of macroeconomic stability, improving of the quality of public institutions and efficiency of public spending (according to the Accounting Chamber for 2020 the total amount of funds audited during the control measures amounted to UAH 1.185 billion, the facts of violations amounted to UAH 26.3 billion [19].

The trap of small GDP was also reflected in the debt policy—the level of public and guaranteed debt is at the limit set by law (60% of GDP). In 2020, Ukraine continued the trend of accumulating debts due to the continuing economic crisis caused by COVID-19 (see Fig. 10). At the same time, low credit rating of Ukraine [20], insufficient level of cooperation with international financial organizations and investor distrust in the state maintained the high cost of government borrowing in both foreign and domestic markets.

There was also a diversion of a significant number of budgetary resources for debt service (due to the existing problems of debt and fiscal policy in recent years) under the circumstances of low efficiency of budget funds and the need to increase security and defence spending and social commitments.

Reaching its maximum in 2015 (14.6% of total amount of state budget expenditures), the level of public debt servicing costs gradually decreased to 9.3% in 2020, increased to 10.4% in 2021 and is planned to increase to almost 12% in 2022. The high cost for internal and external borrowing (in 2009–2010 in some periods reached 30%, in 2021—about 12%) extremely limited fiscal stimulation of the economy, given the prevailing risks to state budget revenues, lack of effectiveness of budget funds and

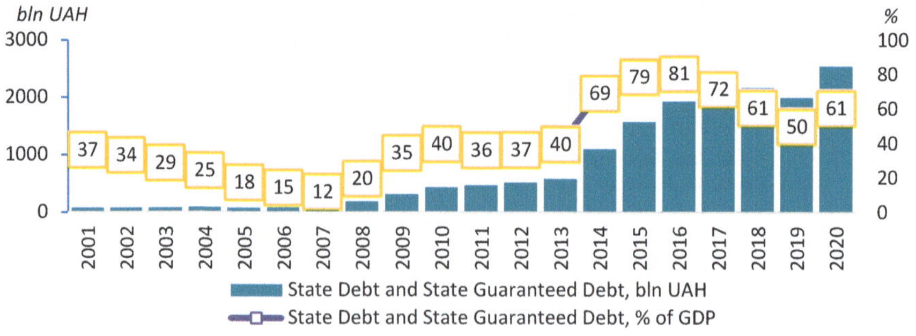

Fig. 10 Dynamics of public and state-guaranteed debt in Ukraine, 2001–2020, % [21]

low opportunities to attract additional financial resources. Thus, in 2020 the cyclically adjusted primary balance in Ukraine amounted to only −1.81% of GDP, while in developed economies it reached −6.78% of GDP. In contrast, in 2014–2019 as part of financial consolidation measures, including keeping the budget deficit low, Ukraine generated a cyclically adjusted surplus [16]. In Ukrainian realities, this indicates the relative rigidity of fiscal policy as a forced measure caused by the accumulated imbalances in fiscal policy and the economy in recent years. In general, this does not contribute to increasing economic growth and causes the accumulation of macroeconomic imbalances in the future, which reduces the economic security of the state.

5 Conclusion

Long-term sustainable non-volatile economic growth on a high-quality competitive basis, provided with sufficient investment and financial resources and focused on strategic structural changes, is a key and most reliable factor in ensuring Ukraine's economic security. The protracted economic recession and the lack of instruments for long-term economic growth arise risks to Ukraine's economic security:

- increasing socio-economic and technological gaps, in particular with the world's leading countries, especially the EU, in the absence of capacity for fixed asset's modernization. The policies of leading countries are focused on encouraging the wider use of advanced technologies, digital reorganization of government, business processes and communications. Other countries have made significant progress in digital and technological transformations in this period, primarily through the financing of infrastructure projects for the production of goods and services on a new basis. Thus, Ukraine is threatened by a double gap due to the lack of funding for the modernization of obsolete fixed assets, as well as increasing the gap from the surge that will make most countries to speed up digital and technological transformation in a COVID-19 crisis;
- critical loss of entrepreneurial potential. Prospects for economic recovery depend on the revised entrepreneurship strategy on the base of creating favourable conditions;
- loss of opportunities owing to the lack of governments incentives for banks which expand lending to priority sectors of the economy, national programs and projects in digital and technological transformations;
- imbalanced public finance system and lack of their significant focus on strategic structural transformations of the economy. The lack of significant prospects for economic growth leads to the risk of reducing the capacity of state and local budgets especially in terms of unstable financial situation of economic entities in increased needs of social spending, the medical sector funding, etc.

For Ukraine the significance of anti-crisis action plan is growing. This plan will help to develop and implement institutional support for investments and entrepreneurship as the most important drivers of long-term economic growth. Clearly defined amounts of financial resources for this plan will increase the opportunities for businesses especially in perception of receiving state financial support and planning their own activities.

Ensuring the conditions for long-term economic growth should be based on the formation of a strategic vision of economic development, effective interaction and coordination of public authorities in the formation and implementation of their own policies and, in particular, include:

- a framework concept to stimulate economic recovery and long-term economic growth (under the new challenges posed by the COVID-19 crisis) to overcome the crisis of low growth, restore consumer and business confidence, increase lending and stimulate investment in the national economy, create jobs and restore entrepreneurship. The creation of this concept will allow to ensure strategic structural transformations, to restore the most affected sectors of Ukraine's economy on an innovative and modernizing basis. Appropriate program to identify the main actions of the Government to increase investment in infrastructure and key sectors; means of overcoming medium- and long-term structural challenges to increase economic growth potential;
- increase of public and private investments in health care, education and infrastructure projects with a high level of return, which will increase the level of economic security of the state. Research spending has the potential to foster innovation and technology, which are key drivers of long-term productivity growth. As well as stimulating the expansion of knowledge, training and retraining of entrepreneurs and employees to ensure their opportunities in the new realities of digital and technological development.

The intensification of long-term economic growth requires a rethinking on the basis of credit expansion. In particular, the approach to the formation of banks' reserves to cover credit risks in Ukraine. This will stimulate the extension of credit terms by introducing a differentiated provisioning rate in accordance with the direction of investment of credit funds. It is necessary to promote the wider usage by banks of savings of individuals as a resource for lending to the economy; to provide state support for the development of small and medium-sized businesses by reforming the banking system to increase the participation of banks in lending to the real sector of the economy.

The formation and implementation of fiscal policy should take into account solving of long-term economic issues. The elements of medium-term budgeting launched in recent years have not yet been supported by the necessary steps in the strategic planning system, and the implementation of budget programs is often reduced to formal implementation of performance indicators rather than effective provision of public services. Therefore, fiscal policy measures should be subordinated to the directions of long-term strategy of economic growth, taking into account the need to increase the efficiency and effectiveness of planning and use of budget expenditures. Strengthening Ukraine's economic security will also help increase the predictability of fiscal and tax policy, increase budgetary discipline, and form medium-term budgetary mechanisms in public–private partnerships to support economically viable businesses and enterprises affected by the crisis.

Attracting significant financial and investment resources to Ukraine's economy should be accompanied by detailed assessments of factors and obstacles that systematically limit the realization of potential for productivity improvements and sustainable long-term growth, as well as identifying ways to prevent and neutralize them.

References

1. European Commission: The European economic and financial system: fostering openness, strength and resilience/communication from the commission to the european parliament, the council, the european central bank, the european economic and social committee and the committee of the regions. 19.1.2021 COM (2021) 32 final. (2021). https://eur-lex.europa.eu/legal-content/EN/TXT/PDF/?uri=CELEX:52021DC0032&from=EN. Accessed 24 Dec 2021
2. Leonard, M., Pisani-Ferry, J., Ribakova, E., Shapiro, J., Wolf, G.: Redefining Europe's economic sovereignty. Issue No 9. Bruegel/Policy Contribution, June 2019, https://www.bruegel.org/wp-content/uploads/2019/06/PC-09_2019_final-1.pdf. Accessed 26 Dec 2021
3. ILO: Definitions: what we mean when we say "economic security". Accessed 20 Dec 2021. https://www.ilo.org/public/english/protection/ses/download/docs/definition.pdf
4. The President of Ukraine: Speech by President Volodymyr Zelenskyy in the Verkhovna Rada with the Annual Address on the Internal and External Situation of Ukraine (2021). Accessed 22 Dec 2021. https://www.president.gov.ua/news/vistup-prezidenta-volodimira-zelenskogo-u-verhovnij-radi-zi-71805
5. Barro, R.J.: Are governments bonds net wealth? J. Polit. Econ. **82**(6), 1095–1117 (1974)
6. Krugman, P.: Will Stagnation Follow the Biden Boom? https://www.nytimes.com/2021/03/08/opinion/biden-economy-stagnation.html. Accessed 10 Jan 2022
7. Aussilloux, V., Baïz, A., Garrigue, M., Martin, P., Mavridis, D.: Fiscal plans in Europe: no divergence but no coordination (2021). https://voxeu.org/article/fiscal-plans-europe-no-divergence-no-coordination. Accessed 14 Jan 2022
8. Gaspar, V., Lam, WR., Raissi, M.: Fiscal Policies for the Recovery from COVID-19 (2020). https://blogs.imf.org/2020/05/06/fiscal-policies-for-the-recovery-from-covid-19/. Accessed 14 Jan 2022
9. Larch, M., Malzubris, J., Santacroc, S.: Economic growth as the ultimate constraint: EU fiscal policies in 2020 (2021). https://voxeu.org/article/eu-fiscal-policies-2020. Accessed 27 Dec 2021
10. Ministry for Development of Economy and Trade of Ukraine: Decree No. 1277 of 29 October 2013 On approval of methodological recommendations for calculating the level of economic security of Ukraine. Accessed 24 Dec 2021. https://zakon.rada.gov.ua/rada/show/v1277731-13
11. The President of Ukraine: Decree No. №347/2021 of 11.08.2021. On decision of the National Security and Defense Council of Ukraine of 11 August 2021. On the Economic Security Strategy of Ukraine for the period up to 2025. https://www.president.gov.ua/documents/3472021-39613. Accessed 22 Dec 2021
12. World Bank Open Data. https://data.worldbank.org/. Accessed 20 Jan 2022
13. The Conference Board: Total Economy Database. https://www.conference-board.org/data/economydatabase/total-economy-database-productivity. Accessed 12 Jan 2022
14. National Bank of Ukraine: Statistics Financial Sector Statistics. https://bank.gov.ua/en/statistic/sector-financial. Accessed 29 Jan 2022

15. IMF: World Economic Outlook Database (2021). https://www.imf.org/en/Publications/WEO/weo-database/2021/October. Accessed 30 Jan 2022
16. IMF: Cyclically adjusted primary balance, % of Potential GDP. https://www.imf.org/external/datamapper/GGCBP_G01_PGDP_PT@FM/ADVEC/FM_EMG/FM_LIDC. Accessed 25 Dec 2021
17. IMF: Fiscal Monitor (October 2021)—Expenditure, % of GDP. https://www.imf.org/external/datamapper/G_X_G01_GDP_PT@FM/ADVEC/FM_EMG/FM_LIDC. Accessed 24 Dec 2021
18. Ministry of Economy of Ukraine: General trends of the shadow economy in Ukraine in January-June 2021. https://www.me.gov.ua/Documents/Download?id=78b79266-3a86-4be0-a528-ef8bb1551809. Accessed 26 Dec 2021
19. The Accounting Chamber: Report of the Accounting Chamber for 2020. https://rp.gov.ua/upload-files/Activity/Reports/2020/ZVIT_RP_2020.pdf. Accessed 29 Dec 2021
20. Ministry of Finance of Ukraine: Credit rating. https://mof.gov.ua/uk/kreditnij-rejting-potochni-rejtingi-zagalna-informacija-istorichni-zmini. Accessed 24 Dec 2021
21. Ministry of Finance of Ukraine: State Debt and State Guaranteed Debt. https://mof.gov.ua/en/derzhavnij-borg-ta-garantovanij-derzhavju-borg

The Impact of Trade Openness on Economic Growth: The Case of Ukraine

Svitlana Tkalenko, Anastasiia Duka, Natalya Sukurova, Liudmyla Kudyrko, and Natalia Litvin

Abstract The first quarter of the XXIst century is characterized by globalization and liberalization trends. For the development of mankind, they are manifested in change, economic, civic and political structures. Today, due to these trends, the economies of the world unite and form a single economic, information space on a global (planetary) scale. Globalization and liberalization are integral attributes of modern development. They help to remove barriers to trade, communication, capital movements, and so on. This is already an objective reality that is difficult to assess and improve; it is an objective process that is conditioned by the development of new types of products, new technologies, means of communication, transport, information, etc. And just in such modern conditions of world economic development countries become more and more open to mutual exchange, and become more and more dependent on each other.

S. Tkalenko
D.Sc. in Economics, Professor, Professor of the Department of European Economics and Business, Kyiv National Economic University named after Vadym Hetman, pr. Peremogy 54/1, Kyiv 03057, Ukraine
e-mail: sv.tkalenko@gmail.com

A. Duka
D.Sc. in Economics, Professor, Professor of the Department of Public Administration and Organizations' Management, Chernihiv Polytechnic National University, Shevchenka Street, 95, Chernihiv 14027, Ukraine
e-mail: apd@online.ua

N. Sukurova (✉)
PhD in Economics, Associate Professor of the Department of Economics, State University of Telecommunications, Solomianska street 7, Kyiv 03110, Ukraine
e-mail: natali11@ukr.net

L. Kudyrko
PhD in Economics, Professor, Professor of the Department of World Economy, Kyiv National University of Trade and Economics, Kyoto str. 19, Kyiv 02156, Ukraine
e-mail: l.kudyrko@knute.edu.ua

N. Litvin
PhD in Economics, Professor, Vice-Rector, KROK» University, Tabirna street, 30-32, Kyiv 03113, Ukraine
e-mail: LitvinNM@krok.edu.ua

A. Hamdan et al. (eds.), *The Implementation of Smart Technologies for Business Success and Sustainability*, Studies in Systems, Decision and Control 216,
https://doi.org/10.1007/978-3-031-10212-7_43

Therefore, in terms of economic growth the questions of influence of openness of economy of our state taking into account world globalization tendencies on economic growth remain actual. This topical research, based on empirical modeling, provides a further conception about the assessment of variables in the constructed model and in the time of Granger's causal relationships in foreign trade openness and economic growth. It is proved that exports have the greatest impact on economic growth. Outpacing of growth rate of imports of goods and services over exports leads to an increase of negative balance of foreign trade, as well as a faster growth of imports of goods and services than exports. The growth of the share of imports in relation to GDP is a negative factor, exacerbated by the imbalance compared to the share of exports in GDP and causes an increase of the share of negative trade balance to GDP. Domestic loans of the financial sector are of great importance for economic growth in current terms of national economic development. Thus, the openness of the economy is directly related with economic growth.

Keywords Open economy · Foreign trade · Real GDP · Domestic credit · Empirical modeling · Globalization · Liberalization

1 Introduction

The process of Ukraine's integration into the world community requires a research of the influence of open trade on the country's economy. The openness of the national economy is associated with the relations that determine the conditions for expansion and the limits of the economic system in the development of international cooperation. The openness of the country's economy, despite its complexity and ambiguity, is mostly considered from the point of view of influence of foreign trade and foreign investments on the economic development of the state.

The purpose of the research is to identify the influence of trade openness on the national economy in the context of globalization of economic development. Based on this goal, the tasks are to research the economic growth of the national economy and identify the impact of indicators of an open economy on it.

An open economy is an economy in which the direction of development is determined by global development trends in the context of globalization, internationalization, the importance of foreign economic relations is growing, and foreign trade turnover reaches the level at which it begins to stimulate overall economic growth. The openness of the national economy plays a significant role in reviving rates of economic growth. Parameters (criteria, indicators) of the level of openness remain a very controversial issue in modern economic science.

Theoretical and applied aspects of openness have been revealed in the works of both foreign and domestic scientists. The impact of trade openness on economic growth remains debatable. Especially for developing countries, such an impact can lead to a decline in economic growth [1].

The relationship between foreign trade and economic growth was researched in the study by Stancheva-Gigov, which showed that open trade is a key component of economic growth at the presence of human capital, investments, institutional quality, education, appropriate level of corruption, population growth rates and government spendings [2]. Such relationship between these determinants is evident. In her research, on the example of an economically developed country, Adeola argues that foreign trade is an economic force of the state, exports and imports are economically beneficial [3]. Such elements as government support, infrastructure, quality of labor, quality of life determine the impact of economic openness on its economic growth. Although international trade contributes to economic growth and global efficiency, it can lead to difficulties for local companies due to the presence of foreign producers.

Daniel Workman in his researches argues that exports, especially export of services, form an increasingly important engine of international trade [4].

The openness of the economy and its relationship with foreign trade has been researched by domestic specialists such as Mazaraki and others in the monograph "Foreign Trade of Ukraine: XXI Century". Foreign trade occupies an important place in the development of Ukraine's economy, and is one of the main components of foreign economic relations of the state [5]. Given the effective usage of comparative advantages, openness is a factor of economic development, which allows to use advanced world technologies and financial resources [6]. From the structure and volume of foreign trade largely depends the possibility of growth of the national economy in general and the well-being of each subject of the economic system in particular [7].

Thus, the research of economic openness through foreign trade in the context of further development of globalization and liberalization tendencies is of great importance. Ukraine has a huge export potential and prospects for economic growth [8–10]. That is why the authors of this article prove the relevance and necessity of empirical research for further economic growth of the national economy.

The relationship between the openness of the national economy and economic growth is based on econometric modeling using the E-Views program.

2 Materials and Methods

This article examines the impact of trade openness on Ukraine's economic growth. Given the use of tools, the usage of E-Views software is relevant to the tasks. Due to this software product, the following tasks can be solved: to analyze scientific information, to model the impact of the most important factors on economic growth, to forecast trends of further economic growth.

A multifactor regression model was built and the hypothesis on the impact of trade openness on economic growth was tested:

$$Y = f(F_1, F_2, \ldots F_n), \quad (1)$$

where, F_n—various influencing factors.

The formation of set of factors influencing on economic growth was carried out on the basis of following principles:

- information accessibility, which means the use of official data (State Statistics Service, Ministry of Finance, National Bank of Ukraine);
- representativeness, which means the includence into the analysis of the most significant and influential indicators on economic growth;
- multiplier, which means that an increase of each factor leades to economic growth. The principle shows the dependence of economic growth from the growth of selected factors for analysis.

The model includes 16 observations, statistics data for analysis were taken and calculated for the period 2005–2020 (Table 1).

The dependent variable of the empirical model is economic growth, which is represented by real gross domestic product (RGDP). The most important independent variables are open trade and financial development. Openness of trade is determined

Table 1 Some indicators of openness of Ukraine economy

Year	Million, USD					% in real GDP	
	Export of goods	Import of goods	Export of services	Import of services	Real GDP	Share of total exports and imports (FT)	Share of domestic loans of financial sector (L)
2005	34,228.4	36,136.3	6443.2	2941.8	27,986.6	110.9	38.9
2006	38,368.0	45,038.6	7791.8	3730.6	48,560.4	97.4	49.8
2007	49,296.1	60,618	9435.1	4995.5	84,528.1	102.7	69.8
2008	66,967.3	85,535.3	12,260.1	6481.5	139,357.2	117.5	95.6
2009	39,695.7	45,433.1	10,129.7	5186.4	92,836.1	93.1	86.0
2010	51,405.2	60,742.2	12,324.2	5467.2	92,346.3	108.6	77.2
2011	68,394.2	82,608.2	11,936.3	5421.6	99,365.4	119.3	70.4
2012	68,830.4	84,717.6	14,180.3	6214.2	101,702.1	106.9	62.5
2013	63,320.7	76,986.8	14,096.2	6650.1	114,752.9	90.6	64.6
2014	53,901.7	54,428.7	14,233.2	7523	85,866.3	113.3	74.7
2015	38,127.1	37,516.4	11,520.8	6373.1	46,253.4	138.8	68.6
2016	36,361.7	39,249.8	9736.6	5523	39,101.6	114.1	49.1
2017	43,264.7	49,607.2	9868	5326.5	38,225.1	117.5	41.6
2018	47,335	57,187.6	10,714.3	5476.1	39,452.6	106.5	34.8
2019	50,054.6	60,800.2	15,618.3	6945.5	37,602.9	93.8	26.4
2020	49,212.9	54,091.3	11,167.0	5209.2	35,180.8	84.5	24.8

Source compiled and calculated by the authors on the basis [15–17]

by the ratio of exports and imports to GDP; financial development is determined by the ratio of domestic loans of the financial sector to GDP. Independent variables are taken into account on the basis of past empirical researches [1, 11–14].

The empirical model is given below:

$$RGDP_t = C(1) * FT_t + C(2) * I_t + C(3) * L_t + C(4)_t \quad (2)$$

where,

RGDP	real gross domestic product;
FT	percentage of foreign trade in real GDP;
L	percentage of domestic loans of financial sector in real GDP;
C	regression coefficients (unknown parameters);
ε	random variable;
t	time, period.

In the empirical model the results of multifactor regression of real GDP growth were analyzed, the presence of autocorrelation of the first and second orders were checked, heteroskedasticity and quality as a linear equation were tested out. The model studied the causal relationships between selected variables and real GDP, analyzed characteristics of model's variables.

Analysis of Table 1 showed that the percentage of foreign trade in real GDP over the period is in average between 90 and 110%. At the same time, the periods 2007–2008, 2010–2012, 2014–2018 have the highest indicators exceeding 100%; the last two years have been characterized by a decline in this indicator and negative trends in the country's foreign trade. The share of domestic loans of the financial sector has declined significantly over the past ten years, reaching the low level of 25%, which is a negative trend for the development of national businesses.

3 Empirical Result and Discussion

We use the multifactor regression method to identify the link between economic growth and open trade. A correlation matrix was constructed on the basis of multifactor regression (Table 2).

The correlation matrix allows to establish the relationship between the selected variables. Based on the matrix data, we have a positive relationship: a strong relationship between the share of domestic loans of the financial sector and real GDP, the correlation coefficient is 84.4%, and a positive weak relationship with a correlation coefficient of 1% between the share of total foreign trade and real GDP. The relationship between the variables is within acceptable limits, which indicates the absence of multicollinearity. It can be assumed that favorable domestic lending is more attractive for economic growth in Ukraine.

The results of the assessment of possible influence of trade openness on economic growth within selected time periods are given in Table 3.

Table 2 Correlation matrix of selected variables

	RGDP	FT	L
RGDP	1	0.0103	0.8443
FT	0.0103	1	0.3302
L	0.8443	0.3302	1

Source authors' development
This table represents a correlation matrix that explains the relationship between selected variables and shows their influence on real GDP. Matrix constructed by us confirms the success of the model

Table 3 The results of multi-factor regression of RGDP

Method: least squares	Sample		Included observations: 16	
	2005	2020		
Variable	Coefficient	Std. error	t-statistic	Prob.
FT	−773.2520	342.4528	−2.257981	0.0418
L	1548.593	218.9299	7.073466	0.0000
C	62,629.85	34,932.86	1.792864	0.0963
R-squared	0.793784	Mean dependent var.		70,194.86
Adjusted R-squared	0.762058	S.D. dependent var.		34,748.39
S.E. of regression	16,950.01	Akaike info criterion		22.48129
Sum squared resid	3.73E+09	Schwarz criterion		22.62615
Log likelihood	−176.8503	Hannan-Quinn criter		22.48870
F-statistic	25.02032	Durbin-Watson stat		1.553718
Prob. (F-statistic)	0.000035			

Source authors' development
This table shows the results of multi-factor regression using the least squares method and different coefficients, which as a whole prove the significance of the equation

Analysis of results in Table 3 multifactor regression of real GDP growth led to the following conclusions:

- the selected variables are statistically significant, as they are within 10% of the significance level: the share of foreign trade in real GDP is 4%, the share of domestic loans of the financial sector in real GDP is 0%; Constanta, in this case it is also significant (9%);
- with a 1% increase of the share of foreign trade, real GDP decreases by 773 million USD. This means that the openness of foreign trade at this level (in most cases more than 100% of real GDP) does not lead to GDP growth. The most effective factor that influences on real GDP growth is the share of domestic loans of the financial sector. With a 1% increase in the share of domestic loans of financial sector, real GDP will grow by 1548 million USD;

- the value of regression R^2 is 79.4% and shows how selected variables are associated with real GDP growth. Adjusted R is 76.2%. This indicates that there is a strong enough link, as there are other quantitative and qualitative factors that affect real GDP growth;
- the probability of accepting the null hypothesis is close to zero (F-statistic = 0.00), which confirms the alternative hypothesis, which indicates the significance of the equation as a whole. According to Fisher's F-statistics, all coefficients of the regression equation do not simultaneously equal zero;
- information criteria, Akaike, Schwarz, Hannan-Quinn are small, confirming the success of the model;
- using the Durbin-Watson test, we check the equation for the presence of first-order autocorrelation. The value of the DW criterion is in the range from 0 to 4. From Table 3 could be seen that this criterion (DW) is 1.5537. Using Durbin-Watson statistics, we determine the critical points d_L and d_U. For the number of observations 16 and 3 variables at a significance level of $\alpha = 1\%$, $0.737 < DW < 1.232$; at the significance level $\alpha = 5\%$, $0.982 < DW < 1.539$. In this model, DW means that there is no reason to reject the null hypothesis (there is no autocorrelation of residues).

The presence of second-order autocorrelation is checked using the Breusch-Godfrey test (Table 4).

Analysis of the table of autocorrelation of the highest order have values of Prob. F (2.11) and Prob. Chi-Square (2) 40.9% and 30.1%, respectively. This is the evidence of absence of higher-order autocorrelation. When using the following lags, we also observe the absence of autocorrelation. Therefore, we accept the null hypothesis.

We test this model on heteroskedasticity using tests: White, Harvey, Glejser, ARCH. The results of the tests are shown in Table 5.

The probability of accepting the null hypothesis according to the White tests is in the range from 5 to 10%, which may indicate the presence of heteroscedasticity.

The next test in the model—Ramsey test, shows the sufficient quality of this model and shows the absence of heteroscedasticity (52.7%). The equation is identified and can be linear (Table 6).

We test this model for explanatory ability. We ascertain in this model to what extent it reflects the dynamics of real GDP growth, i.e. we check it on the explanatory ability (Fig. 1).

The graph shows that simulated values (Fitted) fairly accurately reflect the actual values (Actual), therefore, according to the criterion of explanatory power the model is completely acceptable.

The research of causal relationships between selected variables and real GDP is carried out using the Granger test (Table 7).

The criterion for accepting a hypothesis is the value of the probability of accepting the hypothesis. If Prob. is less than 0.05, then the null hypothesis is not accepted. Feedback is checked at the same time. If two coefficients are statistically significant at the same time, the dependence is two-way or with feedback; may mean the existence

Table 4 Breusch-Godfrey serial correlation LM test

F-statistic	0.969244	Prob. F(2,11)		0.4095
Obs*R-squared	2.397175	Prob. Chi-square(2)		0.3016
Dependent variable: RESID				
Method: least squares				
Sample: 2005 2020				
Included observations: 16				
Presample missing value lagged residuals set to zero				
Variable	**Coefficient**	**Std. error**	**t-statistic**	**Prob.**
FT	123.4989	371.4421	0.332485	0.7458
L	−17.02050	221.4361	−0.076864	0.9401
C	−12,382.48	37,553.23	−0.329731	0.7478
RESID (−1)	0.318602	0.307313	1.036733	0.3221
RESID (−2)	−0.321515	0.286594	−1.121848	0.2858
R squared	0.149823	Mean dependent var.		−8.92E−12
Adjusted R-squared	−0.159332	S.D. dependent var.		15,779.60
S.E. of regression	16,990.25	Akaike info criterion		22.56897
Sum squared resid	3.18E+09	Schwarz criterion		22.81041
Log likelihood	−175.5518	Hannan-Quinn criter		22.58134
F-statistic	0.484622	Durbin-Watson stat		2.062259
Prob. (F-statistic)	0.747020			

Source: authors' development
This table shows the second-order autocorrelation

Table 5 Results of Heteroskedasticity Test

Harvey	F-statistic	3.671213	Prob. F(2,13)	0.0545
	Obs*R-squared	5.775065	Prob. Chi-Square(2)	0.0557
Glejser	F-statistic	4.100895	Prob. F(2,13)	0.0416
	Obs*R-squared	6.189507	Prob. Chi-Square(2)	0.0453
ARCH	F-statistic	0.269649	Prob. F(1,13)	0.6123
	Obs*R-squared	0.304811	Prob. Chi-Square(1)	0.5809
White	F-statistic	2.783989	Prob. F(5,10)	0.0790
	Obs*R-squared	9.311021	Prob. Chi-Square(5)	0.0973

Source authors' development
This table shows the presence or absence of heteroskedasticity

Table 6 Ramsey RESET test

Ramsey RESET test			
Equation: EQ03			
Specification: RGDP FT L C			
Omitted variables: squares of fitted values			
	Value	**df**	**Probability**
t-statistic	0.650822	12	0.5274
F-statistic	0.423569	(1, 12)	0.5274
Likelihood ratio	0.555020	1	0.4563
F-test summary			
	Sum of Sq	**df**	**Mean squares**
Test SSR	1.27E+08	1	1.27E+08
Restricted SSR	3.73E+09	13	2.87E+08
Unrestricted SSR	3.61E+09	12	3.01E+08
Unrestricted SSR	3.61E+09	12	3.01E+08
LR test summary			
	Value	**df**	
Restricted LogL	−176.8503	13	
Unrestricted LogL	−176.5728	12	

Source authors' development
This table shows the quality of the model

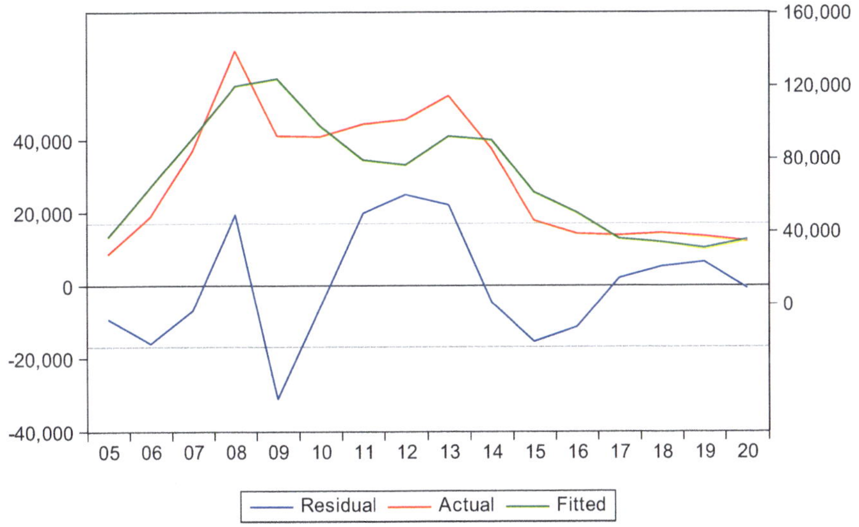

Fig. 1 Explanatory ability of the model. *Source* authors' development

Table 7 Pairwise granger causality test on all the variables, 2005–2020

	The null hypothesis	Lag 2			Lag 3		Lag 4	
		F-Stat	Prob	Conclusion	F-Stat	Prob	F-Stat	Prob
1	FT does not granger cause RGDP	0.3164	0.7365	Accept	0.1473	0.9277	0.2002	0.9229
	RGDP does not granger cause FT	1.1095	0.3709	Accept	0.8831	0.5011	1.0512	0.5039
2	L does not granger cause RGDP	0.2607	0.7761	Accept	0.2650	0.8484	5.3904	0.0989
	RGDP does not granger cause L	0.3425	0.7188	Accept	0.2935	0.8291	7.4028	0.0659
3	L does not granger cause FT	0.6008	0.5689	Accept	0.5025	0.6944	0.2557	0.8896
	FT does not granger cause L	3.1598	0.0913	Accept	1.6896	0.2674	23.616	0.0133

Source: authors' development
This table shows the causal relationships

of a third variable, which is the real cause of changes of those two variables that are represented in the equation.

Descriptive characteristics of the variables are given in the following Table 8.

Table 8 Descriptive characteristics of variables

	RGDP	FT	L
Mean	70,194.86	107.2498	58.43763
Median	66,544.26	107.7464	63.53720
Maximum	139,357.2	138.7918	95.58113
Minimum	27,986.61	84.48186	24.83690
Std. Dev	34,748.39	13.53899	21.17785
Skewness	0.387755	0.367063	−0.071542
Kurtosis	1.871213	3.101170	1.979872
Jarque–Bera	1.250385	0.366118	0.707422
Probability	0.535159	0.832719	0.702078
Sum	1,123,118	1715.997	935.0021
Sum Sq. Dev	1.81E+10	2749.562	6727.519

Source authors' development
This table shows the main characteristics of the selected variables

The largest standard deviation has real GDP, that confirms the influence of external and internal factors on this indicator.

Characterizing the forms of distribution, let's us analyze the coefficients of asymmetry (skewness) and excess (kurtosis). Positive asymmetry coefficients for GDP and foreign trade indicate that in the series of distribution could be values that are higher than the average level (for them the heaviest value is more likely). For the share of domestic loans of the financial sector, in the series of distribution are dominated values lower than the average value (the lowest value is more likely).

The indicator of excess (kurtosis) shows the amplitude of the deviations of variables. If the indicator of excess is greater than 0, then the distribution is acute-vertex and the amplitude is considered significant, if the coefficient of excess is less than zero, the distribution is considered flat-vertex and the amplitude is considered insignificant. The excess of characteristics with a normal distribution is usually in the range from 2 to 4. In our model, all indicators of excess are greater than zero, the distribution is acute-vertex, and are within normal limits. This indicates the compliance of the researched data with the normal distribution law. That is, all the descriptive characteristics of the variables reflect the real GDP with sufficient depth.

We check the model for predictive quality, the MARE criterion is in the range of 10–20, which indicates the normal quality of this model (Fig. 2).

The general form of the model of the dependence of real GDP from the independent variables can be described by the following equation:

Substituted Coefficients:

$$RGDP = -773.252005193 * FT + 1548.59309177 * L + 62629.8549022 \quad (3)$$

The results of the author's model proved its adequacy and high quality.

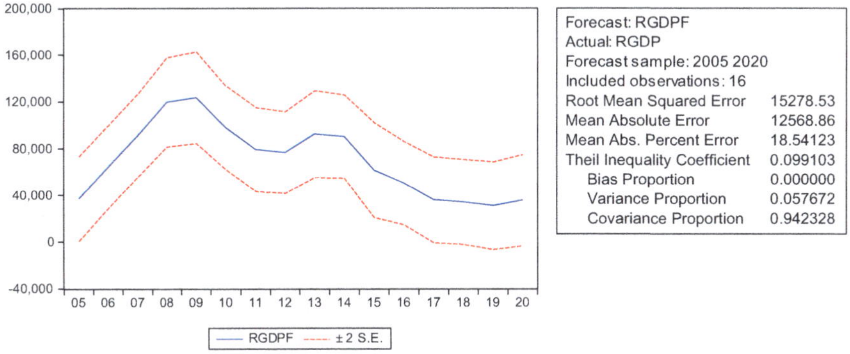

Fig. 2 Forecast. *Source* authors' development

4 Conclusion

In terms of globalization, more and more countries are liberalizing their economies and are entering foreign markets. International trade has the influence on the national economy, which plays a leading role in the country's economic growth. At the same time, the impact of trade openness on economic growth remains controversial, especially for the group of developing countries.

This research analyzed the relationship between economic openness and foreign trade. The relationship between the openness of the national economy and economic growth was more thoroughly made basing on econometric modeling on the example of a country with transition economy—Ukraine. The choice of factors influencing on economic growth, according to the authors, was based on the principles of information accessibility, representativeness and multiplicity. 16 observations were used to build the model. To independent variables, which according to our opinion are the most important, we included trade openness and financial development. In the empirical model, we analyzed the results of multifactor regression of real GDP growth, tested the presence of autocorrelation of the first and second orders, tested for heteroskedacticity and quality as a linear equation.

This model shows the dependence of real GDP from the foreign trade and domestic loans of the financial sector. The coefficients of the equation show the influence of each factor on the resultant indicator with the constancy of other indicators. In foreign trade of Ukraine import is dominated, so an increase in foreign trade under such conditions by 1% leads to a decrease in real GDP by 773 million USD. To eliminate/reduce this value, it is necessary to promote export-oriented sectors of economy. A 1% increase in domestic loans of the financial sector (under normal conditions) contributes to GDP growth of 1548 million USD.

Therefore, by using programs of support and crediting of the real sector of the economy, as well as export-oriented industries in an open economy of Ukraine, we will have economic growth, which will have a positive impact on all economic entities in the country.

References

1. Sook-Yen, L., Mui-Yin, C.: Revising the impact of trade openness on economic growth. Int. Acad. J. Acc. Financ. Manage. **5**(2), 178–188 (2018)
2. Stancheva-Gigov, I.: The impact of foreign trade on the economic growth. J. Sustain. Dev. **5**(11), 52–70 (2014)
3. Adeola, A.A.: The impact of international trade on the economic development of new Hampshire state, USA. Eur. J. Bus. Manage. **12**(20) (2020). Retrieved from file:///C:/Users/User/Downloads/53310-57104-1-PB.pdf
4. Workman, D.: World's Top Export Services (2020). Retrieved from https://www.worldstopexports.com/worlds-top-export-services/
5. Mazaraki, A.A. (ed.): Keiv. In: Ukraine's foreign trade: XXI century. Monograph. 600p (2016). Retrieved from https://knute.edu.ua/file/MjExMzA=/71b98caa3e46dab7dd15e731093dbd4d.pdf

6. Dernova, I.A.: Foreign trade openness of the national economy: a statistical assessment. Eastern Eur.: Econ. Bus. Manage. **2**(19), 410–414 (2019)
7. Yaroshevich, V.: Small open economy in the process of globalization. Belarusian Econ. J. **4**, 37–49 (2016)
8. Romanenko, V.A., Lebedeva, L.V.: Ukraine's foreign trade with the EU: trends, problems and prospects. Econ. Soc. **9**, 71–77 (2017)
9. Lyskova, L.M.: Increasing the export potential of Ukraine as a prerequisite for achieving the goals of regional trade and economic cooperation. Eur. Perspect. **9**, 12–18 (2014)
10. Melnik, T.: Ukraine's export potential: assessment methodology and analysis (2008). Retrieved from http://iepjournal.com/journals/8-9/2008_10_Melnuk1.pdf
11. Tkalenko, S., Melnyk, T., Kudyrko, L.: Macroeconomic factors influencing the development of Ukraine's foreign trade. J. Econ. Stud. **6**(3), 143–156 (2020). https://doi.org/10.30525/2256-0742/2020-6-3-143-155
12. Hafiz, W., Saira, B., Nimrah, H., Arooj, M.: Trade openness and economic growth. Int. J. Econ. Commer. Manage. **4**(3), 180–188 (2016). Retrieved from http://ijecm.co.uk/wp-content/uploads/2016/03/4310.pdf
13. Chernega, O.B., Kozhukhova, T.V.: International trade as a source and factor of sustainable development. Sci. Bull. Kherson State Univ. **12**(3), 32–38 (2015). Retrieved from http://www.ej.kherson.ua/journal/economic_12/115.pdf
14. Tkalenko, S., Sukurova, N., Honcharova, A.: Determinants of the foreign direct investments in terms of digital transformation of the Ukrainian economy. In: Digital Science 2019. DSIC 2019. Advances in Intelligent Systems and Computing, vol. 1114, pp. 148–164. Springer, Cham (2020). https://doi.org/10.1007/978-3-030-37737-3_14
15. State Statistics Service of Ukraine. Retrieved from http://www.ukrstat.gov.ua/
16. Ministry of Finance of Ukraine. Retrieved from https://mof.gov.ua/uk
17. National Bank of Ukraine. Retrieved from https://bank.gov.ua/ua/statistic

Covid-19 and Misinformation Prevalence: A Content Analysis of Fake News Stories Spread in Jordan

Abd Allah Al-Jalabneh, Amjad Omar Safori, and Hatem Shlool

Abstract Since the start of the COVID-19 pandemic social media networks emerged as pivotal means of communication and updates about the pandemic. As the pandemic continued to claim lives, an ample amount of unchecked information was circulating online about the disease, this resulted in citizens facing a wave of fake news not only in Jordan but across the world. This has led governments and health institutions to echo their concerns about the serious threat fake news can cause to citizens' lives. This research study uses a quantitative content analysis to report on the sources and types of fake news spread in Jordan between the period of April 2020 until April 2021 using a data collected by a Jordanian fact checking news website named AKEED. This study is of a vital importance given the scale of concerns resonated by the World Health Organization about facing tsunami of misinformation termed as infodemic. This also comes in a time that is described as "post truth" era. This study concludes that both infodemic and post truth eras hold true in the case of fake news disseminated in Jordan. The study also found that many of these stories are originated from inside of Jordan and that social media networks were the dominant place for these false stories to spread and reach to audiences. The study reveled that fake news stories related to health were the highest number of stories spread in Jordan with the period of this study. Also, the highest percentage of fake news were classified as inaccurate stories with a percentage of 48.40% followed by an intentionally fabricated stories with a percentage of 24.52%.

Keywords Social media · COVID-19 · Fake news · Misinformation

A. A. Al-Jalabneh · A. O. Safori (✉)
Faculty of Media, Zarqa University, Zarqa, Jordan
e-mail: asafori@zu.edu.jo

A. A. Al-Jalabneh
e-mail: ajalabneh@zu.edu.jo

H. Shlool
AL-Ahliyya Amman University, Amman, Jordan
e-mail: h.shlool@ammanu.edu.jo

A. Hamdan et al. (eds.), *The Implementation of Smart Technologies for Business Success and Sustainability*, Studies in Systems, Decision and Control 216,
https://doi.org/10.1007/978-3-031-10212-7_44

1 Introduction

Since the onset of the COVID-19 pandemic in December 2019 millions of deaths occurred worldwide [1, 2]. Informed health information and guidelines from governments and health institutions through various media platforms were in no doubt a key during the pandemic to keep citizens up to date with the fight against the virus. However, this stream of information was also accompanied with a flow of fabricated news stories and misinformation which has not simply misled the public but also claimed peoples' lives [3]. Fake news has also placed a burden on governments and health institutions worldwide [4]. Fake news and misinformation are now reported to be spreading faster than facts; this is what Tedros Adhanom Ghebreyesus, the Director General of the World Health Organization, referred to in one of his televised appearances. He stated, "We're not just fighting an epidemic; we're fighting an infodemic, fake news spreads faster and more easily than this virus and are just as dangerous" [5].

This emphasise the sensitivity and importance of factual information in times of crises when truthful information is being challenged by misinformation. Social media networks and instant messaging applications have facilitated the dissemination of fake news due to their popularity among citizens [1, 6]. Like the rest of the Arab region, Jordan is one of the countries in the Middle East to be socially and economically affected by Covid-19 [7].

Research on fake news is not very recent, nor is the existence of the term itself. False news stories date back to the onset of humans being resided within groups [8, 9]. However, it has recently become a more popular term among politicians and media scholars because of the speed and scale in which false stories can spread due to social media [10]. The term is also frequently used among journalists and flagged up when debating about what is termed as "post-truth" era in which we live in today at the age of social media [11, 12]. The concept of "post-truth" was the 2016 term of the year as chosen by the Oxford dictionaries and usually discussed in the context of politics and mainly become more frequently used during the United Kingdom's referendum regarding its membership in the EU and the presidential elections in the United States [13]. The term post-truth was invented in 1992 by Steve Tesich and essentially means that objective facts are becoming of a secondary importance to people in viewing and weighing information, while subjective opinions are emerging as more relevant in people's lives and the way the view knowledge and distinguish between factual and nonfactual information [14]. Therefore, the concept of "post-truth" becomes more relevant now than ever when discussing about the spread of fake news and misinformation. Fake news can instill misbelieves, confusion and doubt about various aspects of our lives such as swaying elections outcomes or encouraging anti-vaccination beliefs and conspiracies [15–18].

The debate about the existence of fake news and misinformation is not new to media scholars however false information is more concerning at this critical time that humanity lives with the rise of social media usage and the relines on it for news. It is also important to highlight in this regard that the whole world is living in a time of a global pandemic "COVID-19" where factual, reliable and accurate information are at stake. Therefore, the spread of fake news should gain more attention by academics and all actors across the spectrum of media outlets and institutions. This phenomenon needs to be wholly addressed in order to maintain a healthy media environment [19]. Since this research paper is focused on indentifying the various sources of fake news stories spreading in Jordan. The media including its online affiliations such as social media had provided a fertile land where false rumours and misinformation can spread faster than ever before. Thus, this paper provides a quantitative content analysis into the fake news stories collected by a news fact checking websites called *AKEED* in Jordan.

2 Who Makes Fake News During the Pandemic?

Motivations behind the spread of fake news can vary depending on the intentions behind it which could be for example due to ideological or financial gains [13]. However, before initiating the debate on who stands behind creating fake news, a comprehensive definition to fake news is provided in this respect. Allcott et al. [13] defined fake news as "news articles that are intentionally and verifiably false and could mislead readers". Albright [20] has also referred to fake news as a buzzword that emerged during and after the 2016 United States' presidential elections due to the scale of misinformation used in the elections.

Whereas Tandoc et al. [21] have gathered that propaganda, fabrication, advertising, manipulation, news parody, and news satire are all different types of fake news. During the pandemic a video posted on the BBC news platform online presented by Marianna Spring, a reporter and disinformation specialist who underlined the different types of citizens contributing to misinformation online. She highlighted seven misinformation spreader actors in the society which were namely first, the joker who posts funny content, creates memes and shares them with others online. Second, the scammer who utilizes the pandemic to make money through scamming others online such as selling tools, remedies, cures or even provide advice in exchange for money. Third, the politicians who use either their own social media or their state media to deliberately spread false information for their own political gains. Fourth, the conspiracy theorist who does not believe in the pandemic at all and link other factors behind the pandemic such as the 5G theory or even speculating who created the virus. Fifth, the insider who is likely to a hospital worker or a doctor who got things wrong or even do not exist. Sixth, the relative who shares it to their circle of relatives on a social media platform such as WhatsApp. Seventh, the celebrity who amplifies false information to their fans and followers [22].

3 COVID-19 and Fake News on Social Media

As soon as the COVID-19 outbreak stormed the world, people used social media to seek, spread and share timely updates and information about the virus during the pandemic [23, 24]. Social media platforms are open spaces that allow all individuals in the society to participate in the spread of information regardless of their level of expertise or background. As with COVID-19, the harm that fake news can make exceeds creating a panic and fears among the public but extended to reach mistrust in governments, the rise of conspiracy theories, and other forms of misinformation [9, 23].

More evidence suggests that misinformation and falsehoods diffused considerably faster and more broadly than truths on the online sphere. Likewise, people online tend to find false news more novel than true news [25]. Regarding health-related fake news, Chou et al. [26] identified that the lack of scientific evidence in any health-related information is considered misinformation. Cox [27] stated that many people have low health literacy. Thus, information displayed on social media could create a state of information chaos as social media provides an open space for health and non health experts to express their opinions. Alongside the challenges media industries and policy makers confront to curb the spread of fake news on social media, the medical sector had to fight misinformation, disinformation and all sorts of inaccurate health information on social media [23, 28]. Cox [27] indicated that misinformation is not new on social media but what is new is that medical information and science are being politicized during this pandemic over social media. Likewise, health misinformation is aggravated by the effects of echo chambers where information is tailored to individual views, partisan bias, and personal beliefs [26].

4 Method

This study conducted a content analysis method on a data set taken from the Jordanian media credibility monitor *AKEED* within the period from April 2020 to April 2021 which amounted to 624 fake news stories. *AKEED* is considered one of the few reliable fact checking websites in Jordan and the middle east that uses the Arabic language. This fact checking website is run and managed by Jordan Media Institute (JMI). During the period of this study, Jordan has seen many large events happening in the country and worldwide. Jordan also witnessed the closure of many of its private and government sectors due to the pandemic. It has also saw a full curfew and lockdowns on several occasions to contain and curb the spread of the virus. In addition to that, many defense orders were issued and implemented by the government of Jordan to reopen its sectors and economy back to life. The scale of these decisions and events has led to more relines on the media for news updates. The researchers have explored the different types of fake news within the data then classified the data, first, in terms of the sources whether it is coming from inside or outside of Jordan.

Second, the media used in publishing these stories. Third, the type of subjects that these stories are related. Finally, the criteria used in classifying these stories as fake through this research study.

5 Results and Discussion

Table 1 shows that the highest percentage of fake news disseminated during the 12 months period of the study whether from inside or outside sources, was in April 2021 with a total of 80 false story, followed by November 2020 with a total of 60 false stories. Then, September 2020 with 59 false stories. On the other hand, the months that witnessed the least amount of fake news stories where in March 2021 with 35 false stories, then February which witnessed a lower percentage of false stories amounted to 31 stories, then July 2020 which has the least number of false stories amounted to 29 stories. From the data presented in this table we can observe that the majority of fake news disseminated in Jordan are coming from inside of the country with a total of 574 fake news story while the percentage of fake news from outside of Jordan and spreading in Jordan are significantly less than those coming from inside of Jordan amounted to only 50 fake news story.

The month of April 2021 has the highest number of fake news stories spreading from within the inside of Jordan. In contrary, the month of July has the least number of fake stories amounted to only 29 false story. These results pose a concern as many of these fake news stories are produced from the inside of the Jordanian population which reflects the misuse of social media platforms by some Jordanians. Global

Table 1 The source location of fake news

	Inside Jordan	Outside Jordan	Total	Percentage
April 2020	49	0	49	7.85
May 2020	48	3	51	8.17
June 2020	39	4	43	6.89
July 2020	29	0	29	4.65
August 2020	50	3	53	8.49
September 2020	54	5	59	9.46
October 2020	52	1	53	8.49
November 2020	56	4	60	9.62
December 2020	40	4	44	7.05
January 2021	35	2	37	5.93
February 2021	30	1	31	4.97
March 2021	33	2	35	5.61
April 2021	59	21	80	12.82
Total	574	50	624	100

regulations around misinformation online are still a challenge due to the exponential growth of content produced by users online as well as the complex means of communications online which outpaced those of conventional methods of information regulation. Therefore, this may require the Jordanian government to impose some deterrent measures to curb the spread of false news stories from inside of Jordan. Previous research in other countries suggested different actions to be taken to address this phenomenon such as introducing information correction polices or content deletion and blocking or even has gone further to recommend criminality sanction on those who intentionally spread fake news to mislead the public [29, 30]. Other scholarly works of literature signals to introduce media literacy education to tackle the issue of fake news as this solution comes from within citizens self-learning mechanisms to deal with fake news [31, 32].

Table 2 presents the different media sources that published fake news stories divided by traditional media, social media networks and official sources. The table shows that social media networks were platforms that has been used mostly to disseminate fake news during the 12 months of the study period amounted to 421 false story spread on social media. While 198 false news story were disseminated by conventional media such as radio, television, or online newspaper. Whereas only 5 false news stories were spread by official sources. The table also shows that the month of April 2021 has the highest number of fake news stories on both social media with a number of 46 stories while 34 stories were spread by traditional media. The open nature and ecology of social media networks keeps it vulnerable environments for fake news to spread and flourish. Previous research agreed with this result that social media is a more vulnerable place for fake news than any other traditional media platforms [33]. Social media companies have made a noticeable progress to reduce the spread of fake news. However, it is still not enough, and a more holistic approach is still needed to curb the spread of fake news. This approach could map government policy makers, media industry policy makers, social media companies, media educators, as well as online users' feedbacks in this respect.

Table 2 The media source of disseminating fake news

	Traditional media	Social media networks	Published by an official source	Total
April 2020	11	38	0	49
May 2020	13	38	0	51
June 2020	12	31	0	43
July 2020	8	21	0	29
August 2020	20	33	0	53
September 2020	16	41	2	59
October 2020	21	32	0	53
November 2020	16	42	2	60

(continued)

Table 2 (continued)

	Traditional media	Social media networks	Published by an official source	Total
December 2020	14	30	0	44
January 2021	12	24	1	37
February 2021	8	23	0	31
March 2021	13	22	0	35
April 2021	34	46	0	80
Total	198	421	5	624
Percentage	31.70	67.50	0.80	100

Table 3 indicates that the highest number of fake news stories were related to health issues which amounted to 150 story. This result is expected as the country lives through the covid-19 pandemic. The second highest number of fake news stories was related to public affairs with a number of 124 stories. While third, 110 fake news story were related to security, followed by 99 fake news story related to economic affairs, then social 72 and political affairs 69 respectively. This result is expected due to the nature of the period this research was conducted. Many large-scale events related to health policies were implemented in Jordan. The pandemic and its related polices had also touched other fabrics of everyday live and people daily affairs. The concept of fake news was amplified during this time period as fake news affected people on every single level it has even led to misconceptions leading to significant harm [34].

Table 3 The subject area of fake news

	Security	Politics	Health	Economy	Social affairs	Public affairs	Total
April 2020	11	10	11	8	9	0	49
May 2020	8	5	13	6	8	11	51
June 2020	8	6	7	9	5	8	43
July 2020	3	4	2	4	2	14	29
August 2020	8	3	16	9	5	12	53
September 2020	7	4	25	8	6	9	59
October 2020	18	3	14	7	7	4	53
November 2020	15	7	20	9	6	3	60
December 2020	7	5	9	8	9	6	44

(continued)

Table 3 (continued)

	Security	Politics	Health	Economy	Social affairs	Public affairs	Total
January 2021	8	8	3	10	6	2	37
February 2021	7	6	2	9	2	5	31
March 2021	5	2	13	3	4	8	35
April 2021	5	6	15	9	3	42	80
Total	110	69	150	99	72	124	624
Percentage	17.60	11.10	24.00	15.90	11.50	19.90	100

Table 4 presents the criteria used by the researchers to classify the news stories as fake news. It shows that inaccurate information was the criteria in which has the highest percentage of fake news stories (48.40%). Whereas fabricated news stories (24.52%) were the second highest type of fake news stories spread in Jordan. This was followed by fake news that relied on stealing some pieces of information and taking it out of context to support certain agenda (10.90%). After that, news stories that used archival materials (9.46%) which then were refashioned and posted on social media as a new story or event. Lastly, unknown, or unrelated source to the story amounted to 42 fake news story (6.73%). There are no previous studies that have exclusively classified fake news into these categories. Therefore, this categorisation could be useful and used to classify fake news in further research studies as well in order to address this issue in more depth and to highlight the types of fake news that are highly spread in the society so people could be more aware and vigilant towards these types of fake news in the future when they encounter them.

Table 4 The criteria used to classify published news stories as fake news

	Unknown or unrelated source to the story	A piece of information taken out of context	Fabricated information	Inaccurate information	Use of archival materials and reposting it as a new story	Total
Total	42	68	153	302	59	624
Percentage	6.73	10.90	24.52	48.40	9.46	100

6 Conclusion

Fake news crisis reflects several challenges social media networks still struggle to cope with in relation to the authenticity, credibility and accuracy of content posted online. Fake news remains a concern to governments and health institutions worldwide. The challenging danger lies behind the fast nature of the online sphere in disseminating information to a large scale of audiences. Also, citizens' vulnerability to get used to and believe in fake news stories. During the period of this study the scale of fake news was threatening to health institutions as they try to curb the spread of the virus. Health institutions and governments concerns about this phenomenon was exacerbated due to the sensitivity of the time and escalation of many events around the world including inside of Jordan. Jordan has witnessed a wave of fake news stories circulated around on social media and instant messaging applications.

However, the scale of these false stories remains consistent with the global rates of fake news dissemination during the pandemic period. Media literacy and regulations are still required in Jordan to address the problem of spreading fake news and to learn how to deal with it. This research study concludes that social media networks are still the main spreaders of fake news in Jordan. Therefore, this research suggests a comprehensive approach which could link and bring together government officials, decision makers, media institutions, key profession actors such as journalists, social media companies, media scholars, media literacy educators, as well as online users to all participate in finding solutions to deal with fake news.

References

1. Himelein-Wachowiak, M., et al.: Bots and misinformation spread on social media: a mixed scoping review with implications for COVID-19 (Preprint). J. Med. Internet Res. (2021). Available: https://doi.org/10.2196/26933
2. Adam, D.: The pandemic's true death toll: millions more than official counts. Nature **601**(7893), 312–315 (2022). Available: https://doi.org/10.1038/d41586-022-00104-8
3. Wani, A., et al.: Evaluating deep learning approaches for covid19 fake news detection. In: International Workshop on Combating Online Hostile Posts in Regional Languages During Emergency Situation, pp. 153–163 (2021). Available: https://arxiv.org/abs/2101.04012
4. Balakrishnan, V., Ng, K., Rahim, H.: To share or not to share—the underlying motives of sharing fake news amidst the COVID-19 pandemic in Malaysia. Technol. Soc. **66**, 101676 (2021). Available: https://doi.org/10.1016/j.techsoc.2021.101676
5. WHO: Speech by the director-general at the Munich security conference. WHO.int (2022). [Online]. Available: https://securityconference.org/en/msc-2020/speeches/. Accessed: 10-Jan-2022
6. Wang, Y., McKee, M., Torbica, A., Stuckler, D.: Systematic literature review on the spread of health-related misinformation on social media. Soc. Sci. Med. **240**, 112552 (2019). Available: https://doi.org/10.1016/j.socscimed.2019.112552
7. Zeitoun, A.: COVID-19 Pandemic Challenges and Opportunities: The Case of Jordan. euromesco (2020). [Online]. Available: https://www.euromesco.net/publication/covid-19-pandemic-challenges-and-opportunities-the-case-of-jordan/. Accessed: 03 Feb 2022

8. Tandoc, E., Lim, Z., Ling, R.: Defining "Fake News" a typology of scholarly definitions. Digit. Journalism **6**(2), 137–153 (2017). Available: https://doi.org/10.1080/21670811.2017.1360143
9. Al-Jalabneh, A., Safori, A.O.: "Fake News" threats at the time of a global pandemic: Facebook use as a news platform in Jordan. J. Crit. Rev. **7**(19), 9437–9446 (2020)
10. Collins, A., Hoang, D., Nguyen, N., Hwang, D.: Trends in combating fake news on social media—a survey. J. Inf. Telecommun. 1–20 (2020). Available: https://doi.org/10.1080/24751839.2020.1847379
11. Omar Safori, A.: Journalist use of social media: guidelines for media organizations. J. Soc. Sci. Res. 5, 772–779 (2018a). Available: https://doi.org/10.32861/jssr.spi5.772.779
12. McGonagle, T.: "Fake news" false fears or real concerns? Neth. Quart. Hum. Rights **35**(4), 203–209 (2017). Available: https://doi.org/10.1177/0924051917738685
13. Allcott, H., Gentzkow, M., Yu, C.: Trends in the diffusion of misinformation on social media. Res. Polit. **6**(2), 205316801984855 (2019). Available: https://doi.org/10.1177/2053168019848554
14. Grech, V.: Fake news and post-truth pronouncements in general and in early human development. Early Hum. Dev. **115**, 118–120 (2017). Available: https://doi.org/10.1016/j.earlhumdev.2017.09.017
15. Clayton, K., et al.: Real solutions for fake news? Measuring the effectiveness of general warnings and fact-check tags in reducing belief in false stories on social media. Polit. Behav. **42**(4), 1073–1095 (2019). Available: https://doi.org/10.1007/s11109-019-09533-0
16. Germani, F., Biller-Andorno, N.: The anti-vaccination infodemic on social media: a behavioral analysis. PLOS ONE **16**(3), e0247642 (2021). Available: https://doi.org/10.1371/journal.pone.0247642
17. Mamak, K.: Do we need the criminalization of medical fake news? Med. Health Care Philos. **24**(2), 235–245 (2021). Available: https://doi.org/10.1007/s11019-020-09996-7
18. Hernandez, R., Hagen, L., Walker, K., O'Leary, H., Lengacher, C.: The COVID-19 vaccine social media infodemic: healthcare providers' missed dose in addressing misinformation and vaccine hesitancy. Hum. Vaccines Immunotherapeutics **17**(9), 2962–2964 (2021). Available: https://doi.org/10.1080/21645515.2021.1912551
19. Morgan, S.: Fake news, disinformation, manipulation and online tactics to undermine democracy. J. Cyber Policy **3**(1), 39–43 (2018). Available: https://doi.org/10.1080/23738871.2018.1462395
20. Albright, J.: The #Election2016 micro-propaganda machine. Medium (2016). [Online]. Available: https://d1gi.medium.com/the-election2016-micro-propaganda-machine-383449cc1fba. Accessed: 10 Jan 2022
21. Tandoc, E., Lim, Z., Ling, R.: Defining "Fake News". Digit. Journalism **6**(2), 137–153 (2017). Available: https://doi.org/10.1080/21670811.2017.1360143
22. BBC: Who starts viral misinformation? BBC News (2020). [Online]. Available: https://www.bbc.co.uk/news/av/world-52454129. Accessed: 17 Feb 2022
23. Alam, F., et al.: Fighting the COVID-19 infodemic in social media: a holistic perspective and a call to arms. arXiv preprint arXiv (2020)
24. Barua, Z., Barua, S., Aktar, S., Kabir, N., Li, M.: Effects of misinformation on COVID-19 individual responses and recommendations for resilience of disastrous consequences of misinformation. Prog. Disaster Sci. **8**, 100119 (2020). Available: https://doi.org/10.1016/j.pdisas.2020.100119
25. Vosoughi, S., Roy, D., Aral, S.: The spread of true and false news online. Science **359**(6380), 1146–1151 (2018). Available: https://doi.org/10.1126/science.aap9559
26. Chou, W., Oh, A., Klein, W.: Addressing health-related misinformation on social media. JAMA **320**(23), 2417 (2018). Available: https://doi.org/10.1001/jama.2018.16865
27. Cox, S.: COVID-19 era of misinformation: when your family does not trust you, will your patients? J. Am. Pharmacists Assoc. **61**(1), 33 (2021). Available: https://doi.org/10.1016/j.japh.2020.08.020
28. Omar Safori, A.: Social media's impact on a journalist's role. J. Sci. Educ. **19**(1), 148–162 (2018b)

29. Helm, R., Nasu, H.: Regulatory responses to 'Fake News' and freedom of expression: normative and empirical evaluation. Hum. Rights Law Rev. **21**(2), 302–328 (2021). Available: https://doi.org/10.1093/hrlr/ngaa060
30. Hartley, K., Vu, M.: Fighting fake news in the COVID-19 era: policy insights from an equilibrium model. Policy Sci. **53**(4), 735–758 (2020). Available: https://doi.org/10.1007/s11077-020-09405-z
31. Jones-Jang, S., Mortensen, T., Liu, J.: Does media literacy help identification of fake news? Information literacy helps, but other literacies don't. Am. Behav. Sci. **65**(2), 371–388 (2019). Available: https://doi.org/10.1177/0002764219869406
32. Higdon, N.: What is fake news? A foundational question for developing effective critical news literacy education. Democratic Communiqué **29**(1), 1–18 (2020)
33. Yang, J., Tian, Y.: "Others are more vulnerable to fake news than I Am": third-person effect of COVID-19 fake news on social media users. Comput. Hum. Behav. **125**, 106950 (2021). Available: https://doi.org/10.1016/j.chb.2021.106950
34. Patwa, P., et al.: Can self reported symptoms predict daily COVID-19 cases? **2**, 1–8 (2021). arXiv preprint arXiv:2105.08321

Education Management, Technology, Smart Universities, and COVID-19 Impact

Digital Education Technologies and Their Impact on the Quality of Education in Light of the Covid 19 Pandemic of Philadelphia University at Jordan

Marwa Ahmad, Wedad Aqrabawi, and Ahmad Al-Ghoul

Abstract The research aimed to study the effect of using digital educational techniques on the quality of education from the viewpoint of faculty members at Philadelphia University, in light of the Covid19 Pandemic according to academic rank, number of years of experience in teaching at the university, number of training program attended in the field of quality in education and gender. A questionnaire was designed for the study, and this questionnaire consisted of two parts; the first part relates to the personal data of a faculty member, and the second part relates to the importance of using digital educational techniques. Statistical Package for the Social Sciences (SPSS) was used to process the data obtained through means, standard deviations, and percentages. One-way analysis of Variance (ANOVA) was used to study the differences between sample subjects. The study results showed statistically significant differences between the views of the faculty members in the effect of the use of digital educational techniques on the quality of education according to the study's variables. Staff with the rank assistant and associate feel more than the professor that the impact of using digital education techniques on the quality of education is obtained information quickly using Smartphone's while the staff with experience less than five years feel more than the one with experience five years and more than the impact of helps faculty staff to form global relationships. The population comprises all faculties at Philadelphia University members and the sample size was 28 staff members from faculty of Business. It was observed that the female feels the impact of developing study materials that require a specific set of skills compared to the male gender. The study recommended the necessity of paying attention to digital educational techniques and providing the necessary financial requirements for their application.

Keywords Digital learning techniques · Covid19 pandemic · Quality of education · Jordan · Means · ANOVA

M. Ahmad · W. Aqrabawi · A. Al-Ghoul (✉)
Philadelphia University, Amman 19392, Jordan
e-mail: info@philadelphia.edu.jo

A. Hamdan et al. (eds.), *The Implementation of Smart Technologies for Business Success and Sustainability*, Studies in Systems, Decision and Control 216,
https://doi.org/10.1007/978-3-031-10212-7_45

1 Introduction

This century is the century of digital technology par excellence, and despite the development of educational methods, it remained in its main axes: the teacher, the student, and the method of teaching. However, studies have shown that technology in education returns positive results through technology and information dualism, and what the information technology (IT) revolution brought about is an integral part of the educational system.

The digital revolution is transforming the world and helps the flow of information and facilitates taking advantage of new opportunities; digital technology has formed the backbone of daily life for individuals in different places and individuals and "Knowledge" has led to wealth and modernization through "creativity" and "innovation," whether in economic or cultural projects. The value that digital learning using smart devices adds to the educational process the knowledge side and the educational side, and "creativity" has become a competitive industry, which is now called "manufacturing creativity."

We must raise the quality of education to reach the appropriate level of standards in all educational systems by setting a clear vision and goals, providing full responsibility that applies to the quality system so that it can be monitored and controlled through the availability of specific specifications for it. In terms of the importance of using digital education techniques and their effect on the quality of education, this study came to know the extent of using digital technologies on the quality of education in universities [6].

Educational institutions are concerned with building the human personality, as they are a vital tool in human societies, and this is since education is one of the main entrances to the comprehensive development of humanity. These institutions play a critical role in monitoring and enhancing students' engagement in learning, including digital learning materials that have emerged in light of the Corona pandemic. In Jordan, we found that these institutions are unprepared to offer lessons virtually and provide individual support to students. Less than half of the educational institutions have the professional resources needed for faculty members to benefit from digital materials, and this percentage is much lower than the average of the Middle East and North Africa region of 72%.

The results also indicate that most faculty members lack the necessary technical and pedagogical skills to integrate digital resources into their teaching. Therefore, the role of leaders must be reconsidered to achieve integration and harmony with the Covid19 Pandemic and to find new administrative organizations that coordinate between educational institutions and homes. Prepare training plans for digital transformation, determine the skills that must be possessed and practiced, and prepare training plans for digital transformation, in addition to the skills of using communication and digital cooperation applications such as Zoom, Facebook, Skype, and others, verifying digital safety and security, and solving digital problems [7].

2 Literature Review

Aloud and Lina Badr [3] The study aimed to determine the impact of the transition from traditional education to digital.

Altaf [4] explained the procedures followed in applying digital technologies using intelligent devices and their effect on student achievement. The study results concluded that the use of intelligent devices increases the interaction of students and their university achievement.

Amer [6] stated the importance of the study, which lies in the fact that it draws the attention of researchers in educational technology about a solution to problems facing education, and the results of the study were revealed that there is no relationship between the problems of pre-university education and the variables of study in education technology.

Alkhodari [2] attempted to achieve excellence in the education system through digital transformation. Electronic portals must be designed to browse and fast access information and service easily.

Al Shareef and Nayef [1] determined the degree of awareness of digital educational techniques for teachers in Saudi universities, where it was found that there were no statistically significant differences between members of the sample due to the sex factor, and the study recommended the need to employ and use digital educational techniques before the university stage. Al-Zain's [5] study aimed to know the benefits and obstacles of e-learning and showed that teaching methods through digital learning motivate learners to learn and interact with the subject's content well.

3 Significance of the Study

The importance of choosing the subject of research stems from the theoretical and practical importance of digital technologies and their role in the quality of education. This study attempts to know the importance of digital technologies to encourage universities to use and invest in them to achieve comprehensive quality.

The importance of the study emerged from the fact that it relates to a new topic in education through electronic digital education in light of facing the Covid19 Pandemic crisis because it has great importance in confronting and solving the problem faced by the educational process from the impact of the pandemic facing the world.

4 Objectives of the Study

The study's first objective is to answer the question:

What is the impact of using digital education techniques on the quality of education?

The second objective of the study is to know the relationship between the impact of the use of digital technologies on the quality of education and the variables: academic rank, number of years of experience in teaching at the university, number of training programs attended in the field of quality in education and gender.

5 Hypotheses of the Study

1. A statistical indication shows a difference between faculty members in their view on the impact of using digital education techniques on the quality of education in light of the Covid19 Pandemic due to the academic rank at $\alpha \leq 0.05$.
2. A statistical indication shows a difference between faculty members in their view on the impact of using digital education techniques on the quality of education in light of the Covid19 Pandemic due to the number of years of experience at $\alpha \leq 0.05$.
3. A statistical indication shows a difference between faculty members in their view on the impact of using digital education techniques on the quality of education in light of the Covid19 Pandemic due to the number of training programs in the field of quality in education attended at $\alpha \leq 0.05$.
4. A statistical indication shows a difference between faculty members in their view on the impact of using digital education techniques on the quality of education in light of the Covid19 Pandemic due to gender at $\alpha \leq 0.05$.

6 Population and Sample

The study population comprises all faculties at Philadelphia University, and a random sample of faculty members was chosen from the Faculty of Business (28 staff members).

7 Data Collection

The study used two sources of information, the first source is secondary data, and the second source is the primary data collected from the field study through the distribution of the questionnaire. This questionnaire was developed by the researchers and consisted of two parts. The first part included personal information about a teaching staff member, and the second part is 31 variables related to the effect of digital educational techniques on the quality of education.

8 Variables

The independent variables are: Academic rank, number of years of experience in teaching at the university, number of training programs attended in the field of quality in education, and gender. The dependent variable is the impact of Covid 19 Pandemic.

9 Validity and Reliability

To verify the validity and reliability of the questionnaire, we presented the questionnaire to several specialists in Jordanian universities. Their observations were taken, and Cronbach's Alpha was extracted, and it was 0.77; this is acceptable for social and humanitarian studies.

10 Characteristics of the Sample

The characteristics of the sample members are as follows: 10.3% of the samples are teachers, and 53.6% are assistant professors, 50.0% of the sample with experience from 5 years to less than 10 years, 46.4% attended one training program in the field of educational quality, and 85.7% of the sample was male.

11 Results and Discussion

What is the impact of using digital education techniques on the quality of education? To answer this question, means were extracted for all variables, and it was found that the faculty members feel that the impact of using digital education techniques on the quality of education develops mental and cognitive maps of the faculty staff and students, apply smart classes and help to develop knowledge structures, as the mean is 4.8571. On the other hand, the sample members feel low that using digital education techniques on the quality of education develops cognitive skills across the web, as the mean is 2.3214.

12 Test the Hypothesis

1. A statistical indication shows a difference between the faculty members in their view on the impact of using digital education techniques on the quality of education in light of the Covid 19 Pandemic due to the academic rank at $\alpha \leq 0.05$.

To test the hypothesis, we applied ANOVA analysis, and it was found that there are statistically significant differences at $\alpha \leq 0.05$ in staff view of the impact of using digital education techniques on the quality of education in variables related to (enrich digital content through educational media, obtain information quickly using smartphones, use digital and smart resources for exploratory learning, facilitate education for students with special needs, develop faculty staff to become facilitator and monitor and provide information in a digital form). To know which academic rank feel more than others, Scheffe Test is applied, and the result shows that professor and teacher feel more than assistant in variables, (enrich digital content through educational media) Moreover, a teacher, assistant, and associate feel more than a professor in a variable (obtain information quickly using smartphones). Professor feels more than assistant and associate in variables (use digital and smart resources for exploratory learning, facilitate education for students with special needs, and develop faculty staff to become facilitators and monitor and provide information in a digital form).

2. A statistical indication shows a difference between faculty members in their view on the impact of using digital education techniques on the quality of education in light of the Covid 19 Pandemic due to the number of years of experience at $\alpha \leq 0.05$.

To test the hypothesis, ANOVA analysis was applied, and it was found that there are statistically significant differences at $\alpha \leq 0.05$ in staff view of the impact of using digital education techniques on the quality of education in variables related to (apply smart classes, help faculty staff to form global relationships, provide academic and professional development opportunities for faculty staff and gives him a new role and help creative thinking).

To know which years of experience feel more than others, Scheffe Test is applied, and the result shows that the staff with experience less than 5 years feel the importance more than the staff with experience 5 years and more in (apply smart classes, help faculty staff to form global relationships, provide academic and professional development opportunities for faculty staff and gives him a new role and help creative thinking).

3. Statistical indication shows a difference between faculty members in their view on the impact of using digital education techniques on the quality of education in light of the Covid 19 Pandemic due to the number of training programs attended in the field of quality in education at $\alpha \leq 0.05$.

To test the hypothesis, ANOVA analysis was applied, and it was found that there are statistically significant differences at $\alpha \leq 0.05$ in staff view of the impact of using digital education techniques on the quality of education in variables related to (made digital resources available easily through digital libraries, add the teaser on educational patterns and research, use effective communication and smart dialogue via the web and use E-learning platforms in educational activities).

To know which academic rank feel more than others, Scheffe Test is applied, and the result shows that staff who did not attend any training program feel the importance more than the staff attending one training program in (made digital resources available easily through digital libraries, add the teaser on educational patterns and research, use effective communication and smart dialogue via the web and use E-learning platforms in educational activities).

4. A statistical indication shows a difference between the faculty members in their view on the impact of using digital education techniques on the quality of education in light of the Covid 19 Pandemic due to gender at $\alpha \leq 0.05$.

To test the hypothesis, T-test analysis was applied, and it was found that there are statistically significant differences at $\alpha \leq 0.05$ in staff view of the impact of using digital education techniques on the quality of education in variables related to (use to develop study materials which require a specific skill, consider to be an educational medium where the interactive writing boards are used, applied cloud computing as a technical educational environment, increase social and educational interaction between faculty staff and students, facilitate education for students with special needs and connect faculty staff and students with the latest scientific developments available online).

To know the differences between staff, mean was calculated, and it was found that the male feels more than the female that the impact of using digital education techniques on the quality of education is considered to be an educational medium where the interactive writing boards are used, applied cloud computing as a technical educational environment, increase social and educational interaction between faculty staff and students, facilitate education for students with special needs.

On the other hand, females feel more than males that the impact of using digital education techniques on the quality of education is developing study materials that require a specific skill and connecting faculty staff and students with the latest scientific developments available online.

13 Conclusion

Professors and teachers feel more than assistants that the impacts of using digital education techniques on the quality of education are enriching digital content through educational media. Likewise, teachers, assistants, and associates feel more than professors that the impact of using digital education techniques on the quality of education is obtaining information quickly using smartphones.

Staff with experience less than 5 years feel more than the one with experience 5 years and more that the impact of using digital education techniques on the quality of education are applying smart classes, helping faculty staff to form global relationships, providing academic and professional development opportunities for faculty staff and gives him a new role and help creative thinking.

Staff who did not attend any training program feel more than the staff attending one training program that the impact of using digital education techniques on the quality of education is made digital resources available easily through digital libraries, add the teaser on educational patterns and research, use effective communication and smart dialogue via the web and use E-learning platforms in educational activities.

Females feel more than males that the impact of using digital education techniques on the quality of education is developing study materials that require a specific skill and connecting faculty staff and students with the latest scientific developments available online.

Several names and terms have emerged related to the current methods of education, such as e-learning, distance education, and education through the mobile phone, and all of these names are about the use of digital technology in the education process.

14 Limitations of the Study

1. The study is limited to the Faculty of Business at Philadelphia University in Jordan.
2. The study was implemented in a short time during the first semester of the academic year 2021–2022.
3 The study uses a questionnaire designed by the researchers.
4. The study was limited to knowing digital education technologies and their impact on the quality of education in light of the Covid 19 Pandemic.

15 Recommendations

The study recommended the followings:

1. Encouraging faculty and students to use digital education techniques.
2. Conducting training courses for faculty and students in the field of digital educational techniques.
3. Provide technical support for digital learning for staff, such as providing devices and services.
4. Train the staff in educational institutions to gain skills in digital learning.
5. Encourage female teaching staff to employ digital learning techniques effectively.
6. Pay attention to digital educational techniques, and provide the necessary financial requirements for their application.

16 Future Studies

Researchers and academics should continue to study the impact of digital educational techniques on the quality of education to raise the competence of teachers and students.

Expand research to include universities, community colleges, and schools.

References

1. Al Shareef, B., Nayef, M.: The extent of awareness of the digital and intelligent educational techniques of Saudi university faculty members and their attitudes towards them. J. Coll. Educ. Al-Azhar Univ. 179, Part 1 (2018)
2. Alkhodari, B.: The digital technological role in achieving the distinguished leadership of the education system. In: The First Regional Conference for Developing Leadership in the Light of the Digital World, Kuwait (2019)
3. Aloud, M., Lina Badr L.: The impact of switching or the impact of switching from traditional to digital education on the teacher in light of the corona pandemic Covid 19. Jerash Res. Stud. J. **22**(1), Article 14 (2021)
4. Altaf, I.: The effect of digital learning using smart devices on the academic achievement of students in the educational aids course and their direction towards the use of smart devices in learning and education. Umm Al-Qura Univ. J. Educ. Psychol. Sci. **10**(2), Part 1 (2019)
5. Al-Zain, S.: The transformation to the digital learning era: cognitive progress or systematic regression. In: The 11th International Conference of the Center for Generation of Scientific Research on Learning in the Era of Digital Technology—Lebanon (2016)
6. Amer, A.M.: The digital divide in education technology research between reality and expectations (2019). https://www.new-educ.com
7. Putri, M.N., Kristiawan, M., Rohana, R.: Mapping principal's competency of secondary school. Int. J. Educ. Rev. **2**(2), 151–159 p152 (2020)

Online Academic Cheating in the Twenty-First Century

Zorah Abu Kassim and Chiam Chooi Chea

Abstract The issue of poor ethics and integrity among students as well as in the working communities is an alarming issue that needs to be addressed. The value of ethics can be built and instill during the school days especially during the tertiary level because entering the workforce will be their next milestone. Nevertheless, the issue of online academic cheating has been rampant in most universities, mostly due to the advancement of technologies and there are various creative methods to cheating these days and these students have improvised from the usual traditional cheating methods. Traditional classroom learning also has transcend to online learning these days due to COVID-19 pandemic that swept across the world since end of year 2019 till today. This chapter will discuss the concept of online learning, definitions of academic cheating, reasons for online cheating by students, various online cheating methods as well as the ways to mitigate the problems of online academic cheating among students in the twenty-first century.

Keywords Academic online cheating · Ethics and integrity · Technology

1 Introduction: Online Education

UNESCO [1] describes Open Distance Learning (ODL) as a refined form of education that increases accessibility and flexibility. Online education as defined by Allen and Seaman [2], consists of 80% or more of course content being delivered online. On the other hand, hybrid or blended learning has 30–70% of a course content being delivered online.

Z. A. Kassim (✉)
Arab Open University, Bahrain Branch, Building 890, Street 3220, Block 732 A'ali, Manama, Kingdom of Bahrain
e-mail: zorah.kassim@aou.org.bh

C. C. Chea
Open University Malaysia, Block C, Kompleks Centre Point 17, Jalan SS7/19, Kelana Jaya, 47301 Petaling Jaya, Selangor, Malaysia
e-mail: chooi_chea@oum.edu.my

A. Hamdan et al. (eds.), *The Implementation of Smart Technologies for Business Success and Sustainability*, Studies in Systems, Decision and Control 216,
https://doi.org/10.1007/978-3-031-10212-7_46

ODL is an additional option with higher degree of flexibility available for learning to take place. In addition, this has allowed widening accessibility to education specially to working adults and the physically impaired. In other words, online learning takes place over the internet with the use of devices such as mobiles, desktop or laptop computers, tablets and other gadgets. Online learning can be also known as e-learning, one of the distance learning modes of instructions. E-learning is whereby there is a teaching and learning process that takes place regardless of physical distance between the teacher and students. Hence, a classroom is a virtual classroom instead of a physical brick-and-mortar classroom.

1.1 Current Developments in Online Education

Technology advancements with the introduction of the World Wide Web in the late 1970s in particular is the main reason for changing learning pattern in terms of what and how a person can learn. ODL has been stimulated due to the growth of ICT (Information Communication Tools) which has encouraged ODL institutions to integrate these technologies and by using instructional design to further propel and encourage learning. Therefore, this has consequently led learners from a passive mode in traditional learning settings to become active learners. Nevertheless, the success of using such a mode relies heavily on the efficacy of the learning system, stability of the internet as well as effort, willingness and readiness to learn from the learners. Rather, a poor or unstable learning system can hinder and become an obstacle to the learning process rather than providing the support as claimed. Both educators and learners face challenges via ODL as a medium of instruction or delivery.

Due to current pandemic COVID-19 crisis, there have been high expectations from ODL to fill this void as many students are not allowed to go into schools and HEIs due to social distancing and infectious nature of the COVID-19 virus.

1.2 Definitions of Cheating

Examination cheating is an academic dishonesty behavior. A student can intentionally or unintentionally act in such a way as to improve exam conditions compared to what was intended. Cheating can be when one student helps another candidate to obtain an undue advantage at an exam. Cheating is an action or behavior where one committed a dishonest act in order to complete a task or to do something. Cheating in this context will focus in academic cheating, which includes a misconduct such as attempting to use unauthorized information in the completion of an exam or assignment; submitting as one's own the work materials prepared by another person, knowingly assisting another student in obtaining or using unauthorized information or materials, breaking rules of examinations and plagiarism.

E-cheating is digital cheating is a term used to describe students who find a way to cheat using computer technology. One specific form of digital cheating is "e-cheating" which specifically relates to the use of the World Wide Web to assist with cheating [3].

1.3 Types of E-Cheating

There are several types of examination cheating such as, (1) sending a proxy-test taker; (2) using unsanctioned test aids such as notes, books, photos; (3) copying other's work; (4) Obtaining outside help such as video calling friends in the washroom, using of drones; (5) prior knowledge of the test contents and (6) tampering with the examination results. Any attempts of conducting examination cheating are subject to penalty and punishment regardless whether the action carried out is successful or not and also whether the students has the intention or did not have the intention for example accidentally bringing in any unauthorized items to the examination hall. There are specific types of cheating that can be explained in the sections below in the context of E-cheating specifically.

1.4 Examination Cheating

Examination cheating refers to an act of dishonesty committed by students during an exam. Examination cheating occurs when students bring in notes, whether in hardcopy or softcopy, talking or obtaining answers from other students during the examination, peeking and looking at the other students' answers, sitting or taking the exam on behalf of another person and/or having prior knowledge on the examination questions.

1.5 Plagiarism

Plagiarism is an act of abusing the work of an original where using, either in whole or in part, the contents of another person's work in one's own work without acknowledging the source of the original work. Actions such as "copy and paste" and purchase of third-party work are examples of this act. "Copy and paste" is a term used where one copies a sentence, paragraph, diagram, tables, graphs from a certain source and later paste it in their work and later when submitted the write-up claiming to be their own work without any in-text citations and/or proper referencing.

1.6 Using Unauthorized Information in the Completion of an Assignment

There are certain materials with copyrights, which require the permission from the author or the publisher to re-use those materials. Failure to obtain permission is known as using unauthorized information for a report or assignment. This is also a form of plagiarism.

1.7 Submitting as One's Own Work Materials Prepared by Another Person

This is an act where one uses the work of another and claim as their own. This can happen among friends or course mates and senior students. At times the work belongs to a senior who has submitted the work at an earlier semester and then part of or the entire work is used again by another student in a later semester.

1.8 Knowingly Assisting Another Student in Obtaining Answers to Questions

This happen usually when there is an examination, where students would use gadgets such as mobile phones or tablets, smart watches and drones to obtain the answers to the exam questions which are outside the examination venue. This is considered as a type of electronic cheating. Other than that, it can happen in a computer lab where certain software is needed and students can exchange answers using chat box tools.

1.9 Purchase of Academic Work

The service of doing assignments for students due to the high demand for such service that this has developed into a business offering such services to students. Students can pay a certain amount of fees in return for a completed assignment. Hence the purchase of completed assignments done by professional services offered on-line has grown exponentially. Therefore, it is important to understand or to have at least an overall picture on why students tend to cheat.

1.10 Collusion

This type of cheating happens when students share answers and communicate with other students through physical contact or through telecommunications as would happen in an online exam for example. Student tend to deliberately communicate to the students verbally or passing message using various stationeries.

2 Preventive Strategies to Deter Cheating Online

Preventive measures must be taken to detect and prevent such immoral act and it has to be sanctioned at all academic levels. Such strategies are mentioned below.

2.1 Student's Handbook

The awareness about the types of academic dishonesty and the consequences of committing academic dishonesty has to be stated in the students' hand book for every education institution. This information can be disseminated to the students during student orientation at the start of new intake of students for each academic year.

2.2 Conducting Relevant Workshops

Education institutions can organize various relevant workshops for the students to improve their time management skills. Plagiarism workshops can also be conducted. Plagiarism workshops are useful because such workshops give a hands-on approach to students on how they can circumvent plagiarism by proper referencing and in-text citations according to the different style formats adopted by their education institution.

2.3 Counseling

Pressure is one of the main factors for students to conduct academic cheating. At least one licensed and professional counselor should be appointed in each education institution to provide proper guidance to those students in need. The counselor needs to work together with the Student Support Services Department. The counselor can act as a listening ear to many students who under tremendous pressure and offer students strategies on coping with student life. The counselor can also offer advice and guidance to promote the right ethical values.

2.4 Software to Detect Plagiarism

Educational institutions have to invest in anti-plagiarism software to detect such work submitted by the students because students may argue that they did not commit such act and proof of such act is needed in order penalize such students. There are well-known anti-plagiarism softwares in the market such as Turnitin, Vericite, Grammarly, No.plag.com, Dustball, Plagtracker and many others.

2.5 Verbal Warnings

Students can be given strict verbal not to cheat during examinations when exams are being proctored. This to instill fear in students that there will be severe punishment if they are caught cheating. In the case of e-cheating, before examinations begin, where there is online proctoring then this can verbal warnings given on the dire consequences if students are caught.

2.6 Strict Writing Guidelines

Sterngold [4] once used paper strengthening techniques which actually worked well as anti-plagiarism techniques. Sources to writing papers are related to a class discussion or book chapter instead of students being able to pick their own topic [5].

2.7 Guiding Students in Doing Proper Citations

In addition to plagiarism workshops, students need guidance either from academic staff and/or Library on referring to information for their assessments such as project paper, reports, assignments and presentations. Students need to read, understand and then articulate these information and knowledge gained into their written work. Students need to be aware that marks will be deducted for high percentage of plagiarism work submitted by students. This can be usually written at the start of the assignment question and students need to be reminded of this by their tutors.

2.8 A Well-Designed Course

Having a well-designed course will consist of having the assessment structure of two types of assessments; Formative and Summative Assessments. Formative assessments are assessments that require continuous feedback and interaction that happens between the instructor/tutor with the students throughout the course. Examples of formative assessments are in-class discussions, (blogs, forums, WIKIs), observations, weekly quizzes, 1-min reflection assignments, homework assignments, student feedback when students answer tutor's questions in a classroom, conferences and discussions between tutors and students. While summative assessments are assessed at the end of the course in which students are usually evaluated on the whole course contents and materials that they have learnt throughout the semester. Good examples of summative assessment are one major end-of-term exam or final exam, major assignments, term papers, project papers and course portfolios. Most significant is that students need to follow the course throughout consistently which means that the student needs to be able to participate actively in all of the formative assessments.

Formative assignments can be done in such a way that these assessments lead to a progression in learning representing learning blocks that students need to have in order for them to tackle their summative or final exam. Thus, formative assessments act as scaffolding in the learning experience for the student.

With summative assignment or final exam represents the culmination of resources and knowledge obtained and gained from various formative exercises/assessments done previously. So, for instance a student who did not attend or is not actively participating in online forums might not be able to use those resources to answer the summative assignment or final exam. Active participation does mean that students need to post questions, replies and share information that they have collected and read. This active participation requires cooperation, collaboration and an exchange of ideas and information. When this is achieved then there are learning communities formed in which knowledge is not acquired but rather through knowledge is constructed based on student's own experiences and the tutor role is relegated to that of a facilitator rather than a mere disseminator of information. Hence, the use of synchronous tools such as Blackboard, Virtual Classroom and chatrooms where classroom participation is mandatory needs to be combined with asynchronous tools such as online forums. Synchronous tools can use the camera and voice to engage learners and make them participate in classroom discussions. The learner then has the driver's seat. This is another way that ensures the student attends and therefore, is the actual student in the class and not someone else masquerading as the student. This misdemeanor is a syndrome called "ghost students" or imposter students.

Hence, a tutor role is to monitor and guide students. A tutor should have an appropriate number of students allocated to each tutor, for example, no more than 15 students should be allocated for each tutor. This enables monitoring by tutor of postings on forums, blogs, journal entries and participation by students. Tutors are also able to give constructive feedback to individual student especially in essay-based assignments or tutor marked assignments (TMAs).

The ratio of tutors versus students therefore, gives rise to teaching resources in an educational institution needs to be sufficient and adequate for this purpose. This is a major constraint for many educational institutions as teaching resources per ratio to students are not sufficient for the total population of students. There needs to be adequate ratio of tutor versus students and this adheres to many quality accreditation rules and regulations governing many educational institutions globally. A well-designed course is therefore, paramount to minimizing e-cheating.

2.9 Learning Approach by Tutor

Tutor's approach to learning needs to change to constructivism whereby the learner learns from prior knowledge, learning through understanding and social interaction process. This will make students as active and independent learners. Therefore, if the constructivism approach to learning is embraced by tutors this should make assessments not just a tool of evaluation of academic performance but most importantly, a learning experience for students.

2.10 Online Proctoring

Online proctoring can take to try and minimize online cheating by using a camera and webcam recording software. A webcam can be used by a Web-based proctor, for example. Software that is often used for online proctoring is Remote Proctor now (RPNow) and ProctorU. Such software usually requires a computer webcam and an Internet connection.

2.11 Define Cheating and Encourage Honesty

Students should read and sign a policy on honesty at the beginning of the course. For younger students this is seems to be fine but for older students, ethical principles may be difficult to change [6]. Many students are not clear on examination instructions especially if exams are online for example, what resources could be used in take-home exams. Failure to recognize cheating may reflect institutions failure to clarify ground rules with students rather than deliberate acts of sabotage by students [7].

2.12 Time Limits

Firstly, many packaged courseware products, such as Blackboard and WebCT, have the ability to set availability dates and times for all assessments. Time limits and the number of permissible accesses can also be set by the instructor. Tests and quizzes can be set for appropriate time limit that could deter cheating as in a face to face environment. When appropriate time limits are in place students understanding the contents can be successful. For example, an exam set to be completed within 24 h of start time, a student will have to retake the exam within the exam due date limit and not afterward [8].

2.13 Question Banks

Many packaged courseware products have the capability of creating large questions pools for randomized assessments. Randomized questions pools are an excellent tool since they ensure that no two students will take exactly the same assessment [9]. Question banks can be created as a repository of exam questions in which to draw test questions. This is especially true for Multiple Choice Questions (MCQs). A tutor can draw out 50 MCQs to generate a random 20 MCQ quiz which would have a million of unique combinations. Thus, educational institutions need to create Question Banks for course contents for courses offered to students. For example, in Open University Malaysia, the development of question banks have been initiated as assessment structure for many undergraduate courses have been on MCQs and a summative final exam.

2.14 Appropriate Design and Value of the Test

An online assessment worth half of a course grade may entice students to cheat, whereas an online assessment worth a quarter of a course grade may not invoke the same response. Permitting only one question to be viewed at a time and preventing backtracking will prevent students from going back to questions if they finish early and copy them. Having the test questions drawn randomly from a larger pool of questions which are stored in the online test folder is a common practice. Students end up with different overall tests and questions which appear in different orders, making collusion (sitting together to do the test) more difficult [10–13], particularly if there is a time limit on each question. Questions which encourage critical thinking rather than knowledge regurgitation are also more effective as they are harder to cheat on [14]. The benefits of using these measures to increase the security of the online assessments means the test can be made available over a set time period, for example, 24–48 h. This ensures all students, regardless of geography or time zone, are able to find a time to complete the online assessment, an important factor in promoting flexible learning [15].

3 Conclusion

Online cheating is an early sign of poor ethics and integrity among students in their academic life and this could continue to their working lives when they begin work as corruption and bribery has occurred in many corporations. Academic cheating is sign that should not be ignored and needs to be addressed soonest possible by most education institutions. Education is not just about gaining qualifications, but is also a process where students can learn to have good moral values while performing academically well. The perception of education needs to be changed and viewed in a different manner by society at large.

References

1. UNESCO: Open and distance learning trends, policy and strategy considerations. UNESCO, Paris (2002)
2. Allen, I.E., Seaman, J.: Grade level: tracking online education in the United States. Babson Survey Research Group and Quahog Research Group, LLC (2015)
3. Rogers, C.F.: Faculty perceptions about e-cheating during online testing. J. Comput. Sci. Coll. **22**(2), 206–212 (2006)
4. Sterngold, A.: Confronting plagiarism: how conventional teaching invites cyber-cheating. Change: Mag. High. Learn. **36**(3) (2004)
5. Moten, J.Jr., Fitterer, A., Brazier, E., Leonard, J., Brown, A.: Examining online college cyber cheating methods and prevention measures. Electron J. e-Learn. **11**(2), 113–146, (2013). www.ejel.org
6. Rowe, N.C.: Cheating in online student assessment: beyond plagiarism. Online J. Distance Learn. Adm. **7**(2) (2004)
7. Collins, K., Oliver, S.W.: Why students might find it difficult to recognize "cheating". Letters, cheating in medical school exams. BMJ **351**:h4770 (2015). https://doi.org/10.1136/bmj.h4770
8. AlHamdani, W.A.: Online cheating: a prevention procedure. In: Proceedings of International Conference on Computer Science (FCIS'18) (2018)
9. Olt, M.: Ethics and distance education: strategies for minimizing academic dishonesty in online assessment. Online J. Distance Learn. Adm. **5**(3) (2002). https://www.westga.edu/~distance/ojdla/fall53/olt53.html
10. Maguire, M., Ball, S., Braun, A.: Behaviour, classroom management and student "control": enacting policy in the English secondary school. Int. Stud. Sociol. Educ. **20**(2), 153–170 (2010). https://doi.org/10.1080/09620214.2010.503066
11. Prince, D.J., Fulton, R.A., Garsombke, T.W.: Comparisons of proctored versus non-proctored testing strategies in graduate distance education curriculum. J. Coll. Teach. Learn. (TLC) **6**(7) (2009)
12. Trenholm, S.: A review of cheating in fully asynchronous online courses: a math or fact-based course perspective. J. Educ. Technol. Sys. (2007). https://doi.org/10.2190/Y78L-H21X-241N-7Q02

13. Underwood, J.: The impact of digital technology: a review of the evidence of the impact of digital technologies on formal education. BECTA (2009). https://dera.ioe.ac.uk/10491/3/A9RF934_Redacted.pdf
14. Harper, M.: High tech cheating. Nurse Educ. Today **26**(8), 272–679 (2006)
15. Ladyshewsky, R.K.: Post-graduate student performance in 'supervised in-class' vs 'unsupervised online' multiple choice tests: implications for cheating and test security. Assess. Eval. High. Educ. **40**(7), 883–897 (2015). https://doi.org/10.1080/02602938.2014.956683

Distance Learning to Ensure Educational Sustainability in the Era of COVID-19 in Kingdom of Bahrain

Fahad Khaled Alkhaldi and Suad Altaei

Abstract Today, the entire world is experiencing the Covid-19 pandemic known as (Covid-19). The repercussions of this virus struck all countries and affected the economies of the entire world. It is a disease that affects humans, and its symptoms vary on human health, some of them are mild and some are severe, which in some cases may lead to death. However, today we are not only talking about the repercussions of the pandemic in terms of health only, but it has affected all important sectors in our lives, such as social, economic, environmental and education as well. Education is a right that must be guaranteed and always sustained despite all circumstances. It is the basis of renaissance and is one of the most important goals of sustainable development (Goal No. 4). This paper emphasizes the importance of education in times of crisis, and it also presents the unique experience of the Kingdom of Bahrain, which emerged among countries due to its rapid response to the crisis by switching to the distance education process, which reflected the reality of Bahraini digital education. Bahrain experience was analyzed through the application of SWOT Analysis, which exposed the strengths and weaknesses in education. In the end, preparing for crises is a must, by keeping abreast of the latest developments in digital technology and artificial intelligence, and preparing the necessary infrastructure and training to hone the skills of the academic cadre and students as well. Governments should benefit from their experience and the experiences of other countries to better prepare to face future crises to ensure the sustainability of all aspects of human life.

Keywords COVID-19 · Pandemic · Sustainable development · Distance education · Digital education · Kingdom of Bahrain

1 Introduction

In a blink of an eye, our lives have been turned upside down. The present time will be recorded in history and may even be taught in schools to future generations.

F. K. Alkhaldi · S. Altaei (✉)
Environment and Sustainable Development Program, College of Science, University of Bahrain, Sakhir, Kingdom of Bahrain

A. Hamdan et al. (eds.), *The Implementation of Smart Technologies for Business Success and Sustainability*, Studies in Systems, Decision and Control 216,
https://doi.org/10.1007/978-3-031-10212-7_47

Today, all countries around the world are united with one purpose: Control and cure the Corona-virus disease known as (COVID-19). A disease which originated in China and quickly spread to the rest of the world [1]. Moreover, it has attacked the economy of northeast Asia and has affected the rest of the world's economies, without exception [2]. This virus, although for most people causes only minor illness, it can make some people extremely sick, and it can also cause deaths [3]. UNESCO reported that an estimate of more than one billion learners and students has been suffer from the closing of the schools during the pandemic [4].

Bahrain is one of the countries that has been affected by the virus and has gotten its share of human infections. As a result, the Kingdom of Bahrain is suffering the consequences of this disease not only in the health sector, social, economic, but also it has a severe or critical effects on the education sector [5]. There is no doubt that education is one of the rights of all people, and it must be secured and ensured that it continues even in times of crisis. From that point of view, governments should strive to create a resilient learning environment that is safe, encouraging, applicable, and that it is undoubtedly environmentally friendly. In some countries, the lack of advance planning and preparation for crises has caused shock and tension between students and faculty members, whether on a personal or professional level, as the distance education process requires prior preparations and training that goes along with digital education [6]. Therefore, the aim of this study is to explore the governmental efforts in the education sector during the pandemic by delivering friendly learning medium for students and teachers wherever they are and whenever the need is to face any future crises. Also, to shed the light on the importance of education even in the time of crisis and analyze the unique educational experience of the Kingdom of Bahrain within the Coronavirus era by analyzing the current situation.

The authors conducted (SWOT) analysis to provide a comprehensive overview of the current status of the education sector in the Kingdom of Bahrain while taking into consideration the Coronavirus pandemic. Furthermore, it will provide a productive frame-work in order to be able to track the risks that may affect the progress of the educational process and contribute to making the right decisions to support Bahraini society and its government. Therefore, this paper is divided into five parts; the content of each part will be briefly discussed below:

- Part one: gives an overview of the whole study, starting with the background information, statement of the problem, aim, and finally the structure of the study.
- Part two: presents a literature review of education in the Kingdom of Bahrain.
- Part three: presents a literature review of the fourth goal "quality education" listed among the 17 international goals of sustainable development of the United Nations and in the Bahrain's 2030 vision.
- Part four: presents the research methodology used in the study, and shows analytical instrument used to analyze the data collected about education and its continuity during the pandemic.
- Part five: summarizes the findings and the conclusions of the study.

2 Education in the Kingdom of Bahrain

Everyone should have the right to an education. The Kingdom of Bahrain is one of the first countries to provide formal education in its region. The Kingdom of Bahrain is a pioneer in many fields and is considered one of the leading countries in mandatory education and educational development in comparison with other developed countries.

Formal education was established in 1919 when the first school was opened [7]. Education in the Kingdom of Bahrain is compulsory by law for all children aged 6–14 years, and the government is determined to ensure that all children in this age group get the appropriate education, whether in public or private schools. The education system in the country covers nine years of basic education, (primary and intermediate education), followed by three years of secondary education in its various specialties [8].

Currently, MOE has many accomplishments in technological adaptation. In 1999, the ministry established its first website to serve students and their parents (http://www.moe.gov.bh/). This site has seen many updates in line with the latest trends in technology, namely digital education [9]. With technological progress, there is a need to intensify efforts and training to gain new digital skills that guarantee success and resilience in the era of continuous digital transformation [10]. Countries that implement comprehensive strategies for digital skills ensure that their people have the expertise they need to be more employable, productive, innovative, and successful, while staying safe and secure with online communication [11]. It is also worth noting that MOE has the advent of cloud computing, the 365 Microsoft Office system was adopted in all government schools, and it has gradually moved to Amazon cloud. In addition to that, the ministry also contains an electronic archiving system, which is a central store for all ministry documents, to protect them from loss and damage, and, which aims to provide much needed information and documents 24/7 to the public [12].

Many factors enable Bahrain to be an inspiration in digital education and an example to follow in and outside the Arab world. First, Bahrain has a robust ICT infrastructure. Second, Bahrain has qualified human resources in digital education.

This is the result of the project of His Majesty King Hamad for Future Schools, where teachers and students were trained in digital empowerment and using the latest technologies in the service of education [13]. Continuing education by applying e-learning during the Coronavirus quarantine period was an excellent idea and has been a great success and excellent achievement for Bahrain [14]. Therefore, we can say that even before the Coronavirus pandemic period, the MOE in Bahrain, was using digital education to a great extent and has many electronic communication channels in place that have been applied in all government schools and most of the private schools.

3 Education and Sustainability

This refers to the sustainability of teaching and learning, which is in line with the fourth goal of "quality education" listed among the 17 international goals of sustainable development of the United Nations and in the Bahrain's 2030 vision. The Kingdom has achieved many impressive outcomes with regards to the fourth sustainable goal (SDG4), which is providing quality education (the foundation for improving people's lives and achieving sustainable development) [15]. The objective behind SDG4 is to ensure inclusive and equitable quality education and promote lifelong learning opportunities for all [16]. Out of its belief in the "Education for All" principle, the government of Bahrain provides free, high-quality education in public schools for Bahrainis and non-Bahrainis. In the interest of "leaving no one behind", special needs students have been incorporated into schools, and schools have been prepared in terms of awareness and training to equip them with competencies that meet their needs [15].

There is no doubt that this crisis somehow affected the SDGs. Nevertheless, there is always a bright side to any crisis occurring in the world. In distance education, being able to get through all the lessons and exercises to students has been exceptional. After the Coronavirus period, the continued use of tools that have been activated in the distance education will guarantee that no one will be left behind.

Moreover, distance learning has contributed to highlighting the importance of e-books, compared to printed books [17]. Likewise, all activities were implemented through electronic programs in a way that is fun for the students, and this has reduced the consumption of paper, ink, and pens contributing to preserving the environment around us. On another hand, the consumption of petroleum and gasoline has decreased, consequently reducing pollution as well. All of this has contributed in some way to achieving the 17 goals of sustainable development.

This experience should be benefited from and not stopped. Even after we return to everyday life, it should be harnessed in a manner appropriate to the circumstances at the time and strive to achieve the goals of sustainable development and repair what was damaged during this crisis.

4 Distance Learning During the Covid-19 Pandemic: An Analysis

SWOT is a strategic analytical instrument which can be applied at organizational or personal levels for evaluating internal strengths (S) and weaknesses (W) of an organization, as well as exploring the external opportunities (O) available and the threats (T) confronted by the organization [18]. The SWOT analysis is applied in this section to provide a comprehensive overview of the current status of the education sector in Bah- rain while taking into consideration the Coronavirus pandemic. Furthermore, it will provide a productive framework in order to be able to track the

Fig. 1 SWOT analysis. *Source* Authors

risks that may affect the progress of the educational process and contribute to making the right decisions to support Bahraini society and its government (Fig. 1).

4.1 Strength

- **Digitalize the Education**

The ministry is dealing with the current situation with all available resources to ensure the continuation of the learning process for the students of the Kingdom, (including students with special needs), by providing educational services proportionate to their capabilities. Moreover, the biggest asset of the ministry is the unique project of His Majesty King Hamad's for Future Schools. The project began in 2005, working under the framework to provide equal opportunities to all children in Bahrain [13]. Furthermore, the project has succeeded in achieving a transition from traditional education to education based on the employment of information and communication technology [19]. The strengthening of the information society by offering and making the best use of the various sources of knowledge has allowed students to learn according to their abilities and teachers to communicate with students, monitor and cater to their needs more appropriately. Moreover, it has become easier for parents to communicate and follow-up on their children's work at school.

The project has promoted education by advancing technological means that facilitate and develop learning by reinforcing the students with digital and interactive materials in the educational portal, (such as digital copies of books), archiving exams, publishing electronic lessons and interactive electronic enrichment activities, allowing them to interact easily [13, 20]. The MOE has maintained the continuous development of the project and as a result has reached a prominent level of technology in digital education. Therefore, the success of the transition to distance education by employing digital empowerment has been straightforward.

- **Information and Communications Technology (ICT) Infrastructure:**

Despite the small geographical size of the Kingdom of Bahrain, it hosts IT companies that provide an excellent distribution, and data systems management services. Based on Economic Development Board (EDB), Bahrain is ranked first in the Middle East in terms of the availability of information and communications technology, and it provides information and communications technology services at competitive prices [21]. More-over, Bahrain is the first country on a regional level to liberalize the telecommunications market fully. It continues to lead by becoming the first economy in the Gulf Co- operation Council countries to adopt cloud policy. The government is committed to accelerating the transition to the cloud, enabling a more forward-looking approach, has emerged as the best-performing country in the Middle East and ranks 31st globally in the ICT Development Index (released in 2017) [22]. What distinguishes Bahrain is that it ensures a business-friendly regulatory environment with greater access to regional markets. It has a robust infrastructure for information and communications technology, an essential and competitive position, reasonable cost, excellent internet communication systems, and most important of all, the availability of skilled human capital [21]. From this standpoint, we can say that one of the most important reasons behind the success of distance learning is the presence of great information and communication technology infrastructure. Thanks to the strong networks of the internet, students can attend online classes, communicate with their teachers, and download lessons and exercises. There have been outstanding initiatives from communication companies in Bahrain such as (Batelco and Zain) to ensure the continuity of distance education for students [23, 24].

4.2 Weakness

- **Distance Learning Cost**

This does not only involve financial cost. First, several important aspects must be taken into consideration when applying distance education. One issue that has arisen is that the number of students in each family enrolled in different levels of education is not always proportionate to the number of computers, laptops, and TV sets in each home. Another issue is the appropriateness of the home as a venue for online lessons. Many homes have no suitable place for online study or time to prepare a conducive study environment or assist in the continuity of their child's education. There is no doubt that the task has fallen on parents to pursue education with their children from home, in addition to their work responsibilities as a large number are working parents who need their computers and the internet for work purposes. Most often, computers of all kinds and related technology required for effective distance learning are considered expensive. Moreover, internet service is not free and data usage is not limitless [8]. In other words, parents need to pay for the internet. Many lectures and assignments on a range of topics require a high-quality internet connection or a large amount of data usage to ensure that class broadcasts are not interrupted [26].

Distance learning funding is a barrier that parents face. This is in line with a previous study showing that parents who use distance learning classes, those who do not use distance learning, and those who have used distance learning in the past all identified funding as a barrier to distance learning [27]. To sum up, there was not enough time or money for parents to create the appropriate study environment and ready the needed tools to implement distance education from home.

- **Technical difficulties**

The use of modern technology brings with it many technical problems. For Bahraini students, this included: not being able to activate the educational portal, not knowing how to use some of the proposed applications by the ministry (such as Microsoft Teams), and having problems related to the constant high volume of traffic in the portal. Also, one of the obstacles to achieving quality in distance learning during the COVID- 19 pandemic is the poor internet speed in some areas especially in the new residential neighborhoods. Although the MOE has formed a team to address technical problems on the education portal website, distance education faces several problems that are related not only to the educational portal but to adaptability, time management and self-motivation. Pressure on the local network and educational websites have led to some technical difficulties for students, including disrupted online lessons which have made the educational process difficult for students and their parents, as not all parents have sufficient technological expertise, which puts more pressure on them. This is a new experience not only for students but also for their parents. Time, practice, and experience are needed for students and their parents to make this work more efficiently.

4.3 Opportunities

- **Digital Education Hub in the Region**

Many factors enable Bahrain to be an inspiration in digital education and an example to follow. First, Bahrain has a robust ICT infrastructure. Second, Bahrain has qualified human resources in digital education. This is the result of the project of His Majesty King Hamad for Future Schools, where teachers and students were trained in digital empowerment and using the latest technologies in the service of education. More investments must be made to promote digital education and its broader activation in public and private education in all stages and include higher education as well. Continuing education by applying e-learning during the Coronavirus quarantine period was an excellent idea and has been a great success and excellent achievement for Bahrain.

- **Creating a Skilled Human Capital with Digital Technologies**

Preparing a generation capable of using digital technology in all aspects of life, such as education is essential. E-learning and other issues that students are experiencing today, give them the ability to adapt and succeed in distance education and will prepare them to face more significant problems and tackle them more efficiently in the future. Having the strength and knowledge to overcome technical problems and to manage actual crises develops leadership qualities in young people. This generation will be skilled and contribute to the technological renaissance and the transformation to a digital economy. The opportunity that our students have today is unprecedented as the job market they will be part of requires knowledge, skills, and capabilities to help perform any job effectively and innovatively. So, right now, human capital is being built.

4.4 Threats

- **Risks related to the unsafe use of the internet**

The online world is full of threats and dangers, especially for children and students. With distance education, students will engage more and more in the online world. Some threats include cybercriminals to social media posts. These criminals stalk children on social media and through electronic applications that attract children. Some other notable threats are cyber-bullying, cyber-predators, posting personal information, phishing, falling for scams, accidentally downloading malware and posts that can come back to haunt a child later in life. Protecting children on the internet is a matter of awareness. Knowing the underlying risks and how to prevent them is essential when applying distance education because it relies on the internet [25]. Of course, the MOE has played a prominent role in educating students about the dangers of using the internet through holding workshops and seminars in schools. However, as the speed of technology evolves, the problems associated with it develops, accordingly.

5 Conclusion

Dealing with the "Covid-19" crisis was not easy at all, but the Kingdom of Bahrain and the MOE had a great deal of control and a distinguished administration that enabled it to coexist with the crisis by finding quick and alternative solutions that ensured the continuity of all sectors such as the economy, education and health. Undoubtedly, the digital empowerment adopted by the MOE was one of the keys to success in ensuring the continuity of education and the speed of transformation from traditional education in schools to distance learning in homes. Adequate preparedness and planning for crises are crucial, as it provides an opportunity to deal with the crisis in the event of its occurrence and exit from it with minimal losses. As we mentioned before, education is the right of all, so it is imperative to ensure the progress of the educational process even in times of crisis. In the last ten years, the MOE has proven to be highly efficient in over-coming crises and continuing education. Hence, planning for crises and thinking about the various scenarios that Bahrain can face and which could affect education, is essential. Establishing continuity and contingency plans for education, therefore, it is suggested that the role of the crisis team be more active in the ministry, to build perceptions, and create advanced solutions for easy reference to be applied when needed. All schools should be considered and included within the crisis management plan, and it is vital to trial these plans in a real-life scenario. Individuals from the relevant ministries such as Health and Interior can also be added to the team to help get an integrated and coherent plan. Furthermore, no plan will succeed without holding training programs and workshops for employees, teachers, and students as well.

References

1. Nadeem, S.: Coronavirus COVID-19: available free literature provided by various companies, journals and organizations around the world. J. Ongoing Chem. Res. **5**(1), 7–13 (2020)
2. Al-Ubaydli, O.: Understanding how the coronavirus affects the global economy: a guide for non-economists. Derasat Strategic Studies Series, pp. 1–59 (2020)
3. Cucinotta, D., Vanelli, M.: WHO declares COVID-19 a pandemic. Acta Bio Medica **91**(1), 157–160 (2020)
4. Reuge, N., Jenkins, R., Brossard, M., Soobrayan, B., Mizunoya, S., Ackers, J., Jones, L., Taulo, W.: Education response to COVID 19 pandemic, a special issue proposed by UNICEF: Editorial review. Int. J. Educ. Dev. **87**(102485), 1–4 (2021)
5. Elizabeth Noor Coutts, C., Buheji, M., Ahmed, D., Abdulkareem, T., Buheji, A., Buheji, B., Eidan, S., Perepelkin, N.: Emergency remote education in Bahrain, Iraq, and Russia during the COVID-19 pandemic: a comparative case study. Hum. Sys. Manage. 1–21 (2020)
6. Lassoued, Z., Alhendawi, M., Bashitialshaaer, R.: An exploratory study of the obstacles for achieving quality in distance learning during the COVID-19 pandemic. Educ. Sci. **10**(9) (2020)
7. Shaker, H.R.: Distance education in Bahrain: situation and needs. Open Learn. J. Open Distance e-Learn. **15**(1), 57–70 (2000)
8. Shirawi, M.A.A.: Education in Bahrain-1919–1986 an analytical study of problems and progress. Doctoral dissertation, Durham University, Durham, UK (1987)
9. Al-Banki, H.: Evaluation of the digital empowerment project in education from the middle school teachers and principals in the Kingdom of Bahrain. Doctoral dissertation, University of Bahrain, Sakhair, Kingdom of Bahrain (2018)
10. Casalino, N., Żuchowski, I., Labrinos, N., Munoz Nieto, Á.L., Martín, J.A.: Digital strategies and organizational performances of SMEs in the age of Coronavirus: balancing digital transformation with an effective business resilience. Law Econ. Yearly Rev. J. 347–380 (2020)
11. Van Laar, E., Van Deursen, A.J., Van Dijk, J.A., De Haan, J.: The relation between 21st- century skills and digital skills: a systematic literature review. Comput. Hum. Behav. **72**, 577–588 (2017)
12. Microsoft News Center: Bahrain's MoE accelerates e-learning across the Kingdom with Microsoft Teams as an educational platform. Microsoft Inc. (2020). https://news.microsoft.com/en-xm/2020/05/18/bahrains-moe-accelerates-e-learning-across-the-kingdom-with-microsoft-teams-as-an-educational-platform/. Accessed 30 Dec 2021
13. Daradkah, H.M.: Degree of educational portal use in the schools of the future from its teachers' perspective. Int. J. Educ. Technol. Learn. **8**(1), 1–15 (2020)

14. Alalawi, A.: A survey on e-learning methods and effectiveness in public Bahrain schools during the COVID-19 pandemic. In: 2021 International Conference on Innovation and Intelligence for Informatics, Computing, and Technologies (3ICT), pp. 574–579 (2021)
15. Ministry of Education, Kingdom of Bahrain: Education in sustainable development goals
16. del Cerro Velázquez, F., Morales Méndez, G.: Augmented reality and mobile devices: a binominal methodological resource for inclusive education (SDG 4). An example in secondary education. Sustainability **10**(10) (2018)
17. Gueval, J., Tarnow, K., Kumm, S.: Implementing e-books: faculty and student experiences. Teach. Learn. Nurs. **10**(4), 181–185 (2015)
18. Helms, M.M., Nixon, J.: Exploring SWOT analysis–where are we now? J. Strat. Manage. (2010)
19. Ministry of Education: His Majesty King Hamad project for schools of the future. http://www.moedu.gov.bh/DigitalEmp/Docs/khsfp.pdf. Accessed 30 Dec 2021
20. Ministry of Education: Digital empowerment map. http://www.moedu.gov.bh/DigitalEmp/Docs/DigitalEmpowermentMap.pdf. Accessed 30 Dec 2021
21. Economic Development Board: 2017 Annual Report, Economic Development Board, Kingdom of Bahrain 2017
22. International Telecommunication Union: Measuring the Information Society Report, Geneva, Switzerland (2017)
23. Alkhaleej, A.: Educational initiatives by Zain 2020. http://www.akhbar-alkhaleej.com/news/article/1203910. Accessed 30 Dec 2021
24. Batelco: Free browsing on educational sites for Batelco's fixed home broadband customers 2020. https://batelco.com/news-media/free-browsing-on-educational-sites-for-batelcos-fixed-home-broadband-customers/. Accessed 30 Dec 2021
25. Shillair, R., Cotten, S.R., Tsai, H.Y.S., Alhabash, S., LaRose, R., Rifon, N.J.: Online safety begins with you and me: convincing Internet users to protect themselves. Comput. Hum. Behav. **48**, 199–207 (2015)
26. Abuhammad, S.: Barriers to distance learning during the COVID-19 outbreak: a qualitative review from parents' perspective. Heliyon (2020)
27. Tien, J.M.: Toward the fourth industrial revolution on real-time customization. J. Syst. Sci. Syst. Eng. **29**(2), 127–142 (2020)

Acceptance of E-Learning Post-COVID-19: A Case Study on *Pondok* School Institution

Azira Hanani Ab Rahman, Mohd Rushdan Yasoa, Siti Fariha Muhamad, Azwan Abdullah, Siti Rohana Mohamad, Nur Syafiqah A. Samad, and Norzalizah Bahari

Abstract During the pandemic of COVID-19, the education sector has been heavily hit, forcing the educational institutions to close. Throughout the COVID-19 pandemic, Malaysia's education system placed heavy emphasis on online learning to substitute direct conventional instruction. The aim of the research is to determine student acceptability of e-learning implementation following COVID-19 in a *Pondok* school in Kelantan. This study intended to ascertain the implications of students' characteristics and technology acceptance models on their future behavioural intentions to use e-learning continuously. This quantitative research included 100 students from a *Pondok* school in Kelantan. Self-administered questionnaires were used to gather data and Partial least square structural equation modelling (PLS-SEM) was used in the data analysis. Outcomes from this study showed that students' characteristics are positively affected by students' motivation, students' mindset and computer competency. Perceived ease of use and perceived usefulness positively affect technology adoption. On the other hand, economic deprivation negatively affects technology

A. H. Ab Rahman (✉) · M. R. Yasoa · S. F. Muhamad · A. Abdullah · S. R. Mohamad · N. S. A. Samad · N. Bahari
Faculty of Entrepreneurship and Business, Universiti Malaysia Kelantan, 16100 Kota Bharu, Kelantan, Malaysia
e-mail: hanani@umk.edu.my

M. R. Yasoa
e-mail: rushdan.y@umk.edu.my

S. F. Muhamad
e-mail: fariha@umk.edu.my

A. Abdullah
e-mail: azwan.a@umk.edu.my

S. R. Mohamad
e-mail: rohana@umk.edu.my

N. S. A. Samad
e-mail: nursyafiqah.as@umk.edu.my

N. Bahari
e-mail: norzalizah.b@umk.edu.my

A. Hamdan et al. (eds.), *The Implementation of Smart Technologies for Business Success and Sustainability*, Studies in Systems, Decision and Control 216,
https://doi.org/10.1007/978-3-031-10212-7_48

adoption. Furthermore, students' characteristics and technology adoption positively affect the behavioural intent to continuously engage in e-learning in the future.

Keywords COVID-19 pandemic · e-learning · Acceptance · *Pondok* students

1 Introduction

E-learning is a subset of online education that relies entirely on the internet and digital technology to provide instruction to prospective students [1]. Computer and software programmes also facilitate the learning process [2]. E-learning denotes an information system that integrates a variety of educational aspects to provide students with a digital reality in which they can participate in activities such as discovery and audio-visual involvement in multiple subjects, as well as encourage interaction with peers and educators [3, 4]. It was initially designed for working professionals who were incapable of undertaking formal education on a full-time basis [2]. Nonetheless, e-learning capabilities have expanded to support all categories of learners in tertiary education, such as full-time, part-time, or distance learners [5]. Due to the COVID-19 pandemic, e-learning has lately become a necessary component of all educational institutions around the world, including schools, colleges, and universities.

In general, the COVID-19 outbreak has upset human life and work in every part of the globe. In addition to other sectors, the education is severely impacted. Nonetheless, education must resume, so e-learning is the most effective method in lieu of the conventional classroom environment [6]. The COVID-19 pandemic has forced educational institutions to respond by swiftly transitioning to distance and online learning as a viable option [4, 7, 8]. E-learning technologies have become more crucial following COVID-19. E-learning systems may aid learning providers in organizing, planning, executing, and monitoring their learning and teaching initiatives. Additionally, it aims to support instructors as well as schools and universities in promoting student learning during school and university shutdowns. Moreover, the bulk of these programmes are complimentary, which will provide ongoing education throughout the COVID-19 pandemic [7].

Despite the fact that e-learning is the most appropriate strategy in the current environment, its implementation is fraught with difficulties. Inadequate internet access, a lack of technical skills, low administrative support, insufficient content design, decreased students' motivation, reluctance to self-study, low availability of the required services and infrastructure such as high-speed internet access and computer literacy skills, along with the potential danger of test/examination misconduct, and inexperienced educators are just a few of the e-learning hurdles [4, 9–11]. The Malaysian government has improved educational services by updating information and communication technology infrastructure and offering ongoing training for teachers to improve their information technology skills. The benefits, however, are exclusively available to the government and government-aided schools. As a non-governmental school, Pondok students face significant e-learning problems because they do not have access to government-provided infrastructure or teacher training. Pondok schools, as private educational institutions controlled by individuals, non-profit organizations, or Islamic foundations with limited financial resources, rely heavily on student fees and donations from stakeholders to operate and flourish. Regardless of the constraints, Pondok schools must provide e-learning to their students in order to adapt to the current circumstances.

While the majority of e-learning studies in Malaysia focus on higher learning institutions and teaching colleges [3, 5, 9, 12], just a few research has been carried out to assess the acceptability of e-learning in schools. This research addresses that gap by examining the e-learning system acceptance and adoption by students in Pondok schools throughout the COVID-19 crisis.

This study intends to link students' characteristics and the Technology Acceptance Model to Pondok school students' acceptance and preparation for e-learning. This research intends to measure the students' characteristics (students' motivation, mindset, and computer competency) and the Technology Acceptance Model (perceive usefulness, perceive ease of use, and economic deprivation) to the students' acceptance and readiness (behavioural intention to continue using) of e-learning implementation in a Pondok school in Kelantan.

The research is divided into multiple sections. To begin with, this research examines an overview of e-learning, as well as e-learning throughout the COVID-19 crisis and e-learning problems. The research methodology, data collection, and data analysis process are followed by a discussion of related theories and previous studies. Finally, as a wrap-up, the findings are discussed.

2 Literature Review

Several studies have been conducted to investigate students', teachers', and schools' acceptance and preparedness for online teaching and learning. Acceptance and readiness of students is a critical aspect of successfully implementing online learning [9]. Teachers' qualities [13–15], facilitating conditions [11, 14–16], self-directed learning [17] and computers competency [6, 18] are among the factors contributing to e-learning acceptance.

2.1 Students' Motivation

In educational research and practice, motivation has had a significant impact on learners' attitudes and learning habits [17]. Motivation has been proven to play a key role in technology adoption and e-learning in previous studies [6, 18, 19]. Because the nature of the online curriculum is mainly automatic, more personalized, and independent, learners must be sufficiently motivated to engage in successful online learning [18]. Self-efficacy is a critical factor in determining students' motivation, which is a significant aspect of their qualities. A motivated student will actively participate in classroom discussions and activities [6]. Students' inclination to register in e-learning classes may be influenced by the convenience and versatility of e-learning [20] On the other hand, a lack of motivation for online learning has a negative impact on students' mental well-being or characteristics [21]. As a result, with regard to online learning, students' motivation is an imperative feature of the students' profile. As a result, the following hypothesis should be tested:

H1a Students' motivation in e-learning positively influences students' characteristics.

2.2 *Students' Mindset*

Dweck [22] defined mindset as how people think about the mutability of intelligence and ability. Two categories of mindsets have been distinguished: fixed and growth mindsets. Those with a fixed mindset believe that the majority of their abilities are dictated by their intrinsic abilities. Individuals with a growth mindset believe that their capabilities change as a result of learning and experiences that promote skill development. Mindset refers to how much people believe their ability can be improved via hard work and perseverance versus how much they believe their ability is fixed and unchangeable regardless of how much effort they put in. Students with a "growth" mindset believe that they may improve their ability via effort, hard work, and perseverance, whereas those with a "fixed" mindset feel that their ability is predetermined and unchangeable regardless of effort or perseverance [23]. Students' mindset is a reflection of how students perceive and learn from an instructor. Baber's research discovered a significant favorable influence on students' mindsets and characteristics, showing that a student's thinking about perceived learning is a critical element of students' characteristics. As a result, the following hypothesis must be confirmed:

H1b Students' mindset in e-learning positively influences students' characteristics.

2.3 *Computer Competency*

For effective use of ICT in learning, students require experience and some level of skills in ICT systems [24]. According to [25], students with insufficient competency or a lack of exposure to educational technology will not make effective use of such resources since they lack the necessary skills. As a result, the findings imply that students' computer skills have a beneficial effect on their acceptance of an e-learning system. Additionally, in a study to ascertain the key indicators for accepting e-learning, [26] revealed that computer competency demonstrated high levels of consistency, with students' personal computer experience being the most influential determinant of students' characteristics pertaining to the acceptance of e-learning. Consequently, the following hypothesis has been offered in this study:

H1c Students' computer competency positively influences students' characteristics.
H3 Students' characteristics in e-learning positively influence e-learning acceptance.

2.4 *Perceived Usefulness*

As stated by [27], perceived usefulness is the range to which an individual feels that using a specific system would augment individual work performance. One of the most important indicators of e-learning intention is perceived usefulness, which has been verified as a construct to quantify the success of e-learning systems in several studies [3, 14, 28]. In a research of 377 students at three public institutions in Thailand, [15] revealed that perceived usefulness has a favorable influence on students' desire to utilize a specific system (e.g., e-learning). Students are more inclined to accept e-learning if they feel it will assist them in learning more effectively, being more adaptable, and achieving higher academic performance. For example, [29] identified several positive aspects of students' experiences with emergency remote learning throughout the COVID-19 pandemic, including the capacity to process classes for future references, the presence of Learning Management Systems to stimulate the growth of electronic databases, the ability to register in online courses rather than conventional classes, and location adaptability. Moliner et al. [30] discovered a steady reduction in the number of students who favored the first choice (online live sessions), with students gradually shifting to pre-recorded lectures since they preferred to see the teacher's explanations numerous times and regarded this technique as superior to online sessions. As a result, the following theory has been developed:

H2a Perceived usefulness of the learning management system positively influences e-learning acceptance.

2.5 *Perceived Ease of Use*

Davis [27] characterizes perceived ease of use as an individual's confidence that using a specific technology is effortless and simple [31]. The simplicity of using an e-learning system enables students to focus on studying the course materials rather than making additional efforts to learn the instrument [32]. In their research, [28] discovered that perceived ease of use has a substantial positive effect on behavioral intent and practical usage of e-learning. Chang and Tung [33] demonstrated that perceived usefulness and perceived ease of use have a substantial effect on the behavioral intent to employ online learning course webpages. Furthermore, the study discovered that perceived ease of use increases the behavioral intent to utilize websites for online learning courses. As a result, the following hypothesis is proposed in this research:

H2b Perceived ease of use of the learning management system positively influences e-learning acceptance.

2.6 Economic Deprivation

Economic deprivation is defined by [34] as financial difficulties at home and the difficulty of parents' paying for basic needs. The pandemic COVID-19, as well as the Malaysian government's movement control order, forced numerous businesses to close or scale back their operations. People's incomes were impacted as a result of the circumstances, with some losing their jobs and others receiving lower pay as a result of the salary drop. This circumstance has exacerbated the financial pressure on some families, who must spend money on the internet and gadgets in order for their children to participate in e-learning. Financial constraints and students' economic position have been identified in a few studies as impediments to successful e-learning implementation in higher education institutions [35–37]. Remarkably, research conducted by [38] among Malaysian educators on e-learning difficulties throughout the COVID-19 crisis discovered that students from low-income households with guardians who lacked access to digital devices and had poor networks, as well as those who resided in the suburbs and rural areas, were more prone to failure in the e-learning scenario. This argument is in line with [39], reporting that 83.0% of students said the COVID-19 problem had a financial influence on their education. This is due to the high cost of internet connectivity in Zambia, which is in addition to ordinary tuition rates. Accordingly, their e-learning experience suffered due to this predicament. Thus, the following theory is put forth:

H2c Economic deprivation will negatively affect e-learning acceptance.
H4 TAM of e-learning will positively influence e-learning acceptance.

3 Research Methodology

3.1 Participants

The participants in this study were students from Pondok Sungai Durian in Kelantan, Malaysia. This school was chosen because it is one of the few Pondok schools to employ e-learning during the school closure due to the COVID-19 pandemic. During the data collecting process, the school has roughly 75 students staying at hostels, while 275 others were staying at home for online learning. The survey was disseminated to the whole population, where 75 sets of questionnaires were posted to the students at their hostel, while the other students were surveyed through an online platform. 75 sets of questionnaires were returned but only 37 students responded to the link given. Out of 112 responses, 12 questionnaires were identified as having a suspicious pattern. Therefore, only 100 questionnaires were found to be relevant and usable.

3.2 Instrument and Procedures

The designed questionnaire is divided into four sections: (1) demographics; (2) behavioral intention to use e-learning constantly; (3) students' characteristics; and (4) technology acceptance. The first section asks about demographics and e-learning device. In section two, five items question the students' behavioral intention to use e-learning via 5-point Likert scale statements. Part three consists of 11 items that assess the characteristics of the students. Five of the elements were adopted from [26], while the remaining items were adapted from [6]. Part four has 12 items that assess technology acceptance, adapted from [6] and [34].

Initially, the questionnaire was developed in English before being translated into Malay. Three experts from the management of Pondok schools, as well as a senior lecturer from Universiti Malaysia Kelantan, were consulted for content validity. The school's administration has assisted in circulating a paper-based questionnaire for the students staying at the hostel. For the students staying at home, the questionnaire was distributed through WhatsApp. The request to participate was made by the respective teachers. The data was collected within 2 months, between early November 2021 and the end of December 2021.

The partial least square (PLS) structural equation modelling (SEM) technique was applied for data analysis in this research. In the field of behavioral science, partial least squares (PLS), a subset of structural equation modelling (SEM), is an effective technique for understanding predicted behavior. This method was chosen because it allows for the simultaneous examination of a number of dependent relationships, which is especially useful when the model contains both first-order and second-order latent variables [40].

4 Data Analysis and Results

4.1 Descriptive Analysis

Most of the students are between 16 and 18 years old (52%), followed by those between 13 and 15 years old (25%), and the remaining (23%) are between 19 and 21 years old. For economic background, most of the students come from a low-income group, with 72% of their parents' income being less than RM2,000 a month. 22% of the parents have an income of RM2,001–RM4,850 a month, while only 5%

are from the middle-income and high-income groups. Only 37% of the students share devices used during online learning with other family members.

4.2 Measurement Model

Anderson and Gerbing [41] proposed that the analysis be conducted in two phases. After validating the measurement model, the structural model, or hypothesis testing, was done. The loadings, composite reliability (CR), and average variance explained are all utilized to measure construct validity in Table 1. (AVE). According to [42], the loading cut-off should be 0.5 with an AVE of 0.5 and a CR of 0.7. All results in Table 1 are more than the literature's minimal threshold, showing that the measurement model is not a concern in this research (Table 1).

Table 1 Measurement model

Construct	Item	Loading	CR	AVE
SM	SM1	0.735	0.875	0.636
	SM2	0.853		
	SM3	0.776		
	SM4	0.822		
MS	MS1	0.741	0.822	0.606
	MS2	0.764		
	MS3	0.827		
CC	CC1	0.829	0.895	0.634
	CC2	0.835		
	CC3	0.850		
	CC4	0.819		
PU	PU1	0.857	0.926	0.758
	PU2	0.902		
	PU3	0.903		
	PU4	0.817		
POU	POU1	0.860	0.896	0.685
	POU2	0.853		
	POU3	0.882		
	POU4	0.703		
ED	ED1	0.873	0.928	0.764
	ED2	0.878		
	ED3	0.915		
	ED4	0.828		

Table 2 Discriminant validity

	CC	ED	INT	MS	POU	PU	SM
CC	0						
ED	0.219	0					
INT	0.655	0.169	0				
MS	0.857	0.173	0.828	0			
POU	0.662	0.359	0.575	0.567	0		
PU	0.649	0.201	0.579	0.559	0.835	0	
SM	0.573	0.18	0.634	0.9	0.663	0.696	0

Note: *CC* Computer competency, *ED* Economic deprivation, *FWT* Familiarity with technology, *INT* Behavioural intention, *MS* Students' mindset, *POU* Perceived ease of use, *PU* Perceived usefulness, *SM* Students' motivation

After establishing the measurement model, the discriminant validity must be confirmed. Discriminant validity quantifies how different a construct is from another and also how much an indication reflects a single construct [43]. Due to criticism levelled at [44] criteria by [45], this research utilized the heterotrait-monotrait (HTMT) ratio at this stage. When the HTMT value exceeds 0.90, a severe problem with discriminant validity is indicated. As shown in Table 2, the HTMT for all constructs are below 0.90, which demonstrate adequate discriminant validity.

4.3 Structural Model

The next phase in the data analysis process was to obtain the path coefficients in order to determine the relevance of the research model's hypothesized correlations. To evaluate the structural model (Table 3), [42] proposed examining standard beta, t-values using a bootstrapping process with 5000 resampling, predictive relevance (Q2), and effect sizes (f2). The results revealed that students' motivation ($\beta = 0.374$, $t = 11.955$, $LL = 0.318$, $UL = 0.423$, $p < 0.01$), students' mindset ($\beta = 0.301$, $t = 15.281$, $LL = 0.275$, $UL = 0.336$, $p < 0.01$), and computer competency ($\beta = 0.495$, $t = 12.702$, $LL = 0.435$, $UL = 0.571$, $p < 0.01$) have a significant positive correlation with students' characteristics, thus, agreeing with hypotheses H1a, H1b, and H1c. In terms of the technology acceptance model (TAM) of an e-learning system, both perceived usefulness ($\beta = 0.501$, $t = 13.229$, $LL = 4.430$, $UL = 0.554$, $p < 0.01$) and perceived ease of use ($\beta = 0.456$, $t = 14.074$, $LL = 0.399$, $UL = 0.501$, $p < 0.01$) have a significant positive effect. Meanwhile, economic deprivation ($\beta = -0.231$, $t = 2.935$, $LL = 0.061$, $UL = 0.336$, $p < 0.01$) has a negative effect. Hence, hypotheses H2a, H2b, and H2c are also accepted.

Following that, all latent variables of second order were positively associated with the dependent variable of behavioral intention, which was affected by students' characteristics ($\beta = 0.601$, $t = 6.361$, $LL = 0.443$, $UL = 0.757$, $p < 0.01$), and the

Table 3 Path coefficient assessment

Hypotheses	Relationship	Beta	Se	*t*-Value	*p*-Value	LL	UL	Decision
H1a	SM → SC	0.374	0.031	11.955	0.000	0.318	0.423	Supported
H1b	MS → SC	0.301	0.020	15.281	0.000	0.275	0.336	Supported
H1c	CC → SC	0.495	0.039	12.702	0.000	0.435	0.571	Supported
H2a	PU → TAM	0.501	0.038	13.229	0.000	0.430	0.554	Supported
H2b	PEOU → TAM	0.456	0.032	14.074	0.000	0.399	0.501	Supported
H2c	ED → TAM	−0.231	0.079	2.935	0.002	0.061	0.336	Supported
H3	SC → INT	0.601	0.094	6.361	0.000	0.443	0.757	Supported
H4	TAM → INT	0.160	0.098	1.617	0.053	0.014	0.322	Supported

technology acceptance model (TAM) ($\beta = 0.160$, $t = 1.617$, $LL = 0.014$, $UL = 0.322$, $p < 0.05$). Thus, hypotheses H3 and H4 are accepted.

4.4 *Evaluating the Effect Size*

The coefficient of determination (R2), the impact size (f2), and the predictive relevance (Q2) of external factors on endogenous variables are all shown in Table 4. The coefficient of determination (R2) value is the proportion of variation in the endogenous construct that is explained by all the exogenous constructs in the research model. The study's R2 value is 0.484. This indicates that the study's exogenous factors, which include students' characteristics and the technological adoption model, account for 48.4% of the variation in intention. Hair et al. [40] stated that Q2 is a metric of the model's and its parameter estimations' ability to recreate observed values accurately. If Q2 > 0, the model is predictively relevant. The Q2 value for this model is 0.406, greater than Hair et al.'s cut-off value (2013). Cross-validation of redundancy metrics demonstrates that the structural model used in this work is predictive in nature. The effect size (f2) parameter indicates the change in the value of R2 when a particular construct is excluded from the model. The f2 coefficients are 0.02, 0.15, and 0.35, respectively, to indicate the small-to-medium, medium-to-large, and large effects [46]. Each supported exogenous factor (SC, f2 = 0.38; TAM, f2 = 0.03) has a strong influence on the endogenous variable and a small-to-medium effect size.

Table 4 Coefficient of determination, effect size, and predictive relevance

Construct	f^2	Decision	R^2	Q^2
SC	0.38	Large effect	0.484	0.406
TAM	0.03	Small-to-medium		

5 Discussion and Conclusion

The aim of this research is to ascertain the elements that affect students' behavioural intentions at a Pondok school in Kelantan to resume the use of e-learning after COVID-19. This study's findings indicate that students' characteristics are significantly associated with future intent to resume the use of e-learning. Additionally, this study discovered that students' characteristics had a significant impact on their desire to resume the use of e-learning in the future. Thus, it is proposed that students' characteristics are critical to success and play a critical role in online education, as shown by this research, which is in line with the outcomes of past research by [6, 9]. This indicates that the more motivated students are, the more positive their outlook is, and the more proficient they are with computers, the more likely they are to resume the use of e-learning in the future. Thus, highly motivated students with a positive attitude and proficient computer skills will guarantee the success implementing e-learning.

It is critical to assess the acceptability of e-learning by measuring the TAM components. Perceived usefulness (PU) and perceived ease of use (PEOU) are regarded as significant predictors of behavioral intention, either directly or indirectly. The outcomes of this research demonstrate that both PEOU and PU appear to have a beneficial impact on the acceptance model for e-learning, as other studies have revealed [6, 47, 48]. Following that, the external variable is coupled with the model in order to determine its effect on the intents. These were used to evaluate the model's performance in a variety of online education scenarios [49, 50]. COVID-19 has impacted the unemployment rate in Malaysia with an increase of 4.5% in 2020 as compared to 3.3 percent in 2019 [51]. Thus, this study added economic deprivation as the external variable that may influence technology adoption among students. This is due to the current studies by [35, 38] that found financial constraints and students' economic position had hindered the success of e-learning among students, as well as students from low-income families were more likely to be unsuccessful in the e-learning situation. Findings from this study revealed that economic deprivation does have a negative impact on technology acceptance, thus supporting the previous studies by [35, 38]. The more severe the economic deprivation faced by the students' parents, the less likely they are to adopt new technology in learning. This is due to the fact that online learning will be a financial burden on this group as their parents will need to spend extra money on devices and the internet for online learning. During the Parliamentary session in February 2022, it was reported that 21,316 students had dropped out of schools from March 2020 to July 2021. Based on a study done by one of the local universities, it is found that the school dropout rate among students is due to poverty factors. Some of these students had to work during the COVID-19 pandemic to help their parents, and this situation had made the students demotivated to continue their studies [52]. Thus, it is suggested that the authorities pay greater attention to this matter and provide the necessary aids, such as providing suitable devices for online learning and internet data, in order to prevent students from low-income groups from being left behind in today's education system. Other than that, a proper plan needs to be executed for these students to ensure that they continue their studies after the COVID-19 pandemic recovers.

Acknowledgements This research project is funded under the Universiti Malaysia Kelantan Society Impact Research (UMK-SIR) grant.

References

1. Al-Busaidi, K.A.: An empirical investigation linking learners adoption of blended learning to their intention of full e-learning. Behav. Inf. Technol. **32**, 1168–1176 (2013)
2. Moore, J.L., Dickson-Deane, C., Galyen, K.: E-Learning, online learning, and distance learning environments: Are they the same? Internet High. Educ. **14**, 129–135 (2011)
3. Al-Rahmi, W.M., et al.: Use of E-learning by university students in Malaysian higher educational institutions: A case in Universiti Teknologi Malaysia. IEEE Access **6**, 14268–14276 (2018)
4. Siron, Y., Wibowo, A., Narmaditya, B.S.: Factors affecting the adoption of E-learning in Indonesia: lesson from covid-19. J. Technol. Sci. Educ. **10**, 282–295 (2020)
5. Azhari, F.A., Ming, L.C.: Review of e-learning practice at the tertiary education level in Malaysia. Indian J. Pharm. Educ. Res. **49**, 248–257 (2015)
6. Baber, H.: Modelling the acceptance of e-learning during the pandemic of COVID-19-A study of South Korea. Int. J. Manag. Educ. **19**, 100503 (2021)
7. Almaiah, M.A., Al-Khasawneh, A., Althunibat, A.: Exploring the critical challenges and factors influencing the E-learning system usage during COVID-19 pandemic. Educ. Inf. Technol. **25**, 5261–5280 (2020)
8. Callo, E.C., Yazon, A.D.: Exploring the factors influencing the readiness of faculty and students on online teaching and learning as an alternative delivery mode for the new normal. Univers. J. Educ. Res. **8**, 3509–3518 (2020)
9. Adams, D., Sumintono, B., Mohamed, A., Noor, N.S.M.: E-learning readiness among students of diverse backgrounds in a leading Malaysian Higher Education Institution. Malays. J. Learn. Instr. **15**, 227 (2018)
10. Aina, A.Y., Ogegbo, A.A.: Teaching and assessment through online platforms during the COVID-19 pandemic: benefits and challenges. J. Educ. e-Learn. Res. **8**, 408–415 (2021)
11. Ngampornchai, A., Adams, J.: Students' acceptance and readiness for E-learning in Northeastern Thailand. Int. J. Educ. Technol. High. Educ. **13** (2016)
12. Ahmad Shakir, N.S.B., Adnan, N.H.B.: Kebolehgunaan massive open online course (MOOC) Sebagai E-Pembelajaran dalam Pengajaran Pengaturcaraan di Sekolah Menengah. Malays. J. Soc. Sci. Humanit. **5**, 33–41 (2020)
13. So, T., Swatman, P.M.C.: e-Learning readiness of Hong Kong teachers. Hong Kong IT Educ. Conf. 6–8 (2006)
14. Taat, M.S., Francis, A.: Factors influencing the students' acceptance of e-learning at teacher education institute: An exploratory study in Malaysia. Int. J. High. Educ. **9**, 133–141 (2020)
15. Teo, T., Luan, W.S., Thammetar, T., Chattiwat, W.: Assessing e-learning acceptance by university students in Thailand. Australas. J. Educ. Technol. **27**, 1356–1368 (2011)
16. Pham, Q.T., Tran, T.P.: The acceptance of e-learning systems and the learning outcome of students at universities in Vietnam. Knowl. Manag. E-Learn. **12**, 63–84 (2020)
17. Hung, M.L., Chou, C., Chen, C.H., Own, Z.Y.: Learner readiness for online learning: Scale development and student perceptions. Comput. Educ. **55**, 1080–1090 (2010)
18. Rafiee, M., Abbasian-Naghneh, S.: E-learning: development of a model to assess the acceptance and readiness of technology among language learners. Comput. Assist. Lang. Learn. **34**, 730–750 (2021)

19. Tarhini, A., Deh, R.M., Al-Busaidi, K.A., Mohammed, A.B., Maqableh, M.: Factors influencing students' adoption of e-learning: A structural equation modeling approach. J. Int. Educ. Bus. **10**, 164–182 (2017)
20. Baber, H.: Determinants of students' perceived learning outcome and satisfaction in online learning during the pandemic of COVID19. J. Educ. e-Learn. Res. **7**, 285–292 (2020)
21. Lai, C.-S., Au, K.-M., Low, C.-S.: Beyond conventional classroom learning: linking emotions and self-efficacy to academic achievement and satisfaction with online learning during the COVID-19 pandemic. J. Educ. e-Learn. Res. **8**, 367–374 (2021)
22. Dweck, C.S.: Mindsets and math/science achievement. Carnegie Corporation of New York-Institute for Advanced Study Commission on Mathematics and Science Education, vol. 4, pp. 122–126 (2008). https://www.growthmindsetmaths.com/uploads/2/3/7/7/23776169/mindset_and_math_science_achievement_-_nov_2013.pdf
23. Gutshall, C.A.: Student perceptions of teachers' mindset beliefs in the classroom setting. J. Educ. Dev. Psychol. **6**, 135 (2016)
24. Ouma, G., Awuor, F., Kyambo, B.: E-learning readiness in public secondary schools in Kenya. Eur. J. Open Distance e-Learn. **16**, 97–110 (2013)
25. Alasmari, N.: Is internet reciprocal teaching the remedy for Saudi EFL learners' reading difficulties during the Covid-19 pandemic? J. Educ. e-Learn. Res. **8**, 324–332 (2021)
26. Selim, H.M.: Critical success factors for e-learning acceptance: confirmatory factor models. Comput. Educ. **49**, 396–413 (2007)
27. Davis, F.D.: Perceived usefulness, perceived ease of use and user acceptance of information technology. MIS Q. **13**, 319–340 (1989)
28. Ratna, P.A., Mehra, S.: Exploring the acceptance for e-learning using technology acceptance model among university students in India. Int. J. Process Manag. Benchmarking **5**, 194–210 (2015)
29. Quispe, M.D.C.A., Alecchi, B.A.: Business school student satisfaction towards emergency remote teaching. J. Educ. e-Learn. Res. **84**, 375–384 (2021)
30. Moliner, L., Valentin, G.L., Alegre, F.: E-learning during the covid-19 pandemic in Spain: a case study with high school mathematics students. J. Educ. e-Learn. Res. **8**, 179–184 (2021)
31. Nikou, S., Maslov, I.: An analysis of students' perspectives on e-learning participation—the case of COVID-19 pandemic. Int. J. Inf. Learn. Technol. **38**, 299–315 (2021)
32. Sun, P.C., Tsai, R.J., Finger, G., Chen, Y.Y., Yeh, D.: What drives a successful e-Learning? An empirical investigation of the critical factors influencing learner satisfaction. Comput. Educ. **50**, 1183–1202 (2008)
33. Chang, S.C., Tung, F.C.: An empirical investigation of students' behavioural intentions to use the online learning course websites. Br. J. Educ. Technol. **39**, 71–83 (2008)
34. Bernburg, J.G., Thorlindsson, T., Sigfusdottir, I.D.: Relative deprivation and adolescent outcomes in Iceland: A multilevel test. Soc. Forces **87**, 1223–1250 (2009)
35. Aini, Q., Budiarto, M., Putra, P.O.H., Rahardja, U.: Exploring E-learning challenges during the global COVID-19 pandemic: a review. J. Sist. Inf. **16**, 57–65 (2020)
36. Kibuku, R.N., Ochieng, D.O., Wausi, A.N.: E-learning challenges faced by universities in Kenya: a literature review. Electron. J. e-Learn. **18**, 150–161 (2020)
37. Kisanga, D., Ireson, G.: Barriers and strategies on adoption of e-learning in Tanzanian higher learning institutions: lessons for adopters. Int. J. Educ. Dev. using ICT **11** (2015)
38. Lukas, B.A., Yunus, M.M.: ESL teachers' challenges in implementing e-learning during COVID-19. Int. J. Learn. Teach. Educ. Res. **20**, 330–348 (2021)
39. Chola, R., Kasimba, P., George, R., Rajan, R.: COVID-19 and E-learning: perception of freshmen level physics students at Lusaka Apex Medical University. Int. J. Acad. Res. Dev. **5**, 67–76 (2020)
40. Hair, J.F., Hult, G.T.M., Ringle, C.M., Sarstedt, M.: A Primer on Partial Least Squares Structural Equation Modeling (PLS-SEM). SAGE Publication Inc. (2013)
41. Anderson, J.C., Gerbing, D.W.: Structural equation modelling in practice: a review and recommended two-step approach. Psychol. Bull. **103**, 411–423 (1988)

42. Hair, J.F., Hult, G.T.M., Ringle, C.M., Sarstedt, M.: A Primer on Partial Least Squares Structural Equation Modelling (PLS-SEM). SAGE Publications Inc. (2017)
43. Gholami, R., Sulaiman, A., Ramayah, T., Molla, A.: Senior managers' perception on green information systems (IS) adoption and environmental performance: results from a field survey. Inf. Manag. **50** (2013)
44. Fornell, C., Larcker, D.F.: Evaluating structural equation models with unobservable variables and measurement error. J. Mark. Res. **18**, 39–50 (1981)
45. Henseler, J., Ringle, C.M., Sarstedt, M.: A new criterion for assessing discriminant validity in variance-based structural equation modelling. J. Acad. Mark. Sci. **43**, 115–135 (2015)
46. Cohen, J.: Statistical power analysis for the behavioral sciences. Lawrence Erlbaum Associates (1988). https://doi.org/10.4324/9780203771587
47. Farahat, T.: Applying the technology acceptance model to online learning in the Egyptian Universities. Procedia Soc. Behav. Sci. **64**, 95–104 (2012)
48. Teo, T., Luan, W.S., Sing, C.C.: A cross-cultural examination of the intention to use technology between Singaporean and Malaysian pre-service teachers: an application of the technology acceptance model (TAM). Educ. Technol. Soc. **11**, 265–280 (2008)
49. Saadé, R., Bahli, B.: The impact of cognitive absorption on perceived usefulness and perceived ease of use in on-line learning: an extension of the technology acceptance model. Inf. Manag. **42**, 317–327 (2005)
50. Scherer, R., Siddiq, F., Tondeur, J.: The technology acceptance model (TAM): a meta-analytic structural equation modeling approach to explaining teachers' adoption of digital technology in education. Comput. Educ. **128**, 13–35 (2019)
51. Department of Statistic Malaysia. Sustainable Development Goal Indicators, Malaysia 2020 (2021). https://www.dosm.gov.my/v1/index.php?r=column/cthemeByCat&cat=474&bul_id=SHNMV2RNMmhDS2lxS2F3RTdCMkoxQT09&menu_id=WjJGK0Z5bTk1ZElVT09yUW1tRG41Zz09
52. Adnan, A.F.: 21,316 murid berhenti sekolah: Segera lakukan Pelan Pemulihan Pendidikan. Utusan Malaysia (2021)
53. Campbell, C., Al-harthi, A.S., Karimi, A.: Teachers cloud-based content creation in light of the TPACK framework : implication for teacher education. In: Australasian Society for Computers in Learning and Tertiary Education (ASCILITE), pp. 399–403 (2015)
54. Ministry of Education Malaysia: 2018 Annual Report, Malaysian Education Blueprint 2013–2025. Ministry of Education Malaysia (2018)
55. Aboagye, E., Yawson, J.A., Appiah, K.N.: COVID-19 and E-learning: the challenges of students in tertiary institutions. Soc. Educ. Res. **2**, 109–115 (2020)

The Impact of the Corona Pandemic (Covid-19) on the Level of Academic Achievement of Jordanian University Students from the Point of View of Their Parents

Khaled Jumah, Nader Abusheikha, and Marwa Ahmad

Abstract This study aims to determine the impact of the Corona pandemic (Covid-19) on the level of educational attainment of Jordanian university students from the point of view of their parents as staff members. The study sample consisted of 204 teaching staff members who are also parents of students enrolled in public and private Jordanian universities. Data were collected from the sample members through a questionnaire designed for this purpose, and the reliability coefficient of Cronbach reached (0.80). The study results found that faculty staff members from the College of Informatics and Computer feel that e-learning has contributed to the principle of "self-reliance" among students. Results also showed that the level of seriousness of students in following the electronic platform is weak. Assistant professors feel that e-learning has contributed to the development of interaction between the parents and the students, and that e-learning has contributed to creating a family atmosphere. Furthermore, results indicate that the staff from public universities feel that e-learning has contributed to enriching the knowledge and capabilities of parents in electronic technology.

Keywords Corona pandemic · Jordanian universities

1 Introduction

The Corona epidemic cast a heavy shadow on all aspects of our lives, even extending the study of university students, confusing their accounts after the closure of their

K. Jumah (✉) · N. Abusheikha
Faculty of Business, Zarqa University, Amman, Jordan
e-mail: kjumah@zu.edu.jo

N. Abusheikha
e-mail: naderabusheikha@zu.edu.jo

M. Ahmad
Faculty of Business, Philadelphia University, Amman, Jordan
e-mail: Mar_ahmed@philadelphia.edu.jo

A. Hamdan et al. (eds.), *The Implementation of Smart Technologies for Business Success and Sustainability*, Studies in Systems, Decision and Control 216,
https://doi.org/10.1007/978-3-031-10212-7_49

universities. E-learning has become an alternative to face-to-face learning in various fields of specialization in all faculties. It is less capital intensive and fits the need to improve students' digital competence.

Thus, the transition to e-learning (learning via the internet) has shown noticeable impacts on students, and there is a lot to be done to reduce these direct effects through distance learning strategies. Wealthier countries are better prepared to transition to e-learning strategies, even if faculty members and parents face a great deal of effort and challenges. Still, conditions in middle-income and poorer countries are not the same; therefore, if these countries do not act appropriately, this inequality of opportunity—which is appalling and unacceptable in the first place—will only get worse.

Also, some faculty members are skeptical about the results of electronic tests because there were no tangible indicators of students' commitment to the instructions of the tests, which generates doubts about the effectiveness of e-learning among university students despite the widespread use of the internet in the region [1]. Arab experiences are still very modest, and the successful ones are concentrated in some rich countries. Some problems that have emerged in the application of e-learning include the weak employment of some software for e-learning and the weak infrastructure that requires the adoption of specific software and the provision of internet networks, smartphones, and computers for each student.

The fundamental purpose of the success of e-learning was due to the university's rapid response to the pandemic. In a short time, the university prepared an emergency plan. It provided teachers with all the necessary support and motivation to ensure the interactive lectures were delivered on the expected due dates based on the university's annual academic schedule [2].

There is an urgent need to identify the views of parents in terms of the effectiveness of e-learning on the achievement of the students, determine and evaluate the effectiveness of e-learning, and the extent to which e-learning achieves the objectives of education. There is a need to identify whether e-learning can meet students' needs and create an interactive environment that dispenses with face-to-face learning.

E-learning has a significant role in the success of the teaching and learning processes, especially when computer-mediated communication became widely used due to the pandemic. The post-pandemic situation forced educational institutions to rely heavily on technology use by shifting to e-learning to ensure the continuity of the teaching and learning process.

2 Literature Review

The study examined the difference between distance education in emergency situations and online education. The study concluded that online learning experiences differ from learning in emergency situations in terms of the quality of planning and the courses offered online. The study recommended that colleges and universities should work to preserve education during the COVID-19 pandemic [3]. The study aimed to identify the reality of e-learning in light of the Corona pandemic in Algeria and the study conclude that e-learning is one of the best learning methods [4].

In addition, a study indicated that the COVID-19 pandemic challenged the universities to continue to overcome the difficulties facing both students and professors [5]. Furthermore, study relied on the descriptive approach and concluded that there are challenges and problems related to the educational process and its management during the Corona pandemic [6].

The study aimed at knowing the impact of university closures due to the Corona Virus (COVID-19) on the education and mental health of students and faculty. The research highlighted the potential impact of the spread of COVID-19 on students' education and mental health. They believe that students and staff should receive information through email, and the health and safety of students and staff should be a top priority. They assert that faculty must pay close attention to technology to make students' learning experiences rich and effective [7]. Moreover, study aimed to clarify the ways in which the Corona pandemic affected and reshaped education in Indonesia, where the traditional method of education declined to spread instead of learning through the internet because it supports learning from home and reduces the mixing of individuals with each other [8].

3 Theoritical Framework

There are many benefits offered by e-learning, which makes it superior to traditional education methods. E-learning is environmentally friendly, and it costs less than face-to-face courses. Furthermore, it provides flexible learning opportunities to students in terms of time, geographical place, and pace of learning. Thus, it is available and accessible to all learners [9].

E-learning is effective if the educational content is organized in a manner that commensurate with students' individualized learning. It meets students' educational needs and different learning styles and provides teachers with the necessary assessment tools and appropriate teaching aids. However, the teachers are also required to improve their electronic competencies and raise the level of their readiness to use technology adequately and efficiently. Parents play an important role in their children's learning process; thus, it is necessary to investigate the importance of integrating families into their children's educational journey. Parents have an important role in ensuring the progress of the educational process, and educational administrations must clearly communicate what they expect from parents in order for the learning process to be successful. However, this will remain a challenge because not all parents have the same degree of willingness or ability to participate. Despite the difficulty of facing this challenge, educational institutions should not stop communicating with parents to increase their feelings of frustration and helplessness. They should also avoid focusing on negative news related to their children's level of achievement. Parents tend to be interested in knowing what their children are good at, so institutions should involve students in the process of communicating with parents, as this will reflect positively on their learning process. Educational institutions should ask parents to express their opinion on some important matters related to their children's academic achievement instead of arguing with them, and to cooperate with them to achieve common goals and overcome some of the difficulties that might hinder the learner's path and communicates with parents to encourage them to participate in their children's learning. Positive communication about learners' educational process and involving parents in learners' decision-making process, as well as responding to the learners' learning preferences, all these factors will build bridges of trust between the educational institution and students' parents [10].

4 Research Question

This study is derived from the main research question: What is the impact of the Corona pandemic (Covid-19) on the level of achievement of Jordanian university students from the point of view of their parents as staff members?

From the above main question, the following sub-questions emerge:

1. To what extent is parents' reaction affected by the impact of the Corona pandemic (Covid-19) on the level of achievement of Jordanian university students?
2. What are the skills of the faculty members affected by the impact of the Corona pandemic (Covid-19) on the level of achievement of Jordanian university students?
3. To what extent is students' commitment and participation is affected by the impact of the Corona pandemic (Covid-19)?
4. What is the adequacy of the infrastructure for e-learning affected by the impact of the Corona pandemic (Covid-19)?

5 Objectives of the Study

This study aims to reveal the impact of the Corona pandemic (Covid-19) on the level of educational attainment of Jordanian university students from the point of view of their parents as staff members. It also aims to measure the extent of the parents' interaction, the skill level of faculty members, students' commitment and participation, and the adequacy of the infrastructure of e-learning in Jordanian universities.

6 Limitations

This study was applied in public and private Jordanian universities and was applied to (204) faculty members who have enrolled children in university at the time of the Corona pandemic. This study was applied in the first semester of the academic year

2021/2022. This study dealt with the impact of the Corona pandemic (Covid-19) on the level of educational achievement of Jordanian university students and determined the extent of the success of the e-learning experience in light of the Corona pandemic from the point of view of the students' parents as faculty members. The questionnaire is designed by the researchers; therefore, the results are confined to its validity and reliability.

7 Study Instrument

The questionnaire used in this study consisted of (37) paragraphs that were developed by researchers specifically for the purpose of the study. The questionnaire paragraphs were divided according to the five-Likert scale. Researchers used the descriptive-analytical method.

To judge the paragraphs of the questionnaire, it was presented to a group of specialists and experienced people on the validity and clarity of the paragraphs. The researchers modified the paragraphs of the questionnaire based on the opinions of the arbitrators. The reliability coefficient of Cronbach's alpha was extracted, which amounted to (0.80), and this value is acceptable for the purposes of the research. The questionnaire was also distributed to 20 faculty members who were not members of the sample to ensure the validity of the tool.

8 Statistical Analysis

The statistical program (SPSS 24) was used to analyze the data collected from the study tool. Furthermore, descriptive statistics were used to calculate the mean and standard deviation for all paragraphs of the questionnaire. The t-test was also used to identify the differences in the answers of the sample members in the two-level variables. ANOVA test and the Scheff posttest were also used.

9 Population and Sample Size

The study population consisted of 204 faculty members who are also parents to students enrolled in any public or private university in Jordan. The sample was randomly selected from the available study population during the first semester 2021/2022.65.7% of the sample is from public universities, 33.3% are assistant professors, and 20.1% are specialized in Basic.

10 Hypotheses

H1: there is a statistically significant effect at ($\alpha \leq 0.05$) of the impact of the Corona pandemic (Covid-19) on the achievement of university students from the point of view of their parents as a staff member due to staff member specialization.

H2: there is a statistically significant effect at ($\alpha \leq 0.05$) of the impact of the Corona pandemic (Covid-19) on the achievement of university students from the point of view of their parents as a staff member due to staff member academic rank.

H3: there is a statistically significant effect at ($\alpha \leq 0.05$) of the impact of the Corona pandemic (Covid-19) on the achievement of university students from the point of view of their parents as a staff member due to the university (public and private).

11 Results

To answer the main question of the study, which is what is the impact of the Corona pandemic (Covid-19) on the level of achievement of Jordanian university students from the point of view of their parents as staff members?

The mean and standard deviations were calculated for all dimensions of the questionnaire (Table 1).

The results shows that the mean of the skills of the faculty members that affected the level of achievement of Jordanian university students reached (4.6353) with a standard deviation of (0.36703) with a high degree.

The skills that faculty members need in their profession are among the important topics, as faculty members try to achieve the teaching objectives and communicate effectively with students. Results showed that creativity in teaching is an important requirement to achieve good quality University teaching. Good quality of teacher is acquired through faculty members who achieve creative climate within the classroom environment. Creative instruction pushes to activate the foundations of students' academic success. Parents' reactions remained with a mean of (4.6310) and a standard deviation of (0.33848) in a high degree, followed by the field of the students' level of educational attainment with a mean of (4.5763) and a standard deviation of (0.28236)

Table 1 Mean and std. deviation for all dimensions

Dimensions	N	Mean	Std. deviation	Degree
First—parent's reaction	204	4.6310	0.33848	2
Second—skills of the faculty member	204	4.6353	0.36703	1
Third—the student's level of educational attainment	204	4.5763	0.28236	3
Fourth—infrastructure for e-learning	204	4.5605	0.44640	4
Corona	204	4.5362	0.40938	5

Table 2 Mean and std. deviation for parent's reaction dimension

Parent's reaction dimension	N	Mean	Std. deviation	Level
I think that there is a disproportion between computers and the number of students in the same family	204	4.7451	0.54703	1

with a high degree as well. These results mean that the parents as faculty members feel that faculty member skill have strong impact on the level of achievement of Jordanian university students during Covid-19.

11.1 Results of the First Sub-question

To what extent is parents' reaction affected by the impact of the Corona pandemic (Covid-19) on the level of achievement of Jordanian university students?

To answer this question, the mean and standard deviations of the parent's interaction with the impact of the Corona pandemic (Covid-19) was calculated, and the results are as follow: the variable wich state "I think that there is a disproportion between computers and the number of students in the same family." has the highest level with a mean of (4.7451). This is due to that some Jordanian families suffer from financial issues, which do not allow them to provide needed devices to their children (Table 2).

11.2 Results of the Second Sub-question

What are the skills of the faculty members affected by the impact of the Corona pandemic (Covid-19) on the level of achievement of Jordanian university students?

To answer this question, the mean and standard deviations of the skill of a faculty member with the impact of the Corona pandemic (Covid-19) were calculated, and the results are as follow: The variable states that "I think that e-learning have contributed to the development of the skills of the student and the faculty members." with a mean of (4.8676). So, e-learning improves the level of students' self-learning skills (Table 3).

Table 3 Mean and std. deviation for skills of the faculty member dimension

Skills of the faculty member dimension	N	Mean	Std. deviation	Level
I think that e-learning have contributed to the development of the skills of the student and the faculty member	204	4.8676	0.45174	1

Table 4 Mean and std. deviation for student's commitment and participation dimension

Student's commitment and participation dimension	N	Mean	Std. deviation	Level
I think that the level of seriousness of students in following the electronic platform is weak	204	4.9853	0.12067	1

11.3 Results of the Third Sub-question

To what extent is students' commitment and participation is affected by the impact of the Corona pandemic (Covid-19)?

To answer this question, the mean and standard deviations of the impact of the Corona pandemic (Covid-19) on students' commitment and participation was calculated, and the results are as follow: variable states that "I think that the level of seriousness of students in following the electronic platform is weak" has the highest level with of a mean of (4.9853). The seriousness of the follow-up is, weak according to students. This result may be attributed to the fact that e-learning is a new way of learning to them (Table 4).

11.4 Results of the Forth Sub-question

What is the adequacy of the infrastructure for e-learning affected by the impact of the Corona pandemic (Covid-19) on the level of achievement of Jordanian university students?

To answer this question, the mean and standard deviations of the adequacy of the infrastructure for e-learning with the impact of the Corona pandemic (Covid-19) were calculated, and the results are as follow: variable states that "I think that the Internet has worked to create a discrepancy between the abilities of students in educational attainment" has the highest level with a mean of (4.8578) (Table 5).

Table 5 Mean and std. deviation for the adequacy of the infrastructure for e-learning dimension

The adequacy of the infrastructure for e-learning dimension	N	Mean	Std. deviation	Level
I think that the Internet has worked to create a discrepancy between the abilities of students in educational attainment	204	4.8578	0.53855	1

12 Testing Hypotheses

H1: there is statistically significant effect at ($\alpha \leq 0.05$) of the impact of the Corona pandemic (Covid-19) on the achievement of university students from the point of view of their parents as staff members due to staff members specialization:

To test the first hypothesis, we conducted ANOVA test and the results show that there is statistical significance due to the specialization of the staff members in variables 2, 4, 6, 14, 16, 24, 26 and 29.

To find out the differences between the specialization of the staff members, "Scheff Test" was used as a posttest, and the results as follows: Staff members from Economic and Management sciences agree more than staff members from Pharmacy and Basic sciences that there is a disproportion between computers and the number of students living in the same family. The closure of universities and the quality of the internet has negatively impacted students' academic attainment. Results also showed that staff members from Legal sciences and Economic and Management sciences agree more than staff members from Pharmacy and Basic sciences that learning, and e-learning have contributed to the development of positive interactions between faculty members and parents of students. In addition, staff members are more interested in learning about everything new in e-learning. Results revealed that staff from Informatics and Computer agrees more than staff from Engineering sciences that e-learning have contributed to the principle of "self-reliance" among students. The level of seriousness of students in following the electronic platform is weak.

H2: there is statistically significant effect at ($\alpha \leq 0.05$) of the impact of the Corona pandemic (Covid-19) on the achievement of university students from the point of view of their parents as staff members due to staff members' academic rank:

To test the second hypothesis, we conducted ANOVA test and the results show that there is statistical significance due to the rank of the staff members in variables 1, 3, 7, 13, 21, 23, 25, 27, 28 and 30.

To find out the differences between rank of the staff members, post-test "scheff Test" was used, and the results as follows:

Professors agree more than assistant professors that it is difficult to communicate with the concerned faculty members. They feel that the level of interactions between students outside the times of lectures has become weak compared to what was the case in face-to-face education. There is also a noticeable lack of educational resources that support or help students to understand the course content and the low level of educational attainment for students is due to the lack of experience universities have in using e-learning. E-learning is an unfamiliar technology in the Jordanian society and there are no alternative plans for the e-learning in Jordanian universities. The scarce of adequate educational technology tools and the unfamiliarity of e-learning has led to a discrepancy in the level of educational attainment of students. In addition, assistant professors agree more than associate professors that e-learning have contributed to the development of interactions between parents of students. However, in some instances, using e-learning have contributed to creating a family atmosphere tainted by tension, isolation and self-reflection. Therefore, the focus and follow-up by students during the learning and e-learning process is weak.

H3: there is statistically significant effect at ($\alpha \leq 0.05$) of the impact of the Corona pandemic (Covid-19) on the achievement of university students from the point of view of their parents due to the University. To test the third hypothesis, we conduct T-test and the results show that there is statistical significance due to the university (public, private) in variables 4, 5, 7, 12, 16 and 29.

Results show that the staff from public university agree more than the staff from private university that learning, and e-learning have contributed to the development of positive interactions between faculty members and parents of students. Results also revealed that e-learning have contributed to enriching the knowledge and capabilities of parents in electronic technology and that e-learning have contributed to creating a family atmosphere tainted by tension, isolation and self-reflection and there is a difficulty in teaching scientific subjects electronically. Also, the level of seriousness of students in following the electronic platform is weak and it is difficult to follow up with the students to attend the lectures.

13 Conclusion

The results conclude that there is an impact of the Corona pandemic (Covid-19) on the level of academic achievement of Jordanian university students from the point of view of their parents. The skills of faculty members that affected students' level of achievement, reached a mean of (4.6353) with a standard deviation of (0.36703) with a high degree. Parent's reaction remained with a mean of (4.6310) and a standard deviation of (0.33848) in a high degree, followed by the field of the students' level of educational attainment with a mean of (4.5763) and a standard deviation of (0.28236).

The variable stated "I think that there is a disproportion between computers and the number of students in the same family" has the highest level with a mean of (4.7451). The variable stated "I think that e-learning have contributed to the development of the skills of the student and the faculty member" has the highest level with a mean of (4.8676). The variable state "I think that the level of seriousness of students in following the electronic platform is weak" has the highest level with a mean of (4.9853). The variable state that "I think that the Internet has worked to create a discrepancy between the abilities of students in educational attainment" has the highest level with a mean of (4.8578). Staff members from Economic and Management sciences agree more than staff members from Pharmacy and Basic sciences faculties that there is a disproportion between computers and the number of students living in the same family.

14 Recommendation

Based on the findings of this study, one can assert that e-learning is beneficial, and it is essential to encourage faculty members and students to use e-learning. Additionally, students must have an awareness about educational technology tool and e-learning. This kind of ed-technology awareness will help students use ed-technology properly and enjoy using it. Universities must provide learners and educators with appropriate educational training to facilitate the process of implementing e-learning. These kind of offered training must also encourage faculty members to communicate with students through electronic pages and e-mails to prepare them for interactions that take place virtually. Further studies need to be conducted to investigate the effectiveness of e-learning during the Covid-19 pandemic.

References

1. https://www.ecdc.europa.eu/en/covid-19
2. Draissi, Z., Yong, Q.Z.: COVID-19 outbreak response plan: implementing distance education in Moroccan Universities. School of Education, Shaanxi Normal University(2020). https://papers.ssrn.com/sol3/papers.cfm?abstract_id=3586783
3. Affouneh, S., Salha, S., Khlaif, Z.N.: Designing quality e-learning environments for emergency remote teaching in coronavirus crisis. Interdiscip. J. Virtual Learn. Med. Sci. **11**(2), 1–3 (2020)
4. Aljaser, A.: M The effectiveness of e-learning environment in developing academic achievement and the attitude to learn English among primary students. Turkish Online J. Distance Education-TOJDE **20**(2), 176–194 (2019)
5. Bashir, K.: Modeling E-learning interactivity, learner satisfaction and continuance learning Intention in Ugandan higher learning institutions. International Journal of Education and Development using Information and Communication Technology (2019)
6. Basilaia, G., Kvavadze, D.: Transition to online education in schools during a SARS-CoV-2 coronavirus (COVID-19) pandemic in Georgia. Pedagogical Res. **5**(4), em0060 (2020). https://doi.org/10.29333/pr/7937
7. Draissi, Z., Yong, Q.Z.: COVID-19 outbreak response plan: implementing distance education in Moroccan Universities. School of Education, Shaanxi Normal University (2020). https://papers.ssrn.com/sol3/papers.cfm?abstract_id=3586783
8. Yulia, H.: Online learning to prevent the spread of pandemic corona virus in Indonesia. ETERNAL (English Teaching Journal) **11** (2020)
9. Hodges, C., Moore, S., Lockee, B., Trust, T., Bond, A.: The Difference Between Emergency Remote Teaching and Online Learning (2020)
10. Basilaia, G., Kvavadze, D.: Transition to online education in schools during a SARS-CoV-2 coronavirus (COVID-19) pandemic in Georgia. Pedagogical Res. **5**(4), em0060 (2020). https://doi.org/10.29333/pr/7937

Evaluating the Students' Attitude Toward the Use of Moodle Mobile Application at Zarqa University in the COVID-19 Pandemic

Mohammad Musa Al-Momani and Muneer Yousef Hasan

Abstract This study aimed to evaluating the students' attitude toward the use of Moodle mobile application at Zarqa University (ZU) in the covid-19 pandemic. The study used the descriptive analytical approach, and the study sample consisted of (368) students who were randomly selected of undergraduate studies in ZU, and to achieve the objectives of the study, the researchers designed a questionnaire as a tool for collecting data from the members of the study sample. Among undergraduate students in Jordanian universities, There are statistically significant differences in the use of the mobile phone "Moodle mobile application" in distance learning for bachelor's degree students at private Jordanian universities at the significance level (=0.05). This research focuses on the elements that influence students' adoption of the Moodle mobile application and its ability to be used effectively. As a conceptual research framework of Moodle mobile application adoption, the technology acceptance model (TAM) is used. This paper aims to evaluate factors affecting the student's attitude toward the use of the Moodle mobile application at Zarqa University (ZU) in Jordan. The following recommendations were made based on the findings of the study, which showed that students are interested in and satisfied with the Moodle mobile application: Encouraging faculty members and undergraduate students in Jordanian universities to use Moodle mobile application in teaching, design of educational technology lessons for undergraduate and postgraduate subjects to used it in teaching through the Moodle mobile application in order to achieve the goals of distance teaching.

Keywords Online learning · Distance learning · Moodle application · Technology acceptance model · COVID-19 · Zarqa University

M. M. Al-Momani (✉) · M. Y. Hasan
Faculty of Economics and Business Administrative Sciences, Management Information Systems, Zarqa University, Zarqa, Jordan
e-mail: m.momani@zu.edu.jo

M. Y. Hasan
e-mail: mhasan@zu.edu.jo

A. Hamdan et al. (eds.), *The Implementation of Smart Technologies for Business Success and Sustainability*, Studies in Systems, Decision and Control 216,
https://doi.org/10.1007/978-3-031-10212-7_50

1 Introduction

Today, the global community is affected by rapid changes in the field of information and communication technology, which has become a feature of this era, which in turn has been reflected on the institutions of society, including educational institutions. This technological change led to a knowledge revolution that abolished the boundaries of time and space. This led to the need to move from traditional education to technological education through the use of new educational strategies that help keep pace with the current development in the technological system. Distance education is one of these strategies, which relies on the use of modern communication tools such as computers, networks, multimedia of sound, image, graphics, and electronic libraries, whether remotely or in the classroom, that is, the use of technology in all its forms (computer, telephone, mobile, and educational television) all work to deliver information to the learner in the shortest time, with the least effort, and with the greatest benefit.

Conversely, with the emergence of COVID-19 pandemic, the trend to deal with e-learning and the use of modern communication has become an urgent matter. The activation of e-learning and distance education is a real response by educational institutions to keep pace with technological progress to confront the COVID-19 pandemic. Technology plays an important role in students' lives; as it helps them to teach more effectively and stimulates their desire for discovery and experimentation, so we find that most students are interested in technology and in all its forms and tools. With the emergence of the pandemic, the university education system has become unable to meet the needs of large numbers of students due to the discontinuation of the prevailing educational practices associated with its methods and strategies. Therefore, these reasons prompted those in charge of educational administration to search for new educational methods such as e-learning and distance education, which helped to overcome the barrier of place and time in learning, and to provide effective learning based on multimedia elements.

On the surface, the emergence of the idea of distance learning and e-learning as a result of rapid technological changes and the emergence of the Corona pandemic created challenges regarding the need to provide additional educational opportunities without increasing the budget, so many educational institutions have begun to address this challenge by developing distance learning programs [1]. Hence, the importance of this study stems from the importance of the topic it deals with, which is related to knowing the effectiveness of the use of the mobile phone "Moodle application" in distance learning among undergraduate students in private Jordanian universities and in the educational process. Consequently, this study aimed to evaluating the students' attitude toward the use of Moodle mobile application at Zarqa University (ZU) in the covid-19 pandemic.

2 Literature Review

According to quasi-experimental study [2] which aimed to evaluate the effect of task-based learning with the help of mobile devices on students' writing competence and students' writing motivation. The study sample consisted of (40) male and female students divided into two groups, the experimental group was taught through task-based learning with the help of mobile devices, while the control group was taught through task-based learning without the help of mobile devices. The study's findings revealed that (1) task-based learning with the aid of a mobile phone had a significant effect on students' writing efficiency; (2) task-based learning with mobile-assisted learning had a significant effect on students' writing motivation; and (3) task-based mobile-assisted learning had a concurrent effect on students' writing competence and motivation. The study concluded that employing mobile devices as a useful tool in the effective learning process can help students improve their writing skills and increase their drive to write.

In addition, Saleh research [3] aimed to identify the extent of the quality of electronic services in the distance education system at the Faculty of Education in Suhag from the point of view of faculty members and students and their relationship to some variables. The study used the descriptive analytical approach, and a questionnaire was applied to the quality of e-services for distance education on a sample of (50) faculty members and a group of students of the General Educational Diploma Program (200) male and female students. The study showed that the quality level of e-services for distance education from the point of view of students and faculty members came to a medium degree, and there were no differences in the responses of faculty members due to the variable of gender and the level of different computer courses. There are differences due to the academic rank level variable in favor of assistant teachers, there are no differences in students' responses due to gender, and there are differences due to specialization in favor of literary majors.

Moreover, Shuqqura study [4] aimed to identify the degree of mobile phone use in acquiring English language skills from the point of view of Jordanian university students. The study used the descriptive analytical approach, and to achieve the objectives of the study, the questionnaire was used as a tool for collecting data from the study sample members, where the study sample consisted of (400) students from the University of Jordan, and the results showed: From their perspective, there is a high degree of use of mobile phones by University of Jordan students in developing English language abilities. The study also revealed that there were statistically significant differences in the responses of the sample member's terms of the degree of using the cell phone to acquire English language skills based on gender, academic level, and duration of using the cell phone to acquire English language skills.

In addition, one study [5] at Najran University looked into the impact of mobile learning, a sort of e-learning that utilizes mobile devices, on the improvement of academic accomplishment and conversational abilities of English language students. The study employed a quasi-experimental approach. The participants are (50) male and female students enrolled in the Department of Curriculum and Teaching's

"Teaching and Learning Strategies" course for students specializing in English at the Department of English Language at the College of Education. The pupils that took part were placed into two experimental and control groups, each with twenty-five individuals. For students in the course, an electronic version of the course content was posted to the university's blackboard system. At the first meeting, the students in the control group were given another physical copy of the course content. An academic achievement exam and a speaking skills evaluation scale were used to collect data. Mobile learning has a substantial impact on students' academic progress and communication abilities, according to the findings. Faculty members are advised to tailor the usage of mobile learning in their lectures based on these findings.

The researchers [6] constructed an integrated system called (IMELS) which related to evaluate the impact of Blended Learning on students' academic progress, interaction, and satisfaction in higher education. As a treatment group, a group of students who took an Artificial Intelligence course at the University of Jordan's IT College employed for this method. Another group of students who attended the class studied the identical materials in a traditional manner. The findings revealed that blended learning, which makes use of new learning technology, can improve students' academic performance, interactions, and satisfaction. At the same time, [7] employed the descriptive approach to perform a research at Zarqa University on the effectiveness of students' attitudes regarding using E-exams. The findings revealed that Zarqa University students approve and are interested in using E-exams as part of an E-learning system like Moodle to assist them improve their educational attainment. As a result, the researcher recommends that E-exams be used as a permanent component of the educational process because this kind of exam has a clear impact on increasing student interaction and academic achievement, particularly in the COVID-19 pandemic.

3 Research Framework

This research uses Technology Acceptance Model (TAM) as shown in Fig. 1. The model of study is included two independent factors which are technology factors and student factors in order to show the relation between its characteristics and the attitude towards the use of Moodle mobile application.

The technology factors variable is used to study the relationship between perceived usefulness and perceived ease of use in terms of information quality, service quality and system quality. According to Roca et al. [8], the hypotheses are based on the research as following:

H1: "Technology factors will have significant influence on perceived usefulness towards Moodle mobile application."

H2: "Technology factors will have significant influence on perceived ease of use towards Moodle mobile application."

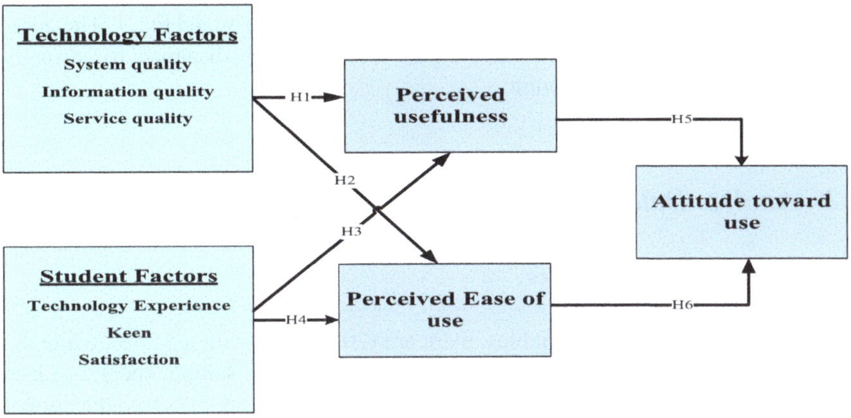

Fig. 1 Research model

The student factors variable is used to study the relationship between perceived usefulness and perceived ease of use in terms of technology experience, keen and satisfaction. According to Roca et al. [8], the hypotheses are based on the research as following:

H3: "Student factors will have significant influence on perceived usefulness towards Moodle mobile application."

H4: "Student factors will have significant influence on perceived ease of use towards Moodle mobile application."

Users' attitudes toward technology acceptance were found to be influenced significantly by perceived ease of use and perceived usefulness [8]. The hypothesis is shown below:

H5: "Perceived usefulness will have significant influence on the attitudes toward use of Moodle mobile application."

H6: "Perceived ease of use will have significant influence on the attitudes toward use of Moodle mobile application."

4 Research Methodology

4.1 Sampling

Around (368) undergraduate students at ZU from more than 20 departments answered the final version of the questionnaire. The online questionnaire distributed between participants to fill it in order to measuring the variables of the model Fig. 1. The questionnaire used five-point Likert scale ranging from "strongly disagree" to "strongly agree" with a middle neutral point.

4.2 Analyze Data

4.2.1 Sample Characteristics

According to demographic variables, there are (201) males with a 54.6%, while the females are (167) with a 45.4%. In addition, 48.1% of the sample spent 1–3 h at the computer per week for non-educational purposes while 33.7% for educational purposes. Also, it is found that (60.1%) of the sample is between 20–25 years old.

4.2.2 Testing of Reliability

"*Cronbach Alpha* (α) *test*" was used to find out instrument reliability. According to [9] the values are accepted if it's more than 60%. The value was = 96.3% for the questionnaire and alpha for each variable is greater than accepted percent 0.60, as shown in Table 1:

Table 1 Cornbach alpha

Variable	α
"Technology factor"	0.93
"Student factor"	0.86
"Perceived usefulness"	0.94
"Perceived ease of use"	0.92
"Attitude toward use"	0.91

4.2.3 Statistical Results

Collinearity Statistics Test

The collinearity statistics, which included the Tolerance and Variance Inflation Factor (VIF), were used to check for multicollinearity between the two independent variables. The VIF's maximum allowable cutoff value is (10). To demonstrate that there is no concern with multicollinearity, the tolerance value should not be less than 0.10, and the VIF value should not be greater than (10) (Table 2).

Table 2 Statistics of collinearity

Model	"Statistics of collinearity"	
	"Tolerance"	VIF
"Technology factor"	0.39	2.617
"Student factor"	0.39	2.617

Hypothesis Testing

All hypotheses are tested using the Statistical Package for Social Sciences (SPSS) software.

Hypothesis #1: "Technology factors will have significant influence on perceived usefulness towards Moodle mobile application."

Simple regression was employed by the authors to test the first hypothesis. Table 3 shows that at the (0.01) level, F value is calculated significant. Thus, with a moderate Pearson correlation (r) of 0.59, technology factors will have a significant impact on perceived usefulness of the Moodle mobile application; additionally, technology factors will have a significant impact on perceived usefulness of the Moodle mobile application, accounting for 35% of the variance (R^2).

Hypothesis #2: "Technology factors will have significant influence on perceived ease of use towards Moodle mobile application."

The authors employed simple regression to test the second hypothesis. Table 4 shows that at the (0.01) level, F value is calculated significant. Thus, with a moderate Pearson correlation (r) of 0.65, technology factors will have a significant impact on perceived usefulness of the Moodle mobile application; additionally, technology factors will have a significant impact on perceived ease of use towards Moodle mobile application, accounting for 43% of the variance (R^2).

Hypothesis #3: "Student factors will have significant influence on perceived usefulness towards Moodle mobile application."

Table 3 Hypothesis #1 testing

"R^2"	"R"	"Sig t"	"T"	"Sig F"	"F"
(0.35)	(0.59)	(0.00^{**})	(8.33)	(0.00^{**})	(69.4)

**"Significant at (0.01) level"

Table 4 Hypothesis #2 testing

"R^2"	"R"	"Sig t"	"T"	"Sig F"	"F"
(0.43)	(0.65)	(0.00^{**})	(9.77)	(0.00^{**})	(95.4)

**"Significant at (0.01) level"

Simple regression was employed by the authors to test the third hypothesis. Table 5 shows that at the (0.01) level, F value is calculated significant. As a result, student factors will have a significant impact on perceived usefulness of the Moodle mobile application, with a moderate Pearson correlation (r) of 0.74; additionally, student factors will have a significant impact on perceived usefulness of the Moodle mobile application, accounting for 54% of the variance (R^2).

Hypothesis #4: "Student factors will have significant influence on perceived on perceived ease of use towards Moodle mobile application."

Simple regression was employed by the authors to evaluate the fourth hypothesis. Table 6 shows that at the (0.01) level, F value is calculated significant. As a result, student factors will have a significant impact on perceived ease of use of the Moodle mobile application, with a moderate Pearson correlation (r) of 0.62; additionally, student factors will have a significant impact on perceived ease of use of the Moodle mobile application, accounting for 38% of the variance (R^2).

Hypothesis #5: "Perceived usefulness will have significant influence on the attitudes toward use of Moodle mobile application."

Simple regression was employed by the authors to examine the fifth hypothesis. Table 7 shows that at the (0.01) level, F value is calculated significant. As a result, perceived usefulness will have a significant impact on attitudes toward using the Moodle mobile application, with a moderate Pearson correlation (R) of 0.72; additionally, perceived usefulness will have a significant impact on attitudes toward using the Moodle mobile application, accounting for 51% of the variance (R^2).

Table 5 Hypothesis #3 testing

"R^2"	"R"	"Sig t"	"T"	"Sig F"	"F"
(0.54)	(0.74)	(0.00^{**})	(12.34)	(0.00^{**})	(152.25)

** "Significant at (0.01) level"

Table 6 Hypothesis #4 testing

"R^2"	"R"	"Sig t"	"T"	"Sig F"	"F"
(0.38)	(0.62)	(0.00^{**})	(8.92)	(0.00^{**})	(79.6)

** "Significant at (0.01) level"

Table 7 Hypothesis #5 testing

"R^2"	"R"	"Sig t"	"T"	"Sig F"	"F"
(0.51)	(0.72)	(0.00^{**})	(11.6)	(0.00^{**})	(133.9)

** "Significant at (0.01) level"

Table 8 Hypothesis #6 testing

"R^2"	"R"	"Sig t"	"T"	"Sig F"	"F"
(0.23)	(0.47)	(0.00**)	(5.94)	(0.00**)	(35.3)

**"Significant at (0.01) level"

Hypothesis #6: "Perceived ease of use will have significant influence on the attitudes toward use of Moodle mobile application."

Simple regression was employed by the authors to examine the last hypothesis. Table 8 shows that at the (0.01) level, F value is calculated significant. Thus, with a moderate Pearson correlation (R) of 0.47, perceived ease of use will have a significant influence on attitudes toward using the Moodle mobile application; also, with 23% of the variance, perceived ease of use will have a significant influence on attitudes toward using the Moodle mobile application (R^2).

5 Discussion and Conclusion

The influence of perceived usefulness, perceived ease of use, technology characteristics, and student factors on students' attitudes on using the Moodle mobile application at ZU during the covid-19 pandemic were evaluated in this study. Based on the results that supported the hypotheses of the study, which showed that there is a strong influence of technology factors and student factors on the perceived usefulness. These findings were in line with previous research, which revealed that system quality and student satisfaction have an impact on the perceived usefulness of using Moodle mobile applications in the covid-19 pandemic [10, 11].

Furthermore, the findings revealed that perceived usefulness has a high significant impact on attitudes toward using the Moodle mobile application in the covid-19 pandemic, but perceived ease of use has a moderate significant impact. However, the findings revealed that perceived usefulness and ease of use had a significant impact on attitudes toward using the Moodle mobile application in the covid-19 pandemic, which is in line with previous research [12, 13].

Consequently, it can be concluded that the students of ZU are attracted to use the Moodle mobile application in the covid-19 pandemic, hence, it can be said that the use of these applications helps in providing valuable information, and that these applications enhance students' productivity and speed up their completion of the tasks assigned to them. Accordingly, students' attraction and satisfaction with the use of the mobile application for the e-learning platform Moodle increased the awareness of faculty members and those in charge of educational institutions of the need to think seriously about integrating the educational process, which in turn will be reflected in the performance of students who have a passion for dealing with modern technological systems and dealing With applications that have become an essential part of their daily lives. Thus, the student becomes able to obtain knowledge and information at any time and any place. Finally, the researchers recommended to activating the role of mobile applications that are related to e-learning, because of their importance in increasing student interaction, especially with the wide spread of these mobile phones among students.

References

1. Al-Momani, M.M.: Evaluating students' attitude toward the use of e-learning: a comparative study between a private and public universities in Jordan. Inf. Knowl. Manag. **7**(1), 17–24 (2017)
2. Dewi, P.A., Ratminingsih, N.M., Santosa, M.H.: Mobile-assisted task-based language learning, writing competency, and motivation. Jurnal Pendidikan Indonesia (JPI) **9**(1), 119–130 (2020)
3. Saleh, S.: Evaluating the quality of e-services for distance education at the Faculty of Education in Sohag from the point of view of faculty members and students. Educ. J. Faculty Educ. Sohag Univ. **74**(1), 205–252 (2020)
4. Shaqqura, T.: The degree of cell phone use in acquiring English language skills from the point of view of Jordanian university students. Master Thesis, AlMafraq, Al al-Bayt University (2019)
5. Elfeky, A.I., Masadeh, T.S.: The effect of mobile learning on students' achievement and conversational skills. Int. J. Higher Educ. **5**(3), 20–31 (2016)
6. Al-Momani, M.M., PILLI, O.: The impact of integrated mobile system on students' academic achievement, interaction and satisfaction. TEM J. **10**(3), 1395–1403 (2021)
7. The Effectiveness of Students' Attitude towards the Use of E-Exams at Zarqa University. In: GAI International Academic Conferences Proceedings, Turkey (2020)
8. Roca, J.C., Chiu, C.-M., Martínez, F.J.: Understanding e-learning continuance intention: an extension of the Technology Acceptance Model. Int. J. Hum. Comput. Stud. **64**(8), 683–696 (2006)
9. Malhotra, N.: Marketing Research, An Applied Orientation. Prentice Hall, New Jersey (2004)
10. Saikat, S., Dhillon, J.S., Ahmad, W.F., Jamaluddin, R.A.: Systematic review of the benefits and challenges of mobile learning during the COVID-19 pandemic. Educ. Sci. **11**(459), 2–14 (2021)
11. Salloum, S.A., Akour, I., Alshurideh, M., Al Kurdi, B., Al Ali, A.: Using machine learning algorithms to predict people's intention to use mobile learning platforms during the COVID-19 pandemic: machine learning approach. JMIR Med. Educ. **7**(1) (2021)
12. Biswas, B., Roy, S.K., Roy, F.: Students perception of mobile learning during COVID-19 in Bangladesh: university student perspective. Aquademia **4**(2), 1–9 (2020)
13. Alhumaid, K., Habes, M., Salloum, S.A.: Examining the factors influencing the mobile learning usage during COVID-19 Pandemic: an Integrated SEM-ANN Method. IEEE Access **9**, 102567–102578 (2021)

The Effectiveness of Digital Communication in the Education System in Jordan in Light of the Covid-19 Pandemic

Tahseen Mohammad Anis Sharadga and Amjad Omar Safori

Abstract In light of the COVID-19 pandemic, the goal of the study was to determine the effectiveness of digital communication in Jordan's educational system. The study sample consisted of 42 faculty members and was conducted using the descriptive-analytical method. The researchers adopted it as a sample for the study in the universities of Yarmouk, Zarqa, and the Middle East, and within the faculties of media, as they were considered experts, and they were selected from the study population according to what is available for collection by random method. The questionnaire was also used to collect data about the study's subject. The study came to some conclusions, one of which was that the effectiveness of digital communication in Jordan's education system in light of the Covid 19 pandemic was low, as the arithmetic means of the total degree of the scale was (2.489), with a standard deviation of (0.592), indicating a low degree of dispersion between the responses of a sample of the study on the total mean, indicating homogeneity of the responses.

Keywords Digital communication · Jordanian education system · COVID-19 pandemic

1 Introduction

Those who closely follow the current era's technological advancements will easily notice the tremendous progress in digital technologies, as well as their widespread adoption by most segments of society and at all levels until the era of digitization was coined; The rapid transition to digital data management was coined as a result of the digitization of many processes [1].

T. M. A. Sharadga · A. O. Safori (✉)
Faculty of Media, Zarqa University, Zarqa, Jordan
e-mail: asafori@zu.edu.jo

T. M. A. Sharadga
e-mail: tsharadqa@zu.edu.jo

A. Hamdan et al. (eds.), *The Implementation of Smart Technologies for Business Success and Sustainability*, Studies in Systems, Decision and Control 216,
https://doi.org/10.1007/978-3-031-10212-7_51

It is important to note that digital communication is primarily based on the use of digital systems and technologies as the sole and primary means of communication between its parties, which is similar to computer-based communication [2].

According to Al-Dahabi and Maamari [3], digital communication is a social process in which participants communicate at a distance by exchanging roles in transmitting and receiving various communication messages via digital systems and their means to achieve specific goals. Digital communication, according to Amroush [4], is communication that takes place through digital technology and relies on the digitization of information. Digital communication, according to Belonovskaya et al. [5], is based on communication technology and economic exchange between participants on the one hand, and Web technologies and services on the other, and thus is mostly virtual, facilitating the process of communication between individuals.

Digital communication and its obligations were placed on planners in general and educational leaders in particular, Demeshkant [6], who were required to consider digitalization and its implications when designing existing and future programs. In the same vein, Al-Bayati [7] stated that digital problems have become a reality and that they must be learned, mastered, and used to increase creativity in administrative, educational, and educational processes. The digital communication processes in the educational system consist of sending ideas and information to the educational system using digital systems, which leads to unifying efforts to fulfill the educational system's goals and mission [8].

It's worth noting that digital communication uses web technology to build an interactive educational system environment by integrating the benefits of electronic content management systems among school leaders, teachers, and educational system personnel [9]. According to Al-Buhairi [10], digital communication facilitates the interchange of ideas, information, knowledge, and opinions among educational leaders, teachers, and workers, as well as the sharing of scientific content, all of which contribute to high-quality educational outcomes.

It's worth noting that the Covid 19 pandemic came on suddenly, prompting the World Health Organization to declare a state of emergency in all aspects of human life, including education, and as a result, health experts recommended several measures to prevent the virus from spreading. Remote work, social distancing, quarantine, and even the complete closure of educational institutions were among the measures implemented [11]. As a result, in March 2020, educational platforms such as Google Classroom, WebQuest, Microsoft Teams, and other online sites made the transition to teaching via digital communication on the Internet. However, due to a lack of training and experience dealing with this type of communication, the majority of faculty members were not well versed in using the requirements of digital communication or dealing with e-learning [12].

Based on the foregoing, the research into the effectiveness of digital communication in the Jordanian education system in light of the Covid-19 pandemic through a field study from the perspective of faculty members at Zarqa University's Faculty of Mass Communication has its justifications, in terms of understanding the current reality of digital communication in the Jordanian education system, as well as attempting to identify the faculty's ability to identify the ability of the faculty.

2 Problem of the Study

What is currently happening in the world is classified as a pandemic, which means that managing the crisis of the Covid 19 spread necessitates preparedness, thorough preparation, scientific planning, and ongoing training to respond to the crisis and minimize its dangers [13]. This means that decision-makers in Jordan's educational system, as well as administrative and operational systems, are always responsible for effective planning, organization, and coordination during the Coronavirus spread.

However, it should be highlighted that, in addition to electronic organization, the use of digital communication necessitates human, administrative, and technical needs, as well as attention to the educational system's internal and external environments [14]. One of the manifestations imposed by digital technology and its challenges is the transformation of educational systems toward digital communication, which has led to changes in the structure of the educational system, performance standards, and digital services within which this system operates in one way or another [15].

The researcher's feelings about the problem arose from his job as a lecturer at Jordan's Zarqa University, where he saw the need for a study on the reality of the Covid 19 outbreak and controlling its growth, which is afflicting the world in general and Jordan in particular. As a result, the study's difficulty occurred, and a series of concerns arose in the researcher's mind about the processes employed, methods employed, and possible scenarios employed by the Jordanian state, represented by the Ministry of Education, in combating the spread of the Coronavirus. This is in the context of diagnosing and understanding the context of the current situation in Jordan and the context of the current global situation, analyzing, interpreting, and providing answers to the following main study question.

What is the effectiveness of digital communication in the education system in Jordan in light of the COVID-19 pandemic through a field study from the point of view of faculty members at the Faculty of Mass Communication at Zarqa University?

3 Significance of the Study

- Shedding light on the importance of digital communication in the education system in Jordan in light of the Covid-19 pandemic.
- Demonstrating the mechanism of confronting the spread of Covid 19 by employing the requirements of digital communication.
- Determining the requirements of digital communication, its mechanisms, and its implications for limiting the spread of Covid 19.

Delimitations of the Study

The goal of the study was to determine the effectiveness of digital communication in Jordan's educational system in light of the COVID-19 pandemic by conducting a field study in 2022 from the perspective of faculty members at Zarqa University's Faculty of Mass Communication.

4 Terms of the Study

Digital communication: Digital communication is a method of dealing with information on an analog basis in addition to the digital method and digitization of information as well, i.e. converting information from text, image, and sound to digitized waves, which facilitates the process of sending, storing, processing and compressing them [16].

Digital communication in this concept falls within the broader framework of the concept of communication and its elements, forms, and models because, in the end, it is a human contact between individuals to achieve certain goals [17].

Covid 19: It is a large family of viruses that may cause illness among animals and humans, and it is known that some coronaviruses cause respiratory infections among humans, ranging in severity from the common cold to more severe diseases such as Middle East Respiratory Syndrome and Severe Acute Respiratory Syndrome (SARS). The newly discovered coronavirus causes COVID-19 [18].

5 Methodology of the Study

By studying a phenomenon to describe it and collect accurate information about it, the study took a descriptive-analytical approach, which is based on collecting, classifying, and categorizing data and facts. The survey included 42 academics from the faculties of mass communication at the universities of Yarmouk, Zarqa, and the Middle East. They were considered experts, and they were chosen at random from the research population based on the available data.

The researcher employed one of the most frequent methods of data collection, the questionnaire, as an instrument for gathering data relating to the study's subject by developing a form for the respondent to fill out to meet the field study's objectives. The researcher gathered comments from faculty members at Zarqa University's Faculty of Mass Communication, who were asked to select the response that best expressed their opinion for each item; responses were limited to the following options: (never, rarely, sometimes, often, always).

The scale, in its final form, after verifying its validity and stability, consisted of (14) items, each of which is listed on a scale of five levels of response to the Likert

Table 1 The key to correcting the study instrument

Cell length	Corresponding relative weight	Estimation
From 1.00–1.80	From 20%–36%	Very low
Greater than 1.80–2.60	Greater than 36%–52%	Low
Greater than 2.60–3.40	Greater than 52%–68%	Medium
Greater than 3.40–4.20	Greater than 68%–84%	High
Greater than 4.20–5.00	Greater than 84%–100%	Very high

scale: (never, rarely, sometimes, often, always) and the following scores are given, respectively: (1, 2, 3, 4, 5). An evaluation judgment was given for the degree of influence for each of the scale items, and Table 1 shows that.

Psychometric Properties of the Study Instrument

6 Validity of the Scale

The validity was calculated using the constructive validity method, where the researcher calculated the correlation coefficient between the items and the total score of the scale to infer the homogeneity between the items of the scale and the total score of the scale. This was done by extracting the values of the (Point-Biserial) correlation coefficients between the item and the total score of the scale, and Table 2 shows these correlations through the data of the pilot sample.

It is clear from Table 2 that the correlation coefficients between the item and the total score of the scale were good, and these results reinforce the constructive validity of the scale.

Table 2 Item correlation coefficients (rpb) with the total score of the scale

Item	rpb. correlation coefficient	*P*-value	Item	rpb. correlation coefficient	*P*-value
1	0.497	0.000*	8	0.646	0.000*
2	0.412	0.000*	9	0.543	0.000*
3	0.628	0.000*	10	0.463	0.000*
4	0.643	0.000*	11	0.294	0.000*
5	0.548	0.000*	12	0.432	0.000*
6	0.558	0.000*	13	0.578	0.000*
7	0.452	0.000*	14	0.667	0.000*

*Statistically significant at ($\alpha \leq 0.05$)

Table 3 Cronbach's alpha

Dimension	Number of items	Alpha reliability coefficien
The effectiveness of digital communication in the education system in Jordan in light of the COVID-19 pandemic	14	0.847

The researcher followed the Cronbach's Alpha method: a method that requires calculating the correlation of items with each other. Table 3 shows the scale reliability coefficients using Cronbach's Alpha.

It is clear from the table that the reliability coefficient of the scale was (0.847), which is a high coefficient of reliability that indicates that the scale has a high degree of stability and can be relied upon in the field application of the study.

7 Results and Discussion

The Study Main Question: What is the effectiveness of digital communication in the education system in Jordan in light of the COVID-19 pandemic through a field study from the point of view of faculty members at the Faculty of Mass Communication at Zarqa University?

To answer this question, the arithmetic means and standard deviations of the answers of the sample members were found about the effectiveness of digital communication in the education system in Jordan in light of the COVID-19 pandemic, as shown in Table 4.

The arithmetic means of the total score of the scale consisting of (14) items according to the five-point Likert scale were (2.489) and a standard deviation of (2.489), indicating that the effectiveness of digital communication in the education system in Jordan in light of the Covid 19 pandemic was low (0.592). It found arithmetic means with responses ranging from medium to low, indicating a lesser degree of dispersion between the study sample's replies than the total mean, which implies response homogeneity.

The previous table also shows that three items with arithmetic mean received a medium mean level: (Current and future programs in the Jordanian education system are being planned around digital connectivity.), (The Jordanian education system relies on communication technology and web services), and (The process of digital communication in the Jordanian education system is to transfer information from school leaders to staff). This result can be explained by the fact that the Jordanian education system has, on average, plans and programs to implement digital communication mechanisms, and the learning system is working hard during this period of Covid 19's spread to apply digital communication between educational institutions of all kinds, as well as the Internet and other means of communication, to transfer information among educational institutions, schools, and universities.

Table 4 Arithmetic means and standard deviations of digital communication effectiveness

No.	Items	Means	Standard deviations	Estimation	Rank
8	Current and future programs are being planned around digital connectivity in the Jordanian education system	2.693	0.615	Medium	1
14	The Jordanian education system relies on communication technology and web services	2.642	0.535	Medium	2
6	The process of digital communication in the Jordanian education system is to transfer information from school leaders to staff	2.616	1.301	Medium	3
4	Digital communication in the Jordanian education system leads to faster achievement	2.591	0.792	Low	4
10	The Jordanian education system employs electronic content technology to limit the spread of COVID-19	2.572	0.544	Low	5
13	Digital communication in the Jordanian education system is based on the use of digital systems and their innovations	2.554	0.972	Low	6
7	Roles are exchanged in transmitting and receiving communication messages through digital systems in the Jordanian education system	2.529	0.716	Low	7
1	The Jordanian education system applies digital communication to limit the spread of COVID-19	2.494	1.333	Low	8
9	The Jordanian education system is working to create an interactive educational system environment to limit the spread of COVID-19	2.453	1.221	Low	9
5	There is communication between school leaders, teachers, supervisors, and administrative staff in the educational system	2.417	0.903	Low	10

(continued)

Table 4 (continued)

No.	Items	Means	Standard deviations	Estimation	Rank
11	Digital communication has an important role in improving productivity and efficiency in the Jordanian education system	2.383	0.950	Low	11
12	Digital information systems in the Jordanian education system can limit the spread of COVID-19	2.336	1.065	Low	12
3	Digital communication seeks positive change within the Jordanian educational system	2.297	1.049	Low	13
2	The Jordanian education system is concerned with the information technology industry related to digital communication	2.274	1.191	Low	14
Total		2.489	0.592	Low	

According to the previous table, the remainder of the items received a low mean score. This indicates that there is a flaw in the e-learning system's application, use, and employment of digital communication. The inability to create an interactive educational system environment targeted at limiting the spread of Covid 19, as well as the lack of use of electronic content for social distance and the non-spread of the Covid 19 pandemic, according to the researcher. This result could also be attributable to Jordan's educational system's failure to use digital information technologies to effect positive change.

8 Recommendations

1. To limit the spread of COVID-19, employees in Jordan's education system are being trained to prepare educational plans using digital communication.
2. The importance of extensive media efforts to raise awareness of Covid-19's spread through Ministry of Higher Education decisions and circulars published on university websites.
3. Emphasizing the importance of using digital communication to limit the spread of Covid 19 in light of the global crisis that is afflicting the world and Jordan in particular.
4. To limit the spread of Covid 19, researchers are focusing on remote administrative practices in terms of correspondence and decisions related to planning and organizing in the educational process.

5. Focusing on university professors' coordination and follow-up practices to achieve high-quality e-learning to avoid mixing in these critical times.
6. Restructuring courses for e-learning and making them available on college websites, which, of course, limits mixing and encourages people to stay at home to prevent the spread of Covid 19.
7. Establishing additional channels of communication between Jordan's educational stakeholders.
8. Using digital communication to provide information to auditors in universities, schools, and educational departments.
9. Increasing the use of digital communication to close the digital divide between developing and developed countries.
10. Qualification, professional development, training, and continuous learning of university and school teaching staff on digital communication requirements.

References

1. Storey, V.C., Woo, C.C.: Data challenges in the digitization era. Workshop on Information Technology and Systems, Santa Clara. Proceedings 28th Workshop on Information Systems and Technologies (WITS), Santa Clara, California, 16–19 Dec 2018, pp. 1–15 (2018)
2. Mutlaq, K.: Measuring the levels of digital competencies for general education teachers in the field of digital transformation. J. Educ. Sci. **87**, 1301–1353 (2020)
3. Al-Dahabi, M., Maamari, A.: Digital communication in higher education institutions, a field study of a sample of professors and administrators at Adrar University, Master's Thesis, Adrar University (2018)
4. Savori, A.O.: Journalist use of social media: guidelines for media organizations. J. Soc. Sci. Res. **5**, 772–779 (2018a). Available: https://doi.org/10.32861/jssr.spi5.772.779
5. Belonovskaya, I.D., Matvievskaya, E.G., Saitbaeva, E.R., Ksenofontova, A.N., Usmanov, S.M., Zatsepina, M.B., Bakshaeva, E.: V: Digital communication in educational process: development trends and new opportunities. Online J. Commun. Media Technol. **10**(2), e202008 (2020)
6. Demeshkant, N.: Future academic teachers' digital skills: polish case-study. Univers. J. Educ. Res. **8**(7), 3173–3178 (2020)
7. Al-Bayati, K.: Digital Communication Rising Nations and Surprised Nations. Dar Al-Bidaya Publishers, Jordan (2015)
8. Al-Baqami, A: A proposed conception of electronic training for educational supervisors in the light of some contemporary experiences. Master's Thesis, Colleges of the Arab East, Saudi Arabia (2016)
9. Al-Hoshan, A.: Evaluating the distance education experience in light of the Corona pandemic using the multiple decisions model (CIPP) a field study on Saudi university students. J. Educ. Psychol. Sci. **4**(8), 61–81 (2020)
10. Al-Buhairi, S.: Attitudes of faculty members towards the application of electronic platforms in the educational process. J. Media Res. **51**(1), 262–288 (2019)
11. Tanveer, M., Bhaumik, A., Hassan, S., Haq, I.: U: Covid-19 pandemic, outbreak educational sector and students online learning in Saudi Arabia. J. Entrepreneurship Educ. **23**(3), 1–14 (2020)
12. Toquero, C.M.: Challenges and opportunities for higher education amid the COVID-19 pandemic: the Philippine context. Pedagogical Res. **5**(4) (2020)
13. Huang, C., Wang, Y., Li, X., Ren, L., Zhao, J., Hu, Y., et al.: Clinical features of patients infected with 2019 novel coronavirus in Wuhan, China. Lancet **395**(10223), 497–506 (2020)

14. Al-Anazi, N., Obaid, A.: Educational leadership strategy in light of the COVID-19 pandemic in the Kingdom of Saudi Arabia. J. College Educ. **37**(5), 375–396 (2021)
15. Ramadan, M.: The role of distance education in solving the problems of the emerging corona epidemic. Educ. J. **20**(7), 1531–1543 (2020)
16. Hammouche, M: Digital communication and its impact on job performance, Master's Thesis, Mohamed Boudiaf University (2020)
17. Portillo, J., Garay, U., Tejada, E., Bilbao, N.: Self-perception of digital competence of educators during the COVID-19 pandemic: A cross-analysis of different educational stages. Basel, Switzerland Basel, Basel: MDPI AG1 (2020)
18. Zhu, N., Zhang, D., Wang, W., Li, X., Yang, B., Song, J., et al.: A novel coronavirus from patients with pneumonia in China, 2019. New Engl. J. Med. (2020)

Dimensions of National Education Policy in Ukraine

Olga Nezhyva and Natalia Teslenko

Abstract This paper examines the degree of development of public policy issues in Ukrainian education. The national policy in the field of Ukrainian education is exercised by the executive authorities and local governments on the basis of the concept of education accepted by authorities and approved by the public opinion. Ukrainian education policy is a tool of influence of the political elite of Ukrainian society for the younger generation. Ukrainian officials, who have all the levers of influence, make certain transformation in the concept of development education, independently, without regard to public view or control of the society. The national policy in the field of Ukrainian education is based on strategic documents of education; constitutional norms, international treaties ratified by the supreme legislative body; legislation; Executive Orders of the President and so on. A preliminary analysis of modern Ukrainian education policy reveals its ideological eclecticism, which creates a favorable environment for the formation of internal contradictions, the slowdown of reforms and the impossibility of transformations. The key assumptions which determine the model of "rational state—sovereign" is determined as: the function of higher education; evaluation of higher education is based on the definition of political effectiveness; terms of interaction based hierarchical subordination and submission, participation of society in education is limited; developing solutions rigidly centralized from top to bottom; hierarchical control system; comprehensive nature of public exposure; autonomy of any of limited scope solution "technical issues"; changes in higher education are the result of changes in political leadership.

Keywords Policy · National policy in the field of education · Education · State · Ukraine

---

O. Nezhyva
National University of Food Technologies, Volodymyrska str. 68, Kyiv 01601, Ukraine

N. Teslenko (✉)
State University of Trade and Economics, Kioto str. 19, Kyiv 02156, Ukraine
e-mail: n-teslenko@ukr.net

A. Hamdan et al. (eds.), *The Implementation of Smart Technologies for Business Success and Sustainability*, Studies in Systems, Decision and Control 216,
https://doi.org/10.1007/978-3-031-10212-7_52

1 Introduction

Modern mankind is experiencing a global educational crisis which deepens in the face of transitive post-Soviet societies such as Ukrainian. In this context, the study of education policy and the development of possible ways of reforming and developing it as an instrument for building a civil society becomes extremely relevant in the scientific and extremely important in the field of praxis. After all, qualitative and effective education is a prerequisite for ensuring sustainable democratic development of society, humanization of socio-economic relations, consolidation of all social institutions, and formation of new-ideological and value-orientated personalities.

There was a common European educational space in the conditions of internationalization of world space, political, economic and cultural convergence of countries. Accordingly, the process of European integration is significantly reflected in every sphere of life of the country, which has also touched on higher education. Consequently, the Bologna system for the reform of higher education has become one of the elements of European integration. After all, Ukraine seeks to enter a common European educational space. Furthermore, in order to achieve this, it is pursuing a strategic policy of integration into the European Union. Thus, considering the common borders of Ukraine with the European Union, as well as the cultural and civilization, historical and territorial proximity of Ukraine, it is appropriate to examine the processes of European integration and their prospects for the development of Ukraine.

Therefore, before the national system of higher education, complex tasks are formulated every day, the solutions of which seem possible, if there is theoretical information on the experience of solving the contradictions of the modernization of education in developed countries. The prospect of creating European spaces for education and science necessitated the formulation of an "agenda" for changes in the political, economic and cultural spheres related to higher education. We include the following features: the inconsistency of the development and implementation of the education policy of Ukraine, the formal and declarative nature of the pro-European course, the total dependence of the state education policy on the current political situation, the eclectic nature of the state educational policy, and the excessive centralization of the education system of Ukraine.

The aim of this paper is to implement a comprehensive analysis of education policy of Ukraine and to examine the national policy in the field of education in Ukraine.

This paper structure consists of such parts as literature review, methodology, results and conclusions.

2 Literature Review

The basis of reflection on educational policy were classical works by domestic and foreign scientists. After all, today Ukrainian philosophers, scientists and educators are studying various aspects of education in Ukraine. The works of other authors are quite diverse. The diversity of research topics and approaches is due to the goals set by specific authors. Let's name only those of them which were the most heuristic for the given dissertation work. Thus, the state education policy is fruitfully studied by well-known foreign researchers. Ukrainian scientists such as Adrushchenko (2004), Korzak [6], Nezyva [8], Kraskov [7] and others pay attention to the philosophical and educational principles of personality development, problems of the education system and its development.

Moreover, Korsak [6] in his work *Education, society, man in the twenty-first century: integral philosophical analysis* defines that national educational policy in the broadest sense as a set of priorities and goals formulated by the government or other higher body to implement measures to improve and develop the system education or its elements.

Besides, Nezhyva Olga in her paper *Modernization of educational policy in the context of modern civilization processes* (2000) presents a systematic analysis of the phenomenon of educational policy in the context of the deployment of a modern civilizational process. Furthermore, she shows that the use of the obtained results in the future can be applied in the development and formation of the state educational policy of Ukraine. Moreover, the other work *The Aspects of Smart Education in The World* (2021) by Olga Nezhyva demonstrates education based on three educational activities, namely, arithmetic, reading and writing, transforms into education which improves the following seven skills: creativity and innovation; communication; cross-cultural understanding; collaboration and leadership; critical thinking and problem solving; ICT literacy; career and life skills.

Furthermore, Andrushchenko and Savelev [1] in the work *Educational policy (review of the agenda)* notes that having considered only the main trends of the world research space of educational policy, we still have some grounds for determining its scientific status among the social sciences.

The analysis of studies about the modern Ukrainian education policy shows its ideological eclecticism, which creates a favorable environment for the formation of internal contradictions, slow reforms and the impossibility of transformational change.

3 Methodology

The methodological framework for the study of the phenomenon of Ukrainian educational policy is based on national ideas about education. In particular, interdisciplinary and transdisciplinary approaches were used in this study. Moreover, it was possible to involve not only traditional but also new methods and principles in the analysis of education. The principles of consistency, integrity, objectivity, historicity and development were applied among the classical approaches to scientific knowledge. The nonclassical approaches such methods as deconstruction and genealogy, which allowed to reveal at a deeper level the essence of the processes which take place in Ukrainian education in the modern society, should be mentioned. Besides, general scientific methods such as analysis, synthesis, abstraction, comparison, systematization are widely used in this study.

4 Results

The term "national education policy" or "public education policy" is used in the scientific literature since 1860s, when the role of education at the state level has been perceived as an important factor of economic development and social progress, as a sphere which requires special care at the national level.

We will briefly consider the degree of development of this issue in the scientific literature. Hence, modern researchers of public policy in education note that a big problem for many scientists is the inability to conceptually define "politics" in their research. Understanding of the policy as "self-evident" leads to weakness in the "analytical study of the structure". Thus, the choice of categorical apparatus, key concepts, working definition of politics is extremely important for the study.

A lot of people think that the positivist view of politics as a product of the state is insufficiently grounded conceptually and methodologically limited. In recognition of this criticism, poststructuralist approach considers the policy as being beyond the scope of official (government) agencies and combines both material and discursive contexts in which policy is made. Jenny Ozga confirms that the policy includes not only political direction, but "negotiations, challenge or struggle between different groups which may lie outside the formal techniques of official policy-making" [12, p. 113]. Moreover, Jenny Ozga confirms that education policy is not limited to formal relations and processes of governance, legislative strengthening relations and procedures in the field of education. A broad definition of policy requires understanding in political, social and economic contexts and the appropriate study how these contexts shape the education policy [12, p. 113].

American researcher Jean Anyon emphasizes the same, too. She confirms that the definition of education policy should be expanded to include consideration of economic policy. Jean Anyon notes that the impact of economic policies, which has its index of the minimum salary, has a significant and often ignorable impact on the experience of urban students. Jean Anyon emphasizes that even small improvements of annual salary can have a direct impact on the experience of urban families living in poverty, and, in particular, on the educational experience of children in these families. Ultimately, Jean Anyon posits the necessity of synchronous coordination of economic policy and education policy, because their separation creates artificial distinction. You cannot hope to influence on the urban schools without the prior decisions of the economic needs of families who visit these schools [5].

Stephen Ball emphasizes that of dual conceptualization of national education policy is necessary i.e. policy is as text and policy is as discourse. Regarding politics as text, Stephen Ball notes that in this case the policy is to be understood as representations which are encoded in many ways (through struggle, compromise, government, public interpretation and reinterpretation) and decoded many ways (using actors, interpretations and meanings in relation to its history, experience, skills, resources and contexts).

Most texts are products of several development programs and compromises. However, Stephen Ball considers this requirement does not mean a pluralistic approach to politics, because alternative views or approaches are excluded in the initial stages of the policy. In addition, the problem of interpretation of policy is difficult because "at all stages of the political process as we are faced with different interpretations" and "interpretations of interpretations". Fazal Rizvi and Stephen Kemmis confirm that politics as text reflects the understanding of politics as a product of compromises between various development plans and interests. In addition, the policy is never complete. Therefore, the researcher always deals with a specific part of the policy which should be considered in connection with other political contexts and historical retrospective of their development [3].

Olena Fimyar notes that understanding of politics as the text is rather heuristic. Her definition is preventing us comprehend that Jenny Ozga calls the "bigger picture" which includes not only what policy thinking and include plans, agendas and programs, but that they do not think or intentionally excluded from the view [2]. Given these criticisms, Stephen Ball provides that the policy is not only the text but also the relation of power, in which power is exercised by "truth and knowledge production, as discourse" [12, p. 21]. Stephen Ball understands politics as discourse which is based on the popular definition of Michel Foucault: "Discourses are practices which systematically form the objects of which are spoken. ... Discourses do not describe objects; they do not identify the objects they construct them, and this practice hide their own intentions" [3, p. 49]. According to Michel Foucault, discourses are combined with "desire and power" and "unconscious manifestations of language and speech" [3, p. 49]. Stephen Ball describes the relationship between discourses and objects which are designed these discourses as follows: "We do not create the discourse. The discourse creates us. We are subjectivities, voices, knowledge, power relations which are constructed and permitted by the discourse. ... We speak through policy; we take a position created for us in politics" [3, p. 22].

According to Ball's approach, the effect of policy is primarily discursive because it changes and excludes the possibility of thinking differently, thereby limiting our response on changes. However, Michel Foucault considers not only the power and domination in the discourse, but also the opportunity to resist, because the discourse can be both instrument of power and influence, and hurdle, a stumbling block, and the starting point for the opposite strategy [3, p. 101].

The term of state education policy emerged in the 60s and 70s of the twentieth century as a social phenomenon. According to A. Mironov, the main requirements for state institutions to start shaping the state education policy are [10]:

- transformation of humanity, i.e. the transition from industrial to post-industrial society; rationalization of human professional activity;

- strengthening the relationship between upbringing and education, culture and education and mentality and education as a means of preserving and developing the civilization of individual states;
- the alternative nature of the development of the educational process, its complexity, systemic and structured content and organizational order;
- the need to meet the universal requirements of educational activities, enrich the experience of providing educational services;
- informatization of public life in general; innovative nature of system development.

The development of state education policy in any society is determined and correlated by the interaction of the triad: power—politics—education system. V. Halperin emphasizes that the nature of the relationship between politics and education, as functional subsystems of society, is determined by all the phenomena embodied in the concept of "social interests" and "power" [9]. The researcher notes: "Education policy is the personification and the result of the interaction of existing social forces, political institutions, individuals, both among themselves and with the environment—economic, social, legal systems, political culture, its various segments" [9, p. 15].

The fact is that the term "education policy" integrates the internal characteristics of such concepts as "policy" (organizational, regulatory and control sphere of society, focused on achieving certain goals, maintaining what has been achieved and its adequate implementation), state policy "(official actions of the state)," education "(the process and result of mastering an individual system of sciences, knowledge, practical skills; socio-cultural institution, through which the social, cultural functioning and improvement of society)" [11].

Thus, education policy is defined as "a field of interaction of different social groups, individuals on the use of government institutions to realize their interests and needs in one of the most important areas of society—education".

Moreover, education policy establishes on the basis of social harmony the fundamental goals and objectives of educational development, guarantees their implementation through concerted action of the state and society. National education policy in its objective sense is equal to its two components: state and public, i.e. public policy. Educational policy is a field of active interaction between the state and society in the implementation of social values, goals and priorities in education.

Based on the analysis of the above definitions of educational policy, we can conclude that its main features are:

- systematicity;
- purposefulness;
- sequence;
- effectiveness;
- focus on the realization of civil human rights.

Other, no less important, characteristics of education policy are given in the work of the research team "Educational policy and the course for change" [9]. The main ones are as follows:

- Policy is more than text.
- Policy has many dimensions.
- The policy is based on certain values.
- Policy exists in context.
- Political decisions are a state affair.
- Educational policy interacts with policy in other areas.
- Policy implementation is never direct.
- Policy entails intentional and unintentional consequences.

One of the key issues in the study of education policy is to determine the role of the state and the functions of government in this context. In the context of education policy, there is always a vertical relationship of power: there are those who govern, and those who govern, no matter how conditional this relationship may be.

Such specialists as Victor Andrushchenko, Leonid Vashchenko, Vasyl Kremen, Sergii Klepko, Victor Lugovyi, Natalia Protasova, Larysa Parashchenko, Vladimir Saveliev, etc. deal with formation and implementation of state policy in the field of education.

In our point view, determining research in the field of education policy of Ukraine is currently Victor Andrushchenko and Vladimir Saveliev's monograph "Education Policy" [1]. The history of research problems were discussed on a professional level in the monograph. It impresses coverage of foreign sources including ideas and opinions of prominent political and public figures in Europe. The author's vision of features of Ukrainian education policy is traced in this work. The monograph of content and structural and functional features of modern education policy is considered in the light of globalization, internationalization and Europeanization which are the main trends of contemporary social development. In our point view, the depth and scope of international experience in authors' scientific work deserves a professional respect. However, the objective and impartial analysis of Ukrainian education policy was not formed by the authors in this work. For the reason that there exists too many minor generalizations and there is no criticism and concrete proposals for the action. Perhaps, this shortcoming occurred in the monograph because one of the authors worked as Deputy Minister of Education and Science of Ukraine for a long time. In addition, analysing that time we can safely say that between thoughts and ideas laid down in the book (edition of 2010) and the actions of the author Victor Andrushchenko as a senior official of Ministry of Education and Science of Ukraine (1995–1999) are observed a significant difference. After all, we can underline that the experience of leading countries in the field of education policy is very different from actions, which was realized by Ukrainian officials in this field of education.

Ukrainian researcher Evgen Kraskov notes that the analysis of the existing legislative and other regulatory and legal framework of education allows us to formulate the concept of "public education policy of Ukraine" [7]. It is a "part of (subsystem) national policy which includes a set of specific goals of the system, objectives, principles, programs and main activities of education authorities, which are aimed at the organizing scientific and methodological support and implementing strategies for the development of the education system" [6, p. 56]. State policy in education and public administration education are related. The governance is a means (mechanisms, tools) of implementation of state policy in the field of education.

The national policy in the field of education of Ukraine is determined by the highest legislative authority and is exercised by the executive authorities and local governments based on the approved authorities and public opinion approved the concept of education. It can be confirmed that education policy of Ukraine is a tool of influence of the political elite of Ukrainian society for the younger generation. In conditions of Ukrainian statehood, which is only just formed, the ideological orientation and motivation of senior officials have brought Ukrainian education system to the point where it is now. Furthermore, there are not any other reservations and excuses, because Ukrainian officials, who had all the levers of influence, make certain transformation in the concept of development education, independently, without regard to public view or control of the society.

The national policy in the field of Ukrainian education is based on strategic documents of advancement of education (programs, doctrines, concepts, etc.); constitutional norms, international treaties ratified by the supreme legislative body; legislations; Executive Orders (Decrees) of the President; Resolutions of the Verkhovna Rada of Ukraine (Parliament) and the Cabinet of Ministers of Ukraine; orders of specially authorized central executive body in the field of Education and Science, other central executive bodies governing educational institutions and so on [8].

The implementation of the national policy in the field of Ukrainian education relies on a specially authorized central executive body in the field of Education and Science and other central executive bodies governing educational institutions, local executive bodies and local government.

Well-known Dutch researchers Åse Gornitzka and Peter Maassen offered four models of higher education management, which was based on the typology of Olsen, in due time. So, the state education policy in Ukraine practically coincides with the criteria model "rational state—sovereign" (the sovereign, rationality-bounded state), which is associated, as the authors note, the "interventionist state", hard "state control" of Education Administration, in which "higher education is seen as a means to achieve economic and social objectives" [4, p. 270].

It turns out that despite all the assurances of the four presidents of Ukraine and other officers with statutory authority, and judging by the features of Ukrainian education policy, several decades "independence" of Ukraine were actually far of democratic reforms. Ukraine with all institutions of governance continued to develop toward authoritarian (presidential-parliamentary) rule. All declared and promised steps on Ukraine's rapprochement with Europe were purely formal. In addition, Ukraine's state education policy includes such priorities as quality of education policy, education policy effectiveness, equality and social justice in education policy. These priorities include strategic education documents and the state policy in educational sphere.

The excessive centralization of the educational system in Ukraine is a factor which essentially hinders democratic processes both in the field of education and science, and in the state as a whole. The political leadership of the country is often criticized in the studies of domestic and Western authors due to the slowdown in the pace of reforms and the selective application of existing policies. Many critics describe the Ukrainian educational system as structurally "too centralized" and, from the point of view of institutional practice, "too Soviet" [10].

After all, the consideration of the peculiarities of the development of Ukrainian education policy, in our opinion, gives a thorough view that in the relatively short time, Ukraine was almost the only post-Soviet country that actually formulated the legislative base for education. Nevertheless, unfortunately, Ukraine did not create mechanisms for the implementation of this base in life. This raises the question of active work on the preparation of the necessary regulations, which would facilitate the implementation of the legislation on education. The National Doctrine of Education Development implies access to quality education for all citizens of Ukraine, which became the main prerequisite for the implementation of external independent evaluation in Ukraine. Therefore, the main goal of the Ukrainian educational system was to provide favorable conditions that enabled the development and self-realization of each individual, since any person has the ability to learn throughout life, to form and develop the values of civil society. Thus, one of the priorities of state policy in the development of education is to create the same and achievable opportunities for children and young people to acquire high-quality education.

5 Conclusions

Thus, the concept of "education policy" means a comprehensive interaction of the state, business and civil society in the field of education, the essential features of which are systematic, purposeful, consistent, effective, focused on the implementation of the principle of human-centeredness. There are two main approaches to the basic concept. First, the positivist view, which considers educational policy exclusively as a product of state activity and, secondly, the poststructuralist approach, which interprets policy as one that goes beyond the work of official (state) institutions and combines both material and discursive contexts in which policies are developed.

It has been found that educational policy is never complete, so a researcher, who seeks to be objective in the analysis, must take into account the fact that he deals only with a certain part of policy. Moreover, the researcher must include in his analysis the entire political context of another decision, taking into account also the history of the decision of a particular issue. According to modern approaches, the results of education policy are often indirect, discursive, because the implementation of a political decision changes the original idea. The prevailing discourse of education policy directs new decisions, defines their context, on the one hand limiting opportunities for operational management of the situation, and on the other—creating long-term effects and new growth points for the opposite (existing) strategy.

Our conclusion is supported by the key assumptions which determine the model of "rational state—sovereign":

- The function of higher education;

- Evaluation of higher education is based on the definition of political effectiveness (effectiveness);
- Terms of interaction based hierarchical subordination and submission, participation of society in education is limited;
- Developing solutions rigidly centralized (single point of control) from top to bottom (top down);
- Hierarchical control system;
- Comprehensive nature of public exposure;
- Autonomy of any of limited scope solution "technical issues";
- Changes in higher education are the result of changes in political leadership.

Thus, we examined the degree of development of public policy issues in Ukrainian education in this paper. The preliminary analysis of modern Ukrainian education policy reveals its ideological eclecticism, which creates a favorable environment for the formation of internal contradictions, the slowdown of reforms and the impossibility of transformations.

References

1. Andrushchenko, V., Savelev, V.: Obrasovatelnaia politika (obzor povestki dnia) [Education Policy (Review the Agenda)]. MP "Lesia", Kyiv, Ukraine (2010)
2. Fimyar, O.: What is policy? In search of frameworks and definitions for non-Western contexts. Educate **14**(3), 6–21 (2014)
3. Foucault, M.: Archaeology of Knowledge. Routledge Classics. Routledge, London (2002)
4. Gornitzka, A., Maassen, P.: Hybrid steering approaches with respect to European higher education. High Educ. Pol. **13**, 265–272 (2000)
5. Gutek, G.L.: Historical and Philosophical Foundations of Education: A Biographical Introduction, 5th edn. Pearson, Boston, MA (2011)
6. Korsak, K.V.: Education, Society, Man in the XXI Century: An Integrated Philosophical Analysis, 210 p. Kyiv-Nizhyn (2004)
7. Kraskov, Y.: The National Policy in the Field of Education of Ukraine: Lessons and Prospects (2012). Retrieved June 11, 2021, from http://kno.rada.gov.ua/komosviti/control/uk/publish/article;jsessionid=29976B8D9EB90F621D61E2DB852DAB7F?art_id=53491&cat_id=44731
8. Nezhyva, O.: The aspects of smart education in the world. Khazar J. Human. Soc. Sci. **24**(3), 62–72 (2021). https://doi.org/10.5782/2223-2621.2021.24.3.62
9. Nezhyva, O.: Modernization of educational policy in the context of modern civilization processes. Int. Sci. J. Univ. Leadersh. № 9 (2020)
10. Nezhyva, O. Future Ukrainian's image: modern vision. Philosophy Cosmol. **14**, 175–179 (2015)
11. Nezhyva, O.: Features of formation of Ukrainian education policy. Future Hum. Image **3**(6), 178–189 (2016)
12. Ozga, J.: Policy Research in Educational Settings: Contested Terrain. Open University Press, Buckingham (2000)

Faculty's Satisfaction Degree of Distance Training on Using Blogs in Teaching During the Pandemic

Asmaa Radi Khanfar

Abstract Within the framework of the effort to prevent infection with the Corona virus and as a precautionary measure to preserve the safety of both sides of the educational process, many Arab countries have resorted to distance education. Saudi universities, including Imam Abdulrahman Bin Faisal University, have responded to the developments caused by the pandemic by moving towards electronic training, by employing all infrastructure and strategies to shift to e-learning and distance education. This is due to the knowledge of these universities of the importance of continuing training, which is a major factor in the development of the university's inputs; this reflects positively on the quality of the outputs. This study came to examine the degree of satisfaction of IAU members with the online training provided to them on the use of blogs in education during the pandemic, and whether there are statistically significant differences in the degree of IAU members' satisfaction with the training as a whole due to a number of variables, using the descriptive approach, and the distribution of a questionnaire to (35) members who attended the training, and the data was analyzed using the SPSS program. The study found that the degree of members' satisfaction with the training provided to them online on the use of blogs in education during the pandemic is very high in all fields, and the fields were arranged in descending order as follows: the field of the trainer, the content. There were statistically significant differences in the degree of members' satisfaction with training in favor of the trainer. The study recommended the adoption of training for all members of the university so that the benefit would prevail.

Keywords Distance training · Higher education · The pandemic · Saudi universities · Electronic blogs

A. R. Khanfar (✉)
Deanship of Academic Development, Imam Abdulrahman Bin Faisal University, Dammam, Saudi Arabia
e-mail: arkhanfar@iau.edu.sa

A. Hamdan et al. (eds.), *The Implementation of Smart Technologies for Business Success and Sustainability*, Studies in Systems, Decision and Control 216,
https://doi.org/10.1007/978-3-031-10212-7_53

1 Introduction

In the context of the effort to prevent infection with the Corona virus and as a precautionary measure to preserve the safety of both sides of the educational process, many Arab countries have resorted to distance education [1], and these repercussions also affected the allocations for training and development by reducing or canceling, which led to the suspension of training activities, conferences, forums and seminars in all Arab countries. This resulted in dependence mainly on electronic training platforms and digital content. Saudi universities have responded to the developments caused by the pandemic, by moving towards e-training, including Imam Abdulrahman bin Faisal University which employed all infrastructure and strategies to shift to e-learning and distance education, where multiple systems have been used to manage e-learning [2]. This is due to its knowledge of the importance of continuing training as a major factor in developing the university's inputs, which reflected positively on the quality of outputs [3], and given the speed of changes around us, especially the labor market's human capital requirements. The required development of training must be in line with these changes, flexible and continuous, and in all aspects of teaching–learning process, the most important of which is a faculty member [4]. Accordingly, before designing any training program, the needs of the trained members must be accurately determined, because the success of any training is related to the demands and needs of the trainees [5]. The members' satisfaction with the training as a whole reflects the extent to which they benefit from it in employing it in education and teaching processes, which is the most important.

In addition to the above, the Deanship of Academic Development (DAD) at (IAU) constantly tracks the members' training needs by directly asking them at the end of each training workshop about the most important topics they wish to be trained on. In its final report [6], the Quality Assurance Committee mentioned a number of topics proposed by faculty such as: practical means to activate e-learning and delivery of content in interactive ways, virtual education and the use of virtual means to enjoy the conduct of the distance learning process, interactive distance teaching applications, modern virtual teaching entrances. So, in response to the trainees' needs that were mentioned in the report, the workshop on using blogs in teaching and learning was one of the workshops offered by the Deanship. And because there is an urgent need to develop the workshops offered for training; the satisfaction of faculty members with the various elements of training is one of the most important indicators of the success of training and also contributes significantly to achieving the goals of the institution [7, 8]. This study came to examine the degree of satisfaction of Saudi university faculties about their distance training on using blogs in teaching during the pandemic.

1.1 Statement of the Problem

The results of many local, Arab and foreign studies showed that the priority areas in need of training from educational fields are the use of alternative teaching methods, the use of information and communication technology in teaching and learning [4] and the results of some studies of the current study sample showed the need for faculty members at IAU for continuous training in several areas, the most important of which is training in the use of technology in education [4, 9] and accordingly, DAD has offered several training workshops in these areas, including the workshop on the use of blogs in teaching and learning, in addition to what was mentioned in the [9] study of the importance of two main dimensions in the level of faculty members' satisfaction with development and training programs, What is offered to members? How does it present? As these two dimensions contribute to expressing approximately (70%) of their level of satisfaction, and in order to research the quality of training and the extent to which it achieves the goals for which it was set in developing members' skills in using blogs in education, and because members' satisfaction with the elements of the training provided to them is one of the most important indicators of the success of training [5, 7] This study came to investigate the degree of satisfaction of the faculties of Imam Abdulrahman bin Faisal University (IAU) about their distance training on the use of blogs in teaching during the pandemic.

1.2 Study Questions

The study questions focused on the following main question:

- What is the degree of satisfaction of faculty members at IAU Imam Abdul Rahman bin Faisal about their distance training on the use of blogs in teaching and learning during the pandemic?

  The main question is divided into the following sub-questions:

- What is the degree of satisfaction of IAU members with the outputs and content of distance training on the use of blogs in education?
- What is the degree of satisfaction of the members of the (IAU) with the performance of the trainer (the event presenter) remotely on the use of blogs in education?
- Are there statistically significant differences in the degree of satisfaction of IAU members with training in favor of one of the two areas (training outputs and content, trainer)?

1.3 Objectives of the Study

The study aimed to reveal the degree of satisfaction of IAU members with the training provided to them remotely on the use of blogs in the educational process during the pandemic in the areas of training output and content, and trainer performance, and to search whether there are statistically significant differences in the degree of their satisfaction with training for one of the two areas.

1.4 The Importance of the Study

The importance of the study also shows that officials at the University of (IAU) and at (DAD) benefit from the results of the study in developing the performance of the university's employees by updating the training provided to them in line with their renewed training needs.

1.5 Study Limits

The study was limited to faculty members at Imam Abdul Rahman bin Faisal University who enrolled in distance training on the use of blogs in education in the academic year (2020–2021).

1.6 The Study Terminology

The study dealt with the following terms, which can be defined idiomatically and procedurally as follows:

Online training: The interactive process through which training is transferred despite the geographical distance, and the presence of means for transmitting information such as printed materials, computers, the Internet and electronic applications such as Zoom, and Teams [10].

Blogs: A web page on the Internet in which content and educational activities are presented in an interesting and attractive manner, allowing users to publish and coordinate text, image, audio, video clips and links to blogs and other sites and comment on them, students and the teacher can exchange discussions during it, and is designed to achieve specific educational and educational goals [11].

2 Previous Studies

Among the studies that examined the role of blogs in higher education, Zwain's study [11] aimed to reveal the effectiveness of using blogs in teaching geography on developing students' digital citizenship skills at Menoufia University. The results indicated the effectiveness of using blogs in teaching geography on developing Digital citizenship skills. At King Saud University, the study of Bin Khanin [12] sought to know the impact of the use of educational blogs on the achievement of female students, using the experimental approach, and the results showed an increase in the level of achievement of the experimental group students who used the second generation tools for education. The most difficult difficulties that faced by the students were related to the use of computers.

As for the studies that dealt with measuring the extent of faculty members' satisfaction with the training provided to them, such as the Darawshe study [7] at An-Najah National University in Palestine, which examined the degree of members' satisfaction with the electronic training provided to them in light of the variables of gender, academic specialization, academic qualification, and years of experience. The study found that the degree of members' satisfaction with the training was high, and there were no statistically significant differences between the degrees of their satisfaction with the training due to any of the study variables. At Princess Nourah University, Al-Amr's study [5] examined the effectiveness of training programs in meeting the needs of faculty members. It used the comprehensive social survey approach for faculty members in the Faculty of Social Work, and its most important results were: the most important means and procedures used in determining training needs (opinion poll and workshops Discussion and workshops), as for the most important obstacles that limit the effectiveness of training programs: overload on the faculty member, the speed in implementing the programs, and the focus on current programs rather than future ones. At Shaqra University, Al-Sadhan's study [2] aimed to give explanation of the effectiveness of a distance training program in developing the skills of using e-learning management among faculty members using the experimental research based on the design of one group, the most important results of the study are the presence of statistically significant differences between the mean scores of the members of the research group in the pre and post applications of research tools in favor of the post, which indicates the effectiveness of the distance training program to develop the skills of using the e-learning management system. It is noted from the previous studies that the degree of faculties' satisfaction with their distance training on a specific topic was not searched, but rather the search for their satisfaction with electronic training in general, and this is what distinguishes this study from previous studies, which came to reveal the degree of faculties' satisfaction with their distance training on the use of blogs in education.

3 Philosophy of Using Blogs in Teaching and Learning

According to the constructivist theory, the learning process is based on the idea of active learning, problem solving and cooperation among learners, as learning occurs when learners initiate to build their own knowledge rather than present it to them, as constructivism emphasizes the occurrence of learning in social contexts that support initiative in learning and the independence of the learner, and constructivists believe that the use of learning, the electronic and its tools enhance personal and collective knowledge through dialogue with others, the learner's interactions with his peers and his undertaking of new activities that stimulate his learning. Technology is used in the structural learning environment as a tool with which the learner learns and not from [13, 14]. Accordingly, technology is seen as a means that provides learners with opportunities to build their own understanding and knowledge, and as supportive of collaborative learning, depending on the quality of the programs offered that will provide an environment for cooperation [15].

Blogs, as one of the technology tools, have added new dimensions to the educational and learning process, as they are characterized by high interactiveness and high flexibility that transforms the learner from a future of information to a producer and participant in it [11]. More space for creativity outside the classroom at the time and place chosen by the student [16], and this was confirmed by Vise [17] when he referred to the ability of blogs to make students interact virtually in a manner similar to their face-to-face interaction between them and their teacher with complete comfort that he accepts and loves students.

Among its characteristics is what was reported by Omran [8] and Mahmoud [18]: Everyone can access the contents of the blogs at the right time and place, and they have flexible ready-made templates, simple with graphical interfaces characterized by flexibility, blogs allow the teacher to communicate directly and continuously with his students, and they are a means of disseminating information, ideas and various experiences, and blogs facilitate the process of reviewing topics In the old days, most blogs are created free of charge on free sites and therefore their cost is low. Blogs are easy to use by the teacher and student, as well as easy to set up and update. Blogs guarantee everyone the freedom to express what they want. Blogs preserve the information written in them, and when needed, they are easy to refer to. By obtaining it, the blog encourages students to be creative by providing them with tools that enable them to reach the highest degree of creativity in the educational product. The blogs also encourage students to participate actively through interesting and attractive learning.

Benefits of using blogs in the educational learning process [8]: Educational blogs help create an atmosphere of social interaction between students on the one hand and between students and their teachers on the other hand, and contribute significantly to raising the level of students' interaction with each other and exchanging ideas among them. In which peer guidance is provided, a group of older and more experienced students can help a group of younger students in developing their various skills, and blogs increase the level of internal motivation of the student towards

Table 1 Distribution of study members according to the study methodology: the study used the descriptive method for its relevance to the research topic

The semester	No.	Percentage
The first	12	34.3
The second	23	65.7
Total	35	100.0

learning, and raise the level of self-confidence and self-esteem, especially learners who feel ashamed to participate in the classroom. As it allows learners to express their ideas freely without fear of criticism, as the learner remains active in thinking preoccupied with the ideas addressed in the blog, and increases the degree of concentration of the learner when studying. It helps learners to organize their thoughts when writing and to compare their thoughts with those of others.

4 Study Methodology

The study used the descriptive method for its relevance to the research topic.

4.1 Study Community and Sample

The study community represented the faculty members at the University of **(IAU)**, and an intentional sample was chosen and represented by all the faculty members who joined the distance training program on the use of blogs in education, and their number is **(35)** members, which illustrates this variable of the semester (Table 1).

4.2 Study Tool

The study used a questionnaire of (15) items divided into two areas: training outcomes and content, trainers, where the five-point Likert scale was used to determine the degree of members' satisfaction with training, and Table 2 illustrates this.

Validity and stability of the tool: To verify the internal consistency of the questionnaire, the Pearson correlation coefficient was used to measure the relationship between each statement and the total degree of the dimension to which it belongs, as well as between each dimension and the total degree of the questionnaire, and Tables 3 and 4 explain this.

It is clear from Table 3 that all the correlation coefficients between each phrase and the dimension belonging to it were positive and statistically significant at the

Table 2 Arithmetic mean and degree of satisfaction

Mean	Degrees of satisfaction
Above 4.2	Very high
From 3.4 to less than 4.2	High
From 2.6 to less than 3.4	Medium
From 1.8 to less than 2.6	Low
Less than 1.8	Very low

Table 3 The correlation coefficients of the expressions of each dimension with its total degree

The dimension	Phrase	Correlation coefficient	Phrase	Correlation coefficient
Satisfaction with output and content	1	0.825**	4	0.803**
	2	0.938**	5	0.834**
	3	0.843**	6	0.854**
Satisfaction with the presenter(s) of the event	1	0.954**	6	0.956**
	2	0.821**	7	0.902**
	3	0.896**	8	0.963**
	4	0.945**	9	0.901**
	5	0.854**		

**Function at (0.01)

Table 4 Correlation coefficients for each dimension with the total score of the questionnaire

The dimension	Correlation coefficient
Satisfaction with output and content	0.906**
Satisfaction with the presenter(s) of the event	0.900**

**Function at (0.01)

level **(0.01)**. Table 4 shows that the correlation coefficients for each dimension with the total score of the questionnaire were all positive and statistically significant at the level **(0.01)**, and this indicates that all the questionnaire statements were true and measure the goal for which it was set.

Reliability: To verify the stability of the resolution, the stability coefficient of Cronbach's alpha was found for the dimensions of the resolution, and the results were as shown in Table 5.

Table 5 shows high values of Cronbach's alpha coefficients for the dimensions of the resolution, which reassures that the resolution has a high degree of stability.

Table 5 Values of stability coefficients for the dimensions of the resolution

The dimension	Cronbach's alpha coefficient
Satisfaction with output and content	0.914
Satisfaction with the presenter(s) of the event	0.975
The resolution as a whole	0.974

4.3 *Statistical Processing Methods*

To achieve the objectives of the study, the Statistical Package for Social Sciences (**SPSS**) program was used to analyze the data and obtain the results as follows: frequencies and percentages to describe the characteristics of the study sample, arithmetic averages and standard deviations to identify the responses of the sample members to each of the phrases, coefficient Pearson Correlation to verify the internal consistency of the questionnaire, Cronbach's Alpha to verify the stability of the resolution, Paired Samples T test to indicate the differences between two related variables.

5 Results and Discussion

5.1 *Main Question*

"What is the degree of satisfaction of faculty members at IAU with their distance training on the use of blogs in teaching and learning during the pandemic?" To answer this question, means and SDs were calculated for the participants' responses, and the results were as shown in Table 6.

Means and its value (4.62) and a very high degree of satisfaction, followed by the satisfaction with the outputs and content and obtained a mean of (4.49) and a very high degree of satisfaction. The table also shows that the total dimensions obtained a mean of (4.49) and a very high degree of satisfaction, and this indicates that the satisfaction of faculty members at IAU with their distance training on the use of blogs in teaching and learning during the pandemic was very high. Where the following

Table 6 Means and SDs of the responses of faculty members about their degree of satisfaction with their remote training on the use of blogs in teaching and learning during the pandemic

The dimension	Mean	SD	Degree of satisfaction	Ranking
2 satisfaction with the output and content	4.49	0.609	Very high	2
1 satisfaction with the presenter(s) of the event	4.62	0.573	Very high	1

Table 7 Arithmetic mean and degree of satisfaction

Mean	Degree of satisfaction
4.2 above	Very high
From 3.4 to less than 4.2	High
From 2.6 to less than 3.4	Medium
From 1.8 to less than 2.6	Weak
Less than 1.8	Very weak

grading (Table 7) was used to indicate the average responses of the sample members to the degree of satisfaction.

Their high degree of satisfaction with the training provided to them is explained by their urgent need for this training, which was online and at the beginning of the pandemic, that allowed them to identify many technical tools during the training, as it will be clear from the upcoming results that the trainers have used them in their training, which raised the degree of satisfaction. The trainees expressed their trainers' performance, as the trainer's field achieved the highest degree of satisfaction, as this is also explained by the importance of the topic presented at that stage of the pandemic, which dealt with electronic blogs, the importance of which in the educational process was demonstrated by many studies [11, 12] which agrees with the results of the current study, as well as with the results of the study of both Darawshe [7] and Al-Sadhan [2] in the high degree of members' satisfaction with the training provided to them online.

5.2 *Second Question; "What is the Degree of Satisfaction of the Members of the (IAU) with the Outputs and Content of the Distance Training on the Use of Blogs in Education?"*

The means and SD's of the responses of faculty members at IAU about their degree of satisfaction with the outputs and content of distance training on the use of blogs in education were calculated, and the results were as shown in Table 8.

Table 8 shows that the mean of the responses of faculty members at IAU about the degree of their satisfaction with the outputs and content of distance training on the use of blogs in education, their values ranged between (4.29–4.63), all with very high satisfaction levels. No. (3), which states that "the presented concepts are clear and related to the topic of effectiveness" has the highest mean and its value is (4.63), while statement No. (1) states that "the agenda (agenda)/program topics met my expectations and needs" got the lowest. A mean and its value is (4.29). The table also shows that the total of the phrases obtained a mean of (4.49) and a very high degree of satisfaction, and this indicates that the satisfaction of IAU members with the outputs

Table 8 Means and SD's of the responses of faculty members of IAU about the degree of their satisfaction with the outputs and content of distance training on the use of blogs in teaching

No.	Phrase	Mean	SD	Degree of satisfaction	Ranking
3	The concepts presented are clear and related to the topic of the event	4.63	0.547	Very high	1
6	Event tools (presentation, worksheets/if available) are well prepared and professional	4.57	0.558	Very high	2
4	Effectiveness and themes are closely related to practical realism	4.51	0.742	Very high	3
5	The activities of the event reflected the practical application of the scientific content	4.49	0.781	Very high	4
2	The scientific content that was presented achieves the outcomes of the event	4.46	0.780	Very high	5
1	Agenda/program topics met my expectations and needs	4.29	0.893	Very high	6
	Overall average	4.49	0.609	Very high	

and content of distance training on the use of blogs in education was very high. This result is explained by the importance of the topic presented, which helped members to use blogs in their teaching as an effective tool in enriching students' learning and increasing their interaction by integrating them in learning, thus increasing their achievement. This is also confirmed by the fact that the "Agenda" paragraph received a very high degree of satisfaction despite its last ranking among paragraphs.

5.3 The Results of the Answer to the Third Question; Which States: "What is the Degree of Satisfaction of the Members of the (IAU) with the Performance of the Trainer (the Event Presenter) Remotely on the Use of Blogs in Education?"

To answer this question, means and SD's of the responses of the faculty members at IAU about the degree of their satisfaction with the performance of the trainer (the event presenter) were calculated remotely on the use of blogs in education, and the results were as shown in Table 9.

Table 9 shows that the mean of the responses of the faculty members at IAU about the degree of their satisfaction with the performance of the trainer (the event presenter) from a distance on the use of blogs in education, their values ranged between (4.46–4.69), all of them with very high satisfaction levels, where statement

Table 9 Means and SD's of the responses of faculty members at IAU about the degree of their satisfaction with the performance of the trainer remotely on the use of blogs in teaching

No.	Phrase	Mean	SD	Degree of satisfaction	Ranking
4	Interact with the participants with enthusiasm	4.69	0.530	Very high	1
6	He/she was able to successfully manage the discussions and balance among the participants	4.69	0.631	Very high	2
2	Accept and respect the opinions of the participants	4.68	0.589	Very high	3
3	Effectively present information and ideas	4.66	0.684	Very high	4
1	Demonstrate a good and organized preparation for the presentation of the event	4.65	0.691	Very high	5
8	Create/create a positive, interactive environment among the participants	4.63	0.646	Very high	6
9	Employ the technology in a way that adds value to the event	4.59	0.657	Very high	7
5	Use/use a variety of interesting activities to present the content	4.57	0.558	Very high	8
7	He/she was able to manage time and present the agenda appropriately	4.46	0.701	Very high	9
	Overall average	4.62	0.573	Very high	

No. (4), which states "interact with the participants with enthusiasm," got the highest mean and its value is (4.69), while statement No. (7) states that "they were able/were able to manage time and present the agenda (agenda) in a better way suitable" at the lowest mean and its value (4.46). The table also shows that the total of the phrases obtained mean of (4.62) and a very high degree of satisfaction, and this indicates that the satisfaction of IAU members with the performance of the trainer (the event presenter) from a distance on the use of blogs in education was very high, and this is explained by understanding the trainers for the current stage of the pandemic by meeting the needs of the trainees from the members, and their use of clear, easy and effective tools in distance training, which helped to show their enthusiasm for training and encourage the trained members for training. The result is in agreement with the results of Al-Amr [5] and Al-Sadhan [2] study.

Table 10 Of the results of the paired samples *T* test to indicate the differences in the degree of satisfaction of IAU members with training in favor of one of the two areas (training outputs and content, trainer)

The field	Mean	SD	*T*-values	Freedom degrees	Significance level
Training outcomes and content	0.609	4.49	0.012	34	2.656
The trainer	0.573	4.62			

5.4 *The Results of the Answer to the Fourth Question; Which States: "Are There Any Statistically Significant Differences in the Degree of Satisfaction of IAU Members with Training in Favor of One of the Two Areas (Training Outputs and Content, Trainer)?"*

To answer this question, a *t*-test of correlated samples was used, as shown in Table 10.

Table 10 shows that the significance level value amounted to (0.012), which is less than (0.05), and this indicates that there are statistically significant differences in the degree of satisfaction of IAU members with training in favor of one of the two areas (training outputs and content, the trainer), and the means show that these differences were in favor of the trainer's field. This indicates that the IAU members' satisfaction with the trainer was higher than their satisfaction with the training outcomes and content. This result is explained by the result of the previous question, which showed a very high degree of satisfaction with the trainers' performance in all paragraphs of the trainer's field, and this is explained by the fact that the trainer is the one who chooses the training subject based on a prior survey of the trainees' needs, and prepares and reviews it with a qualified and trained professional team to reach the highest levels of professionalism.

6 Summary of Results

This study came to examine the degree of satisfaction of IAU members with the online training provided to them on the use of blogs in education during the pandemic, and whether there are statistically significant differences in the degree of IAU members' satisfaction with the training as a whole due to a number of variables. The study found that the degree of members' satisfaction with the training provided to them online on the use of blogs in education during the pandemic is very high in all fields, and the fields were arranged in descending order as follows: the field of the trainer, the content. There were statistically significant differences in the degree of members' satisfaction with training in favor of the trainer.

Recommendations: Based on the results of the study, the researcher recommends the importance of training faculty members in general and members of the University (IAU) especially remotely in cases that call for remote training, and adopting distance training as a good option using the available technology, which has proven its effectiveness in many topics through its Positive impact on the educational process and the integration of students in education and increase their achievement. The researcher also recommends the adoption of electronic blogs in university education, as it contributes significantly to raising the level of students' interaction with each other and exchanging ideas among them, and works to develop and develop learning communities.

References

1. Al Lily, A., Ismail, A., Abunasser, F., Alqahtani, R.: Distance education as a response to pandemics: coronavirus and Arab culture. Technol. Soc. **63**, 101317 (2020). online first. https://doi.org/10.1016/j.techsoc.2020.101317
2. Al-Sadhan, A.: The effectiveness of a distance training program in developing the skills of using the e-learning management system Moodle among faculty members at Shaqra University. Umm Al-Qura Univ. J. Educ. Psychol. Sci. **13**(2) (2021). https://uqu.edu.sa/jep/App/FILES/116012
3. Al tuwaiygri, A.: Training needs and their impact on educational services quality marketing in light of knowledge management. Int. Interdisc. J. Educ. **6**(1), 1–16 (2017)
4. Muammar, O., Khanfar, A.: Towards high quality and distinct educational practices for the academic staff at Imam Abdul Rahman bin Faisal University. Umm Al-Qura Univ. J. Educ. Psychol. Sci. **13**(2) (2021). https://uqu.edu.sa/jep/App/FILES/116012
5. Al-Amr, B.: The effectiveness of training programs in meeting the needs of faculty members an applied study on faculty members at Princess Nourah Bint Abdul Rahman University. J. Soc. Work **58**(2) (2017). https://egjsw.journals.ekb.eg/article_173359.html
6. Roca, J., Martines, F.: Understanding e-learning continuance intention: an extension of the Technology Acceptance Modle. Int. J. Human Comput. Stud. **64**(8), 683–696 (2006)
7. Darawshe, B.: The extent of faculty members' satisfaction with the training and technical support provided by the E-Learning Center at An-Najah National University. An unpublished master's thesis, An-Najah National University, Nablus, Palestine (2016)
8. Omran, K.: The effectiveness of using blogs in teaching geography on cognitive achievement in developing geographical research skills and motivation for learning among first year secondary students. Educ. J. **31**, 353–425 (2012)
9. Muammar, O., Alkathiri, M.: What really matters to faculty members attending professional development programs in higher education. Int. J. Acad. Dev. (2021). https://doi.org/10.1080/1360144X.2021.1897987
10. Al-Essa, G., Al-Omran, A.: E-training (distance training): its justifications, requirements, and obstacles from the point of view of female and male trainers. Arab J. Manag. **41**(2), 355–374 (2021)
11. Zwain, S.: The effectiveness of using blogs in teaching geography on developing digital citizenship skills for student-teachers in the College of Education. Sci. J. **33**(9) (2017)
12. Bin Khanin, A.: The effect of using educational blogs on the achievement of students of King Saud University in Riyadh. J. Educ. Psychol. Sci. Islamic Univ. Gaza **25**(1), 248–267 (2017)
13. Al-Saleh, B.: Technology and the school of the future, myths and facts. Knowl. J. (103) (2003)
14. Hashem, R.: E-learning, its philosophy and values from the point of view of students and faculty members of the Arab Open University. J. Specif. Educ. Res. Mansoura Univ. **26**, 83–122 (2012). https://journals.ekb.eg/article_145418.html

15. Nichols, M.A: Theory for e-learning educational technology & society. Learn. Consultant Ucol, Palmerstone North, N. Z. **6**(2) (2003)
16. Kajder, S., Bull, G.: A space for "writing without writing": blogs in the language arts classroom. Learn. Lead. Technol. **31**(6), 32–35 (2004)
17. Vise, D.: The effect of wiki and blog-technologies on the student' performance when learning the preterite and imperfect aspects in Spanish. West Virginia: Unpublished Ph.D. thesis, College of Human Resources and Education, West Virginia University (2007)
18. Mahmoud, R.: The effectiveness of electronic educational blogs in developing some technical skills for middle school students. Sci. J. **37**(1), 151–174 (2021)
19. The Quality Assurance Committee Report Internal Report: Imam Abdulrahman Bin Faisal University (2021)

COVID-19—Reality and Challenge's in The Northern Governorates of Palestine

Heba Manasrah

Abstract This study aims to identify the reality of the outbreak of the Corona epidemic in Palestine and the challenges and difficulties to confront it and limit its spread. The study followed the descriptive analytical approach due to its relevance to the nature of this study, through qualitative and quantitative tools, namely, the interview and the questionnaire. Interviews were conducted with the heads of the emergency committee, which consisted of governors, mayors and chambers of commerce, health directors, and heads of financial committees in the governorates. The questionnaire dealt with the point of view of Palestinian citizens in the northern governorates. And the study sample amounted to (1064). The results of the interviews reflected the commitment and response of citizens in the first months of the epidemic, as well as a full awareness and response to dealing with safety and prevention measures, after which the interest decreased and the response became like ebb and flow, and therefore there has become a complacency in behavior, especially in wearing masks and physical distancing, and the large number of gatherings, Which is considered a negative indicator in dealing with the emerging corona virus. From the results of the questionnaire, it was found that the impact of the Corona epidemic on citizens in the northern governorates in terms of health was a high percentage of 74%, and in terms of social terms, the percentage was high, amounting to 78%. On the economic side, it was average, with a percentage of 69%. The education rate is also high, at 78%. The study recommended to Establishing an economic committee in each governorate and evaluating the previous stage through mechanisms and field studies, to monitor the current economic situation on the one hand, and what will happen to matter in the future.

Keywords COVID-19 · Reality · Challenges

H. Manasrah (✉)
An-Najah National University, Nablus, Palestine
e-mail: Heba.jetawi@gmail.com

A. Hamdan et al. (eds.), *The Implementation of Smart Technologies for Business Success and Sustainability*, Studies in Systems, Decision and Control 216,
https://doi.org/10.1007/978-3-031-10212-7_54

1 Introduction

Corona pandemic spread within months in most countries of the world, and it has already caused a negative decline in the activities of trade, industry, education, and tourism. In view of the reality that happened to the Palestinian people in light of this crisis, it varied between the division of people's opinions and ideas on social media into two separate branches, the first being convinced of the existence of the virus and practicing prevention as a daily routine, and dealing with Corona as the most dangerous thing that could be infected with it, and the second saying that the whole thing is a conspiracy resulting from political differences or economic wars, and it does not kill a large percentage of people, so there is no need to fear it or take any precautionary measures. This reality, accompanied by the continuous increase in cases infected with the Corona virus, and the view of the majority of citizens that the virus is a global conspiracy, and the urgent need to find practical and scientific solutions to confront the virus now and in the future, this study came to shed light on the challenges and difficulties facing the West Bank governorates in the face of the Covid-19 epidemic.

2 The Study Problem

At different times, the world witnessed many infectious diseases and epidemics, some of which were epidemics confined to countries, and some were global epidemics, or what is called a "pandemic".

With regard to the Corona virus, it was declared a pandemic by the World Health Organization on March 11, 2020, and from this point of view, most governments, including the State of Palestine, have taken preventive measures to limit the spread of the virus, the most important of which was the imposition of home quarantine, and many people did not adhere to the prevention guidelines that The World Health Organization and governments have stressed it, and they insist on breaking the quarantine and roaming the streets, considering that the virus is a global political conspiracy, but in view of the economic reality, which has declined significantly due to the pandemic, and with reference to the continuous statistics of the Ministry of Health on the number of HIV infections, and citizens' frequent breach of the curfew, the problem of the study revolved around the researcher's desire to learn about the reality of the outbreak of the Corona epidemic in Palestine and the challenges and difficulties to confront it and limit its spread.

3 Study Questions

1. What is the reality of the Corona epidemic in the West Bank governorates, and what are the challenges and difficulties to confront it?
2. Do the responses of the study sample differ from the impact of the Covid-19 pandemic on the health, economic, social, distance learning aspects according to gender, educational qualification, and governorate variables?

4 Hypotheses of the Study

1. There are no statistically significant differences at the significance level ($\alpha \leq 0.05$) of the corona virus on citizens in the West Bank governorates according to the gender variable.
2. There are no statistically significant differences at the significance level ($\alpha \leq 0.05$) of corona virus on citizens in the governorates of the West Bank according to the educational qualification variable
3. There are no statistically significant differences at the significance level ($\alpha \leq 0.05$) of the corona virus on citizens in the West Bank governorates according to the governorate variable.

5 Background

5.1 The Palestinian Reality with Covid-19

The first cases of corona virus infection were recorded in Palestine in the city of Bethlehem, in early March 2020, and the Palestinian Authority immediately responded with a strict response to the discovery of the first infections, as President Mahmud Abbes announced on March 5, 2020, a presidential decree declaring a state of emergency in all Palestinian territories for a period of 30 days to confront the virus and prevent its spread. Under the declared state of emergency, the competent authorities were authorized to take the necessary procedures and measures to protect public health and achieve security and stability, and the Prime Minister was empowered with the necessary powers to achieve the goals of this declaration [1], and after discovering new cases in other governorates, the Palestinian government quickly took a number of strict measures to counter the outbreak of the virus, represented by quarantine and comprehensive closures. With the emergence of the epidemic, emergency committees were formed to deal and cooperate in confronting the epidemic and to know the health, economic and social repercussions, as a higher emergency committee was formed headed by the Prime Minister, Muhammad Shtayyeh, and this committee consisted of (Health Committee, Security Committee, Provincial Committees, Economic Committee) [2] In general, this pandemic came to cause great harm to the Palestinian citizen in terms of economic, social, educational and health terms, to impose a new reality and a radical change on the individual and the official and popular society.

With regard to the health sector, the Ministry of Health has made great efforts to combat the epidemic, including planning and following up the epidemiological maps, updating them on a daily basis, examining samples in all governorates of the country.

	West Bank	Gaza	Palestine
Number of hospitals	52	32	85
Number of beds	3950	2485	6435
Number of beds per 1000 people	1.3	1.3	1.3

Palestinian Central Bureau of Statistics, 2019 [3]

The table shows the number of hospitals and beds in Palestine. The number is limited to 1.3 beds per 1000 people. Reading and interpretation of this number shows the limited ability and readiness of the Palestinian health system in the West Bank and Gaza to face health pandemics and disasters caused by natural conditions, accidents, or wars. However, the Palestinian Authority adopted A plan dated March 14, 2020, aimed at increasing examination capabilities to 20,000 samples, raising the capacity of hospitals to 200 additional beds, expanding the capabilities to treat respiratory infections, and providing intensive care to support 100 additional beds. At the beginning of the pandemic, examination samples were going to Israeli laboratories, then it is now being tested in the central laboratories of the Ministry of Health, and today the Ministry of Health has seven accredited laboratories to conduct a corona examination [4]. As for the medical staff, the World Health Organization has set some standards for the adequacy of medical staff in any country, where the organization recommended that there should be at least one doctor per thousand people as an acceptable minimum (WHO 2010).

As for the issue of distance education, educational institutions began to search for an alternative to regular education, here is the idea of distance education and the use of technology in education [5].

6 Methods

The descriptive-analytical approach was used, being the most appropriate for this type of research, "and it is an approach that is based on the study of reality, or on presenting the phenomenon as it is, without increase, decrease or modification, by providing an accurate description of that reality or that phenomenon."

6.1 Study Community

The study community is made up of citizens living in the northern governorates (West Bank) who are 18 years old and over (2017) with a total of (776,611), Appendix 1 shows the distribution of community by governorate and educational qualification for the year 2017 [6].

6.2 Study Sample

The study sample consisted of (1064, 421 = male, 643 female) citizens aged 18 years and over from citizens residing in the northern governorates (West Bank), and it was calculated according to Robert Mattson's equation to determine the sample size. The questionnaire was distributed electronically and on paper, depending on the governorate classes and the educational qualification, and it was distributed randomly after choosing the two classes. Appendix 2 shows that.

6.3 Study Tools

Questionnaire tool: The study tool consisted of (37) items. It is divided into (4) areas (the health aspect, the social aspect, the economic aspect, the distance education aspect) of the impact of the Corona epidemic on the citizens in the northern governorates of Palestine.

6.4 Validity of the Tool

The tool was presented to a group of experts and specialists to express their opinion on the content of the scale's paragraphs and its effectiveness towards the target group. Some of the paragraphs were modified and reformulated to suit the reality of our Palestinian society. The five-point (Likert) scale was used to measure the level of the Corona epidemic, reality, and challenges. The items for this scale were evaluated as follows: (strongly agree (5) points, agree (4) points, neutral (3) points, disagree (2) point, strongly disagree one point).

6.5 *Reliability of the Tool*

After applying the study tool to the sample, the reliability coefficient of the tool was calculated by using the (Cronbach's alpha) equation for internal consistency, where the value of the reliability coefficient of the tool was (0.89), which is a statistically acceptable value.

6.6 *The Interview Tool*

The researcher adopted the semi-regular interview, and the questions were semi-open to achieve the objectives of the study.

7 Results of the Study

This part dealt with a presentation and discussion of the results of the study according to the order of its questions, the questionnaire questions and its hypotheses, the interview questions and then the general results.

The first question: What is the impact of the Corona epidemic on the citizens in the northern governorates of Palestine (the West Bank) in the aspects of (health, social, economic, distance education)?

To interpret the results, the following statistically approved arithmetic means were used for answering the items: (80% or more, a very high impact scores) (70%–79.99% a high impact score) (60%–69.99% a medium impact score) (from 50%–59.99% a moderate impact score) Low effect) (less than 50% the degree.

To answer this question, it was necessary to display the arithmetic means, standard deviations, and the degree of impact for each of the fields (health, social, economic, distance education)?

We notice through the data in Table 1 that the percentage of the arithmetic means for the areas of impact that the Corona epidemic had on citizens in the West Bank ranged between 69%, with a moderate impact degree for the field of economic aspects, so that it obtained the lowest percentage. But, the field of distance education and social aspects obtained the highest percentages of the arithmetic mean, reaching 78%, with a significant degree of impact. We also notice that the level of the total degree of the impact that the Corona epidemic had on citizens in the governorates was high, with a percentage of 75%, meaning that the respondents believed that the Corona epidemic had a significant impact on their lives in all aspects of life despite the disparity in this impact.

Table 1 Arithmetic means, standard deviations, and the degree of impact for the stress fields

Area	Arithmetic mean	Standard deviation	Percentage (%)	Degree of impact
Heath aspects	0.58	3.70	74	High
Social aspects	0.56	3.88	78	High
Economic aspects	0.56	3.43	69	Moderate
Distance education aspects	0.52	3.89	78	High
Total degree	0.43	3.73	75	High

The second question: Is there a statistically significant difference in the impact of the Corona epidemic on citizens in the northern governorates of Palestine (West Bank) according to the variables (gender, educational qualification, governorate)?

A set of hypotheses emerged from the study question, which will be examined as follows:

Results related to the first hypothesis: There are no statistically significant differences at the significance level ($\alpha \leq 0.05$) of the corona virus on citizens in the northern governorates of Palestine (West Bank) according to the gender variable.

To test the hypothesis, an independent t-test of two independent groups was used. As shown in Table 2.

We reject the null hypothesis, as evidenced by the data in the previous Table 2, meaning that there are statistically significant differences at the significance level ($\alpha \leq 0.05$) for the impact of the Corona epidemic on citizens in the northern governorates of Palestine (West Bank) according to the gender variable, according to The total degree

Table 2 Results of the T-test for two independent samples to test the level of differences for the Corona epidemic according to the gender variable

Area	Male		Female		T value	Significant
	Arithmetic mean	Standard deviation	Arithmetic mean	Standard deviation		
Health aspects	3.60	0.64	3.76	0.53	−4.21	0.00
Social aspects	3.76	0.61	3.95	0.51	−5.74	0.00
Economic aspects	3.38	0.59	3.46	0.54	−2.24	0.03
Distance learning aspects	3.88	0.53	3.90	0.51	−0.86	0.39
Total degree	3.66	0.48	3.78	0.39	−4.26	0.00

*Statistically significant at the level ($\alpha \leq 0.05$)

Table 3 Arithmetic averages of areas of the impact of the corona epidemic on citizens in the northern governorates of Palestine (West Bank) according to the educational qualification variable

Area	Secondary school and below	Diploma	Bachelor	Post graduate
Health aspects	3.70	3.66	3.70	3.67
Social aspects	3.88	3.82	3.90	3.78
Economic aspects	3.46	3.53	3.44	3.27
Distance learning aspects	3.91	3.84	3.88	3.94
Total degree	3.75	3.75	3.74	3.68

and on the areas (health aspects, social aspects, economic aspects), its significance level ranged between (0.00–0.03) and these values are all less than the significance level (0.05) and this effect was on females. The hypothesis was accepted for the aspects of distance education so that the significance level for this field was 0.39, which is greater than the significance level 0.50.

Results related to the second hypothesis: There are no statistically significant differences at the significance level ($\alpha \leq 0.05$) of the corona virus on citizens in the northern governorates of Palestine (West Bank) according to the educational qualification variable.

To test the second hypothesis, the arithmetic means of the impact of the Corona epidemic on citizens in the northern governorates were extracted, as well as the standard deviations, then the one-way analysis of variance test was used to examine the differences between the means according to the variable of educational qualification as shown in Tables 3 and 4.

We accept the null hypothesis, meaning that there are no statistically significant differences at the significance level ($\alpha \leq 0.05$) of the corona virus on citizens in the governorates (West Bank) according to the educational qualification variable based on the total degree and areas (health aspects, social aspects, distance education aspects), its significance level ranged between (0.27–0.89) and these values are all greater than the significance level (0.05).

As for the economic aspects, we reject the null hypothesis in the sense that there are statistically significant differences at the significance level ($\alpha \leq 0.05$) for the impact of the Corona epidemic on the citizens in the northern governorates of Palestine according to the educational qualification variable, in favor of those who hold a secondary educational qualification.

The results related to the third hypothesis: There are no statistically significant differences at the significance level ($\alpha \leq 0.05$) of the corona virus on citizens in the northern governorates of Palestine (West Bank) according to the governorate variable.

To test the third hypothesis, the arithmetic means of the impact of the Corona epidemic on the citizens of the northern governorates of Palestine were extracted, as well as the standard deviations, then the one-way analysis of variance test was used

Table 4 Results of the one-way analysis of variance test, to test the significance level of differences according to the educational qualification variable

area	Contrast source	Sum of squares	degrees of freedom	mean squares	*F* value	Significance level
Health aspects	Between groups	0.21	3	0.07	0.21	0.89
	Within groups	358.38	1060	0.34		
	Total	358.59	1063			
Social aspects	Between groups	1.02	3	0.34	1.09	0.35
	Within groups	331.55	1060	0.31		
	Total	332.57	1063			
Economic aspects	Between groups	2.83	3	0.94	2.99	0.03
	Within groups	334.42	1060	0.32		
	Total	337.25	1063			
Distance learning aspects	Between groups	0.60	3	0.20	0.74	0.53
	Within groups	286.38	1060	0.27		
	Total	286.98	1063			
Total degree	Between groups	0.73	3	0.24	1.32	0.27
	Within groups	194.31	1060	0.18		
	Total	195.04	1063			

to examine the differences between the means according to the governorate variable as shown in Tables 5 and 6.

We reject the null hypothesis, which means that there are statistically significant differences at the significance level ($\alpha \leq 0.05$) of the corona virus on the citizens in the northern governorates of Palestine (West Bank) according to the governorate variable, the total score and all aspects, its significance level for all aspects reached (0.00) and all these values are less than the significance level (0.05).

This means that the impact of the Corona epidemic on citizens in the northern governorates of Palestine, according to the governorate variable, was the most in the governorates of Jericho, Hebron, Tulkarm, and Nablus.

Table 5 Arithmetic means of areas of the impact of the corona epidemic on citizens in the northern governorates of Palestine (West Bank) by governorate variable

Area	Jericho	Hebron	Jerusalem	Bethlehm	Jenin	Ramallah	Salfit	Tubas	Tulkarm	Qalqilya	Nablus
Health aspects	3.97	3.84	3.63	3.63	3.38	3.58	3.40	3.69	3.85	3.63	3.74
Social aspects	3.96	3.96	3.78	3.89	3.78	3.87	3.61	3.03	3.95	3.81	3.91
Economic aspects	3.51	3.42	3.41	3.43	3.58	3.28	3.16	3.58	3.69	3.45	3.53
Distance learning aspects	4.14	3.92	3.71	3.71	3.88	3.95	3.81	3.83	3.93	3.83	3.98
Total degree	3.90	3.79	3.64	3.64	3.67	3.69	3.51	3.86	3.86	3.69	3.80

Table 6 Results of the one-way analysis of variance test, to test the significance level of differences according to the governorate variable

Area	Contrast source	Sum of squares	Degrees of freedom	Mean squares	*F* value	Significance level
Health aspects	Between groups	15.37	10	1.54	4.71	0.00
	Within groups	343.22	1053	0.33		
	Total	358.59	1063			
Social aspects	Between groups	8.03	10	0.80	2.60	0.00
	Within groups	324.55	1053	0.31		
	Total	332.57	1063			
Economic aspects	Between groups	13.33	10	1.33	4.33	0.00
	Within groups	323.92	1053	0.31		
	Total	337.25	1063			
Distance learning aspects	Between groups	8.21	10	0.82	3.10	0.00
	Within groups	278.76	1053	0.27		
	Total	286.98	1063			
Total degree	Between groups	7.12	10	0.71	3.99	0.00
	Within groups	187.92	1053	0.18		
	Total	195.04	1063			

8 General Results of the Interviews

1. The results of the study showed that the commitment and response of citizens in the first months of the epidemic was comprehensive and complete. There was also awareness, acceptance, and complete response to dealing with all safety and prevention measures from all aspects, but after that this interest decreased and the response became like ebb and flow and thus there became complacency in behavior, especially wearing masks, physical distancing, and staying away from gatherings, which is a negative indicator in dealing with the emerging corona virus.
2. It was found through the answers that there is no specialized economic committee in each governorate separately.

3. The results of the study showed the diversity of challenges and difficulties facing the governors of the West Bank between the presence of the occupation, which is the biggest challenge because it often prevented field missions from reaching areas that needed to work as security and medical agencies, and the difficulty of convincing citizens of the existence of the epidemic and the need for social distancing and adherence to the health protocol, as well as the harmonization between the needs of citizens and their provision and the continued rotation of the economy wheel on one hand, and the requirements and decisions related to confronting the pandemic on the other hand, and the decline in the economic situation, which was accompanied by a high rate of unemployment, and the emergence of some problems with the security establishment as a result of the current situation.
4. As for the obstacles of the heads of the chambers of commerce, they varied between the forms of closure and its nature on the ground. The city center was completely closed, while in the countryside and the camps it was working normally, or because of the presence of settlements that were operating inside the governorate with the presence of a high movement of Israeli security within the governorate and therefore it was Israeli merchants are allowed to work, while the Palestinian merchant is not.

9 Recommendations

Based on the findings of the study, the researcher recommends the following:

- Despite the commitment of several the Palestinian community to the instructions of the Ministry of Health and the competent authorities, there is a large number who do not believe in this pandemic as a matter of conspiracy, denial and disbelief, so the researcher recommends that educational programs be created to enable the citizen to understand the pandemic and its consequences.
- The researcher also recommends evaluating the previous stage and making the necessary recommendations.
- Establishing an economic committee in each governorate and evaluating the previous stage through mechanisms and field studies, to monitor the current economic situation on the one hand, and what will happen to matter in the future.
- The researcher considered that the competent authorities that dealt with the pandemic were flexible in dealing with citizens without enacting laws that oblige them to adhere to the discipline.
- Managing a media and advertising plan with the aim of positively dealing with the upcoming vaccines in order to overcome our health, economic and social problems and activate the role of audio-visual media and social networking sites, and link most of these means within a collective plan.

List of Sources and References

1. Palestinian News and Information Agency Wafaa: The President issues a decree declaring a state of emergency in all Palestinian territories for a month (2020). http://www.wafa.ps/ar_page.aspx?id=QVYJ3sa871529739630aQVYJ3s. Last accessed 24/11/2020
2. Orabi, S.: Corona-confrontation committees in the West Bank ... between challenge and politicization, Raya Center for Political Development (2020). https://vision-pd.org/archives/508911. Last accessed 25/11/2020
3. Palestinian Central Bureau of Statistics, Number of hospitals, hospital beds, and beds per 1000 citizens in Palestine by region (2019) http://www.pcbs.gov.ps/Portals/_Rainbow/Documents/%D8%B9%D8%AF%D8%AF%20%D8%A7%D9%84%D9%85%D8%B3%D8%AA%D8%B4%D9%81%D9%8A%D8%A7%D8%AA%20%D9%88%D8%A3%D8%B3%D8%B1%D8%A9%20%D8%A7%D9%84%D9%85%D8%B3%D8%AA%D8%B4%D9%81%D9%8A%D8%A7%D8%AA.html. Last accessed 29/11/2020
4. Salama, A.: Corona Pandemic: Palestinian Performance in a Volatile Global Scene, A Study in the Journal of Palestine Politics Between the Corona and Normalization Pandemic, volume (49–50) Al-Ayyam Foundation for Printing Ramallah, Palestine
5. Palestinian News and Information Agency, Wafaa, distance education in Palestine (2021). https://info.wafa.ps/ar_page.aspx?id=c2xypma27836871744ac2xypmlast. Accessed 28/11/202
6. Palestinian Central Bureau of Statistics, 2017 Palestinian population (10 years and over) in Palestine by governorate, gender, and educational status, Ramallah, Palestine (2017)

The Effectiveness Degree of Distance Learning at Zarqa University in Light of Corona Pandemic from Faculty Members' and Students' Point of View

Nidal Alramahi, Rada Almwdiah, Alaeddin Sadeq, Khaled Alzeaideen, Laith Hadla, Maram Al-Safarini, Amer Abu Salem, Mahmoud Odeh, Khaled Al-qawasmi, and Aliaa Burqan

Abstract This study aims at finding out the effectiveness degree of distance learning at Zarqa University in light of Corona Pandemic from faculty members' and students' point of view. The descriptive survey methodology is used. The study is applied to a stratified random sample of faculty members and students at Zarqa University, whose number was (210) faculty members and (1261) male and female students. The questionnaire, which consists of (40) items, distributed into three domains, is used as a tool to collect data, after ensuring its validity and reliability. The findings show that the mean of the effectiveness degree of distance learning at the university is medium from faculty members' point of view with a mean (3.67) and a standard deviation of (0.47). The study found that the effectiveness degree of distance learning at the university is also medium from the students' point of view. The mean is (3.36) with a standard deviation of (0.62). The findings indicate that there are significant differences at ($\alpha \leq 0.05$) between the views of faculty members and students' views on the overall degree and the three domains in favor of faculty members. There are no significant differences at ($\alpha \leq 0.05$) of the degree of effectiveness of distance learning at the university from the faculty members' point of view due to the academic rank and gender variables. While there are significant differences attributed to the variable of the faculty, in favor of the humanities faculties. There are significant differences in the degree of effectiveness of distance learning at the university from the students' point of view in the total score and the three domains attributed to gender, in favor of males. There are significant differences due to the variable of the faculty in the domains of (Distance learning effectiveness and Distance learning system) in favor of humanities faculties. There were no significant differences in the domain of (Distance learning obstacles).

Keywords Degree of effectiveness · Distance learning · Zarqa University · Corona pandemic (Covid-19)

N. Alramahi · R. Almwdiah · A. Sadeq · K. Alzeaideen · L. Hadla · M. Al-Safarini · A. A. Salem · M. Odeh (✉) · K. Al-qawasmi · A. Burqan
Zarqa University, Zarqa, Jordan
e-mail: Modeh@zu.edu.jo

A. Hamdan et al. (eds.), *The Implementation of Smart Technologies for Business Success and Sustainability*, Studies in Systems, Decision and Control 216,
https://doi.org/10.1007/978-3-031-10212-7_55

1 Introduction

The world is facing the Corona crisis, which has put it in front of many challenges that cannot be avoided or overlooked, and imposed on countries all over the world to coexist with this crisis in light of different and unstable circumstances, and perhaps other data that it has not experienced before. Among the most prominent and important of these data is the issue of distance learning, or what is called "e-learning", which is considered direct education, as the educational content is presented through electronic means [1, 2].

As a result of the outbreak of the Corona pandemic around the world during the past year (2020), this damage affected the countries of the Middle East and Africa region, and to limit the spread of the epidemic infection, governments in this region began to restrict the movement of citizens and close schools and universities in the beginning of March of the year (2020) [3]. These schools and universities are still closed to the present time [4]. There are more than (100) million students outside the classroom. The numerical modeling predictions indicated a decline in learning outcomes according to what was documented in school analyses, and to mitigate the negative impact of this closure of schools and universities in the Middle East and North Africa region, learning in this region has become remotely via the Internet [5].

E-learning is a form of distance learning that relies on the capabilities and tools of the International Information Network, the Internet, and computers in studying specific educational content through continuous interaction between the teacher, the learner, and the content [6]. It is, as described by some researchers, an educational system to provide educational or training programs to learners or trainees at any time and in any place, by employing interactive information and communication technologies such as the Internet, local or satellite channels for television, magnetic and cellular disks, e-mail, computers, and teleconferences to provide a multi-source interactive environment in a simultaneous or asynchronous way inside the classroom, without committing to a specific location, depending on the self-learning and the interaction between the learner and the teacher.

2 Study Questions

The study questions included the following sub-questions:

Q1: What is the degree of effectiveness of distance learning at Zarqa University in light of the Corona pandemic from point of view of faculty members?

Q2: What is the effectiveness of distance learning at Zarqa University in light of the Corona pandemic from the point of view of students?

Q3: Are there statistically significant differences at the level of ($\alpha \leq 0.05$) in the degree of effectiveness of distance learning at Zarqa University between the views of faculty members and the views of students?

Q4: Are there statistically significant differences at the level of ($\alpha \leq 0.05$) in the degree of effectiveness of distance learning at Zarqa University from the point of view of the teaching staff due to variables: academic rank, gender, and faculty?

Q5: Are there statistically significant differences at the level of ($\alpha \leq 0.05$) in the degree of effectiveness of distance learning at Zarqa University from the viewpoint of students due to the variables of sex and faculty?

3 Study Objective

This study aims to identify:

1. The degree of effectiveness of distance learning at Zarqa University in light of the Corona pandemic from the point of view of faculty members.
2. The degree of effectiveness of distance learning at Zarqa University in light of the Corona pandemic from the point of view of students.
3. Statistically significant differences in the degree of effectiveness of distance learning at Zarqa University between faculty members and students.
4. The statistically significant differences in the degree of effectiveness of distance learning at Zarqa University from the point of view of faculty members are due to variables of academic rank, gender, and faculty.
5. The significant statistical differences of the degree of effectiveness of distance learning at Zarqa University from the students' point of view is due to the variables of gender and faculty.

4 Importance of Study

The importance of this study stems from the importance of the topic it addresses, which is represented in distance learning in light of the Corona pandemic and its degree of effectiveness and importance in facing this pandemic, according to what Zarqa University did in this regard. It is expected that this study would benefit:

- Administrative leaders at Zarqa University, by knowing the degree of effectiveness of distance learning.
- Teaching staff and students, by knowing the importance and necessity of distance learning in light of the Corona pandemic.
- Researchers in this scope in light of the theoretical literature and previous studies presented in this study and a tool whose validity and reliability have been verified, so that it can be a starting point for other studies.

5 Theoretical Literature

Education and e-learning is one of the methods of distance learning through the use of computers, its networks, and its multimedia, including sound, image, graphics, search mechanisms, libraries, and Internet portals [7]. The following is a review of previous Arab and foreign studies related to distance learning or e-learning according to the date of their publication from the oldest to the most recent. The study of [8] aimed to know the reality of e-learning at Tishreen University from the point of view of faculty members and students. Two questionnaires were used, the first was applied to a sample of the teaching staff at Tishreen University, whose number reached (113) members, and the second was applied to a random sample of fourth-year students at Tishreen University, the number of its members was (774) students. The results showed that there were no statistically significant differences between the average grades of the teaching staff according to the variables of scientific rank and specialization on the axis of "the extent of using e-learning and its pros, cons, and obstacles," and the existence of teams on the axis of "negatives" according to the variable of specialization and in favor of the literary specialization. There was no statistically significant difference between the average scores of students on the e-learning axis and its obstacles according to the specialization variable. The interest of faculty members and students in e-learning was minimal. The sample members emphasized the role of this type of education in increasing computer skills. The most disadvantages of it are that it reduces the burden of teaching staff, and that sitting in front of the computer causes many diseases [9].

6 Research Methodology

In this study, the descriptive survey methodology is used as the appropriate method for the study topic and its problem. Surveying and reviewing the relevant studies are used to clarify the nature of the topic, the foundations and the theoretical perspectives on which the subject of the current study is based. The questionnaire was used to collect data after confirming its validity and reliability.

7 Study Sample

The sample is chosen from the study population using the random stratified method, as the number of teaching staff reached (210) faculty members, (156) males and (54) females, and (1261) male and female students, (682) male students and (579) female students, from Zarqa University. Table 1 shows the distribution of the study sample according to the study variables.

Table 1 The study sample individuals are distributed according to the study variables

Category	Variable	Variable categories	Number	Percentage (%)
Faculty members	Academic rank	Assistant Professor	106	50.5
		Lecturer	43	20.5
		Associate Professor	40	19.0
		Professor	21	10.0
		Total	210	100
	Gender	Male	156	74.3
		Female	54	25.7
		Total	210	100
	Faculty	Humanities	108	51.4
		Scientific	102	48.6
		Total	210	100
Students	Gender	Males	682	54.1
		Female	579	45.9
		Total	1261	100
		Humanities	780	61.9
		Scientific	481	38.1
		Total	1261	100

8 Study Tools

To achieve the objectives of the study by answering its questions, a questionnaire was developed for this purpose, referring to the relevant theoretical literature and previous related studies. The questionnaire included, in its initial form, (40) items distributed in three areas, namely: the effectiveness of distance learning, distance learning system, and barriers to distance learning. Table 2 shows the distribution of the items of the questionnaire (the study tool) among the three scopes.

The questionnaire is designed according to the quintuple Likert-scale. The alternatives of the answer were: strongly agree and give five degrees, agree and give four degrees, neutral and give three degrees, while the alternative do not agree was given two degrees, and the alternative strongly disagreed with one degree.

Table 2 The items of the questionnaire (the study tool) are distributed among the three scopes

Scope	Number of items
Effectiveness of distance learning	14
Distance learning system	18
Barriers to distance learning	8
Total	40

9 Validity of the Study Tools

The validity of the study tool is verified by presenting it to a number of experienced and competent referees who are members of the teaching bodies in Jordanian universities. They were asked to express their opinions regarding the items of the tool in terms of their belonging to the scope that they fell under, the accuracy of their linguistic formulation, and the extent of their clarity, and adding, modifying, or deleting what they deem appropriate of the items. The approval percentage of (80%) or more of the opinions of the arbitrators was adopted to maintain the item, and to delete or amend the item if it obtained less than this percentage, and in light of all that, the arbitrators' observations and suggestions were taken into account, and the required amendments were made to the items that some arbitrators requested to amend, and no items were deleted, and the questionnaire remained as it is, consisting of (40) items, in their final form, distributed over the three areas of the tool. Taking into account the opinions, observations, and suggestions presented by the arbitrators, regarding the items of the tool, is considered evidence of the validity of the tool, which increases the degree of reliability of the results reached by this study.

10 Validation of Tool's Construct

Validity of the tool's construct was verified using the internal consistency method, which is one of the methods used to find construct validity. The correlation coefficient between each scope of the study tool and the overall degree of the tool was found. Table 3 shows the results.

It can be seen from Table 3 that all the values of the Pearson correlation coefficient between each scope of the study tool and the total score of the tool were statistically significant at the level ($\alpha \leq 0.05$), which indicates the validity of the internal consistency and hence the construct validity of the tool, as this shows the values of the Pearson correlation coefficient. (Pearson), which ranged between (0.69) for the scope of distance learning system and (0.77) for the scope of distance learning barriers.

Table 3 Pearson correlation coefficient values for the study instrument domains with overall score

Scope	Domain is the value of the Pearson correlation coefficient
Effectiveness of distance learning	0.73*
Distance learning system	0.69*
Barriers to distance learning	0.77*

*Statistically significant at ($\alpha \leq 0.05$)

Table 4 Values of the stability parameters using the test–retest, and internal consistency methods

Scope	The stability parameter evaluated by using the test–retest method	Values of the stability parameter using the internal consistency method using the Cronbach-alpha equation
Efficacy of distance learning	0.86	0.84
Distance learning system	0.88	0.87
Barriers to distance learning	0.91	0.88
Total	0.92	–

11 Stability of Study Tools

The stability of the study tool (the questionnaire) was confirmed by applying the tool to a sample consisting of (30) members of the study community, including faculty members and students, and from outside the study sample, and the tool was re-applied again to the same sample members with an interval of two weeks using the test and retest method, and the Pearson correlation coefficient was calculated between the two applications, and the value of the correlation coefficient was (0.92) for the overall score of the tool.

The stability of the tool was also confirmed by finding its internal consistency parameter, using the Cronbach Alpha equation for each of the three domains. Table 4 shows the results of stability using the two methods.

The results in Table 4 show that the values of the stability parameter using the test–retest method ranged for scopes between (0.86–0.91) and for the overall score (0.92), and with the internal consistency method, the values of the Cronbach-Alpha equation ranged between (0.84–0.88). And all the values of the two methods are statistically significant at ($\alpha \leq 0.05$). These values are acceptable for the purposes of the present study.

12 Presentation and Discussion of Results and Recommendations

The results of the first question, which states: "What is the degree of effectiveness of distance learning at Zarqa University in light of the Corona pandemic from the faculty members' point of view?"

To answer this question, averages and standard deviations were calculated and the ranks were determined for the degree of effectiveness of distance learning at Zarqa University in light of the Corona pandemic from the point of view of the faculty members, according to each of the three scopes of study (the effectiveness of distance learning, the distance learning system, and obstacles of distance learning) and the overall degree. Table 5 shows the results.

Table 5 Arithmetic means, standard deviations, ranks and degree of effectiveness of distance education at Zarqa University in light of the corona pandemic from the viewpoint of faculty members arranged in descending order

No.	Scope	Arithmetic mean	Standard deviation	Rank	Efficacy
3	Barriers to distance learning	3.85	0.60	1	High
2	Distance learning system	3.71	0.63	2	High
1	Effectiveness of distance learning	3.53	0.61	3	Medium
	Overall score	3.67	0.47	–	Medium

It can be seen from Table 5 that the arithmetic average of the degree of effectiveness of distance learning at Zarqa University in light of the Corona pandemic from the viewpoint of the faculty members was medium, as the arithmetic mean was (3.67) with a standard deviation (0.47). The scopes of "distance learning obstacles" and "distance learning system" came with a high degree of effectiveness, with two arithmetic averages (3.85) and (3.71), ranking first and second respectively. As for the scope of "effectiveness of distance learning", it came in the third and final rank, as it reached an arithmetic average (3.53) and a medium degree.

This result may be attributed to the awareness of faculty members at Zarqa University about the essence of distance learning, its importance, and degree of effectiveness in the current circumstance through what they have seen from the results of students and their responses to the programs offered by this type of education, especially since the current reality that Jordan and all countries are experiencing in light of the epidemic, the rapidly spreading Corona pandemic, which requires protection, prevention, and social distancing, and as a result, universities and schools were closed, and this imposes the continuation of the educational process by using distance learning to ensure the isolation of students from each other and their stay in their homes to keep them from contracting this deadly disease. Perhaps this result is due to the high level of obstacles facing the process of implementing remote education, which faculty members cannot overcome or surpass, which in turn affected the degree of effectiveness of this type of education. This result may be attributed to technical difficulties in the distance learning system, parents' lack of conviction about the effectiveness of distance learning, and the lack of computers for a good proportion of student families, in addition to the financial cost of the Internet, which is difficult, or even impossible, for poor families to pay for [10].

13 Recommendations

In light of the results of this study, the researchers made a number of recommendations and proposals as follows:

- Increasing the interest of those responsible for public and private Jordanian universities, including deans of faculties and heads of academic departments, in the importance of distance learning and working to provide it for students in all specialties, especially at the present time as the Corona epidemic invaded all countries of the world.
- Providing support, encouragement, and material and moral motivation for faculty members to practice e-learning and distance learning, and encourage technicians and administrators and motivate them to provide the required services and support them in the process of practicing this type of education.
- Conducting a study similar to the current study on Jordanian public universities and comparing its results with the results of the current study.
- Conducting an evaluation study for the experience of distance learning in public and private Jordanian universities.

Acknowledgements This research is funded by the Deanship of Research in Zarqa University, Jordan.

References

1. Al-Ramahi, N., Odeh, M.: The impact of innovative technology on the quality assurance at higher education institutions in developing countries: a case study of Jordan. Int. J. Inform. Educ. Technol. **10**(11) (2020)
2. Al-Owaid, M.S., Al-Hamid, A.b.A.: "E-learning in the Faculty of Communication and Information in Riyadh: a case study", a working paper submitted to the E-learning seminar in King Faisal Schools/Riyadh in the period 19–21 Safar 1424 AH corresponding to 04/23/21/2003 (2003)
3. Azzi, H.K., Blom, A., Bend, M.: Innovation in Responding to Coronavirus Could Pave the Way for Better Learning Outcomes in MENA (2020)
4. World Health Organization (2020). http://www.whoint/ar
5. Kaliope, A.-H., Tigran, S.: Managing the Impact of COVID-19 on Education Systems around the World: How Countries are Preparing, Coping and Planning for Recovery (2020)
6. Hussamo, S., Al-Abdullah, F.I.: The reality of e-learning at Tishreen University from the point of view of both faculty members and students. Damascus Univ. J. **27**(Suppl), 243–278 (2011)
7. Al-Shomali, K.: What is the Meaning of Distance Learning? (2017). mawdoo3.com. Smadi, O.: Corona and E-Learning (2020). ammon news
8. Miqdadi, M.A.: Perceptions of high school students in public schools in Jordan for the use of distance learning in light of the Corona Crisis and its developments. Arab J. Sci. Publishing (AJSP) **16**, 97–113 (2020)
9. Salem, A.M.: Teaching Technology and E-Learning. Al-Rashed Library, Riyadh (2004)
10. Nasser, R.: Distance Learning. mawdoo3.com (2019)

The Extent of Meeting Global E-Learning Standards in Jordanian Universities from the Perspective of Faculty Members

Khaled Alzeaideen, Ali M. Alelaimat, Reda Al-mawadieh, and Issa abdulwahab Al Tarawneh

Abstract This study explored the extent of meeting global e-learning standards in Jordanian universities from the perspective of faculty members. Sample: (58) faculty members from faculty of educational sciences at Hashemite University and Zarqa University during (2020–2021). The validity and reliability of the questionnaire were checked. It targets three areas: (the process of designing the e-content, the support and assistance provided by supervisors, and administrative services). Results: The extent of meeting global e-learning standards in Jordanian universities from the perspective of faculty members is high. The means of the areas are either higher or moderate. The mean of the process of designing the e-content is high and ranked first. The mean of the support and assistance provided by supervisors is high and ranked second. The mean of the administrative services is moderate and ranked third. There isn't any significant difference between the sampled faculty members' attitudes towards the extent of meeting the aforementioned level which can be attributed to university. However, there is a difference between the sampled faculty members' attitudes towards the extent of meeting the aforementioned level which can be attributed to academic rank for the favour of the ones holding the higher rank. Recommendations: Providing more attention to e-content designing process through holding workshops & symposiums for faculty members.

Keywords E-Learning · Jordanian Universities · E-learning standards

K. Alzeaideen (✉)
Zarqa University, Zarqa, Jordan
e-mail: khaledz@zu.edu.jo

A. M. Alelaimat
Queen Rania Faculty for Childhood, Hashemite University, Zarqa, Jordan
e-mail: aliolaimat@hu.edu.jo

R. Al-mawadieh
Faculty of Educational Science, Zarqa University, Zarqa, Jordan
e-mail: rmawadieh@zu.edu.jo

I. Al Tarawneh
Zarqa, Jordan
e-mail: spss_esa@yahoo.com

A. Hamdan et al. (eds.), *The Implementation of Smart Technologies for Business Success and Sustainability*, Studies in Systems, Decision and Control 216,
https://doi.org/10.1007/978-3-031-10212-7_56

1 Introduction

Covid-19 lockdown enforced delivering electronic education through educational and academic institutions. This crisis forced 1,576,021,818 students worldwide (91.3% of the overall students worldwide) to set in their homes. Distance education is no longer an additional option instead it's essential and can't be dispensed by educational institutions today. However, they aren't ready for making this digital shift.

Students using e-learning platforms maximized 200 times. That led to raising the severity of the competition between the owners of e-learning platforms. Platforms features were developed to meet the requirements of e-learning environment. Six months later, those platforms played a role that couldn't have been played during the past ten years [1].

Technology has been developing fast and rapidly. The teaching process is more effective when shifting to electronic means during such conditions. During such conditions, the electronic educational pattern is considered the only suitable teaching pattern during the lock down period. It's expected that this pattern shall become soon the best pattern [2]. Some people claimed for employing the efforts of teachers in the aim of overcoming the obstacles. Some people claimed for shifting the concentration from concentrating on the educational content into educational technologies. That's because the significance of the teaching methods is equivalent to the significance of the educational content.

Other challenges that are still facing the educational sector: technological illiteracy, and the absence of a suitable infrastructure [1]. Students and faculty members today are in need for being developed in order to identify the way of dealing and adapting themselves with the shift that occurred to the pattern of delivering education during the COVID 19 crisis. There is no wonder that some teachers who adopt conventional teaching methods are questioning the effectiveness of delivering distance education. It was expected to have a conflict between the teachers who excel in using technology and the teachers who adopt conventional teaching methods. Such conflict is represented in debates over students' scores, learning outcomes, and educational flexibility. The novice teachers believe that the virtual classrooms provide students with independency that can't be enjoyed in the conventional classroom [3].

Meeting the e-learning goals mentioned in the plans set by the Jordanian government requires comprehensive and well-planned development. This development must target all students and the educational system. It must be accompanied with an assessment for the outcomes of the educational system. It shall lead to meeting the changing needs of society and the changing demands of the labour market. That shall be done through developing the educational system to benefit from the knowledge expansion and technical developments. The modern technical developments, the communication development and the emergence of the web led to having great expansion in information in various areas. All people today seek benefiting from the modern technologies. That applies to the ones working in the education sector. The latter ones benefit from the modern technologies in the e-learning area [4].

According to [5], during the last couple of years, the use of internet increased significantly. That affected various aspects in life. Such aspects include: the educational aspects. For instance, the e-learning term emerged. E-learning has many merits. For instance, it reduces the incurred costs, and provides learning opportunities that are more flexible. The number of the learners enrolled in the electronic training programs increased. Many educational institutions in developed countries provide distance educational programs for their learners who are located in various countries. Those programs deliver synchronous and asynchronous education.

There are concepts that are associated with e-learning, such as: computer-based learning, internet-based learning, distance education, web-based education, mobile learning, resources-based education, and technology-based education. There are similarities between the latter concepts and the e-learning concept. However, there are differences between them. There are correlative relationships between those terms. In all cases, all those concepts are types of e-learning.

E-learning are terms that emerged recently. It emerged due to the development of ICTs. It involves several types of education that don't require having a trainer nor buildings. It's enough to have a teacher providing guidance and providing assistance and training bodies running and supervising the educational program. Al-Hameedy [6] adds that this education serves as a reform for the conventional education rather than an alternative for it.

The Arabic language association defines the word (electronic) as a trait for everything related to the electronic machines and tools and systems. Such tools include Magnetic pumps and transistors. There are various definitions for e-learning. E-learning may be defined as the learning that aims at creating interactive environment that's rich with computer-based applications that allow students to access learning resources at any time and place [7]. It can be defined as the illustration of an educational content with providing exercises and following up the things illustrated in classroom through using advanced programs [8]. It refers to the use of an educational system that involves using information technologies and the computer network for expanding the score of the teaching process. It may involve using computer, internet and electronic programs by experts in the ministry of companies [21].

The developments that occurred to ICTs contributes significantly in delivering information to learners in various places fast without exerting much effort [9]. The latter researchers added that the goal-oriented use for ICT in the field of education facilitated the delivery of education to learners regardless of their place and the providing of learners with knowledge, information and skills without having those learners to attend in person at the higher educational institution.

Through the e-learning application, there are two types of education that can be delivered (i.e., synchronous and asynchronous education). Both of those patterns of educations take the conditions of learners and the individual differences between learners into consideration. They provide learners with learning opportunities any time. They enable students to save information and retrieve it at any time they want [10]. The latter researchers add that both of those types of education are essential for delivering education successfully. Delivering asynchronous education only isn't enough. That's because doing that shall make the learner play an inactive role. To solve this problem and achieve the best results, it is necessary to deliver synchronous education.

Al-Leqani [11] carried out a study. He identifies the characteristics of the teachers delivering e-learning. Such characteristics include the following ones:

- The teachers must be having much general knowledge
- The teachers must be capable of running the educational context in the light of having a modern technology
- The teachers must be capable of managing the interaction inside the classroom and outside it
- The teachers must be capable of creating and acquiring knowledge
- The teachers must encourage learners to exchange expertise and benefit their peers
- E-learning offers learning opportunities that fit with the time schedule of learners and educational institutions
- E-learning facilitates the access to the educational websites
- E-learning provides learners with equal access to the learning resources

Pilkington et al. [9] found that there is a relationship between the increase of collaboration between learners through using the e-learning platforms and the increase of achievement.

Several studies suggest that it's significant to deliver e-learning. They include: the study of [12]. The latter researcher reviewed and assessed the role of e-learning in improving the educational climate. Such studies include: the study of [13]. The latter researcher adds that it's significant to assess the e-learning practices and methods for developing e-learning.

Coombs and Rodd [14] add that students believe that e-learning offer opportunities to learn at any place and time the learners want. They add that e-learning takes the individual differences between learners into consideration. E-learning provides learning opportunities that fit with the speed that the learners need to perform the tasks that fit with their capabilities.

Atan et al. [15] found that students have positive attitudes towards e-learning. They found that students believe that e-learning is enjoyable and facilitates cooperation and communication between them and their peers. It's clear that the demand on e-learning has been increasing due to its merits. E-learning provides opportunities to learn and train. E-learning is beneficial for the learners and the academic institution. However, some people can't benefit from e-learning efficiently. Thus, it's necessary to identify the way one ought to benefit from e-learning. The demand on e-learning by various categories of people in society has been increasing. E-learning is characterized by being flexible and effective on the economic level. It eliminates the spatial and temporal barriers. Due to such merits, there is a great demand on the e-learning services by various categories of people [16].

The most important factor that positively affect the increasing use of e-learning technologies worldwide [8] include the ones below.

- There is a constant need for training and education due to the development in various areas of knowledge

- There is a constant need for training and education in the suitable time and place
- E-leaning provides learning opportunities that are feasible. It reduces the costs of teaching and training employees and learners worldwide.

The increasing demand on the use of e-learning platforms made many users pay attention to the defects in those platforms. That forced the owners of those platforms to make reforms and fix those defects in order to win the competition during the COVID 19 crisis. They made reforms to ensure that the user admire their platforms and ensure that the use of their platforms increase. However, e-learning requires making upgrades in an ongoing manner. It requires setting the fundamentals of the elements of the teaching process (i.e., students, educators and content).

2 Statement of the Problem

The number of the universities and the educational institutions delivering e-learning has been increasing worldwide. The significance of e-learning can be seen in various majors and areas. There are various attempts in Jordan for delivering e-learning. Those attempts don't meet the basic e-learning requirements. They don't comply with the total quality standards in this area. That's why those attempts fail and the learning outcomes don't reach the required level by the ones in the labor market.

The problem is represented in failure to meet the basic e-learning requirements. For instance, the IT infrastructure is poor. The computer devices are not sufficient in homes, schools and universities. The number of users of modern technologies increased recently. The requirements of e-learning management systems aren't met. Thus, the researchers in this study aimed to explore the degree of meeting the e-learning standards In Jordanian universities in the light of the global e-learning standards.

3 The study's Questions

This study offers answers to the questions below:

Q.1. What's the degree of meeting the e-learning global standards In Jordanian universities from the perspective of faculty members?

Q.2 Is there any significant difference—at the significance level of ($a \leq 0.05$)—between the sampled faculty members' attitudes which can be attributed to university or academic rank?

4 Significance of the Study

This study is significant because it provides decision makers in educational institutions with ideas and suggestions about e-learning. It enables decision makers to make long-term future plans and educational policies that improve the teaching process.

5 Definition of Terms

E-learning: It refers to a training educational program that delivers education through using electronic means. Such means allow learners to acquire information and expertise in a synchronic manner or not. Such means ensure providing learners with an attractive educational environment [22].

Standards: They refer to a set of indicators that are used for assessing a performance or the way of performing a task. They are used to describe the extent of mastering a task or the performance level in the light of various contexts [13].

6 Methodology and Procedures

The researchers presented above the theoretical framework. This framework involves scientific contributions and the relevant previous studies and research. This part sheds a light on the methodology used in this study. The researchers aimed to meet the goal of this study. This part presents the approach, population, sample, and data collection methods. It presents data about the way of checking the validity and reliability of the instrument. It presents data about the steps carried out to carry out this study. It presents data about the statistical analysis methods. Data is shown below.

6.1 Methodology and Procedures

The researchers presented above the theoretical framework. This framework involves scientific contributions and the relevant previous studies and research. This part sheds a light on the methodology used in this study. The researchers aimed to meet the goal of this study. This part presents the approach, population, sample, and data collection methods. It presents data about the way of checking the validity and reliability of the instrument. It presents data about the steps carried out to carry out this study. It presents data about the statistical analysis methods. Data is shown below.

6.2 The Study's Methodology

The descriptive cross-sectional analytical approach is adopted. It involves making a review for the relevant resources to draft the theoretical framework. The researchers carried out a field investigation for the data through using a survey. This survey was drafted. It was passed to the members of the sample to analyse the data in a statistical manner and answer the study's questions. That was done to meet the gaols, reach results and offer recommendations.

6.3 Population

The population consists from all the faculty members in the educational sciences faculty at Zarqa University and Hashemite University during the year 2020/2021. It consists from 92 faculty members.

6.4 Sample

After excluding the pilot sample, the researchers passed the questionnaire forms to the rest of the members of the population (62 faculty members). The researchers used the comprehensive survey method. 58 forms were retrieved and deemed valid for analysis. The retrieved forms represent 94% of the distributed surveys. Table 1 presents the distribution of the members of the sample in accordance with variables.

Table 1 Distribution of the sample in accordance with variables

Variable	Category	Frequency	Percentage (%)
University	Zarqa University	17	29
	Hashemite University	41	71
	Overall	58	100
Academic rank	Professor	15	26
	Associate professor	20	34
	Assistant professor	18	31
	Lecturer	5	9
	Overall	58	100

6.5 Instrument

To meet the intended goals, the researchers carried out a review for the relevant literature and previous studies. Based on this review, they developed a survey that consists from the parts below:

Part one: It collects data about the demographic and personal characteristics of the respondents (i.e., university and academic rank).
Part two: It involves 30 items that aim to measure the quality of e-learning:

Items No. (1–10): They are related to the process of designing the e-content
Items No. (11–20): They are related to the support and assistance provided by supervisors
Items No. (21–30): They are related to the administrative services

The five-point Likert scale was used. It consists from the following categories: to a very great extent, to a great extent, to a moderate extent, to a little extent and to a very little extent. Those categories stand for the following scores: 5, 4, 3, 2 and 1.

6.6 Validity

The researchers checked the validity of the instrument through the following methods:

Content validity: The initial version of the questionnaire was passed to 10 experts who possess much expert. Those experts are faculty members. They were asked to provide their view about the clarity, relevancy and language of the items. Changes were made based on the comments of the experts. Some items were deleted. Some items were added. The final version of the questionnaire consists from 30 items.

Internal consistency validity: The researchers checked the internal consistency validity of the survey was measured through calculating the values of the coefficient of correlation between each item and the area it belongs to. The latter values range between 0.61–0.82. They are statistically significant at the significance level of ($a \leq 0.05$). The values of the coefficient of correlation are calculated for the items with the overall degree. They range between 0.74–0.92. Those values indicates that the instrument is valid and the results are reliable.

6.7 Reliability of Instrument

The reliability of instrument is measured through calculating the internal consistency coefficient values. That is done through calculating the Cronbach alpha coefficient values for the areas. Those values are high and indicate that the reliability of the instrument is high. Table 2 presents the values of the reliability coefficient.

Table 2 The values of the internal consistency coefficient for each area

Area	Cronbach alpha
The process of designing the e-content	0.89
The support and assistance provided by supervisors	0.86
Administrative services	0.90
Overall	0.93

Based on Table 2, the values of the reliability coefficient range between 0.86–0.90. The overall value is 0.93. Those values are accepted for the goals of the study.

6.8 Procedures

After checking the validity and reliability of the instrument and drafting the final version of the instrument, the following procedures were carried out:

- The study's instrument was used to collect data from the respondents. The respondents' queries were answered while collecting data through the instrument. The respondents were provided with data about the significance of the study and the importance of answering the items objectively. Then, the researchers retrieved the survey forms.
- Data was administered to the SPSS program to analyze it and reach results.
- The researchers used specific criteria for classifying means. Those criteria are shown below:

  Less than 2.33: **Low**
  2.33–3.66: **Moderate**
  More than 3.66: **High**

6.9 Variables

The study used two variables:

1. **Independent variables**: University and academic rank
2. **Dependent variable**: The extent of meeting global e-learning standards in Jordanian universities from the perspective of faculty members. It's measured through the survey developed for this goal.

6.10 Statistical Processing

To answer the questions in this study, the researchers used SPSS program version No. 21. They used descriptive analytical methods. They used the following methods:

- *Descriptive statistical methods*: They include: percentages to describe the characteristics of the sample.
- *Two-way analysis of variance analysis*: It was used to identify whether there is a significant difference between attitudes which can be attributed to university and academic rank.
- *Pearson correlation coefficient* values are calculated to check the consistency validity.
- *Cronbach coefficient* values are calculated to check the reliability of the instrument.

7 Results and Discussion

This result presents the results in a detailed manner with providing a discussion.

7.1 Results Related to Question One:

Q.1. What's the degree of meeting the e-learning global standards In Jordanian universities from the perspective of faculty members?

To answer this part, means and standard deviations are calculated for the dimensions jointly and separately. Table 3 presents the results.

Based on Table 3, the overall mean is high which is 3.68. That indicates that the extent of meeting the e-learning global standards in Jordanian universities is high. The overall standard deviation is 0.43. The mean of the administrative services is 3.82 which is high. The mean of the support and assistance provided by supervisors is 3.71 which is high. The mean of the process of designing the e-content is 3.53 which is high. Table 1 presents the perceptions of the faculty members towards the degree of meeting the e-learning global standards in Jordanian universities.

Table 3 The means and standard deviations for the degree of meeting the e-learning global standards in Jordanian universities

Rank	Area	Mean	Std	Level
1	Administrative services	3.82	0.49	High
2	The support and assistance provided by supervisors,	3.71	0.48	High
3	The process of designing the e-content	3.53	0.54	High
	Overall	3.68	0.43	High

Based on the results related to first question, the extent of meeting the e-learning global standards in Jordanian universities is high. The latter result is attributed to the fact that Jordanian universities made many preparations for delivering e-learning. Jordanian universities today are obliged in the light of the COVID 19 crisis to deliver distance education to fight against the spread of this virus among students and carry out the social distancing measures. That motivated educational institutions to improve the distance education methods and meeting the global e-learning standards. The minor differences between the extent of meeting the e-learning global standards in Jordanian universities are attributed to the fact that e-learning is still being developed and changes are still made to it. For instance, there are still changes being made to computer programs, and e-content.

In terms of the quality of administrative services and the support and assistance provided by supervisors, they show high means. The latter result is attributed to the fact that Jordanian universities provided much support in this regard. For instance, Jordanian universities have clear policies in providing e-curricula. They provide students with all the required communication means for delivering e-learning. The costs incurred for delivering such education are less than the costs needed for delivering conventional education. The burdens assigned to supervisors in the e-learning programs are clear. In terms of the process of designing the e-content, it shows a moderate mean that's ranked last. That may be attributed to the nature of this process which requires having much skills and capabilities that aren't possessed by the ones. This process requires making ongoing developments and assessments in order to take the individual differences between learners into consideration. It requires providing various learning resources and assessment and measurement mechanism.

7.2 *Results Related to the Second Question*

Q.2 Is there any significant difference—at the significance level of ($a \leq 0.05$)—between the sampled faculty members' attitudes which can be attributed to university or academic rank?

The two-way analysis of variance was carried out to explore the respondents' attitudes in accordance with university or academic rank. The results of this analysis are presented in Table 4.

Based on the data in Table 4, there isn't any significant difference—at the significance level of ($a \leq 0.05$)—between the sampled faculty members' attitudes which can be attributed to university. However, there is a significant difference—at the significance level of ($a \leq 0.05$)—between the sampled faculty members' attitudes which can be attributed to academic rank. To identify the ones who's the latter difference is for their favor, Sheffeh test was carried out. Table 5 presents the results of the latter analysis.

Based on Table 5, the differences are significant for the favour of the ones holding higher academic rank. The latter differences are for the favour of the professors, followed by associate professor, assistant professor and lecturer respectively.

Table 4 The results of the two-way analysis of variance for exploring the respondents' attitudes in accordance with university or academic rank

Source of variance	Sum of squares	Df	Mean square	*F* value	Sig
University	1350	1	0.1350	0.8480	0.3610
Academic rank	18.477	3	6.159	38.636	0.000*
Error	8.449	53	0.1590		
Overall	541.885	58			
Corrected overall	26.928	57			

*This sign means that the value is significant at the significance level of (a ≥ 0.05)

Table 5 The results of Sheffeh test for identifying who's the academic rank-related differences are for their favour

Means	Academic rank	Differences			
		Professor	Associate professor	Assistant professor	Lecturer
3.81	Professor	–	0.11*	0.23*	0.26*
3.7	Associate professor	0.11–*	–	0.12*	0.15*
3.58	Assistant professor	0.23–*	0.12–*	–	0.03*
3.55	Lecturer	0.26–*	0.15–*	0.03–*	–

[*] : statistical significance

There isn't any significant difference—at the significance level of (a ≤ 0.05)—between the sampled faculty members' attitudes which can be attributed to university. The researchers attribute the latter result to having similar conditions and potentials inside the Jordanian universities. Jordanian universities have similar equipment and resources that support the delivery of e-learning. That's because Jordanian universities are affiliated with the Ministry of Higher Education that enforces control over their performance. There is a significant difference between the sampled faculty members' attitudes which can be attributed to academic rank for the favour of the ones having higher rank. The researchers attribute the latter result to the fact that professors have less academic burden than others. That enable professors to show much competency. They attribute the latter result to the fact that professors are provided with more support. The faculty members with lower rank have more teaching burden. Thus, they aren't provided with many opportunities to access the e-content like others.

8 Recommendations

In the light of the results, the researcher recommends:

- Promoting a suitable e-learning culture inside Jordanian universities

- Providing more attention to the process of designing the e-content through holding workshops and symposiums for faculty members
- Developing the faculty members digital skills in order to raise their performance level in delivering e-learning.

Acknowledgements This research is funded by the Deanship of Research in Zarqa University, Jordan.

References

1. Najm Aldeen, H.: The reality of using madrasati platform in the light of the Coronavirus crisis from the perspective of the female teachers teaching social studies in the kingdom of Saudi Arabia. The Virtual International Conference for learners in the Arab World: Problems and Solutions. Enriching Knowledge for Conferences and Research, pp. 205-222 (2021)
2. Kiryakova, G.: Review of distance education. Trakia J. Sci. **7**(3) (2009)
3. Amro, H.: Faculty perceptions of student performance in the online classroom. E-Learn. Digital Media. **10**(3) (2013)
4. Shaheen, M.: Distance Education: Challenges, Future Aspirations and Success Opportunities. Al-Quds Open University (2011). Salah Aldeen, S.: The role of e-learning in developing education in Egypt. J. Middle East Res. **45**, 646–596 (2018)
5. Kanpper, K., Cropley, A.: Lifelong Learning in Higher Education, 3rd edn. Clays Ltd., UK (2000)
6. Al-Hameedy, A.: Distance Education: Theory and Application. Al-Farazdaq Al-Tejareyeh Printing House, Reyad (1997)
7. Sugges, S., Cissell, W., McIntyre, C., Ward, M.: Adoption of communication technologies in a Texas health setting. Educ. Technol. Soc. **2**(5) (2002)
8. Al-Ghareb, Z.: Information Technology and Upgrading Education, p. 129. Alam alketab., Cairo (2001)
9. Pilkington, R., Bennett, C., Vaughan, S.: An evaluation of computer mediated communication to support group discussion in continuing education. Educ. Technol. Soc. **3**(3) (2000)
10. McInnerney, J., Roberts, T.: Online learning: social interaction and the creation of sense of community. Educ. Technol. Soc. **7**(3) (2004)
11. Al-Leqani, A.: Characteristics of the teachers in electronic school. The Tenth Scientific Conferences: The Egyptian Association for Educational Technology: Series for Peer Reviewed Studies and Research. Cairo. New initiatives in e-learning in the Arab world: Arab PC Journal (2005). http://arabic.pcmag-arabic.com/news/news.php?id=EplEplZypytMhEvcq
12. Al-Mana'y, S.: The impact of e-learning on developing students' creative capabilities. Educ. Stud. J. (2014)
13. Al- Alakloby, S.: Evaluating educational programs based on multimedia in arabic language subjects at the primary stage in The Kingdom of Saudi Arabia in light of educational and technical standards. Al-Azhar J. Educ. **163**(4), 421–460 (2015)
14. Coombs, S., Rodd, J.: Using the internet to deliver higher education. Cautionary Tale About Achieving Good Practice: Comput. Sch. **17**(3/4), 67 (2001)
15. Atan, H., Rahman, Z., Idrus, R.: Characteristics of the Web-Based Learning Environment in Distance Education: Students' Perceptions of Their Learning Needs. Penang, Malaysia (2002)
16. Al-Baghdadi, M.: Teaching and Learning Technologies, p. 127. Dar alfeker Alarabi, Cairo (1998)

Impact of Coronavirus on People's Life, Education and Socio-economic Implications—A Review

Maram Y. Al-Safarini, Reda Second, Fatima Haimour, Luma Fakhir, Ghassan Samara, and Siham Haimour

Abstract There have been about 4.3 million confirmed cases and over 290,000 fatalities as a consequence of this viral outbreak. Concerns about an impending economic slump have also been raised as a result. Reduced employment across all economic sectors as a consequence of people becoming socially and personally isolated, as well as travel restrictions, has resulted in many jobs being lost. Products and manufactured items have seen a decrease in popularity after the closure of schools. Alternatively, the demand for medical goods has risen dramatically. The food industry is also seeing an upsurge in demand that results in an economic downturn. Food goods are being purchased in a hurry and stockpiled. A list of socioeconomic elements has been prepared to assess the effect of COVID-19 on the global economy in the wake of this worldwide epidemic. The study warns against depending on pandemic-driven advantages to achieve sustainable growth areas, and highlights the necessity for a significant, basic structural shift in how we live. It contends that the recent global market growth paradigm, which is shaped by a linear economy, has to be rethought exploiting and energy-guzzling trade techniques, in support of more maintainable one. The study has been re-calibrated to fit the circular economy (CE) type. Building

M. Y. Al-Safarini (✉) · F. Haimour · G. Samara
Computer Science Department, Zarqa University, Zarqa, Jordan
e-mail: maram@zu.edu.jo

F. Haimour
e-mail: fhaimour@zu.edu.jo

G. Samara
e-mail: ghsamarah@zu.edu.jo

R. Second · L. Fakhir
Faculty of Educational Science, Zarqa University, Zarqa, Jordan
e-mail: rmawadieh@zu.edu.jo

L. Fakhir
e-mail: lfakhir@zu.edu.jo

S. Haimour
Faculty of Science, Zarqa University, Zarqa, Jordan
e-mail: shaimour@zu.edu.jo

A. Hamdan et al. (eds.), *The Implementation of Smart Technologies for Business Success and Sustainability*, Studies in Systems, Decision and Control 216,
https://doi.org/10.1007/978-3-031-10212-7_57

on the sign that CE is a good thing, the mechanism for matching the complicated equation of achieving turnover while causing the least amount of environmental harm. The study lays out precise sector-specific proposals for CE-based solutions are a world accelerator. Economic-based advancement in a post-viral world that is resilient. As a result, critical theory should lead the lessons learnt, and circular seeking to conceive innovations must also shift their focus from giving a competitive advantage, profitability, or development of enterprises and economic systems to focusing on the general well-being of the public instead. Our study, we summarized an impacts of coronavirus on normal peoples, education and socio-economical aspects.

Keywords COVID-19 · Impacts · Socio-economic and learning

1 Introduction

People throughout the world's mental health have been negatively affected since December 31, 2019, when Wuhan (China) was struck by a new coronavirus infection (COVID-19). Local and international governmental organizations (like as the World Health Organization) have issued data and advice on the rapid spread of the virus, as have epidemiologists, researchers, and public opinion pollsters all across the world [20]. Since this virus is dangerous to patients, as well as people in the general population and government officials and healthcare organizations, its influence on global psychological well-being has been disregarded so far. There has been an outbreak of febrile respiratory syndrome linked to a newly unknown COVID-19 since December 2019 in the Chinese city of Wuhan, which has been linked to a wholesale seafood store in Huanan [36].

A wide range of public health measures, such as rigorous monitoring and epidemiological investigation, were taken by Chinese health authorities as a precautionary measure. In 2020, the date is January 1. It was rapidly determined that acute severe Coronavirus-related respiratory syndrome (SARS-CoV), Coronavirus causes Middle Eastern Respiratory Syndrome (MERS-CoV), avian influenza, and other breathing illnesses (CDC year 2020) were not present.

Since it was isolated from a single individual for less time, the genome of COVID-19 will be sequenced on this day in 2020. WHO formally acknowledged the genetic sequence of SARS-CoV-2 on January 12th, 2020, paving the path for the establishment of polymerase chain reaction-based investigative testing in various nations to detect the new sickness. An unnamed beta-coronavirus formerly known as SARSCoV-2 has been found as the 7th coronavirus that infects individuals. Humans are infected with SARS-CoV-2, a member of COVID-19 that is distinct from MERS-CoV and SARS-CoV. Food from wild or confined animals may be a contributing factor to this outbreak [25].

The World Health Organization has given this new respiratory illness the designation COVID-19. The epidemic began to impact men first, with fewer than half of them having preexisting diseases including diabetes, hypertension, and heart disease, which contributed to their vulnerability. During the onset of the disease, fever, coughing, and myalgia, or fatigue, were all prevalent symptoms. Symptoms including coughing up mucus, having a headache, having hemoptysis, and having diarrhea are becoming less prevalent. Secondary significant contagiousness and acute respiratory distress disorder were among the effects of the incident [41]. In this review, we have summarized the effects of COVID-19 in normal and corporate people life issues, learning and economic issues in worldwide [29].

2 Socio-economic Impacts of Historical Outbreaks

A pandemic at the very least causes a two-fold catastrophe, placing pressure on the healthcare system and the economy. There has long been a consensus that any sort of infectious sickness outbreak—dependent on the disease's latency, transmission rate, and regional expansion—is a daunting route of localized economic threats. In the kind of endemic species, waves, plagues, and catastrophic events, history is riddled with such breakouts [13].

There have been numerous instances when these illnesses contributed to the downfall of empires, flooded hospital institutions, led to civil upheaval, and impacted a broad variety of sectors. A major concern for public health and global finances arose early on in the COVID-19 pandemic when it became evident that any natural, inadvertent, or planned biological danger or eruption may occur anywhere [14].

Numerous pandemics have occurred in the last 100 years, according to Saunders-Hastings and Krewski. The Black Death, the Third Plague Pandemic, the Spanish Flu, and HIV/AIDS are all examples of the Great Plague. Each outbreak is distinct in terms of its strength. As a result, various retrospective evaluations have found that their economic effects varied. For example, the Ebola pandemic in West Africa cost $53 billion in sociopolitical expenses between 2013 and 2016, reduced Sierra Leone's GDP by 20% in 2015 and Liberia's by 8% in 2013 and 2014, and resulted in a decrease in death rates over the same time [32].

Even if the environment we live in today is very different from that of earlier epochs, certain old lessons from past epidemics are still relevant as the world reaches the present tipping point. Numerous variables distinguish the present situation. COVID-19's socioeconomic crisis differs from earlier ones that allows for straightforward evaluations with previous worldwide epidemics are inconceivable. The point that COVID-19 is a world epidemic, as well as the fact that it is a highly contagious virus, are some of the distinguishing features [15]. Given that the globe has become more globalised, this has resulted in knock-on impacts across supply chains. As a result of globalisation and technological improvements, people have grown considerably more integrated in terms of technology. Furthermore, scientific, medicinal, and technical improvements have occurred around the globe. During earlier pandemics, the virus's transmission was slowed by a limited number of air travelers, but worldwide travel has increased considerably in recent years. Interest rates were at all-time lows in terms of expenditure, and there was a big disparity between market pressures for goods. More importantly, many of the countries most affected by the current pandemic are not just low- and middle-income countries, but also those at the top of the engineering and global supply chain pyramids [16].

3 Global Economy

COVID-19 has evolved during the course of the pandemic's severity and longevity, which was marked by a scarcity of resources. COVID-19 has a particularly harsh impact on China, the USA, Korea, Italy, and the United Kingdom, which account for 65% of global production and exports. Due to worldwide shortages of health-care equipment such as personal protective equipment (PPE) and ventilators, various policy focuses have shifted to medical apparatus procurement such as PPE and ventilation systems. In order to maximize the use of ventilators, outlines for rationing them, as well as bed spaces, have to be devised [18]. Other industries, such as CE-sensitive constituents harvesting and mining, have been impacted by disruptions in processes and worldwide commodity prices (Fig. 1).

From the Fig. 1, it was clear that the gaia's global systems comes under four different categories. It includes social features where the people can able to face the consequences of coronavirus. Also, the world economic growth also went very low which results in shorter life span. In particular, the people who is living in below poverty line were completely destroyed because of this coronavirus consequences. Moreover, our environment has deteriorated because of this current virus. Eventually, we are understand that all these features plays a major role in human life.

One of COVID-19's psychological consequences, as mentioned in the following parts, is fear purchasing, which arises as a result of national factors such as insufficient

Fig. 1 Gaia's world systems

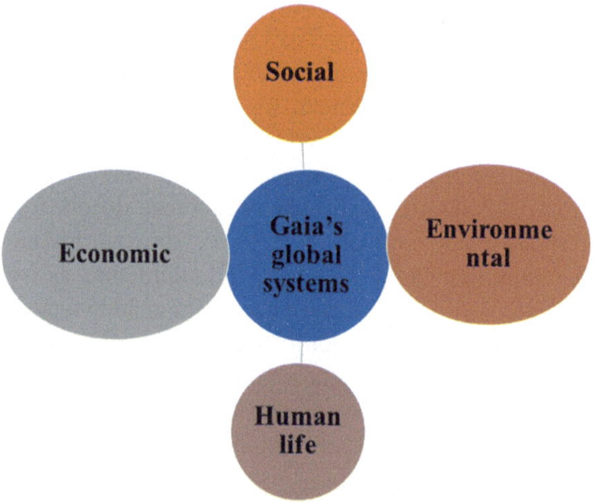

equipment or different difference uncertainties. The instability, exploitation, and wastefulness of the present supply chain architecture have been exposed in both situations. Sarkis questioned if the global economy has the money to return to the just-in-time (JIT) supply chain plan favored by the health-care sector [35], citing an evident fault in crucial commodities' ability to cope. Following that is a sub-section that looks at few macro and micro-economic consequences of COVID-19 [7].

3.1 Macro-economic Impact

World productions, exports, as well as imports have a macroeconomic impact. In countries like the United States, best practices like the JIT macroeconomic blueprint don't encourage the hoarding of critical therapeutic equipment, which is a problem for the healthcare sector. Many nations (including the UK, India, and the US) had to make extraordinary steps to protect their supply chains, leading Ford and Dyson to join the ventilator design or production market, despite massive budgets [19]. With the Defense Production Act, the United States pushed vehicle manufacturers to shift their focus to ventilators because of their high cost and unavailability. Hospitals and suppliers in the United States were forced to join the global market because of a recurrent lack of N95 masks and the necessity to acquire lower-cost equipment. Global production of these special masks is expected to be led by China, where the virus started out, with EU supplies mostly coming from Malaysia and Japan. The gap was so acute that the US was accused of 'pirating' medical supplies earmarked for EU states from Asian countries [28].

France and Germany following with comparable inward-looking policies, as well as the EU implemented export limitations on PPEs, putting many previously reliant countries in jeopardy. Naturally, China and the EU look suitable to decrease or eliminate import tariffs on fresh substances and PPE. A life-threatening repercussion of logistical catastrophes will become more common in the future [11]. The misallocation of critical equipment as well as products might give the industry a fresh lease on life blockchain, RFID, as well as the Internet of Things (IoT) are giving novel life and impetus to technologies including blockchain, RFID, followed by the Internet of Things (IoT), and transparency. Collaboration on a world scale and scenario development are always required to round out the picture. Despite trade deals and other measures, nevertheless, for crucial healthcare equipment, the world supply chain can't rely on imports–or donations due to cooperative arrangements. This realization opens the way for production to be localized enhancements in ecological and social sustainability are expected as a result. Baldwin and Evenett [22]. This is demonstrated in the circumstance of N95 masks that became so popular overnight that airfreights were unable to keep up with the growing (Fig. 2).

They were delivered through private and profitable planes in contrast to conventional container shipping these innovations. In this favor, the EU devised a cooperative obtaining strategy while creating a structure to limit competition amongst member nations throughout the United States. Although there was a major reduction

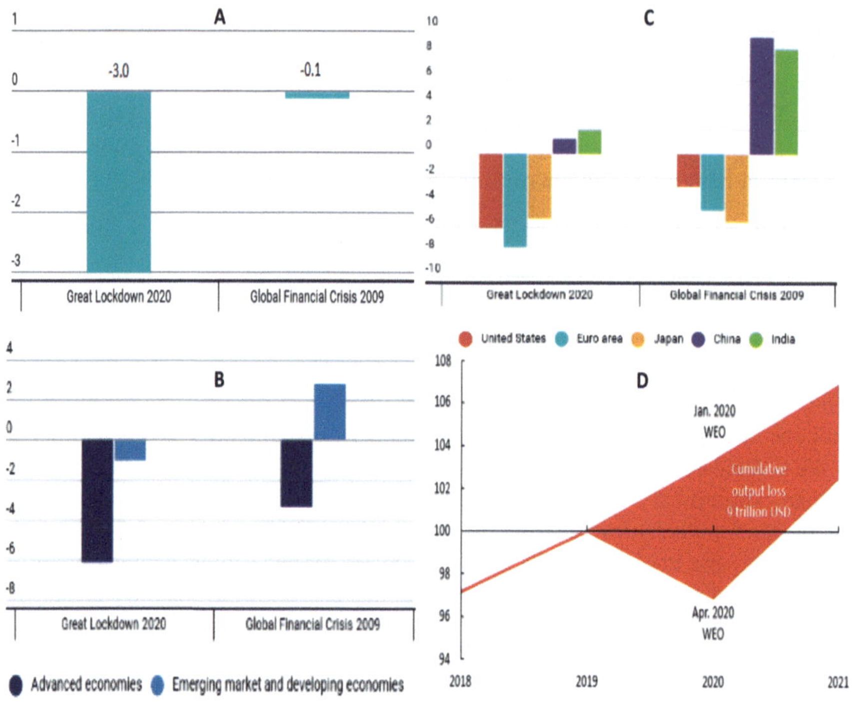

Fig. 2 Socio-economical effect of COVID-19 lockdown. (Have to take permission from Ibn-Mohammed et al. [23])

in emissions associated with traditional shipping, due to desperation and urgency of demand, there was a growth in the consumption of airfreighting. Many nations have been compelled to rethink their global value chains as a consequence of the COVID-19 outbreak. As a consequence of the COVID-19 outbreak, as well as the fact that numerous eastern European/Mediterranean countries have a competitive advantage over Chinese products, national interests and protectionism were at play [39].

3.2 *Microeconomic Effects: Consumer Behavior*

There has been a long-term disconnect between the wants of consumers and the realities of the biophysical world. To put it another way: The effects of one's actions on society have been brought into sharper focus because to COVID-19. Anxiety-fueled shopping sprees for food and feminine hygiene products were common in many countries. There has also been a movement in consumer sentiment on an individual level. Consumers have been forced to rethink their purchasing habits and aspirations because of limited availability to essential items and services [38]. Spash argued that the linear market model is likely to be impacted by technological ineffectiveness of modern products as a result of rapid advancement and personal consumerism, as evidenced by the fact that, for example, mobile phones have an average lifespan of four years, assuming that their manufacture or repair services were forced due to economic closures and lockdowns. However, in a subject like health, where mass manufacturing and consumption of key equipment may benefit, the patent system is a concern. The copyright breach of a 3D printer in Italy led demands for "Open Source Ventilators" and "Good Samaritan Laws" to help in the reaction to global health disasters like COVID-19 [37].

4 Learning Impacts

Schools, training facilities, and other educational resources have been forced to close as a consequence of the COVID-19 pandemic in the majority of nations. An educational paradigm shift had occurred, as educators were now giving high-quality instruction through a variety of internet mediums. To address these unprecedented worldwide pandemics, educators and students alike have found that online, distance, and continuous training are excellent alternatives [2]. Learning online may be an entirely new experience for both learners and educators, but they have few or no other alternatives. Due to a wide range of online platforms, educators and the educational system as a whole were dubbed "Education in Emergency" and had to set up a structure for what they had developed [5].

During the epidemic, e-learning tools were critical in supporting schools as well as universities in facilitating student erudition during the shutdown of universities followed by schools. Staff and management are adjusting to the new developments. Student willingness should be assessed and assisted as needed. The students' learners by a secure mindset find it difficult to acclimate and modify, while those by an open mindset find it easier [9]. A development mindset adapts readily to novel learning situations. A one-size-fits-all solution did not exist. There was no such thing as a one-size-fits-all solution for OL. There were several themes from which to choose, each with its own set of requirements. OL is required in a variety of professions and for different age groups. Physical handicapped pupils can also profit from OL since it allows them to involve in learning in a virtual area with negligible crusade [17]. Figure 3 depicts the possible losers due to COVID-19 infection.

Scholars, parents, especially educators near the global has feel the unanticipated rippling effect of viral epidemic as schools were locked to pact by the world epidemic. All kids are given the opportunity to get a high-quality education, regardless of the efforts of governments and public health authorities in the fight against the spread of the epidemic. At home or in their living quarters, many students struggled to express themselves and were unable to communicate effectively. It's not yet known which online homeschooling approaches are the most successful [33].

Effective online education may be influenced by teachers' and students' knowledge of and comfort with data and communication technology (ICT). Now, teachers are using platforms like Microsoft Teams, Google Classroom (GCR), Canvas, and

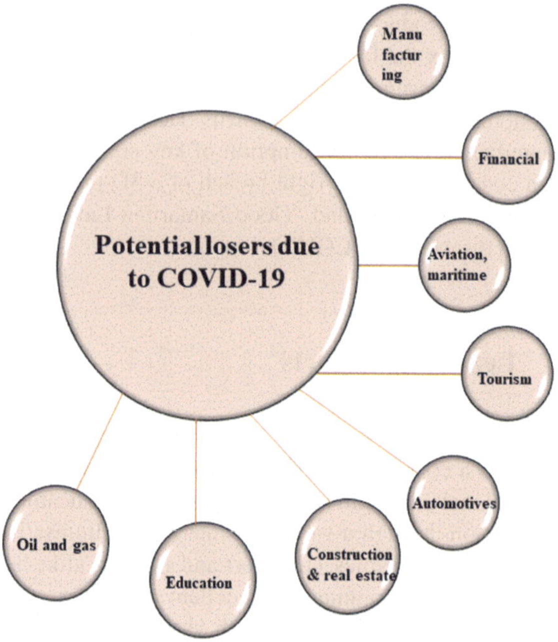

Fig. 3 Outline of potential losers due to COVID-19

Blackboard to create educational courses and programs for training and skill growth. Work chat, video meetings, and file storage are just some of the features that help keep courses organized and running well. Allows for a wide variety of file types to be separated (such as Word and PDF), as well as video. Quizzes and other forms of acquiesced work-based evaluations make it possible to measure student teaching and progress [26].

The flipped CR, a basic method for delivering learning resources including as articles, verified movies, and YouTube relatives previous to the beginning of class. Conversations with instructors and other students on the online CR allowed students to make better use of their CR-related time. This is a great way to get students thinking critically, solving problems, and taking responsibility for their own education. More and more virtual classrooms were relying on video conferencing and cloud-based learning management systems like Elias and Moodle [34].

4.1 Last Learning

For at least part of the first half of 2020, no lessons were conducted in schools for many children and teenagers. Predictions of days of work are shown in (Fig. 4). An OECD/Harvard study found that schooling is being lost due to closure. In mid-May 2020, a survey will be performed. In most cases, the school year had not yet ended these nations. While the specifics of closures necessitate. Further analysis reveals

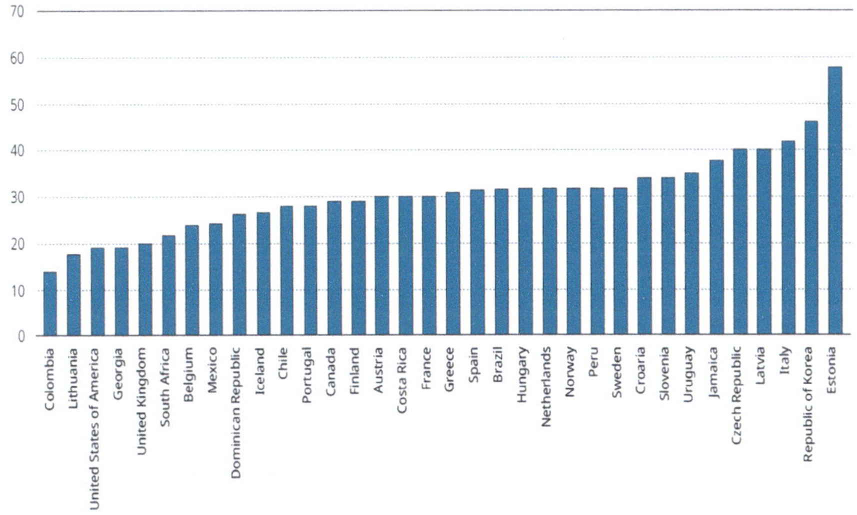

Fig. 4 Days of schooling lost by mid-May 2020. Source: OECD/Harvard University as a source (2020 [5]) Rapid Assessment of the COVID-19 Education Response by the Harvard Global Education Innovation Initiative and the OECD (Have to take permission)

that these preliminary estimations are correct. In most countries, there have been significant losses [4]. Additional setbacks since the survey, and also anticipated losses to the future. In most cases, the future may be predicted to be far greater nations. Little was recognized about the success of home learning for the total student populace, or what this means for skill growth. Nevertheless, there were several exceptions there are signs from a number of nations that many people. There was a lack of adequate instruction for the pupils. For a significant amount of time a percentage of students who learn during school closures appeared to be nearly non-existent [30].

Early tracking statistics from an online mathematics tool utilised in a number of US school districts previous to corona, for example, propose that kids' learning performance have experienced a significant reduction during the disaster, particularly in lower-income schools. According to a study of parents of schoolchildren in Germany, the amount of time spent on school-related actions each day was cut in half during the virus. From 7.4 to 3.6 h of school shutdown. Indeed, 38% of students agree [24]. I spent little more than 2 h per day studying for school. For no more than 4 h, at 74%. The time, on the other hand, spent seeing television, playing computer games, and speaking on the phone. The amount of time spent on passive activities has climbed to 5.2 h each day. Children with better-educated parents had the same decrease in school activities as other kids, however an increase in passive behavior was somewhat smaller. Passive pursuits such as hobbies were more common among students who were struggling academically [12]. Less than half of all students took group online courses on a weekly basis, with just 6% taking them daily. The amount of one-on-one time students get with their professors has decreased. Typical instructional aids were weekly assignments for students to complete. Overall, learning opportunities were considerably curtailed as a result of the school closures, with the greatest decreases occurring among underprivileged students [40].

This year's school year was also canceled in a number of countries, in addition to the complete closure that had already taken place. Educational districts that are experimenting with new school innovations, such as asynchronous presentations and video work, are likely to underestimate the amount of time students have lost in school [21].

4.2 Tasks in Teaching and Learn

Users, both educators and students, regularly encounter difficulties while accessing or referring to the different platforms and digital educational resources available. Many experts have noted and stressed the following issues as being of particular concern: Accessibility, affordability, flexibility, teaching methods, lifelong learning, and educational reform are all problems with e-learning. Maintaining an Internet connection and gaining access to digital gadgets may be difficult in many countries. Students in developing nations who cannot afford online learning gadgets might nevertheless benefit from online education since it increases their media consumption options [8].

Because they need little monitoring and help, intrinsically driven learners were relatively untouched in their learning, but students who are having to learn inadequate have obstacles. A small number of financially underprivileged students who are academically eligible are yet unable to attend and pay for OL [42]. Due to decreased contact hours for beginners as well as a shortage of communication by professors when sessions are conducted for both year-end and internal examinations, pupils' academic achievement is prone to suffer having difficulty understanding/learning. Online grading necessitated a lot of trial and error, as well as skepticism and misunderstanding among instructors, students, and families. How an online test is administered depends on both the examiner's suitability as a teacher and how well the students are able to demonstrate their appropriateness as students. Numerous schools and institutions have yet to execute plagiarism detection systems. Mostly because of the high population of students [31]. Figure. 5 exemplifies the challenges faced during post COVID-19 period.

As a result of the closures of schools and colleges, key public credentials like the General Certificate of Secondary Education (GCSE) have been adversely affected (GCSE), but also external calculations and inspections for the major public qualifications including the GCSE but whole cohort in the UK has had their A levels annulled. Depending on the length of the lockdown, the lockdown may be postponed or cancelled. It's possible that the whole test valuation will be a nightmare. Numerous state-level board-, recruiting-, university-level-exams as well as other tests are available. Because of its COVID-19 epidemic, all entrance tests in India have been delayed as well as a state of emergency [27].

Fig. 5 Post COVID-19 challenges

Additionally, school time aids in children's social and emotional development. There are financial, social, and psychological ramifications for children who miss school. Because so many students are now taking courses online and spending more time on technology than ever before, they're at greater danger of being victimized by scammers. In addition to being exposed to potentially hazardous and violent material, youngsters who spend more time on OL also face a heightened risk of cyberbullying as a result of their unstructured use. More families were turning to technology and digital solutions to keep their children occupied, interested, and connected as a consequence of school closures and rigorous confinement measures, but not all children had access to the information, services, or resources they needed to do so.

Many of Bhutan's internet users are from rural areas where their parents are largely farmers who cannot read or write. It's common for students to help out their parents with agricultural chores including farming, caring for livestock, and more. There were some students who specifically asked that their tests be rescheduled [10]. So it was in the afternoon since they had to work in the field all morning. In other cases, children claimed that they were compelled to accompany their ill grandparents, parents, or other relatives, even if it meant driving them to the hospital. It becomes more difficult for them to keep up with the teachings when they get home at night.

Parents of children in lower grades think that allowing their children to repeat the school year is the best option for them. There is also a lack of Internet access at home, thus the majority of students are unable to use cellphones or televisions [29]. Many people are left with no or very little money as a result of the decline in commercial activity. When compared to typical profits, the data package (costs) is too expensive for farmers to maintain. Though the majority of students prefer video-based CRs, some have expressed dissatisfaction with the face-to-face online class's need for more information correspondences. In the end, teachers were caught between who to follow and what methods to utilize. People seldom consider the possibility of using pre-recorded films, although this might restrict face-to-face engagement. It's difficult to come up with a plan that covers all of the educational needs of children while also being practical [3].

5 Impact of COVID-19 in UAE Nations

To combat the COVID-19 pandemic, the UAE has implemented a number of steps, including converting schools and universities to online platforms, restricting public meetings, adjusting work-from-home arrangements, and requiring social separation. The epidemic has also caused a shift in the way healthcare is delivered. The adoption of telemedicine has increased, and the shift is already taking place. To complete its transfer to online platforms, services relied on an already strong technology infrastructure [6]. These changes suggest that "return to normal" measures will most likely not return us to the world we knew before COVID-19. Recently, foreign governments have begun debating the best strategies for reopening their economies, and the United Arab Emirates is no exception. Local economic departments were given the duty of looking into the prospect of reopening business centers, as long as all necessary precautions and preventative measures were taken. The major issue here is how to strike a balance between development, economic progress, and people's well-being. There is no risk to one's health or well-being. Never before have governments needed to rely on the scientific community as much as they do now. To address these issues, the Mohammed Bin Rashid School of Government suggests using its collective knowledge, expert knowledge, and facilities to host leaders from the UAE's research world to brainstorm evidence available and talk about policy options in order to effectively meet key questions about the UAE's post-pandemic long term [1].

6 Conclusion

COVID-19 has shown the ecological absurdity of the 'extract-produce-use-dump' economic paradigm. There is no guarantee that short-term remedies to the pandemic's severity would last in the long run. However, they do raise crucial issues, such as the relationship between pollution and transportation that needs to be recognized.

Unrestricted air travel plays a critical role in the spread of pandemics, particularly those caused by influenza viruses, and the tourism and aviation industries would feel the effects of any reduction in the number of passengers (some airlines may not ever recoup or return to viability for a longer time). There is a strong and decisive leadership in healthcare, industry, government, and society at large required because of fears of a fresh recession and financial catastrophe. We must take immediate action to alleviate the situation and make necessary modifications to ensure that no one is left behind. When it comes to rebalancing and revitalizing the economy, long-term and medium-term planning are necessary. It is also necessary for individuals with long-term company models to have a comprehensive socioeconomic growth strategy that includes sector-by-sector initiatives as well as an entrepreneurial environment. To ensure that the "whatever it takes" promise is honored, governments and financial institutions should constantly review and reevaluate the present situation. Supply and demand necessitated that cargo transportation be curtailed in favor of air freights, whose transatlantic cost per kilogram rose overnight. Loss of employment, income inequality, an increase in the global poverty levels, and changes in the economy across sectors and supply networks all contribute to this. For various industries, the practicality of remote work and the resulting implications on worker mobility have been thoroughly investigated and evaluated. Telemedicine and remote work are no longer thought unfeasible since they have been effectively applied over the greater part of a 4-month globe shutdown duration. The decrease in primary energy consumption that has occurred as a result of the slowdown and closure of commercial and industrial activities means that distance schooling is no longer a problem. More people realize the benefits of cloud computing analytics in speeding up vendor selection procedures and managing supplier relations and logistics, as well as in boosting manufacturing processes. Because of the necessity to enhance ventilation, such as in hospitals, offices and commercial buildings are underutilized during the lockdown, resulting in higher energy usage. Structures that can be disassembled and reassembled to meet changing needs, as shown in China, are becoming more common because to the growing use of modular technology for fast construction. Instead than just enhancing the competitiveness, profitability, or expansion of businesses and economies, circular thinking breakthroughs should focus on the fundamental well-being of the people in the future. If government agencies promise economic recovery, they should include post-COVID-19 assets required for the transition to more resilient, lower carbon, and circular markets in the incentive packages for economic recovery claims made by them.

References

1. Ajmal, M. M., Khan, M., Shad, M.K., AlKatheeri, H., Jabeen, F.: Socio-economic and technological new normal in supply chain management: lessons from COVID-19 pandemic. Int. J. Logistics Manage. (2021)
2. Andrew, A., Cattan, S., Costa Dias, M., Farquharson, C., Kraftman, L., Krutikova, S., Sevilla,

A.: Inequalities in children's experiences of home learning during the COVID-19 lockdown in England. Fiscal Stud. **41**(3), 653–683 (2020)
3. Azevedo J.P., Hasan, A., Goldemberg, D., Iqbal, S.A., Geven, K.: Simulating the potential impacts of COVID-19 school closures on schooling and learning outcomes: a set of global estimates. World Bank Policy Research Working Paper (9284) (2020). https://openknowledge.worldbank.org/handle/10986/33945. Accessed 9 Feb 2021
4. Bacher-Hicks, A., Goodman, J., Mulhern, C.: Inequality in household adaptation to schooling shocks: Covid-induced online learning engagement in real time. J. Public Econ. **193**, 104345 (2021)
5. Bansak, C., Starr, M.: COVID-19 shocks to education supply: how 200,000 US households dealt with the sudden shift to distance learning. Rev. Econ. Household **19**(1), 63–90 (2021)
6. Cheikh Ismail, L., Mohamad, M.N., Bataineh, M.A.F., Ajab, A., Al-Marzouqi, A.M., Jarrar, A.H., Abu Jamous, D.O., Ali, H.I., Al Sabbah, H., Hasan, H., Stojanovska, L., Hashim, M., Shaker Obaid, R.R., Saleh, S.T., Osaili, T.M., Al Dhaheri, A.S., Al Dhaheri, A.S.: Impact of the coronavirus pandemic (COVID-19) lockdown on mental health and well-being in the United Arab Emirates. Front. Psychiatry **12**, 265 (2021)
7. Dargaville, T., Spann, K., Celina, M.: Opinion to address the personal protective equipment shortage in the global community during the COVID-19 outbreak. Polym. Degrad. Stab. **176**, 109162 (2020)
8. Dhawan, S.: Online learning: a panacea in the time of COVID-19 crisis. J. Educ. Technol. Syst. **49**(1), 5–22 (2020)
9. Dietrich, H., Patzina, A., Lerche, A.: Social inequality in the homeschooling efforts of German high school students during a school closing period. Eur. Soc. **23**(sup1), S348–S369 (2021)
10. Dreesen, T., Akseer, S., Brossard, M., Dewan, P., Giraldo, J.P., Kamei, A., Mizunoya, S., Ortiz, J.S.: Promising practices for equitable remote learning: emerging lessons from COVID-19 education responses in 127 countries (2020)
11. Friant, M.C., Vermeulen, W.J., Salomone, R.: A typology of circular economy discourses: navigating the diverse visions of a contested paradigm. Resour. Conserv. Recycl. **161**, 104917 (2020)
12. Fuchs-Schündeln, N., Krueger, D., Ludwig, A., Popova, I.: The long-term distributional and welfare effects of Covid-19 school closures, no. w27773. National Bureau of Economic Research (2020)
13. Gates, B.: Responding to Covid-19—a once-in-a-century pandemic? N. Engl. J. Med. **382**(18), 1677–1679 (2020)
14. Geissdoerfer, M., Savaget, P., Bocken, N.M., Hultink, E.J.: The circular economy—a new sustainability paradigm? J. Cleaner Prod. **143**, 757–768 (2017)
15. Gondi, S., Beckman, A.L., Deveau, N., Raja, A.S., Ranney, M.L., Popkin, R., He, S.: Personal protective equipment needs in the USA during the COVID-19 pandemic. Lancet **395**(10237), e90–e91 (2020)
16. Govindan, K., Hasanagic, M.: A systematic review on drivers, barriers, and practices towards circular economy: a supply chain perspective. Int. J. Prod. Res. **56**(1–2), 278–311 (2018)
17. Grätz, M., Lipps, O.: Large loss in studying time during the closure of schools in Switzerland in 2020. Res. Soc. Stratification Mobility **71**, 100554 (2021)
18. Guan, D., Wang, D., Hallegatte, S., Huo, J., Li, S., Bai, Y., Lei, T., Xue, Q., Davis, S.J., Coffman, D.M., Cheng, D.: Global economic footprint of the COVID-19 pandemic (2020)
19. Guerrieri, V., Lorenzoni, G., Straub, L., Werning, I.: Macroeconomic implications of COVID-19: can negative supply shocks cause demand shortages? National Bureau of Economic Research (2020)
20. Hamzah, F.B., Lau, C., Nazri, H., Ligot, D.V., Lee, G., Tan, C.L., Shaib, M.K., Zaidon, U.H., Abdullah, A.B., Chung, M.H.: CoronaTracker: worldwide COVID-19 outbreak data analysis and prediction. Bull. World Health Organ. **1**(32), 1–32 (2020)
21. Hanushek, E.A., Woessmann, L.: The economic impacts of learning losses (2020)
22. Hellani, H., Sliman, L., Samhat, A.E., Exposito, E.: On blockchain integration with supply chain: overview on data transparency. Logistics **5**(3), 46 (2021)

23. Ibn-Mohammed, T., Mustapha, K. B., Godsell, J., Adamu, Z., Babatunde, K. A., Akintade, D.D., Acquaye, A., Fujii, H., Ndiaye, M.M., Yamoah, F.A., Koh, S.C.L.: A critical analysis of the impacts of COVID-19 on the global economy and ecosystems and opportunities for circular economy strategies, Resour. Conserv. Recycl. **164**, 105169 (2021)
24. Kaffenberger, M.: Modelling the long-run learning impact of the Covid-19 learning shock: actions to (more than) mitigate loss. Int. J. Educ. Dev. **81**, 102326 (2021)
25. Lai, C.C., Shih, T.P., Ko, W.C., Tang, H.J., Hsueh, P.R.: Severe acute respiratory syndrome coronavirus 2 (SARS-CoV-2) and coronavirus disease-2019 (COVID-19): the epidemic and the challenges. Int. J. Antimicrob. Agents **55**(3), 105924 (2020)
26. Lewis, K., Kuhfeld, M., Ruzek, E., McEachin, A.: Learning during COVID-19: reading and math achievement in the 2020–21 school year. Center for School and Student Progress (2021). https://www.nwea.org/research/publication/learning-duringcovid-19-reading-and-math-achievement-in-the-2020-2021-school-year
27. Manca, F., Meluzzi, F.: Strengthening online learning when school are closed: the role of families and teachers in supporting students during the COVID-19 crisis (2020)
28. McKibbin, W., Fernando, R.: The global macroeconomic impacts of COVID-19: seven scenarios. Asian Econ. Pap. **20**(2), 1–30 (2021)
29. Montacute, R., Cullinane, C.: Learning in Lockdown. Research Brief. Sutton Trust (2021)
30. Olaimat, A.N., Aolymat, I., Shahbaz, H.M., Holley, R.A.: Knowledge and information sources about COVID-19 among university students in Jordan: a cross-sectional study. Front. Public Health **8**, 254 (2020)
31. Pham, T.T.T., Le, H.A., Do, D.T.: The factors affecting students' online learning outcomes during the COVID-19 pandemic: a Bayesian exploratory factor analysis. Educ. Res. Int. (2021)
32. Qureshi, A.I.: Economic and political impact of Ebola virus disease. Ebola Virus Dis. 177 (2016)
33. Rashid, S., Yadav, S.S.: Impact of Covid-19 pandemic on higher education and research. Indian J. Hum. Dev. **14**(2), 340–343 (2020)
34. Rose, S., Twist, L., Lord, P., Rutt, S., Badr, K., Hope, C., Styles, B.: Impact of School Closures and Subsequent Support Strategies on Attainment and Socio-Emotional Wellbeing in Key Stage 1: Interim Paper 1. National Foundation for Educational Research (2021)
35. Sarkis J., Cohen M.J., Dewick P., Schröder P.: A brave new world: lessons from the COVID-19 pandemic for transitioning to sustainable supply and production. Resour. Conserv. Recycl. (2020)
36. Sharifi, A., Khavarian-Garmsir, A.R.: The COVID-19 pandemic: impacts on cities and major lessons for urban planning, design, and management. Sci. Total Environ. **749**, 142391 (2020)
37. Sim, K., Chua, H.C., Vieta, E., Fernandez, G.: The anatomy of panic buying related to the current COVID-19 pandemic. Psychiatry Res. **288**, 113015 (2020)
38. Spash, C.L.: 'The economy' as if people mattered: revisiting critiques of economic growth in a time of crisis. Globalizations **18**(7), 1087–1104 (2021)
39. Vallance, P., Norman, J.: A Time of Unprecedented Change in the Transport System. Government Office for Science, London, UK (2019)
40. Von Hippel, P.T.: How will the coronavirus crisis affect children's learning? Unequally. Educ. Next (2020)
41. Wang, H., Li, X., Li, T., Zhang, S., Wang, L., Wu, X., Liu, J.: The genetic sequence, origin, and diagnosis of SARS-CoV-2. Eur. J. Clin. Microbiol. Infect. Dis. **39**(9), 1629–1635 (2020)
42. Zalat, M.M., Hamed, M.S., Bolbol, S.A.: The experiences, challenges, and acceptance of e-learning as a tool for teaching during the COVID-19 pandemic among university medical staff. PLoS ONE **16**(3), e0248758 (2021)

Sustainable Finance, Innovation, and Business Uncertain Situations

The Impact of Blockchain Technology on Internal Auditing in the Financial Sector

Chi Zhang and Sabarina Mohammed Shah

Abstract Blockchain technology is claimed to lead innovations to many business paradigms by reducing fraud and providing safe transactions. This study examined whether the adoption of blockchain technology will impact their internal audit function (IAF) and the role of IAF in blockchain smart contracts in the financial sector. The study referred to the related literature review and followed the international standards for the professional practice of internal auditing (IPPF). The study found that internal audit function and relative associate with the basics of blockchain technology and reached a continuous audit model with blockchain smart contracts. Considering the significant benefits and also potential challenges of applying blockchain technology in the modern accounting system, the study filled a gap in knowledge on the impact of blockchain on internal auditing.

Keywords Blockchain technology · Internal audit function · Blockchain smart contract · IPPF

1 Introduction

Industry 4.0 creates enormous impact and opportunity, and businesses must find a new path to innovate for competitive advantage that disrupts technology, innovations, and business modes [29]. As an integral part of daily life, these radical developments will inevitably impact the financial sector. In the meantime, globalization and technological advancements are interconnected forces that internal audit function will be a significant impact [11]. The most cutting-edge technology in finance technology is now blockchain technology [4]. It has attracted the attention of government entities, financial institutions, and investors in a number of countries. While the drawbacks of today's credit mechanism are apparent and in crisis, the blockchain-based distributed cryptographic ledger solves the credit insufficiency problem that has a tremendous impact on the current industry ecology and way of operation [23].

C. Zhang (✉) · S. M. Shah
Universiti Putra Malaysia, 43400 UPM Serdang, Selangor Darul Ehsan, Malaysia
e-mail: gs60174@student.upm.edu.my

A. Hamdan et al. (eds.), *The Implementation of Smart Technologies for Business Success and Sustainability*, Studies in Systems, Decision and Control 216,
https://doi.org/10.1007/978-3-031-10212-7_58

In a questionnaire of global banking executives conducted in May 2016, McKinsey discovered that around half of the executives thought blockchain would have a significant influence within three years, and some predict that this will happen during the next 18 months. Another survey of 200 worldwide banks indicated that 15% would be using blockchain technology substantially by the following year. According to IBM, 66% of banks will have commercial blockchain at scale in four years [15]. In 2016, blockchain representatives from the Big Four accounting firms gathered at the American Accounting Association to explore the creation of a distributed ledger alliance. The Australian Securities Exchange has revealed its desire to use blockchain technology to rebuild its clearing and settlement systems [3]. As the incomplete statistics, according to the website of HACKERNOON, by 2020, approximately 400 banks and financial institutions will have used or will be utilizing blockchain technology [2].

As highly innovative business technology, blockchain technology has disrupted the audit function [8]. Blockchain, essentially an accounting technology, might make asset transfers and ledgers of financial and non-financial data more efficient and accurate. Because all transaction data is stored in up-to-date, immutable, and historical distributed ledgers, the potential of real-time audit is high. Blockchain might finally put an end to auditors' random sampling by allowing them to review every transaction [10]. As a result, it offers increased audit efficiency [8, 31]. Although scholars have affirmed the various advantages of blockchain technology, applying this technology in the financial field is still a relatively new research topic.

This study examined that the adoption of blockchain technology in financial institutions will impact their internal audit function (IAF). In particular, it looked whether and if so, blockchain smart contracts can reach continuous auditing in real-time. To address the research question of the impact of blockchain technology on internal audit in financial sector, the study referred to the related literature review and in accordance with the international standards for the professional practice of internal auditing (IPPF). Considering the significant benefits and also potential challenges of applying blockchain technology in the modern accounting system. The study filled a gap in knowledge on the impact of blockchain on internal auditing.

The remainder of the paper is divided into six sections. Section 2 review relative literature, Sect. 3 illustrate the basics of blockchain technology and internal audit, and Sect. 4 discuss blockchain technology drove internal audit function change. Section 5 discusses the continuous auditing with blockchain smart contracts, and Sect. 6 summarizes of the findings and recommendations for further research.

2 Review of Related Literature

In relatively early blockchain studies, researchers highlighted the benefits of reducing operational costs. Lazanis [20] analyzed peer-to-peer transactions, automated settlements, and updated ledgers on blockchain networks, arguing that corporate fund payments can be made through blockchain networks, thereby reducing commission costs by cutting across settlement intermediaries. This study highlights the advantages of blockchain and looks more closely at the fact that if peer-to-peer payments are made using the blockchain network's own digital currency, the reduction in intermediate cost fees is more pronounced, and it has the potential to greatly streamline the transaction process and lower operational finance expenses. Liu and Shi [21] suggest that accounting in blockchain mode will be transferred from "peer-to-peer" to "peer-to-network." This study found that the common audit of each node will record transactions and maintain updated data, which can reduce the influence of personal subjective factors of bookkeepers to a minimum. This is also one of the most significant benefits of utilising blockchain technology in the accounting profession.

Yoon et al. [36] analyze blockchain from another dimension, and the author argues that the blockchain comes with an audit system that will transform the accounting ecosystem. The process of blockchain information writing is an automated audit of accounting transactions, review, confirmation, bookkeeping, and other bookkeeping activities, in which participants jointly confirm whether the accounts are true and complete, reducing falsification and satisfying stakeholders' requirements for accurate and authentic financial data. Sun et al. [32] investigated blockchain from the standpoint of financial security, simulating a transparent distributed financial system on a private blockchain, with the financial system's customers being a corporation or a collection of companies. Each record node maintains a comprehensive and tamper-proof ledger, allowing any node to query all data at any time; at the same time, visitors' access is limited by cryptographic encryption technology, and external personnel can only access with authorization, ensuring financial data privacy. As a result, the blockchain financial system provides accounting data integrity and openness while preserving financial data safely and securely.

Tapscott [34] suggests that blockchain auditing can be used to replace third-party auditing services, resulting in more secure and trustworthy "self-auditing." All participating nodes in the blockchain distributed ledger network are auditors and recorders of information and data. In data auditing, the decentralised auditing mechanism has significant advantages, and the blockchain ledger can guarantee the independence, objectivity, and impartiality of auditing work compared with the third-party auditors, and the information data of enterprises on their own blockchain financial system can guarantee the objectivity and accuracy of financial reports. Zhong and Jia [38] also studied blockchain from the perspective of data reliability. According to the findings, the characteristics of blockchain and the benefits of big data may be used to tackle the problem of data silos in a targeted manner by evaluating the common difficulties of traditional Internet banking. The application of blockchain technology can improve accounting data structures and meet societal demands for online finance correctness and validity.

Through the analysis of previous studies, as a unique distributed ledger technology, blockchain's own structure can realize the functions of transparency, sharing, and non-tampering, which has a natural fit with accounting bookkeeping and auditing, and it is predictable that blockchain will be the best way to store data and share information. With the exploration and application of technology, the transformative role of blockchain in the audit industry is emerging step by step. In the future internal audit business, a lot of work can be handed over to blockchain-based smart contracts to complete. After a lot of research and analysis, scholars currently agree more on the great advantages of blockchain technology, which could be the Internet technology with the greatest impact on auditing in the future. However, to the best of my knowledge, there is relatively little research literature on the adoption of blockchain technology in the auditing industry, and more research is needed to deepen our knowledge of blockchain. Overall, the current research on blockchain internal auditing is far from adequate.

3 Basics of Blockchain Technology and Internal Audit

3.1 Definition of Blockchain Technology

In 2008, Satoshi Nakamoto created Bitcoin, the first cryptocurrency designed for peer-to-peer trade. Bitcoin transactions in digital assets are now safe and secure, primarily to the development of blockchain technology [30].

As a distributed ledger technology, blockchain technology is a data structure that uses cryptography principles to record the flow of transaction funds. It uses its own distributed nodes for network data storage, verification, transmission, and exchange, providing safe and stable, transparent, traceable, and efficient technical solutions for information recording and data interaction. On a blockchain, when a transaction is performed, it is broadcast to all nodes in the network. This transaction is verified by miners that the transaction is valid and called this process mining. Cryptographically signed transactions are organised into blocks once they have been validated by the miners. After validation and consensus, each block is cryptographically connected to the one before it. Each block contains a cryptographic procedure called a "hash" of the previous block, as well as a timestamp and transaction data. New blocks are distributed across network copies of the ledger, and any conflicts are resolved automatically according to pre-determined procedures [27]. As new blocks are added, the transactions in a block cannot be modified (creating tamper resistance). When transactions are recorded, the data in any one block cannot be changed retroactively without affecting all following blocks since a different hash value will be generated, and other blocks related to it will not be able to confirm the hash value, necessitating network majority consensus [26]. Various data is recorded on a list of records, which are linked together like chains. Every node on the blockchain network may see every transaction on that block, as well as all previous blocks, resulting in a comprehensive transaction database that is constantly updated and accessible to all blockchain participants [37].

For an illustration of how does blockchain technology work to prevent any unauthorized data changes, draw an example (see Fig. 1.)

As Fig. 1 explain that there is a hash value in Block 2. This hash value is a 256-byte digital code randomly generated by an encryption algorithm. The authentication of

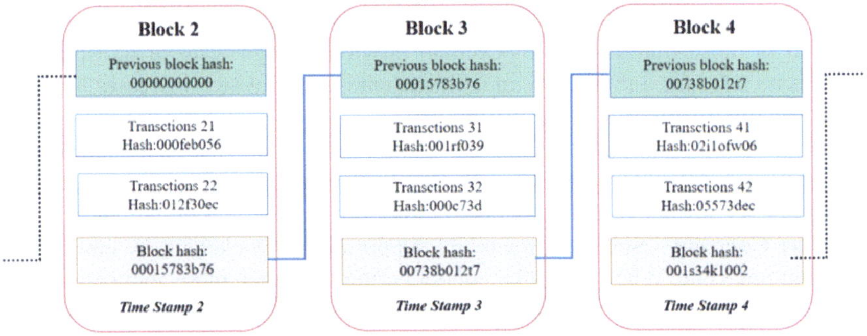

Fig. 1 A figure to explain the way of encryption between blocks

the next block is based on the hash value of the previous block. Once the hash value of a certain block is changed, then every subsequent block cannot be authenticated [14]. Due to the length of the hash value and the number of blocks, it becomes difficult to recalculate the hash after the change. Therefore, it will cause huge modification costs and essentially cannot cause data tampering.

On a peer-to-peer network, blockchain technology allows a digital ledger of transactions to be recorded, preserved, and verified. Decentralized blockchain provides the same function as many intermediaries in our society in establishing trust and maintaining integrity between transacting parties by authenticating and documenting immutable transactions [22]. As a distributed ledger in which groups of transactions or events are recorded and kept in a chain-like data structure, blockchain eliminates the need for costly middleman validation and verification methods [5, 8].

3.2 Characteristics of Blockchain Technology

Blockchain can be thought of as a decentralised architecture with built-in security that improves transaction trust and integrity [9, 22]. This section seeks to provide an overview of the blockchain's most frequent traits as well as its disadvantages.

Decentralisation—In contrast to centralised architecture, which has issues such as single point of failure and scalability, the blockchain uses a decentralised and distributed ledger to take advantage of the processing capabilities of all users in the blockchain network, reducing latency and eliminating the single point of failure [1].

Immutability—One of the most important properties of the blockchain is its capacity to create immutable ledgers, which ensures transaction integrity. Databases can be tampered with in typical centralised designs, necessitating the establishment of trust with a third party to ensure data integrity. Because each block in the distributed ledger is linked to the preceding block in a chain of blocks, the blocks in blockchain technology are permanently recorded and never modified as long as the network is maintained by the participating user [1].

Transparency—By sharing transaction details among all participating users involved in those transactions, blockchain provides a high level of transparency. There is no need for a third party in a blockchain setting, which increases business friendliness and ensures a trusted workflow [1].

Real-time—Because transactions are recorded and reconciled on the blockchain almost immediately after they occur, blockchain technology delivers near-real-time transaction records and account reconciliation [22].

Smart Contract—Smart contracts are supported by blockchain by embedding programming code. When particular contract criteria are met, these applications can conduct transactions and make related ledger entries. In a decentralised setting, self-executing smart contracts allow for the timing of ownership transfers from one party to another [17, 22].

Better Security—Although security is a concern for most new technologies, blockchain is more secure because it has a public key infrastructure that guards against unwanted data manipulation. Participants in the blockchain network put their faith in the consensus mechanism's integrity and security characteristics [1].

High Energy Consumption—Keeping a real-time ledger necessitates the use of energy. Every time a new node is created, it connects with the other nodes at the same time. Transparency is achieved in this manner. Miners on the network are attempting to solve a large number of solutions each second in order to validate transactions. They're putting a lot of processing power into it. Every node provides great fault tolerance, guarantees zero downtime, and assures that data recorded on the blockchain is immutable and censorship-resistant forever. However, these processes use both electricity and time [14].

3.3 The Relationship Between Blockchain Technology and Internal Audit

As known, Internal auditing is an independent, objective assurance and consulting activity designed to add value and improve an organization's operations [7]. It assists a company in achieving its goals by implementing a systematic, disciplined approach to evaluating and improving the efficacy of risk management, control, and governance [25]. Internal auditors are expected to apply and uphold the Internal Audit Standards and IPPF. The Code of Ethics interprets internal auditor should be followed as Integrity, Objectivity, Confidentiality, and Competency. However, the audit risk caused by personal factors is also possible.

Independence—Audit independence is threatened by self-interest, self-evaluation, over-promotion, intimate relationship, and external pressure. Technology can avoid the judgmental influence of such human emotions.

Integrity—Lack of integrity is a flaw in human nature; they may turn a blind eye when auditors find fraud and errors. However, the capacity to create immutable ledgers ensures the integrity of transactions, which is a key aspect of blockchain technology [1].

Objectivity—With the development of the legislation and International Standards on Auditing, auditors' professional ability will be deviation whether they can catch it in time. Suppose we added the Standards in the protocol layer of blockchain that can be reduced audit mistakes caused by the lack of auditors' professional ability.

Confidentiality—When auditing, many trade secrets can be touched by auditors. By using blockchain technology, it may be protected. Such as, Deloitte used blockchain technology to integrate with the Irish banking system.

Competency—Blockchain technology is more trustworthy than auditors. For example, the blockchain audit application—Perma Rec of Deloitte's Rubix platform. This programme establishes a real-time connection between the audit firm and the audited firm's financial system, monitors and verifies the audited firm's transactions and financial entry, and keeps track of business operations. Through this application, established distributed accounts. This is an innovation in auditing, which enables help to do audits by using blockchain and smart contracts.

4 Blockchain Technology Drove the Role of IAF Changed

Internal audit is an independent supervision and evaluation activity based on the company's internal economic activities and serving the management department. At the same time, it also supervises the authenticity, completeness, and legality of the company's accounting information and plays a role in inspecting, supervising, and evaluating the safety of the company's assets and the company's own business performance. With the development of the market economy, many enterprises have become larger and larger, and their fiduciary and liability relationships have become more and more complicated. Great changes have taken place in the content and main structure, leading to corresponding changes in a wider range of internal audit services and service targets. Internal audits are conducted at many levels, from management to the company, with the goal of improving governance, risk management, and control systems. Internal audit functions that are most effective will reflect each organization's priorities and values. Each organization's senior managers and audit committees will have different expectations of the internal audit role [25].

The impact of blockchain on auditing is still in its early stages [24]. In a world where new technologies, such as blockchain advancements, are thriving but standards are behind [13], the function of internal audit needs to be continuously developed to include the ability to verify the regular operation of the various components of the blockchain, which provides for verification of access rights, encryption, and encryption codes, and proof of smart contract transaction codes, functions, and security [16]. Internal audits will need to build procedures for evaluating the operation of blockchain systems as a result.

Auditors could consider using relevant data analytic in blockchain and expanding consulting services such as control design, change management, and blockchain governance with resources freed up from traditional evidence collection and testing [22]. Internal auditors will also need to update their understanding of both internal and external risks associated with blockchain processes, as well as develop relevant monitoring procedures [16], as the internal audit function has changed as a result of blockchain technology, necessitating the development of a conceptual framework to envision the function of internal audit.

5 Continuous Auditing with Blockchain Smart Contract

Blockchain has a lot of potential as a tamper-proof audit trail because of its cryptography and consensus methods that ensure transaction integrity. Smart contracts on the blockchain can be used for a variety of things, including the automatic settlement of financial derivatives and the secure transfer of property titles [6, 12]. When combined with smart contracts [33], which are computer programme that perform tasks on behalf of a human user, blockchain has the potential to drastically change current business procedures. If a financial institution's database architecture is a blockchain platform, blockchain-enabled smart contracts can generate agile supply chains and financial organisations by automatically monitoring and executing the terms of bills of lading and financial derivatives [3, 28]. Then according to the encryption and consensus mechanism, the execution of many businesses is realized through the blockchain smart contracts, including internal audits. Therefore, based on blockchain technology, the roles of internal audits of financial institutions can get a conceptual framework.

Continuous auditing was pioneered by Vasarhelyi and Halper [35], and continuous monitor the internal control was implemented. Also can achieve continuous data assurance, continuous risk, monitoring and assessment, and continuous auditing in management systems [19, 28]. Internal auditing has been associated with commonly continuous auditing because they are less constrained by legislative obligations [18]. Rozario and Vasarhelyi [28] stated that audit data analytic advance to the natural progression ADA3.0 with blockchain smart contracts. Fundamentally, the blockchain platform enables smart audit procedures' developing that run prediction models autonomously, discover critical items, and offer real-time audit data. Smart auditing models can improve audit quality and reduce the gap between auditors' and stakeholders' expectations [28].

However, with the application of blockchain platforms by financial institutions, internal audits only need to design corresponding smart contracts for the audit tasks to be carried out. Smart audit methods are now included in the definition of blockchain smart contracts (e.g., the analyses of audit evidence) [28]. Through data analysis and mining, risk assessment and abnormal early warning of supervision and special audits can be realized in a timely manner. Find audit doubts or audit trails. For example, commercial banks have added loan usage analysis rules, credit risk early warning rules, and internal control monitoring rules to the blockchain platform.

In addition, when internal auditors design automated audit procedures-smart contracts when smart business contracts executed by financial institutions violate the rules and cannot be executed, internal auditors need to audit smart business contracts. For example, bank deposit and loan pledge third-party supervision and credit business. Inventory pledgee applies to the bank for inventory pledge third-party supervision and credit business, including pledge price, quantity, description, pledge verification, etc. These phrases will be embedded in business logic and implemented on the blockchain. The blockchain enables banks and pledgers to authenticate the data given by the pledger and the actual pledge through smart contracts. An error message is generated if one or more of the smart contract's rules are broken, and the transaction cannot be completed. The internal auditor must rectify such breaches in order to authenticate the transaction's authenticity.

6 Conclusion

The idea of adopting blockchain technology in internal auditing is investigated in this study. Internal audits demand decentralisation and tamper-proof characteristics, which blockchain technology provides. Traditional internal audits will be drastically altered by the concept of blockchain technology internal audits. However, data security difficulties, timeliness issues, and legal issues remain obstacles to using blockchain technology in internal audits. In this paper, a conceptual framework for the application of blockchain technology to internal audits is presented. The results will be more generalised if a questionnaire is given to a large number of internal auditors at financial institutions. Conducting comparisons with empirical research as a future study option might be intriguing.

References

1. Atlam, H., Wills, G.: Technical aspects of blockchain and IoT. In: Advances in Computers, pp. 1–39 (2019)
2. Comprehensive list of banks using blockchain technology in 2020. https://hackernoon.com/comprehensive-list-of-banks-using-blockchain-technology-in-2020-revised-and-updated-uq493yrb. Accessed 11 Jan 2020
3. Yermack, D.: Corporate governance and blockchains. Rev. Financ. **21**(1), 7–31 (2017). https://doi.org/10.1093/rof/rfw074
4. De Filippi, P.: Blockchain and the Law: The Rule of Code. Harvard University Press, England (2018)
5. De Oliveira, L., Simoyama, F., Grigg, I., Bueno, R.: Triple entry ledgers with blockchain for auditing. Int. J. Auditing Technol. **3**, 163 (2017)
6. DeCovny, S.: Chips off the old blockchain. CFA Inst. Mag. **26**, 24–25 (2015)
7. Definition of internal auditing—Institute of Internal Auditors—Australia. https://www.iia.org.au/about-iia-australia/WhatIsInternalAudit/DefinitionOfIA.aspx
8. Dyball, M., Seethamraju, R.: The impact of client use of blockchain technology on audit risk and audit approach—an exploratory study. Int. J. Auditing **25**, 602–615 (2021)

9. El Bassam, N.: Blockchain. In: Distributed Renewable Energies for Off-Grid Communities, pp. 447–450 (2021)
10. Ernst & Young (EY). https://assets.ey.com/content/dam/ey-sites/ey-com/en_gl/topics/digital/ey-reporting-building-blocks-of-the-future.pdf?download. Accessed 15 Jan 2021
11. Ernst & Young (EY). https://eyfinancialservicesthoughtgallery.ie/wp-content/uploads/2014/11/Internal-Audit-Insights-thought-leadership-for-Internal-Auditors-in-banking-and-capital-markets.pdf?cv=1. Accessed Sep 2014
12. Fanning, K., Centers, D.: Blockchain and its coming impact on financial services. J. Corp. Acc. Financ. **27**, 53–57 (2016)
13. Gauthier, M., Brender, N.: How do the current auditing standards fit the emergent use of blockchain? Manag. Audit. J. **36**, 365–385 (2021)
14. Golosova, J., Romanovs, A.: The advantages and disadvantages of the blockchain technology. In: 2018 IEEE 6th Workshop on Advances in Information, Electronic and Electrical Engineering (AIEEE) (2018)
15. Guo, Y., Liang, C.: Blockchain application and outlook in the banking industry. Financ. Innov. **2**, 24 (2016). https://doi.org/10.1186/s40854-016-0034-9
16. Kloch, R., Little, S.: Blockchain and internal audit. Intern. Audit Found. (2019)
17. Kosba, A., Miller, A., Shi, E., Wen, Z., Papamanthou, C.: Hawk: the blockchain model of cryptography and privacy-preserving smart contracts. In: 2016 IEEE Symposium on Security and Privacy (SP) (2016)
18. Kuenkaikaew, S., Vasarhelyi, M.: The predictive audit framework. Int. J. Digit. Acc. Res. **13** (2013)
19. Kuhn, J., Sutton, S.: Continuous auditing in ERP system environments: the current state and future directions. J. Inf. Syst. **24**, 91–112 (2010)
20. Lazanis, R.: How technology behind bitcoin could transform accounting as we know it. TechVibes Jan 22 (2015)
21. Liu, F.M., Shi, W.: Blockchain-based defense method of ARP spoofing attack. J. Netw. Inf. Secur. **004**(009), 36–43 (2016)
22. Liu, M., Wu, K., Xu, J.: How will blockchain technology impact auditing and accounting: permissionless versus permissioned blockchain. Curr. Issues Auditing **13**, A19–A29 (2019)
23. Liu, R.: A preliminary study of the impact of blockchain technology on internal auditing. In: 2020 2nd International Conference on Applied Machine Learning (ICAML) (2020)
24. Lombardi, R., de Villiers, C., Moscariello, N., Pizzo, M.: The disruption of blockchain in auditing—a systematic literature review and an agenda for future research. Acc. Auditing Accountability J. (2021)
25. Pitt, S.: Internal Audit Quality
26. Raval, S.: Decentralized Applications. O'Reilly, Beijing (2016)
27. Rooney, H., Aiken, B., Rooney, M.: Q&A. Is internal audit ready for blockchain?. Technol. Innov. Manag. Rev. **7**, 41–44 (2017)
28. Rozario, A., Vasarhelyi, M.: Auditing with smart contracts. Int. J. Digit. Acc. Res. 1–27 (2018)
29. Salminen V., Ruohomaa, H., Kantola, J.: Digitalization and big data supporting responsible business co-evolution. In: Kantola, J., Barath, T., Nazir, S., Andre, T. (eds.) Advances in Human Factors, Business Management, Training and Education. Advances in Intelligent Systems and Computing, vol. 498. Springer, Cham (2017)
30. Nakamoto, S.: Bitcoin: a peer-to-peer electronic cash system (2008). https://bitcoin.org/bitcoin.pdf
31. Schmitz, J., Leoni, G.: Accounting and auditing at the time of blockchain technology: a research agenda. Aust. Account. Rev. **29**, 331–342 (2019)
32. Sun, Y.P., Jiang, T., Lan, M.L.: Financial crowdsourcing model application based on blockchain technology. Financ. Acc. **000**(017), 48–50 (2017)
33. Szabo, N.: Formalizing and securing relationships on public networks. First Monday **2** (1997)
34. Tapscott, D., Tapscott, A.: Blockchain Revolution: How the Technology behind Bitcoin Is Changing Money, Business, and the World. Penguin, New York (2016)

35. Vasarhelyi, M., Halper, F.: The continuous audit of online systems. Auditing J. Pract. Theory (1991)
36. Yoon, K., Hoogduin, L.A., Zhang, L.: Big data as complementary audit evidence. Acc. Horiz. **29**(2), 150227131434007 (2015)
37. Yoon, S.: A study on the transformation of accounting based on new technologies: evidence from Korea. Sustainability **12**, 8669 (2020)
38. Zhong, W., Jia, Y.Z.: Outlook on the application of blockchain technology in accounting. Friends Acc. **17**, 122–125 (2016)

Cyberphysicality: Toward a Conceptual Framework for Studying the Fourth Industrial Revolution and its Implications on Business, Communication and Learning

Ibrahim Subeh

Abstract Blockchain, artificial intelligence and the internet of things are a few examples of the infrastructural innovations of the cyberphysical paradigm of technological development. These innovations are ubiquitously researched in social sciences, yet they are studied separately, in fragmented projects, and, in isolation from the overarching context they are part of. Cyberphysical systems are technologies that are capable of continuous interaction with the physical world by facilitating connections with, and between, real world objects. Cyberphysicality is the essence of the fourth industrial revolution, which realization is accelerated by the Covid-19 restrictions on movement, socialization and commerce. The paradigm of cyberphysicality is enabled by a variety of new technologies, including virtual and augmented reality, neural interfaces and haptic feedback, which are integral to implementing the metaverse and other engaging and immersive experiences that will transform the way humans communicate, transact and learn. This paper seeks to demonstrate the importance of taking a holistic approach to the study of technology. This paper also aims to construct a transdisciplinary frame of reference for social scientists seeking to study technological change in context, by borrowing from the approaches and concepts of cultural studies, sociology and communication.

Keywords Covid-19 · Transdisciplinarity · Cyberphysicality · Convergence · Fourth industrial revolution · Internet of things · Blockchain · Artificial intelligence · Smart environments

1 Introduction

The fourth industrial revolution is "an era where people are using smart, connected and converged cyber, physical and biological systems and smart business models to define and reshape the social, economic and political spheres" (PC4IR, 2020). It is understood as technology-enabled paradigm shift in various aspects of human

I. Subeh (✉)
University of Business and Technology, JCA, Jeddah 23435 21361, Saudi Arabia
e-mail: i.subeh@ubt.edu.sa

A. Hamdan et al. (eds.), *The Implementation of Smart Technologies for Business Success and Sustainability*, Studies in Systems, Decision and Control 216,
https://doi.org/10.1007/978-3-031-10212-7_59

life [53]. This shift gained visibility after the Covid-19 pandemic as the adoption of immerging technologies has accelerated and transformed many aspects of daily life. For example, the use of artificial intelligence in tracking the spread of disease changed the way public spaces are accessed. The implementation of Covid-19-related technology is a process that requires a behavioral adjustment and compliance at a societal scale, leading to a range of unexamined implications.

Educational, commercial and governmental institutions alike are developing content and strategies for addressing and researching individual cyberphysical technologies. However, the interaction of these technologies with one another and with the society remains largely under-researched, especially at the level of theorization. This gap in research is discussed by Lee and Lim [45] in their systematic review of the technologies of the fourth industrial revolution:

> As a multidimensional agenda, Industry 4.0 continues to attract interest from practitioners and academics across a variety of disciplines. While many studies have reported technologies and applications about the agenda. (e.g., Liao et al. [46]; Frank et al. [29], theorizing and conceptualization efforts on Industry 4.0 per se are still relatively scarce (2021)

This study contributes to theory, practice and directions for further research in three ways. First is by highlighting the potential of transdisciplinarity as an approach to this multifaceted topic. Second is by highlighting relevant concepts from various disciplines. Third is by engaging with the topic at multiple levels, including the technical and conceptual aspects, the effects and implications, the various dimensions of context and the levels of socio-technical interaction.

This study offers a comprehensive discussion of cyberphysicality as a paradigm for technological development. It aims to propose a transdisciplinary frame of reference for social scientists seeking to study technological change in context. The paper is structured as follows: Sect. 1 introduces the problem. Section 2 discusses pandemics as catalysts for technological and social change. Section 3 provides an overview of the social, economic and political implications of the four industrial revolutions leading to the cyberphysical paradigm. Section 4 provides a technical overview of cyberphysicality. Section 5 proposes a conceptual framework for social science research of technology. The last part of the study presents conclusions and directions for further research.

2 Pandemics as Catalysts for Technological and Social Change

2.1 The Black Death

The black death was one of the most significant pandemics in world history. It accelerated dramatic political, economic, social and cultural changes [19]. The pandemic caused a demographic collapse while natural and physical resources were untouched, resulting in a resource share increase and a shortage of labor, leading to a reduction in income inequality and various social transformations [35]. The social transformations resulting from the black death pandemic are regarded as contributing factors to the end of feudalism in Europe and the emergence of capitalism [25]. The black death has been the subject of numerous studies in economics, sociology and cultural studies. In his discussions of disease and social regulation, Michael Foucault compares two distinct models resulting from black death and leprosy, each with its own power-knowledge and technology [49]. The Foucauldian analysis of pandemics could provide important insights into Covid-19 as a catalyst for technological and social change.

2.2 *Covid-19*

Studies on Covid-19 and technology are ubiquitous. They cover a wide range of topics, including healthcare, education, business and communication. Particular topics related to this paper are health status tracking and remote interaction. The latter is a recurring theme that focuses on overcoming the challenges resulting from social distancing on work, healthcare, learning, shopping, socialization and entertainment.

Health Status Tracking. Healthcare literature on technology and Covid-19 discusses a wide range of infrastructural technologies of the cyberphysical paradigm. Mbunge et al. [51] for example, discuss the use of geospatial technology, artificial intelligence, big data, blockchain and the internet of medical things for the purposes of detecting, monitoring, diagnosing, screening, surveilling, mapping, tracking and creating awareness (2021). Two important themes emerge in the context of Covid-19 health status tracking: the tracking of infection patterns through the continuous checking of individuals' health status and the tracking of vaccination status. These potential benefits, however, are not free of potential risks; as Brough and Martin [7] argue: "response to the outbreak has threatened privacy by reducing consumer control over the collection, sharing and protection of some of the most sensitive types of personal information" (2021). In Foucauldian terms, the medical uses of technology under the apparatus of urgency and exception could be associated with serious irreversible societal implications that should be the subject of further inter- and transdisciplinary research.

Remote Interaction. The need to overcome lockdown and social distancing challenges resulted in vast social and infrastructural changes. According to Koumpouros [43], after Covid-19, global internet peak traffic has increased by 47%. Global internet capacity was enhanced and broadband was subsidized (2022). Attitudes toward remote work, healthcare, education, shopping, socialization and entertainment have changed as well. As various studies suggest, this shift in attitudes toward remote and hybrid interaction is potentially permanent [8, 32]. Leaps in communication and logistic solutions opened the door for new possibilities and paved the way for the adoption of the technologies of the fourth industrial revolution [54]. The centrality of remote interaction to post-Covid-19 life brings to prominence concepts such as digital poverty. As Seah [70] explains, new levels of access to basic digital needs are leading to a range of different inclusions and exclusions that "reflect, reproduce and amplify divides which exist between socio-economic class, ethnicity and gender, to name a few" (2020). Such concepts are of tremendous value to the study of cyber-physicality due to the centrality of connectivity as a necessary tool to preform basic tasks and access basic services.

3 Technology as an Agent of Global Social, Economic and Political Change

An industrial revolution is a process by which radical technological advancements occur in a very short time [18], leading to major social, political and economic transformations. Four major stages in the history of modern industrialization are identified: mechanization, mass production, digitization and cyberphysical interaction [22]. Each era is associated with a set of convergent technologies that collectively constitute a distinct technological paradigm with its own social, political and economic implications. The following paragraphs discuss the global effects of the four stages of industrialization in an attempt to contextualize cyberphysicality.

3.1 First Industrial Revolution: The Steam Age

The first industrial revolution began in Europe. It started from the middle of the eighteenth century and lasted until the middle of the nineteenth century [76]. It is signified by mechanization. Incremental technological advancements centered around steam power led to a significant increase in efficiency and productivity [80]. Economic models, divisions of labor, patterns of human settlement, governance and international relations were profoundly impacted by the technological changes that marked the transition from agriculture to industrialization. The first industrial revolution model that started in Britain produced a set of challenges at national and global scales as it was adopted in various parts of the Western world. The resulting demand for raw materials was compensated by an increase in the centrality of the role of slavery [10] as a social institution, and thus, changing the demographic, economic, cultural and political landscape of the affected parts of the world, especially Africa. The first industrial revolution was marked by rapid technological advancements that paved the way for the electric age.

3.2 The Second Industrial Revolution: The Electric Age

The second industrial revolution occurred between the late nineteenth century and early twentieth century [9]. It is signified by the use of electricity and the emergence of assembly lines which enabled mass production [71]. Mass production changed society in profound ways, from encouraging hyperconsumerism to changing perceptions of art, as discussed by Benjamin [5]. The second industrial era was marked by substantial leaps in all sectors of technology in the West [20]. Politically and economically, this further materialized Eurocentric worldviews which manifested itself in the form of global colonialism; "it created not only the need for Europe to expand, but the power to successfully take and profitably maintain so many colonies overseas" [63]. The second industrial revolution "transitioned the Western world from the modern into the contemporary era" [44]. It did not only transform industrialized societies internally; it also restructured their relationship with the world, leading to an array of dependencies and international conflicts. In this context, technology can be understood as an agent in social, political and economic change by materializing culture and ideology.

3.3 The Third Industrial Revolution: The Digital Age

The third industrial revolution came about in the 1960s [77]. It was enabled "by the development and availability of computers and communication technologies, such as networked computing, the internet and wireless communications" [3]. The convergent technologies of the digital era led to the formation of what Stein Braten describes as network societies [24]. The concept became a central theme in the work of Castells [11] who offers a useful framework for discussing the social, political and economic implications of the digital age. According to Castells [11], a network society can be understood as a "society whose social structure is made up of networks powered by micro-electronics-based information and communications technologies" [57]. Unlike the preceding stages of industrialization, in this era, the nature of society changes as human action and social relations are uprooted from local contexts and cultures and reconstituted within distant networks free of geographical constraints [23]. The development of network societies signifies the emergence of the information economy where data is understood as the new means of production [75] and control over information is understood as the chief form of power [11]. This era is marked by the centrality of data and mediated communication, a trend that will continue in the following stage of modern industrialization as new domains are digitized.

3.4 The Fourth Industrial Revolution: The Cyberphysical Age

The fourth industrial revolution started in the early 2000s [81]. It is signified by the blurring of the boundaries between the cyber, the physical, and the biological domains [59]. As explained by Raj et al. [65] the fourth industrial revolution can be understood as the convergence of several emerging concepts and new technologies leading to the restructuring of production systems and the integration of humans, data and machines (2020). The fourth industrial revolution encompasses a wide range of infrastructural technologies that enable cyber and physical interaction. According to Schwab [68], who coined the term in 2016 [67], the fourth industrial revolution will disrupt jobs and skills, innovation and productivity, ethics and identity, security and conflict and various aspects of human life [68]. Cyberphysicality is the essence of these disruptions.

Cyberphysicality. Cyberphysicality can be understood as a paradigm of technological development characterized by the convergence of the technologies of the fourth industrial revolution in ways that enable the continuous interaction between the cyber and the physical world. It is an evolution of the digital age that is marked by quantitative and qualitative developments in digital technology and technological infrastructure. Social, political and economic transformations resulting from cyberphysicality can be analyzed at three different levels: human-to-human, human-to-machine and machine-to-machine.

Human-to-Human level. As remote interaction becomes an integral aspect of work, healthcare, learning, shopping, socialization and entertainment, and as communication technologies leap to higher levels of immersion, the nature of virtuality evolves from virtual reality to real virtuality:

> Real virtuality is defined as a true high-fidelity multi-sensory virtual environment that evokes the same perceptual response from a viewer as if he/she was actually present, or "there", in the real scene being depicted... such environments are interactive and based on physics. All five senses are concurrently stimulated to deliver real-world modalities naturally and in real time. [13]

Sensory data is one of the multiple layers of user data that can be potentially mined in the cyberphysical age, which rases important ethical and regulatory challenges. This is especially relevant to discussions of the metaverse as real virtuality is at its core.

The augmented nature of data in the cyberphysical age and its significance as the new means of production are disrupting power dynamics among the stakeholders of technological infrastructure. Today, software development communities are arguably more impactful than legislators as the concept of regulation by design shifts from the physical to the cyber world. This critical change is a recognized development as reflected in the axiom "code is law", which is a theme in numerous digital studies [47, 60]. This phenomenon is enabled by the increasing dependency on digital solutions and the emergence of blockchain as a disruptive, disintermediating and decentralizing technology [75]. On human-to-human level, decentralizing and disintermediating technologies disrupt traditional organizational hierarchies at the institutional, national and global scales.

Human-to-Machine level. The fourth industrial revolution opens new frontiers for human-to-machine interaction. This leads to the emergence of three important topics: first is the configuration or coding bias as technology assumes grated regulatory impact [27]; second is the human-to-machine competition as more jobs are placed by machines [48]; third is the transformations in the field of interface design as digital experiences become multi-sensory and immersive [55]. All three topics constitute important directions for further exploration and research.

Machine-to-Machine level. Distinct technologies are converging, and previously distant systems are becoming interoperable as they connect with the internet and with one another [26]. Internet of things and cyberphysical systems will open new frontiers for machine-to-machine interaction; incidents of all kinds would trigger instant specialized notification and response [72]. Smart alarm clocks would ensure the timely arrival of everyday commuters by adjusting to external factors [56]. Smart fridges become mindful of quantities, nutritional values and expiration dates [42]. Metaphorical consciousness highlights the power and significance of cyberphysical interaction.

All three levels discussed should be taken into consideration in reference to their social, political and economic implications.

4 Technological Overview

Cyberphysical interaction is enabled by several convergent and interoperable technologies. This section discusses a number of relevant technologies and their convergence.

4.1 Relevant Technologies

Relevant technologies include mesh connectivity, neural interfaces, blockchain, artificial intelligence and the internet of things.

Mesh Connectivity. "A wireless mesh network (WMN) is a communication technology suitable for cyberphysical applications, such as healthcare devices, smart grids and the internet of things" [78]. It is a proposed solution that is capable of linking billions of physical objects to one another and to the internet [36]. Mesh networks are self-organizing, self-healing, dynamic, and inexpensive to implement

[1]. Their wireless and decentralized topology that relies on peer-to-peer connectivity makes them potentially resistant to centralized censorship and control, and thus, disrupting existing forms of internet governance.

Neural Interfaces. "Devices such as keyboards and touchscreens allow humans to communicate with machines. Neural interfaces, which can provide a direct, electrical bridge between analogue nervous systems and digital man-made systems, could provide a more efficient route to future information exchange" [82]. The significance of neural interfaces, also known as brain-computer interfaces, exceeds the mere replacement of keyboards and touch screens; it opens the door for new fields of inquiry such as bioelectronic medicine [17]. Its applications cover a wide range of domains, including transportation, art and entertainment [58]. As a new form of cyberphysical interaction, neural interfaces revolutionize human-to-machine interaction.

Blockchain. In simple terms, "blockchain can be understood as a distributed database connected in a decentralized manner" [12]. It is a new way for organizing, sharing, storing and perceiving data, leading to the development of asset classes and new industries such as FinTech, DeFi and RegTech [15]. "For the first time in history, trust no longer requires a third party. Unrelated individuals are able to perform a spectrum of collaborative operations on a global scale without institutional oversight" [75]. In addition to its various financial applications, blockchain can be applied to smart cities and smart manufacturing, governance, agriculture, transportation, education and supply chain [16]. Through disintermediation and decentralization, blockchain radically disrupts organizational hierarchies and economic models.

Artificial Intelligence. The term artificial intelligence broadly "refers to the idea of machines being capable of performing human tasks" [69]. According to Chen et al. [14] artificial intelligence encompasses the development of machines that have some level of intelligence, with the ability to perform human-like functions, including cognitive, learning, decision making and adapting to the environment (2020). According to Walther and Weigold [79], artificial intelligence encompasses two related concepts: machine learning and deep learning (2021). Each of these concepts has its own subfields, processes and applications, from computer vision and pattern recognition to robotics and knowledge management.

Internet of Things. The internet of things refers to the idea of connecting billions of objects and machines to the internet and to one another [37]. Connected 'things' continuously and autonomously interact to perform a wide range of activities without the need for human interactions. According to Alkhammash et al. [2] the internet of things is at the heart of concepts such as the smart home, the smart campus and the smart city (2020). The internet of things revolutionizes both machine-to-machine interaction and human-to-machine interaction.

Relevant technologies include, but are not limited to, haptic feedback, augmented, virtual and mixed reality, big data and cloud computing.

4.2 Cyberphysical Convergence

Convergence, as the idea of distinct entities or domains merging, is the subject of numerous studies in communication [61], political science [21] and technology [34]. Convergence is understood as a phenomenon with various dimensions, including technological, social, cultural, industrial and political aspects, that provide insight into future societal changes. Three consequences of convergence are discussed: the smart environment, the prosumer and the cyborg.

The Smart Environment. As Park [62] explains, technological convergence refers to the phenomenon where two or more independent technologies integrate and form a new outcome, as can be seen in smart phones which integrate the telephone, the computer, the camera and wide range of tools (2019). Technological convergence in the cyberphysical age takes more complex forms as smart technologies, devices and platforms converge to form smart environments, such as smart homes, smart campuses, and smart cities. This type of convergence is a subject of numerous studies [38, 73, 74]. The metaverse is the ultimate embodiment of this concept.

The Prosumer. The concept of prosumer refers to convergence of roles of producer and consumer [52]. As Khan [40] points out, the concept was discussed by Marshall McLuhan in the context of media ecology (2020). It gained popularity with the advent of new media as barriers to entry to the industry are diminished. Today, the concept is discussed in the context of blockchain-enabled distributed energy systems. As Hua et al. [33] points out, smart grid allows households to transact excess solar power between them without the need for utility companies (2022). The emergence of prosumerism in the utility sector is a direct consequence of cyberphysical convergence.

The Cyborg. This concept is understood and approached in various ways. It is discussed as a metaphor and a vehicle for countercultural discourse, a recurring theme in science fiction literature and as a conceptualization of a spectrum of human machine integration. In their book, *Modified: Living as a Cyborg,* Gray et al. [30] travel along this spectrum and introduce real-world examples of Cyborgism from surgical implantation of sensors and devices to accounts of living in the cyberspace (2021). The development of smart environments and neural interfaces and the emergence of synthetic biology will open new possibilities for cyborgism in the cyberphysical age.

Other consequences worth of studying include digital scarcity, the primary engine behind the immerging fintech industry [4] as well as the new forms of poverty that may develop due to increased digital dependency.

5 Toward a Framework for Social Science Research

A good starting point for researching technology is to ask the following three questions. First, what is the researcher's fundamental philosophical position on the role of technology as an agent in social and political change? Second, what aspects of technology must be considered? This question encompasses an array of topics including technical knowledge, conceptual knowledge, technology ethics and regulations, perceptions and attitudes towards technology, technology effects, the dimensions of context and the levels of socio-technical interaction. Finally, what approaches can be used to answer these questions?

5.1 *Philosophical Positions*

Philosophical positions on the relationship between society and technology are grouped into two main categories: technological determinism and social determinism. According to Khan [41], technological determinism refers to the notion that technology is the key force in society (2021). Social determinism, on the other hand, refers to the role of users in shaping technology [50]. Views within both camps fall at a wide spectrum. This study adopts a softer form of technological determinism that acknowledges the role of technology as an important agent in social and political change without ignoring the important role of social, economic and cultural forces. According to this view, every technology encompasses a set of inherent qualities that is socially consequential; to choose a technology is to enable particular modalities of power relations [75]. Defining the philosophical position provides structure and sets the background for the conceptual and theoretical frameworks.

5.2 *Technology as Multifaceted Subject*

The following aspects of technology are important to developing a nuanced understanding of the subject. First is the technical knowledge which entails developing a basic understanding of the studied technologies, its uses and application and its interoperability with other technological solutions. Second is the conceptual knowledge which encompasses an understanding of concepts such as smart innovations and smart environments. Third is technology ethics and regulations. The emergence of new technologies is associated with a set of ethical and legal challenges. These challenges might be more visible in some fields than others, such as synthetic biology for example; however, they are relevant to all newly emerging sectors. Fourth is perceptions and attitudes toward technology, which are useful sources of insights for secondary research. Fifth is technology effects. These may include a combination of a wide range of inherent qualities of technology, such as centralization or decentralization, intermediation, or disintermediation. Sixth is technology in context. This includes its relation to other technologies, the paradigm in which it functions and the social, cultural, historical, economic and political contexts. The last aspect is the levels of socio-technical interaction which include the human-to-human level, human-to-machine level and the machine-to-machine level.

5.3 Approaches to the Study of Technology

As a multifaceted subject, the study of technology requires the use of various approaches from a wide range of disciplines. This study aims to highlight the role transdisciplinarity in relation to this subject.

Transdisciplinarity. According to Renn [66], transdisciplinarity aims to establish effective links between a variety of truth claims, social goals and policy choices with the aim of solving real problems by involving multiple stakeholders (2021). Under this view, the benefits of transdisciplinarity include overcoming the limitations of individual disciplines as well as overcoming the challenges of both disciplinary and stakeholder bias.

While the utility and methods of transdisciplinary research may vary, their aims remain the same. According to Hashmi et al. [31] "transdisciplinary system approach has the advantage of a self-improving model that drives realization of digital transformation in evolutionary cycles" (2021). According to Broo [6], transdisciplinary approaches are important in dealing with the challenges related to the cyberphysical systems (2021). Fitzek et al. [28], on the other hand, directs his focus on the role of transdisciplinarity in tackling subjects related to human–machine interaction. This emerging field of transdisciplinary research aims to promote next generation digitalized human–machine interactions in perceived real time (2021). The following disciplines present a range of synergizing tools and approaches.

Relevant Disciplines. The field of cultural studies is a good example of the utilization of transdisciplinary approaches in the study of culture. It is a model that can be followed in the study of technology and its societal implications.

> Cultural studies operate with a transdisciplinary conception that draws on social theory, economics, politics, history, communication studies, literary and cultural theory, philosophy and other theoretical discourses—an approach shared by the Frankfurt school, British cultural studies, and French postmodern theory. Transdisciplinary approaches to culture and society transgress borders between various academic disciplines. [39]

Cultural studies are also understood as a futuristic discipline, as Powers [64] demonstrates (2020). Their futuristic orientation and ability to transcend disciplinary boundaries are compatible with the nature of cyberphysicality as a subject of inquiry.

As a revolution in communication technology, scholars researching cyberphysicality could benefit from a wide range of media and mass communication tools and approaches, such as content analysis, and a wide range of conceptual frameworks, including media convergence and media ecology. Useful sociological concepts include the technology adoption life cycle, conflict theory and dialectics. The work of Michael Foucault embodies this approach as it falls at the intersections of sociology, literature, philosophy and political science. It offers a wide range of concepts and tools useful and relevant to the study of technology and society. Relevant Foucauldian concepts include power-knowledge, heterotopia and discourse on technology.

6 Conclusion

In conclusion, this study fills an important gap in the social science research of technology by addressing the recognized problem of the lack of conceptual frameworks for studying the fourth industrial revolution. It takes a unique scope and perspective that recognize cyberphysicality as a phenomenon and a paradigm of technological development. It explores various aspects of cyberphysicality and discusses the internal and external contextual complexities influencing its development. It discusses the catalyzing role of pandemics, such as black death and Covid-19, in the context of technological and societal development. It maps the position of cyberphysicality in relation to consecutive industrial revolutions and highlights its social, political and economic implications in reference to the three levels of sociotechnical interaction: human-to-human, human-to-machine and machine-to-machine. It provides a technical overview of relevant technologies and highlights the consequences of its convergence. These consequences include the emergence of the prosumer, the smart environment and the cyborg. Furthermore, this study recommends a three-dimensional framework for studying cyberphysicality by highlighting three important elements: the philosophical position, the scope and the approach. The proposed approach acknowledges the significance of technology as one of the various forces dialectically shaping the society. It emphasizes the multifaceted nature of technology as a subject of inquiry and highlights a number of these facets. Finally, this study underlines the value of transdisciplinarity and the relevance of various disciplinary approaches and concepts borrowed mainly from sociology, communication and cultural studies. This study achieved its purpose of developing a basic understanding of the concept of cyberphysicality in the context of contemporary societal and technological changes; however, an important question remains unresearched and is worthy of further studying: How would cyberphysicality impact specific sectors such as advertising, marketing, education, business and healthcare? Cyberphysicality is no longer a subject of science fiction. It is the essence of the fourth industrial revolution which has become the subject of growing educational, commercial and governmental interest.

References

1. Al-Juaifari, M.K.R., Hussain, M.A., Khammas, N.H.A.: Power enhancement based link quality for wireless mesh network. Int. J. Electr. Comput. Eng. **11**(2), 1388 (2021)
2. Alkhammash, M., Beloff, N., White, M.: An internet of things and blockchain based smart campus architecture. In: Science and Information Conference, pp. 467–486. Springer, Cham (2020)
3. Arden, N.S., Fisher, A.C., Tyner, K., Lawrence, X.Y., Lee, S.L., Kopcha, M.: Industry 4.0 for pharmaceutical manufacturing: preparing for the smart factories of the future. Int. J. Pharm. **602**, 120554 (2021)
4. Bao, H., Roubaud, D.: Non-Fungible token: a systematic review and research agenda. J. Risk Financ. Manage. **15**(5), 215 (2022)
5. Benjamin, W.: The Work of Art in the Age of Mechanical Reproduction. Penguin, UK (2008)
6. Broo, D.G.: Transdisciplinarity and three mindsets for sustainability (2021)
7. Brough, A.R., Martin, K.D.: Consumer privacy during (and after) the COVID-19 pandemic. J. Public Policy Mark. **40**(1), 108–110 (2021)
8. Buchanan, N.D., Aslaner, D.M., Adelstein, J., MacKenzie, D.M., Wold, L.E., Gorr, M.W.: Remote work during the COVID-19 pandemic: making the best of it. Physiology **36**(1), 2–4 (2021)
9. Campa, R.: Fourth industrial revolution and emotional intelligence: a conceptual and scientometric analysis. Changing Soc. Personalities **4**(1), 8–30 (2020)
10. Candido, M.P.: The expansion of slavery in Benguela during the nineteenth century. Int. Rev. Soc. Hist. **65**(S28), 67–92 (2020)
11. Castells, M.: A critical theory of informational society. In: Critical Education in the New Information Age, vol. 37 (1999)
12. Castro, R.Q., Au-Yong-Oliveira, M.: Blockchain and higher education diplomas. Eur. J. Invest. Health Psychol. Educ. **11**(1), 154–167 (2021)
13. Chalmers, A., Howard, D., Moir, C.: Real virtuality: a step change from virtual reality. In: Proceedings of the 25th Spring Conference on Computer Graphics, pp. 9–16 (2009)
14. Chen, L., Chen, P., Lin, Z.: Artificial intelligence in education: a review. IEEE Access **8**, 75264–75278 (2020)
15. Chen, C.C., Huang, H.H., Takamura, H., Chen, H.H.: An Overview of Financial Technology Innovation (2022a)
16. Chen, Y., Lu, Y., Bulysheva, L., Kataev, M.Y.: Applications of blockchain in industry 4.0: a review. Inf. Syst. Front. 1–15 (2022b)
17. Cho, Y., Park, J., Lee, C., Lee, S.: Recent progress on peripheral neural interface technology towards bioelectronic medicine. Bioelectron. Med. **6**(1), 1–10 (2020)
18. Damayanti, D.: Human resources competency in the era of industrial revolution 4.0. Econ. Bus. Solutions J. **5**(1), 13–27 (2021)
19. Deming, D.: Science and Technology in World History, volume 3: The Black Death, the Renaissance, the Reformation and the Scientific Revolution, vol. 3. McFarland (2014)
20. Dogaru, L.: The main goals of the fourth industrial revolution. Renewable energy perspectives. Procedia Manuf. **46**, 397–401 (2020)
21. Dowd, R.: EU policy convergence—supranational data policies in the EU (Mid 1990s–2016). In: The Birth of Digital Human Rights, pp. 159–209. Palgrave Macmillan, Cham (2022)
22. Duan, L., Da Xu, L.: Data Analytics in Industry 4.0: a survey. Inf. Syst. Front. 1–17 (2021)
23. Elliott, A., Lemert, C.C.: Introduction to Contemporary Social Theory. Routledge (2022)
24. ESCOBAR, A., SACHS, W.: Post development theory–development as discourse. Theorizing Dev. Globalization **86** (2021)
25. Fang, G.W.: The effects of the black death on international relations during the renaissance period compared with the Covid-19 in 2020. In: 2021 International Conference on Public Art and Human Development (ICPAHD 2021), pp. 919–922. Atlantis Press (2022)

26. Farahani, B., Firouzi, F., Luecking, M.: The convergence of IoT and distributed ledger technologies (DLT): opportunities, challenges, and solutions. J. Netw. Comput. Appl. **177**, 102936 (2021)
27. Ferrer, X., van Nuenen, T., Such, J.M., Coté, M., Criado, N.: Bias and discrimination in AI: a cross-disciplinary perspective. IEEE Technol. Soc. Mag. **40**(2), 72–80 (2021)
28. Fitzek, F.H., Li, S.C., Speidel, S., Strufe, T., Seeling, P.: Frontiers of transdisciplinary research in tactile internet with human-in-the-loop. In: 2021 17th International Symposium on Wireless Communication Systems (ISWCS), pp. 1–6. IEEE (2021)
29. Frank, A.G., Dalenogare, L.S., Ayala, N.F.: Industry 4.0 technologies: Implementation patterns in manufacturing companies. Int. J. Prod. Econ. **210**, 15–26 (2019)
30. Gray, C.H., Figueroa-Sarriera, H., Mentor, S. (eds.) Modified: Living as a Cyborg. Routledge (2021)
31. Hashmi, M.A., Mo, J.P., Beckett, R.C.: Transdisciplinary systems approach to realization of digital transformation. Adv. Eng. Inform. **49**, 101316 (2021)
32. Hern, A.: Covid-19 could cause permanent shift towards home working. Guardian **13** (2020)
33. Hua, W., Chen, Y., Qadrdan, M., Jiang, J., Sun, H., Wu, J.: Applications of blockchain and artificial intelligence technologies for enabling prosumers in smart grids: a review. Renew. Sustain. Energy Rev. **161**, 112308 (2022)
34. Hussain, A., Jeon, J., Rehman, M.: Technological convergence assessment of the smart factory using patent data and network analysis. Sustainability **14**(3), 1668 (2022)
35. Jedwab, R., Johnson, N.D., Koyama, M.: The economic impact of the black death. J. Econ. Lit. **60**(1), 132–178 (2022)
36. Jiang, X., Zhang, H., Yi, E.A.B., Raghunathan, N., Mousoulis, C., Chaterji, S., Peroulis, D., Shakouri, A., Bagchi, S.: Hybrid low-power wide-area mesh network for Iot applications. IEEE Internet Things J. **8**(2), 901–915 (2020)
37. Kagita, M.K., Thilakarathne, N., Rajput, D.S., Lanka, D.S.: A detail study of security and privacy issues of internet of things. arXiv preprint (2020). arXiv:2009.06341
38. Kaur, M.J., Mishra, V.P., Maheshwari, P.: The convergence of digital twin, IoT, and machine learning: transforming data into action. In: Digital Twin Technologies and Smart Cities, pp. 3–17. Springer, Cham (2020)
39. Kellner, D.: Cultural Marxism, British cultural studies, and the reconstruction of education. Educ. Philos. Theory 1–13 (2021)
40. Khan, A.: Understanding Mcluhan: the times and relevance. Int. J. Knowl. Manage. Practices **8**(1), 41 (2020)
41. Khan, J.: CRISPR, like any other technology: shedding determinism and reviving Athens. Can. J. Law Technol. **19** (2021)
42. Kore, U., Akre, P., Mashayak, U.: Smart refrigerator. In: 2nd International Conference on Communication and Information Processing (ICCIP) (2020)
43. Koumpouros, Y.: Innovative learning technologies for children with autism and special educational needs during COVID-19. In: Quality of Healthcare in the Aftermath of the COVID-19 Pandemic, pp. 95–111. IGI Global (2022)
44. Ladd, M.: Music and the second industrial revolution ed. by Massimiliano Sala. Notes **78**(3), 439–442 (2022)
45. Lee, C., Lim, C.: From technological development to social advance: a review of Industry 4.0 through machine learning. Technol. Forecast. Soc. Change **167**, 120653 (2021)
46. Liao, Y., Deschamps, F., Loures, E.D.F.R., Ramos, L.F.P.: Past, present and future of Industry 4.0-a systematic literature review and research agenda proposal. Int. J. Prod. Res. **55**(12), 3609–3629 (2017)
47. Liu, S.: Code is law. In: Rights Limitation in Digital Age, pp. 117–122. Springer, Singapore (2021)
48. Magnani, M.: The technological revolution: professions at risk and new jobs. In: Making the Global Economy Work for Everyone, pp. 53–71. Palgrave Macmillan, Cham (2022)
49. Marinković, D., Major, S.: COVID-19 and the genealogies of biopolitics: a pandemic history of the present. Sociologija **62**(4), 486–502 (2020)

50. Martin, K.: Value-laden biases in data analytics. In: Ethics of Data and Analytics, pp. 1–5. Auerbach Publications (2022)
51. Mbunge, E., Akinnuwesi, B., Fashoto, S.G., Metfula, A.S., Mashwama, P.: A critical review of emerging technologies for tackling COVID-19 pandemic. Human Behav. Emerg. Technol. **3**(1), 25–39 (2021)
52. Mehdinejad, M., Shayanfar, H., Mohammadi-Ivatloo, B.: Decentralized blockchain-based peer-to-peer energy-backed token trading for active prosumers. Energy **244**, 122713 (2022)
53. Mhlanga, D.: Stakeholder capitalism, the fourth industrial revolution (4IR), and sustainable development: issues to be resolved. Sustainability **14**(7), 3902 (2022)
54. Mhlanga, D., Moloi, T.: COVID-19 and the digital transformation of education: what are we learning on 4IR in South Africa? Educ. Sci. **10**(7), 180 (2020)
55. Mishra, A., Shukla, A., Rana, N.P., Dwivedi, Y.K.: From "touch" to a "multisensory" experience: the impact of technology interface and product type on consumer responses. Psychol. Mark. **38**(3), 385–396 (2021)
56. Modoni, G.E., Caldarola, E.G., Sacco, M., Wasielewska, K., Ganzha, M., Paprzycki, M., Szmeja, P., Pawłowski, W., Palau, C.E., Solarz-Niesłuchowski, B.: Integrating the AAL CasAware platform within an IoT ecosystem, leveraging the INTER-IoT approach. In: Proceedings of First International Conference on Computing, Communications, and Cyber-Security (IC4S 2019), pp. 197–212. Springer, Singapore (2020)
57. Moll, I.: The fourth industrial revolution: a new ideology. tripleC: communication, capitalism and critique. Open Access J. Glob. Sustain. Inf. Soc. **20**(1), 45–61 (2022)
58. Mridha, M.F., Das, S.C., Kabir, M.M., Lima, A.A., Islam, M., Watanobe, Y.: Brain-computer interface: advancement and challenges. Sensors **21**(17), 5746 (2021)
59. Mustafa, N.A., Shah, N.M., Hashim, N.W., Desa, M.M.: An overview of stem education and industry 4.0 for early childhood education in Malaysia. J. Positive Sch. Psychol. 53–62 (2022)
60. Oster, J.: Code is code and law is law—the law of digitalization and the digitalization of law. Int. J. Law Inf. Technol. **29**(2), 101–117 (2021)
61. O'Sullivan, J., Fortunati, L.: Media convergence. In: Digital Roots, vol. 41 (2021)
62. Park, S.E.: Technological convergence: regulatory, digital privacy, and data security issues. Congressional Res. Serv. Tech. Rep. (2019)
63. Parvanova, D.: The industrial revolution was the force behind the new Imperialism. Essai **15**(1), 30 (2017)
64. Powers, D.: Towards a futurist cultural studies. Int. J. Cult. Stud. **23**(4), 451–457 (2020)
65. Raj, A., Dwivedi, G., Sharma, A., de Sousa Jabbour, A.B.L., Rajak, S.: Barriers to the adoption of industry 4.0 technologies in the manufacturing sector: an inter-country comparative perspective. Int. J. Prod. Econ. **224**, 107546 (2020)
66. Renn, O.: Transdisciplinarity: synthesis towards a modular approach. Futures **130**, 102744 (2021)
67. Ross, P., Maynard, K.: Towards a 4th industrial revolution. Intell. Buildings Int. **13**(3), 159–161 (2021)
68. Schwab, K.: The Fourth Industrial Revolution. Currency (2017)
69. Schwendicke, F.A., Samek, W., Krois, J.: Artificial intelligence in dentistry: chances and challenges. J. Dent. Res. **99**(7), 769–774 (2020)
70. Seah, K.M.: COVID-19: exposing digital poverty in a pandemic. Int. J. Surg. (London, England) **79**, 127 (2020)
71. Shahbakhsh, M., Emad, G.R., Cahoon, S.: Industrial revolutions and transition of the maritime industry: the case of Seafarer's role in autonomous shipping. Asian J. Shipping Logistics **38**(1), 10–18 (2022)
72. Sharma, A., Awasthi, Y., Kumar, S.: The role of blockchain, AI and IoT for smart road traffic management system. In: 2020 IEEE India Council International Subsections Conference (INDISCON), pp. 289–296. IEEE (2020)
73. Shi, F., Ning, H., Huangfu, W., Zhang, F., Wei, D., Hong, T., Daneshmand, M.: Recent progress on the convergence of the internet of things and artificial intelligence. IEEE Netw. **34**(5), 8–15 (2020)

74. Singh, S., Sharma, P.K., Yoon, B., Shojafar, M., Cho, G.H., Ra, I.H.: Convergence of blockchain and artificial intelligence in IoT network for the sustainable smart city. Sustain. Cities Soc. **63**, 102364 (2020)
75. Subeh, I.: Blockchain as a technological imaginary: media framing and the views of blockchain professionals in the arab world. (2020)
76. Toan, N.T.: Industrial revolution 4.0 and global citizenship education in Vietnam today. Adv. Soc. Sci. Res. J. **7**(3) (2020)
77. Tri, N.M., Hoang, P.D., Dung, N.T.: Impact of the industrial revolution 4.0 on higher education in Vietnam: challenges and opportunities. Linguist. Cult. Rev. **5**(S3), 1–15 (2021)
78. Vijayanand, R., Devaraj, D.: A novel feature selection method using whale optimization algorithm and genetic operators for intrusion detection system in wireless mesh network. IEEE Access **8**, 56847–56854 (2020)
79. Walther, J., Weigold, M.: A systematic review on predicting and forecasting the electrical energy consumption in the manufacturing industry. Energies **14**(4), 968 (2021)
80. Wardley-Kershaw, J., Schenk-Hoppé, K.R.: Economic growth in the UK: the inception. World **3**(2), 162–174 (2022)
81. Yusuf, B., Walters, L.M., Sailin, S.N.: Restructuring educational institutions for growth in the fourth industrial revolution (4IR): a systematic review. Int. J. Emerg. Technol. Learn. **15**(3) (2020)
82. Zhang, M., Tang, Z., Liu, X., Van der Spiegel, J.: Electronic neural interfaces. Nat. Electron. **3**(4), 191–200 (2020)

Do Service Quality and Satisfaction Affect the Intention of Using Application-Based Land Transportation? A Study on Generation YZ in Jakarta

Michael Christian, Suryo Wibowo, Eko Retno Indriyarti, Sunarno Sunarno, and Yustinus Yuniarto

Abstract Many researchers have been investigating the intention of Generations Y and Z, who are known to be picky customers, to use various services. The purpose of this study is to investigate the impact of application-based land transportation service quality on usage intentions as mediated by satisfaction. The sample size for this quantitative study is 200 users of application-based transportation from Generations Y and Z in Jakarta. Data are analyzed using structural modeling with partial least squares using SMART PLS 3.0. The analysis includes reliability (composite reliability and loading factors) and validity (average variance extracted) tests, as well as hypothesis testing (t-value and *p*-value). This study's findings support all proposed hypotheses. The satisfaction and intention to use application-based land transportation are significantly influenced by service quality. Meanwhile, user satisfaction influences the intention to use application-based land transportation. Furthermore, satisfaction successfully mediates the effect of service quality on the intention to use application-based land transportation. It is evident that to cultivate users' intention to reuse, user satisfaction must be achieved; and to shape user satisfaction, quality services must be offered to users.

Keywords Intention to use · Satisfaction · Service quality

M. Christian (✉) · Y. Yuniarto
Universitas Bunda Mulia, Jakarta, Indonesia
e-mail: michaelchristianid@gmail.com

S. Wibowo · S. Sunarno
Universitas Persada Indonesia YAI, Jakarta, Indonesia

E. R. Indriyarti
Universitas Trisakti, Jakarta, Indonesia

A. Hamdan et al. (eds.), *The Implementation of Smart Technologies for Business Success and Sustainability*, Studies in Systems, Decision and Control 216,
https://doi.org/10.1007/978-3-031-10212-7_60

1 Introduction

In regard to consumer behavior, young people's loyalty frequently changes, and today's younger generation is more fickle and swiftly switches to other options for a product or service. This is due to changes that happen and the popularity of brands, both of which have an impact on the lifestyle of this group. According to Valentine and Powers [1], millennials lives are closely linked to the usage of technology, and social and economic developments, which influence purchase behavior. This is consistent with Jackson et al. [2] stance that social, political, and economic events can influence a group's values, beliefs, expectations, and behavior. According to Ordun [3], younger generations, such as millennials, are open-minded, technologically savvy, innovative, and ambitious. Millennials have variously been defined as those born from 1980 through the 1990s [4], from 1980 to 2000 [5], and from 1990 through the 2000s [3, 6]. Petra [7] uses a specific label, Generation Y, to represent the group born between 1980 and 2000, while Bednall et al. [8] defines Generation Y as those born between 1980 and 1994. The next generation, which was born after 1995, is known as Generation Z [9].

As previously stated, these generations are familiar with the usage of technology, such as social media platforms and applications. This influences these generations' voter characters even more. The causes of this phenomenon include the ease with which information may be accessed and the availability of a diverse range of products and services. This is supported by Omar et al. [4] who argue that the younger generations are inextricably linked to the usage of smartphones, the internet, and social media. Services offered by way of technology, such as applications, are widespread in the younger generation's daily life. One of the more interesting aspects of this is application-based land transportation (ALT). Currently, in Jakarta, users now have several ALT options, mainly Gojek and Grab; although, over time, new players such as Blujek, TopJek, and Ladyjek, have entered the market. However, fierce rivalry and customers' tendency to switch ALT services pose a significant barrier for new businesses with minimal capital, human resources, and technology. Thus, understanding the user's intentions in using a specific ALT service proves to be a difficult task. This occurrence is consistent with Christian et al. [10] that market orientation, which emphasizes the importance of focusing on market-responsive strategies, competition strength, and customers' satisfaction in determining business performance.

Previous studies, such as ALT Grab [11] and Gojek [12], explain the critical significance of service quality in predicting behavioral intentions. However, these studies do not incorporate real usage behavior, which is a crucial aspect in describing the experience of utilizing ALT. Furthermore, these studies assess ALT users from non-specific age generations. According to other study [13], satisfaction is an endogenous variable, with user behavior characteristics acting as exogenous variables. Because of the dynamic rivalry in the ALT market, satisfaction cannot be considered as the final measure of ALT use. More research is needed to quantify the intention to use ALT so that the intention to stick with a brand or switch to a different brand can be discovered. Sutia et al. [14] research explains the extension from assessing satisfaction to the intention to use ALT but does not include the service quality element, which is a crucial identifier of service delivery for ALT.

Using structural modeling and satisfaction as a mediator, this study aims to measure the intention to use ALT of Generations Y and Z in Jakarta, highlighting the phenomenon and research gaps described above. The study's novelty comes from measuring the intention to use ALT specifically with satisfaction as a mediator while focusing on Generations Y and Z. This angle has not been directly investigated since prior studies have focused on satisfaction as an endogenous variable associated with the characteristics of ALT users in general. The results of this study are likely to expand and supplement the analyses of previous studies, particularly in understanding consumer behavior while using ALT across generations.

2 Literature Review and Hypothesis Development

Several factors can influence one's decision to use ALT. According to the theoretical concept, intention can be defined as an action to be taken when a perception of usefulness has been created [15]. In addition to perceived usefulness, the ease-of-use factor influences intention to use [16]. Attractive services provided by ALT service providers that can meet and exceed users' expectations can be one of the most important aspects when consumers decide to use these services. Price is a major concern for the younger generations who use ALT services. This will have a more effective impact when delivered through advertising media [17]. The lower the price in effect at the time, the more consumers will be interested in selecting a particular ALT service.

Human resources are another factor that can influence users' intention to use ALT services. The availability of human resources, particularly drivers, is critical in the transportation business. Given present-day Jakarta's size and population, the availability of ALT drivers must be proportional to the number of requests. This continues to be a challenge for service providers. If the number of requests is extremely high, especially during peak hours, some users may not be served. As a result, service providers frequently charge higher prices during peak usage hours, sometimes charging double the standard price. For price-conscious users, this is a drawback for service providers seeking to retain loyal customers. In this scenario, the user can quickly switch to a competitor or another mode of transportation.

The human resource aspect also includes the driver's ability. User complaints against drivers include things like lack of knowledge of road routes and driving at fast speeds. Avoiding traffic congestion is frequently the primary motivator for consumers to use ALT services, particularly two-wheeled vehicles [18]. Their small and slim frames can provide reliable travel time and the ability to travel down local roads to reduce travel time. ALT services also offer four-wheeled modes of transportation, which some users may find to be more comfortable to use than two-wheeled vehicles.

The advancement of information and communications technology is also an important factor in the growth of the ALT industry. The ability of a company to provide simple-to-use applications, attractive displays, and continuing improvements is vital for users' intention to use an ALT service. Satisfaction when using smartphone applications is a crucial element in offering quality services. Aspects of service quality, such as tangible, reliability, responsiveness, empathy, and assurance, influence satisfaction and intention to reuse [19]. Christian et al. [20] noted that the concept of service, service encounter, and service delivery are essential parts that determine the service strategy, emphasizing the concept of services provided to customers.

Reuse, which is theoretically a measure of loyalty, is heavily influenced by characteristics of service quality. In addition, Fauzi [21] reveals in his research that service quality affects customer satisfaction in ALT services in Indonesia. This study looks at the quality of electronic services, namely information quality, application design, payment methods used, security, and client privacy. Although overall service quality and satisfaction are good predictors of user loyalty, there is no clear association between safety aspects and user loyalty [22]. Thus, this study proposes the following hypotheses (H):

H1: Service quality affects the intention to use application-based land transportation.

H2: Service quality affects the satisfaction of using application-based land transportation.

Given the complexities of the problem of service quality in relation to usage intentions, it is worth emphasizing that the quality of services supplied by ALT providers can have an impact on user satisfaction. On the other side, user dissatisfaction almost always occurs in terms of a driver's ability to provide services in accordance with customers' standards. Things such as effective communication, polite attitudes, or providing a sense of security and comfort while traveling can help create a positive experience from the driver's service side. Ensuring that users' enjoy a positive experience is vital since satisfaction with the user experience influences ALT reuse [19].

In a similar vein, Marisse and Dewi [23] explain in their research that the quality of services in systems and information that provide a pleasant form of use might influence satisfaction and the intention to continue using ALT services. The intention to switch to utilizing ALT is also influenced by several factors, including functional, social, and psychological benefits. As a result, knowing ALT users' wishes cannot be observed solely from the existing functional apps. Based on the explanations provided above, the following hypotheses are proposed:

H3: Satisfaction affects the intention to use application-based land transportation.

H4: Service quality affects the intention of using application-based land transportation mediated by satisfaction.

In summary, the objective of this study is to examine the existence of factors that cause satisfaction and the intention to use ALT services in Jakarta. The combined group of Generations Y and Z is an interesting target to study because it is a large population with substantial global purchasing power and high consumption [24] and is rarely used as a study sample. In addition, in the generally accepted concept of consumer behavior, satisfied customers will automatically become repeat users of products or services. However, several studies prove this does not always hold true. This study's research will help deepen the current knowledge of consumer behavior, especially of that pertaining to users of ALT services.

3 Methods

This quantitative research consists of one exogenous variable (service quality), one mediator variable (satisfaction), and one endogenous variable (intention of use) as shown in Fig. 1.

The operational variables used in this study consist of three variables. Service quality in this study is explained as aspects related to services provided by ALT service providers to users. The aspects of this variable are determined only on the main aspects related to ALT services such as a sense of security and comfort, suitability, and reasonability in terms of service time [25, 26]. The user satisfaction variable in this study is described as an output in the form of satisfaction from users after using ALT services. This variable includes aspects of satisfaction in delivery services and satisfaction in using applications [25]. The third variable in this study is the intention of use, which is described as a behavior to plan to use ALT services in the future with indicators, namely the desire to continue using ALT services, a feeling of confidence in continued use of ALT services and plans to change to using ALT services instead of other types of transportation [26, 27].

Data for this study were collected during November 2020, via an online questionnaire designed with a Likert 1–5 scale where 1 = strongly disagree, 2 = disagree, 3 = neutral, 4 = agree, and 5 = strongly agree. Users of ALT services (Gojek and/or Gojek) in Jakarta were surveyed. The study sample is restricted to female and male users who have used ALT (two-wheeled and/or four-wheeled) within one year of data collection and are from the age groups (generations) Y (1980–1994) and Z (1995 and later). To obtain representation from each of the predetermined criteria, this study employs the average proportional cluster method consisting of generation, gender, and area groups (North Jakarta, Central Jakarta, West Jakarta, East Jakarta, and South Jakarta).

Fig. 1 Research paradigm

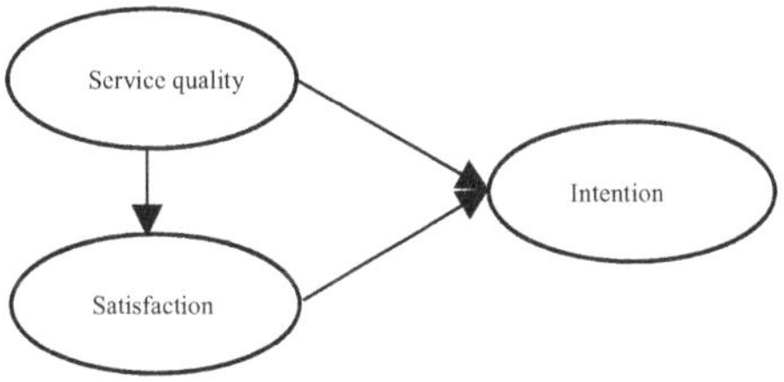

The total sample population consists of 200 individuals and is distributed as follows: Generation Y and Generation Z—100 each, male—100, female—100, and 5 area groups—40 each. According to Hair et al. [28], the size of unknown populations can be determined from the number of items multiplied by 5 (minimum) to 10 (maximum) to determine the number of samples needed. Thus, the number of samples determined still meets the requirements in the number of samples. This study uses structural equation modeling (SEM) on partial least squares with SMART PLS 3.0 as an analysis tool. Tests are carried out on the inner and outer models. The analysis of research results is discussed based on the results of hypothesis testing that have been determined in this study.

4 Results and Discussion

This study involves 200 participants consisting of 100 participants from Generation Y and Generation Z, each. In each generation category, the gender of participants (female and male) was evenly distributed. The regional representation of ALT users was also distributed evenly, using 40 participants from each area (North Jakarta, Central Jakarta, West Jakarta, East Jakarta, and South Jakarta). Thus, this research provides equal representation for each determined criterion.

By using structural equation modeling (SEM), the reliability and validity test are determined by the scores of composite reliability (CR), loading factor (LF), and average variance extracted (AVE). The expected results for CR and LF are ≥ 0.7 while AVE is ≥ 0.5. In the first test, several items do not meet the expected conditions. For the service quality variable, SQT2, SQT5, SQT8, and SQT10 are determined to be less than 0.7, so these items must be eliminated. In addition, for the intention variable, IUT3 must be deleted because the result is also below 0.7. The retesting process is carried out after these items are removed.

Table 1 shows that all the criteria—composite reliability (CR), loading factor (LF), and average Variance Extracted (AVE)—meet the conditions with the expected results. These overall results explain that the construct is reliable (CR > 0.7) and valid (AVE > 0.5, OL > 0.7) [29]. Thus, the subsequent testing and analysis process can continue.

To obtain the significance result (t-value), the bootstrapping process was carried out on the data. Table 2 shows that the significance value can be explained by the t-value obtained after testing. The expected t-value must be greater than 1.96. The t-value on service quality → intention is 4.501, service quality → satisfaction is 23.847, satisfaction → intention is 4.030, and the indirect effect of service quality → intention is 3.853. These results indicate that all paths show significance based on the t-values obtained.

The test results of each hypothesis in this study can be determined based on the results of Table 2. Service quality → intention of 4.501 (>1.96) with a p-value of 0.000 (<0.05) indicates that service quality has a significant effect on the intention of using ALT services. Therefore, **Hypothesis 1 is accepted**. These results are in line

Table 1 Reliability-validity test

Variable	Item	LF	CR	AVE
Service quality	SQT1	0.705	0.885	0.562
	SQT3	0.733		
	SQT4	0.756		
	SQT6	0.773		
	SQT7	0.773		
	SQT9	0.756		
Satisfaction	SFT1	0.888	0.904	0.758
	SFT2	0.847		
	SFT3	0.877		
Intention	IUT1	0.914	0.920	0.852
	IUT2	0.932		

Table 2 Hypothesis tests

Path	Original Sample	T-Value	P-value	Remark
Service quality → Intention	0.567	4.501	0.000	H1: accepted
Service quality → Satisfaction	0.749	23.847	0.000	H2: accepted
Satisfaction → Intention	0.322	4.030	0.000	H3: accepted
Mediation: Service quality → Intention	0.241	3.853	0.000	H4: accepted

with study conducted by Sumaedi et al. [28]. Service quality → satisfaction of 23.847 (>1.96) with a *p*-value of 0.000 (<0.05) shows that service quality has a significant effect on satisfaction in using ALT services; thus, **Hypothesis 2 is accepted**. This result is supported by the research of Mugion et al. [25].

Quality of services by ALT service providers—that is, to feel safe and comfortable in using both two-wheeled and four-wheeled vehicles—have been proven to have an impact on the desires of Generations Y and Z users in Jakarta. Therefore, the ability of a driver to operate a vehicle safely is a critical factor. The reliability of the ALT application system also factors into the service quality variable. Another aspect that has an impact on the intention to use ALT services for Generations Y and Z is time. The reasonableness of a driver's waiting time to pick up users at the designated point is an important factor. Furthermore, the suitability of the travel time is equally important. Drivers whose knowledge helps in determining the shortest route possible are preferred by users.

The results of satisfaction → intention of 4.030 (>1.96) with a *p*-value of 0.000 (<0.05) indicate that satisfaction has a significant effect on the intention of using ALT; therefore, **Hypothesis 3 is accepted**. These results support research conducted by Hussein and Hapsari [30]. The satisfaction obtained by users should form a desire to reuse services. For ALT in Jakarta, Generation Y and Z users are very dependent on the smoothness of the services provided. In this group of users, the desire to change brands/companies happens quite readily and change is carried out from a momentary or brief assessment. In this variable, the satisfaction aspect can be observed from the satisfaction in the driving service and the satisfaction in using the application.

From the driving service aspect, the factors in service quality that have been described previously are important in shaping the level of user satisfaction. Satisfaction with the user experience in getting a driver who drives at a reasonable speed so that the travel time is as expected is the key factor to be considered by drivers and ALT service providers. Comfort, sense of courtesy, and proper speech can be determinants of satisfaction for users of transportation mode services. Application system reliability is essential as applications that often experience errors can damage the level of user satisfaction, which in the long run can damage the desire to use the ALT service.

The results of this study also explain the indirect effect of service quality on the intention to use the mode of transportation mediated by satisfaction. The indirect effect of service quality → intention of 3.853 (>1,96) with a *p*-value of 0.000 (<0.05) shows that the intention to use ALT is significantly affected by service quality mediated by satisfaction in using ALT; therefore, **Hypothesis 4 is accepted**. These results are in line with research conducted by Ismail et al. [31] where usage satisfaction successfully mediates service quality towards usage intentions. In this study, the user intention variable focuses on the aspects of shaping the desire of users to continue using ALT services.

The results of this study indirectly provide a formula to form the intention of using ALT services, namely service quality coupled with user satisfaction. Together, these two factors can shape users' desire to reuse existing ALT services. Even in the longer term, it can shape users' perceptions to be more willing to use ALT than other types of land transportation in Jakarta. Looking at the original sample, this study succeeds in explaining that the intention to use ALT is influenced more by service quality (0.567) than satisfaction (0.322). These results further reinforce the concept that to form the intention to reuse ALT services, it is necessary to make efforts to continue to maintain and improve aspects of the quality of services of ALT to keep users satisfied.

5 Conclusion

In this study, service quality has been shown to have a significant effect on the intention of using ALT for Generations Y and Z users in Jakarta. To increase the sense of security for ALT users regarding driving speed, ALT service providers can implement a speedometer feature for their applications. The results of this study also indicate that service quality has a significant effect on satisfaction in using ALT for users of Generation Y and Z in Jakarta. The functionality or reliability of existing applications is highly vulnerable to the development of technology itself. This study also shows that the intention of using ALT is significantly affected by usage satisfaction. In order to attract and maintain fickle users by building deeper relationships with them, understanding aspects that can increase user satisfaction is vital. Finally, the results of this study indicate that service quality has a significant effect on intentions mediated by satisfaction in using ALT for Generation Y and Z users in Jakarta. It is undeniable that to form reuse intentions, it is necessary to achieve user satisfaction, and to shape user satisfaction requires quality services provided to users. The sample criteria applied in this study, especially generations Y and Z, have limitations. Involving participants from generation X, for example, can improve the research's comparison and analysis. Secondly, the limited area for adopting application-based land transportation, particularly Jakarta, must be expanded to other locations, whether or not they are big cities.

References

1. Valentine, D.B., Powers, T.L.: Generation Y values and lifestyle segments. J. Consum. Mark. **30**(7), 597–606 (2013)
2. Jackson, V., Stoel, L., Brantley, A.: Mall attributes and shopping value: differences by gender and generational cohort. J. Retail Consum. Serv. **18**(1), 1–9 (2011)
3. Ordun, G.: Millennial (Gen Y) consumer behavior, their shopping preferences and perceptual maps associated with brand loyalty. Can. Soc. Sci. **11**(4), 40–55 (2015)
4. Omar, F.I., Hassan, N.A., Sallehuddin, I.S., Abdullah, N.H., Gen, Y.: A study on social media use and outcomes. J. Manag. Muamalah [Internet] **6**(1), 53–64 (2016). Available from: https://jmm.kuis.edu.my/index.php/jurnal/article/view/98/80
5. Lee, N.R., Kotler, P.: Social Marketing: Changing Behaviors for Good. 5th ed. SAGE Publications, Inc. (2016)
6. Moore, M.: Interactive media usage among millennial consumers. J. Consum. Mark. **29**(6), 436–444 (2012)
7. Petra, K.K., Generation, Y.: Attitudes towards shopping: a comparison of the Czech Republic and Slovakia. J. Competitiveness **8**(1), 38–54 (2016)
8. Bednall, D.H., Valos, M., Adam, S., McLeod, C.: Getting Generation Y to attend: friends, interactivity and half-time entertainment. Sport Manag. Rev. **15**, 80–90 (2012)
9. Bassiouni, D.H., Hackley, C.: 'Generation Z' children's adaptation to digital consumer culture: a critical literature review. J. Customer Behav. **13**(2), 113–133 (2014)
10. Christian, M., Dewi, D., Rembulan, G.D., Indriyarti, E.R., Wibowo, S., Yuniarto, Y.: Business performance determinants of salted fish distribution in Kapuk during the COVID-19. J. Distrib. Sci. **19**(6), 29–39 (2021)
11. Septiani, R., Handayani, P.W., Azzahro, F.: Factors that affecting behavioral intention in online transportation service: case study of GO-JEK. Procedia Comput. Sci. **124**, 504–512 (2017)
12. Septiani, R., Handayani, P.W., Azzahro, F.: An empirical examination of factors affecting the behavioral intention to use online transportation service: case study of GRAB. In: 9th International Conference on Advanced Computer Science and Information Systems, ICACSIS 2017. Jakarta: Institute of Electrical and Electronics Engineers Inc., Jakarta, Indonesia (2018)
13. Sanny, L., Larasathy, K., Claudia, R., Widarma, B.: The customer satisfaction of online transportation in Indonesia. J. Phys. Conf. Ser. **1175**, 1–7 (2019)
14. Sutia, S., Adha, S., Fahlevi, M.: Why do customers intend to repurchase transportation online in Indonesia? In: The 4th International Conference on Energy, Environment, Epidemiology and Information System (ICENIS 2019). Semarang: E3S Web of Conferences, pp. 1–5 (2019)
15. Wang, Y., Wang, S., Wang, J., Wei, J., Wang, C.: An empirical study of consumers' intention to use ride-sharing services: using an extended technology acceptance model. Transportation (AMST) (2018)

16. Widianto, Y.F., Haryadi, H., Indyastuti, D.L.: Technology acceptance model on Gojek applications in Purwokerto. J. Akuntansi Manaj dan Ekon. **22**(1), 12–20 (2020)
17. Christian, M., Pardede, R., Indriyarti, E.R.: Generation Z in Jakarta's attitude towards COVID-19 ad distribution on YouTube. J. Distrib. Sci. **20**(3), 13–22 (2022)
18. Sambara, K.: Quality of Service and customer satisfaction in grab transportation service business. J. Ekon. Paradig. [Internet] **21**(1), 29–39 (2019). Available from: https://journal.uniba.ac.id/index.php/PRM/article/view/154
19. Santoso, A., Aprianingsih, A.: The influence of service and e-service quality to repurchase intention the mediating role of customer satisfaction. Case study: online transportation by Go-Jek Indonesia. J. Bus. Manag. [Internet] **6**(1), 32–43 (2017). Available from: https://journal.sbm.itb.ac.id/index.php/jbm/article/view/2138/1112
20. Christian, M., Haris, K., Indriyarti, E.R., Wibowo, S., Sunarno, S.: Service distribution strategy on business performance of Padang restaurants in North Jakarta. J. Distrib. Sci. **19**(12), 57–69 (2021)
21. Fauzi, A.A.: Electronic service quality on mobile application of online transportation services. J. Manajemen Indonesia **18**(1), 13–27 (2018)
22. Su, D.N., Nguyen-Phuoc, D.Q., Johnson, L.W.: Effects of perceived safety, involvement and perceived service quality on loyalty intention among ride-sourcing passengers. Transportation (AMST) **48**, 369–393 (2021)
23. Marisse, K., Dewi, R.H.: The effect of system quality, information quality, and perceived playfulness on the use of transportation online application. Russ. J. Agric. Socio-Econ. Sci. **108**(12), 112–119 (2020)
24. Moreno, F.M., Lafuente, J.G., Carreón, F.Á., Moreno, S.M.: The characterization of the millennials and their buying behavior. Int. J. Mark. Stud. **9**(5), 135–144 (2017)
25. Mugion, R.G., Toni, M., Raharjo, H., Di, P.L., Sebathu, S.P.: Does the service quality of urban public transport enhance sustainable mobility? J. Cleaner Prod. **174**, 1566–1587 (2018)
26. Sumaedi, S., Bakti, I.G.M.Y., Yarmen, M.: The impact study of public transport passengers behavioral intentions: the roles of service quality, perceived sacrifice, perceived value, and satisfaction (case study: paratransit passengers in Jakarta, Indonesia). Int. J. Traffic Transp. Eng. **2**(1), 83–97 (2012)
27. Pavlina, P.: The factors influencing satisfaction with public city transport: a structural equation modelling approach. J. Competitiveness [Internet] **7**(4), 18–32 (2015). Available from: https://jmm.kuis.edu.my/index.php/jurnal/article/view/98/80
28. Hair, J.F., Anderson, R.E., Black, W.C.: Multivariate Data Analysis. 7th ed. Pearson Education Limited, Essex (2014)
29. Rahadianto, N.A., Maarif, S., Yuliati, L.N.: Analysis of intention to use TransJakarta bus. Independent J. Manag. Prod. **10**(1), 301 (2019)
30. Hussein, A.S., Hapsari, R.: How quality, value and satisfaction create passenger loyalty: an empirical study on Indonesia bus rapid transit passenger. Int. J. Account. Bus. Soc. [Internet] **22**(2), 95–115 (2014). Available from: http://sabilfeb.lecture.ub.ac.id/files/2014/07/262-311-1-SM.pdf
31. Ismail, A., Rose, I.R., Tudin, R., Dawi, N.M.: Relationship between service quality and behavioral intentions: the mediating effect of customer satisfaction. Etikonomi **16**(2), 125–144 (2017)

Ownership Structure and the Quality of Corporate Social Responsibilities Disclosures in Malaysian Companies

Abdul Wahid Bin Mohd Kassim, Raman Bin Noordin, Junainah Jaidi, and Sulaiman Bin Tahajuddin

Abstract This paper investigates the linked between Ownership structure and the quality of Corporate Social Responsibilities Disclosures (CSRD). Four variables namely managerial ownership, foreign ownership, government ownership and Malay ownership are investigated in Malaysian companies' annual reports. This study used content analysis and a dataset from Bursa Malaysia public listed companies. Results based on the full regression model with the four variables indicated that except for Managerial ownership, three of the independent variables government ownership, foreign ownership and Malay ownership showed association with quality of CSRD in the predicted directions. In addition, all these three variables government ownership, foreign ownership, and Malay ownership are highly significant at the 1% level. On the other hand, the variable managerial ownership, Malay ownership is found to be insignificant and showed association with quality of CSRD not in the predicted directions.

Keywords The quality CSRD · Ownership structure · Annual reports

1 Introduction

This research investigates the quality of Corporate Social Responsibilities Disclosures (CSRD) in the annual reports of the public listed companies (PLCs) in Bursa Malaysia. Conventionally, companies only inform their business activities in financial reports to their major stakeholders—the shareholders and investors with stressing on its overall financial health. Due to public pressure and raising awareness on environmental and social responsibility issues, companies began reporting on economic, social, and environmental issues [2, 9, 45, 49]. This new type of reporting is known as CSRD, CSR reporting, Corporate Social and Environmental Disclosure (CSED) or sustainability reporting practices [2, 9, 15, 49].

A. W. B. Mohd Kassim (✉) · R. B. Noordin · J. Jaidi · S. B. Tahajuddin
Faculty of Business, Economics and Accountancy, Universiti Malaysia Sabah, 88400 Kota Kinabalu, Sabah, Malaysia
e-mail: abdulwah@ums.edu.my

A. Hamdan et al. (eds.), *The Implementation of Smart Technologies for Business Success and Sustainability*, Studies in Systems, Decision and Control 216,
https://doi.org/10.1007/978-3-031-10212-7_61

This concept argued and highlighted whether a corporation shall be responsible to its stakeholders, such as shareholders, customers, employees, suppliers, and the society. The situation further exacerbated by worldwide corporation scandals. These scandals heightened public awareness globally including Malaysia that more stakeholders will suffer if the importance of CSR is not recognized [31, 36, 37, 50]. In short, assuring the public trust is the main theme or order of business of corporate players and leaders rebuilding [8]. Company's disclosure such as CSRD can be a vital means to realise this objective [8].

Therefore, as suggested by several authors (such as Rahman and Ismail [41]; She and Michelon [44]; Xu et al. [51]) further examination of the determinants of CSRD is still relevant and warranted as evidenced by previous studies. Studies done by several scholars such as Andrew et al. [10]; Ahmad and Sulaiman [3]; Elijido-Ten [17]; Ahmad et al. [4]; Thompson and Zakaria [48]; Othman et al. [40]; Haji [26]; Esa and Ghazali [19]; Fatima et al. [20]; Sundarasen et al. [47]; Anuforo et al. [11], found that the level of CSRD made by Malaysian PLCs are still considered low and need further improvement.

In addition, it was also recognised that previous studies gave over emphasis on investigating on the type and the extent of disclosures [13, 25]. Yet, there has been lack of study that considered the quality of CSRD practices [13].

As such, the objective of this study is to find out the link between ownership structure and with the quality CSRD in Malaysian. The remains of the paper is organised as follows. It begins with reviewing of the literature and hypothesis development. The paper continues with the methodology that utilised in this study. Next section will provide the findings and discussion of the study. The final section of this paper will provide the conclusion of this study.

2 Literature Review and Hypothesis Development

2.1 CSR and CSR Disclosure

Neu et al. [39] claimed that CSRD could be recognised as a symbolic impression or constructed image of itself that an organisation is expressing to the entire world. Gelb and Strawser [21] a positive linked between CSR and CSRD. Simply put, organisations that involve in CSR activities offer further helpful and/or comprehensive disclosures than do other organisations that participate in fewer CSR activities. The reasoning behind is that, when an organisation is carrying socially responsible activities, it has the motivations to offer comprehensive and useful disclosure on the activities to make an impact on their stakeholders.

2.2 Managerial Ownership

Coffey and Wang [16], Leung and Horwitz [35], and Nasir and Abdullah [38] found that managerial ownership has a significant and positive relationship with voluntary disclosure. Quite the opposite, there are past studies that found either negative relationship [33, 34] or insignificant [5, 32]. Eng and Mak [18] alternatively found that increased voluntary disclosures are positively linked to lower managerial shareholding [14]. Thus, it is hypothesised that:

H1: There is a positive relationship between managerial ownership and the quality of CSRD.

2.3 Government Ownership

Most researchers such as Albawwat and Basah [7], Alhazaimeh et al. [8], Al Amosh and Khatib [5], Eng and Mak [18], Khan et al. [33], Lagasio and Cucari [34], Nasir and Abdullah [38], Rudyanto [42], and Said et al. [43] found that government shareholding was related with the amount and level of voluntary disclosures. They also found that there is positive linked between government ownership on disclosure practices. Thus, it is predicted that government ownership will promote more CSRDs because of increase transparency among PLCs in Malaysia. Therefore, it is hypothesised that:

H2: There is a positive relationship between government ownership and the quality of CSRD.

2.4 Foreign Ownership

According to Al Amosh and Mansor [6] the participation of foreign shareholders is important because they promoted transparency and trust between firms and stakeholders. Researchers such as Alhazaimeh et al. [8], Al Amosh and Khatib [5], Grassa et al. [23], and Guo and Zheng [24], also believed that foreign ownership has a positive consequence on the extent of disclosure and disclosure practices of firms, Thus, Said et al. [43] believe that the participation of foreign ownerships in Malaysian PLCs will improve the level of CSRD in Malaysia. This is confirmed by Haniffa and Cooke [28]. They found a significant linked between foreign ownership and CSRD in Malaysia PLCs. Malaysian PLCs use CSRD as an affirmative legitimating strategy satisfy ethical investors and to encourage continuous FDI [28]. Thus, it is hypothesised that:

H3: There is a positive relationship between foreign ownership and the quality of CSRD.

2.5 Malay Ownership

The disclosure practice of a company will be greatly influenced by the race of the majority owners the company. If companies are dominated by Malay shareholders, they also will implement several strategies for legitimacy purposes, either to divert or to attract attention to issues that are deemed important to them. Malays are Muslims and their ethical values are influenced by the Islamic principles. Thus, issues related to Islam, such as the companies' business operations are permitted by Islam, i.e., Halal or praying halls are provided for the companies' employees are among Malay shareholders' agendas [28]. As such, transparency is part of the Islamic business ethics and as such, the majority Malays being Muslims are anticipated to be more transparent in their disclosure practice [27]. Thus, it can be hypothesised that:

H4: There is a positive relationship between Malay ownership and the quality of CSRD.

3 Research Methodology and Sample Selection

This study is carried out among the Malaysian PLCs. The data for this paper consists of 347 firms listed in Bursa Malaysia and was mainly secondary in nature. In line with prior studies on CSR disclosures, a content analysis method is utilised (for instance, Abbott and Monsen [1]; Belal [12]; Imam [30]).

3.1 The Dependent variable–the Quality of CSRD

Several methods are available to measure the volume of disclosure in firms' annual reporting. One of the most common methods is counting words, sentences, paragraphs, pictures, and pages.

However, as mentioned before those previous studies over emphasis on investigating the type and extent of disclosures [13, 25] only. Yet, there has been lack of study that considered the quality of CSRD practices [13]. Therefore, in this study, the authors will be using the disclosure index approach. The quality of reporting is assessed on the disclosure index approach. The disclosure index was derived from a checklist developed by Ghazali [22].

Table 1 The independent variables and measurement

No	Variable	Measurement
1	Managerial ownership	The proportion of executive directors to total shareholders
2	Government ownership	Proportion of government shareholders to total shareholders
3	Foreign ownership	Proportion of foreign shareholders to total shareholders
4	Malay ownership	The proportion of Malay shareholders to total shareholders

3.2 The Independent Variables

The independent variables and their measurement are shown in Table 1.

4 Results and Discussion

Table 2 shows multiple regression results for the samples of 347 firms. The model reported an F-test value (F-test = 65.169, $p < 0.001$) which confirms the significance and sufficiently robustness of the model. The R-squared coefficient suggested that 43.3% of the variation in dependent variable is explained by the variation in independent variables. Thus, this model has reasonable explanatory power. The VIF values that ranged between 1.077 and 2.330 which do not exceed the value of 10.000. This indicates that multicollinearity is not an issue in explaining the regression results.

All the independent variables Malay ownership [MALOWN (+)], government ownership [GOVOWN (+)], and foreign ownership [FOROWN (+)] showed association with quality of CSRD in the predicted directions. All the three variables Malay ownership (+), government ownership [GOVOWN (+)], and foreign ownership [FOROWN (+)] are highly significant at the 1% level. While managerial ownership

Table 2 Multiple Regression (N = 347)

Variable and predicted sign	Regression	VIF	t-value	*p*-value
MANOWN (+)	0.131	1.259	−1.436	0.076
GOVOWN (+)	0.01	2.282	3.162	0.001
FOROWN (+)	0.14	1.077	4.996	0.000
MALOWN (+)	0.161	2.330	7.599	0.000
Summary Statistics:				
Intercept	0.283			
R^2	0.433			
F-Test	65.169***			

*, **, *** = Significant at the 10%, 5% and 1% level respectively using a two tailed test

[MANOWN (+)] is found to be not statistically significant and showed association with quality of CSRD not in the predicted direction.

It was found three of the four investigated variables to be significant and positively related to the quality of CSRD. Government ownership [GOVOWN (+)] also found to be significant and positively linked to the quality of CSRD (t-statistics = −3.162; *p*-value = 0.001). This result does support H2, which claims that the quality of CSRD will be improved the higher the number of government ownerships. This finding confirming the results of past studies by Eng and Mak [18], Haniffa and Cooke [28] and Juhmani [32].

5 Conclusion

This paper findings underline the distinction of between various types of shareholdings is vital in establishing the influence of ownership structure on CSRD. The insignificance of managerial ownership indicate high proportion of managerial shareholding does not influence CSRD. However, the negative relationship between managerial ownership with CSRD indicate that their linked with CSRD are in opposite directions as indicted by prior studies (e.g., Eng and Mak [18], Huafang and Jianguo [29]) trend of CSRD. Secondly, the study only covers on the quality of CSRD. In future study, it would be beneficial to examine the quantity and quality of CSRD together. Thirdly, the suggestions by Sisaye [46] could be considered to enrich and better understanding of CSR and CSRD by considering the elements of ecosystems, environments, natural resources, demography, population, culture, political systems, and history.

References

1. Abbott, W.F., Monsen, R.J.: On the measurement of corporate social responsibility: self-report disclosure as a method of measuring social involvement. Acad. Manag. J. **22**, 501–515 (1979)
2. Abernathy, J., Stefaniak, C., Wilkins, A., Olson, J.: Literature review and research opportunities on credibility of corporate social responsibility reporting. Am. J. Bus. **32**(1), 24–41 (2017)
3. Ahmad, N.N., Sulaiman, M.: Environmental disclosures in Malaysian annual reports: a legitimacy theory perspective. Int. J. Commer. Manag. **14**(1), 44–58 (2004)
4. Ahmad, N.N., Sulaiman, M., Siswantoro, D.: CSR disclosure in Malaysia: an analysis of annual reports of KLSE listed companies. IIUM J. Econ. Manage. **11**(1), 1–37 (2003)
5. Al Amosh, H., Khatib, S.F.A.: Ownership structure and environmental, social and governance performance disclosure: the moderating role of the board independence. J. Bus. Socioecon. Dev. 2635–1374 (2021). https://doi.org/10.1108/JBSED-07-2021-0094
6. Al Amosh, H., Mansor, N.: Disclosure of integrated reporting elements by industrial companies: evidence from Jordan. J. Manage. Gov. **25**(1), 121–145 (2021)
7. Albawwat, A., Basah, M.: Corporate governance and voluntary disclosure of interim financial reporting in Jordan. J. Public Adm. Gov. **5**(2), 100–127 (2015)

8. Alhazaimeh, A., Palaniappan, R., Almsafir, M.: The impact of corporate governance and ownership structure on voluntary disclosure in annual reports among listed Jordanian companies. Procedia Soc. Behav. Sci. **129**, 341–348 (2014)
9. Ali, W., Rizwan, M.: Factors influencing corporate social and environmental disclosure (CSED) practices in the developing countries: an institutional theoretical perspective. Int. J. Asian Soc. Sci. **3**(3), 590–609 (2013)
10. Andrew, B.H., Gul, F.A., Guthrie, J.E., Teoh, H.Y.: A note on corporate social disclosure practices in developing countries: the case of Malaysia and Singapore. Br. Acc. Rev. **21**, 371–376 (1989)
11. Anuforo, P.U., Aripin, N., Mohamed, M.I.: The extent of corporate social responsibility reporting among Malaysian firms: evidence from past studies. Int. J. Manage. Res. Rev. **8**(5), 8–23 (2018)
12. Belal, A.R.: A study of corporate social disclosures in Bangladesh. Manag. Audit. J. **1615**, 274–289 (2001)
13. Buniamin, S.: The quantity and quality of environmental reporting in annual report of public listed companies in Malaysia. Issues Soc. Environ. Acc. **4**(2), 115–135 (2010)
14. Buniamin, S., Alrazi, B., Johari, N.H., Rahman, N.R.: An investigation of the association between corporate governance and environmental reporting in Malaysia. Asian J. Bus. Acc. **1**(2), 65–88 (2008)
15. Cerin, P.: Communication in corporate environmental reports. Corp. Soc. Responsib. Environ. Manag. **9**, 46–66 (2002)
16. Coffey, B.S., Wang, J.: Board diversity and managerial control as predictors of corporate social performance. J. Bus. Ethics **17**(14), 1595–1603 (1998)
17. Elijido-Ten, E.: Determinants of environmental disclosure in a developing country: an application of stakeholder theory. In: Asia Pacific Interdisciplinary Research in Accounting (APIRA) (2004)
18. Eng, L.L., Mak, Y.T.: Corporate governance and voluntary disclosure. J. Account. Public Policy **22**, 325–345 (2003)
19. Esa, E., Ghazali, N.A.M.: Corporate social responsibility and corporate governance in Malaysian government-linked companies. Corp. Gov. **12**(3), 292–305 (2012)
20. Fatima, A.H., Abdullah, N., Sulaiman, M.: Environmental disclosure quality: examining the impact of the stock exchange of Malaysia's listing requirements. Soc. Responsib. J. **11**(4), 904–922 (2015)
21. Gelb, D.S., Strawser, J.A.: Corporate social responsibility and financial disclosures: an alternative explanation for increased disclosure. J. Bus. Ethics **33**, 1–13 (2001)
22. Ghazali, N.A.: Ownership structure and corporate social responsibility disclosure: some Malaysian evidence. Corp. Gov. Int J. Bus. Soc. **7**(3), 251–266 (2007)
23. Grassa, R., Moumen, N., Hussainey, K.: Do ownership structures affect risk disclosure in Islamic banks? International evidence. J. Financ. Reporting Acc. **19**(3), 369–391 (2020)
24. Guo, M., Zheng, C.: Foreign ownership and corporate social responsibility: evidence from China. Sustainability **13**(508), 1–22 (2021)
25. Guthrie, J., Cuganesan, S., Ward, L.: Industry specific social and environmental reporting: the Australian food and beverage industry. Acc. Forum **32**, 1–15 (2008)
26. Haji, A.A.: Corporate social responsibility disclosure over time: evidence from Malaysia. Manag. Audit. J. **28**(7), 647–676 (2013)
27. Haniffa, R.M., Cooke, T.E.: Culture, corporate governance and disclosure in Malaysian corporations. Abacus **38**(3), 317–349 (2002)
28. Haniffa, R.M., Cooke, T.E.: The impact of culture and governance on corporate social reporting culture, corporate governance and disclosure in Malaysian corporations. J. Account. Public Policy **24**(5), 391–430 (2005)
29. Huafang, X., Jianguo, Y.: Ownership structure, board composition and corporate voluntary disclosure: evidence from listed companies in China. Manag. Auditing J. **22**(6), 604–619 (2007)
30. Imam, S.: Corporate social performance reporting in Bangladesh. Manag. Auditing J. **15**(3), 133–41 (2000)

31. Jaturat, M., Dampitakse, K., Kuntonbutr, C.: The effect of corporate governance on the board of directors' characteristics and sustainability disclosure: an empirical study from Thailand. J. Asian Financ. Econ. Bus. **8**(12), 0191–0201 (2021)
32. Juhmani, O.: Ownership structure and corporate voluntary disclosure: evidence from Bahrain. Int. J. Acc. Financ. Reporting **3**(2), 133–148 (2013)
33. Khan, A., Muttakin, M.B., Siddiqui, J.: Corporate governance and corporate social responsibility disclosures: evidence from an emerging economy. J. Bus. Ethics **114**(2), 207–223 (2013)
34. Lagasio, V., Cucari, N.: Corporate governance and environmental social governance disclosure: a meta-analytical review. Corp. Soc. Responsib. Environ. Manag. **26**(4), 701–711 (2019)
35. Leung, S., Horwitz, B.: Director ownership and voluntary segment disclosure: Hong Kong evidence. J. Int. Financ. Manag. Acc. **15**(3), 235–260 (2004)
36. Majdi, S., Rahman, R.A.: Labour market consequences after fraud and lawsuit revelation. Int. Res. J. Financ. Econ. **63**, 189–198 (2011)
37. Melliniawati, F., Prasetyo, A.B.: The analysis of the effect of corporate governance on corporate risk disclosure (Empirical study of manufacturing company listed in the Indonesia stock exchange in the period of 2018–2019). Int. J. Arts Soc. Sci. **4**(5), 290–291 (2021)
38. Nasir, N.M., Abdullah, S.N.: Voluntary disclosure and corporate governance among distressed firms in Malaysia. Financ. Reporting Regul. Gov. **3**(1) (2004). Available from http://www.cbs.curtin.edu.au/files/nasir-abdullah.pdf
39. Neu, D., Wasame, H., Pedwell, K.: Managing public impressions: environmental disclosures in annual reports. Acc. Organ. Soc. **23**(3), 265–289 (1998)
40. Othman, S., Darus, F., Arshad, R.: The influence of coercive isomorphism on corporate social responsibility reporting and reputation. Soc. Responsib. J. **7**(1), 119–135 (2011)
41. Rahman, I.M.A., Ismail, K.N.I.K.: Do representation of women directors and their demographic characteristics affect CSR disclosure? Evidence from Malaysia. Bus. Manage. Econ. Res. **2**(4), 81–89 (2016)
42. Rudyanto, A.: State ownership, family ownership, and sustainability report quality: the moderating role of board effectiveness. Acc. Financ. Rev. (AFR) **2**(2), 15–25 (2017)
43. Said, R., Zainuddin, Y., Haron, H.: The relationship between corporate social responsibility disclosure and corporate governance characteristics in Malaysian public listed companies. Soc. Responsib. J. **5**(2), 212–226 (2009)
44. She, C., Michelon, G.: Managing stakeholder perceptions: organized hypocrisy in CSR disclosures on Facebook. Crit. Perspect. Account. **61**, 54–76 (2019)
45. Shinde, J.S., Wacker, R., Shinde, U.S., Hou, Z.J.: Superfluity and murkiness in the understanding of social responsibility: a content analysis. J. Leadersh. Account. Ethics **8**(3), 30–42 (2011)
46. Sisaye, S.: The organizational ecological resource framework of sustainability reporting: implications for corporate social reporting (CSR). J. Bus. Socioecon. Dev. (2021). https://doi.org/10.1108/JBSED-05-2021
47. Sundarasen, S.D.D., Tan, J.Y., Rajangam, N.: Board composition and corporate social responsibility in an emerging market. Corp. Gov. Int. J. Bus. Soc. **16**(1), 35–53 (2016)
48. Thompson, P., Zakaria, Z.: Corporate social reporting in Malaysia. J. Corp. Citizsh. **13**, 125–136 (2004)
49. Tschopp, D., Huefner, R.: Comparing the evolution of CSR reporting to that of financial reporting. J. Bus. Ethics **127**(3), 565–577 (2015)
50. Uwuigbe, U., Uwuigbe, O., Ajayi, A.O.: Corporate social responsibility disclosures by environmentally visible corporations: a study of selected firms in Nigeria. Eur. J. Bus. Manage. **3**(9), 9–17 (2011)
51. Xu, S., Chen, X., Li, A., Xia, X.: Disclosure for whom? Government involvement, CSR disclosure and firm value. Emerg. Markets Rev. **44**, 100717 (2020)

Are Intrapreneurs the Right Innovation Champions in Organisations?

Nomkhosi Radebe and Helen Duh

Abstract Intrapreneurs are autonomous strategic actors within organisations who depart from customary ways of doing things, exploit opportunities and leverage resources to achieve extraordinary goals. Despite being flexible and daring to take on new business ventures for organisational renewal, insights into intrapreneurs' roles as champions of innovation within their organisations are not fully understood. This study used qualitative methods (interviews and systematic observations) and collected data from 43 mining sector intrapreneurs to gather insights into the roles of intrapreneurs in the adoption of innovation in organisations in Africa. Data analysed using the Atlas ti software revealed that intrapreneurs are critical to any organisation in fostering innovation. Furthermore, intrapreneurs exist at the different hierarchical levels of the organisation as opposed to being situated in one reporting level.

Keywords Intrapreneurship · Innovation

1 Introduction

The importance of innovation has never been more critical than in the current fourth industrial revolution (4IR) era we live in. Today, most organizational tasks are automated and robotics point towards artificial intelligence as the future. These need innovative minds in organizations. Intrapreneurs in organisations are seen as custodians and the cornerstones for organizational innovations by bringing about renewal strategises to the business [18]. Pellegrini et al. [19] view intrapreneurs as autonomous strategic actors within organisations who depart from customary ways of doing things, exploit opportunities and leverage resources to achieve extraordinary goals. They recommend the investigation of intrapreneurs' roles in sustainable innovation within organisations. This is particularly important in Africa which is projected to drive the next phase of global economic growth. In an African organizational setting therefore, this study investigated whether intrapreneurs are the ideal

N. Radebe (✉) · H. Duh
Witwatersrand University, 1 Jan Smuts Ave, Braamfontein, Johannesburg 2000, South Africa
e-mail: Khosi.radebe@wits.ac.za

A. Hamdan et al. (eds.), *The Implementation of Smart Technologies for Business Success and Sustainability*, Studies in Systems, Decision and Control 216,
https://doi.org/10.1007/978-3-031-10212-7_62

innovation champions. It asks the charateristics, types and roles of intrapreneurs in fostering innovations within organisations. Already, and characterising intrapreneurs as being autonomous, innovative, risk taker, proactive, and competitively aggressive, and considering their ability to apply and implement renewal strategies in organisations, Pellegrini et al. [19] posit that intrapreneurs are key catalysts or champions of innovations within organisations. Ganzer et al. [10] also contend that intrapreneurs bring about novelty in organizations. With the notion that the characteristics and roles of entrepreneurs are contingent on the industry, Marques et al. [16] conducted a study in the health industry and found that its intrapreneurs produce innovative outputs because they have rational cognitive leadership styles. In the mining sector which is capital intensive, face frequent market fluctuations and are urgently required to use green exploration methods and be environmentally friendly [15], it is questionable as to the characteristics, types and roles intrapreneurs play?

This study starts by refining these questions, it then sets the objectives, after which literature reviewed is reported, followed by the research methodology, results, discussions and conclusions.

2 Research Questions and Objectives

Four research questions are to be answered in this study:

Research question 1: Which innovation types exists in the organisation, especially a mining organization?
Research question 2: What are the impediments to innovation in the organisation?
Research question 3: Aligned to innovation, who are intrapreneurs, what are their characteristics and roles?
Research question 4: In a mining sector, are intrapreneurs innovation champions as literature suggest?

Research objectives

The current study aimed to investigate if intraprneurs are best placed to drive innovation.

To identify to innovation types that exist in the organisation.

To recognise the barriers to innovation.

To establish the understanding of what an intrapreneur is and the traits they have inline wit innovation.

3 Literature Review

3.1 What is Intrapreneurship?

Intrapreneurship can be defined as "a process whereby employee(s) recognize and exploit opportunities by being innovative, proactive and by taking risks, in order for the organization to create new products, processes and services, initiate self-renewal or venture new businesses to enhance competitiveness and performance of the organization" [18, p. 551]. Seshadri and Tripathy [7] provide defininition of intrapreneurship as a process upon which an individual or people within an organisation create a new organisation or initiate renewal or innovate within same organisation. According to Al-Obadi et al. [1], intrapreneurship in a large corporation are propellers promoting research and development, improving products, services and processes for immediate economic and environmental benefits. Using intrapreneurship for innovations is also less costly because inherent expertise is used [1]. There is consistency across board on what intrareneurship is. One of the founders of the concept of entrepreneurship who extended the understanding of entrepreneurship to intrapreneurship through the incorporation of innovation was Joseph Schumpeter, the Austrian economist. He viewed an entrepreneur as someone who "carries out **new combination**," he further mentions that "the essence of Entrepreneurship lies in the perception and exploitation of **new opportunities** in the realm of business". Hartman [12]. The notion of "**exploiting new opportunities**" resonated with the meaning of innovation, where innovation is "the action of innovating and creating processes which promoted disruption on the economic system while allowing the emergency of novelty" [10: 322]. Foss et al. [9], in Schumpeter's opinion, view entrepreneurs only when they were carrying out **new combinations**, not when they are running already established businesses. The authors further emphasised that entrepreneurship only occurred when **new products, processes or strategies got introduced**.

3.2 Characteristics of Intrapreneurs/Corporate Entrepreneurs that Make Innovation Possible

Intrapreneurs are employees of an organisation who have eyes on opportunities, make risky decisions and use company resources for implementation [5]. They however are guided by pre-existing rules and regulations Mariluz and Sousa-Ginel [4]. Cadar et al. (2015) indicated that an intrapreneur's activity has a restoring character. Like entrepreneurs, intrapreneurs are innovative individuals. They both believe in adding value to products and services as well as formulate new ways and proposals. They undertake risks in different degrees as asserted by Moriano et al. [17]. The entrepreneur takes a risk with his own money whilst the intrapreneur risks the company's money allocated to new products and processes [3] (Cadar et al. 2015). Moriano and Levy Mangin (2014) assert that intrapreneurs should be innovative, pro-active in their approach at all times. Whether these characteristics are found in intrapreneurs in the mining companies of South Africa, needs investigation.

3.3 The Concept of Innovation

Ganzer et al. [10: 322] describe innovation as "the action of innovating and creating processes which promote disruption on the economic system while allowing the emergency of novelty". Razavi et al. [20] establish that one cannot speak about competition nor growth without innovation, asserting that innovation is key for organisational success. Capitanio et al. (2009) indicate how innovation was paramount to companies as it helped firms to challenge competition both from national and international companies through attaining competitive advantage. In the context of international organisations such United Nations, Ambos and Tatarinov [2: 92] define innovation as "developing new products or processes to serve the organisations' beneficiaries and to address grand challenges more effectively, such as developing new ways to grow food in slum areas, giving refugees access to high-quality education, or predicting migration patterns". In the mining industry, which is the context of this study, innovation takes the form of clean sources of energy supply, discovery of new and finite deposits of minerals and metals and better and faster ways of mining agreements [15]. In the health sector, Reinius et al. [21] define innovation in the sense of "identifying new or improved health policies, systems, products and technologies, or services and delivery methods that improve people's health and well-being". They assert that the innovation should be aimed at adding value in terms of improved efficiency, quality, effectiveness, safety, sustainability, and/or affordability. Some innovations are for benefit of customers, environment, social good and for the organisation itself (e.g., cost effectiveness, efficiency and more competitive products). According to Reinius et al. [21], innovations are customer driven or to satisfy unmet needs of the innovators. While various industries see innovation in different lights, there are some general categorizations of innovation.

3.4 Benefits of Innovation in Corporate Organisation

Innovation has a number of benefits as alluded by D'Attoma and Ieva [6]. These include environmental benefits as the world moves towards a greener domain with minimal negative impact to the environment particularly within the mining industry. The authors specifically look at marketing innovation as the driver for environmental benefits both internally (production process) and externally (consumption stage). Hernandez-Vivanco et al. [13] acknowledge the urgent need to accelerate the implementation of innovation through the integration of management systems and open innovation to achieve the maximum innovation benefits which lead to increased sales productivity. Furthermore, the authors mention how investing in innovation positively influences the productivity and commercialisation of products.

3.5 Innovation Barriers

Despite the benefits of innovation, organisations are often unware or ignore the barriers to innovation that could potentially affect innovation in an organisation. Gluszczuk [11] elaborates on the factors contributing to restrictive innovation for a product, process, organisational and marketing innovations. They include but not limited to cost factors, knowledge factors, market factors, institutional factors, cognitive barriers, structural barriers, and other reasons. *Cost factors* include high expenses, lack of finance, and riskiness. *Knowledge factors* entail lack of qualified personnel and management attitude. *Market factors* involve uncertain demand for innovative goods or services. *Institutional factors* cover aspects such as lack of infrastructure, legislation, and regulations. *Cognitive barrier* is when managers who use a specific business model find it hard to understand new business models. *Other reasons* include factors such as no need to innovate because of lack of demand. In the mining sector, risk-averse mindset, frequent market fluctuations, capital intensiveness, and bureaucratic structures are some of the barriers that prevent innovation [15]. The mitigation of some of these barriers may emerge from the presence of intrapreneurs in organisations. For example, while [14, 15] suggest that managers should create a conducive environment and good organisational culture for innovation to thrive, Pellegrini et al. [19] the presence of intrapreneurs in organisations as one key solution.

Addressing the case study followed a qualitative research method with an exploratory research design to assist in unpacking the role and position of intrapreneurs within an organisation that promotes innovation. The collection of data was through interviews and systematic observations. A total forty-three (43) interviews were conducted onsite. Data validation was achieved through triangulation—**data triangulation** (participants from the different managerial levels) and **methodological triangulation** ((interviews and observations), Farquhar et al. [8].

3.6 Data Collection Method and Sample

The sampled number of participants for the study was 50 participants. There were 43 out of 50 interviews that were successfully conducted and subsequently analysed. Therefore, the response rate was 86% of the study sample. According to Zvonkovic et al. [22] this sample size is high enough to produce valid data and ultimately credible research results. The characteristics of the sample were representatives from both the technical and support departments. The interviewees consisted of the executive committee, senior managers, middle managers and supervisory. The researcher could not interview the lower level as they were unionised and were thus rebuked from participating. However, there was a 50% representation from the executive level inclusive of the General Manager who is at the apex of the mine. There was a 47% representation of senior managers both from the core and support departments. They constituted a bigger portion of the target population. There were about 35% of middle managers and only 2% from lower management as the unions refuted access to that employee band.

4 Results

The subsection provides a presentation of the qualitative data collected on the field. The results are presented in the same sequence and logic as the research questions that were asked to the participants to maintain the same flow of understanding.

4.1 The Reasons Provided for Innovation in the Mining Sector

Five respondents suggested that innovation involves the generation of new solutions to improve operations to mitigate both internal and external challenges. Interviewee P19 pointed out that: "*...innovation is vital as a way of* ***developing solutions*** *to deal with both internal and external challenges.*" This view provided a link between innovation and the formulation of strategies in the organisation. Another IT analyst in middle management indicated that: "G*enerally,* ***employees come up with new ideas and provide the business case*** *and submit requests for capital approval to take the ideas further for implementation.*" [P26].

4.2 Types of Innovation

The interview data that was analysed showed that there were two types of innovation that were introduced in the different operations in the mining company. The two types of innovation supported by the data from the current study were process innovation and technological innovation.

Technological innovation

Thirteen participants stated that acquired technologies were used across the different departments in the company to improve the quality of decisions and services, reduced costs and safety and security of the stakeholders. Interviewee 17, mentioned different technological acquisitions that include: "*An integrated software known as GEMS used* ***for modelling*** *, and we're moving into 3D-experience, which has analytics capability.*" P27 [Mining Engineer] shared a similar view that: "*…we have introduced new* ***software package*** *that is used to perform 3D modelling of our plans*".

Process innovation

The reconfiguration of production processes was targeted at lowering costs, increasing output and improving safety. The ore processing management included the improvement of the ore resource management. Six participants suggested how they improved plant processing, mining, and fieldwork processes. Data from a Training Manager at middle management confirmed that: "*There are a lot of* ***process improvements at the process plant*** *in terms of addressing issues of driving technology advancement for the diamond recovery plant.*"

4.3 Existing Barriers to Innovation

The implementation of innovation is not without impediments. The following are the barriers to innovation that were uncovered at data collection stage.

4.3.1 Resistance to Change

Fourteen interviewees shared the same sentiments about being resistant to change. A middle manager in HR postulated that "*If you bring change to an environment the change is a risk, because you would need* ***buy-in from the business leadership*** *and the people you are serving because people do not want change. R* ***esistance to change*** *needs to be managed properly because it is a reality*".

4.3.2 Increased Risk Exposure

The participants noted that employee safety was compromised because with a new system comes uncertainty, cyber-crime may increase as initially there would be lack of enough knowledge or research on how to mitigate any unwarranted attacks and security breach like in the case of flying in a drone to unauthorized areas. The introduction of new IT would be exposed to the cyber risks that include malware, application attacks and hackers.

4.4 Understanding the Concept of an Intrapreneur

The interviewees described the term intrapreneur using different attributes associated with the person. Six participants associated intrapreneurship with senior managers who were responsible for identifying business opportunities and making key decisions that make the organisation profitable. One Mining Manager alluded that "*An intrapreneur is someone very innovative to ensure that the organisation that he/she is working for operates* ***profitable and sustainable*** ." [P22] *The* interviewees also confirmed that managers operated their departments like mini business areas (MBA) in a sustainable manner like how the owners (entrepreneur) would push the corporate agenda. One Business Improvement Manager corroborated that: "*I think being an intrapreneur,* ***fits the mini business area (MBA)*** *philosophy; this is your business now. It involves sustaining the MBA as my own business because* ***the owner outside*** *would do the same.*" [P40] From the field data, an intrapreneur was also described as person who contributes innovative ideas and organises the factors to exploit the opportunities. A Medical Superintendent supported that: "*Intrapreneurship is when a person is enterprising and innovative* ***inside an organisation*** ." [P19]. This was similar to the results of Sharma and Chrisma (1999) which state that intrapreneurship is a process that involves internal people who create new organisations, or initiate renewal in the same organisation.

Camelo-Ordaz et al. [5] echo the view that an intrapreneur is an architect of innovation within an organisation. The typical examples of intrapreneurs were given as the **chief executive offic**er, **general manager**, **executive committee team**, **middle managers,** and **frontline managers**. A senior manager (ore processing engineer) said that: "*In corporate language by intrapreneurs we refer to the* ***CEO, GM and Exco team*** *, if they are not intrapreneurs, they will run down the company. I give you a cut-off up to the level of* ***foreman*** ." [P43] The view excluded operatives or ordinary employees from being intrapreneurs.

An IT senior manager associated intrapreneurship with executive management when she commented that: "*Intrapreneurs are c* ***orporate champions and trailblazers who*** *receive bonus schemes when the company performs well. They make* ***impactful decisions*** *that take the company to* ***new heights*** ." [P27] The data showed that an intrapreneur operates inside an organisation and receives performance incentives when the company meets the set targets. Similarly, an Asset Manager corroborated that the intrapreneur focuses on accomplishing the long-term goals of the organisation. He was quoted as follows: "*I think* ***intrapreneurs are committed to pushing the corporate agenda*** *and are seen by some employees as slave drivers.*" [P41].

Three other participants explained the concept of intrapreneur by focusing on the limitations faced by the intrapreneurs that include making cushioned decisions inside the organisation with limited decision rights. The issue of limited decision rights was highlighted by P19 who stated that: "*As an intrapreneur you might face risks, but you have a* ***limited decision rights*** ." The other constraint raised by some interviewees related to shortage of resources to pursue investment opportunities. This view is confirmed by a senior security manager who shared that: "*A creative employee can be* ***constrained by resources*** *such that the person has to rely on or report to someone who might not be interested in what you want to do, but the owner of a business* ***allocates resources independently*** ." [P37] The field data also showed that an intrapreneur may have limited capacity to exploit new business opportunities.

4.5 Intrapreneurship Spectrum

Interviewee P39 shared that: "*The renewal energy is* ***radiated by everyone*** *even the employee I give small tasks they do it in a manner that brings value to the business.*" Another Training Manager at senior management levels substantiated that: "*I think managers are just the drivers who can strategise and come up with tolls, coordinate and even educate the general workforce on innovation. I think innovation must be* ***spread out to everyone*** *in the organisation.*" [P16] A Senior Mining Manager also responded that: "*We are pushing the pillars of innovation and digital innovation down to the lower levels so that* ***each and every single employee is integrated in the strategy***." [P22] This notion is in line with the beliefs of (Marion and Levy 2014) as quoted on the literature review that at individual level, intrapreneurship must be a bottom-up setting, work related initiatives by the individuals within the organisation where any employee has the same drive as the entrepreneur. At the same time, the inclusion of everyone into thinking like an intrapreneur supported by the field data above contradicts who [22] categorises as an intrapreneur. According to the scholar, intrapreneurship befits certain managers (not all) who are then targeted or identified to be intrapreneurs.

4.6 The Characteristic of an Intrapreneur

Intrapreneurs within an organisation assume certain traits which allows them the status of being categorised as intrapreneurs. The following qualities were uncovered on the field:

One Senior Project Manager suggested that: "*Entrepreneurs are always* ***looking for opportunities*** *and are* ***visionaries and goal-getters*** . *They are creative* ***thinkers who are open-minded*** ."

The remainder of the respondents (middle managers and supervisors) possessed attributes like **self-driven, bold, open-minded, interpersonal skills** and **inquisitive**.

5 Discussion

5.1 Types of Innovation

Findings from the previous chapter uncovered two types of innovation from mining and supporting departments: **process innovation** and **technological innovation**. Some participants linked innovation to the corporate strategy where the corporate strategy plan promoted experimentation with new things. This linkage was further substantiated by the structure of the new corporate strategy, which embedded an innovation pillar known as the **Innovation & Digital Transformation** pillar. Some aspects of how innovation manifests itself were mentioned, for example, when business processes and systems were reconfigured to enhance the quality of work out.

5.2 Technological Innovation

From the findings, technological innovation referred to acquisitions of technology or software packages made to improve the quality of decisions and reduce costs, safety and security of stakeholders. Some of the acquired software that the participants mentioned included the use of software such as WAVE used for assessment and approval of innovative ideas. GEMS, on the other hand, was used for modelling, leapfrog, which was an improvement of GEMS, IT software for modernisation of application packages, industrial computer systems for controlling the environment from a central point, a procurement system was known as SAP Ariba procurement system to name just a few. In essence, the mine spent many funds purchasing systems and software that we describe as acquired technology.

5.3 Process Innovation

The operational processes were reconfigured to increase output, reduce costs, and improve safety. Some of the improvements included introducing a new process called Dynamic Mine Design, where information was collected to have flexible plans that changed as the mining process. Another operation process improvement was on the Deep Slope Mining (DSM). DSM is a process that involves mining across the bedding planes following the gradient of the planes. Traditionally, the company used to mine across the area of bedding planes of the pit, yet the new process fostered mining along the bedding planes. A senior manager alluded that the ***ISO 3100 risk management standard*** was implemented to revamp the operating model to align with the requirements of the risk management international standards.

5.4 Barriers to Innovation

The overall findings revealed that some of the barriers to innovation include resistance to change and increased risk exposure. These constitutes internal barriers of which the organisation has the capability to overcome. Some of the suggested ways to overcome these barriers are having a visionary and supportive leadership who understand the importance of innovation to the organisation. The introduction of change management was suggested as a mitigating factor to help overcome the resistance that employees had to change.

Table 1 Comparison of characteristics of an intrapreneurs from field research findings and literature review

Field research findings	Literature review
Risk-taker	Risk-taker/risk-conscious
Innovative	Innovative
Self-driven	Proactive
Interpersonal	
Open-minded	
Inquisitive	
Goal-getter	

5.5 Intrapreneurship

The findings revealed that there were varying views about the understanding of who an intrapreneur was however, what all participants agreed on was that an intrapreneur is someone who is innovative within the organiation. Furthermore, a critical outcome of the field study was when the participants from the three managerial levels stated that any employee in the organisation could be viewed as an intrapreneur. The appreciation of intrapreneurship was not restricted to a particular group of people since all employees had an equal chance of providing innovative ideas (distributed intrapreneurship). This consistency was in line with Martiarena (2013), who stated that intrapreneurs were employees of an organisation without explicitly mentioning their level.

5.6 Qualities/Chracteristics of an Intrapreneur

The qualities of an intrapreneur in the literature review compared to the field research findings are presented in Table 1.

6 Study Limitations

6.1 Meso-Organisational Level

Interview participants were called for interviews at their convenient time during working hours. The arrangement might have compromised the work schedules prepared during the field data collection period.

6.2 Micro-Individual Level

The interviews scheduling was a challenge for some participants during the second interview cycle as the focus was on winding up their work in preparation for the forthcoming festive season.

6.3 Macro-External Level

The case study was conducted in a mining company in Botswana. Therefore, the researcher had to schedule interviews from South Africa and then travel to Botswana to conduct the interviews. As a result, logistical challenges were straining and were overcome to stay ahead of interview schedules.

7 Implications for Future Research and Conclusion

The present research design opened research gaps for other similar research studies to use different research methodologies such as the quantitative approach and mixed methods to expand the knowledge about intrapreneurship and innovation. Furthermore, the research must be extended to other industries other than mining which is a product and technical industry. Service industries with soft skills must also be considered.

Even though there are pending limitations to this study, a tangible contribution has been made regarding the responsibility that intrapreneurs have in driving innovation within the organisation. The two (2) types of innovation that are in existence at the mine and implemented by the intrapreneurs are technological and process innovation. They are implemented through inclusive / distributed intrapreneurship. Common intrapreneurship traits that are found in the literature review and observed on the field are innovative and a risk-taker. The current research requires further investigation through other research methods (quantitative and mixed methods research) in a different industry or company setup.

References

1. Al-Obadi, M., Ayad, H., Pokharel, S., Ayari, M.A.: Perspectives on food waste management: prevention and social innovations. Sci. Dir. 190–208 (2022)
2. Ambos, T.C., Tatarinov, K.: Building responsible innovation in international organizations through intrapreneurship. J. Manage. Stud. 92–125 (2022)
3. Badulescu, C.: Entrepreneur, entrepreneurship and Intrapreneurship. A literature review. Econ. Sci. Series 658–664 (2015)

4. Camelo-Ordaz, C., Fernández-Alles, M., Ruiz-Navarro, J., & Sousa-Ginel, E. (2012). The intrapreneur and innovation in creative firms. *International Small Business Journal, 30*(5), 513–535
5. Camelo-Ordaz, C., Fernández-Alle., Ruiz-Navarro, J.: The intrapreneur and innovation in. Int Small Bus. J. 513–535 (2011)
6. D'Attoma, I., Ieva, M.: The role of marketing strategies in achieving the environmental benefits of innovation. 1–14 (2022)
7. D.V.R., Tripathy, A.: Innovation through Intrapreneurship: The road less travelled, Vikalpa: The Journal for decision makers, **31** (1), 17–29 (2006)
8. Farquhar, J., Michels, N., Robson, J.: Triangulation in industrial qualitative case study research: widening the. Ind. Mark. Manage. 160–170 (2020)
9. Foss, N., Klein, P.: Organizing entrepreneurial judgement: a new approach to the frm. Cambridge University Press (2012)
10. Ganzer, P.P., Chais, C., Olea, M.P.: Product, process, marketing and organizational innovation industries of the flat knitting sector. Revista de administracao e inovavacao 321–332 (2017)
11. Gluszczuk, D.: Barriers to innovation activities in industrial enterprises by Polish Regions in 2004–2006 and 2008–2010. Research Papers of the Wroclaw University of Economics, pp. 181–189 (2013)
12. Hartman, H.: Manager and entrepreneurs: a useful distinction? Adm. Sci. Quartely **3**(3), 429–451 (1959)
13. Hernandez-Vivanco, A., Cruz-Cazares, C., Bernardo, M.: Openness and management systems integration: pursuing. 76–90 (2018)
14. Ibrahim, S., Fallah, M.: Drivers of innovation and influence of technological clusters. Eng. Manage. J. 33–41 (2005)
15. Kashan, A.J., Wiewiora, A., Mohannak, K.: Unpacking organisational culture for innovation in Australian mining industry. P1–13(2021)
16. Marques, C., Lopes, C., Braga, V., et al.: Intuition and rationality in intrapreneurship and innovation outputs: the case of health professionals in primary health care. Int Entrep Manag J (2021). https://doi.org/10.1007/s11365-021-00761-7
17. Moriano, J.A., Molero, F., Topa, G., Mangin, J.L., The influence of transformational leadership and organizational identification on intrapreneurship. Int. Entrepreneurship Manage. J. **10**(1), 103–119 (2014)
18. Neessen, P. C., Caniels, M. C., Vos, B., de Jong, J. P.: The intrapreneurial employee: toward an integrated model. Int. Entrepreneurship Manage. J. 545–571 (2019)
19. Pellegrini, C., Nnunziata, E., Rizzi, F., Frey, M.: The role of networks and sustainable intrapreneurship as interactive drivers catalyzing the adoption of sustainable innovation. Corp. Soc. Responsib. Environ. Manage. 1026–1048 (2019)
20. Razavi, S., Nargesi, G., Hajihoseini, H., Akbari, M.: The impact of technological innovation capabilities on competitive performance of Iranian ICT firms. Iran. J. Manage. Stud. 855–882 (2016)
21. Reinius, M., Mazzocato, P., Riggare, S., Bylund, A., Jansson, H., Ovretveit, J., Hasson, H.: Patient-driven innovations reported in peer-reviewed journals. A scoping review 1–10 (2022)
22. Zvonkovic, A., Goldberg, A., Sharp, E., LaRossa, R.: Sampling richness and qualitative integrity. J. Marriage Fam 243–250 (2015)

The Role of Open Innovation in New Product Development and Its Effect on Firm Performance During an Economic Crisis: A Case Study COVID-19 Vaccine

Intesar Saleem, Tariq Yaseen, Mohammad Najjar, and Aiman Daifallah

Abstract This study aims to explore the role of open innovation in developing a new product by relying on two main dimensions (collaboration, digital technology) of open innovation during an economic crisis and its effect on firm financial performance. A conceptual framework has been developed to explain the role of open innovation on new product development by considering open innovation's collaboration and digital technology dimensions and their effect on firm financial performance. Hence, literature related to these aspects has been reviewed. COVID-19 vaccine case study sheds light on applying the open innovation approach to product and service development. The results show the relationship between open innovation through collaboration and digital technology, positively affecting firm financial performance. One limitation lies in that this paper lacks exploratory or survey-based data. Future research can investigate the case of each company separately to get a deeper understanding of the role of open innovation in producing the COVID-19 vaccine under uncertainty. Also, future research can study different dimensions of open innovation (e.g., R&D and intellectual property) on the development of the COVID-19 vaccine. The paper suggests ways in which firms could use open innovation for the efficient exploration, acquisition, integration, and development of valuable knowledge.

Keywords Open Innovation (OI) · New product development (NPD) · Collaboration · Digital technology (DT) · Firm performance · COVID-19

1 Introduction

According to the rising rate at which technology and consumer tastes are evolving, the shorter innovation cycles, and the increasing cost of research and development, firms are prompted to head toward open innovation. Chesbrough (2003) argued that the traditional innovation model, in which the innovation process takes place totally inside a company's walls—with no collaboration with external stakeholders until a

I. Saleem (✉) · T. Yaseen · M. Najjar · A. Daifallah
An-Najah National University, Nablus, Palestine
e-mail: Intesar.saleem@hotmail.com

A. Hamdan et al. (eds.), *The Implementation of Smart Technologies for Business Success and Sustainability*, Studies in Systems, Decision and Control 216,
https://doi.org/10.1007/978-3-031-10212-7_63

product is introduced into the marketplace—is no longer productive [1]. As a result of this shift in marketplace dynamics, there has been an expansion in interest in the "open innovation" approach, emphasizing the inflow and outflow of knowledge to speed up internal innovation while also expanding the markets for the external application of invention [2]. In previous research, openness has been defined in various ways [3]. Open innovation is defined as the act of developing with others to generate mutual advantages via shared risk and reward, which has grown dramatically over the previous two decades [4]. The existing literature on open innovation (OI) has demonstrated a variety of ways that firms can use inbound knowledge from their relationships with others to develop new products and services and improve their overall performance [1, 5–7]. The concept of "openness" in innovation activities refers to incorporating outside ideas and routes to market into a company's innovation processes and initiatives [1]. Where openness has been explained due to the number of partners involved, as stated by Lauren and Salter (2006) and Leiponen and Helfat (2010), the main focus has been on sourcing ideas and knowledge from a diverse range of sources [6, 8]. Others define it as partnerships between a wide range of life sciences organizations that are becoming ever more numerous, cross-disciplinary, open (in terms of data and knowledge sharing) and are reaching further down the value chain and the process of development [9]. According to studies, firms that keep investing in their innovative capabilities during the recession have a better chance of surviving when growth resumes [10]. That is how the United States' chemical industry surpassed the United Kingdom's after World War I, how Sears surpassed Montgomery Ward as the leading U.S. retailer after World War II, and how Japanese semiconductor manufacturers outpaced U.S. companies following the economic downturn in the early 1980s [11].

The history of science searches full of studies on open innovation and its significant role in developing creativity and its impact [12, 13]. Chesbrough and Garman (2009) stated that outbound open innovation through spinning off a part of the firms' internal projects and intellectual property to external partners during a recession would accelerate the development process and cut-down costs [11]. On the other hand, Chesbrough (2020) focused on the role of open innovation in managing crisis and how this contributes to vaccine production. He emphasized that openness is essential in times of crisis because good ideas may come from anybody [14]. Opening up will speed up your internal innovation process and enable you to benefit from the expertise of people in your organization (from the outside in) while also allowing others to contribute [14]. While Puslecki et al. (2021) discussed the formation of new partnerships, research and development collaborations in the wake of the COVID-19 epidemic and emphasized the role the COVID-19 pandemic played in the formulation of new strategic initiatives [15].

Hence, this research attempts to develop a conceptual framework that explores how open innovation can assist companies in producing new products (e.g., COVID-19 vaccine) during an uncertain situation and economic crisis. It also investigates the impact of open innovation on firms' financial performance while taking into account two main dimensions in open innovation: collaboration and digital technology. More specifically, the current research seeks to answer the following research questions:

RQ1: What is the impact of open innovation in developing a new product (COVID-19 vaccine) during an economic crisis?
RQ2: How is open innovation affected by successful network collaboration?
RQ3: How open innovation during economic crisis will affect firms' financial performance?

The research has addressed significant literature gaps that delimit our current understanding of open Innovation During COVID-19. This research attempts to develop a conceptual framework explaining the role of open innovation in developing new products, using the COVID-19 vaccine as a case study and its effect on the firm's financial performance under an economic crisis.

The research structure is as follows. After this introduction, the second section discusses the methodology. The third section of this research is a literature review about the role of open innovation in new product development, collaboration, and technology dimensions of open innovation. Section four is the discussion, and finally, the conclusion in section five.

2 Methodology

In this research, we aimed to develop a conceptual framework to explain the role of OI on NPD by considering the collaboration and digital technology dimensions of OI and their effect on firm financial performance. Hence, literature related to these aspects was reviewed. First, literature related to OI was reviewed to explore its role in NPD. Second, we investigated literature related to collaboration to gain a broader understanding of its impact on NPD. Thirdly, we examined literature related to digital technologies to see how they can support OI and NPD. Finally, we explored literature that studied how OI can affect firms' financial performance during economic crisis. To gain enough knowledge on these topics, relevant and well-cited papers were searched across different journals (e.g., Research Policy, R&D Management) using online databases such as Wiley Online Library, Science direct, and Emerald Group. The following keywords were utilized to collect relevant research papers: "Open Innovation," "New Product Development," "Firm Performance," "Open Innovation Firm Performance," "Open Innovation COVID-19," and "Open Innovation economic crisis."A list of highly cited and peer-reviewed papers has been categorized and explored in more detail (initial abstract investigation), filtering the articles based on their relevance and added value to the research topic in this study. Finally, the list of selected papers was then screened again to remove articles that provided identical results and link the different papers to the different research questions proposed in this study.

3 Literature Review

3.1 The Role of Open Innovation in New Product Development

Open innovation significantly improves the complicated process of new product development, as it enhances the gathering and integration of complex and fragmented knowledge from different fields [16, 17]. Furthermore, Frankort (2016) found that R&D alliances led to better knowledge acquisition and application, which enhanced new product development performance [18]. According to Leonard and Sensiper (1998), today's complexity necessitates matching opinions and expertise from different disciplines in the process of developing new products and services [19]. Open innovation facilitates new product development through inbound and outbound directions of open innovation [20]. Inbound open innovation is known as the acquisition of knowledge, ideas, and projects from external partners to incorporate it with the internal development process. This step usually takes place at the early stage of new product development [1]. On the other hand, outbound open innovation is defined as exporting internal ideas and projects to external partners, usually at the late stages of the development process [21]. Despite that open innovation breadth and depth accelerated the process of new product development [22], other research found an inverted U-shape relationship between research depth, exploratory and exploitive learning, and new product development [23, 24].

According to this part of the literature review, this research studies the relationship between open innovation and new product development during an economic crisis.

3.2 Collaboration as a Part of Open Innovation

Collaboration between two or more companies in new product development will assist companies in introducing new products to the market. Collaboration between companies has two forms: direct tie and indirect tie. Direct collaboration improves the spread and transition of knowledge, resources, and complimentary services between companies [5, 25]. Research showed that direct collaboration carries many benefits to the process of new product development. First, new product development is considered a knowledge-intensive, heterogeneous process requiring knowledge acquisition from different sources [26]. According to the knowledge-based view (KBV), knowledge is a core resource of the firm [27]. Hence, direct collaboration facilitates firms' accessibility to the necessary high-value knowledge, which will lead to better knowledge distribution, resulting in speeding up the process of new product development [5, 25, 28]. Secondly, new product development is a process that requires a firm's complementarity and experience [29]. Collaboration seems to allow firms to achieve this complementarity [30], as the direct collaboration between partners will enable firms to use each other's talents, tacit knowledge, expertise, and complementary technologies interchangeably [5, 25]. Consequently, firms will meet the complementarity and experience needed for new product development. Finally, new product development is associated with a higher degree of risk due to shorter product lifespans and rapid technological changes [31]. Hence, direct collaboration lowers these risks by dividing risks across different partners. This fact can be achieved through resource sharing between partners, obtaining newer technologies, new market entry, and performing economies of scale; all would participate in reducing risks associated with the process of new product development [5, 32].

From the indirect perspective, indirect connections' benefits for new product development mainly focus on knowledge [5, 32, 33]. Two reasons indirect collaboration benefits new product development. The first reason is that firms use indirect connections to look for knowledge outside their boundaries [23]. As technological trajectories and experiences of other firms in their field are available on social networks, firms may strengthen their indirect linkages and gain a deeper understanding of other firms' experiments' success and failure to raise the chances of success of their new product development [33]. The second reason is that these linkages may act as information selecting, classifying, absorbing, and processing tools, which will incorporate the acquired knowledge into firms' internal capabilities [5, 25, 33].

According to this part of the literature review, this study will clarify the effect of network collaboration as a part of open innovation to produce new products during an economic crisis.

3.3 Technology as a Part of Open Innovation

Firms increasingly adopt the open innovation approach in their new product development due to the growing need for speed and accuracy in the development process, which entails incorporating digital technologies in their development process to create newer products, processes, business models, and ecosystems, that will lead to creating more excellent value due to enhance creativity gained by open innovation [34, 35]. Chesbrough (2003) focused in his definition of open innovation procedures on the inflow and outflow of information to boost internal developmental capabilities and commercialization opportunities, respectively [1]. Thus, digital integration through disruptive technologies led to growing radical changes in the development process [36]. Integrating digital technologies with the business processes brought several advantages to the product development process. These advantages transform actual items into virtual ones, improving their transferability and accessibility [37]. Digital technologies unleashed creativity for stakeholders' management and competitive advantages, creating newer ways for innovation, development, customer involvement, and information examination [38]. Digital Innovation is defined as combining digital technology in the innovation process for developing new products and services [39]. Digital Innovation is becoming more evident in many firms that integrate digital technologies with their innovation process [40]. In the battle against the COVID-19 pandemic, digital technologies -as a part of open innovation- played an essential role in the development of COVID-19 vaccines (e.g., Pfizer and Moderna) through collaboration platforms that facilitated partner selection, collaboration, and knowledge and experience sharing among various partners, which contributed to speeding up the development process. In addition, the collaboration between medical and technological sectors facilitated the development of diagnostic tools relying on modern advanced technologies such as Artificial Intelligence, Machine Learning, the Internet of Things, Blockchain, and Virtual Reality [15]. Openness in the healthcare system is critical during crises that lead to better performance and quicker response for public health challenges [14]. Digitalization of the healthcare system shifts us toward a more patient-centric approach, which allows healthcare professionals to collaborate more efficiently for the common good and public health issues, as using digital technologies facilitates non-equity R&D alliances between medical and biopharmaceutical companies to develop monitoring, diagnostics, and therapeutics for current and future pandemics [15, 41].

According to this part of the literature review, this study will clarify the role of digital technology as a part of open innovation to produce new products during an economic crisis.

4 Discussion

Innovation is considered a source of competitive advantage that allows firms to outperform their competitors. Thus, leading to superior performance [42, 43]. The open innovation approach is increasingly seen as an alternative approach to the closed approach. OI cooperatively takes place with external partners like external companies, research institutions, and the technology sector. Open innovation enhances a firm's competitive position by broadening its innovative capabilities. Firms will be exposed to newer ideas and technologies from external sources and thus will not fall into the trap of core rigidities or path-dependency [44, 45]. By doing so, firms will be able to develop new products that carry out the potential to be commercially successful [46]. The development process of a new product takes place at R&D departments [47], which is considered a complex, interrelated, risky process and carries a higher potential for failure [48, 49].

Economic crisis challenges the firm's ability to continue its innovative projects. Therefore, it obstructs its ability to grow and expand by forcing companies to cut down a part of their initiative that could be promising in good times due to lack of resources and markets stagnation. However, firms that keep their innovative pace during tough times will perform better when growth comes back. Open innovation will assist firms to keep their innovation initiatives during a recession by exchanging their capabilities (ideas, intellectual property, and technologies) beyond their boundaries through strategic partnerships with external parties. Thus, firms will be able to save time and costs and boost relationships with partners. As a result, it may harvest better financial awards [11]. As shown in Fig. 1, collaboration has facilitated NPD through knowledge, resources, risk-sharing, and knowledge acquisition, processing, and exploitation. According to our case study, the collaboration between biopharmaceutical companies, academia, governments, and non-profit organizations, as seen in "ACTIV" and "COVAX" alliances, speeded up the introduction of COVID-19 vaccines to the world. Collaboration between Pfizer/BioNTech, Moderna/NIAID, and AstraZeneca/Oxford University resulted in the early introduction of COVID-19 vaccines to the market in a record time by early 2021 [50]. Figure 1 also demonstrates that Digital technologies are becoming a key enabler of OI and NPD through cutting-edge technologies (like AI, IoT, and Blockchain). Also, external technology acquisition and exploitation supported OI, which in turn supports NPD and firm performance [51] by facilitating the gathering of distributed knowledge to support OI [52] through OI platforms that assisted firms in partner selection, finding financial support, and knowledge co-creation [41]. OI platforms helped pharmaceutical companies in their R&D processes and drug discovery by facilitating knowledge exchange with external partners while protecting their intellectual property and tracking new ideas to incorporate them with their internal development process in a quick and flexible way [53]. Open innovation has a positive effect on NPD, and NPD has a positive impact on firm performance (in terms of sales and return on sales) in a consequential relationship. [54]. As Open Innovation helps firms to overcome problems related to the NPD process and maximizes the chance of the development process success through R&D partnerships, which supply the firm with complementary external resources [27], reduce development cost and risk [55], and adapt to the dynamic environment [56], which will lead to better financial performance and higher revenues [57]. According to Du et al. (2014), OI through scientific-based partnerships between firms, universities and research institutions is inexpensive, linked to a lower degree of risk, and results in better financial performance and higher returns [58]. The scientific-based partnership will enrich the firm with the latest scientific knowledge and accessibility to scientific facilities and equipment, which will give the company a first-mover advantage and new market space [55, 59]. Chesbrough and Garman (2009) stated five strategic inside-out open innovation moves that could help firms during a recession by spinning off a part of their projects and intellectual property to an external partner. The first move is to become a customer or supplier of its existing project. The firm will benefit from the spin-off project that external partners have developed as being only a customer or supplier of that project while saving costs and time of development and keep their participation in that project. The second move is to spin off non-strategic initiatives to external investors, maintaining scarce funds for more critical projects. Thirdly, transfer unused intellectual properties to external partners and generate licensing income. This might also help start-up companies during a crisis, where financing is scarce, collaborating with capital-rich firms to get funds for their projects. In the fourth move, firms grow and expand relationships with their ecosystem. Firms that have active projects during a recession will find opportunities in their ecosystem to collaborate in commercializing these projects and get access to venture capital. Finally, transferring internal knowledge into an open domain and letting others participate in the development process of existing projects will

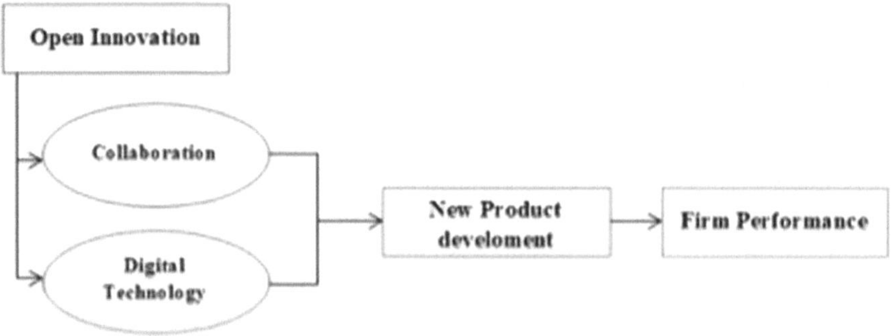

Fig. 1 Illustrates the relationship between open innovation (and its dimensions: collaboration and digital technology), new product development, and firm performance

accelerate the development process by exchanging ideas and facilities, shifting costs outside, and turning R&D departments from cost centers to profit centers [11].

5 Conclusion

Open innovation is increasingly taking place in the new product development, which entails exchanging knowledge and resources with external parties through inbound and outbound ways. As the COVID-19 pandemic negatively affected all aspects of life worldwide, pharmaceutical companies are in a race against time to develop vaccines, which raised the necessity to adopt the OI approach to accelerate the development process and save costs during the tough time. In this research, we developed a conceptual framework to explain the role of OI on NPD and their effect on firm performance, considering two dimensions of OI: collaboration and digital technologies and taking the COVID-19 vaccine as a case study. Our findings are demonstrated in Fig. 1. The role of OI in developing the COVID-19 vaccine supported the link between OI and NPD during economic crisis. Findings of collaboration in this case study also supported the link between collaboration and NPD. Although the role of digital technologies is significant in NPD, its role in developing the COVID-19 vaccine is unclear. And lastly, firm financial performance is enhanced by OI during economic crisis. This has implications for management to deal with the uncertainty that arises due to crisis to consider the OI approach as it allows firms to keep their innovative pace and reduce risks and costs through collaborating with others instead of putting promising projects and initiatives on the shelf. This will enhance their financial performance during economic crisis and allow them to better financially perform when growth returns. This research provides an understanding of the role of OI in developing COVID-19 vaccines; it lacks exploratory or survey-based data. Future research can bridge this gap. Moreover, future research can investigate the case of each company separately to get a deeper understanding of the role of OI in producing COVID-19 vaccine under uncertainty. Also, future research can study different dimensions of OI (e.g., R&D and intellectual property) on the development of the COVID-19 vaccine.

References

1. Chesbrough, H.: Open Innovation: the new imperative for creating and profiting from technology. **21**, (2003)
2. Chesbrough, H., Vanhaverbeke, W., West, J.: Open Innovation: Researching a New Paradigm. OUP Oxford (2006)
3. Dahlander, L., Gann, D.: How open is innovation? Res. Policy **39**, 699–709 (2010)
4. Yu, H.: Achieving proof of concept in drug discovery and development. In Bibliography. Edward Elgar Publishing (2016)
5. Ahuja, G.: Collaboration networks, structural holes, and innovation: a longitudinal study. Adm. Sci. Q. **45**(3), 425–455 (2000)
6. Laursen, K., Salter, A.: Open for Innovation: the role of openness in explaining innovation performance among U.K. manufacturing firms. Strateg. Manag. J. **27**(2), 131–150 (2006)
7. Powell, W.W., Koput, K.W., Smith-Doerr, L.: Interorganizational collaboration and the locus of Innovation: networks of learning in biotechnology. Adm. Sci. Q. **41**(1), 116–145 (1996)
8. Leiponen, A., Helfat, C.E.: Innovation objectives, knowledge sources, and the benefits of breadth. Strateg. Manag. J. **31**(2), 224–236 (2010)
9. Pigott, R., Barker, R., Roberts, M.: Shaping the future of open Innovation: a practical guide for life science organisations. Wellcome Trust, London (2014)
10. Am, J.B., et al.: Innovation in a crisis: Why it is more critical than ever. McKinsey & Company (2020)
11. Chesbrough, H.W., Garman, A.R.: Use open innovation to cope in a downturn. Harv. Bus. Rev. **87**(12), 68–76 (2009)
12. Lee, Y., et al.: Why do experts solve complex problems using open innovation? Evidence from the U.S. pharmaceutical industry. Calif. Manag. Rev. **62**(1), 144–166 (2019)
13. Liao, S., Fu, L., Liu, Z.: Investigating open innovation strategies and firm performance: the moderating role of technological capability and market information management capability. J. Bus Ind. Mark. **35**(1), 23–39 (2020)
14. Chesbrough, H.: To recover faster from Covid-19, open up: managerial implications from an open innovation perspective. Ind. Mark. Manag. **88**, 410–413 (2020)
15. Puślecki, Ł, Dąbrowski, M., Puślecki, M.: Development of innovation cooperation in the time of COVID-19 pandemic. Eur. Res. Stud. J. **24**(1), 1049–1073 (2021)
16. Enkel, E., Gassmann, O., Chesbrough, H.: Open R&D and open innovation: exploring the phenomenon. R&D Manag. **39**(4), 311–316 (2009)
17. Gassmann, O., Enkel, E., Chesbrough, H.: The future of open innovation. R&D Manag. **40**(3), 213–221 (2010)
18. Frankort, H.T.W.: When does knowledge acquisition in R&D alliances increase new product development? The moderating roles of technological relatedness and product-market competition. Res. Policy **45**, 291–302 (2016)
19. Leonard, D.A., S. Sensiper, The role of tacit knowledge in group innovation. In: Managing Knowledge Assets, Creativity and Innovation. pp. 301–323
20. Grönlund, J., Sjödin, D., Frishammar, J.: Open innovation and the stage-gate process: a revised model for new product development. Calif. Manag. Rev. **52**, 106–131 (2010)
21. Lichtenthaler, U.: The drivers of technology licensing: an industry comparison. Calif. Manag. Rev. **49**, 67–89 (2007)
22. Zhu, X., et al.: The fit between firms' open innovation and business model for new product development speed. Contingent Perspect. **86–87**, 75–85 (2019)

23. Katila, R., Ahuja, G.: Something old, something new: a longitudinal study of search behavior and new product introduction. Acad. Manag. J. **45**(6), 1183–1194 (2002)
24. Li, C.-R., Chu, C.-P., Lin, C.-J.: The contingent value of exploratory and exploitative learning for new product development performance. Ind. Mark. Manag. **39**(7), 1186–1197 (2010)
25. Vanhaverbeke, W., Gilsing, V., Duysters, G.: Competence and governance in strategic collaboration: the differential effect of network structure on the creation of core and noncore technology. J. Prod. Innov. Manag. **29**(5), 784–802 (2012)
26. Goffin, K., Koners, U.: Tacit knowledge, lessons learnt, and new product development. J. Prod. Innov. Manag. **28**(2), 300–318 (2011)
27. Grant, R.M., Baden-Fuller, C.: A knowledge accessing theory of strategic alliances. J. Manag. Stud. **41**(1), 61–84 (2004)
28. Pallotti, F., Lomi, A.: Network influence and organizational performance: the effects of tie strength and structural equivalence. Eur. Manag. J. **29**(5), 389–403 (2011)
29. Eric, F.: The effect of strategic alliance knowledge complementarity on new product innovativeness in China. Organ. Sci. **22**(1), 158–172 (2011)
30. Arranz, N., de Arroyabe, J.C.F.: The choice of partners in R&D cooperation: an empirical analysis of Spanish firms. Technovation **28**(1), 88–100 (2008)
31. Droge, C., Calantone, R., Harmancioglu, N.: New product success: is it really controllable by managers in highly turbulent environments? J. Prod. Innov. Manag. **25**(3), 272–286 (2008)
32. Karamanos, A.: Leveraging micro- and macro-structures of embeddedness in alliance networks for exploratory innovation in biotechnology. R&D Manag. **42**, 71–89 (2012)
33. Salman, N., Saives, A.-L.: Indirect networks: an intangible resource for biotechnology innovation. R&D Manag. **35**(2), 203–215 (2005)
34. Chiaroni, D., Chiesa, V., Frattini, F.: The open innovation journey: how firms dynamically implement the emerging innovation management paradigm. Technovation **31**, 34–43 (2011)
35. Urbinati, A., et al.: The role of digital technologies in open innovation processes: an exploratory multiple case study analysis. R&D Manag. **50**(1), 136–160 (2020)
36. Kekäle, T.H., Stanisauskaite, P., Lammi, V., Miia. Design of a VR/AR environment for service prototyping. In: Proceedings of the 3rd International Scientific and Practical Forum 'Russia, Europe, Asia: Global Environment for Digitalisation". (2020). Russia: Sekvoyija
37. Rayna, T., Striukova, L.: From rapid prototyping to home fabrication: how 3D printing is changing business model innovation. Technol. Forecast. Soc. Chang. **102**, 214–224 (2016)
38. Brunswicker, S., Bertino, E., Matei, S.: Big Data for open digital innovation—a research roadmap. Big Data Res. **2**(2), 53–58 (2015)
39. Fichman, R.G., Santos, B.L.D., Zheng, Z.: Digital innovation as a fundamental and powerful concept in the information systems curriculum. MIS Q. **38**, 329–343 (2014)
40. Chen, P.: The impact of collaboration network on new product development: evidence from the automobile industry of China. J. Ind.-Univ. Collab. (2019). ahead-of-print.
41. Abbate, T., Codini, A.P., Aquilani, B.: Knowledge co-creation in open innovation digital platforms: processes, tools and services. J. Bus. Ind. Mark. **34**(7), 1434–1447 (2019)
42. Ferrier, W.J., Smith, K.G., Grimm, C.M.: The role of competitive action in market share erosion and industry dethronement: a study of industry leaders and challengers. Acad. Manag. J. **42**(4), 372–388 (1999)
43. Helfat, C.E., et al.: Dynamic Capabilities: Understanding Strategic Change in Organizations. Wiley (2009)
44. Leonard-Barton, D.: Core capabilities and core rigidities: a paradox in managing new product development. Strateg. Manag. J. **13**(S1), 111–125 (1992)
45. Rosenkopf, L., Nerkar, A.: Beyond local search: boundary-spanning, exploration, and impact in the optical disk industry. Strateg. Manag. J. **22**(4), 287–306 (2001)
46. Prabhu, J.C., Chandy, R.K., Ellis, M.E.: The Impact of Acquisitions on innovation: poison pill, placebo, or tonic? J. Mark. **69**(1), 114–130 (2005)
47. Cassiman, B., Di Guardo, M.C., Valentini, G.: Organizing links with science: cooperate or contract?: A project-level analysis. Res. Policy **39**(7), 882–892 (2010)

48. Clark, K.B., Wheelwright, S.C.: Managing New Product and Process Development: Text and Cases. Free Press (1993)
49. Cooper, R.G., Edgett, S.J., Kleinschmidt, E.J.: benchmarking best NPD practices—II. Res. Technol. Manag. **47**(3), 50–59 (2004)
50. Richardson, B., Bentley,S., Dillon,S. A., Abuaf, L., Mohanna, N. The race for the COVID-19 vaccine: a story of innovation and collaboration. (2020)
51. Zanjirchi, S.M., Jalilian, N., Shahmohamadi Mehrjardi, M.: Open innovation: from technology exploitation to creation of superior performance. (2019)
52. Dodgson, M., Gann, D., Salter, A.: The role of technology in the shift towards open innovation: the case of procter & gamble. R&D Manag. **36**(3), 333–346 (2006)
53. Carroll, G.P., et al.: Measuring the effectiveness and impact of an open innovation platform. Drug Discov Today **22**(5), 776–785 (2017)
54. Rubera, G., Chandrasekaran, D., Ordanini, A.: Open innovation, product portfolio innovativeness and firm performance: the dual role of new product development capabilities. J. Acad. Mark. Sci. **44**, (2015)
55. Leten, B., et al.: IP Models to Orchestrate innovation ecosystems: IMEC, a public research institute in NANO-electronics. Calif. Manag. Rev. **55**, 51–64 (2013)
56. Dittrich, K., Duysters, G.: Networking as a means to strategy change: the case of open innovation in mobile telephony. J. Prod. Innov. Manag. **24**, 510–521 (2007)
57. Faems, D., Van Looy, B., Debackere, K.: Interorganizational collaboration and innovation: toward a portfolio approach*. J. Prod. Innov. Manag. **22**(3), 238–250 (2005)
58. Du, J., Leten, B., Vanhaverbeke, W.: Managing open innovation projects with science-based and market-based partners. Res. Policy **43**, 828–840 (2014)
59. Fabrizio, K.: Absorptive capacity and the search for innovation. Res. Policy **38**, 255–267 (2009)

The Impact of the Change in the Elements of Cash Flows on the Capital Structure of Pharmaceutical Companies Listed on the Amman Stock Exchange in the Shade of the Covid-19 (Analytical Study)

Abdelwhab Alrwashdeh, Riham Alkabbji, Ala' Alrazim, and Farah Hamdan

Abstract The aim of this study is to clarify the impact of the change in the elements of the cash flow statement on the capital structure. However, this study was conducted on six pharmaceutical companies listed on the Amman Stock Exchange during the period between 2018 and 2021. Meanwhile, the descriptive analysis of the study variables was used, besides each of the correlation tests, simple regression, and analysis of changes in financial ratios. Thus, for the purpose of showing the impact of cash flows elements on the capital structure of pharmaceutical companies and the extent to which it is affected by the Coronavirus (COVID-19) pandemic. Whereas, the current study concluded that there is a statistically significant impact (reject the null hypothesis) of cash flows on the capital structure as well as having an impact of cash flows on indebtedness and that there is no statistically significant impact (rejection of the null hypothesis) of cash flows on the shareholders' contribution to property rights. One of the most important conclusions of the current study is that the change in cash flows negatively affects the indebtedness rate, and this is a good thing, as the increase in cash flows leads to a decrease in the indebtedness ratio and this is desirable by the companies' management, owners and creditors, but the decrease in these flows leads to an increase in the indebtedness rate. Meanwhile, the current study recommended the possibility of studying the company's ability to sustain in

A. Alrwashdeh · R. Alkabbji (✉) · A. Alrazim
Accounting Department, Zarqa University, Zarqa, Jordan
e-mail: rkabbji@zu.edu.jo

A. Alrwashdeh
e-mail: arawashdeh@zu.edu.jo

A. Alrazim
e-mail: aalrazim@zu.edu.jo

F. Hamdan
Zarqa University, Zarqa, Jordan
e-mail: 20209033@zu.edu.jo

A. Hamdan et al. (eds.), *The Implementation of Smart Technologies for Business Success and Sustainability*, Studies in Systems, Decision and Control 216,
https://doi.org/10.1007/978-3-031-10212-7_64

light of the high indebtedness' circumstances and ways to survive from insolvency and bankruptcy.

Keywords Cash flow statement's elements · Capital structure · Amman stock exchange · Pharmaceutical companies

1 Introduction

The cash flow statement is the fourth and basic list that relied on the cash basis in clarifying a detailed summary of each of the inflow and outflow cash, which means, the company's ability to generate cash in order to pay its obligations and finance its activities, also it is considered one of the statutory financial statements as illustrated in the Financial Accounting Standards Board FASB No. 9 of the year 1987, and the following action was the issuance of the international accounting standard No. 7.

It can be said that companies focus on the capital structure, for the purposes such as financing, the company's sustainability in its main activities, and the company's ability to meet its obligations. Because of its importance for owners, lenders, and investors. Also, the company depends on debt or ownership when it is about financing; Companies that rely on indebtedness have a stronger capital structure than companies that rely on the issuance of shares. However, coronavirus (COVID-19) pandemic appeared in December 2019, where the pandemic caused a severe economic shock. Knowing that at the beginning of March 2020, Jordan besides many other countries' economies suffered from a collapse, as there were closures in all sectors, which led to many accounting considerations on these sectors.

Hence, this study came to shed light on the Jordanian pharmaceuticals' industry sectors, because of their significant role in Jordan's economy. As it clarifies the impact of the cash flow statement's elements alongside its dimensions (operational, investment, financial), on the capital structure dimensions (short-term debts, long-term debts, and ownership), and the impact of the Coronavirus (COVID-19) pandemic on it.

2 Previous Studies

Many researchers have studied the impact of the change in the cash flow statement's elements on the capital structure. Muhammad and Al-Sleihat [1] said that there is an impact of cash flows on the realized earnings per share concerning the Jordanian industrial companies. Also, Mohamed Mahmoud [2] indicated that there is a statistically significant impact of the operational cash flows, financial, investment and company size on the speed of adjusting the capital structure, as well as there is a negative impact of the return on assets and equity on adjusting the speed of the capital structure. Meanwhile, Awadallah and Gilani [3] recommended that Jordan

Steel Group must reconsider its capital to finance its assets and reduce the debt ratio due to the negative impact of financial leverage on its financial performance. Additionally, Rahim [4] agreed with the previous one. Whereas [5] found that there is a positive and statistically significant relationship between profitability and debt ratio, which forms part of the financial structure.

However, Al-Monem and Kishik [6] study showed that financial leverage does not have an important impact on capital investment in Jordanian industrial companies, the study pointed out that this could be due to the fact that debts in industrial companies do not play an important role in the capital investment decision and companies depend on the issuance of shares and retained earnings in the process of financing its capital investments. Moreover, Ahmad [7] found that there might be a growth opportunity relying on the policy of dividend distribution and the policy of reliance on debt.

3 Study Methodology

The current study used the descriptive-analytical method to achieve the objectives of the study, namely, to answer the following questions:

What is the impact of the change in cash flows on the long-term indebtedness of the pharmaceutical companies listed on the Amman Stock Exchange in light of the Coronavirus (COVID-19) pandemic?

What is the impact of the change in cash flows on the capital to equity ratio in pharmaceutical companies listed on the Amman Stock Exchange in light of the Coronavirus (COVID-19) pandemic?

3.1 Hypotheses

This study is based on the Pecking Order Theory, which was developed based on the work of [8]. According to this theory, there are three sources of financing for institutions, internal financing based on funds generated from the company's profits, financing with debt, and financing with new shares. As for the priorities of funding sources for institutions, according to this theory, internal financing comes first, then debt financing, and finally financing with new shares. Institutions must abide by this hierarchy in choosing financing sources, where equity financing is a last resort, because it brings new owners to the institution, which affects decision-making [9].

This study is concerned with determining the impact of cash flows on financing sources, where indebtedness represents the second option for financing, which is debt financing, while capital-to-Equity represents the first option for financing based on internal financing based on funds generated from the company's profits.

The consequences of the Coronavirus (COVID-19) pandemic affected the economic situation in Jordan; Thus, because of the closures policy implemented by the country, in order to reduce the spread of the pandemic, therefore, these closures had many side effects on the cash flow in Jordan, where buying and selling decreased significantly, and even the ability of companies to pay their obligations decreased as a result of the depreciation in the value of cash inflows, and this had a significant impact on changing the financing structure of companies listed on the Amman Stock Exchange. Notably, the pharmaceutical companies were affected by the Coronavirus (COVID-19) pandemic. The present study also follows up cash flows and their impact on their financing structure to examine the change in the pharmaceutical companies'. Based on the above, the hypotheses were formulated as follows:

Ho1: There is no statistically significant impact of the change in cash flows on the indebtedness of pharmaceutical companies in light of the Coronavirus (COVID-19) pandemic.
Ho2: There is no statistically significant impact of a change in cash flows on the capital to equity ratio in pharmaceutical companies in light of the Coronavirus (COVID-19) pandemic.

3.2 Sources of Data Collection

There are two types of data collection sources that were relied upon in the process of preparing this study, namely:

Primary sources: The data that will be collected using the study tool, such as the annual financial reports, financial statements, and private information in pharmaceutical and medical industries published on each of the Amman Stock Exchange and the companies' websites.

Secondary sources: The data that will be collected from books, researches, literature reviews, and previous studies related to the research topic, besides a number of internet.

3.3 Study Population and Sample

The study population consists of all listed pharmaceutical and medical industries companies; which have shares in the Amman Stock Exchange during the period between 2018 and 2021, they are six companies. The following companies are ordered respectively concerning the capital, where the highest capital for Dar Al Dawa for Development and Investment Company with a capital of 35,000,000 Jordanian dinars, the average capital of these companies is 14,760,513 Jordanian dinars, and the least of which is Middle East Pharma. & Chemical Ind. & Medical Appliances Company with a capital of 6,250,580 Jordanian dinars. Taking into account that all the entire previous quarters were relied upon, except for the last quarter of 2021 because the study was conducted before it, notably, the six companies were selected to represent the study sample.

Table 1 Methods of measuring study variables

Variable name	Variable type	Definition
Net cash flow	Dependent	The value of the cash flows was extracted from the published financial reports of pharmaceutical companies, as this list is prepared within the mandatory financial statements in the financial reports
Debt to equity	Independent	This variable is measured by dividing the total liabilities by the total equity
Capital to equity	Independent	This variable is measured by dividing the share capital stocks by the total equity

3.4 Study Variables

To test the hypotheses of the study, net cash flows were used as an independent variable, while the capital structure was measured through the debt ratio and the capital to equity ratio. Whereas the financing structure is divided into internal and external financing, the internal financing constitutes (capital stocks, reserves, and retained earnings). While the external financing represents long and short-term debts, and therefore in order to determine the percentage of external financing's contribution to the formation of the financing structure, the indebtedness ratio was used for the pharmaceutical companies listed on the Amman Stock Exchange. Also, the capital to equity ratio was used to measure the extent of the shareholders' contribution to the capital, as well as to measure the impact of the Coronavirus (COVID-19) pandemic as a mediating variable by introducing the impact before and during the Coronavirus (COVID-19) pandemic. The following table shows an explanation of the concerned ways to measure these variables (Table 1).

3.5 Statistical Methods of the Research

Descriptive analysis of the study variables through the measures of central tendency (maiden, standard deviation, and the lowest and highest value). In addition, testing the study's hypotheses by applying the multiple regression test.

Descriptive Statistics

It can be seen from the Table 2 the descriptive statistics of the study variables, as there is a discrepancy in the size of the paid-in capital since the highest value reached 35 million with a variation of 30 million compared to the minimum capital, and therefore the standard deviation is high, and this will certainly be reflected in the results of the study, as the volume of the cash flow will vary depending on the variance level in the capital, so the capital will be used as an intermediate variable to adjust this variance.

Also, it can be noted that the total equity is less than the capital for some companies, due to the retained losses suffered by the pharmaceutical companies listed on the

Table 2 Descriptive statistics for the study variables

Variable	Mean	Maximum	Minimum	Std. Dev.
Net cash flow	135,029	9,092,000	−6,558,657	3,244,116
Capital	14.67 million	35 million	5 million	12.4 million
Equity	9211870.2	31,117,620	−10,963,035	14,310,534
Debt to equity	0.933	23.15	−8.15	6.67
Capital to equity	0.6528	15.8	−5.774	4.446

Amman Stock Exchange, and this is clearly reflected in the debt ratios, as the debt ratio for some companies is very high. So, it constitutes 2315% (23.51) at the highest limit, which means that debts are greater than the equity with a large variation, hence, the equity is not sufficient for paying these obligations, while the minimum indebtedness is (−815%), whereas the minus sign indicates that the value of the total equity is a loss due to the accumulated and constantly increasing retained losses, and this is evident from the previous variable, equity, which indicates the minus equity value.

It can be observed from the capital to equity ratio, and here it refers to the shareholder's contribution to equity, i.e. the value of the paid-in capital from the total equity. We note that there is a large discrepancy in values, where the highest value of this ratio is 1580% (15.8). Whereby the additional capital is higher than the value of equity and this value is due to the company that has the highest value of capital, while the lowest value is minus −577.4% (−5.774). The reason behind the minus sign is that the total equity has a negative value therefore the capital is eroded.

3.6 Empirical Estimation Method

The following equations are needed to measure the hypotheses of the study:

$$DE = a_0 + B_1 NCF + \in \tag{1}$$

$$CE = a_0 + B_1 NCF + \in \tag{2}$$

where as, the symbols are abbreviations for the study variables, as DE stands for the dependent variable indebtedness, NCF stands for the independent variable cash flows, and CE stands for capital to equity.

3.7 Empirical Results and Analyses

Results of the hypothesis test

The (Statistical Package for the Social Sciences) SPSS software was utilized to find the results of the simple regression model, and the results were as follows:

The first hypothesis

Ho1: There is no statistically significant impact of a change in cash flows on the indebtedness of pharmaceutical companies in light of the Coronavirus (COVID-19) pandemic.

This hypothesis was initially measured by simple regression eq. (1), which measures the impacts of the change in the net cash flows on the debt ratio for the period between 2018 and 2019, which is the period before the Coronavirus (COVID-19) pandemic.

And since the indebtedness ratio indicates the extent of the company's reliance on external debt in financing assets and operational operations. However, it is better for the creditors when this ratio is low, as this ratio gives an indication of the creditors' ability to recover their money in the event of the company's liquidity. In short, the higher the ratio, the greater the risk to the creditors.

This indicates a lot of reliance on long-term debt. In contrast, a lower ratio reveals a high margin of safety for the creditors. Also, Table 3 shows the results of the first hypothesis test before the Coronavirus (COVID-19) pandemic.

The Simple Regression Test was used to test the first hypothesis, and Table 3 shows the results of testing the impact of the change in the net cash flows on the company's debt ratio before the Corona pandemic, where it was P-value sig. = 0.003, which is a value less than the level of significance $\alpha = 0.05$, thus rejecting the null hypothesis and accepting the alternative hypothesis.

That is, the change in the net cash flows affects the change in the indebtedness ratio, and since the value of β is negative, and therefore the increase in cash flows leads to a decrease in the indebtedness ratio, and this means that the higher the value of the cash flows, the higher the company's ability to pay its obligations, and thus the decrease in the value of the financing external assets and resources of the company.

Consequently, the company's financial structure has improved and its reliance on the external financing is reduced. The table also shows that the value of adjR2 is (0.332), which means that net cash flows explain (33%) of the variance in the debt ratio of the companies under study. The results showed that the net cash flow variable has a negative impact on the debt ratio.

Table 3 Results of the first hypothesis test before (COVID-19) pandemic

Statistical significance	β	Value of t
0.003	−0.583	4.297
R^2	$adjR^2$	
0.339	0.332	

Table 4 Results of the first hypothesis test after (COVID-19) pandemic

Statistical significance	β	Value of t
0.002	−0.522	8.297
R^2	adjR^2	
0.349	0.322	

Measuring of the first hypothesis for the period between 2020 to 2021, which constitutes the period during the Coronavirus (COVID-19) pandemic, and Table 4 shows the results of testing the hypothesis after the Coronavirus (COVID-19) pandemic:

Table 4 shows the results of testing the impact of the change in net cash flows on the company's debt ratio after the Coronavirus (COVID-19) pandemic, where the P-value sig. = 0.002, which is less than the level of significance $\alpha = 0.05$. Additionally, it can be noted that the value of β is negative before and during the Coronavirus (COVID-19) pandemic and that the variance between before and during the pandemic is almost non-existent in the value of β, and therefore we conclude that there is no impact of the Coronavirus (COVID-19) pandemic on the results of the hypothesis measurement. Whereas, the results were computable whether before or after the Coronavirus (COVID-19) pandemic.

The second hypothesis

Ho2: There is no statistically significant impact of a change in cash flows on the ratio of capital to equity in pharmaceutical companies in light of the Coronavirus (COVID-19) pandemic.

The second hypothesis was measured through simple regression eq. (2), which measures the impact of the change in net cash flows on capital to equity for the period between 2018 to 2019, notably during the Coronavirus (COVID-19) pandemic, and since the capital to equity ratio refers to the shareholder's contribution in equity, then it is the second part of the financing structure. As a result, the higher this ratio, the better for the owners, as this ratio indicates the extent to which the shareholders' equity contribute the total long-term funds expressed as a percentage. Accordingly, the higher the percentage, the company reliance on (external financing) is less. Table 5 Indicates the results of the second hypothesis test.

The simple regression test was used to test the second hypothesis, and Table 5 shows the results of testing the impact of the change in net cash flows on the capital to equity ratio for the period before the Corona pandemic, where P-value sig. = 0.001, which is a value that is less than the level of significance $\alpha = 0.05$. And as a

Table 5 Results of the second hypothesis test before (COVID-19) pandemic

Statistical significance	β	Value of t
0.001	0.448	9.320
R^2	adjR^2	
0.519	0.510	

Table 6 Results of the second hypothesis test after (COVID-19) pandemic

Statistical significance	β	Value of t
0.004	0.408	7.320
R^2	$adjR^2$	
0.50	0.509	

consequence, rejecting the null hypothesis and accepting the alternative hypothesis. That is, the net cash flows affect the capital to equity ratio in the pharmaceutical companies listed on the Amman Stock Exchange, and since the value of β is positive, therefore the increase in cash flows raises the company's reliance on internal financing from shareholders. Also, the table shows that the value of $adjR^2$ is (0.510), which means that the change in net cash flows explains (51%) of the variance in the capital to equity ratio.

However, measuring the second hypothesis for the period between 2020 and 2021, which constitutes the time period during the Coronavirus (COVID-19) pandemic, and Table 6 Shows the results of the hypothesis testing after the Coronavirus (COVID-19) pandemic:

4 Summary of the Study Results

This study discussed the impact of the change in cash flows on the financing structure of pharmaceutical companies listed on the Amman Stock Exchange and reached the following results:

1. The pharmaceutical companies listed on the Amman Stock Exchange suffer from accumulated revolving losses before the beginning of the Coronavirus (COVID-19) pandemic, and therefore there was a big variance between the capital stocks and the total equity as a result of these losses, which led to an erosion in the capital until the value of the total equity became negative.
2. Pharmaceutical companies suffer from high inflation in indebtedness, as the total debts were adopted in calculating the indebtedness ratio, but these debts mostly consist of short-term debts, and this matter is more serious for the company compared to the long-term debts, and this means higher rates of financial leverage, thus, it is consistent with [3] study findings, where these findings negatively affect the evaluation of the companies' financial performance.
3. The study concluded that the net cash flow affects the formation of the capital structure, this is consistent with [1] conclusions, too.

5 Study Conclusions and Recommendations

5.1 Conclusions

1. The change in cash flows negatively affects the indebtedness ratio, which is a good thing, as the increase in cash flows leads to a decrease in the indebtedness ratio and this is desirable by the companies' management, owners and creditors, but the decrease in these flows leads to an increase in the indebtedness ratio.
2. There is no impact of the Coronavirus (COVID-19) pandemic on the change in the results of studying the impact of cash flows on the debt ratio whether before or during the Coronavirus (COVID-19) pandemic, where the impact of the change was consistent.
3. The change in cash flows positively affects the shareholder's contribution to equity, where the greater the cash flows, the company's ability to pay its obligations increases, and consequently the value of the internal financing in the financing structure increases, namely, the increase in the value of the shareholders' contribution to equity.
4. There is no impact of the Coronavirus (COVID-19) pandemic on the change in the results of studying the impact of cash flows on the capital to equity ratio whether before and during the Coronavirus (COVID-19) pandemic, as the impact of the change was consistent.

5.2 Study Recommendations

1. The study recommends that pharmaceutical companies reconsider their capital to finance their assets and reduce the debt ratio because of the negative impact of financial leverage on their financial performance.
2. The study recommends the possibility of studying the company's ability to continue with high indebtedness and finding particular ways to survive from insolvency and bankruptcy.

References

1. Muhammad, T.Y., Al-Sleihat: The impact of cash flows on the shares returns of Jordanian industrial public shareholding companies, an unpublished master's thesis, Amman Arab University, Amman—Jordan (2021)
2. Mahmoud, A.-A.: The impact of cash flows on the speed of adjusting the optimal capital structure in Jordanian industrial companies: a study on Jordanian industrial companies for the period between 2006–2017, thesis. Al al-Bayt University, Mafraq—Jordan (2019)
3. Awadallah, A., Gilani, S.: The impact of the return on assets and the degree of financial leverage on the return on equity of Jordan Steel and its subsidiaries: an analytical study. J. Econ. Sci. Zarqa Univ Zarqa Jordan (2016)

4. Muhammad, R.: The cash flow statement and its role in evaluating the financial performance of the Jordanian banking sector, unpublished master's thesis. Al al-Bayt University, Mafraq—Jordan (2016)
5. Negasa, T.: Impact of capital structure on corporate profitability (Evidence from Ethiopia) preprints 201607.0013/v1 (On-Line), (2016), available
6. Samih, A.A.M, Kishik, A.: Impact of financial leverage on capital investment: case of Jordanian Industrial firms for the period 2008–2002, (Unpublished Master Thesis), Yarmouk University, Irbid, Jordan (2010)
7. Ahmed, H.J.: The impact of financing decision, dividend policy, and corporate ownership on firm performance at presence or absence of growth opportunity: a panel data approach evidence from kualalumpor stock exchange. Corp Ownership Control **6**(1), 485–491 (2008). https://www.preprints.org/manuscript/201607.0013/v1
8. Myers, S.C., Majluf, N.: Corporate financing and investment decisions when firms have information investors do not have. J Financ Econ **13**, 187–221. (1984)
9. Brealey, R.A. Myers, S.C.: Principles of Corporate Finance, 7th edn, McGraw Hill (2003)

Effect of Operating Cash Flows on Company Financial Performance Before and During Corona Pandemic. "Evidence from Insurance Company Listed on the Amman Stock Exchange"

Husni K. Al-Shattarat

Abstract The current study aims to know effect of operating cash flows on company financial performance represented by the ratios of quick ratio and current ratio on insurance sector for the company listed on the Amman Stock Exchange (ASE), before and during COVID 19 pandemic during the period (2019–2020), which represents before the spread of the Corona pandemic and the period in which the pandemic spread and all were closed Aspects of economic life, the study sample consisted of 17 public shareholding insurance companies. Where the main hypotheses were formulated in light of the study problem, and to achieve the purpose of the current study, financial data were obtained from the financial statements of insurance companies, and these data were analyzed through SPSS. The most prominent results were: There is a positive statistically significant effect between operational cash flows on the financial performance represented by the ratios of circulating liquidity and quick liquidity during the period of the Corona pandemic, as a result of the decrease in accidents and sickness cases, and the decrease in operational costs and public expenses due to the commitment of citizens and remote work, while the study recommended The government taking financial measures to reduce the additional burden on these companies, such as exempting these companies from financial and customs fees and reducing the percentage of taxes related to these companies as a result of the operational activity.

Keywords Operating cash flows · Financial performance · Current ratio · Quick ratio · COVID-19 pandemic

1 Introduction

Liman and Mohammed [7] Reported that financial performance assessment is the most essential concerns of all companies employing various financial resources to complete project successfully in order to maximize profit and income. In this regard,

H. K. Al-Shattarat (✉)
Economics and Administrative Sciences, Zarqa University (ZU), Zarqa, Jordan
e-mail: hshattarat@zu.edu.jo

A. Hamdan et al. (eds.), *The Implementation of Smart Technologies for Business Success and Sustainability*, Studies in Systems, Decision and Control 216,
https://doi.org/10.1007/978-3-031-10212-7_65

the economic success of any company engaged in product or service operations is dependent on effective and efficient cash flow managing both inside and outside the company.

As the epidemic of Covid-19 has created new obstacles in the business field. The Covid-19 is placing pressure on companies specially the insurance sectors, including revenues, profit, and liquidity, the companies are facing a probable economic closure. Many businesses are facing less revenue as a result of the pandemic Covid-19, leading in reduced cash flow and postponed receivables collection. Companies should handle cash in the best feasible way to ensure liquidity during epidemic days since cash flow is critical. Governments have recognized the significance of cash flow and have acted with a variety of policies, particularly in the area of liquidity, to assist businesses in overcoming the Covid-19 crisis. The major concern is whether there is effect of operating cash flow on financial performance for insurance companies during and after Covid-19 pandemic. So the aim of this current study was to examine the consequence of operating cash flows on company financial performance represented by the ratios of quick ratio and current ratio on insurance sector for the company listed on ASE before and during COVID 19 pandemic.

To best of our knowledge this is the first study in Jordan carried out to investigate the effect of operating cash flows on company performance from 2019 to 2020, and by examining the influence of liquidity ratio on cash flows from operations on company performance. The current study differs from previous studies, especially those conducted in the Jordanian environment that most of them tended to study the impact of cash flows and capital structure on financial performance and measured through return on assets and return on equity. However, our addition to accounting literature may be represented by studying the effect on financial performance through Liquidity ratios before and during the pandemic.

2 Theoretical Background

2.1 Liquidity and Financial Performance

EY [4] claimed that in order to decrease the demand, challenges in obtaining receivables, in addition to needing to pay obligations to competitive vendors, government forced measures to avoid the expansion of corona pandemic have made it extremely complex for business to handle their operating capital.

These issues may have an impact on the company's liquidity situation, as the liquidity ratio evaluates the company's capacity to handle its short-term commitments. Companies have a stronger inclination to keep healthy liquidity before the onset of the corona virus than throughout the onset of pandemic [12]. Businesses that have run effectively their liquidity are anticipated to benefit more than any other. Justified this for the reason that they are able to make enhanced investment judgments whereas still satisfying their short-term obligations.

Li et al. [6] stated that having possessing greater liquidity indicates inefficiently utilizing assets to create profits. Companies were urged that they needed a strong and balanced liquidity position in maintaining their operation and make suitable investment choices.

2.2 *Operating Cash Flows and Financial Performance*

Cash flows from operating activity have been a critical component of a business's existence, and investment decisions would be made depending on whether or not the enterprise has enough money to invest, which will affect its success. Shafi et al. [9] suggest that if enterprises do not sustain robust cash flows through the Covid-19 epidemic spread, they risk going bankrupt and losing company. As a result, firms must improve their cash flow situation in order to limit the corona various spread's protracted detrimental impacts [9]. Consistent with [7], Operating cash flows, according to the literature, have a considerable positive link with corporate success. Additionally [3] applied that companies with net cash inflows from operations may have few liquidity problems since they can earn cash from their core business. As having a large amount of cash might suggest ineffective cash handling governance [2].

2.3 *Related Literatures (Previous Study)*

According to [10] The Covid-19 is expected to have a significant influence on company's performance on Bursa Malaysia because of the movement restriction order. As a result, the aim of the research is to look at how leverage Ratio, liquidity Ratio, and cash flows from activities affect company performance throughout Covid-19. It is discovered that liquidity in addition to cash flows from operations have a substantial effect on firm performance using secondary data from public listed for two financial quarters in the fiscal year 2020.

Al Maani [1] The objective of this report is to look at the influence of profitability as well as liquidity parameters on the market value of stocks insurance businesses in Jordan over the course of five years (2015–2019). The following variables were used; net-operating capital (NOC), current trading ratios (CTR), and profitability ratios (PR), which include return on equity (ROE), return on assets (ROA), and operational net profit margin, to collect data from yearly reports (OPM). The study found that NOC, CTR, ROA, and OPM had no effect on the market value of the shares in Jordanian insurance businesses, but the ROE did.

Rahman and Sharma [8] suggested that appropriate cash flow administration may help a company enhance its financial performance, but failure to manage cash flow in an optimal and appropriate manner leads to financial disaster. This study explores on the impact of (CFOs) on the financial performance of Saudi Arabian insurance and industrial firms. The profitability ratio represented by the following ratios return on equity (ROE) also return on assets (ROA) were used as dependent variables, As for the dependent variable, it was represented by CFOs, leverage (LEV) in addition to firm size (SIZE) as control factors, and an industry dummy were used to collect annual data for the period. They found that the financial performance (ROE & ROA) and operating cash flows (CFOs) have a significantly positive relationship, but SIZE and LEV have a negative relationship.

Yahiaoui and Mahdi [11] Attempted to investigate the influence of several internal and external variables on the financial performance of Algerian non-life insurance businesses over a nine-year period from 2010 to 2018. They concluded that the size of the company, its liquidity ratio, and its Solvency Ratio all have a significant influence on insurance companies' financial performance when assessed by Return on Assets (ROA). They found that the size of the company and its liquidity ratio had a positive influence. The solvency ratio, on the other hand, suggested a negative influence.

Khresiat [5] Who founded that insurance company' investment decisions had a significant influence on their earnings. And reported that when further insurance companies concentrate their investments to financial assets, loans, and real estate investments, their profits will have an influence on their profitability.

2.4 *Hypotheses Development*

To study the effect of operating cash flows on company financial performance represented by the ratios of current ratio and quick ratio on insurance sector has the following hypotheses:

Ho1: operating cash flows hasn't positively effect on company financial performance before and during Corona pandemic of Insurance sector listed on the Amman Stock Exchange (ASE).

The main hypothesis leads to the following mainly two sub-hypothesis:

Ho1-1: operating cash flows hasn't positively effect on current ratio before and during Corona pandemic of Insurance sector listed on ASE.

Ho1-2: operating cash flows hasn't positively effect on quick ratio before and during Corona pandemic of Insurance sector listed on ASE.

3 Study Design

3.1 *Population and Sample Study and Data Collection*

The study population contains Twenty-three companies' where the financial statements of seventeen insurance companies listed on ASE were studied and analyzed, which in total represent the sample of the study with time interval from 2019 to 2020, where this period included the absence and presence of the virus. Six companies were excluded for the following reasons. The following companies were excluded from the Arab Union International Insurance Company, Philadelphia Insurance Company and Medgulf Company due to the realization of losses in the year 2020; Al Safwa Insurance Company and Al Arab Insurance Company were excluded as they are not listed in the financial market and finally Because of the difficulty in obtaining the financial statements for the year 2020 of the Holy Land Insurance Company.. The current study's sample size is determined by the availability of financial data. Data was gathered from company financial statements and used to calculate several financial ratios, which were employed as dependent and independent variables were tested. Correlation, descriptive statistics was implemented. Moreover, test of normality was evaluated.

3.2 *Measure of Variables*

Two types of variables were used in this study. First, the dependent variables are the current and quick ratios. Second, the independent variable is Cash Flow from Operations (CFO).

3.2.1 Independent Variables

Operating Cash Flows (CFO) Net cash from operating activities divided by Total Current Liabilities.

3.2.2 Dependent Variables

Current Ratio = Current Assets divided by Current Liabilities.

Quick Ratio = (Current Assets − Inventory − Prepaid Expense) divided by Current Liabilities.

Table 1 Correlation analysis of study variables

	CFOs	CUR ratio	QUI ratio
CFOs	1		
Curratio	0.260	1	
Quratio	0.291	0.893	1

3.3 Empirical Models

The following two regression models are used to test hypotheses and interpret for this study:

$$\text{CUR Ratio} = \alpha_0 + \alpha_1(\text{CFO}) + e$$
$$\text{CUR Ratio} = \beta_0 + \beta_1(\text{CFO}) + \varepsilon$$

where

CUR Ratio	Current Ratio
QUI Ratio	Quick Ratio
CFO	Cash Flows Operating
β_1, β_2	Regression Coefficient
e, ε	error term

3.4 Statistical Analysis

The correlation between CFOs and Current, Quick Ratio variables is shown in Table 1. The correlation is between −1 and 1. CFO are positively connected with (CUR) according to the correlation analysis of Model 1, where (CUR) is the dependent variable (CUR). Similarly, Model 2's correlation analysis, where (QUI) is the dependent variable, reveals that CFOs is positively connected with (QUI).

4 Empirical Results and Analysis

Table 2 illustrates the effect of operating cash flows on financial performance that measured by current ratio before and during coved-19 pandemic, as we note from the data contained in this table that there is a negative effect which means that the higher the operating cash flow, the lower the liquidity as a result, and the sig value reached a significance level of (0.841), this indicate that the effect is not accepted from the statistical point of view, and accordingly we accept the null hypothesis that states that "operating cash flows hasn't positively effect on current ratio before and during Corona pandemic of Insurance sector listed on ASE". However, for the year

Table 2 Results of analysis to examine the effect of operating cash flows on current ratio

Covid19	Model		"Unstandardized coefficients"		"Standardized coefficients"	t	Sig.	R^2
			B	Std. error	Beta			
Before	1	(Constant)	1.871	0.145		12.938	0.000	0.003
		CFOs	−0.063	0.306	−0.053	−0.204	0.841	
After	1	(Constant)	1.779	0.155		11.495	0.000	0.125
		CFOs	0.646	0.288	0.501	2.241	0.041	

[a]Dependent variable: qui ratio

2020, the results showed that there is a positive effect in the sense that the higher the operational cash flows, the higher the liquidity ratio, and this is what led to value of sig 0.041 which is less than (0.05), so it was concluded that there is an effect. Which means accept the alternative hypothesis, which states "operating cash flows has positively effect on current ratio before and during Corona pandemic of Insurance sector listed on ASE".

Table 3 shows the effect of operating cash flows on financial performance that measured by quick ratio before and during coved-19 pandemic, as we note from the data contained in this table that there is no a significant and the sig value reached a significance level of (0.985), this means that the effect is not accepted from the statistical point of view and accordingly we accept the null hypothesis that states that "operating cash flows hasn't positively effect on quick ratio before and during Corona pandemic of Insurance sector listed on ASE". However, for the year 2020, the results showed that there is a significant effect between the operating cash flows and quick ratio, where the value of sig 0.050 which is less than (0.05) so it was concluded that there is an effect. Which means accept the alternative hypothesis, which states "operating cash flows has positively effect on quick ratio before and during Corona pandemic of Insurance sector listed on ASE".

Table 3 Results of analysis to examine the effect of operating cash flows on quick ratio

Covid19	Model		"Unstandardized coefficients"		"Standardized coefficients"	t	Sig.	R^2
			B	Std. error	Beta			
Before	1	(Constant)	2.137	0.131		16.253	0.000	0.000
		CFOs	0.005	0.279	0.005	0.019	0.985	
After	1	(Constant)	2.088	0.163		12.832	0.000	0.232
		CFOs	0.645	0.303	0.482	2.128	0.050	

[a] Dependent variable: qui ratio

5 Conclusion and Recommendation

The present study investigates at how operating cash flows effect on financial performance in the insurance business in ASE. Before and during covid19, the operating cash flow has a positive effect on the financial performance for insurance industries, as a result of the decrease in accidents and sickness cases, and the decrease in operational costs and public expenses due to the commitment of citizens and remote work according to the data provided. This shows that increasing cash flows in this industry would improve their financial performance. Furthermore, the companies keep enough assets and use them properly.

The study recommends applying it to other sectors of enterprise listed on the Amman Stock Exchange, taking into consideration the financial ratios most related to the nature of the work of these companies, and the nature of their activity, both in terms of solvency, efficiency, liquidity and efficiency of the financial performance of these companies. Future studies can be expanded in several ways; first, being to increase the sample size, second, being to conduct a comparison between the insurance and industrial sectors, and finally being to include another dependent as variable Net working capital.

References

1. Al Maani, A., Alawad, A.S., Karaki, A.: Impact Of liquidity and profitability ratios on the stock market value of Jordan insurance companies. Acad. Acc. Finan. Stud. J. **25**(2), 1–14 (2021)
2. Alshammari, T.: Cash level and corporate performance: evidence from the gulf cooperation council countries. Investment Manage. Finan. Innovations **17**(4), 14–24 (2020)
3. Alslehat, N., Al-Nimer, M.: Empirical study of the relationship between cash flow management and financial performance of the Jordanian insurance companies. Int. Bus. Manag. **100**(3), 776–782 (2017)
4. EY.: Covid-19 business impact survey—understanding the now, preparing for the next and beyond. Build. A Better Working World **8**(7) (2020)
5. Khresiat, O.: Impact of investment decisions on the profits of Jordanian insurance companies listed on the Amman stock exchange. Acad. Acc. Finan. Stud. J. **23**(5), 1–12 (2019)
6. Li, K., Musah, M., Kong, Y., Adjei Mensah, I., Antwi, S.K., Bawuah, J., Donkor, M., Coffie, C.P.K., Andrew Osei, A.: Liquidity and firms' financial performance nexus: panel evidence from non-financial firms listed on the Ghana stock exchange. SAGE Open **10**(3), 1–20 (2020)
7. Liman, M., Mohammed, A.: Operating cash flow and corporate financial performance of listed conglomerate companies in Nigeria. J. Humanit. Soc. Sci. **23**(2), 1–11 (2018)
8. Rahman, A., Sharma, R.: Cash flows and financial performance in the industrial sector of Saudi Arabia: with special reference to insurance and manufacturing sectors. Investment Manage. Finan. Innovations **17**(4), 76–84 (2020)
9. Shafi, M., Liu, J., Ren, W.: Impact of COVID-19 pandemic on micro, small, and medium-sized enterprises operating in Pakistan. Res. Globalization (2020)
10. Shaharuddin, H., Mahmud, R., Azhari, M., Perwitasari, W.: Company performance during Covid-19: impact of leverage, liquidity and cash flows. Environ.-Behav. Proc. J. **6**(17), 11–16 (2021)

11. Yahiaoui, S., Mahdi, O.: The determinants of financial performance of non-life insurance companies In Algeria. Revue Des Sci. Commerciales Et De Gestion **16**(1), 60–77 (2020)
12. Zimon, G., Dankiewicz, R.: Trade credit management strategies in SMEs and the COVID-19 pandemic-a case of Poland. Sustainability **12**(6114), 1–16 (2020)

An Evaluation Model of a Disruptive Technology During Covid-19 Pandemic

Mohammad Omar Sabri, Mahmoud Odeh, Khaled Al-Qawasmi, and Nidal Al-Ramahi

Abstract Disruptive Technologies have created radical transformations in different industry fields. They also changed principles governing market and competitions. This role was considered an advantage in different organizational aspects including performance, effectiveness and competitiveness. The attendance of these technologies was not bounded by these aspects alone, it has also flourished in crises and specifically Covid-19 pandemic. One of these disruptive technologies that caught interest in this crisis is cloud collaboration platforms (CCPs). CCPs were essential and critical in a key sector during the pandemic, which was education. The pandemic outbreak caused a mandatory conversion towards online education leaded by CCPs. Demonstration of these platforms during the pandemic was accompanied with their evaluation using different criteria. Still, these criteria were neither clearly justified in their reasons of use nor classified according to their importance during the pandemic. In this paper, a model of CCPs evaluation criteria during the pandemic is presented. This model is based on previous research criteria and applies inductive approach through discovering users experience regarding these criteria during the pandemic. Functionality, reliability and usability in addition to their sub-standards were the main criteria that have been assessed and ranked by users. The users who rated the criteria were academics and experts of a university in Jordan and they were engaged in online education using CCPs. Evaluation was conducted using a quantitative approach using an online questionnaire which was distributed to these academics. Findings show that security and understandability are the most significant criteria that we should look

M. O. Sabri · M. Odeh (✉)
Business Information Technology Department, Zarqa University, Zarqa, Jordan
e-mail: Modeh@zu.edu.jo

M. O. Sabri
e-mail: msabri@zu.edu.jo

K. Al-Qawasmi
Cyber Security Department, Zarqa University, Zarqa, Jordan

N. Al-Ramahi
Accounting Department, Zarqa University, Zarqa, Jordan

A. Hamdan et al. (eds.), *The Implementation of Smart Technologies for Business Success and Sustainability*, Studies in Systems, Decision and Control 216,
https://doi.org/10.1007/978-3-031-10212-7_66

after during the pandemic. Fault tolerance, accuracy, suitability and maturity standards have the second priority and finally, compliance, recoverability, learnability and interoperability have the third priority.

Keywords Disruptive technology · Cloud collaboration platforms · Online education · ISO 9126–1 standards

1 Introduction

Implications and role of advanced technologies have increased during Covid-19 pandemic. The key of them have been nominated as disruptive since they cause massive changes and result in substantial advantages in business [1]. They also enable digital transformation and change conventional business into smart one. A disruptive technology is defined as a "technology that changes the bases of competition by changing the performance metrices along which firm compete" [2, 3]. Such disruptive technologies include pillars of industry 4.0 such as big data, cyber security, 3D printing, cloud technology and internet of things (IOTs) [4].

Disruptive technologies have unfolded new vision in acting towards pandemic challenges. They revealed power in controlling, monitoring and constricting pandemic outbreaks, in addition to addressing business needs in critical situations. While these benefits were discussed by different authors, previous research is still limited in deciding what metrics are the most appropriate in order to assess disruptive technologies in pandemic situations.

In this paper, we have selected an important disruptive technology that has significantly affected the educational sector in Jordan during Covid-19 pandemic. This disruptive technology comprises all cloud-based collaboration platforms [5]. Different criteria were presented to a number of academics and technology experts in order to assess these criteria and arrange their significance during pandemic in relation to cloud-based collaboration platforms (CCPs).

2 Disruptive Technologies

Disruptive Technologies are related to emerging technical interferences which create new markets or implementations. These technologies change the dominant ones since they are usually more helpful, valuable and economic [6]. They gradually replace existing practices with their new value propositions and grow till they become prevalent [7].

Disruptive technologies have generated new tracks and opportunities for entrepreneurship and innovations in various industrial fields. Due to their indispensable role in the industrial evolution, they were appealing to be reviewed and evaluated according to these fields [8]. Such fields include education, risk management, clinical trials, learners experience and real estate.

In addition to their role in industrial evolution, disruptive technologies have participated in the restriction of Covid-19 outbreak. Cloud platforms were one of the main disruptive technologies that have been utilized in Covid-19 pandemic. Furthermore, they were common and necessary in embracing and activating other disruptive technologies such as Artificial Intelligence (AI), Internet of things (IOT) and Internet of Medical Things (IOMT) [9].

3 Cloud-Based Collaboration Platforms

The arise of cloud computing has allowed different users to accomplish their work remotely. Its role in distance working was not limited to personal tasks alone. It has expanded to cover collaboration platforms that engage team groups and audience overseas. Collaboration platforms are one of the main disruptive technologies that have changed the rules of work in the new era. These platforms have enabled better communication and demonstrated an effective utilization in various areas such as integrated programming, crowdsourcing and online education.

Online Education is an urgent field that was a demand during Covid-19 pandemic. CCPs have revealed a sustainable approach to embrace online education. They support a collaborative sharing and creation of documents, in addition to learning process through the use of cloud computing resources [10].

Adoption of CCPs in online education has unleashed a set of challenges that students and academics has experienced. Many of these challenges were emphasized and highlighted during Covid-19 pandemic. Accordingly, some analyses were conducted to CCPs in terms of interaction and communication, reliability, timely transmission of video information, access speed, course management [11]. However, a model that ranks these criteria and determines significant ones in the assessment of CCPs during the pandemic is not identified.

4 CCPs Pandemic Assessment Criteria

Different criteria were suggested and listed for CCPs assessment during Covid-19 pandemic. These criteria were evaluated by users but lacked to be reviewed and ranked regarding their importance by academics and IT experts. In addition, the criteria that are chosen for assessment should correspond to the environment where the CCPs exist. In Jordan, online education during pandemic has adopted three main CCPs; Moodle, Microsoft Teams and Zoom platforms. The authors used specific criteria which are considered essential while using these CCPs. These criteria are based on features that seem significant and are considered as indicator of proper performance and functioning of an organization's information system (IS) [12].

Features include supporting tasks that an IS was designed for, running without breaks or unpredictable interruptions, and ease of use with supportive services for users. With the previous features, we proposed three sequential corresponding criteria to be discussed and ranked regarding CCPs assessment; functionality, reliability and usability. The preceding criteria match the defined features that we have mentioned and present three of six quality characteristics introduced by ISO 9126–1 [13].

4.1 Functionality

Functionality of an IS refers to a set of capabilities that affect the existence of certain business functions and their specific properties; these functions address the declared needs [14]. It is basically the goal for which an IS designed for. Functionality of IS could be measured through [15]: (i) evaluation of the set of attributes of the program, (ii) the generality of system functions that are provided, and (iii) the security of the system.

Functionality include sub-criteria [13]; compliance, suitability, accuracy, interoperability and security. Compliance is adhering to standards in relation to governmental or industrial regulations. Suitability is the consistency between business activities and system functionalities. Accuracy is the correctness of IS functions according to specification. Interoperability is the capability of integrating with other IS components. Finally, security is safeguard and defending from any threats or unauthorized access.

4.2 Reliability

Reliability refers to the extent of which could an IS work under certain environment conditions and during a specific period of time. A reliability model of an IS include three components according to Tworek and others [12]; reliability of supporting services delivered for IS, reliability of the Information that the system offers, and reliability of the system independently.

Reliability has the following sub-criteria [13]; fault tolerance, recoverability, and maturity. Fault tolerance is system capability to afford failure and remain active. Recoverability is system ability to requite without loss of data, and lastly maturity which means rareness of system failure.

4.3 Usability

The assessment of IS or any CCP regarding its ease of use and installation is defined as the usability of this IS. Usability indicates to user experience in managing system

Table 1 A summary of suggested criteria for CCPs assessment during Covid-19 pandemic

Criteria	Sub-Criteria/standard
Functionality	*Compliance*
	Suitability
	Accuracy
	Security
Reliability	*Fault tolerance*
	Recoverability
	Maturity
Usability	*Understandability*
	Learnability
	Operability

services. It also measures the application of an IS in relation to efficiency, effectiveness, and satisfaction by certain users in order to achieve different organizational goals [16].

Usability refers to three sub-criteria in ISO 9126-1 quality model [13]; understandability, learnability, and operability. Understandability is user capability to understand system functions in relation to business activities. Learnability is about time and effort that is necessary to learn and understand the application and its services. The last sub-criteria, operability is system capability to run in work and function within few practice sessions.

Based on the preceding criteria and its sub-standards, an evaluation of CCPs would be accomplished in order to extract the most demanding ones in Covid-19 pandemic. A summary of these suggested criteria is presented in Table 1. The evaluation will not be restricted to these criteria alone and responders can have the opportunity to add any additional criteria that should have given attention during the pandemic.

5 Research Methodology and Data Analysis

After the determination of the criteria, a quantitative approach using an online survey questionnaire was conducted in order to extract a pandemic evaluation model. This model would be proposed according to the rating of the criteria in relation to the main CCPs that were used during Covid-19 pandemic. 46 academics from engineering and IT departments were selected as a sample from a university in Jordan in order to fill the survey. The sample was specifically chosen from academics with IT and engineering background since they were expected to have clearer understanding of the suggested criteria in addition to their academic experience. The questionnaire was built using google forms and its link was sent directly by email to all members of both IT and engineering faculties. The responders were asked to rate the criteria using 3-levels scale (high, medium and low). High level responses present criteria that require more attention and are given priority while using CCPs during pandemic. On the other hand, medium and low levels are successively given less priority. After each rating of the sub-criteria that belong to the major one, the responders were asked to arrange these criteria from high to low priority. Moreover, the responders have the opportunity to suggest any missing criteria that should be added to the list as well as its level of rating. Following the collection of participants' responses, data was exported to be analyzed using SPSS statistical analysis tool. Descriptive statistics were used to present the results of the responses. These statistics include the mean, minimum, maximum and standard deviation.

Depending on the responses of the participants that we have received in the survey questionnaire, we report the main following findings:

- Largest number of responses rated *security* number one in its importance among other standards in *functionality* criteria. *Security* standard has also largest number of rating responses of high scale level between all standards of main criteria with 73.3% of the respondents (see Table 2).
- 60% of the sample reported high scale level of *suitability* and *accuracy* standards of *functionality* criteria. *Accuracy* and *suitability* were also rated sequentially the second and third important standards among *functionality* sub-criteria (see Fig. 1).
- *Fault tolerance* is the most significant standard of *reliability* criteria followed by *maturity* then *recoverability* (see Fig. 2). However, *recoverability* has the largest number of rating responses of high scale level with 56.7% followed by *fault tolerance* with 50%.
- *Understandability* standard follows *security* in the largest number of rating responses of high scale level among all sub-criteria with 66.7% of the sample. *Understandability* was also considered the most significant standard out of all

Table 2 Descriptive statistics of all standards of main criteria

	Min	Max	Mean	Std. deviation
Security	2	3	2.73	0.450
Understandability	2	3	2.67	0.479
Compliance	1	3	2.63	0.556
Learnability	2	3	2.60	0.498
Accuracy	1	3	2.57	0.568
Suitability	1	3	2.57	0.568
Recoverability	1	3	2.53	0.571
Operability	2	3	2.47	0.507
Fault tolerance	1	3	2.47	0.571
Maturity	1	3	2.43	0.568
Interoperability	1	3	2.27	0.583

Note 1 = Low; 2 = Medium; 3 = High

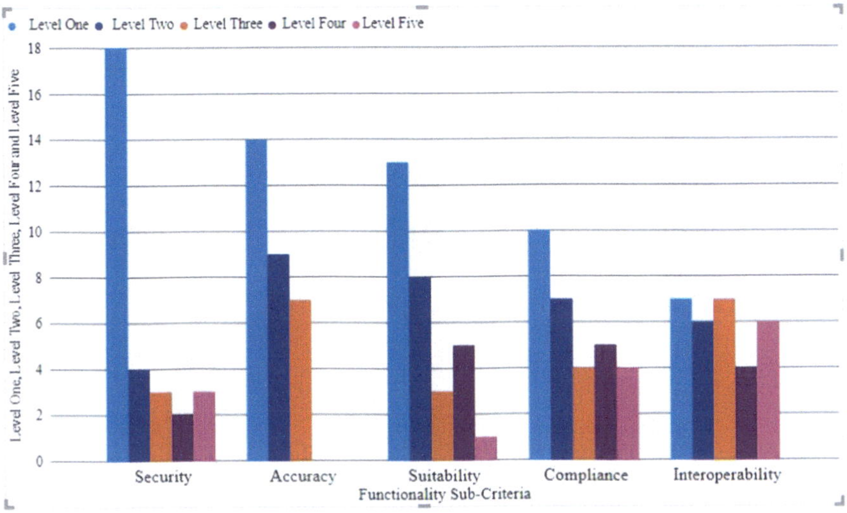

Fig. 1 Distribution of respondents ranking of functionality sub-criteria from highest to lowest levels (1–5)

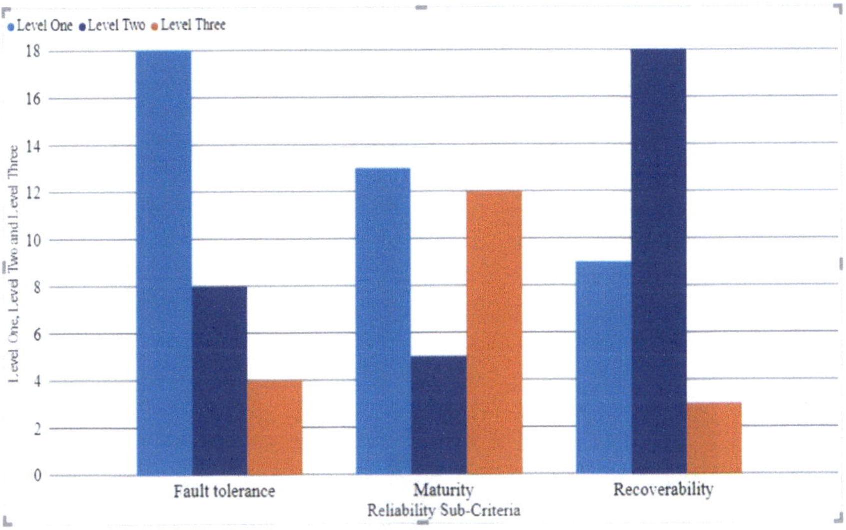

Fig. 2 Distribution of respondents ranking of reliability sub-criteria from highest to lowest levels (1–3)

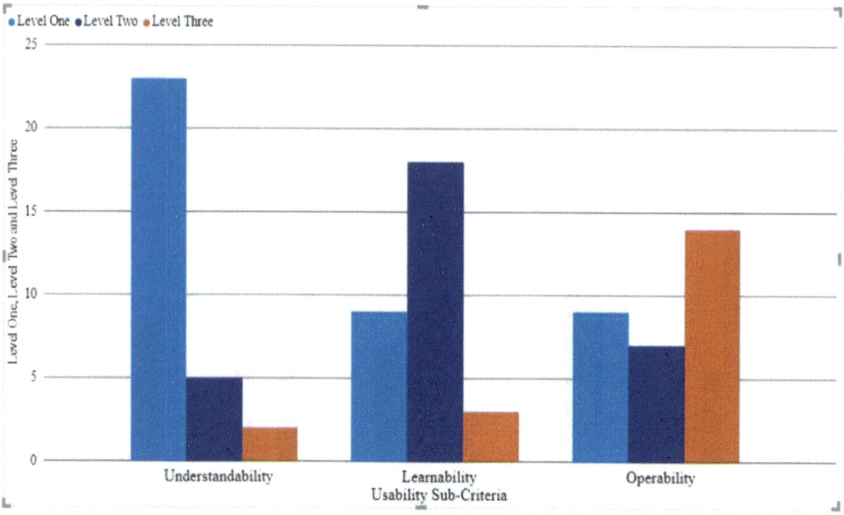

Fig. 3 Distribution of respondents ranking of usability sub-criteria from highest to lowest levels (1–3)

usability sub-criteria (see Fig. 3). In addition, it has the largest number of agreements on classification one rating among all sub-criteria including *security*.

- Finally, an overall rating of the main criteria shows that *functionality* is the most important criteria for respondents with highest priority followed by *reliability* then *usability* (see Fig. 4).

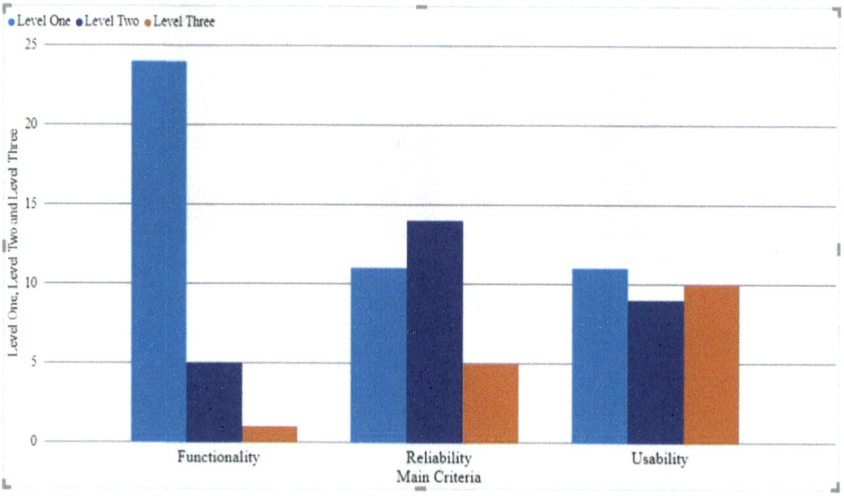

Fig. 4 Distribution of respondents ranking of main criteria from highest to lowest levels (1–3)

With reference to former analysis, a pandemic evaluation model of CCPs is proposed in Fig. 5. This model prioritizes sub-criteria according to its importance on the basis of the findings we have presented, with consideration to both main criteria and sub-criteria rating and ranking by respondents. *Security* and *understandability* were placed at first priority since they were given highest means with classification one ranking among their sub-criteria. *Fault tolerance*, *accuracy*, *suitability* and *maturity* had the second priority according to each one rating and ranking in their sub-criteria in addition to the ranking of the three main criteria (*functionality*, *reliability* and *usability*). Remaining sub-criteria which include *compliance*, *recoverability*, *learnability*, *operability* and *interoperability* were graded as third priority.

We suggest to apply the proposed model while using CCPs during urgent cases and specifically in educational sector. The prioritizing of criteria in this model could facilitate the work of IT technicians and make them understand what CCPs' users essentially need in a time-limited situation. It also highlights what CCPs' users of different sectors may give attention to while using these platforms in critical conditions. Furthermore, the sub-criteria indicate to what users usually worry about and rank first while shifting to a new technology.

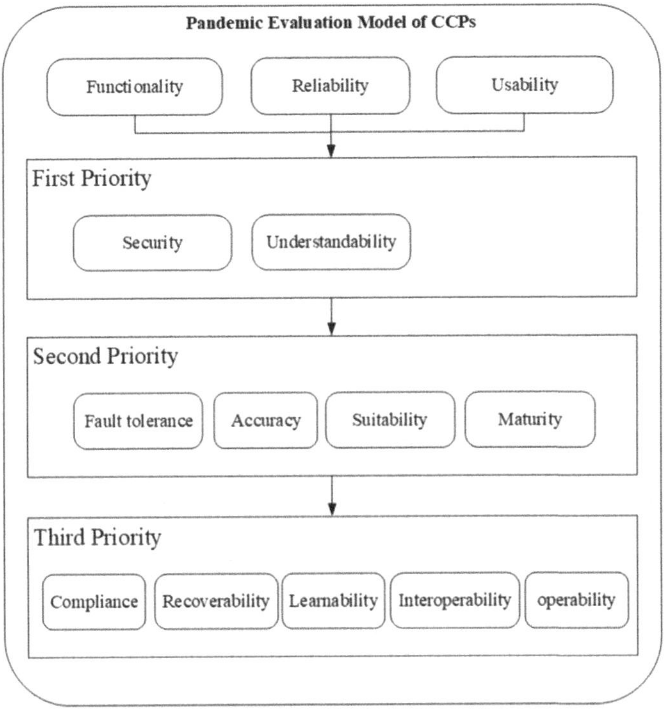

Fig. 5 Pandemic evaluation model of CCPs

6 Conclusion

Demonstration of a disruptive technology during pandemics requires prompt evaluation and consideration of criteria priorities due to time and pressure that organizations run across in such crises. CCPs are one of these technologies that countries utilized effectively for online education during Covid-19 pandemic. Different criteria were used to evaluate CCPs in the pandemic without determination of most imperative ones. In this paper, we present different critical criteria that are necessary to consider in CCPs evaluation. We have also proposed a pandemic evaluation model based on preceding research criteria in addition to academics' views of a Jordanian university by using an inductive approach method. Functionality, reliability and usability criteria besides their standards are used to develop this model. The model is divided into three levels of priorities which are sequentially arranged from first to third level according to their importance. Each priority level includes different standards or sub-criteria which are imported from the main criteria. The criteria distribution among these priorities is related to their rating and ranking in their sub-criteria as well as the ranking of the main criteria.

In conclusion, the pandemic evaluation model shows that security and understandability sub-criteria are the most significant standards to consider during the pandemic. The second priority are recommended to be assigned to fault tolerance, accuracy, suitability and maturity standards. Remaining standards are compliance, recoverability, learnability, operability and interoperability.

Further studies are suggested to conduct a comparison with these criteria significance in normal or no pandemic situations. We also recommend different case studies to be applied in the development of the pandemic evaluation model such as schools and training centers.

References

1. Siebel, T.M.: *Digital transformation: survive and thrive in an era of mass extinction.* RosettaBooks (2019)
2. Bower, J.L. Christensen, C.M.: Disruptive technologies: catching the wave (1995)
3. Li, M., Porter, A.L., Suominen, A.: Insights into relationships between disruptive technology/innovation and emerging technology: a bibliometric perspective. Technol. Forecast. Soc. Chang. **129**, 285–296 (2018)
4. Forcina, A., Introna, V., Silvestri, A.: Enabling technology for maintenance in a smart factory: a literature review. Procedia Comput. Sci. **180**, 430–435 (2021)
5. Gejendhiran, S., Anicia, S.A., Vignesh, S., Kalaimani, M.: Disruptive technologies-a promising key for sustainable future education. Procedia Comput. Sci. **172**, 843–847 (2020)
6. Sengupta, T., Narayanamurthy, G., Moser, R., Pereira, V., Bhattacharjee, D.: Disruptive technologies for achieving supply chain resilience in COVID-19 era: an implementation case study of satellite imagery and blockchain technologies in fish supply chain. Inf. Syst. Front. 1–17 (2021)
7. Flavin, M.: Disruptive conduct: the impact of disruptive technologies on social relations in higher education. Innov. Educ. Teach. Int. **53**(1), 3–15 (2016)

8. Ullah, F., Sepasgozar, S.M.E., Wang, C.: A systematic review of smart real estate technology: drivers of, and barriers to, the use of digital disruptive technologies and online platforms. Sustainability **10**(9), 3142 (2018)
9. Abdel-Basset, M., Chang, V., Nabeeh, N.A.: An intelligent framework using disruptive technologies for COVID-19 analysis. Technol. Forecast. Soc. Chang. **163**, 120431 (2021)
10. Kurniawan, D., Suganda, L.A.: Cloud collaboration: its effect toward writing achievement and impact toward attitude to learning. IRJE Indonesian Res. J. Educ. 466–482 (2020)
11. Chen, T., Peng, L., Jing, B., Wu, C., Yang, J., Cong, G.: The impact of the COVID-19 pandemic on user experience with online education platforms in China. Sustainability **12**(18), 7329 (2020)
12. Tworek, K., Walecka-Jankowska, K., Zgrzywa-Ziemak, A.: Information systems reliability and organizational performance. In: International Conference on Information Systems Architecture and Technology, pp. 201–210 (2018)
13. Kumar, V., Vidhyalakshmi, R.: Reliability Metrics. In: Reliability Aspect of Cloud Computing Environment, pp. 51–77. Springer (2018)
14. Behkamal, B., Kahani, M., Akbari, M.K.: Customizing ISO 9126 quality model for evaluation of B2B applications. Inf. Softw. Technol. **51**(3), 599–609 (2009)
15. Rawashdeh, A., Matalkah, B.: A new software quality model for evaluating COTS components. J. Comput. Sci. **2**(4), 373–381 (2006)
16. Jeddi, F.R., Nabovati, E., Bigham, R., Farrahi, R.: Usability evaluation of a comprehensive national health information system: a heuristic evaluation. Informatics in Medicine Unlocked **19**, 100332 (2020)

The Impact of Covid-19 Pandemic on the Financial Performance Indicators for Palestinian Firms Listed at PEX

Ghassan Daas and Khaled Zedan

Abstract The literature examining the impact of the Covied-19 pandemic on the financial performance through the most popular financial performance indicators. Six indictors were used as dependent variables, where the Covid-19 considered as independent dummy variable measures by (0, 1) for the periods, 2019, and 2020 responded to the before and after the pandemic measurement. 44 companies listed at PEX were searched with its five main industry categories as sectoral classification (industrial, investment, service, insurance, and banking). Data were extracted from PEX web-page and the published financial statements for the listed companies. The research findings indicates that the industrial sector is the less impacted of the pandemic consequences for the ROA, ROE, EPS, and PE indicators, whereas the service sector is the most painful with an impact of the pandemic consequences. The investment sector is also has its portion of painful results with an impact of the pandemic consequences with dark future of investment opportunities. The insurance sector is still fighting with a moderate impact of the pandemic consequences supported by the demand for insurance service as a hedge for the expected losses. Finally, the banking sector affected mainly by the ROA and ROE indicators, as the results of the closer periods, the customers' liquidity problems, and the restrictions from the regulatory bodies.

Keywords Financial performance indicators · Covied-19 · ROA · ROE · EPS · PE · Tobin's Q · Debt Ratio · PEX

G. Daas (✉) · K. Zedan
An-Najah-National University, Nablus, Palestine
e-mail: Daas@najah.edu

K. Zedan
e-mail: kl_zedan@najah.edu

A. Hamdan et al. (eds.), *The Implementation of Smart Technologies for Business Success and Sustainability*, Studies in Systems, Decision and Control 216,
https://doi.org/10.1007/978-3-031-10212-7_67

1 Introduction

Coronavirus (COVID-19) has rapidly spread all over the world; it continues to spread uncontrollably and seriously affecting all economies, especially the operations of listed companies [1]. Therefore, since the outbreak of the COVID-19 pandemic in December 2019, various waves of infections have occurred in all parts of the world, it is a global pandemic. Threatening people's health, the pandemic has also seen ongoing lockdown measures that limit personal movement and negatively affect the global economy [2]. COVID-19 pandemic is not considered only as a health problem but also disturbs the global economies and corporations either listed or not listed at financial markets. The governments fight against it including mobility restrictions, stay-at-home orders, social distance policies, and community lockdowns [1].

Covid-19 exponentially accelerated the process, stressing the need to remove obstacles to digital transformation and foster innovative growth. During the current Covid-19 pandemic crisis, the digitalization went from being a strategic option to a necessary requirement for the survival of the mains industries [3].

With the emergence of COVID-19 at the end of 2019, the cost of the epidemic was catastrophic, and the governments attempt to overcome the pandemic, and to take all the precautions to stop or reduce the spread of the virus. The consequences of the COVID-19 pandemic are significant; this created great uncertainty regarding the economic consequences of the crisis, and as COVID-19 continued to spread around the world. The listed companies as part of the economy mainly those who has the influence in the great economy harmed a lot. The results were summarized in its financial statements and reflected in its financial indicators.

Based on the foregoing, it became clear to us the importance of shedding light on the impact through the financial indicators both before and after to the pandemic even it still spread and the virus still takes new shapes and affect overall the world.

2 Problem Statement

The ability and speed of companies to adapt to changed business is determined by many factors [4]. The aim of the article is to illustrate the impact of the pandemic on the financial results of the listed firms at PEX and to seek answers to the following questions:

1. What is the behavior of the financial performance indicators before and after (still) the pandemic?
2. Is the impact of the pandemic on the financial situation of enterprises within one industry in a uniform and one-way?
3. Is there any difference of the for the financial performance indicators before and after (still) (through) the pandemic?

3 The Importance of Research

The importance of research emerges from the impact of the pandemic all over the world. The developing countries with its ability to overcome the consequences of the pandemic vary from those developed ones. Therefore, the consequences in the pandemic on the listed companies for developing countries, as the case of Palestine will also vary. So, the impact of the pandemic on the listed companies on those developing countries worth to be studied.

4 Research Objectives

In light of the research problem, this research try to achieve the following objectives:

1. Identifying the behavior of the financial performance indictors before and after (still) the Corona pandemic.
2. Identifying the impact of the pandemic on the financial performance indicators.
3. Identifying the impact of the pandemic on the financial performance indicators through sectoral analysis for the five sectors identified in PEX.

5 Limitations Research

Based on the real situation of the listed companies at PEX, and the available data demonstrated at PEX and the published financial statement of the listed companies, the following limitations were addressed:

1. Comparing the published financial statements and its calculated financial performance indicators for the available data of 2019, and 2020.
2. The period of the research represent years of 2019 and 2020 before the Corona pandemic after (through) the Corona pandemic.

6 Theoretical Framework and Hypothesis Development

Financial measures have long been recognized as the basis for corporate performance measurement [5]. Researchers measure firm performance using indicators such as Return on Asset (ROA), Return on Equity (ROE) and Earning per Share (EPS) as well as Debt Ratio (DR) [6]. Whereas others using three indicators; Internal–based performance measured by Return on Assets, Market-based performance measured by Tobin's Q model (Price/Book value of Equity) and Economic-based performance measured by Economic Value add [7].

Following to that, the most of the used of financial indicators concentrated on the return on assets (ROA), return on equity (ROE), earning per share (EPS), price earning (PE), Tobin's Q ratios for the performance measurements [8]. ROA reflects the efficiency of utilizing available assets in creating profits [9].

ROA: is one of the profitability indicators used to measure the financial performance, which is Return on Assets (ROA) or Return on Investment (ROI), with the following formula [10]. ROA = Net Income/ Total Assets).

H01: There is no statistically significant impact at the level of significance ($\alpha \leq 0.05$) of the Corona pandemic on the financial performance indicator (ROA) for the Palestinian companies listed at PEX on sectoral categorization.

ROE: represents return generation on common stocks of shareholders and is recognized as an important financial indicator for owners [9]. Return on equity (ROE) is a measure of financial performance calculated by dividing net income by shareholders' equity. Because shareholders' equity is equal to a company's assets minus its debt, ROE is being considered the return on net assets [11].

H02: There is no statistically significant impact at the level of significance ($\alpha \leq 0.05$) of the Corona pandemic on the financial performance indicator (ROE) for the Palestinian companies listed at PEX on sectoral categorization.

EPS: Earning Per Share (EPS) is the top most importance as compared to Market share, company's prestige and liquidity [12]. Maximizing EPS involving maximizing the shareholding wealth. Maximizing share price and EPS not only important for firms, but also it will create corporate image in investors' mind [13]. Basic EPS is calculated by dividing profit or loss attributable to ordinary equity holders of the parent entity (the numerator) by the weighted average number of ordinary shares outstanding (the denominator) during the period. The earnings numerators (profit or loss from continuing operations and net profit or loss) used for the calculation should be after deducting all expenses including taxes, minority interests, and preference dividends [14].

H03: There is no statistically significant impact at the level of significance ($\alpha \leq 0.05$) of the Corona pandemic on the financial performance indicator (EPS) for the Palestinian companies listed at PEX on sectoral categorization.

PE: Price Earnings Ratio (PE): the Price Earning (P/E) ratio is indicator of the investment performance. Stock market analysts especially those who believe in value investing have used price earnings (P/E) ratio as a tool to measure how cheap or expensive a stock is [15]. P/E ratio calculated as the stock price divided by the earnings per share. PE is calculated as follow [16]: PE = Market Value per Share/ Earnings per Share.

H04: There is no statistically significant impact at the level of significance ($\alpha \leq 0.05$) of the Corona pandemic on the financial performance indicator (PE) for the Palestinian companies listed at PEX on sectoral categorization.

Tobin's Q: James Tobin first mentioned Tobin Q in 1977. The Q ratio indicator was defined as the market value of a firm divided by the replacement cost of the firm's assets. The Tobin's Q ratio indicator is mostly used to calculate a company's assets in comparison to the market value of the assets of the company. The greater the value of Tobin's Q encourages companies to inject more funds to the firm. As it was indicated in the organization that the funds invested yield a great deal of return compared to the cost incurred in relation to the capital acquired. For Tobin's Q indicator the formula is [16]: Tobin's Q = Total Market Value of Firm/Total Asset Value of Firm.

H05: There is no statistically significant impact at the level of significance ($\alpha \leq 0.05$) of the Corona pandemic on the financial performance indicator (Tobin's **Q**) for the Palestinian companies listed at PEX on sectoral categorization.

Debt Ratio: the debt ratio is the indictor of the total debt of the company-to-company assets. The lower the debt indicator, the lower the source of financing through debt. The higher the debt ratio, the higher the source of financing through debt [17].

H06: There is no statistically significant impact at the level of significance ($\alpha \leq 0.05$) of the Corona pandemic on the financial performance indicator (Debt ratio) for the Palestinian companies listed at PEX on sectoral categorization.

7 Method and Procedures

Secondary sources were relied on books, periodicals, magazines related to the subject of the research. As well as relying on published financial statements for the listed companies as well as the data extracted from PEX, mostly for the financial indicators for 2019 and 2020.

7.1 Population and Sample of the Research

The research population consists of all Palestinian companies listed on PEX. With all the companies that have its data published and indicators were extracted from PEX. The following table summarizes the study population and sample (Table 1).

Table 1 Research population/sample

Sector	# of Companies
Industrial	13
Service	9
Banks	5
Insurance	7
Investment	10
Total	44

7.2 Research Variables

Dependent Variables: The dependent variable in (the financial performance indicators): through a set of financial ratios that work on evaluating the financial performance of the listed companies at PEX and based on prior research and the problem statement consisted of the (Return on Asset (ROA), Return on Equity (ROE) and Earning per Share (EPS) as well as Debt Ratio (DR)).

Independent variable: Using COVID as dummy for comparing purposes, and created, where the value of "1" is assigned to all observations during the year 2020 and "0" to the prior COVID-19 period (the year 2019). The practice of assigning dummy Impact of COVID-19 on financial performance variables within the time series data [8].

Control variable: the listed companies at PEX categorized for five main categories, (industrial, banking, investment, insurance, and service) [18]. The performance of the Palestinian companies listed at PEX may vary according to the sectorial categorization. Therefore, this research will use the sector as control variable, and will be tested.

7.3 The Statistical Methods Used

1. Descriptive analysis: which aims to describe the data of the research sample through the measures of central tendency represented (arithmetic mean).
2. Analysis of financial ratios: this aims to clarify the importance of each ratio and the pandemic impact in the research sample.
3. Paired Samples t Test compares the means of two measurements 2019, and 2020.

8 Testing Hypotheses

The hypotheses of the research were tested through the SPSS program using the Paired Samples t Test compares the means of two measurements taken from the same individual, object, or related units. These "paired" measurements can represent things like: A measurement taken at two different times (e.g., pre-test and post-test score with an interventionin order to achieve the objectives of the research and answer its assumptions, and the results were as follows:

First : H01: There is no statistically significant impact at the level of significance ($\alpha \leq 0.05$) of the Corona pandemic on the financial performance indicator (ROA) for the Palestinian companies listed at PEX on sectoral categorization.

By investigating the results at Table 2 it indicates the rejection of the hypothesis and there is significant impact of the pandemic on the ROA, and through the table the

Table 2 Impact of COVID-19 pandemic on ROA

Paired differences 95% confidence interval			T	df	Sig. (2-tailed)
ROA	Lower	Upper			
	−1.58325	−1.36812	−27.269	87	0.000
Sector			ROA		Impact
			2019 (%)	2020 (%)	
Industrial			5.01	5.03	+
Service			2.81	1.87	−
Banks			0.75	0.49	−
Insurance			2.50	2.62	+
Investment			0.50	−0.60	−

industrial and insurance sectors were not influenced, where the other sectors were influenced and in decreased indicator.

Second: H02: There is no statistically significant impact at the level of significance ($\alpha \leq 0.05$) of the Corona pandemic on the financial performance indicator (ROE) for the Palestinian companies listed at PEX on sectoral categorization.

By investigating the results at Table 3, it indicates the rejection of the hypothesis, there is significant impact of the pandemic on the ROE, and through the table the Industrial, and insurance sectors were not influenced, where the other sectors were influenced and in decreased indicator.

Third: H03: There is no statistically significant impact at the level of significance ($\alpha \leq 0.05$) of the Corona pandemic on the financial performance indicator (EPS) for the Palestinian companies listed at PEX on sectoral categorization.

By investigating the results at Table 4 it indicates the rejection of the hypothesis and there is significant impact of the pandemic on the EPS, and through the table the

Table 3 Impact of COVID-19 pandemic on ROE

Paired differences 95% confidence Interval			t	Df	Sig. (2-tailed)
ROE	Lower	Upper			
	−1.56608	−1.34574	−26.267	87	0.000
Sector			ROE		
			2019 (%)	2020 (%)	Impact
Industrial			6.12	5.37	+
Service			3.68	1.35	−
Banks			9.26	6.05	−
Insurance			8.65	9.51	+
Investment			0.95	−2.62	−

Table 4 Impact of COVID-19 Pandemic on EPS

Paired differences 95% confidence interval			t	Df	Sig. (2-tailed)
EPS	Lower	Upper			
	−1.44664	−1.20298	−21.614	87	0.000
Sector			EPS		
			2019 (%)	2020 (%)	Impact
Industrial			28.33	28.33	+
Service			10.56	7.95	–
Banks			24.16	28.64	+
Insurance			21.40	24.32	+
Investment			5.07	3.93	–

Industrial, Banks and insurance sectors were not influenced, where the other sectors were influenced and in decreased indicator.

Fourth: H04: There is no statistically significant impact at the level of significance ($\alpha \leq 0.05$) of the Corona pandemic on the financial performance indicator (PE) for the Palestinian companies listed at PEX on sectoral categorization.

By investigating the results at Table 5 it indicates the rejection of the hypothesis and there is significant impact of the pandemic on the PE, and through the table the industrial, service, banks and investment sectors were not influenced, where the other sectors were influenced and in decreased indicator.

Fifth: H05: There is no statistically significant impact at the level of significance ($\alpha \leq 0.05$) of the Corona pandemic on the financial performance indicator (Tobin's Q) for the Palestinian companies listed at PEX on sectoral categorization.

By investigating the results at Table 6 it indicates the rejection of the hypothesis and there is significant impact of the pandemic on the Tobin's Q, and through the

Table 5 Impact of COVID-19 Pandemic on PE

Paired differences 95% confidence interval			t	Df	Sig. (2-tailed)
PE	Lower	Upper			
	6.19654	148.05156	2.161	87	0.033
Sector			PE		
			2019 (%)	2020 (%)	Impact
Industrial			7078.88	8278.27	+
Service			14,876.15	32,685.49	+
Banks			786.32	4736.90	+
Insurance			1910.77	1137.32	–
Investment			−62.46	1411.66	+

Table 6 Impact of COVID-19 Pandemic on Tobin's Q

Paired differences 95% confidence interval			t	Df	Sig. (2-tailed)
Tobin's Q	Lower	Upper			
	−1.08572	−0.80290	−13.273	87	0.000
Sector			Tobin's Q		
			2019	2020 (%)	Impact
Industrial			85.65	80.62	−
Service			67.32	57.26	−
Banks			5.88	6.48	+
Insurance			34.47	28.23	−
Investment			51.40	57.56	+

table the banks and investment sectors were not influenced, where the other sectors were influenced and in decreased indicator.

Sixth: H06: There is no statistically significant impact at the level of significance ($\alpha \leq 0.05$) of the Corona pandemic on the financial performance indicator (Debt Ratio) for the Palestinian companies listed at PEX on sectoral categorization.

By investigating the results at Table 7 it indicates the rejection of the hypothesis and there is significant impact of the pandemic on the Tobin's Q, and through the table the industrial sector was not influenced, where the other sectors were influenced and in decreased indicator.

Table 7 Impact of COVID-19 pandemic on debt ratio

Paired differences 95% confidence interval			t	Df	Sig. (2-tailed)
Debt ratio	Lower	Upper			
	−1.12551	−0.88701	−16.772	87	0.000
Sector			Debt ratio		
			2019 (%)	2020 (%)	Impact
Industrial			33.28	33.17	+
Service			40.58	41.52	−
Banks			91.92	92.05	−
Insurance			71.90	72.38	−
Investment			39.98	41.52	−

9 Conclusions

Through the analysis of the research findings, the following conclusion be addressed (Table 8)

Table 8 Findings summary

Sector	Financial performance indicators (Impact)					
	ROA	ROE	EPS	PE	Tobin's Q	Debt ratio
Industrial	+	+	+	+	–	+
Service	–	–	–	+	–	–
Banks	–	–	+	+	+	–
Insurance	+	+	+	–	–	–
Investment	–	–	–	+	+	–

1. The industrial sector is the pioneer fighter with the less impact of the pandemic consequences for the ROA, ROE, EPS, and PE indicators.
2. The service sector is the most painful with an impact of the pandemic consequences for all indicators except for the RE ratio supported by the telecommunication companies performance as the result of the high intensive pressure of using the interment lines.
3. The investment sector is also has its portion of painful results with an impact of the pandemic consequences for all indicators except for the PE, and Tobin's Q ratios with dark future of investment opportunities.
4. The insurance sector is still fighting with a moderate impact of the pandemic consequences for all indicators except for the RE, debt and Tobin's Q ratios ratio supported by the demand for insurance service as a hedge for the expected losses.
5. The banking sector also affected mainly by the ROA and ROE indicators, as the closer periods, the customers' liquidity problems, and the restriction from the regulatory bodies.

10 Recommendations

To reduce the impact and the consequences of the pandemic, and based on the findings of the research, it is recommended to enhance the operational efficiency, and dig for the new market opportunities to raise-up the net income with acceptable margins. This is, as for all, the performance indictors rely on the net income as main component in calculation the main performance indicators. In addition to balance the opportunities with the debt financing. In addition more dependence on the information technology and use the digital services. Moreover, PEX shall develop the trading process to continuous working in such conditions both technical, administrative and legal.

References

1. Khatib, S.F.A., Nour, A.-N.: The impact of corporate governance on firm performance during the COVID-19 pandemic: evidence from Malaysia. J Asian Finance Econ. Bus. **2**(8), 943–952 (2021)
2. Giulia, L., Lai, A., Riccardo, S., Steccolini, I., Brammer, S., Linnenluecke, M., Demirag, I.: The pervasive role of accounting. Acc. Auditing Accountability J **1**(35), 1–19 (2022)
3. Cicchiello, A.F.: Digital transformation in South Africa: opportunities and challenges. Inf. Technol. Innovations Econ. Finance Acc. Law **1**(7), (2020)
4. Spoz, A., Fabrowska, I.S., Kotliński, G., Żukowska, H.: The impact of the Covid-19 pandemic on the financial. Eur. Res. Stud. J. **XXIV**(4), 955–976 (2021)
5. Hegazy, M.A.: The development of key financial performance indicators for U.K construction companies. Account. Accountability Perform. **1&2**(17), 49–77 (2014)
6. Nassar, S.: The impact of capital structure on financial performance of the firms: evidence from Borsa Istanbul. J. Bus. Financ. Aff. **5**(2), (2016)
7. Alkhatib, A., Harasheh, M.: Financial performance of Palestinian commercial banks. Int J Bus. Soc. Sci. **3**(3), (2012)
8. Atayah, O.F., Najaf, K.: Impact of COVID-19 on financial performance of logistics firms: evidence from G-20 countries. (2021)
9. Al-Musali, M.A., Ismail, K.N.I.K.: Cross-country comparison of intellectual capital performance and its impact on financial performance of commercial banks in GCC countries. Int. J. Islam. Middle East. Financ. Manag. **9**(4), 512–531 (2016)
10. Yuniarti, Analysis of financial performance and services performance before and during the Covid-19 pandemic (case study at Bayu Asih Hospital Purwakarta). Turkish J. Physiother. Rehabil. **32**(3), 6103–6112 (2021)
11. Investopedia, "Investopedia;," (2021). (Online). Available: https://www.investopedia.com/terms/r/returnonequity.asp. Accessed 7 Feb 2022
12. Joshi, M., Bayra, Y.: Factors affecting Investment decision in stock market. In: Business and Competitive Advantage: Creating and Sustaining Ideas to solve Contemporary Issues in Recent Era, pp. 204–209 (2018)
13. Subramaniam, V., Murugesu, T.: Impact of earning per share (EPS) on share price (listed manufacturing companies in Sri Lanka). Int. J. Innovative Res. Stud. **2**(12), (2013)
14. Deloitte, "Deloitte," Deloitte, 2 (2020). (Online). Available: https://www.iasplus.com/en/standards/ias/ias33.
15. Lakshi, V., Roy, B.: Price Earning ratio effect: a test of the semi strong form of efficient market hypothesis in Indian stock market. In: XI Capital Markets Conference (2012)
16. Fu, L., Singhal, R., Parkash, M.: Tobin's Q ratio and firm performance. Int. Res. J. Appl. Finance **7**(4), 1–10 (2016)
17. Husna, A., Satria, I.: Effects of return on asset, debt to asset ratio, current ratio, firm size, and dividend payout ratio on firm value. Int. J. Econ. Financ. Issues **9**(5), 50–54 (2019)
18. PEX (2022). (Online). Available: https://web.pex.ps

Delivery Service via Electronic Applications and Its Impact on Customers Satisfaction at Retail Stores in Amman City/Jordan

Mustafa Said Al-Shaikh and Iyad A. A. Khanfar

Abstract The aim of this study is to understand the impact of the delivery service through E applications and its impact on customers satisfaction during the Covid-19 pandemic at retail stores in Amman city by selecting the independent variable which is delivery service includes (Speed and response, product safety, quality of service, price, employees).A dependent variable is customer satisfaction and the mediating role is purchasing behavior. The research population is customers of retail stores in the city of Amman. The researchers distributed (480) questionnaires to the study population, (432) questionnaires were retrieved from the respondents from the sample with complete answers and valid for analysis. The researchers adopted the random sample method in distributing the questionnaires. The research results showed that the delivery service dimensions have a high contribution on customer's satisfaction about (56.2%). Also the employee's variable had the highest contribution among the delivery service on customer satisfaction. The study recommended that the price set based on the service quality, and restudy the pricing system to encourage the customer.

Keywords Delivery service · Customers satisfaction · Purchasing behavior · Covid-19 pandemic · Retail stores and Jordan

1 Introduction

Electronic orders delivery applications witnessed a great demand from users during the pandemic Covid-19, and enabled them to get their daily needs while they are at home. And many electronic applications have emerged that have provided the services of purchasing and delivering food, catering, medicines, medical products

M. S. Al-Shaikh (✉) · I. A. A. Khanfar
Zarqa University, Zarqa, Jordan
e-mail: malshaikh@zu.edu.jo

I. A. A. Khanfar
e-mail: i.khanfar@zu.edu.jo

A. Hamdan et al. (eds.), *The Implementation of Smart Technologies for Business Success and Sustainability*, Studies in Systems, Decision and Control 216,
https://doi.org/10.1007/978-3-031-10212-7_68

and other items, some of which offer the service of purchase and delivery, and others that provide delivery service only [1].

Based on what was previously mentioned, this study aims to emphasize the importance of delivery services and its impact on customer's satisfaction during the period of the Covid-19 pandemic at retail stores in Amman city.

2 Problem of the Study

Going shopping and contact with others carries the risk of contracting the virus. The retail stores are an ideal medium for the transmission of the virus, where a lot of people touch, check, and change goods, as well as dealing with visa cards, parking tickets, ATM buttons and receipts etc., as well as the closeness of people in those shops. Based on this, the problem of this study lies in the ability of the retail stores to provide services that suit the needs and desires of customers during this pandemic. Moreover, this study answered main question: the first one is what are the important of the delivery service and its impact on customer's satisfaction during the Covid-19 pandemic at retail stores in Amman city? And the second question refers to is purchasing behavior as mediating variable play role between delivery service and customer satisfaction.

3 Importance of the Study

The scientific importance of this study is to identify the impact of delivery service on customer's satisfaction during the Covid-19 pandemic at retail stores in Amman city. This research attempts to enrich the Arabic library to the research scholars with a topic that is considered one of the modern marketing topics in our time. Practically, the importance of the study stems from providing consumers with relevant information about delivery service, customer satisfaction and purchasing behavior. Moreover, this research can help to identify how to improve the performance of retail stores in the Jordanian market, especially in the city of Amman during covid-19.

4 Objectives of the Study

The objectives of this study is to identify the delivery service through E applications and its impact on customer satisfaction during the Covid-19 pandemic at retail stores. This basic objectives can only be achieved by the following:

1. To examine the impact of speed and response on the customers satisfaction.
2. To find out the impact of product safety on the customers satisfaction.

3. To analysis impact of quality of service on the customers satisfaction.
4. To study impact of price on the customers satisfaction.
5. To know the impact of employees on the customers satisfaction.

5 Studymodel

This model was prepared based on previous studies [1–3]:

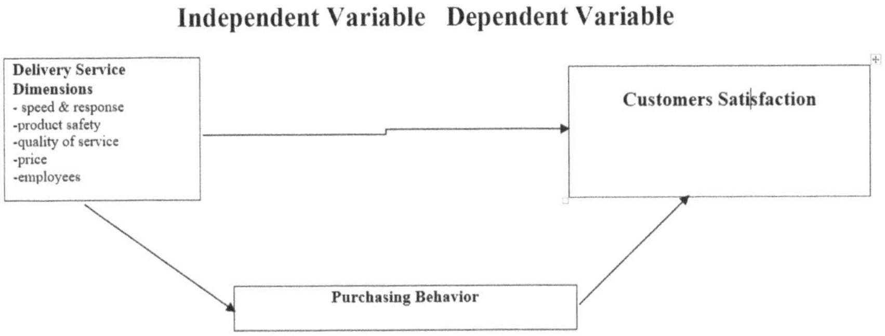

6 Study Hypotheses

This study includes the following main hypotheses:

HO_1: There is no statistically significant impact at the level of significance of α) ≤ **0.05)** for the cofactors that assist the delivery service through E applications and customers satisfaction during the Covid-19 pandemic at retail stores in Amman city. Through this main hypothesis that the following sub-assumptions can be derived:

$HO_{1.1}$: There is no statistically significant impact at the level of significance α) ≤ 0.05) of the Speed and response on the delivery service and customers satisfaction.

$HO_{1.2}$: There is no statistically significant impact at the level of significance of α) ≤ 0.05) of the product safety on the delivery service and customers satisfaction.

$HO_{1.3}$: There is no statistically significant impact at the level of significanceα) ≤ 0.05) of the quality of service on the delivery service and customers satisfaction.

$HO_{1.4}$: There is no statistically significant at the level of significance α) ≤ 0.05) of the price on the delivery service and customers satisfaction.

$HO_{1.5}$: There is no statistically significant at the level of significance α) ≤ 0.05) of the employees on the delivery service through E applications and customers satisfaction.

H02: There is no statistically significant impact at the significant level ($\alpha \leq 0.05$) for delivery service dimensions on customer's satisfaction due to purchasing behavior as a mediating variable.

7 Operational Definitions

Delivery Service: distribution services provided by business organizations to deliver consumer demand in a specific location within a specified time by using different transportation bearing the logo for the licensing of this service [4].

Customers Satisfaction: the difference between a customer's prior expectations for service performance and the actual performance of that service [5].

Purchasing behavior: It is the actions of individuals or groups that use products, services, experiences or ideas to meet their needs and the impact of these processes on the consumer and society [6].

Transportation: A vehicle equipped with a special equipment for use in the delivery of foodstuffs, whether hot, cold, or frozen, in order to reach the consumer in the required condition, or be a type of closed shipping vehicle that is tightly sealed by the manufacturer and can be equipped with tools [7].

8 Previous Studies

Researchers are interested in the subject of delivery service and its impact on customer satisfaction because of its importance in the field of contemporary marketing, we will mention some of them as follows:

Some author discuss about how purchasing behavior impact on using social media through electronic word of mouth about clothing sector in Jordan [8]. Other article mention about the customer satisfaction in the telecommunications sector might happen due to how the customers deal with the telecommunications company [5]. Moreover, short delivery time is a feature that can influence consumers' purchasing decisions [9]. On other paper for [10] Individual economic and social features are considered among the most important factors that affect and contribute to electronic delivery services in terms of spending and frequency.

Meanwhile, other study for [1] fond that there is directly impact between the customer attitude and the behavioral intention for delivery services price and employees dealing. This is in addition to multichannel service delivery quality positively impacts customer experiences, which in turn influence their continued engagement intentions [2]. Other study results show significant spatial clustering in customer demand; whilst the delivery of products to customers, in contrast, is spatially dispersed. There is a substantial gap between revenue generated and costs

[11]..Last but not least the study of [12] revealed that emotional labor moderated the relationship between service actions and customer buying behavior. Moreover, the enjoyment and trust are also key factors determining behavior intention toward customer attitude using online food delivery services, and there is appositive relationship between customer attitude and behavior intention in the online food delivery service context [13]. The delivery service, time, security, privacy, and price have significant relationship with satisfaction of customer, with delivery service has the most significant relationship [11].

9 Theoretical Framework

Nowadays customers want to enjoy their shopping with convenience and maximum value for their money. On a large scale, online retailers' self-service parcel delivery as one of the solutions to problems in the last mile [3]. It must be said that the delivery service is one of the important tasks of online purchasing related to logistics, the performance of which directly affects the level of consumer satisfaction. Moreover, this helps to save time, resources, fuel and efforts for people, which is one of the key reasons why this industry is thriving and will tend to do so [14].

[15] Al so [16] showed that the factor of e applications affects customer satisfaction through the delivery services systems. Where customer satisfaction is an important element for all services, in addition to that, it has an important role in online purchasing [9, 17], It is clear that the marketing orientation of the relationship aims to establish successful and strong relationships between customers and companies. Where this relationship must be satisfactory. In addition, relationship marketing is one of the tools that the company uses to improve customer satisfaction because customer satisfaction is among the main goals that companies seek to reach. Other authors have mentioned five important factors in terms of online customer satisfaction [13]: website design, reliability, time saved, product diversity, and delivery performance.

10 Methodology

The descriptive analytical method was used in addition to the statistical analysis within this research. Data were collected from primary and secondary sources. Where secondary sources included magazines, books and statistics. For the purpose of collecting the primary data, a questionnaire was designed and developed for this purpose.

Table 1 Reliability of the independent and dependent variables (Cronbach's alpha)

Number	Variables	Number of items	Cronbach's alpha
1	Speed and response	5	0.808
2	Product safety	4	0.795
3	Quality of service	5	0.857
4	Price	5	0.838
5	Employees	5	0.854
6	Purchasing Behavior	5	0.842
7	Customers satisfaction	6	0.889
	Over all	35	0.959

10.1 Population Size

The study population size representatives all delivery service users whatever the products. The researchers couldn't count the population, and based on based on [3], the representatives sample size shouldn't be less than (386), (480) questionnaires were distributed. Just (432) of the questionnaires were valid for actual analysis, which respondents sample size.

10.2 Reliability

The reliability of an instrument is the degree of consistency which measures the attribute; It is supposed to be measuring [18]. Cronbach's alpha coefficient method was selected among the many statistical methods to measure reliability for this research [19]. The lower limit for Cronbach's alpha value is 0.70, although according to [19]. Table 1.

We note from the following table that all Cronbach's alpha values related to the study variables are more than (0.7). Therefore, the results of the study can be accepted.

10.3 Level of Delivery Service

To measurement, the level of delivery service based on the study sample individuals point view, the researcher depend on arithmetic mean and the Table 2 show the result.

Based on the result in Table 2 the most important variable is the product safety with mean (3.60). About the lowest mean belong to the price (3.45).

Table 2 Means for delivery service dimension

Number	Variables	Mean	Importance	Level
1	Speed and response	3.64	3	Medium
2	Product safety	3.66	1	Medium
3	Quality of service	3.60	4	Medium
4	Price	3.45	5	Medium
5	Employees	3.65	2	Medium

Table 3 Variance inflation vector and the tolerance

Variables	Tolerance effect	VIF
Speed and response	0.444	2.252
Product safety	0.391	2.556
Quality of service	0.275	3.632
Price	0.571	1.752
Employees	0.367	2.721

10.4 The Multicollinearity

It is important to check the multicollinearity problem for continuous and dummy variables before running the model, we can indicates the multicollinearity problem by variance inflation factors (VIF) and tolerance impact. As stated by [20], lower tolerance (say, below 0.1) and the larger variance inflation factors (say, above 10) specify the attendance of multicollinearity, the Table 3 show the results.

From Table 3, we can see all independent variables' tolerance effect is more than (0.1), on the other hand, all VIF less than (10), indicating a lack of multicollinearity within exogenous constructs.

10.5 Hypothesis Test

Multiple linear regression was used to test the first main and sub hypotheses at the significance level ($\alpha \leq 0.05$). The Table 4 show the results.

The Multiple correlation coefficient $R = 0.750$ and $R^2 = 0.562$. This means (56.2%) of the customer satisfaction is explained by all of the independent variables together "delivery service dimension. The analysis of variance for the regression model = 109.497, Sig. = 0.000, so there is a significant impact delivery service dimension on customer satisfaction. For the variables (product safety, quality of service, price, employees), the sig. which is less than (0.05), so there significant impact on customer satisfaction. About the variable (Speed and response), the sig. is greater than (0.05), then there isn't significant impact on customer satisfaction. The

Table 4 Results of multiple linear regressions of the first main and sub hypotheses

Variables	B	t	Sig	R	R^2	f	Sig
Speed and response	0.077	1.529	0.127	0.750	0.562	109.497	0.000
Product safety	0.181	3.404	0.001				
Quality of service	0.146	2.409	0.016				
Price	0.152	4.119	0.000				
Employees	0.329	5.914	0.000				

Table 5 Indicators goodness of fit for study model

Chi square $\chi 2$ calculated	Chi square $\chi 2$ table	DF	Sig	RAMSEA
735.112	7.81	3	0.000	0.752

results indicates the employee's variable has the highest contribution on the customer satisfaction about (47%).

In order to test the second main hypothesis, the researchers used path analysis through sequence equation modeling to ensure that purchasing behavior will mediate the relationship between delivery service and customer satisfaction.. In the beginning should examinations of the goodness of fit study model. Table 5 shows the result.

GFI: Goodness of fit index

RMSEA: Root Mean Square Error of Approximation

All the indications of goodness-of-fit (GOF) indices show an adequate fit. From the Table 5 we observed that there is a significant impact of delivery service on customer's satisfaction through purchasing behavior. The calculated Chi^2 was (735.112) at level ($\alpha \leq 0.05$), which is greater than tabled Chi^2 (7.81). In addition, Root Mean Square Error of Approximation (RMSEA) is (0.752) less than (0.80) [19], which indicates the model is fit. To examine the purchasing behavior that will mediate the relationship between delivery service and the customers satisfaction, the researcher used a path analysis test, the Table 6 observed direct, indirect, and total impacts.

Based on Table 6, only one indirect effect appears, this indirect effect is (0.348), and belongs to delivery service indirect impact on customer's satisfaction through

Table 6 Direct, indirect, total impacts in path analysis

Variables	Direct effect		Indirect effect		Total effect	
	Delivery service	Purchasing behavior	Delivery service	Purchasing behavior	Delivery service	Purchasing behavior
Purchasing behavior	0.889	0.000	0.000	0.000	0.889	0.000
Customers satisfaction	0.534	0.392	0.348	0.000	0.882	0.392

purchasing behavior. But, the direct impact on customer's satisfaction is greater than the indirect impact; therefore, the researcher refused the second main hypothesis.

11 Conclusion

- The study investigated the impact of delivery service dimensions on customer satisfaction and there is strong relationship between the variables. That indicates any improvement in delivery service terms positively affect the customers satisfaction. Also, the result indicated that the delivery service dimensions have a high contribution on customer's satisfaction about (56.2%). Also the employee's variable had the highest contribution among the delivery service on customer satisfaction. The results appear the variables (product safety, quality of service, price, and employees), have significant impact on customer satisfaction while Speed and response hasn't significant impact on the dependent variable.
- The delivery service impact on purchasing behavior, on the same time purchasing behavior also has significant impact on customer's satisfaction. But purchasing behavior doesn't play as a mediate role between delivery service and customers satisfaction.
- The product safety has the most important variables among to delivery service dimensions, most introducer delivery service give the product service high importance. While the lowest mean belong to the price.

12 Recommendations

- The service delivery company raise an interested in Improving employees performance through the training course and capacity building.
- Continuous studying customer purchasing behavior to identify the changing on the customer needs and recognize the potential market.
- The price set based on the service quality, and restudy the pricing system to encourage the customer.

13 Contribution of the Study

Based on the above the study identify the delivery service through e applications and its impact on customers satisfaction during the Covid-19 pandemic at retail stores in Amman city that help retails stores to adapt new strategies to face new situation of Covid-19 pandemic.

14 Future Studies

This study discussed the delivery service through e applications and its impact on customer satisfaction during the Covid-19 pandemic at retail stores in Amman city. It must be noted that there are other points of view that can be preferred to be studied in the future from other aspects, which in turn can serve retail stores such as after-sales services, in addition to studying other factors.

References

1. Chhetri, P., Kam, B., Lau, K.H., Corbitt, B.: France cheong: improving service responsiveness and delivery efficiency of retail networks: a case study of melbourne. Int. J. Retail Distrib. Manage. **45**(3), (2017)
2. Home Delivery Service Manual: Royal commission for Jubail public services sector—property management. Kingdom of Saudi Arabia (2017)
3. Sabagh, N.: The impact of E-marketing on customer's satisfaction in the telecommunications sector (a field study). Master Published Thesis, Faculty of Economics, Syrian Virtual University, Syria (2016)
4. Hau, L., Ngo, L.: Relationship marketing in vietnam: an empirical study. Asia Pac. J. Mark. Logistics **24**(2), (2012)
5. Rudansky-Kloppers, S.: Investigating factors influencing customer online buying satisfaction in gauteng, South Africa. Int. Bus. Econ. Res. J **13**(5), (2014)
6. Chen, J.-S., Tsou, H.-T., Chou, C.Y., Ciou, C.H.: Effect of multichannel service delivery quality on customers' continued engagement intention: a customer experience perspective. Asia Pac. J. Mark. Logistics **32**(2), (2019)
7. Seger-Guttmann, T., Medler-Liraz, H.: Does emotional labor color service actions in customer buying? Journal 6 (2020)
8. Alyaseen, M.: An impact of using social media on purchasing behavior through electronic word of mouth: empirical study of clothing sector in Jordan. Master Published Thesis, Faculty of Economics, Middle East University, Jordan (2017)
9. Belanche, D., Flavián, M., Pérez-Rueda, A.: Mobile apps use and WOM in the food delivery sector: the role of planned behavior perceived security and customer lifestyle compatibility. Sustainability **12**, 42–75 (2020)
10. Chen, Y., Yu, J., Yang, S., Wei, J.: Consumer's intention to use self-service parcel delivery service in online retailing: an empirical study. Internet **28**(2), (2018)
11. Pallant, J.: SPSS survival manual: a step by step guide to data analysis using SPSS: Open University Press (2010)
12. Sekaran, U.: Research Methods for Business, 4th edn. Willey, New York (2003)
13. Jun, K., Yoon, B., Lee, S. Lee, D.-S.: Factors influencing customer decisions to use online food delivery service during the COVID-19 pandemic. J. Foods **11**(64), (2022)
14. Kumar, R., Sachan, A., Gupta, P.: Different stages of the E-service delivery system process: belief attitude–intention-framework. Int. J. Retail Distrib. Manage. **48**(7), (2020)
15. Coşar Corvinus, C., Panyi, K., Varga, A.: The importance of delivery service in online shopping. Organ. Mark. Emerg. Econ. **8**,**2**(16), (2017)
16. Ramesh, R., Prabhu, S., Sasikumar, B., Devi, B.S.K., Prasath, P., Kamala, P.R.: An empirical study of online food delivery services from applications perspective. Elsevier (2021)
17. Hair, J.F., Money, A.H., Samouel, P., Page, M.: Research Method for Business. Wiley, West Sussex, England (2008)
18. Marino, G., Zotteri, G., Montagna, F.: Consumer sensitivity to delivery lead time: a furniture retail case. Int. J. Phys. Distrib. Logistics Manag. **48**(6), (2018)

19. George, D., Mallery, P.: SPSS for windows step by step. A simple guide and reference. Allyn and Bacon, Boston, MA, USA, (2006), p. 231
20. Yasin, M.N., Alam, S.S.: An investigation into the antecedents of customer satisfaction of online shopping. J. Mark. Dev. Competitiveness **5**(1), (2010)

Websites

21. https://lusailnews.net/article/society/qatari/19/03/2020/12:08pm

Judicial Rulings Issued by Bahraini Courts in Litigations Related to the Lease Ending with Ownership "Ijarah Muntahiya Bitamleek": Pathway of Innovation Finance

Hamad Farooq Alshaikh

Abstracts The research sheds light on the judicial rulings issued by the Bahraini civil courts, the Court of Cassation, and the Judicial Disputes Chambers, with respect to the Ijara Muntahiya Beltamleek disputes, which is used by Islamic banks as a mode of financing to help customers acquire real estate property. Considering the principle of commitment to the donation that Islamic banks use to deter the procrastinating debtor, the research also aims to clarify the reasons that prompted the Kingdom of Bahrain to issue legislation to adopt the sharia standards issued by AAOIFI and use it as a reference to the judiciary when any conflict arises, and the researcher seeks to show the most important gaps that may exist because of this change. The research concludes with some suggestions and recommendations on the subject and ways to reduce the gaps in the future. The researcher has analyzed the reasons why the judiciary re-adapted Islamic finance contracts, and the reasons for the emergence of those gaps, and concluded that the application of Sharia standards as a reference in disputes will contribute to the stability of Islamic banks, but it does not guarantee that other problems will not arise because of this new legislation. More studies on the most important differences between local laws and Sharia standards, in addition to ensuring the training of a sufficient number of judges, lawyers and arbitrators on Islamic financial transactions, and a comprehensive study of Sharia standards to examine the gaps and weaknesses through which Islamic banks and their contracts may be affected.

1 Introduction

Praise be to the God, Lord of the Worlds, and prayers and peace be upon the one who was sent as a mercy to the worlds, our master Muhammad, and his family and

H. F. Alshaikh (✉)
Institute of Islamic Banking and Finance—International Islamic University of Malaysia, Jalan Gombak 53100, Kuala Lumpur, Malaysia
e-mail: hamadfarooq@hotmail.com

A. Hamdan et al. (eds.), *The Implementation of Smart Technologies for Business Success and Sustainability*, Studies in Systems, Decision and Control 216,
https://doi.org/10.1007/978-3-031-10212-7_69

all his companions, and on behalf of their followers in goodness and reform until the Day of Judgment.

The lease contract applied by Islamic bank is characterized by its name (called "Ijarah Muntahyia Bittamleek"), which was given to differentiate between it and the ordinary (operating) lease. The latter only provides a benefit or usufruct for a temporary period, in return for a specified rent, and then the leased property returned back to its original owner. While in the case of the "Ijarah Muntahyia Bittamleek" the ownership of the leased asset is transferred to the same tenant.

The idea of this product goes back to1864 AD, when conventional banks, particularly in the United Kingdom, wanted to develop installment sales applications that were prevalent at that time, as they felt that it didn't fulfill neither their requirements nor that of customers. An additional condition was added to that kind of lease agreement, where the ownership of the leased property would be transferred to the tenant automatically after settlement of the installments. Then came in the leasing contract which resolved many problems left by its predecessor, in the sixties of the last century, particularly in France, where this product was distinguished by being more obvious in terms of applying the principles of leasing with containment of a mixture of contracts, in addition to the intervention of a third party in between the two parties, as a financier. And it is worth to mention here that some Arab legislations have introduced this contract in their laws and called it as "financial lease". And in this financial lease, the lessee (buyer) is required to guarantee, insure, and provide the basic maintenance to the leased asset, while some other legislations did not address this type of lease, including the Kingdom of Bahrain.

After the emergence and development of Islamic banks, and in their pursuant to meet customers' needs, many of them began to apply the idea of financial leasing, after removing all non-Shariah compliant items, the most important of which included the separation of the sales contract from the lease agreement, especially when the lease contract entails transferring the ownership of the asset to the lessee, and applying Islamic lease provisions during the period of the lease contract, in addition to the condition of bearing the costs of maintenance, insurance and guaranteeing of the leased asset by the lessor.

Several Fatwas and decisions have been issued regarding the lease ending with ownership (Ijara Muntahyia Bittamleek), the most important of which was that stated in the decision of the International Islamic Fiqh Academy No. 44, 1988 AD, which paved the way for this Ijarah (lease) without giving details of it or addressing its forms. Then came the decision that allowed some forms of it, and stipulated Shariah control measures, as per the Council Resolution No. 110 issued in 2000, Which was a prelude to the issuance of the Shariah standards by the Accounting and Auditing Organization for Islamic Financial Institutions (AAOIFI) in 2002, specifically for the Ijarah Muntahyia Bittamleek, and its amendment in 2015.

After the passage of decades in the application of lease contract (Ijarah Muntahyia Bittamleek), by Islamic banks, some judicial and legal disputes began to pop-up between Islamic banks and their customers, especially after the global financial crisis that contributed to the bankruptcy or failure of many businesses. But what was unexpected were the judicial procedures and rulings issued from Bahrain courts regarding disputes related to the lease contracts ending with ownership (Ijarah Muntahyia Bittamleek). Surprisingly, these rulings adapted in their judgement the lease contract (Ijarah Muntahyia Bittamleek) either as being a loan, or in some cases as an installment sale. Therefore, Islamic banks were forced to deal with the cases brought by them or against them with other considerations.

Given the seriousness of these rulings, and as some of them were issued by the Court of Appeal, it was inevitable for Islamic banks to raise attention to save their banking experience, demonstrate the validity of their contracts, and relabel the concept of the Ijara Muntahyia Bittamleek. And as a result of this turmoil, there was a proposal to issue an urgent legislation, by Law No. (14) of 2020 AD, so as to amend some provisions of the Central Bank of Bahrain Law, specifically in Article (39) Paragraph (C), which should stipulate that "All transactions and contracts concluded by Islamic financial institutions shall be subject to the Shariah standards determined by a decision issued by the Central Bank of Bahrain, and no provisions other than those standards shall be taken into account." This legislation has interrupted any possible provisions that may be issued to change the qualification of contracts or rule out their invalidity.

This stage was considered as one of the most sensitive stages for the Islamic banks, here in Bahrain, and could have led to a destabilization in customers and employee's confidence as well, to Islamic banking experience. However, this experience, with its pros & cons, has not yet entered the stage of analysis, collection, documentation, and extensive discussion to identify the reasons behind issuing such rulings, which would have opened the way to probe the depths of this problem, and revealed the huge gap between the judicial system and that of the Islamic Shariah.

Therefore, and throughout this research, we will try to analyze and identify the reasons behind issuing such rulings, and what are the legal loopholes that have contributed to the issuance of such rulings, which were inconformity with Islamic Shariah principles. Perhaps this experience is worthy of being researched and analyzed to clarify and document what happened for the benefits of all people inside and outside the Kingdom of Bahrain. It is an attempt to understand the nature of those judicial rulings that have changed the adaptation of the Islamic financial contracts to other considerations, such as installment sale or loan agreements, and thus had a detrimental effect over the Islamic financial industry, and subsequently led to the interference of the Central Bank of Bahrain along with a number of Islamic banks and issuance of a prominent legislation requiring application of Shariah standards as a reference for all Islamic financial disputes.

In addition to the above, even after the issuance of the modern legislation concerning Islamic Banks, the problem has not yet ended, and Islamic banks are still facing many challenges represented in reviewing conformity of their applications of the Ijarah as an instrument of financing with the Sharia standards pertaining to this mode, and whether Islamic banks are properly applying it and appropriately verifying their contracts and products, or otherwise. Another challenge is represented in the judicial system which is still lacking specialized courts to consider the Sharia standards issued by the Accounting and Auditing Organization for Islamic Financial Institutions, which shall need as a first step to recognize the differences between civil laws and the Sharia standards pertaining to the Ijarah contract. Moreover, there is a gap that still exists between the Shariah standards and the judicial laws, especially when drafting and implementing such contracts pertaining to the Islamic modes of financing.

Our research, in this respect, shall follow the qualitative approach, and focuses in collecting data and information via open communication and conversation, and determine the reason behind the phenomenon of issuing judicial rulings based on the re-adaptation of the Ijara Muntahyia Bittamleek contract as a sale or a loan contract, and further to explore and analyze the causes of this phenomenon. We shall collect a sample of judicial rulings issued in this respect for discussion and citation purposes, and propose a number of ideas which attempt to interpret the reasons behind the adaptation of the Islamic financial financing contracts, such as lease contracts, which are originally based on the Ijara Muntahyia Bittamleek mode of financing.

2 The Adaptation Made by the Judicial System in Bahrain to the Ijara Muntahyia Bittamleek Contracts

2.1 Judicial Verdicts Issued in Disputes Pertaining to Lease Contracts Ending with Ownership (Ijara Muntahyia Bittamleek)

2.1.1 The First Case (Issued by the Courts of First Instance)

The court concluded that the agreement between the plaintiff and the defendant in its entirety is an ordinary commercial loan according to the intention of the parties, and not a lease contract in the Islamic sense, and that the sale contracts are nothing but guarantees of the amount of financing, and that the lease contract and its amendment is nothing but a cover for an agreement on a clear commercial loan, and whenever was that, the description of the Ijara Muntahyia Bittamleek was negated from the contract subject to the dispute, and that the relationship between the plaintiff and the defendant in reality is an ordinary loan relationship, and therefore the plaintiff's lawsuit is unfounded *(Bahrain court, invitation list, 2017).*

2.1.2 The Second Case (a Judgment of Cassation Issued in a Dispute Related to an Ijara Muntahyia Bittamleek Transaction)

It is well known that the Court of Cassation is the highest court in the Kingdom of Bahrain and that its judiciary and rulings are binding to lower courts. Therefore, the ruling issued by this court has a paramount importance in determining the course of future disputes related to the Ijara Muntahyia Bittamleek. And from the text of the ruling, we may quote the following: "The previous rulings shall apply even if the contracting parties called the sale as a lease agreement… It does not matter whatever the two parties has given as a name to the contract, by calling it as an Ijarah Muntahyia Bittamleek. And if the contested ruling agreed with this consideration, concluding that the rent stipulated in the contract is in fact the installments of the sale price, and thus naming of the underlying contract as an Ijara Muntahyia Bittamleek is incorrect." *(Bahrain court, invitation list, 2017).*

2.2 Effects of Judgments Issued

There is no doubt that the above mentioned two rulings issued and others similar ones are considered unfair and damaging to the reputation of Islamic banks, especially since the Court of Cassation supported such rulings, which are considered binding to the lower courts. In fact, there are many consequences to such verdicts issued by different courts, including exposure of all Ijara Muntahyia Bittamleek transactions to the contest and re-adaptation, which would have damaged effect on the reputation of Islamic banks, contracts in force, and the Shariah supervisory bodies governing compliance of such banks to the Islamic Shariah principles.

3 The Legal and Shariah Rules Governing the Ijarah Muntahia Bittamleek Contract

3.1 The Contract and the Modus Operandi of the Ijarah Muntahia Bittamleek

3.1.1 The Definition of the Ijarah Muntahia Bittamleek Contract

The Ijarah Muntahuia Bittamleek contract does not depart from being a lease agreement upon which all the provisions of the lease are applicable, and it is usually accompanied by a promise of ownership of the underlying asset at the end of its term. And Its legitimacy has been confirmed by the decision of the International Islamic Fiqh Academy." The Leasing is included in Islamic Fiqh within the named contracts. (AAOIFI, Sharia Standard).

3.1.2 Forms of the Ijarah Muntahyia Bittamleek

- **The legitimate forms of Ijarah, from Shariah perspectives**.

The first form: Initial leasing of the underlying asset(s) and then the gift.

The second form: leasing and then sale by a symbolic price.

The third form: leasing and then sale of the underlying asset by way of a non-symbolic price, which represents the agreed upon price.

The Fourth form: Leasing through sale of the underlying asset with the rest of the installments.

The fifth form: the lease and then the gradual sale.

The sixth form: sale and then leasing, using both operating and Ijara Muntahyia Bittamleek methods. *(AAOIFI, Sharia Standard—Hammad , Nazih. (2008). The Dictionary of Financial and Economic Terms in the Language of Jurists,).*

- **The legal forms of Ijarah:**

The first form: a lease that conceals a sale arrangement, and this may be realized if the intention of the two contracting parties is to sell by installments, and sale as such is considered a normal sale contract, and accordingly, the provisions of the sale shall apply.

The second form: the lease associated with a promise to sell, but the intention involves a promise to sell, and the two contracting parties initially require selling by installments, but it is contingent on the full payment of the tenant to the agreed installments, and thus this form in its legal terms is a sale, and not a lease contract.

The third form: the serious lease agreement, which is associated with a promise to sell, and in which the rental amount is suitable compared to the underlying asset's price, and the promised sale price at the end of the term is an actual price, and not a symbolic one. *(Al-Qara Daghi, Ali. (2013). the Student's Economic portfolio, Research in Islamic Financial Transactions).*

3.1.3 The Legal Relationship Between the Ijarah Muntahyia Bittamleek and the "Ownership Lease"

We found that the Ijarah Muntahyia Bittamleek differs from the lease sale in the following aspects:

1. The Islamic bank owns the underlying asset of the Ijarah Muntahyia Bittamleek, and guarantees the liability of its destruction, while the ownership of the underlying asset is transferred to the buyer in the case of a deferred installment sale, and accordingly the guarantee is transferred to the buyer.
2. The buyer is considered as a debtor if he defaulted in payment in an installment sale contract, whereas the tenant's right to stay is forfeited if he fails to pay the rental, in case of Ijarah Muntahyia Bittamleek.
3. The landlord is committed in case of the Ijarah Muntahyia Bittamleek contracts to bear the cost of maintenance and insurance on the leased property, unlike the lease sale where he is not obligated to do so.
4. The process of an Ijarah Muntahyia Bittamleek involves a lease contract and then a sale or a gift contract to transfer the ownership of the underlying asset to the

lessee at the end of the term, whereas an installment sale contract only involves one contract.

5. All the Shariah provisions of the lease agreement shall be applied to the Ijarah Muntahyia Bittamleek, whereas the provisions of a sale contract shall be applied to the installment sale. *(AAOIFI, Sharia Standard-(Al-Qara Daghi, Ali. (2013). the Student's Economic portfolio, Research in Islamic Financial Transactions)*

3.2 The Legal and the Legislative System in the Kingdom of Bahrain

3.2.1 Sources of Law

The Bahraini legislator divides legislation into official sources of law, represented in the following: legislation, customary, Islamic law, principles of natural law and rules of justice, approval, analogy, deduction, and application of personal discretion by the judge. However, the Islamic Shariah is regarded as one of the reserve sources of law. *(Ahmed, Khaled Jamal (2010). Introduction to the Principles of Bahraini Law, University of Applied Sciences).*

3.2.2 Applicable Laws to the Operations of Islamic Banks

The rules of civil and commercial laws are currently applied to the Islamic banking business. The judge usually resorts on the sources of commercial law when he issues any verdict pertaining to an Islamic banking transaction. These commercial laws include agreement, custom, and civil law.

3.2.3 The Legal Value of the Shariah Standards Issued by the Accounting and Auditing Organization for Islamic Financial Institutions (AAOIFI)

The Kingdom of Bahrain is considered one of the first countries that obliged Islamic financial institutions to operate in accordance with the Shariah standards of AAOIFI. The Central Bank of Bahrain has issued its regulations in this respect, in the year 2014, giving these regulations a legal force, and exposing its violator to financial and administrative penalties. However, despite of this clear legal obligations, controversy arose between jurists, on the status of the Sharia standards in Bahrain legislation, and the extent of their mandatory and legal force. *(AAOIFI, Sharia Standard).*

We may extract from the following opinions to determine the position of the Sharia standards in Bahraini legislation, as follow:

1. The Shariah standards are regarded as a general custom.

2. The Shariah standards are only binding rules for Islamic banks. Which means that they are binding only the concerned parties (i.e., Islamic banks), but not other parties, which include the judiciary system.
3. The Shariah standards are considered as a general regulatory framework for the Shariah provisions that would be applied only to the Islamic banks, without conflicting with the Law of the Central Bank of Bahrain and other applicable laws in the Kingdom.
4. the Sharia standards are within the original sources of the law and not among the indicative sources.

(Bu Ghammar , Haitham, the comprehensive legal interface for the operations of commercial and Islamic banks- Ahmed, Khadija, The Legal Value of Shariah Standards in the Kingdom of Bahrain).

Through the foregoing opinions, we may conclude that the obligation to apply Sharia standards is no more than being general rules binding only Islamic banks, because they were issued by the executive authority and not issued as a legislation or a law by the legislative authority, and it is known that the legislation is regarded as the first original source in Bahrain law, and that means it is not mandatory for Litigations or when applying laws. This opinion was valid prior to the issuance of the urgent legislation, by Law No. (14) in 2020, and amending some provisions of the Central Bank of Bahrain Laws, particularly Article (39) Paragraph (C) which stipulates that "all transactions and contracts concluded by Islamic financial institutions are subject to the provisions of the Islamic Shariah, which shall be determined and decided by the Central Bank of Bahrain, and no other provisions other than those standards shall be taken into consideration." Thus, the Bahraini legislator has settled the controversy overwhelmed the Sharia standards, and accordingly became an original legislation, since they were issued as a legislative provision.

4 The Reasons Behind the Re-Adaptation of the Ijarah, and the Treatment of the Gap Between the Judiciary Laws and the Islamic Banks Contracts

By extrapolating the judicial verdicts and legal defenses, we see that the reasons for re-adapting lease contracts (Ijarah Muntahyia Bitamleek), are as follows:

1. Mandatory Civil laws: Islamic banks are obligated to apply civil and commercial laws as companies listed under Bahraini law. The Bahraini legislator considers that Ijarah Muntahyia Bittamleek is an installment sale contract, in accordance to the Article No. 391.
2. There is a defect in the application of lease contract by the Islamic banks, and they are not adopting the underlying instrument in a proper manner.
3. There is a defect or omission in the claims submitted to the judiciary.
4. The judiciary is not familiar with the jurisprudence of Islamic financial transactions and Sharia standards.

(Bahrain Islamic Bank. (2018). Sessions of the Second Bahrain Islamic Bank Forum, Islamic Banks between Shariah Standards and Applicable Laws, Bahrain).

5 Treatment of the Gap Between the Judiciary Laws and the Financing Contracts Adopted by Islamic Banks

5.1 The Role of Islamic Banks in Resolving the Gap Between the Judiciary, Contracts, and Its Application

In general, the Islamic banks ought to do the following:

1. Remedy of defects and shortages in the contracts.
2. Putting pressure on the regulatory authorities to fill the legislative gap.

5.2 The Role of the Central Bank of Bahrain in Resolving the Gap Between the Judiciary, Contracts, and the Application

We would not be exaggerating if we said that the Central Bank of Bahrain role, in this respect, is the most important one, considering its responsibility as a regulator, and as a representative of Islamic banks in communicating with other government channels, including legislative, executive, and judicial authorities. However, the Central Bank of Bahrain has played a remarkable and efficient role in remedying the legislative deficiency.

5.3 The Role of the Legislature in Resolving the Gap Between the Judiciary, Contracts and Application

The legislature has an obvious role in controlling legislations and addressing gaps, including:

1. Restricting the role of the judiciary so that it is only work in reviewing the procedures and granting the executive capacity.
2. Strengthening the role of the central Shariah Board.
3. Working to qualify lawyers, jurists, judges, and arbitrators.
4. Establishing specialized Shariah courts for disputes pertaining to Islamic banks.
5. Issuance of a separate law for Islamic banks.
6. Reviewing of civil and commercial laws and amending them to comply with the works of Islamic banks.

7. Issuance of a new legislation, which specifies the legal reference when resorting to the judiciary.

Bahrain has chosen to issue the Law No. (14), in 2020, to amend some provisions in the law of the Central Bank of Bahrain, as we have previously referred to.

(Bahrain Islamic Bank. (2018). Sessions of the Second Bahrain Islamic Bank Forum, Islamic Banks between Shariah Standards and Applicable Laws, Bahrain).

5.4 Foresight for the Post-Approval Phase of the Sharia Standards

It can be said that most critical problems of the Islamic banks pertaining to the judicial disputes have been addressed, and therefore, the Bahrain judiciary has become obligated to apply what is stated in the Sharia standards issued by the Accounting and Auditing Organization for Islamic Financial Institutions. Application of these Shariah standards shall result in the stability of the Islamic banks and their legal bodies. However, it seems that we are still at the beginning of the path and need to pave our way forward, as the effects of this decision has not yet reflected upon the ground, and it is expected emergence of some issues and gaps that need to be resolved, such as:

1. It seems to be difficult for the judiciary members to comprehend the large number of the Shariah standards, pertaining to each mode of financing, and properly interpret them, given that Bahraini judiciary is still suffering from lack of specialists, so the best solution for that is to refer the transactions to the Central Shariah Board for the purpose of interpretation and recommendation, or Furthermore, an independent body from Shariah experts may be established, provided no one of them has a membership in the existing Shariah boards, and they can give their collective opinions on disputes related to Islamic banking instruments of financing. In addition to that, providing training courses in the field of the Shariah standards to the judges, arbitrators, and jurists is vital.
2. The failure of the Islamic banks to ensure a proper application of Shariah standards, even with the presence of Sharia boards. Therefore, the best solution that we propose is to establish a body affiliated from either the Accounting and Auditing Organization or the Central Bank of Bahrain, so as to issue an assessment report for extent of appropriateness of application of Sharia standards by each Islamic bank.
3. The Sharia standards have been formulated and expressed in a Shariah form, but not on a legal ground, hence, there might be some legal defects in the wording of these standards, and we believe that the best solution in this regard is to reformulate the Sharia standards in a legal form so as to fill the gaps that may exist.
4. The most effective and optimal solution that the researcher sees remains the formulation of an integrated law for banks and money exchangers in general,

in which there will be a section for conventional banks, and another for Islamic banks, in which the circulars of the Central Bank of Bahrain are collected and what is mentioned in the laws related to banking work with the adoption of Sharia standards as a main reference.

6 Conclusion

One of the most crucial and complex stages that ever faced by Islamic banks, was due to the inappropriate judicial rulings issued against some modes of financing would have almost bombarded the work of Islamic banks. However, the proposal that was raised for the purpose of issuance of a law that refers disputes to Sharia standards, has contributed to the strengthening and stability of Islamic banks, through adopting such a solid reference. However, Islamic banks are still facing many challenges. One of them is represented in reviewing their applications with the Sharia standards, and whether they are committed to them or otherwise, as well as verifying compliance of their contracts and products to such standards. Another challenge is represented in the judicial side, which still lacks specialized courts in the Shariah standards, to investigate those Sharia standards which were issued by the Accounting and Auditing Organization for Islamic Financial Institutions, and which will need as a first step to identify the differences between civil laws and Sharia standards. In addition, there is still a gap existing between the Shariah standards and the law, especially when drafting and implementing contracts. Therefore, the concerned supervisory authorities have many tasks to perform, so as to limit the negative effects of the new law.

The researcher benefited from the following research presented in the forum:

- Al Mahmoud, Abdul Latif, The Role of Shariah Supervisory Boards in Addressing the Legislative Deficiency.
- Ahmed, Khadija, The Legal Value of Shariah Standards in the Kingdom of Bahrain.
- Bu Ghamar, Haitham, Leasing ending with ownership (Ijara Muntahyia Bittamleek) in the Bahraini Civil Law.
- Al-Khuwaildi, Abdul-Sattar, Experiments in the Linkage between the Shariah and Law in the Business of Islamic Banks.
- Al-Khuwaildi, Abdul-Sattar, The Most Important common Legal and Shariah Notes in Islamic Finance—Models.
- Al-Anazi, Essam, Islamic Finance between Shariah rulings and legal systems.
- Ghamid, Salem, the extent of the need for a specialized judicial system to settle the disputes of Islamic banks.
- Yassin, Salah, reviewing examples of some judgments and judicial disputes related to Islamic banks in the Kingdom of Bahrain.
- Accounting and Auditing Organization for Islamic Financial Institutions. (2012).

Shariah Board Conference:

- Abu Ghaddah, Abdul Sattar, Legal Issues in the Islamic Financial Institutions.
- Al-Khuwaildi, Abdul-Sattar, Legal Issues in the Islamic Financial Institutions.

References

The books

1. Ahmed, K.J.: Introduction to the Principles of Bahraini Law, University of Applied Sciences. (2010)
2. Ismail, M.: The Origins of Civil Law
3. Bu Ghammar, H.: the comprehensive legal interface for the operations of commercial and Islamic banks.
4. Bahrain Islamic Bank: Fatwas of the Shariah Supervisory Board (2015)
5. Nazih, H.: The Dictionary of Financial and Economic Terms in the Language of Jurists, Damascus, Dar Al-Qalam (2008)
6. Abdel Razzaq Ahmed, A.S.: Mediator in Explaining Civil Law, Beirut, Arab Heritage Revival House (2007)
7. Abdel Moneim Faraj, A.S.: The Origins of the Law, Beirut, Dar Al-Nahda Al-Arabiya
8. Ahmed, A.W., Sayed Ahmed, I.: The lease contract in the light of the opinions of scholars, legislation and judicial rulings, Cairo, House of Justice (2018)
9. Eman, A.Q.: Financing through lease ending with ownership (Ijara Muntahyia Bittamleek), a critical reading of the Saudi financial leasing system
10. Daghi, A.Q., Ali.: The Student's Economic portfolio, Research in Islamic Financial Transactions, Beirut, Dar Al-Bashaer, 2nd edn. (2013)
11. Al-Najida, A.: The Summary of the Sale Contract. Dar Al-Nahda Al-Arabiya, Cairo (2003)
12. Hadi, W. The Origins of Contemporary Transactions Control, Qatar (2015)
13. The Accounting and Auditing Organization for Islamic Financial Institutions (2015). Shariah Standards, Manama, Bahrain
14. The Accounting and Auditing Organization for Islamic Financial Institutions (2015). Accounting Standards, Manama, Bahrain
15. Yusriya, Abdul Jalil. (2011). Mediator in the Bahraini Commercial Companies Law, Fakhrawi Foundation, Bahrain
16. Yakan, Zuhdi, Lease Contract, Beirut, Al-Asriya Library

Periodicals

17. The Official Gazette, Kingdom of Bahrain
18. Real Estate Regulation Law, Kingdom of Bahrain
19. Central Bank of Bahrain, Central Bank of Bahrain decisions and circulars for financial institutions in the Kingdom of Bahrain
20. Department of Legal Affairs, Constitution of the Kingdom of Bahrain, and its explanatory note, 2002 AD
21. Ministry of Justice, Islamic Affairs and Endowments, Bahraini Civil Law (2002)

Conferences and Seminars

22. Bahrain Islamic Bank. (2018). Sessions of the Second Bahrain Islamic Bank Forum, Islamic Banks between Shariah Standards and Applicable Laws, Bahrain.

Standardizing the Shariah Regulation of the Forensic Audit Profession of Islamic Financial Institutions

Abdul Naser Omar Al Mahmood

Abstract A juristic frame of reference is considered a crucial building block for Islamic banks, Sharīʻah boards and both internal and external Sharīʻah audit. It is aalso important at the level of regulatory and supervisory authorities. In this regard, many of those interested in the Islamic banking industry and those practicing the profession of Sharīʻah audit are looking forward to crystallizing the idea of unifying the frame of reference for the Sharīʻah pronouncements (fatwā) that are issued by the Sharīʻah supervisory boards of regional and international Islamic financial institutions. This study discusses the topic of establishing a juristic frame of reference, which refers to unifying the juristic frame of reference by relying on all the credible fatwā of Islamic banks that have been issued by various Sharīʻah supervisory boards in a country or a jurisdiction and all the Sharīʻah standards that have been issued by the Accounting and Auditing Organization for Islamic Financial institutions (AAOIFI). The topic of unifying the juristic frame of reference does not refer in any way whatsoever to the issue of unifying the juristic opinions, since it is completely different from the concept of unifying juristic opinions. Based on the problem statement, the objectives of the study has been determined, which aim to identify the concept of juristic frame of reference, its nature and relationship with Islamic financial institutions. Moreover, the study will analyze the obstacles of unifying the juristic frame of reference within the Islamic banking system and its effectiveness in any Islamic financial institution. In this context, the study reached a number of conclusions and recommendations, namely the need for the regulatory and supervisory authorities to impose the implementation of Sharīʻah and accounting standards issued by AAOIFI and in the event a Sharīʻah board of an Islamic financial institution did not follow a certain standard, disclosure should be made about such an incident. There is also the need to establish a supreme Sharīʻah board in the central bank that is tasked with providing guidance and identifying the preponderant view when different Sharīʻah boards of Islamic financial institutions differ on an issue not to unify the fatwā, but to support it with juristic evidence with the aim of disseminating knowledge among the new students.

A. N. O. Al Mahmood (✉)
Khaleeji Commercial Bank, Manama, Kingdom of Bahrain
e-mail: aalmahmood@khcbonline.com

A. Hamdan et al. (eds.), *The Implementation of Smart Technologies for Business Success and Sustainability*, Studies in Systems, Decision and Control 216,
https://doi.org/10.1007/978-3-031-10212-7_70

Keywords Frame of reference · Unifying the frame of reference · Sharī'ah audit · Separation between fatwā and audit · Sharī'ah standards · Shariah regulation · Forensic audit profession · Islamic financial institutions

1 Introduction

Islamic financial banks are institutions that adhere to Shariah compliance grounded in the tenets of the Islamic faith as they relate to commercial transactions regarding financial rights and duties. In fact, Islamic banks are required to comply with Shariah by adopting the legal code of Islam (based on the teachings of the Quran)and following the rules that govern commercial transactions following Shariah rules such as Fatwa, opinions of Shariah bodies or Shariah standards, and prohibited practices issued by (AAOIFI).

In this regard, the profession of the forensic audit of Islamic Financial institutions that heavily depends on Shariah Fatawi needs to refer to a standard set of rules. Yet, the primary purpose is not to standardize legitimate decisions; it is indeed intended to foster credibility and transparency and a fair basis for competition to enable Islamic banks to be rated on international credit rating scales. Accordingly, this will ensure the Shariah banking provisions become the legal repository recourse for litigation before civil courts and a medium to legalize the work of Shariah committees.

For instance, the governance standard No. (1) consists of "the provisions and principles of Islamic law as stated in the Fatawi's opinions and guidelines issued by the Shariah Supervisory Authority," which promulgates the laws related to the creation of the Shariah Supervisory Authority and the appointment of respective members. In addition, the external forensic audit standard no. (6), issued by the Auditing Authority of Islamic Financial Institutions (AAOIFI), stipulated that the Islamic financial institutions should refer first to AAOIFI standards, second to the local country law, third to the central Shariah authority if the country has a supreme legal body, and finally to Fatawi and decisions of the Shariah body. Therefore, the preferential arrangement will have a significant impact on the scope of the work of legitimate audit in the institutional evaluation process in terms of compliance with the provisions and principles of Islamic law, Fatawi, and decisions of the Shariah authority, and the laws and regulations stipulated by the regulatory and governance entities.

On the other side, the financial audit of conventional banks reached a steady stage with regard to the international accounting standards and Basel requirements. Moreover, financial audits for Islamic financial institutions were also required by central banks in some countries such as Bahrain and Sudan, where the financial system allows Islamic financial institutions to operate. To mention, there was a rising interest of forensic auditors to adopt the financial audit procedures. Yet, there was no actual in-depth scientific study to standardize the forensic audit rules to regulate the profession and facilitate the conflicts and incidents of Islamic financial institutions' stakeholders.

2 Research Concepts and Definitions

2.1 Definition of Shariah Rules

The Shariah authority is a specialized body that legislates the fundamental provisions and holistic rules from an Islamic financial perspective. In some instances, Islamic financial institutions can rely on additional resources like the Fatawi and decisions stipulated by the Shariah body of the Islamic Financial Institution, legislations of jurisprudential councils, and the Shariah standards issued by the Accounting and Audit Authorities related to the Islamic financial institutions.

Therefore, the control standard no. (2) stated that the forensic monitoring should benefit from the support of the authority of the forensic audit coupled with the Fatawi and resolutions stipulated by the institution's legislative body or the authority of Fatwa and Supreme Shariah control, if any.

2.2 Definition of the Fatwa

Fatwa is a device that plays a vital role in ensuring that Islamic financial principles are followed in the true spirit of Shariah. It is issued by a recognized religious authority in Islam whose rulings are based on knowledge and wisdom. They need to supply evidence from Islamic sources for their opinions regarding emerging incidents to correct people's words, actions, and behaviors. The legal standard (29) defined it as an Islamic religious ruling and a scholarly opinion that can resolve past or expected conflict or incident. This judgment should not be based on presumption or referendum ruling requests related to past or future matters.

2.3 The Concept of Standardizing Legislative References

The concept of standardizing and consolidating Shariah rules requires the reliance on a unique set of legal repository recourse extracted from previous settlements between the industry and regulators—(it is not intended to unify the doctrinal opinions)—so that this standard Shariah becomes a reliable reference for Islamic banking transactions to control their activities and prohibit illegal acts as it is the case for some countries such as Sudan and the Kingdom of Bahrain. In both countries, the central banks have imposed on the financial institutions to adopt Shariah and accounting standards issued by the Accounting and Audit Authority of Islamic Financial Institutions (AAOIFI). At the same time, the controlling entities did not cancel the Fatawi and Shariah decisions in case of violated standards by the Islamic financial institutions, provided that they will disclose such matters in their annual reports.

2.3.1 The Experience of the International Federation of Islamic Banks Related to the Inception Decision of the Federation Board of Directors

The Supreme Shariah Oversight Authority was first established in late 1982. It was formed by 20 members of the heads of the Banks' Shariah oversight bodies. The Board of Directors has the right to elect members from Shariah jurists and scholars. This authority specializes in the following:

- Follow-up of the association members of Islamic banks to ensure their compliance with the Shariah provisions. To perform its mission, it has the right to ask for any additional data from the General Secretariat of the Union from member banks.
- Review and consider any Muslim citizen's proposal regarding the legitimate activities of any of the member banks.
- Assess previous Fatawi issued by the Shariah oversight bodies of the union members in an effort to unite opinion.
- Converge perceptions and applications of Shariah oversight bodies in various banks to ward off suspicions raised about their main activities and avoid conflicts and inconsistencies in implementing Fatawi decisions.

This attempt was the first initiative to unify opinions. Yet, it didn't last for long as its activities were frozen after several meetings and upon the issuance of specific Fatawi that aim to converge perceptions and applications of Shariah oversight bodies and prevent suspicions from appearing on the bulletin of the Supreme Authority, issued by Dubai Islamic Bank in 1984.

2.3.2 Alternative Ways to Reach a Unique Legitimate Reference for Islamic Financial Banks

The standardization of the ruling sets of the forensic audit profession and the need of Islamic financial banks to refer to a single set of legitimate references becomes an urgent institutional requirement. The topic gained the interest of scholars and was addressed during important conferences and seminars, but the process of unifying Fatawi appeared an arduous and impractical task. Appropriate control and full coordination are required to narrow disagreement about Fatawi, avoid contradiction or conflict, and achieve legitimate purposes for Islamic entities' sake to prevent corruption and thefts.

Accordingly, we propose the following research question: "What is requested to standardize Shariah rules in Islamic financial banks? The answer is as follows:

1. Communication and coordination between Fatwa bodies in the Muslim world to consult opinions and decisions related to past incidents and conflicts.
2. Appropriate atmosphere to standardize legitimate references such as laws, regulations, and resolutions to realize the objective.

3. A decision by regulators such as Central Banks to make mandatory the Shariah and accounting standards issued by the Accounting and Audit Authority of Islamic Financial Institutions as the legal entity or Islamic audit companies and financial institutions.
4. Establish a higher Shariah body related to the Central Bank in the country of origin with the purpose to gather and classify all issued Fatawi and opinions in order to analyze them and match their compliance with existing standards. The work will then be compiled in an encyclopedia distributed to all Islamic financial institutions located in the country.
5. Establishing a regulatory mechanism to be agreed upon by the heads of the Shariah bodies, the Supreme Shariah body of the Central Bank, and the Shariah Council of the Shariah Standards Authority. The objective is to analyze and discuss new issues, modify previous opinions, or add a recent opinion that has not been considered.
6. Establishing the Association of Forensic Examiners to organize and defend all certified auditors in the country of origin and provides them with the latest rules issued by the supervision and legitimate review.

2.3.3 The Positive Benefits of Legitimate Standard Reference

There are many benefits to standardizing the legitimate Shariah references. Some of them are listed below:

1. Creating consensus among professionals in referring to legitimate standards, Fatawi, and decisions of the Supreme Shariah Authority when practicing the audit profession to promote the application of forensic audit of Islamic financial institutions professionally and adequately.
2. Unifying the mechanisms and procedures of internal or external forensic auditors for Islamic financial institutions.
3. Strengthening communication and coordination between Islamic financial institutions and external auditing companies.
4. Alleviating potential risks for Islamic financial institutions and their clients in cases of uncontrolled Fatawi and fostering quality control and compliance with Shariah rules.
5. Standardize and articulate the task of the internal forensic auditor in all Islamic financial institutions to improve auditors' work and efficiency.
6. A regulatory mechanism between the legitimate bodies of Islamic financial institutions, the Supreme Authority of Central Shariah, and the Shariah Council of the Accounting and Audit Authority to foster communication, address critical issues facing Islamic banks, and provide solutions and legal standards.

7- Review all previous regulations and principles issued by Islamic financial institutions and the Supreme Authority of Central Shariah and use them as a reference to control and monitor banks' activities and as a medium to solve conflicts and incidents based on the agreed-upon Fatawi.

8. Circulating all Fatawi and resolutions adopted by all Islamic financial institutions periodically to facilitate the implementation process by the regulations and procedures specified under each Fatwa.
9. Representation of Shariah bodies in Shariah Council to keep updated about new legal issues related to forensic and accounting standards.

2.3.4 Proposal to Standardize Sharia Regulations

It is recommended to develop an appropriate plan or draw up a "common legitimate policy" between the Shariah Council through the Accounting and Audit Authority and Shariah bodies through Islamic financial institutions to achieve desired objectives and goals.

The former plan should be implemented with careful consideration and due diligence for the sake of all members. Thus, it is entitled to prepare a conceptual framework to converge the different methodologies of Fatawi related to Islamic financial institutions based on the approximation principle that will account for the Fatawi methodology and priorities. We can divide principles and priorities into two groups according to the following:

First: Key and fundamental principles of the conceptual framework/ Fatawi methodology:

- Adopt the "Al-Rajeh" principle as a decisive basis for discussions and opinions
- "Al-Iakhayor" principle from the jurists ruling and permissibility sets for odd situations.
- "Al-Maslaha" principle based on maxims related to relaxation and permissibility controls.
- "Al-Ijtihad" principle of collective jurisprudence and permissibility controls.
- "Al-Thabet & Moutaghayer" principle from respective provisions and controls.

Second: Supporting principles for the approximation framework of Fatawi methodology:

- Standardize the required conditions and specifications of the members of Shariah bodies.
- Legalize the Fatawi methodology following any legal form such as law, regulation, decision, and generalization.
- Unify the operational framework of Islamic banking.
- Classify the Islamic financial institutions' operating activities.
- Re-establish the Supreme Authority for Fatwa and the forensic audit of Islamic Financial Institutions and revamp their respective roles.
- Integrate bank, finance, and Islamic investment services under the umbrella of the Islamic bank and Islamic financial institutions and foster the exchange and communication of Fatawi and decisions issued by legal bodies and printing them.

2.3.5 The Impact of the Consolidation of Islamic Rules on the Development of Innovative Islamic Financial Products

Some opponents believe that the consolidation and standardization of Islamic Shariah will negatively impact the development of innovative Islamic financial products. This view is accurate in light of the lack of visibility due to several reasons:

- Many countries where Islamic financial institutions are operating do not have special laws that govern Islamic financial institutions' activities. Scarce are such laws and they do not adequately meet the required purpose.
- There are no deliberate and specialized operational mechanisms to legitimize a legal ruling entity for Islamic banks. There exist individual initiatives that allocate importance to the Islamic economy.
- Central banks - in countries where Islamic financial institutions are operating – are not yet convinced of the importance of Islamic financial banks and their role as a key component of the domestic and international economy.
- The Shariah bodies and the Shariah Council of the Accounting and Audit Authority do not have the appropriate coordination and communication mechanisms. Such mechanisms would foster the gathering of agreed-upon decisions, legal standards, forensic accounting, Fatawi, and decisions to facilitate their access. They will become a requisite source of data that can be consulted, amended, or adjusted to keep pace with the dynamic changes in the banking sector.
- Islamic financial banks lack joint cooperation to introduce novel Islamic financial products that can promote their cohesion and unity in facing local and global economies.
- Some of the administrators of Islamic financial banks are not convinced of adhering to Islamic law in the context of banking transactions. They do not allocate great importance to this subject because of their former experience in traditional banks.

2.3.6 The Role of Regulatory Bodies, Shariah Bodies, and Professional Syndicates to Standardize the Rules and Requirements for the Forensic Audit Profession

The regulatory bodies play a crucial role in standardizing the rules and requirements for the forensic audit profession. A significant effort is needed to achieve such a goal while accounting for the importance of constantly updating and amending the said rules to cope with local and international developments. This can be done through:

- Legislation, guidelines, and regulations that regulators must issue to foster the legislation of Islamic banks rules.
- Segregating Islamic banks' regulations from traditional banks.
- Islamic banks should consult on a mandatory basis the Shariah authority, forensic accounting standards, and the standards of the Islamic Services Council.

- Appoint a supreme legitimacy body for each country and define its duty, course of actions, membership requirements, and guidance about its controlling and monitoring role of Islamic financial institutions.
- The obligation of Islamic financial institutions to establish a committee of internal control that includes professionals with solid expertise and professionalism to work efficiently.

The role of the Shariah bodies is undoubtedly essential to standardize the governing rules of Islamic financial institutions. In fact, scholars' and jurists' experience, competence, and legal knowledge contribute to a major role in clarifying the Shariah purposes and improving the efficiency of Islamic financial institutions and respective economies.

2.3.7 Recommendations

The most important recommendations are:

- Establish the association of forensic auditors and foster its role and effectiveness to develop an appropriate set of regulations, controlling mechanisms, and ethical conducts. The association also facilitates the convergence of discussions and the dissemination of essential rules. It provides adequate models and appropriate support for forensic auditors in the context of Islamic financial institutions.
- The role of supervisory oversight bodies lies in two aspects:
    - o Legislation of Islamic banks' regulations in each country keeps a constant follow-up on significant changes and updates.
    - o Compliance with the Shariah and accounting standards issued by the Audit and Accounting Authority for Islamic financial institutions. In some countries where Shariah rules are non-mandatory, bank stakeholders can refer to them as a guiding tool to resolve conflicts and disputes.
- The Supreme Council's presence in each country under the central bank supervision has the role of coordination and guidance. It also has the power to resolve conflicts between The Shariah boards of Islamic financial banks. The main objective is not to unify Fatawi but to support it with tangible evidence to spread jurisprudence and knowledge in the said matters.

References

1. Ibn Al-Kayem AL Jawziya, Informing the Lord Signatories, the Library of Ibn al-Qayyim Series, Ibn al-Jawzi Publishing House, Saudi Arabia, 1st edn, Rajab 1423 AH.
2. Abu Al-Bassal, Abed Al-Nasser, Fatwa Controls, 12 Nov 2018. www.aliftaa.jo
3. Abu Naeem Ahmed bin Abdullah, The Ornament and Layers of The Esfiyeh, Al-Khanji Library Cairo, Dar al-Fikr for Printing, Publishing, and Distribution, Part 6. (1996)

4. Abu Ghadda, Abdul Sattar, Reasons for the Differing Fatawi of The Shariah Bodies of Islamic Financial Institutions, Doha II Islamic Finance Conference, Oct 10 2011, Doha, Qatar. (2011)
5. Abu Ghadda, Abdul Sattar, Research in Islamic civilization and values p. 308, and from the research of the methodology of comparison between the doctrines of jurisprudence, and from its sources: colleges for the Kafawi 61, approvals for Al-Shatabi 4/214 and 222, the almighty conquest of Ibn al-Hammam 6/294, and Hashiah Ibn Abedin 4/331
6. Al-Husseini, A.B., Mohammed, Prerequisites of Investigation in Forgery and Fabrication, Dar Al-Qadri, Damascus, Beirut, 2nd edn (1997)
7. Hammad, Nazih, The role of legitimate bodies and the need for their existence, Doha II Conference on Islamic Finance, Oct 10 2011, Doha, Qatar. (2011)
8. Al Khafif, The summary of legislative transactions, the light of the lights of the trumpet in Anwah Al fourouk, Karafi, 3rd edn
9. Al-Baali, A.H.: Investment and Legal Control. Publisher Faisal Islamic Bank, Turkish Republic of Cyprus, 1st edn (1991)
10. Al-Baali, Hamid, A.: Legalizing the activities of Shariah bodies, features, and mechanisms, the third conference of the legitimate bodies, Oct 5–7 Manama, Kingdom of Bahrain. (2003)
11. Al-Baali, Hamid, A.: Islamic banking legislation comparative analytical study and mechanism for preparing integrated model legislation, Dallah Al-Baraka Group—Bahrain.
12. Salami, Mohammed: Reasons for the differences of fatawi of Shariah Bodies. In: 2nd Doha Islamic Finance Conference, Oct 10 2011, Doha, Qatar. (2011)
13. Al-Fazih, Mohammed: Standardization of forensic audit profession requirements. In: A Paper Presented at the Third Conference of Forensic Audit, May 18, 2011, Manama, Kingdom of Bahrain p. 176 Shura Encyclopedia for Forensic Audit Research. (2011)
14. The International Islamic Jurisprudence Complex, 1427/2006 in two recommendations attached to resolution 153 (2/17)
15. Bari, M.A.: The safety of implementing the resolutions of the Shariah authority. In: A Working Paper Presented to the Symposium on the Future of Islamic Banking, Al Ahli Commercial Bank, Jeddah. (2008)
16. Bari, M.A.: Integrated general framework for the governance of the Islamic financial industry. In: An Integrated Study Presented at the 14th Conference of The Shariah Bodies of Islamic Financial Institutions of the Islamic Financial Institutions Accounting and Audit Authority, Mar 22–23, 2015, Manama, Kingdom of Bahrain. (2015)
17. Bari, M.A.: Future prospects for legitimate oversight, vision for development. In: Paper presented to the Islamic Finance Forum on the Future of Islamic Finance, Arab Academy for Shariah Research, and Islamic Institute for Research and Training p. 5. (2010)
18. Bari, M.A.: Unifying the legitimate reference of the Islamic financial industry at the international level. Al Salam Mag. Islamic Econ. (1), (2020)
19. Accounting standards, Accounting and Audit Authority of Islamic Financial Institutions (AAOIFI), paragraph (13) governance standard No. (1), Appointment of an oversight body, inception, and composition, and governance standard No. (2) Shariah control, paragraph (2).
20. Accounting standards issued by the Accounting and Audit Authority of Islamic Financial Institutions, standard (2) Shariah supervision p. 16 (2003–2004)
21. External Forensic Audit Standard No. (6), Accounting and Audit Authority for Islamic Financial Institutions
22. Legal Standard No. (29), Fatwa controls and ethical conduct in institutions, paragraph 2/1 and 2/2 p. 747
23. The website of the Islamic Financial Services Council www.ifsb.org
24. The website of the International Islamic Jurisprudence Council
25. The website of the Islamic Jurisprudence of the Association of the Islamic World https://ar.the mwl.org
26. The website of the Accounting and Audit Authority of Islamic Financial Institutions www.aao ifi.com

Interactive Procedures for Group Selection of Economic Mechanisms for Solving System Problems

Fisun Konstantyn, Tkachenko Yurii, Cherkashyn Oleksandr, Chukhlatyi Andrii, and Reznik Nadiia

Abstract The methods of expert assessments, which imply the formalization of the group selection procedure, have been investigated. This takes into account the fact that an interactive approach to conducting expert assessments and the formation of a group choice is possible. The situation when it is impossible to obtain an acceptable solution for the agreement of all members of the expert group on the proposed options for alternatives implies a choice between the conditions of selectivity and efficiency of the general opinion. The situation when it is impossible to obtain an acceptable solution for the agreement of all members of the expert group on the proposed options for alternatives implies a choice between the conditions of selectivity and efficiency of the general opinion. The following concepts are defined: group choice functions, individual choice functions. Individual preferences of experts are determined based on the personal opinion. In this case, the status of the participant does not play a role, however, an option is possible when the significance of the opinions of one or another participant in the approvals is assumed. Finding a compromise solution is an important aspect in the development of the strategic direction of an organization of any profile. Interactive group selection procedures are characterized by three main components. The first step of the procedure assumes an axiomatic approach to the group choice rule. The second solution is determined by expanding the structures of the family of the decision set. To increase the effectiveness of the group's opinion, it is envisaged to appeal to experts in order to clarify their opinion on alternative proposals, taking into account the minimum number of votes. The implementation of such a procedure makes it possible to eliminate the possibility of manifestation of the dictator's rule in conditions of group choice with expert assessments.

Keywords Group choice · Procedures · Interactive coordination · Control system · Efficiency

---

F. Konstantyn · T. Yurii · C. Oleksandr · C. Andrii · R. Nadiia (✉)
Department of Economic, National University of Life and Environmental Sciences of Ukraine, Kyiv, Ukraine
e-mail: nadya-reznik@ukr.net

A. Hamdan et al. (eds.), *The Implementation of Smart Technologies for Business Success and Sustainability*, Studies in Systems, Decision and Control 216,
https://doi.org/10.1007/978-3-031-10212-7_71

1 Introduction

The ambiguity associated with the construction of a formal theory on the behavior of a group of experts within the frame work of the management system for socio-economic objects is one of the problems of methodological support for the coordination of the interests of various parties in the frame work of multi-project planning. In the plane of the problem of collective interaction, essentially different aspects of this problem statement are investigated. The structure of work execution of the process of implementing the methodologies of the theory of collective decisions in the form of interactive procedures with the corresponding definition of the problem of group choice is presented. The schematic nature of this procedure is a reflection of the general procedure for constructing the sequence of actions of the managing entity responsible for the implementation of the project. For the formal relation of the considered procedure, some assumptions are fixed regarding the model, which determines the order of implementation of the search for an agreed choice. The problem is that the more choice alternatives are simultaneously in the control system, the greater the degree of danger that a decision will not be made. A conflict of priorities between interested parties is possible. One of the ways in resolving the conflict is to make an informed decision by a group of experts.

Decision-making in complex economic, technical, administrative systems is characterized by a high degree of uncertainty, which is due to insufficient information or differences in the positions of the parties involved. The concept of decision-making is based not only on quantitative characteristics, but also on factors that do not always have quantitative measurements. The field of application of qualitative data is wide: forecasting scientific and technological development, production management, planning research work, designing new technology, allocating reserves, evaluating complex scientific and technical programs and products of enterprises, developing socio-economic plans.

2 Analysis of Research and Publication, Definition of the Problem

K. Errou made an important contribution to the study of the economic theory of information, uncertainty and risks due to incomplete knowledge of economic agents [1]. The problem of the influence of uncertainty on economic group fit is the focus of modern economists. Nevertheless, where uncertainty is concerned, there is an opportunity to remove it, or at least reduce it.

The issues of efficient economic mechanisms for solving systemic problems have become very urgent in Ukraine now.

Despite the existence of an objective need for joint decisions, this does not prevent different participants in coordinated actions to evaluate differently the options for implementing a common event for them, their collective actions and their results

[2–5]. Such discrepancies in assessments of the preference of various options may reflect really conflicting interests, fundamentally different views, both on strategies for achieving results and the interpretation of the goals of cooperation. The problems of making a group decision are considered from a unified position [19]. This is quite justified, since empirical data indicate that there are general patterns in the behavior of people when making economic, political, social and technical decisions.

Planning for the development of the socio-economic system consists of planning individual projects [19], which have different time intervals and resource provision [8]. At the empirical level, an objectively necessary condition for developing a program for the development of a system of any level is the correspondence between adjacent elements of projects of different functional purposes with different dimensions [9, 10]. Adjacent elements of the system make up the structure of any system, which is changed periodically under the influence of various factors [8].

The problem of non-quantitative group choice is investigated in four different settings. The difference between these statements is how the concepts of "individual opinion" and "group opinion" ("opinions of a collective of individuals") are formally defined [6].

The main content of the theory of group choice is made up of procedures based on the coordination of individual opinions in the group; at the same time, two approaches can be distinguished—the metric approach and the approach based on the principles of rationality, reasonableness, which the considered procedures must satisfy. Within the framework of the second approach, more particular problem of analysis is naturally distinguished, when the principles (axioms) that the investigated matching rule satisfies are analyzed, and more general synthesis problem, when it is required to describe all the matching rules that satisfy a given set of axioms. The relevance of these tasks is due to the fact that they formalize the "natural" requirements for group selection procedures, and help in practice to choose a necessary rule from the set of existing rules taking into account specific requirements [11].

One of the main requirements (conditions) for the structure of group choice in the process of managing organizational systems is the requirement that the final solution is nonempty and non-trivial (when the set of selected options is always less than those required for comparison) of the final solution. The requirement of nonempty is called the selectivity condition, and the requirement of nontriviality is called the efficiency condition. Failure to comply with these conditions of group choice will not allow making any general decisions required for management. Therefore, it seems necessary to study the conditions for observing the selectivity and effectiveness of group choice for a wide class of the procedures most often used in practice for agreeing a set of individual opinions.

There is a noticeable gap between works of a theoretical and applied nature in the scientific literature on the problems of choice. The theory considers the logical foundations of choice, axiomatic, general principles of rationality and agreement, and most of the applied works are devoted to the construction of specific models and procedures, to a large extent heuretic [2, 14]. In this situation, the issues of development and systematization of theoretically substantiated methods that can serve as a methodological basis for solving applied problems are brought to the fore.

3 Group Selection Procedures

3.1 Publication Output

Direct algorithms do not always make it possible to realize the required optimality of the matching procedures, that is, the given set of conditions for the "reasonableness" of the group choice turns out to be inconsistent. Therefore, one has to either refuse to comply with certain conditions, or introduce additional restrictions on the profiles of individual selection functions (SF).

Another possible method for constructing a selective and effective group choice function is the procedure below. The implementation of this procedure allows one to obtain a non-empty and non-trivial group choice with a minimum, in a sense, the number of selected alternatives (the so-called "subtlety" property of group choice is fulfilled). In this case, the algorithm for obtaining a group decision is interactive, and as the basic (initial) rule for group choice, an unmanipulated rule is used that satisfies the conditions of monotony, neutrality, and the absence of a dictator.

The object of the research is the sports complex "Lokomotiv" named after Hero of Ukraine G. Kirpa (JV "Locomotiv"), which is in Kharkov. The main task and function of the SC "Lokomotiv" is to participate in the development and implementation of programs for sports and recreation activities at the place of residence and in places of mass recreation of the population.

The following were chosen as the main indicators of the development of SC Lokomotiv for the examination by the top management of this organization:

(1) Creation of an information databank on issues that meet the main objectives and activities of the sports complex;
(2) Implementation of measures for information, scientific and methodological support of sports, health and sports activities in the social sphere;
(3) Organization and holding of competitions, competitions of demonstration performances, festivals, sports events and other events;
(4) Organization of research activities on modern health technologies for different groups of the population;
(5) Development of direct international relations, including the conclusion, within its competence, relevant agreements on specific cooperation;
(6) Assistance in strengthening the material and technical base.

Let's introduce the following designations: x, y—alternatives; A—many alternatives; M—many significant factors; GPF—group preference function; n is the number of experts; l is a subset of the set of alternatives; R—binary relations expressing individual preferences; DM—a decision-maker; k—majority—group choice rule (GCR).

Let's define three main components of the following procedure:

1. At the first step of the procedure, the group choice function is constructed according to the axiomatically described rule.

2. The selectivity of the choice function is achieved by extending the structure of the family of decision sets.
3. To increase efficiency, it is envisaged to re-appeal to individuals in order to clarify their views on the alternatives included in the "best" group, but with a minimum number of votes.

A repeated appeal to an individual assumes that if the individual considers the alternative $x \in A$ "best", but the alternative $y \in A$ is not, then after repeated appeals, either the situation will not change, or both alternatives will not be "best". Therefore, at each step of the procedure the number of "best" alternatives in function choice FC (if they were) does not increase.

The process stops either when there is only one best alternative in each admissible set $S \subseteq A$, or when, after the next appeal to individuals, their opinions have not changed.

Two options for the implementation of the interactive procedure are considered:

- Individual opinions are set in the form of preference relations in the class of dichotomous rankings, which ensures thecyclicity of group preference;
- Individual opinions are set as FC.

Let us describe the first variant of the procedure implementation.

An arbitrary group preference function (GPF) f allows one to associate any profile [15] with a certain subset of the "best" alternatives $M \subset A$ if at least one individual can compare in terms of preference the alternatives from a given set A. We call M the set of maximal elements, maximal in the sense that these alternatives are recognized as "best" by a large number of individuals compared to the rest.

Let us consider the algorithm for working out a group decision using the example of the weighted majority rule. It is easy to see that without loss of generality, the weights of the experts p1, p2,…, pn can be determined by natural numbers. For any subset, we denote $L = \text{card}\ (I)$ and

$$P(I) = \sum_{i=1}^{L} P_i, \mathrm{P} = \mathrm{P(N)} = \sum_{\mathrm{i}=1}^{\mathrm{n}} \mathrm{P_i}. \quad (1)$$

We define the family Wg as follows:

$$\mathrm{Wg} = \{\mathrm{I} \mid \mathrm{I} \subset \mathrm{N} \xi \mathrm{P(I)} \geq \mathrm{g}\}, \quad (2)$$

where the number g is chosen in accordance with the following statement.

Statement. Let $\mathrm{J} = \left\{\mathrm{I} \mid \mathrm{I} \subset \mathrm{N} \xi \mathrm{P(I)} \leq \frac{\mathrm{P}}{2}\right\}$, and I^* satisfies the conditions:

(1) $\mathrm{I}^* \subset \mathrm{J}$;
(2) $\forall \mathrm{I} \in \mathrm{J} \quad \mathrm{P(I^*)} \geq \mathrm{P(I)}$.

The Wg family satisfies the conditions:

$$\forall I,\ I' \in Wg \quad (I \cap I' \neq \phi), \tag{3}$$

if and only if $g > P(I^*)$.

Evidence. Suppose that $\forall I,\ I' \in Wg \quad (I \cap I' \neq \phi)$ and the set J are also defined according to the condition of the statement. Let's define *g* as $g \leq P(I^*),\ I^* \in J$. Let, $P(I^*) = \frac{P}{2}$, then, if $g \leq P(I^*)$, among $I,\ I^* \in Wg$ you can always find such that $P(I) = P(I') = g$ and $I \neq I'$ while $P(I^*) > P(I)$. Hence, $P(I) < \frac{P}{2}$ and $P(I') < \frac{P}{2}$. By definition $P(I) = \sum_{i=1}^{L} P_i$. So if $\sum_{i=1}^{L} P_i < \frac{P}{2}$, then $L < \frac{n}{2}$ $(n = card(N))$.

If $P(I^*) < \frac{P}{2},\quad I^* \subset J$, then that is all the more there is a contradiction with the statement condition. Therefore the necessity of the statement conditions is proved.

The sufficiency of the assertion condition $g > P(I^*)$ for the fulfillment $I,\ I' \in Wg \quad (I \cap I' \neq \phi)$ is obvious if we take into account that $P(I^*) = \sum_{i=1}^{L} P_i = \frac{P}{2}$ $(i \in N)$. Then $g > \frac{P}{2},\ \ n - L = L'$ and for every $P(I) = \sum_{i=1}^{L} P_i > \frac{P}{2}$.

The statement is proven.

Let f be an arbitrary GPF, $R \in R^n$, $x, y \in A$ and $x \neq y$.

Suppose $I\tilde{x}\tilde{y}(R)$—the set of all individuals i that consider x and y equal, i.e., $x\tilde{R}iy$, and through $Ixy(R)$—the set of all those individuals i that consider x being no worse than y, i.e. $xRiy$.

When the condition $g > P(I^*)$ is satisfied, the family Wg for the basic (initial) rule defines a monotonic GPF fg as follows:

$$x\bar{f}_g(R)y \Leftrightarrow \overline{Ixy}(R) \in Wg \Leftrightarrow P(\overline{Ixy}(R)) \geq g. \tag{4}$$

This expression shows that the larger the number g [16], the more justified the group choice, but the more likely that the group choice of the "best" alternatives will be empty.

Let us introduce the notation: S_1 and S_2 are labels that allow you to control the process when $M = \phi$.

3.2 *The Results of the Study on the Algorithmicization of Procedures for the Agreement of Individual Opinions*

The procedure for agreeing individual opinions in the management process of SC "Lokomotiv" is implemented in the following sequence of steps.

1. Initial opinions of all experts are set in the form of a profile of individual preferences. Let R denote (the set A is divided into two classes of "best" and "worst" alternatives). g should be accepted equal $P(I^*) + I$, or determine according to the direction of the decision-maker (DM). Previous selection set $M_{\Pi p} = \phi,\ S_1 = 0,\ S_2 = 0$.

2. To form an attitude according to the profile of individual preferences. On a computer, this will be reduced to the formation of the incidence matrix of the dimension relation m × m.
3. Find the set M of maximal elements in A relatively preference $\overline{\rho}$.

$$M = \{x \in A \mid \forall y\ P(I\overline{yx}(R)) < g\}. \quad (5)$$

Let us denote the real choice set by Mn.
4. Check whether the set Mn is empty. If $M_H = \phi$, then go to step 5, otherwise go back to step 13.
5. Remember the current value of the individual preference profile $R_H = R$. Remember $S_2 = I$.
6. Select a set of alternatives from the "best" class, the choice of which is not yet sufficiently substantiated. In order to further clarify the preference of these alternatives, find some set of pairs of alternatives D as

$$D_{p*} = \left\{(x, y) \in A \mid \forall x \in M \xi I \tilde{x} \tilde{y} \neq \phi \xi\ P^* = \min P(I\overline{xy})\right\} \quad (6)$$

7. Remember $D_H = D^{p^*}$.
8. Check if the set DH is empty. If $D_H = \phi$, then go back to step 9, otherwise go back to step 15.
9. Check if a new pair has appeared in DH. If $D_H \backslash D_{\Pi p} \neq \phi$, then go back to step 11, otherwise go to step 10.
10. Define a new set of pairs

$$D_{p*} = \left\{(x, y) \in A \mid \forall x \in M \xi I \tilde{x} \tilde{y} \neq \phi \xi\ P^* = \min P(I\overline{xy}) + 1\right\}. \quad (7)$$

Go back to step 7.
11. Each individual from the set $I\tilde{x}\tilde{y} \neq \phi$ is offered to carry out an additional division of the class containing pairs of alternatives DH (to distinguish these alternatives according to their preference). Form a new profile R′ of individual preferences. Remember $D_{\Pi p} = D_H$.
12. Check the condition of profiles coincidence R_H and R′. If $R_H = R'$, then go back to step 15, otherwise—put $R = R'$ and go back to step 2.
13. Check the condition $S_2 = I$. If it is satisfied, then go back to step 14, otherwise—to step 15.
14. Check the condition $S_1 = 0$. If it is working, then assign $S_1 = I$ and go back to step 15, otherwise go back to step 16.
15. Check the condition $g < P$. If it is fulfilled, then put $g = g + I$ and go back to step 2, otherwise go back to step 16.
16. Remember the result set M of the group choice, providing its profile R of individual preferences and the final value g. Finish the procedure.

If experts and DM are equal, then the k-majority rule is used in the procedure. In this case, all p1, p2, ..., pn must be set equal to I.

We will assume that the set $\Re$ of individual FC always satisfies the normality condition: for any alternative $x \in A$ there is an admissible set $S \in W$ and FC $h_1, h_2 \in \Re$ such that and $x \notin h'(S)$. The normality condition ensures for any $x \in A$ and $I \subset N$ existence $H \in \Re$ and $S \in W$ such that

$$I = I_x(H, S).$$

Violation of the conditions of normality for $\Re$ means the existence of such an alternative that experts never choose, or that is included in every admissible set. Each individual is obliged to include it in the set of selected alternatives. A set of selective and effective FC satisfies the normality condition if:

$$\cup S = A,$$
$$S \in W',$$

where W' is obtained from W by discarding all single-element subsets.

If the assertion conditions for the family w_g are satisfied, then v(H) for the basic group choice rule defines a monotone function as follows. For all S, belonging to the set of subsets W of the set A, formed in the process of polling experts:

$$x \in v_g(H) \Leftrightarrow I_x(H, S) \in w_g \Leftrightarrow P(I_x(H, S)) \geq g. \tag{8}$$

The proposed procedure is implemented as follows:

1. Initial opinions of all experts are set in the form of a profile of individual FC on set A. Let's denote this profile by H (the set of alternatives $S \in W$ is divided into two classes of "best" and "worst" alternatives). Take the previously defined to be equal $P(I^*) + I$ (or as instructed by the DM). The previous set of "best" alternatives will be denoted as $M_{\Pi p} = \phi$.
2. Formulate GCR $v = v_g(H)$ according to the previous expression for the profile of individual FC.
3. Find the set M of maximal elements in S regarding GCR.

$$M = \{x \in S \mid (P - P(I_x(H, S))) < g\}. \tag{9}$$

Let us denote this set by M_H.

4. Check if the set Mn is empty. If $M_H \neq \phi$, then assign $S_2 = I$ and go back to step 5, otherwise go back to step 11.
5. Determine a subset of alternatives from the class of "best", the choice of which is carried out with a minimum preponderance of the votes of individuals, i.e. to further improve the efficiency of choice find a subset of alternatives B as:

$$B_{p*} = \left\{x \in M \mid P^* = min P(I_x(H < S))\right\}. \tag{10}$$

6. Remember $B_H = B_{p*}$.
7. Check if a new alternative has appeared in Bn. If $B_H \backslash B_{\Pi p} = \phi$ ($B_{\Pi p}$ is the previous subset), then go back to step 9, otherwise—to step 8.
8. Define a new subset of alternatives:

$$B_{p*} = \left\{x \in M \mid P^* = \min P(I_x(H < S)) + 1\right\}. \tag{11}$$

   Go back to step 6.
9. Determine a group of experts and customers from the total set N according to the rule:

$$I = \{i \in N \mid B_H \subset h_i(S)\}. \tag{12}$$

   Invite individuals from this group to clarify their opinion regarding the alternatives of the project parameters $x \in B_H$. (Do individuals $i \in I$ still believe that alternatives $x \in B_H$ are in the subset of "best" alternatives $x \in h_i(A)$ and $h_i(A) \neq \phi$?). Generate a new profile of individual FC for $i \in I$ $I \subset N$. Designate it $H'(I)$. Remember $B_{\Pi p} = B_H$.
10. Check the coincidence of the profiles H(I) and $H'(I)$, where H(I) is the profile obtained from the profile $H = (h_1, h_2, \ldots, h_n)$ (see step 1 of this variant of the procedure) by discarding individual FC h_j, $j \in N \backslash I$. If $H(I) = H'(I)$, then go back to step 13, otherwise—put $H(I) = H'(I)$ and go back to step 2.
11. Check the condition $S_2 = I$. If it is satisfied, then go back to step 12, otherwise—to step 13.
12. Check the condition $S_1 = 0$. If it is satisfied, then assign $S_1 = I$ and go back to step 13, otherwise—to step 14.
13. Check the condition $g < P$. If it is satisfied, then put $g = g + I$ and go back to step 2, otherwise go back to step 14.
14. Remember the set of group choice M, the final profile of the individual FC and the final value of g. Finish the procedure.

If we assume that the weights of experts and customers are equal, then the GCR is determined by the k-majority rule.

We will show the procedure for obtaining the resulting ranking and determining the coefficients of relative importance using the example of the performance indicators of the work of the expert commission, consisting of the heads of divisions of SC "Lokomotiv" (Kharkov, Ukraine).

The expert commission involved in the work on this project consisted of the following specialists: chief engineer—Stakankov S.V., chief accountant—Muricheva O.N., senior administrator—Dolgikh M.E., leading economist—Chuychenko M.I., 1st category engineer—A.I. Salasin.

Six main indicators of development were investigated, $M = \{1, 2, 3, 4, 5, 6\}$. Their content is given in Sect. 3.

Methods for conducting expert assessments were also tested. These indicators were proposed by the management of SC "Lokomotiv" in the development of guidelines for the strategic development plan.

To form an effective management system and development strategy for SC Lokomotiv, it was necessary to determine the priorities of influence on the main economic indicators selected by experts [16].

The task was to qualitatively analyze the comparability of performance indicators of SC Lokomotiv and its development prospects.

Competition in the information space between various sports and recreation institutions has determined the practical value of solving the problem. The need to solve it is due to the fact that not all performance indicators of SC Lokomotiv can be calculated in quantitative terms.

Six experts determined the following individual preference relationships on the set of indicators $M = \{1, 2, 3, 4, 5, 6\}$:

$$
\begin{aligned}
R_1 &: 6 > 1 > 2 > 3 > 4 > 5; \\
R_2 &: 6 > 2 = 3 > 4 > 1 = 5; \\
R_3 &: 6 > 1 > 4 = 5 > 3 > 2; \\
R_4 &: 6 > 1 = 2 = 3 = 5 > 4; \\
R_5 &: 6 > 4 > 2 > 1 = 3 = 5; \\
R_6 &: 6 > 5 > 1 = 4 > 3 > 2.
\end{aligned}
$$

Due to the fact that all six experts prefer indicator 6 to the rest, we will exclude it from consideration for now, but in the resulting ratio its rank will be equal to 1.

Let n (a, b) denote the number of relations Ri for which aRi b, that is, number of experts preferring indicator a over indicator b. We find the relationship based on the group decision from the principle:

$$aRb \Leftrightarrow n(a, b) \geq f(n), \tag{13}$$

where f (n) is an arbitrary function of a natural argument satisfying the condition $0 \geq f(n) \leq n$.

In this case $f(n) = 2/3$.

As a result of the analysis of these relations, the agreement of the experts was found to be satisfactory. For example, a coefficient of concordance $W \approx 0.37$, that at a significance level of 5% is acceptable.

To find the resulting relationship, the majority rule algorithm was first used. The resulting relation matrix for $n(a, b) > f(n)$ is:

$$R = \begin{pmatrix} 1\,1\,1\,1\,1 \\ 0\,1\,0\,0\,1 \\ 0\,0\,1\,0\,0 \\ 0\,0\,0\,1\,1 \\ 0\,0\,0\,0\,1 \end{pmatrix}$$

This relationship is nonlinear, so a more complex algorithm was used—the algorithm for modifying the transitive closure. For this it was required to find one more relation satisfying the condition n (a, b) ≥ f (n),

$$R = \begin{pmatrix} 1\,1\,1\,1\,1 \\ 0\,1\,1\,0\,1 \\ 0\,1\,1\,0\,1 \\ 0\,0\,0\,1\,1 \end{pmatrix}$$

Since the function f (n) was chosen such that f (n) > 1/2, the ratio matrix R is replaced by the ratio matrix R0 satisfying the relation $R_0 = R \bigcup \left(\overline{R} \bigcap R^{-1}\right)$.

The search for the resulting relation Q by the transitive closure modification algorithm is applicable provided that $R \subseteq Q \subseteq R_0$. The resulting ratio Q was acceptable for further analysis, since satisfies the linearity and transitivity property:

$$Q = \begin{pmatrix} 1\,1\,1\,1\,1 \\ 0\,1\,1\,1\,1 \\ 0\,0\,1\,0\,1 \\ 0\,0\,1\,1\,1 \\ 0\,0\,1\,0\,1 \end{pmatrix}$$

The resulting order of indicators is as follows:

Rобщ.: 6 > 1 > 2 > 4 > 3 = 5

To determine the coefficients of the relative importance of indicators by the priority method, the following scale was chosen:

1—equally important;
3—slightly more important;
5—more important;
7—is much superior in importance.

Intermediate values are accepted in cases of fluctuation between the main estimates.

Based on the judgments of experts about the comparative importance of each pair of indicators, taking into account the fact that aij = 1/aji, aii = 1, the following matrix A was obtained:

$$A = \begin{pmatrix} 1 & 2 & 5 & 3 & 2 & 1/3 \\ 1/2 & 1 & 4 & 2 & 4 & 1/4 \\ 1/5 & 1/4 & 1 & 1/4 & 1 & 1/7 \\ 1/3 & 1/4 & 4 & 1 & 4 & 1/5 \\ 1/5 & 1/4 & 1 & 1/4 & 1 & 1/7 \\ 3 & 4 & 7 & 5 & 7 & 1 \end{pmatrix}$$

Let us denote the vector of the sums of the columns of the matrix A by C, and the vector of the inverse sums by C^{-1}. We have:

$$C = (5.233;\ 7.75;\ 22;\ 11.5;\ 22;\ 2;\ 0.69),$$
$$C^{-1} = (0.19;\ 0.13;\ 0.05;\ 0.09;\ 0.05;\ 0.48).$$

As noted earlier, when measuring accurately, it should be $\omega = c^{-1}$. In this case:

$$\sum_{i=1}^{6} C_i^{-1} = 0.99 < 1.$$

This is because judgments like aik = aij × ajk are violated. The expert group decided to increase the weight of the fourth indicator, and then the vector of coefficients of the relative importance of the indicators of the third section became equal to:

$$\lambda^3 = (0.19;\ 0.13;\ 0.05;\ 0.10;\ 0.05;\ 0.48.$$

The given examples reflect the mathematical essence of the proposed procedures and algorithms.

The proposed procedure was implemented during the examination of selected indicators of the activity of SC "Lokomotiv" (Kharkov, Ukraine). Moreover, it should be noted that the customer can also be an expert as a decision-maker when approving the results of a group choice (according to the selected commission).

The effectiveness of the decisions made on the management of SC "Locomotiv" as a result of the use of the proposed mathematical models in this work is proposed to be determined by an expert method based on the indicators:

(1) The time for making managerial decisions;
(2) Adaptive reliability, by which we mean the ability of the proposed mathematical models to implement their decision support functions during changes in the initial parameters within acceptable limits;
(3) Functional completeness, which should be understood as the share of automated information processing based on the proposed mathematical models in the total amount of information for the functioning of the entire control system of the SC "Locomotiv".

Table 1 Values of the possible result of achieving efficiency

Statement of possible outcome	Numerical equivalent
Slightly worsening	−1
No change	0
Minor improvement	+1
Significant improvement	+2
Significant improvement	+3

The values of the possible result of achieving the quality of management of SC "Locomotiv" as a result of using the proposed algorithmization of procedures for agreeing individual opinions are given in the Table 1.

Under the conditions of the control problem for the SC "Locomotiv", the verification of the choice of the numerical equivalent of the possible result of achieving efficiency (table) was carried out on the basis of using the z-score method. In the course of using this method, normalized values of the numerical equivalent were obtained, converted to values of a scale based on a unit normal distribution with a mean equal to 0 and a standard deviation equal to 1. Deviation with a "−" sign means deterioration, with a "+ " sign improvement. The choice of the z-score method for solving the problem of forming a numerical equivalent of a possible result of achieving management efficiency of SC "Lokomotiv" is due to the standardization (normalization) of the distributions of possible results. That is, all distributions of a possible result of efficiency were reduced by the z-score method to one measurement.

The formulations of the possible result (table) do not provide for significant deterioration and deterioration, to a large extent as a result of using the proposed mathematical models for the following reason. The method of group decision, on which the proposed models are based, does not contradict the limitations of the subject area of SC "Lokomotiv" and, accordingly, its use will not be able to significantly worsen the management processes of SC "Lokomotiv".

To assess the level of efficiency of the decisions made on the management of SC "Lokomotiv" as a result of using the proposed mathematical models for managing the development of SC "Lokomotiv" with the involvement of experts from SC "Lokomotiv", the following scale was developed:

from −3 to 0—low efficiency;
from 1 to 4—satisfactory efficiency;
from 5 to 9—high efficiency.

As a result of expert evaluation of the use of the proposed algorithmization of procedures for agreeing individual opinions in SC Lokomotiv, a value of 8 was obtained. This value corresponds to a high level of efficiency of decisions made on the management of SC Lokomotiv as a result of using the mathematical models proposed in the work.

Thus, the examination made it possible to conclude that the developed procedure makes it possible to obtain a high level of efficiency of the decisions taken on the management of SC Lokomotiv.

3.3 Discussion of the Results of Algorithmicization of Procedures for the Coordination of Individual Opinions in the Management Process of Sc "Lokomotiv"

Within the framework of this study, an algorithmization of the procedures for agreeing individual opinions has been developed. The proposed methodology is a continuation of research [16, 17] on the development of a mathematical apparatus and formalization of interactive processes. Possible areas of practical application of the methodology can be socio-economic systems of various types. Based on the results of the development, the following conclusions can be drawn regarding the possible practical application of algorithmization of procedures for agreeing individual opinions in the management process of SC Lokomotiv:

(1) The axiomatically described rule in interactive group selection procedures should take into account the wishes of the enterprise administration. Otherwise, as a result of group choice, the output will be inadequate data for potential consumers and partners;
(2) When using the metric approach to the group choice problem, the main issue is the question of finding the median in a particular class of preference relations or choice functions.

When matching individual preferences into the group preference function, a certain complexity of the algorithm for determining the median arises. This laboriousness lies in the fact that for the general case of finding a finite solution, it is necessary to enumerate all possible options for group preferences. The Khemming metric is used as the distance.

The advantages of the proposed algorithmization of procedures for agreeing individual opinions in the control process of the SC "Locomotiv" are:

- Formalization of group selection procedures, which creates conditions for processing results in decision support information systems;
- Construction of an algorithm for finding transitive modifications of majority a relationship that allows reaching a common opinion of all experts;
- Consideration of the initial opinions of all experts is set in the form of a profile of individual selection functions. This circumstance makes it possible to maximize the approximation of the created algorithmization of the procedures for agreeing individual opinions to the practice of SC "Lokomotiv".

The disadvantages of the proposed algorithmization of procedures for agreeing individual opinions in the management process of SC "Lokomotiv" should include the fact that individual assessments of experts as the central link in the development can lead to subjectivity of the results obtained.

In the process of using the proposed results, the following restrictions of a subjective nature may be imposed:

- Personal assessments of experts can lead to a change in the configuration of possible variations of the choice functions;

- In the process of software implementation of the proposed algorithmization of procedures for agreeing individual opinions, rankings should be formed. The essence of these rankings is to reflect the individual assessments of experts. When forming them, experts are guided only by their subjective preferences.

Further areas of research can be:

- Assessment of the management efficiency of SC «Lokomotiv» based on the agreement of individual opinions of experts;
- Formalization of the procedure for assessing the degree of manipulability of group choice;
- Development of a methodology for achieving dynamics equilibrium and stability of collective decisions.

In the process of implementing the indicated further directions of research, such problems may arise.

- In the course of assessing the management efficiency of SC «Lokomotiv» based on the agreement of individual opinions of experts, it may be difficult to determine the integral indicator of such efficiency and quantify each component of this indicator.
- Formalization of the procedure for assessing the degree of manipulability of group choice may face a problem selection of parameters reflecting various aspects of such a choice.
- In the process of developing a methodology for achieving the dynamics of equilibrium and stability of collective decisions, it may be difficult to identify the factors of usefulness of various variations of such dynamics for the practice of SC "Lokomotiv".

4 Conclusion

1. The analysis of group selection procedures for managing complex systems has been carried out. The necessary preconditions for the construction of mathematical models of group choice in the control of complex systems are systematized.

   The problem posed during the study was to find effective procedures for coordinating a group decision. The practical application of the proposed methods were tested in the practice of expert assessments on the basis of the SC "Lokomotiv" (Kharkov, Ukraine).

   The analysis showed the applicability of this methodology in the study of complex economic organizational systems. Results have been obtained that show that the first important indicator is the assistance to strengthen the material and technical base.

   The consistency of the approach is determined by the following. Among the large number of methods for analyzing and synthesizing complex control systems, those that take into account the behavior of the most important element

of any control system—a person—are of significant theoretical and practical interest. Within the framework of the problem of collective interaction, substantially different aspects of this problem are investigated. One of these aspects is the problem of group choice.
2. The development of mathematical models for the formation of group decisions in various classes of relations is proposed. As a result, majority relations and their generalizations were obtained, expressing the general opinion of the entire group of experts. As a result, the resulting transitive relation is quite close to the original relation in terms of Khemming distance. The resulting relation is linear and therefore can be taken as the desired solution.
3. The formalization of interactive procedures for group selection has been carried out. The result of this formalization is the clarification of the opinions of experts regarding the alternatives included in the group of the best, but with the minimum number of votes. This formalization creates the necessary conditions for processing the results in decision support information systems. Also, this formalization made it possible to build an algorithm for finding transitive modifications of majority relations, as well as relations of a more general form, in which individual preference relations are linear quasi-orders.
4. Algorithmization of the procedures for agreeing individual opinions in the management process of SC “Lokomotiv” was carried out. The proposed algorithmization makes it possible to take into account the initial opinions of all experts in the form of a profile of individual selection functions. As a result, a sequence of steps has been formed to coordinate individual opinions in the management process of SC “Lokomotiv”. The vector of the relative importance of the indicators turned out in the form:

$\lambda^3 = (0.19;\ 0.13;\ 0.05;\ 0.10;\ 0.05;\ 0.48)$..

The algorithm itself proposes to expand and refine the opinions of experts until they become convenient for making a decision. In this case, the broader concepts of “voter decision” and “list of candidates” are considered. As a result, a methodology for searching for transitive agreement of a group of experts was formed. The interactivity of the procedure makes it possible to eliminate the inefficiency of the work of the expert group, regardless of the object of the applications of its assessments and the rules of coordination. Thus, the results of the examination will not be distorted if the experts are asked again several times.

Author Contributions Conceptualization: Fisun Konstantin, Tkachenko Yurii.
Data curation: Cherkashyn Oleksandr, Chukhlatyi Andrii.
Formal analysis: Fisun Konstantin, Cherkashyn Oleksandr.
Funding acquisition: Cherkashyn Oleksandr.
Investigation: Fisun Konstantin, Tkachenko Yurii, Chukhlatyi Andrii.
Methodology: Fisun Konstantin, Tkachenko Yurii.
Project administration: Fisun Konstantin, Tkachenko Yurii.
Validation: Fisun Konstantin, Tkachenko Yurii.
Visualization: Cherkashyn Oleksandr, Chukhlatyi Andrii.
Writing—original draft: Fisun Konstantin.
Writing—review and editing: Tkachenko Yurii, Chukhlatyi Andrii.

References

1. Arrow, K. I.: Social Choice and Individual Values, p. 124. Wiley, New York (1951)
2. Grabovetsky, B. Y.: Method of Expert Evaluations: Theory, Methodology, Directions of Use: Monograph, p. 171. VNTU, Vinnytsia (2010)
3. Gibbard, A.: Manipulatio of Voting Chemes: A General, p. 587. Econometrica (1977)
4. Danylchuk, H., Chebanova, N., Reznik, N., Vitkovskyi, Y.: Modeling of investment attractiveness of countries using entropy analysis of regional stock markets (June 11, 2019). Global J. Environ. Sci. Manage. **5**(SI), 227–235 (2019). Available at SSRN: https://ssrn.com/abstract=3540484
5. Danelyan, T. Y.: Formal methods of expert assessments [Electronic resource]. Econ. Stat. Inf. 1, 183–187 (2015). Access mode: file: /// D: / Downloads-Web/formalnye-metody-ekspertnyh-otsenok.pdf
6. Kuzmin, V. B.: Construction of Group Solutions in the Space of Clear and Fuzzy Binary Relations, p. 168. Nauka (1982)
7. Litvak, B. G.: Development of Management Decisions, p. 392. Delo (2004)
8. Lyndina, M. I.: Mathematical theory of ratings [electronic resource]. Sci. J. KubGAU **104**(10) (2015). Access mode: http://ej.kubagro.ru/2015/10/pdf/01.pdf
9. Reznik, N. P., Dolynskyi, S. V., Voloshchuk, N. Y.: Retrospective analysis of basic risk as a part futures trading in Ukraine. Int. J. Sci. Technol. Res (2020). Available at: http://www.scopus.com/inward/record.url?eid=2-s2.0-85078765334&partnerID=MN8TOARS
10. Maslennikov, E. V.: Possibilities of using expert knowledge as a source of concepts for the development of organizations. Bulletin of the Moscow University. Series 18, Sociol. Polit. Sci. **23**(2), 229–249 (2017)
11. Mc Shane, B., Chen, C., Anderson, E., Simester, D.: Decision stages and asymmetries in regular retail price pass-through. Mark. Sci. **35**(4), 619–639 (2016)
12. Motalo, V.:. Analysis of Measurement Scales, pp. 21–35. Lviv Polytechnic National University, Lviv (2015)
13. Novosad, V. P.: Scientific and methodological development. In: Novosad, V. P., Seliverstov, R.G., Artim, I. I. K. (eds). Quantitative methods of expert evaluation: method development, p. 36. NADU (2009)
14. Reznik, N. P.: Organizational and economic mechanism of agro-industrial investment: monograph, p 300. «Sochinsky», Uman (2014)
15. Chen, L., Ellis, S., Suresh. N.: A supplier development adoption framework using expectancy theory. Int. J. Operat. Prod. Manage. **36**(5), 592–615 (2016)
16. Semenov, S. S.: Estimation of quality and technical level of complex systems: practice of application of a method of expert estimations, p. 352. LENAND (2015)
17. Reznik, N. P., Popov, V. M., Podplietnii, V. V., Popova, S. P.: Financial support for the development of joint territorial communities. Test Eng. Manage (2020). Available at: http://www.scopus.com/inward/record.url?eid=2-s2.0-85082999646&partnerID=MN8TOARS
18. Shubyn, A. N., Sub, A. D.: Application of methods of expert assessments to build a rating. Yzd-vo LTD «Zhurnal innovatsiy i investiciy» **10**, 208–217 (2014).
19. Fisun, K.A.: Implementation of group selection methods in the theory of system constraint. Cyber Govern. Inf. Technol. **1**, 52–61 (2014)
20. Fisun, K., Naumenko, M., Bortnik. N.: Price of the efficiency of functioning of organizational systems using expert. Baltic J. Econ. Stud. **4**(3), 325–333 (2018) (Riga: Publishing House Baltiya Publishing)
21. Fisun, K., Vavzhenchuk, S., Tkachenko, Y., Naumenko, M.: Algorithmization of Procedures for Coordination of Individual Opinions in the Process of Implementation of Organizational and Managerial Labor Relations, 237–250
22. Yurin, A. Y.: Methods of group choice for the adaptation of decisions obtained as a result of reasoning on the basis of precedents. Artif. Intell. Decis. Making. Moscow: Federal Res. Center Inf. Manage. Russian Acad. Sci. **3**, 78–85 (2013)

23. Yurynets, R. V., Savras, I. Z.: Formation of expert groups using economic and mathematical decision models. Bulletin of socio-economic research. zb. science. pr. ODEU. pp. 413–417 (2008)
24. Reznik, N. P., Oleksandr, G. V.: Strategic planning of the socio-economic development of Ukraine: conceptual aspects. Int. J. Sci. Technol. Res. **9**(3), 7034–7038 (2020). Retrieved from: http://www.scopus.com/inward/record.url?eid=2-s2.0-85082866
25. Reznik, N. P.: Foreign experience of investing in agro-industrial complex. Culture Peoples Black Sea Region **147**, 77–80 (2008). Retrieved from http://dspace.nbuv.gov.ua/handle/123456789/34555
26. Reznik, N. P.: Organizational-economic mechanism of attraction of investments into agro-industrial complex of Ukraine, Ekonomika APK **7**, 70–77 (2013). Retrieved from http://eapk.org.ua/en/taxonomy/term/1032
27. Reznik, N. P.: Social responsibility of business: essential-theoretical aspects. J. Eur. Econ. **13**(3), 296–303 (2014). 364&partnerID=MN8TOARS

Strategies of the Jordan Banking System to Face Corona Pandemic Repercussions

Yazan Abdullah Qasrawi and Ahmed Mohamed Al Saad

Abstract This research aims to provide a brief statement of the Corona pandemic, and how the Jordan banking system which represented by the Central Bank, the Social Security Corporation, and the banks operating in Jordan, were able to confront this pandemic through strategies that enabled through multiple procedures and in coordination with the Social Security Corporation, which the researchers addressed in the research demands by explaining the concept of the Corona pandemic, And this effects On banking sector in Jordan and the role at Confrontation its repercussions, And the role the bank central of Jordan too at Confrontation repercussions and also the role of the Consumer Protection Department in facing these repercussions. This research showed some results, that the Central Bank and the banking sector in Jordan, in coordination with the Social Security Corporation, was able to bypass this pandemic crisis and keep the economy safe. The Consumer Protection Department had a clear role in confronting the Corona pandemic. The researchers recommend the Central Bank, the banking sector, and the Social Security Corporation to develop plans either in the form of measures or legislation that protects institutions and individuals, and the involvement of the private sector in these plans to confront any stage of this epidemic.

Keywords Jordan banking system · Corona pandemic · The impact of Corona pandemic · The role of the government in facing the pandemic · Banks role · COVID-19 · Central Bank of Jordan

Y. A. Qasrawi (✉)
Islamic Economics and Banking, Islamic International Arab Bank, Amman, Jordan
e-mail: yazan_fm@yahoo.com

A. M. Al Saad
Department of Economics and Islamic Banks, Faculty of Sharia, Yarmouk University, Irbid, Jordan

A. Hamdan et al. (eds.), *The Implementation of Smart Technologies for Business Success and Sustainability*, Studies in Systems, Decision and Control 216,
https://doi.org/10.1007/978-3-031-10212-7_72

1 Introduction

The world is facing multi-dimensional crises: financial, economic, health, and others, we have been experiencing health crises for nearly two years. Corona pandemic crisis was followed by uncertainty and declension in commodity prices, it became clear that nearly two years after the crisis, the global economy entered into a state of contraction that is the deepest in decades which, according to the latest forecasts issued by the International Monetary Fund, reached (4.3%), compared to a decline of (3%) from what was expected, and despite the optimism that prevails in many countries of the world after the easing of the restrictions in them, these expectations are still marred by a strong state of uncertainty, and will certainly be subject to continuous adjustment in a manner unprecedented due to the high risks surrounding the global economy [1].

2 Study Problem

The research problem asks the main question: **What are the most economic strategies undertaken by the Jordanian government to confront the Corona pandemic repercussions?** Branch the following questions:

(1) What is the definition of the Corona pandemic?
(2) What is the impact of the corona pandemic on Jordan's banking sector and its role in confrontation repercussions?
(3) What is the Central Bank of Jordan's role in facing the repercussions of the Corona pandemic?
(4) What is the Consumer Financial Protection Department's role in facing the repercussions of the corona pandemic?

3 Study Objectives

(1) Knowing the impact of the corona pandemic on the Jordan banking sector and its role in facing repercussions.
(2) Knowing the role of the Central Bank of Jordan is facing the repercussions of the pandemic.
(3) Knowing the role of the Consumer Protection Department in facing the pandemic repercussions.

3.1 First Requirement: The Impact of the Corona Pandemic on the Jordan Banking Sector and Its Role in Facing the Repercussions

The corona crisis has caused bad effects on Jordan's economy and the other world. In light of the corona pandemic, we cannot isolate Jordan's economy from the situation of the global and regional economies. It is considered as a major driver of external demand for foreign trade or investment and tourism, these are the main external channels that influence the Jordan economy [2].

The national economy has generally been accustomed to facing challenges. Once we go back in time to about a decade ago, we will remember how the results of the repercussions of the global financial crisis in the year 2008, which resulted from the mortgage crisis in the United States of America, and its repercussions on the whole world, and the outputs represented by a downturn During that period, we also recall what our region experienced in terms of wide political turmoil, and severe political and security events during the period of the what called Arab Spring, and the accompanying repercussions on the national economy and a slowdown in growth rates and the effects it had on the economy [3].

The economy of pandemic's repercussions is different from the economy crises, as they arise from the health sector due to its close connection and its great role in combating pandemics, which is one of the infrastructure sectors, and economic and financial crises differ in that they arise from the financial and business sector, so that the first affected in it is the financial sector, thus the impact extends to the productive sectors, then the infrastructure sectors in the medium to long term as for the economic and financial crises resulting from pandemics and epidemics [4, 5].

The economic crises caused by pandemics and epidemics may be similar to the economic and financial crises, but they differ in stimulating aggregate demand through fiscal and monetary policy tools, and thus the economic stagnation associated with the corona pandemic may differ from that stagnation resulting from the financial crisis in 2008, because the slowdown determinants of aggregate demand under stimulus packages offered, as a result of global shutdowns global lockdown which directly affected the financial stability, which central banks aim for and strive to achieve [3, 6].

The corona pandemic had an impact on the banking and financial sector, as a result of the decrease in the revenues of companies in the services and tourism sector. As a result of total and partial closures, these companies were forced to stop their work, which led to an increase in the number of liabilities against assets, and thus the difficulty in keeping their obligations towards the banking sector [7].

Jordan's banking sector, as other global banking sectors, was affected by the corona pandemic, and this was reflected in the decline of about (1.5%) in the economic growth rate in the first three quarters of 2020, the government expects that the decline will reach (3%) in the economy [3].

3.2 *Second Requirement: The Role of the Government and the Social Security Corporation in Facing the Corona Pandemic Repercussions*

2020 was not an ordinary year, full of challenges in light of the unusual economic health crisis, it witnessed the most severe in the modern era caused by COVID-19. The pandemic left behind a shadow cast on the economies of developed and developing countries are alike which led to a contraction in economic activity, high unemployment rates, deteriorating public financial conditions, and high levels of public debt.

During 2020, as for the monetary and banking policy of the Central Bank of Jordan, the policy was on a high level of flexibility, quick and proactive response to the situation of corona pandemic, and to control repercussions' effects on the national economy. In line with the bank's objectives of maintaining monetary and financial stability, and maintaining inflation rates within suitable levels, the stability of the exchange rate of the Jordanian dinar, to maintain the power and the strength of the banking system to protect the rights of depositors and shareholders, and contribute the encouraging economic growth in the kingdom with the relevant official authorities [8, 9].

To face corona pandemic repercussions, the government and the Social Security Corporation, during the year 2020, followed a few measures to mitigate the bad effects on the economy in general and the economic facilities and affected individuals in particular, to face one part of the economic pandemic consequences and generally, to mitigate the downturn in the national economy [10].

The Social Security Corporation, under several defense orders, launched some programs and procedures, aimed at alleviating the corona pandemic repercussions on workers and establishments in the most affected economic activities and sectors. These programs (9):

Suspending work on (Old-Age and Survivors Insurance) program: It gives facilities the option to suspend 'Old-Age and Survivors Insurance' for their employees or some of them, on the other hand, reducing social security contributions to (5.25%) instead of (21.75) for three months from the year 2020. about 23 thousand establishments got benefited from this program.

(Tadamon) programs: were launched to protect workers and establishments. About 106 thousand workers and 11 thousand establishments got benefited from these programs. These programs included:

(Tadamon) Program (1): which supports establishments with activities that have been completely or partially suspended, by resorting to the Social Security Corporation to pay a percentage of wages.

(Tadamon) Program (2): which benefits the facility that is not covered by the provisions of the Social Security Law, and all or some of its workers have stopped working during the pandemic.

(Musaned 1, Musaned 2, Musaned 3) program: which aims to help Jordanian and non-Jordanian social security workers to obtain additional income by allowing them to spend from their Unemployment Insurance saving accounts.
(Himayah) program: It aims to prevent the layoffs of Jordanian workers in the most affected economic sectors (the tourism sector and transportation included by the provisions of the Social Security Cooperation Law).
(Tamkeen 1) program: It includes partial coverage of workers with Old-age and survivors' insurance.
(Eatidamah) program: It was launched between the government and the Social Security Corporation. The program includes support for establishments that are not authorized to work and sectors that were mostly affected by the corona pandemic.

3.3 Third Requirement: Bank's Role in Facing the Repercussions

Banks in Jordan had a role in facing the corona pandemic repercussions, as banks took a wide package of measures and implemented the central bank's programs aimed at supporting and financing the economic sectors and small and medium-sized companies Such as postponing installments, reducing interest rates, providing liquidity and granting financing to various economic sectors and the Jordan government [3].

The various and large measures taken by Jordan's banks to reduce the pandemic repercussions on the national economy came in line with the directives of the Central Bank of Jordan, which enabled the wise and skillful management of monetary policy, which was reflected in the financial and monetary stability in the Kingdom, and the maintenance of comfortable levels of foreign reserves [9, 11].

The most important measures taken by banks to face the corona crisis repercussions [6, 12].

- Postponement of loans and other payments: Since the beginning of the corona crisis, the Central Bank allowed other banks to postpone installments, with a permitting period of three months.
- Observance of health and safety measures: As banks were keen to take decisive and significant measures, to provide the best health safety conditions for their customers and employees, to prevent infection and prevent its spread by reducing the number of workers in their branches, providing the necessary health supplies for customers and employees.
- Technological development of banks: It has allowed banks to provide most of their banking services through various electronic means to reduce the need for customers to visit bank branches.
- Continuous provision of ATMs: At the beginning of the crisis, the Central Bank of Jordan has specialized a plan with other banks that are keen to sustain the

provision of funds through their ATMs to provide them with the necessary cash on an ongoing basis.

- Supporting the economic sectors: Banks continued to provide support to the various economic sectors, including the programs launched by the Central Bank for this purpose.
- Huge value of financial liquidity at banks: By exploiting the additional amount of money that comes from reducing the required reserve ratio, expanding credit grants, lending various economic sectors, affected groups, and major institutions.
- Implementation of the National Program for Financing to Confront the Corona Crisis: According to the conditions specified in the program the new Central Bank program has been fully applied with the infusion of (500) million dinars and directing it quickly to finance the beneficiaries of the program.
- Arranging priorities according to productive sectors: Banks have been keen to work on re-arranging their lending priorities for the first productive sectors, with labor-intensiveness and high added value, and trying to include multiple and many categories of borrowers from various sectors to achieve a well-diversified portfolio.
- The corporation and coordination continuously with the Central Bank: The banks reviewed the instructions and regulatory requirements and coordinated with the Central Bank of Jordan in an attempt to ease the conditions that limit the ability of banks to expand credit.
- The social responsibility of banks: Based on the role 'Good Banking Is Good Citizenship', banks have been taking the initiative within the framework of their social responsibilities, as they rushed to provide donations and support to all accounts and funds established by the government to receive donations, to overcome the economic effects of corona crisis repercussions.

3.4 Fourth Requirement: The Role of the Central Bank of Jordan is Facing the Repercussions of the Corona Pandemic

In 2020, the Central Bank of Jordan took several monetary and precautionary measures that it had taken during the corona pandemic, with a total reached (2.5) billion dinars.

These procedures aimed to provide the necessary financial liquidity for economic activity, enhance the ability of economic sectors to face the negative repercussions of the emerging pandemic, and strengthen the local economy's resilience by preserving jobs and sustaining the business ability to carry out economic activities, regulating the work of banks and strengthening their financial positions, by following several procedures represented by the following (13):

- Reducing interest rates on individuals and companies, as the bank reduced main interest rates on all financial policy instruments twice in a row.

- Enhancing the financial liquidity in the banking system by reducing the mandatory reserve ratio on deposits with banks, liberating additional financial liquidity for banks through open market operations, and stopping the issuance of deposit certificates.
- Meeting the financing needs of the targeted economic sectors within the Central Bank program to finance the economic sectors with a value of (1.2) billion dinars through banks.
- Strengthening the resilience of small and medium-sized companies, professionals, craftsmen, and individual institutions.
- Supporting companies and individuals that have declined their cash flows and protecting them from bank failure.
- Supporting small sector clients.
- Strengthening the resilience of pioneering and emerging companies by allocating some of the funds from the Jordan Fund for entrepreneurship and direct investment in these companies.
- Granting the Agricultural Credit Corporation additional loans, with the aim of re-lending it by the Corporation to farmers for 10 years, including a grace period within two years.
- Enhancing electronic payments to ensure the continuity of the monetary cycle in the economy, enable payment companies to provide the service of opening electronic wallets to customers remotely and without any costs, organizing and encouraging contactless electronic payments, and enabling customers to withdraw and deposit cash from Electronic wallets, through ATMs, without charging customers any commissions for withdrawals.

3.5 *Fifth Requirement: The Role of the Consumer Protection Department is Facing the Repercussions of the Corona Pandemic*

In light of the quick global developments as a result of the impact of the emerging coronavirus, and the challenges it posed to the national economy and its serious effects on individual borrower customers in terms of interruption or decrease in their income, and the impact of this on their ability to pay their debts for banks, the Central Bank takes several measures aimed of reviving as much as possible on financial services consumers. We summarize these measures as follows [8].

- Postponing loan installments: Where the Central Bank issued orders according to which banks were allowed to postpone loan installments for affected customers by the pandemic until the end of 2020, without commissions or delay interests. Moreover, the Central Bank, following a circular issued to banks, affirmed the necessity of clearly disclosing and announcing the costs and benefits that will be paid to be collected from the customer and as a result of the postponement of installments.

- Reducing interest rates: where the Central Bank reduced the interest rate on its monetary policy tools by (1.5%) during March, intending to reduce the costs of financing and debt servicing, for all economic activities and recovery after the crisis, and the Central Bank followed up on banks to ensure that they reduced the interest rate.
- The procedures for checks and their settlement, as the Central Bank decided not to include clients who have rejected checks for financial reasons in the list of clients who defaulted on payment, and not to charge them commissions resulting from checks returned during the crisis, and to give this category of clients an appropriate period to settle their checks, a period of three months instead of Fourteen working days.
- Dealing with customer complaints and inquiries, despite the conditions of the quarantines, the Central Bank continued to receive peoples' complaints by phone. It allocated an integrated team to receive calls on mobile phones, where the Central Bank received more than (5000) calls, most of which were inquiries and complaints regarding the postponement of installments, the bank has dealt with all complaints received from the beneficiaries of the National Aid Fund.
- E-wallets: Where the government has approved mobile payment service companies to support daily workers by opening electronic wallets, under the principle of taking simplified due diligence procedures, in terms of identifying the customer and the real beneficiary to ensure the promotion of access to all society segments to official financial services, and the continuity of By providing electronic payment services that adequately meet the needs of the service recipient. As the Central Bank circulated to mobile payment companies, the necessity of adhering to the provisions and instructions of the Central Bank to deal with customer complaints in a professional, efficient and effective, and in a manner way that ensures their adoption of best practices in this regard, and to preserve the rights of customers and the company's reputation and customers' trust in it.
- Withdrawals from ATMs: In the interest of the Central Bank of Jordan to provide customers' needs with the lowest commissions possible, the Central Bank has decided not to charge customers any commissions in return for making cash withdrawals through ATMs, the Central Bank has also coordinated with banks to equip mobile teller machines for various governorates

4 Findings

(1) The Central Bank of Jordan and the Social Security Corporation provide direct financial support for economic enterprises, especially small and medium ones, and arrange priorities according to the productive sectors with labor intensity and high.

(2) The Social Security Corporation Prepare and implement a plan for a safe return to work, based on a risk assessment plan which faces establishments,

defining economic priorities, and adopting a package of programs and procedures, aimed at mitigating the repercussions of the Corona pandemic on workers and establishments in the most affected economic sectors and activities.

(3) The measure came taken by Jordanian banks, in line with the directives of the Central Bank of Jordan, so by adopting modern technologies and creating new business models To reduce the repercussions of the pandemic on the national economy and those dealing with the banking sector.

5 Recommendations

(1) The necessity of designing general banking policies that include developing the Jordanian banking sector by setting time plans that include benefiting from the Corona pandemic through the most prominent preparations to face the negative repercussions on that sector.

(2) That the Central Bank of Jordan develops supervisory procedures to deal with exceptional circumstances in terms of confrontation, treatment, and recovery.

(3) The need for the Consumer Protection Department to update the regulations to ensure fair treatment and protect financial consumers, especially in light of the exceptional circumstances that were accompanied by the expansion of financial and technological products, which made it difficult for the financial consumer to make the right decision.

References

1. Association of Banks in Jordan: Journal of Banks in Jordan. Amman, Jordan **40**(15) (2020). https://www.who.int/emergencies/diseases/novel-coronavirus-2019
2. Association of Banks in Jordan: Journal of Banks in Jordan. Amman, Jordan **40**(15), 5 (2020)
3. Association of Banks in Jordan: Journal of Banks in Jordan. Analytical reading of the Association of Banks in the draft Budgets Law of Government Units for the Fiscal Year 2021. Amman, Jordan **41**(16), 5 (2021)
4. The central bank of Jordan: (Annual Report 2020), research department, Amman, Jordan, p. 7 (2021)
5. The central bank of Jordan: (Annual Report 2020), research department, Amman, Jordan, pp. 26–27 (2021)
6. Association of Banks in Jordan: Journal of Banks in Jordan. Analytical reading of the Association of Banks in the draft Budgets Law of Government Units for the Fiscal Year 2021 Amman, Jordan **41**(16), 6 (2021)
7. Association of Banks in Jordan: Journal of Banks in Jordan, Analytical reading of the Association of Banks in the draft general budget law for the fiscal year 2021 AD, Amman, Jordan **41**(16) (2021)
8. The central bank of Jordan.: (Cbj's Measures for Protection the Financial Consumer under the Repercussion of Covid-19 Pandemic), financial consumer protection department, Amman, Jordan, pp. 3–6 (2020)
9. Association of Banks in Jordan.: Journal of Banks in Jordan, Amman, Jordan, **40**(15) (2020)

10. Talha, W.A.: The repercussions of the spread of the Coronavirus on the banking and financial sector in the Arab countries. King Abdulaziz University Journal Institute of Islamic Economics **33**(3), 131–140 (2020)
11. Belwafi, A.: The Economic and Social repercussions of the emerging coronavirus covid-19 Islamic point of view. King Abdulaziz University Journal of Islamic Economics **33**(3), 43–74 (2020)
12. The central bank of Jordan: (Annual Report 2020), research department, Amman, Jordan, pp. 35–38 (2021)
13. Association of Banks in Jordan: Journal of Banks in Jordan. Amman, Jordan **40**(15), 7 (2020)
14. Belwafi, A.: The Economic and Social repercussions of the emerging coronavirus covid-19 Islamic point of view. King Abdulaziz University Journal Institute of Islamic Economics **33**(3), 43–74 (2020)
15. The Central Bank of Jordan: (Annual Report 2020), research department, Amman, Jordan, p. 8 (2021)
16. Association of Banks in Jordan.: The expected effects of the Corona Crisis on some indicators of the Jordan economy, what are the most important initiatives and incentive packages have taken and the role of the banking sector. Amman, Jordan, pp. 24–27 (2020)
17. Association of Banks in Jordan: The expected effects of the Corona Crisis on some indicators of the Jordan economy, the most important initiatives and incentive packages that have taken, and the role of the banking sector. Amman, Jordan, pp. 24–27 (2020)

The Impact of the Profits of Jordanian Insurance Companies on Their Financial Performance Indicators During the Corona Pandemic

Ola Mohammad Khresat and Ibrahim Sadiq

Abstract The study aims to investigate the impact of the profits of Jordanian insurance companies on their financial performance indicators in the light Corona pandemic by comparing the financial performance indicators of insurance companies between the years 2019 (before the COVID-19 pandemic) and 2020 (during the COVID-19 pandemic). The study sample consists of (17) general insurance companies that made profits under the Corona pandemic. The study relied on the descriptive analytical approach by relying on quantitative and numerical data to describe the study sample through annual financial reports issued by the insurance industry to measure financial performance indicators for a time series of two consecutive years (2019–2020). The study found that there is an impact of the profits of insurance companies on the financial performance indicators measured by financial leverage ratios (the ratio of debts to assets), and there is no impact for the profits of insurance companies on the financial performance measured by other ratios for the year (2019) before the COVID-19 pandemic, as for the year (2020). During the COVID-19 pandemic, the study found an impact of the profits of insurance companies on the financial performance measured by (quick liquidity), and there is no impact of the profits of insurance companies on the financial performance measured by other ratios. The study therefore recommends decision-makers to seek to improve the financial performance of the insurance sector by taking this study by risk managers and regulatory authorities in order to understand the internal factors represented by financial planning, internal management and administrative efficiency, as well as external factors represented by the legal and economic environment affecting financial performance. Future studies also recommend conducting a study model on other financial sectors and the industrial and service sectors to find out how corporate profits impact the financial performance of these sectors during the COVID-19 pandemic.

O. M. Khresat (✉)
Department of Accounting Science, Zarqa University, Zarqa, Jordan
e-mail: okhresat@zu.edu.jo

I. Sadiq
Income and Sales Tax Department, Zarqa University, Zarqa, Jordan

A. Hamdan et al. (eds.), *The Implementation of Smart Technologies for Business Success and Sustainability*, Studies in Systems, Decision and Control 216,
https://doi.org/10.1007/978-3-031-10212-7_73

Keywords Insurance company profits · Financial performance · Liquidity ratios · Leverage ratios · Jordanian insurance companies

1 First Section

The insurance sector is one of the most important economic sectors in both developing and developed countries, as this sector constitutes an important and main source of savings to finance various economic activities and creates multiple opportunities. Insurance is a means to facilitate credit operations as well as saving means. Also, insurance works on increasing production as it achieves stability for the individual and the family [1, 2].

Financial ratios constitute the most widely used means and the most important model for analyzing financial reports and reaching an accurate assessment of the efficiency and effectiveness of the company [3]. For this reason, the researchers considered studying the impact of insurance companies' profits on their financial performance indicators considering the Corona pandemic, especially after the preliminary aggregated financial statements of the 23 insurance companies on Amman Stock Exchange showed, an increase in net profit by 45%;to record 24.8 million dinars in 2020, compared to 17.1 million dinars in 2019. It also showed that the profit before income tax for the sector increased by 48%, with an increase of 10.6 million dinars in 2020; Profits before tax recorded about 32.8 million dinars, compared to 22.17 million dinars in 2019, and that the operating profit of the companies also showed an increase for insurance sector companies to reach 62.93 million dinars in 2020 compared to 48.2 million dinars in the year 2019, with an increase of 31%.

The financial statements of the companies showed stability in the net written premiums in general, which was offset by a noticeable decline in the net compensation paid in the first nine months of 2020, which was reflected in the rise in the sector's net profits as it led to partial and complete closures, in addition to extending the period of daily ban hours within The government's plan for 2020 to control the spread of corona virus, to essentially reduce vehicle accidents; This was reflected in the decrease in the value of compensation paid by insurance companies. Additionally, the decrease in the number of patients visiting hospitals contributed to a decrease in health insurance costs for insurance companies. The number of companies that achieved profits reached 17 companies, with a total profit of 28.06 million dinars, while 6 companies had losses of amounted to 3.25 million dinars, and it should be noted that 20 insurance companies listed on the regular market in the Amman Stock Exchange achieved profits of 25.75 million dinars, and 3 companies in the "unlisted" stock market collectively recorded net losses of 940 thousand dinars in 2020, compared to a net loss of 3.6 million dinars in 2019 (https://www.almamlakatv.com/news).

As the Jordanian Federation of Insurance Companies mentioned in its report of the year (2020), the total realized premiums for the insurance sector in 2020 and for all insurance branches compared to 2019 decreased by 3.6%. The decrease in the total premiums came because of the decrease in the written premiums of some insurance branches. The premiums of the fire insurance branch increased by 9.7% in 2020 compared to 2019 and the premiums of the life insurance branch increased by 8.1% in 2020 compared to 2019. The premiums of the marine Insurance branch have decreased by (3.7%) in 2020 compared to 2019, and the premiums for the medical insurance branch decreased by (5.7%) in 2020 compared to 2019.

The premiums of the general accident insurance branch decreased by (8.0%) in 2020 compared to 2019, and the premiums of the vehicle insurance branch decreased by (9.4%) in 2020 compared to 2019. Also, the premiums of the credit insurance branch decreased by (49.8%) for the year 2020 compared to 2019. The premiums of the General Accident Insurance Branch include the premiums of the liability insurance branch, which increased by 3.9% for the year 2020 compared to the year 2019, and the insurance branch of the other branches, where its premiums decreased by (13.9%) for the year 2020 compared to the year 2019.

The total insurance premiums in Jordanian companies for the year 2020 amounted to a 3.6% decrease, of which the share of general insurances amounted to 84.4% of the total collected premiums, and a decrease of 5.4% from 2019.

The marine insurance premiums for the year 2020 compared to 2019 amounted to a decrease of 3.7%, while the compensation paid by the branch for the year 2020 compared to 2019 amounted to a decrease of (22.7%) The marine insurance business includes aviation insurance, and the total aviation insurance premiums for the year 2020 written by (8) companies, with a decrease of (39.0%) compared to 2019, and one insurance company paid compensation to the aviation insurance branch for the year 2020 at an increase rate of 164.9% compared to 2019.

(Report of the Jordanian Federation of Insurance Companies, 2020), [4].

2 Literature Review

Several studies have dealt with insurance companies in Jordan and the impact of international accounting standards on their profits, Referring to the study [5, 6]. The study of [5] concluded that there is an impact of the application of IFRS 9 in the year of application of the standard 2018, on the opening balances of retained earnings for the year 2018 for Jordanian insurance companies. Moreover, the study [6] showed that there is a statistically significant effect of applying IFRS 9 on the profits and losses of insurance companies. Bank deposits were the most affected and influential. It was also found that FAAC and TA ranked second, LHLT and PR ranked third in terms of impact. Also, many studies dealt with the investments of insurance companies and their impact on profits, including the study [7]. The study [7] concluded that whenever insurance companies direct their investments in the first place for financial assets, followed by loans and then real estate investments, this will affect their profits by (50.109%), and the more they direct their investments in the first place towards financial assets and real estate investments loans, the returns of these companies will affect their profits with 53.836%. As for [8, 9], Melhem and Diab [8] study showed an impact of external factors, the legal dimension, and the economic aspect, collectively and individually, on the financial performance as measured by both the return on assets and the return on equity and the existence of an impact of the financing structure on the financial performance measured by the return on assets and the return on equity, (while it was found that there was no impact of operational performance on the financial performance measured by both the return on assets and the return on equity. The study [9] found that there is a statistically significant impact of the internal factors represented by (total insurance, administrative expenses, and company size) on the profitability of insurance companies, while the results showed no impact of claims and financial leverage, and the presence of a statistically significant impact of the external factors measured by the interest rate on the profitability of insurance companies, Rahman and Sharma [10] study examined the impact of operating cash flows for companies in the insurance and manufacturing sectors in the Kingdom of Saudi Arabia on financial performance and concluded that there is a positive and significant relationship between financial performance (ROA and ROE) and (CFOs), and a negative correlation between SIZE and LEV. and CFOs, and it was found that the operating cash flows of companies in the insurance and manufacturing sectors in the Kingdom of Saudi Arabia affect the financial performance. On the other hand, the study, Alqaisi et al. [11] found an impact of the return on assets, debt ratio, company age and company size on the market share price in Insurance companies listed on the Amman Stock Exchange, while the results showed that there was no impact of the return on equity on the market share price.

3 Hypothesis and Methodology

To achieve the purpose of this study, the following main hypothesis was formulated:

HO: There is no statistically significant effect at the level of ($\alpha = 0.05$) of the profits of Jordanian insurance companies on their financial performance in light of the Corona pandemic and the following sub-hypotheses are branched from it:
HO 1: There is no statistically significant effect at the level of significance ($\alpha = 0.05$) of the profits of Jordanian insurance companies on liquidity ratios (current ratio) in light of the Corona pandemic.
HO 2: There is no statistically significant effect at the level of significance ($\alpha = 0.05$) of the profits of Jordanian insurance companies on liquidity ratios (quick liquidity ratio) in light of the Corona pandemic.
HO 3: There is no statistically significant effect at the level of significance ($\alpha = 0.05$) of the profits of Jordanian insurance companies on leverage ratios (debts to assets ratio) in light of the Corona pandemic.
HO 4: There is no statistically significant effect at a significant level ($\alpha = 0.05$) of the profits of Jordanian insurance companies on their leverage ratios (debts to equity ratio) in light of the Corona pandemic.

The Study Community and Sample: The study community covers the 23 insurance companies listed in the Securities Depository Center for the year 2020. The study sample consists of 17 public joint-stock insurance companies that made profits under the Corona pandemic. The study depends on the descriptive—analytical methodology through using quantitative and numerical data to describe the study sample through the annual financial reports issued by the insurance sector to measure financial performance indicators and for a time series of two consecutive years (2019–2020), which is the period covered by the study. The Arab Union International Insurance Company, Philadelphia Insurance Company and Medgulf Company were excluded due to losses in the year 2020. Al-Safwa Insurance Company and Al Arab Insurance Company were excluded as they are not listed in the financial market. Besides; the researchers were unable to obtain the financial statements for the year 2020 of the Holy Land Insurance Company.

A number of statistical methods appropriate to the nature of the data available to them which depend on (SPSS) have been used, to achieve the study's objective, which is to measure the impact of insurance companies' profits on their performance before the Corona COVID-19 pandemic in (2019) and during the pandemic in (2020), the test of Linear Regression was used to test all the null hypotheses of the study as follows [12].

> HO 1: There is no statistically significant effect at a significant level ($\alpha = 0.05$) of the profits of Jordanian insurance companies on their measured financial performance indicators (current ratio) in light of the Corona pandemic.

Table 1 shows the effect of the profits of Jordanian insurance companies on the liquidity ratios (current ratio) during the (COVID-19) pandemic, as it was not statistically significant, and the calculated (F) value amounted to (3.55) with a significance level of (0.079), and where the value of The significance level is greater than the value (0.05), it means that the effect is not accepted from the statistical point of view, and the value (R^2) refers to the value of the change or difference that may occur in the value of the dependent variable, which is attributed to the value of the independent variable. If the value of the independent variable increases, it would indicate the ability of this variable to explain this variation and consequently its ability to predict the dependent variable. This value was equal to (0.191). Regarding the size of the effect, it is expressed by (β), which amounted to (−0.438), and this value is considered non-effective and not statistically significant because the level of the function (β) which is (0.079) is greater than (0.05), and with this result, the null hypothesis was accepted, which states **that "there is no statistically significant effect at the level of significance** ($\alpha = 0.05$) **of the profits of Jordanian** insurance companies on the measured financial performance indicators (current ratio) during the Corona pandemic for the year (2020).

As for the pre-pandemic (COVID-19), where the impact of insurance companies' profits on the current ratio was not statistically significant, and the calculated (F) value amounted to (0.218) with a significance level of (0.647), and since the value of the significance level is greater than the value of (0.05), This means that the effect is not accepted from the statistical point of view, and the value of (R^2) indicates that it

Table 1 Results of the first null hypothesis test for the year (2019–2020)

Dependent variables	Model summary		Variance analysis		Transactions		
	R	r^2	F	Sig	Beta	T	Sig
Current ratio 2020	0.438	0.191	3.55	0.079	−0.438	10.659	0.079
Current ratio 2019	0.12	0.014	0.218	0.647	−0.12	−0.467	0.647

was a very low value, as it was equal to (0.014), but as for the size of the effect, it is considered (β), which amounted to (−0.120), and this value is considered not statistically significant. Therefore, the effect is not statistically significant because the level of the function (β), which amounts to (0.647), is greater than (0.05) and we conclude that there is no effect Which means that there is no statistically significant effect at the level of significance ($\alpha = 0.05$) of the profits of Jordanian insurance companies measured financial performance (current ratio) before the Corona pandemic for the year (2019).

HO 2: There is no statistically significant effect at a significant level ($\alpha = 0.05$) of the profits of Jordanian insurance companies on their measured financial performance (quick ratio) in light of the Corona pandemic.

Table 2 shows the impact of the profits of Jordanian insurance companies on their measured financial performance (quick liquidity ratio). It was statistically significant as the calculated (F) value amounted to (4.738) with a significance level of (0.046), and where the value of the significance level is less than the value of (0.05),it means that the effect is accepted from the statistical point of view, and the value of (R^2) is (24%). Statistically, because the level of the function (β), which amounted to (0.046) is less than (0.05), the value of the coefficient of determination β (−0.490) is negative, which indicates an inverse relationship with the profits of insurance companies, and with this result, we conclude that there is an effect. Therefore, we reject the null hypothesis and accept the alternative that states **there is statistically significant effect at the level of significance** ($\alpha = 0.05$) **of the profits of Jordanian insurance companies measured financial performance (quick liquidity ratio) during the Corona pandemic for the year (2020)**. On the other hand, and before the pandemic (COVID-19), it was not statistically significant and the value (F) amounted (0.871) with a significance level of (0.365), and since the value is more than (0.05), it means not accepting the impact from the statistical point of view. The value of (R^2) represents a very low value where it was equal to (0.055).Regarding the impact size (β) which amounted (−0.234), and this value is not statistically significant, and therefore the effect is not statistically significant because the level of the function (β) which is (0.365), is greater than (0.05), which means that there is no significant effect level ($\alpha = 0.05$) of the profits of Jordanian insurance companies, the measured financial performance (quick liquidity ratio) before the Corona pandemic (2019).

Table 2 Results of the second null hypothesis test for the year (2019–2020)

Dependent variables	Model summary		Variance analysis		Transactions		
	R	r^2	F	Sig	Beta	T	sig
Quick liquidity 2020	0.49	0.24	4.738	0.046	−0.49	−2.177	0.046
Quick liquidity 2019	0.234	0.055	0.871	0.365	−0.234	−0.933	0.365

HO 3: There is no statistically significant effect at a significant level ($\alpha = 0.05$) of the profits of Jordanian insurance companies on their measured financial performance (debts to assets) in light of the Corona pandemic.

Table 3 shows the impact of the profits of Jordanian insurance companies on their measured financial performance (debts to assets ratio) during the (COVID-19) pandemic, as it was not statistically significant, and the calculated (F) value amounted to (0.436) with a significance level of (0.519). Since the value of the significance level is greater than the value (0.05), this means that the effect is not accepted from the statistical point of view. The value of (R^2) was low (0.028). Regarding the effect size expressed by (β) which amounted to (0.168), this value is considered not statistically significant, and therefore the effect is not statistically significant because the level of the function (β) which is (0.519) is greater than (0.05). With this result, we conclude that there is no effect, and this means accepting the null hypothesis which states that "there is no statistically significant effect at the level of significance ($\alpha = 0.05$) of the profits of Jordanian insurance companies on the measured financial performance (debts to assets ratio) during the Corona pandemic for the year (2020)".

As for before the Corona pandemic, the impact of the profits of Jordanian insurance companies on their measured financial performance (debts to assets ratio), was statistically significant, as the calculated (F) value amounted to (10.38) with a significance level of (0.006), and where the value of the significance level is less than (0.05), this means that the effect is accepted from the statistical point of view, and this value (R^2) reached (40.9%), as for the size of the effect, it is expressed by (β), which amounted to (0.64), and the final value is considered to be statistically significant m and so the effect is statistically significant because the level of the function (β) amounted to (0.006), which is less than (0.05), and with this result, we conclude that there is a statistically significant effect at the level of significance ($0.05 \geq \alpha$) of the profits of Jordanian insurance companies measured financial performance (debts to assets ratio) before the Corona pandemic for the year (2019).

Table 3 Results of the third null hypothesis test for the year (2019–2020)

Dependent variables	Model summary		Variance analysis		Transactions		
	R	r^2	F	sig	Beta	T	sig
Debts to assets	0.168	0.028	0.436	0.519	0.168	0.66	0.519
Debts to assets 2019	0.64	0.409	10.38	0.006	0.64	3.222	0.006

Table 4 Results of the froth null hypothesis test for the year (2019–2020)

Dependent variables	Model summary		Variance analysis		Transactions		
	R	r^2	F	sig	Beta	T	sig
Debts to equity 2020	0.272	0.074	1.197	0.291	0.272	1.094	0.291
Debts to equity 2019	0.234	0.055	0.866	0.367	0.234	0.931	0.367

HO 4: There is no statistically significant effect at a significant level ($\alpha = 0.05$) of the profits of Jordanian insurance companies on their measured financial performance (debts to equity ratio) in light of the Corona pandemic.

Table 4 shows the impact of the profits of Jordanian insurance companies on their measured financial performance (debts to equity ratio) during the (COVID-19) pandemic, as it was not statistically significant, and the calculated (F) value amounted to (1.197) with a significance level of (0.291). Since the value of the significance level is greater than the value (0.05), it means that the effect is not accepted from the statistical point of view, and the value of (R^2) was low (0.074). Regarding the effect size expressed by (β), it amounted to (0.272). This value is considered not statistically significant, and, therefore, the effect is not statistically significant because the level of the function (β) which is (0.291) is greater than (0.05), and with this result, we conclude that there is no effect, and this means accepting the null hypothesis which states that "there is no statistically significant effect at the level of significance ($\alpha = 0.05$) of the profits of Jordanian insurance companies on the measured financial performance (debts to equity ratio) during the Corona pandemic for the year (2020)".

As for the pre-pandemic (COVID-19), where the impact of insurance companies' profits on its measured financial performance (debts to equity ratio) was not statistically significant, and the calculated (F) value amounted to (0.866) with a significance level of (0.367), and since the value of the significance level is greater than the value of (0.05), it means that the effect is not accepted from the statistical point of view, and the value of (R^2) indicates that it was a very low value as it was equal to (0.057). However, for the size of the effect, it is considered (β), which amounted to (0.234), and this value is considered not statistically significant. Therefore, the effect is not statistically significant because the level of the function (β), which amounts to (0.367), is greater than (0.05). We conclude that there is no effect which means that there is no statistically significant effect at the level of significance ($\alpha = 0.05$) of the profits of Jordanian insurance companies measured financial performance (debts to equity ratio) before the Corona pandemic for the year (2019).

Through the results of the statistical analysis of leverage ratio, the ratio of debts to assets and debts to equity are shown to have decreased. The compensation paid to the owners of insurance services decreased due to the restrictions imposed by the government during the pandemic, and there was a positive reflection on the profits of the insurance companies and consequently on equity rights.

4 Conclusion and Recommendations

After testing the previous hypotheses, the study concluded that there is no impact on the profits of Jordanian insurance companies on the measured financial performance indicators (current and debts to equity ratios) during the pandemic of the year (2020) and before the Corona pandemic of 2019) Also, there is also an impact on the profits of companies Jordanian insurance measured financial performance (quick liquidity ratio) during the pandemic in the year (2020), and there was no effect before the pandemic for the year 2019. In addition, it was found that there is no effect for the insurance companies' profits on the measured financial performance (debts on assets) after the Corona pandemic in 2020 m but there is an effect before the pandemic in 2019. The results of the study conform with study [8] which concluded that there is an effect of the financial structure on the measured financial performance regarding the return of the assets and equity. On the other hand, no effect was found for the operational performance on the measured financial performance with returns of assets and return on equity. The results of the study agree with, [9] study that there is no impact of claims and financial leverage on the profitability of insurance companies, and it contradicts the study [11] that indicated that there is an impact of the debts ratio on the market share price of insurance companies listed on the Amman Stock Exchange.

The study therefore recommends decision-makers to seek to improve the financial performance of the insurance sector by taking this study by risk managers and regulatory authorities in order to understand the internal factors represented by financial planning, internal management and administrative efficiency, as well as external factors represented by the legal and economic environment affecting financial performance. And work on studying the liquidity ratios and financial leverage ratios that guarantees benefiting from this money for the benefit of insurance companies. The future studies also recommend conducting a study model on the other financial, industrial and service sectors to find out the extent of the impact of the breaches. on the financial performance of these sectors during the COVID-19 pandemic.

References

1. Al-Masry, F.: Determinants of the risks of financial hardship for Jordanian insurance companies, unpublished master's thesis (2019)
2. Abu Nassar, MH. Hamidat, JF.: International Accounting and Financial Reporting Standards: Theoretical and Practical Aspects, ed 1, Amman, Dar Wael for printing, publishing and distribution (2020)
3. Shojaee, M. R., Haghbin, A., Saadati, A.: Evaluating the critical success factors for KASRA insurance company and providing the appropriate strategy for company based on the "SWOT" model. J. Basic Appl. Sci. Res. Text Road Publication **2**(3), 2165–2171 (2012).

https://insurancepanorama.com/296. https://altaj.news/business/68641 . https://www.iclclaw.com/ar/. http://www.mit.gov.jo/Default/. https://www.ccd.gov.jo/Default/Ar. http://www.pm.gov.jo/newspaperResult?version=&year=2007&subject-. https://www.sdc.com.jo/arabic/index.php?option=com_content&task=view&id=118&Itemid=80. https://www.ase.com.jo/ar/company_guide/information/247. https://www.almamlakatv.com/news
4. Barros, C. P., Barroso, N., Borges, M. R.: Evaluating the efficiency and productivity of insurance companies with a Malmquist index: a case study for Portugal. Int. Assoc. Study Insur. Econ. The Geneva Papers on Risk and Insurance **30**, 244–267 (2005)
5. Abudallou, M.: The Impact of the Application of International Financial Reporting Standard IFRS 9 (Financial Instruments) On the Financial Statements of Jordanian Insurance. Dissertation Master's degree unpublished (2021)
6. Khersiat, O., ALkabbji, R.: Impact of the application of IFRS 9 standards on the profits and losses of insurance companies listed on the Amman stock exchange. Acad. Strat. Manage. J. **19**(1) (2020)
7. Khersiat, O.: impact of investment decisions on the profits of Jordanian insurance companies listed on the Amman stock exchange. Acad. Account. Fin. Stud. J. **23**(5) (2019)
8. Melhem, O., Diab, M.: Factors affecting the financial performance of dissertation Master;s degree unpublished (2019)
9. Rashid, A., Kamal, U.: Impact of internal (MICRO) and external (MACRO) factors on profitability of insurance companies. J. Econ. Policy Res. **5**(1), P35-57 (2018)
10. Rahman, A., Sharma, R.B.: Cash flows and performance in the industrial sector of Saudi Arabia: with special reference to insurance and manufacturing sectors. Invest. Manage. Fin. Innov. **17**(4), 76–84 (2020)
11. Alqaisi, F., Tahtamouni, A., Alqudah, M.: Factors affecting the market stock price: the case of the insurance companies listed in Amman stock exchange. Int. J. Bus. Soc. Sci. **7**(10), 81–90 (2016)
12. Field, A.: Reliability analysis. In: Field, A. (ed.) Discovering Statistics Using SPSS, 2nd Edition. Sage, London, Chapter 15 (2005)

The Role of the System of Internal Control in Improving the Financial Performance of Water Authority of Jordan in the COVID-19 Pandemic

Hiba Al Zyadat and Eman Al Hanini

Abstract This study aimed to demonstrate the role of internal control in improving the financial performance in Water Authority of Jordan in COVID-19 Pandemic, and dealt with the dimensions of internal control according to COSO framework. (85)E-questionnaires were distributed to all Financial employees for (2021), that be used (SPSS) to analyze data, the study concluded there is a role for internal control in improving the financial performance in Water Authority of Jordan in The COVID-19 Pandemic at a high rate. and recommended working to raise the academic and professional qualification for the top administrations employees and their acquisition of certificate (CPA, CMA, CFA), in addition to They need to work on settling water price tariffs so that it reflects the fairness of water distribution and the fairness of providing water services to customers, which helps in increasing revenue.

Keywords Internal control · Financial performance · Water authority of Jordan · COVID-19 pandemic

H. Al Zyadat
Water Authority of Jordan, Amman, Jordan

E. Al Hanini (✉)
Balqa Applied University, As-Salt, Jordan
e-mail: hanini_bau@yahoo.com

A. Hamdan et al. (eds.), *The Implementation of Smart Technologies for Business Success and Sustainability*, Studies in Systems, Decision and Control 216,
https://doi.org/10.1007/978-3-031-10212-7_74

1 Introduction

COVID-19 pandemic caused a crisis for Jordanian economy where economic growth rates decreased during (2020–2021). Financial performance provides an integrated system of reliable information to compare actual performance of management activities [1]. In Water Authority of Jordan internal control is considered defense to protect public funds, and helps administration to take decisions, helps in planning and evaluating.

1.1 Objectives of the Study

The main objective is to demonstrate the role of the internal control system with all its dimensions (the Risk Assessment, Control Activities, Information and Communication, Control Environment and Monitoring Activities) in improving financial performance (Water Revenue Collection, Sewage Revenue Collection, private wells revenue Collection, and the operating costs Recovery) in Water Authority of Jordan in the COVID-19 pandemic.

1.2 Hypothesis of the Study

The main null study hypotheses there is no role of the internal control system with all its dimensions (which are the Risk Assessment, Control Activities, Information and Communication, Control Environment and Monitoring Activities) in improving financial performance in Water Authority of Jordan in the COVID-19 pandemic and the following null hypotheses:

H01: There is no role of the internal control system with all its dimensions in improving the Water Revenue Collection, in Water Authority of Jordan in the COVID-19 pandemic.

H02: There is no role of the internal control system with all its dimensions in improving the Sewage Revenue Collection, in Water Authority of Jordan in the COVID-19 pandemic.

H03: There is no role of the internal control system with all its dimensions in improving the private wells revenue collection, in Water Authority of Jordan in the COVID-19 pandemic.

H04: There is no role of the internal control system with all its dimensions in improving the operating costs Recovery, in Water Authority of Jordan in the COVID-19 pandemic.

1.3 Problem and Questions of the Study

Study's Problem is: What is the role of the internal control system in improving the financial performance of Water Authority of Jordan in the COVID-19 pandemic?

1.4 The Model of the Study

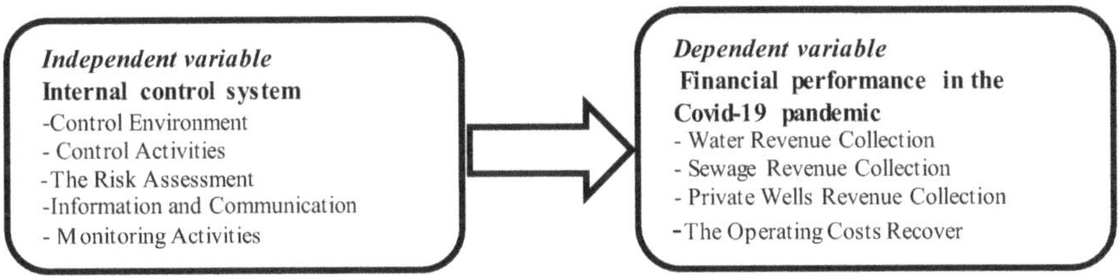

*The model is developed by the researcher using the study [2].

2 Theoretical Framework

2.1 Internal Control System

The role of the internal control system is to lead the organization's internal controls by improving organizational performance and governance, with the aim of changing the attitude of employees towards spending government money; In line with public sector policies [3]. There are five dimensions, according to the (COSO) framework, during which internal control is evaluated, which are the control environment: policies and procedures that reflect managers' attitudes regarding the importance of internal control. And risk assessment: is to identify the risks facing the institution, then analyze them, and know the extent of their impact on the financial statements. And Information and Communication: It is the process of identifying the correct information and communicating it appropriately in a timely manner to the organizational structure of the organization. Control activities: policies and procedures to confirm the implementation of managers' directives and take the necessary measures to address risks and achieve specified goals [4]. As for monitoring: the periodic and continuous evaluation of the internal control components by the management to determine if they are working as required, and to identify weaknesses in the control system followed.

2.2 Financial Performance

The financial performance is the diagnosis of the financial position of the institution, to know its future financial capacity, depending on the budgets [5]. The study presented the most important dimensions of financial performance in the Water Authority of Jordan, which are:

(a) Water Revenue Collection: It is collection money for the price of water, and water reconnection service, the water subscription fees.
(b) Sewage Revenue Collection: It is the money received as a result of providing sewage services, such as the sewage use fees, sewage fees from the building and land tax and sewage connection service fees.
(c) Private Wells Revenue Collection: It is money received in exchange for the prices of water imported from wells belonging to the private property.
(d) Operating Costs Recover: It is realized when revenues cover costs of operation and maintenance and calculating the depreciation rate of fixed assets, based on the fair value.

2.3 Literature Review

There are a number of research studies that fall within the scope of the current study. One of them is Sabah's study [6] entitled "The Reality of the internal Control System in The Income Tax Departments and its Impact on The Income Tax System to Achieve its Goals" which aimed to identify the reality of the internal control system in the income tax departments and its impact on the income tax system achieving its objectives. It concluded that there is a direct positive relationship between the dimensions of the internal control system and the financial income tax objectives, and recommended working on developing the internal control dimensions in the income tax departments and paying attention to training courses for employees.

Albakheet's study [2] entitled "The impact of internal control system on the financial performance of Jordanian civil courts: an empirical study: evidence from Jordan", which aimed to know the impact of the internal control system on the financial performance of the Jordanian regular courts, and concluded that there is an impact of the internal control system on the financial performance of the courts to a medium degree. It also recommended that the Jordanian Ministry of Justice establish a risk management department in the Jordanian regular courts because of its impact on the financial performance.

Adom-Frimpong's study [7] with title "Evaluating The Effectiveness Of Internal Controls Between Public And Private Hospitals", which aimed to evaluate the effectiveness of internal control systems in Ghanaian hospitals, and concluded that there are some policies and procedures to ensure internal control in hospitals, and that the control systems are well effective, and it was revealed that internal control had a positive impact on the necessity of providing hospital management for both policies and procedures. for the necessary training of its employees.

Ahmed and Nganga's study [8] entitled "Internal Control Practices and Financial Performance of County Governments in the Coastal Region of Kenya", which aimed to determine the internal control practices and the financial performance of county governments in the coastal region of Kenya, summarized in the presence of a positive impact between internal control systems and financial performance, and concluded that identifying and mitigating risks contribute to the impact on financial performance, and recommended the need to identify effective strategies Timely risk.

3 Methodology

3.1 Data Collection

The study data were collected through the questionnaire as a study tool. In addition to data collected from books, scholarly journals, and Internet.

3.2 Population and Sample Study

The population of this study was taken from (85) employees for (2021) within the Control and Finance Department of the Water Authority of Jordan, whose job titles are (financial manager, head of the internal audit department, internal auditor, accountant). Using the comprehensive survey method by publishing (85) E-questionnaires to all employees who received (83) questionnaires.

Table 1 shows a description of the study sample, which includes scientific disciplines, where the proportion of accounting majors was 63.9% of the sample size. This is logical because the financial and control departments in the Water Authority of Jordan are based on those who hold the accounting specialization, which is the largest percentage. The study sample included various qualifications focused on bachelor's holders, who accounted for 74.7% of the sample size. The table shows the availability of appropriate practical experiences in the sample. The percentage of respondents with experience (5–10) years was 31.3% of the sample size. The table shows the availability of professional certificate (CPA, CMA, CFA) in the sample. The percentage was 3.6% for CPA certificate and this percentage means the sample is a weak for professional certificates.

Table 1 Description of the study sample

	Description	Frequency	Percentage (%)
Majors	Accounting	53	63.9
	Accounting information systems	10	12
	Business administration	8	9.6
	Others	12	14.5
Qualification	Diploma	13	15.7
	Bachelor	62	74.7
	Master	5	6
	PHD	3	3.6
Experience	Less than 5 years	11	13.3
	5–10 years	26	31.3
	11–14 years	24	28.9
	More than 15 years	22	26.5
Position	Financial manager	5	6
	Head of the internal audit department	8	9.6
	Internal auditor	20	24.1
	Accountant	50	60.2
Certificate	CPA	3	3.6
	CMA	2	2.4
	CFA	2	2.4
	OTHERS	76	91.6
	Total	83	100

4 Statistical Analysis of Data

4.1 Data Analysis and Discussion

Statistical methods were used to analyze the data and test the hypotheses of the study such as Descriptive Statistic Measures: they included frequencies and percentages to describe the study sample namely arithmetic mean and standard deviation for analyzing the sample answers. Table 1 showed a description of the study sample and Table 2 demonstrated the mean and standard deviation of the items of the questionnaire representing the study variables. It is clear from the statistical analysis that all the paragraphs and variables of the study have an arithmetic mean greater than 3 which implies their acceptance. The standard deviation is less than 1, and this indicates the plausibility and acceptance of the dispersion of the data [9].

Inferential Statistic: where (SPSS) was conducted as follows:

(a) Cronbach Alpha test used to measure the internal consistency of the questionnaire, which has reached 96%. Since it is greater than 70%, this means the study

Table 2 The arithmetic mean and standard deviation of the paragraphs of the questionnaire

No	Paragraphs	Arithmetic mean	Standard deviation
Independent variable (Internal control system)			
Control environment			
1	Administration is providing the required professional capabilities for its employees	4.18	0.683
2	Administration is interested with the periodic reports issued by the internal audit committees	4.19	0.788
3	Administration establishes the principle that oversight is the responsibility of all employees	4.20	0.761
Control activities			
4	Administration is keen on having a job description to describe the tasks that each employee performs	4.30	0.761
5	The administration provides physical and electronic protection measures for the Authority's assets	4.22	0.750
6	Administration's control reports include periodic comparisons between the results of different periods	4.10	0.775
The risk assessment			
7	The administration has system for identifying and assessing risks on a regular basis	4.16	0.862
8	Administration controls the levels of risks (financial and operational) that impede the job's performance	4.14	0.813
9	There is a robust and effective contingency plan to face all foreseeable and emergency risks	4.11	0.884
Information and communication			
10	Communication methods within the department deliver information on time and accuracy	4.19	0.876
11	Administration develops systems and keeps track of each update	3.87	0.838
12	The administration provides a secure (financial and operational) information system from hackers	3.94	0.832

(continued)

Table 2 (continued)

No	Paragraphs	Arithmetic mean	Standard deviation
Monitoring activities			
13	Employees in the field of internal control are efficient in the performance of their jobs	4.07	0.852
14	Administration develop business by the results of control and performance reports	3.96	0.890
15	The administration uses the complaints of service recipients as a kind of follow-up and evaluation	3.90	0.835
Dependent variable (Financial performance in the COVID-19 pandemic)			
Water revenue collection			
16	Customer invoices are collected on their due dates	3.98	1.00
17	E-Fawateercom increase the percentage of revenue	4.30	0.837
18	The water price reflects the fairness of water distribution and the provision of water services to costumers	3.84	0.994
Sewage revenue collection			
19	Administration includes sewage fees within the water fees consumed facilitate revenue collection	4.35	0.688
20	Administration provides a preventive maintenance plan for all sewage treatment facilities on a regular basis	4.08	0.752
21	Reusing treated sewage as a secondary water source revenue increases	4.16	0.848
Private wells revenue collection			
22	Owners of private wells use an advance form that enables them to know the cost of their water	4.20	0.808
23	Administration works to reduce water losses through specialized teams to maintain meters	4.17	0.809
24	The administration negotiates a price offers from the owners of private wells to water resources	3.93	0.777

(continued)

Table 2 (continued)

No	Paragraphs	Arithmetic mean	Standard deviation
The operating costs recover			
25	Administration determines how long the costs will be due operational by accuracy	4.36	0.774
26	Administration reduces expenses and operational costs by doing regular maintenance of water lines	4.28	0.856
27	The administration calculates accurately expenses and operational costs for each m^3 of water sold	4.07	0.745

tool is stable, and collected data is appropriate for measuring variables, and can be relied for statistical analysis [10].

(b) Skewness test, torsion factor and One-Sample Kolmogorov–Smirnov Test was used to identify whether the data fall within the normal distribution or not, that mean the data are normally distributed.

(c) Multiple Linear Regression test to test the effects of independent variables on the dependent variable [11].

Table 3 shows results of statistical test for study model that includes the independent variable of internal control with all its dimensions and the dependent variable representing the financial performance of Water Authority of Jordan in the COVID-19 pandemic with all its dimensions. The correlation coefficient $R = (0.894)$ indicates a strong relationship between internal control in its dimensions and the financial performance of Water Authority of Jordan in the COVID-19 pandemic. It was also found that is a statistically significant role of internal control and the dependent variable representing the financial performance of Water Authority of Jordan in the COVID-19 pandemic, through the value of (F. Sig) of (0.00) which is less than (0.05) and also through the value of (F) calculated and its value is (61.180). Latter is greater than Its tabular value (1.989). This implies the significance of this model at the degree of freedom (5/77), and the value of the coefficient of determination ($R2 = 0.799$) indicates that the internal control in its dimensions justified (79.9%) of the financial performance of Water Authority of Jordan in the COVID-19 pandemic.

4.2 *Hypotheses Testing*

H01: There is statistically significant role at $\alpha \leq 0.05$ for the internal control system with all its dimensions in improving the Water Revenue Collection, in Water Authority of Jordan in the COVID-19 pandemic.

Table 3 The multiple regression test

	Independent variable	Unstandardized coefficients	Standardized coefficients	The complete (T) value	T. Sig
	Internal control	Std. error	Beta		
Dependent variable Financial performance in the Covid-19 pandemic	Control environment	0.070	0.232	3.411	0.001*
	Control activities	0.068	0.255	3.706	0.000*
	The risk assessment	0.065	0.194	2.623	0.011*
	Information and communication	0.060	0.255	3.609	0.001*
	Monitoring activities	0.056	0.194	2.891	0.005*
Correlation coefficient R		Determination coefficient R2	The computed F value		F. Sig
0.894		0.799	61.180		0.00*
				Significant at (0.05 ≥ α *)	
Tabular F value = (2.27)		DF = 5/77		Tabular T value = (1.989)	

H02: There is statistically significant role at $\alpha \leq 0.05$ for the internal control system with all its dimensions in improving the Sewage Revenue Collection, in Water Authority of Jordan in the COVID-19 pandemic.
H03: There is statistically significant role at $\alpha \leq 0.05$ for the internal control system with all its dimensions in improving the private wells revenue Collection, in Water Authority of Jordan in the COVID-19 pandemic.
H04: There is statistically significant role at $\alpha \leq 0.05$ for the internal control system with all its in improving the operating costs Recovery, in Water Authority of Jordan in the COVID-19 pandemic.

5 Conclusion

5.1 Results

(1) The study found a statistically significant impact of the internal control system with all its dimensions the financial performance of Water Authority of Jordan in the COVID-19 pandemic by Water Revenue Collection and Sewage Revenue Collection from the viewpoint of the respondents.
(2) There is statistically significant impact of the internal control system with all its dimensions, the financial performance of Water Authority of Jordan in the

COVID-19 pandemic by private wells revenue Collection and operating costs Recovery from the viewpoint of the respondents.

5.2 *Recommendation*

The study recommended and also working to raise the academic and professional qualification for the top administrations employees and their acquisition of professional courses such as (CPA, CMA, CFA) to raise and enhance the Scientific knowledge level in financial management, in addition to They need to work on settling water price tariffs so that it reflects the fairness of water distribution and the fairness of providing water services to customers, which helps increasing revenue.

References

1. Siam, W., Ghunaimat, A.: Factors affecting the effectiveness of the internal control systems in the Jordanian ministries. Jordan J. Bus. Admin. **7**(4), 625–646, 22 (2011)
2. Albakheet, A.: The impact of internal control system on the financial performance of Jordanian civil courts: an empirical study: evidence from Jordan. Jordan J. Appl. Sci. (JJOAS) **19**(2), 1–25 (2016)
3. Jeanne, M.: Internal control components and financial performance in public institutions in Rwanda case study: Rwanda Social Security Board (RSSB). Doctoral dissertation, University of Rawanda. Collage of Business and Economic, pp. 56–89, (2019)
4. Edward, B.: Internal Controls and the Quality of Financial Statements in Local Governments Case Study: Wakiso District. Makarere University (2011)
5. Abdullawi, A.: The role of the financial accounting system in evaluating the financial performance of small and medium enterprises. Damascus University J. Econ. Legal Sci. **25**(2), 889–942 (2017)
6. Sabah, R.: The Reality of the internal Control System in The Income Tax Departments and its Impact on The Income Tax System to achieve its Goals. An-Najah National University, Palestine (2021)
7. Adom-Frimpon, N.Y.: Evaluating the Effectiveness of Internal Controls between Public and Private Hospitals. Doctoral dissertation, University of Ghana (2019)
8. Ahme, S.O., Nganga, P.: Internal control practices and financial performance of county governments in the coastal region of Kenya. Int. J. Curr. Aspects (3), 28–41 (2019)
9. Bani Younes, A.: The beginner guide to general curricula in scientific research (1st ed.). Dar Al- Nafa'as, Amman, Jordan (2017)
10. Sekaran, U., Bougie, R.: Research Methods for Business: A Skill Building Approach, 7th edn. John Wiley & Sons Inc., New York, NY (2016)
11. Hair, J.F., Black, W.C., Babin, B.J., Anderson, R.E., Tatham, R.L.: Multivariate Data Analysis (8th ed.). Cengage Learning EMEA (2018)

The Fourth Industrial Revolution, Tourism and Hospitality, and Healthcare Services

Hospitals in UAE Between Normal and COVID-19 Patients: A Descriptive Study of the Beginning of the Pandemic

Anis Bachta

Abstract The present report is based on describing the impact of the COVID-19 pandemic on the hospitals in the UAE. Like other nations in the world, the UAE is also hardly hit by this epidemic in terms of public health and the nation's economy. This sudden emergence of this pandemic created immense pressure on the hospitals, clinics, and healthcare workers. With the exponential rise in the number of infected persons, the healthcare institutions find it overburdened to handle all the patients at a time. The overloading pressure of treating the COVID-19 positive patients by the healthcare professionals has also increased a high risk on their part. Majority of the countries have also lost their healthcare professionals through COVID-19 infection. To address this problem, the following paper will be organized around three sections. In the first section, we will start by studying and comparing the impact of the COVID-19 in the world and in the UAE. In the second section, we will move to discuss more specifically the impact of COVID-19 on the health institutions in UAE. The last section will be reserved to the empirical investigation in which we will focus on studying how the hospitals managed the pandemic, especially regarding their role with the regular patients and their fight against the COVID-19.

Keywords COVID-19 · Health Institutions · Normal and COVID patients

1 The Impact of COVID-19 in the World and in the UAE

1.1 In the World

The novel coronavirus has affected human health and is slowing down the global economy. In this approach, governments have imposed restrictions on their population by locking and setting strict quarantine for minimizing and controlling the spread of this highly communicable disease. According to Rossi et al. [1], this pandemic has

A. Bachta (✉)
Liwa College of Technology, Abu Dhabi 41009, UAE
e-mail: Anis.bachta@ect.ac.ae

A. Hamdan et al. (eds.), *The Implementation of Smart Technologies for Business Success and Sustainability*, Studies in Systems, Decision and Control 216,
https://doi.org/10.1007/978-3-031-10212-7_75

gravely injured the world's economy, presenting with serious consequences from all individuals and communities. According to the report by Hsiang et al. [2] employers are cutting more than 100 of thousands of jobs, since the COVID-19 pandemic continues to hit the economy. In accordance with the report by the Office for National Statistics (ONS) the unemployment rate from August to October of 2020 is found to be 4.9%. [1]. With the rise in unemployment and restrictions in trade operations, the marginalized population and low-income countries became the hardest hit of the pandemic [3].

1.2 In the UAE

Like other countries, UAE also cannot escape from the COVID-19 and its impact. Being one of the high-income nations and the second-largest economy in the Arab region, it is no exception from the reach of COVID-19. In this country, the total number of people infected from coronavirus till 31 Dec 2020, are 204,369 and out of the 662 people have died [4]. The UAE National Emergency Crisis and Disaster Management Authority (NCEMA) along with the federal and local government are responsible for coordinating the initiatives to reduce the spread of the virus and combat its adverse impact. It is to be noted that the UAE is the first country to detect the virus within the Arab region. According to the report of Rossi et al., UAE became the top 3 countries in terms of testing its most of the population. In February 2020, the demand for UAE oil from China (major buyer) also reduced, reducing the conjunction level by 20% [1]. It is expected that the oil price in the country will continue to experience the downside of recent lows because the level of production is far above the demand [5]. In addition to this, the unemployment rate in the country has increased from 2.35% to 2.45% from 2019 to 2020, which is primarily due to the COVID-19 effect [6].

2 The Impact on Health Institutions in UAE

Despite, fact that UAE has a robust public health system, COVID-19 has increased the pressure on healthcare workers as well as the organization. For health institutions, the year 2020 has been a period of intense pressure. The country has approximately 1.3 hospital beds per 1000 persons, which is relatively less than other countries like South Korea and Russia had 11.5 and 8.2 beds per 1000 persons, respectively [6]. The UAE healthcare professionals must prioritize the urgent life-saving medical care to COVID-19 patients. Moreover, in UAE 82% of physicians and 96% of nurses are expatriates, and the pandemic increases the risk of losing these skilled healthcare professionals, creates a healthcare scarcity [7]. Irrespective of this critical situation, the UAE healthcare system has been successfully able to build field hospitals in minimum time, set up screening facilities, and started to use innovative technologies like telemedicine and other healthcare applications.

2.1 Health Institutions Dealing with Ordinary Patients and COVID-19 Patients

Normal process (before COVID-19)—Like other nations, the UAE healthcare institutions are open for both ordinary patients and chronic patients. In most of the regions, general practitioner doctors are treating patients regularly. However, with pandemic situation, they have either closed their hospitals (or clinics) voluntarily or due to municipality guidelines.

Specific process (for COVID-19)—It is evident that COVID-19 has created a panic situation among the public and healthcare system, which as a result has affected the treatment of ordinary patients. Other instances observed is the fear among healthcare workers to treat ordinary patients, since they might be potential COVID-19 spreaders [8]. In the spirit of medical treatment, these professionals have to learn new practice measures to handle co-COVID-19 cases. The negative consequences of this result in the worst health condition for both practitioners as well as the patients [9].

The changes (impact related to COVID-19)—The major change or impact of COVID-19 on ordinary patients is the cancellation or postponing planned elective operations. According to research by Rossi et al., nearly 42% of routine appointments for medical counselling and surgeries were rescheduled or cancelled due to such reason. Further, with the increase in COVID-19 patients and burden on healthcare professionals' increases, comorbidity cases increased, hence it increases the health risk for those patients. Notably the health care institutions also opted for digital methods of communication like telemedicine, but this account for approximately <5% of cases.

2.2 Difficulty Faced by Health Institutions

Before the advent of the COVID-19 pandemic, the health institution was more focused on individual care. However, in the current outbreak scenario, the healthcare system has shifted its focus to public health. In other words, this indicates that the burden on hospital beds, as well as the shortage of available healthcare professionals, also becomes a serious issue (Massoud 2020). Note that health institutions have adjunct responsibility to provide requisite safety and precautionary training and PPE kit to the physicians, nurses, and other paramedical staff. The practice-related challenge includes lack of evidence-based practices protocol for COVID-19 cases.

Particularly in UAE, more than 1253 physicians were found to be affected while providing the COVID-19 treatment to community members [10].

According to the report of Dwivedi et al. [9], the issue of mental health, tiredness, and lethargy, that were faced by healthcare professionals in these tuff times were seldom neglected and not discussed with serious notes.

3 Empirical Investigations

3.1 Methodology

The empirical investigations in this paper are based on the use of the descriptive research. This enables public health administrators to target a particular segment of the population for education or prevention programs and can help allocate resources more efficiently [11]. In this direction, we will be using quantitative methods where the focus is to collect quantifiable information for statistical analysis of the population sample. A Cross-sectional study is conducted with the use of social surveys based on structured questionnaire. Multiple-choice questions will be asked to the administrative staff in health institutions in UAE who have experience and knowledge regarding the COVID-19 pandemic and its impact on the healthcare institutions. In specific, 20 different hospitals in the UAE will be considered for this survey purpose.

3.2 Hypothesis

The hypotheses which are included in this research study are the following:

H1: Adaptation that are incorporated by the hospitals in the UAE are effective in dealing with COVID-19 situation [12].
H2: Hospitals has incorporated key adaptations to deal with the COVID-19 situation and continue serving the society by providing health services [13].
H3: COVID-19 increases the challenges for the treatment of ordinary patients [1].

4 Results and Discussions

4.1 Demographic Survey Analysis

From the given data it is revealed that there are 26% female participants and 74% male participants in the study.

As far as age is concerned 11% of the participants were below the age of 20 years, 53% of the participants were in between the age of 21–30 years and 36% of the participants were above the age of 30 years.

4.2 *Survey Questions and Findings*

Except 14% of the respondents stayed neutral and 1% of the participants has denied, 85% of the participants stated that their organization have taken step to deal with COVID-19 situation. Therefore, it will be interesting next to assess the importance of the measures taken.

Moreover, 95% of the participants have said that testing facility is available at their hospital. However, 2% respondents disagreed and 3% chose to stay neutral. In this way it is revealed that most of the hospitals has incorporated COVID-19 testing facility at their hospitals.

Regarding the existence of sufficient number of ventilators in the ICUs, the answers are mixed. In one hand, only 48% of the answers state that there are enough ventilators. In the other hand, 28% denied the statement and 25% preferred to stay neutral.

Similarly, 72% of the candidates agrees that their hospital is disinfected periodically. However, 8% of the candidates chose to remain neutral by not giving any answer and 20% of the participants disagreed with the statement. Therefore, it is revealed that majority of the health organizations have incorporated periodic disinfection mechanism as one of their business practices.

In addition, 74% of the participants have said that their hospital do have placed safety measure instructions at various places in their hospital. 19% of the participants have chosen to stay neutral while 7% of the participants have denied the statement. Therefore, it is revealed that majority of the hospitals have placed safety instructions in their hospitals for making people aware about the virus.

Regarding the existence of a periodic cleaning and sanitization of the facility, 62% of the participants strongly agreed with statement, 30% of the participants have also agreed making a total of 92% participants agreeing with the statement. However, 8% of the participants stayed neutral which shows that majority of the hospitals have also undertakes cleaning and sanitization activity at hospital premises.

In the same direction, 48% of the answers stated that the currently prescribed medication on COVID-19 is effective, 49% are less sure and finally, only 3% denied this statement. Therefore, our first hypothesis is confirmed:

H1: Adaptation that are incorporated by the hospitals in the UAE are effective in dealing with COVID-19 situation [12] (Table 1).

The key adaptations engaged by the health system in the UAE can be summarized in three main dimensions explaining 70% of the variance explained. The first dimension is related to **Controlling the contamination**, which includes: the steps taken to deal with the COVID-19 situation, the availability of the testing facilities

Table 1 Component matrix

	Component		
	1	2	3
Your organization have taken several steps to deal with COVID-19 situation at your hospital	0.785	−0.286	0.123
COVID-19 testing facility is available at the hospital premises	0.608	−0.143	−0.416
Air disinfectants are used to make the environment free from virus	0.775	0.064	0.394
Proper safety instructions are made available at strategic locations	0.251	0.817	−0.012
Periodic cleaning and sanitization of the facility is undertaken	−0.063	0.649	0.515
Intensive Care Units (ICUs) in the UAE healthcare facilities have always been equipped with a sufficient amount of ventilators	−0.172	−0.566	0.660

Extraction Method: Principal Component Analysis
[a]3 components extracted

and the use of air disinfectants to make the environment free from virus. The second dimension is more related to **Preventing the contamination**, based mainly on the use of proper safety instructions at the strategic locations and the periodic cleaning and sanitization of the facilities. The third dimension focuses mainly on **Treating the contaminated.**

At another level, 95% of the participants have said that separate COVID-19 wards are available in their organizations. 1% of the respondent disagreed while 4% of the participants chose to remain neutral. The data has revealed that in majority hospitals separate COVID-19 isolation wards are created. Thus, considering the fact that the hospitals has arranged quarantine facilities for COVID patients, 76% of the interviewee denied the statement. Only 23% agreed and 1% preferred to stay neutral. Moreover, 90% of the participants have said that their organization has provided COVID-19 safety kits to their health staff for the protection against COVID-19. However, 10% of the participants chose to stay neutral and not to give any answer. Therefore, it is revealed that majority of hospitals have given COVID-19 safety kits to their staff such as doctors, nurses, and other ward staff for their protection against the virus. Furthermore, 75% of the participants have said that their organization do have systematic mechanism of handling and disposing of biomedical waste. However, 14% of the participants have disagreed with the idea while 11% stayed neutral. Therefore, it is revealed that majority of the hospitals have adapted effective bio-medical waste system to avoid the contamination of the regular patients by those affected by the COVID-19. Regarding the existence of reserved beds for hospital staff in case the staff get infected. 72% of the candidates agreed with the statement that their hospital do have reserved beds for hospital staff. On the contrary, 22% of the participants stayed neutral and 6% of the participants disagreed with the statement.

From the results of data analysis, we can confirm our second hypothesis:

H2: Hospitals has incorporated key adaptations to deal with the COVID-19 situation and continue serving the society by providing health services [13] (Table 2).

Table 2 Component matrix

	Component	
	1	2
Separate wards for COVID-19 patients are created in the hospital premises	0.574	0.531
Measures are taken for the effective handling and disposal of bio-medical waste	0.783	0.308
Hospital has reserved separate beds for hospital staff, in case they get infected	0.854	−0.304
COVID-19 Safety kits are given to the medical staff dealing with COVID-19 positive patients	0.175	0.604
Hospital has arranged quarantine facility for patients so that they will not occupy hospital beds which will be given to serious patients	0.543	−0.723

Extraction Method: Principal Component Analysis
[a]2 components extracted

In this regard, with 67% of the variance explained, two main dimensions were identified. The first one is related to **Preventing the internal contamination** with separating wards for COVID patients and regular one, taking measures for effectively handling the bio-medical waste and reserving separate beds for staff in case they get infected. The second dimension is more likely related to **Preventing the external contamination** by arranging quarantine facilities for the COVID patients so they will not occupy hospital beds with serious patients. In addition, COVID safety kits are given to the doctors dealing with contaminated patients.

Thus, the study of Repici et al. [14] has also revealed that health sector is the most vulnerable sector for COVID-19 pandemic, the reason is that it is a frontline defense of humanity against the virus. Similarly, the outcome of the study undertaken by Wei et al. [15], has also suggested that health organizations have to take extra precautionary measures in order to deal with COVID-19 pandemic because the infected patients are also taken into the hospitals for treatment.

On another hand, only 27% of the interviewed staff considered that their hospitals were understaffed to effectively deal with a large influx of COVID-19 patient. 47% denied this statement and 26% preferred to stay neutral.

In addition, 78% of the answers state that the healthcare system is capable to manage the huge number of the COVID-19 patients. Only 7% of them denied the statement and 14% preferred to stay neutral.

Moreover, 89% of the interviewee stated that their hospitals observed structural changes in terms of departmentalization because of the COVID-19 pandemic. Those changes are made to be able to deal effectively with the COVID-19 patient but also to be able to continue serving the regular patients.

Even with the efforts made to reorganize the health institutions, the workload was very important for the doctors. In facts, 63% of the answers state that the mental health of the doctors and paramedics has been affected because of the pandemic. 8%, denied that and 28% preferred to stay neutral.

In this regard, and to be able to manage effectively the workload engaged by the pandemic, the hospitals in the UAE hired more doctors and paramedics to effectively deal with the affected patients. In facts, 42% of the answers states that they recruited more doctors, 22% stayed neutral and 36% denied the statement.

With enough staff, 72% of the interviewee state that the patient with diseases other than COVID-19 are treated with the same level of urgency as they were prior to the pandemic. Only 11% denied the statement and 17% stayed neutral.

To make the efforts engaged by the healthcare system more effective, a partnership between the government and the hospitals was built through a taskforce team. In fact, 88% of the answers support the existence of such partnership. Only 1% disagreed and 9% stayed neutral (Table 3).

In this direction, three main challenges can be identified. In fact, with 64% of the variance explained, the factor analysis revealed three main dimensions. The first dimension is essentially related to **the need for medical staff** which is positively correlated with the fact that initially, UAE hospitals were understaffed to effectively deal with a large influx of COVID-19 patients and hiring more doctors and paramedics to effectively deal with the influx of patients because of the pandemic. On the other hand, the need for medical staff is negatively correlated with the fact that patients with diseases other than COVID-19 are treated with the same level of urgency as they were prior to COVID-19. The second dimension is mainly related to **the readiness of the healthcare system** which is positively correlated with the ability to manage

Table 3 Rotated component matrix

	Component		
	1	2	3
Initially, UAE hospitals were understaffed to effectively deal with a large influx of COVID-19 patients	0.805	−0.038	−0.156
The healthcare system capable to manage a huge number of COVID-19 patients	−0.186	0.812	−0.058
Hospitals have observed structural changes in terms of departmentalization as a result of the COVID-19 pandemic	−0.326	−0.033	0.735
The mental health of doctors and paramedics has been affected as a result of the COVID-19 pandemic	0.088	0.178	0.811
Hospitals in the UAE have hired more doctors and paramedics to effectively deal with the influx of patients as a result of the pandemic	0.751	0.064	−0.168
Patients with diseases other than COVID-19 are treated with the same level of urgency as they were prior to COVID-19	−0.597	0.257	−0.297
The government and healthcare institutions are working in coordination to deal with this challenge	0.089	0.796	0.211

Extraction Method: Principal Component Analysis
Rotation Method: Varimax with Kaiser Normalization
[a]Rotation converged in 6 iterations

a huge number of COVID-19 patients and the partnership between the government and the healthcare institutions. The third dimension suggested by the factor analysis is related to **the human and organizational challenges** for the hospitals which is positively correlated with the structural changes observed by the hospitals during the COVID pandemic and the affected mental health of the doctors. In the light of the results above, we can confirm our third hypothesis stipulating that:

H3: COVID-19 increases the challenges for the treatment of ordinary patients [1].

To face those challenges, a specific action plan was made possible essentially under the tutelage of the UAE government. The considered actions focused mainly on educating the community, motivating the healthcare system, and managing the post COVID transition. Regarding the first set of measures, 68% of the interviewee stated that the public campaign was the main tool used for educating the community and assessing the mental health. 22% indicated the use of the door-to-door checkup and only 9% stated the use of the campaign in school. Concerning the second set of measures, with 40% each the interviewee stated that the acknowledgment through increased salary and additional benefits to the healthcare professional were the most used strategies to motivate the healthcare system. Only 19% asked for more funding. Finally, regarding the post COVID situation, 45% of the answers stressed the importance of the routine free checkup. At a lower level with 28% the interviewee stated the importance of the education related to nutrition, medication, and physical activities.

5 Recommendations and Conclusions

Results—Through the present study we found out that health sector of UAE has taken certain measures due to the pandemic situation. These measures are taken to continue business practices by effectively controlling the impact of COVID-19 situation. In fact, the health organizations such as hospitals and other health institutions have taken several effective measures to deal with COVID-19 situation. For example, health institutions have ensured the availability of COVID-19 testing facility at the hospital with the help of which the hospital staff is periodically tested to identify the effected staff members and provide necessary treatment to those staff members. In addition to that, majority of health institutions have also established separate COVID-19 isolation wards for COVID-19 positive patients to make sure the containment of pandemic. Likewise, majority of the hospitals of UAE have also arranged all the necessary devices and equipment such as X-ray machine, MRI system, CT Scanning devices etcetera in their hospital to deal with emergency situations. Further, every hospital has provided COVID-19 safety kits to the hospital staff and made wearing at mandatory for the staff that deals with COVID-19 patients. Furthermore, Hospital has also made sure periodic cleaning of floor and disinfection of air twice a day so that the spread of virus will be minimized and contained. Additionally, the management of health organizations have also ensured the effective handling and disposal of biomedical waste so that spread of virus would not be possible through biomedical waste produced at hospitals. Moreover, the management has placed safety instructions at several important spots in the hospitals so that people visiting the hospital become aware about the virus and take proper measures.

Suggestions—Besides focusing on in-person care, the healthcare should emphasize delivering a common service to all the COVID-19 patients, so that the treatment procedure is fastened. It is the responsibility of the healthcare settings and the staff to follow infection prevention and control guidelines. The telehealth service can be used for screening patients having the potential COVID-19 symptoms. The healthcare institution should also organize mobile services for those patients who have difficulty in accessing care, especially those who live in rural settings, older population, and people with limited mobility. The mildly ill patients with COVID-19 can be managed and treated at home by prescribing certain medications, to avoid the rush in the hospital. The healthcare staff are required to monitor these patients with regular check-in using telephone calls, or patient portals.

To curb the rising peak of COVID-19 positive cases, the country should identify and prioritize the essential services to avoid any form of difficulty for the general public. All the non-essential services should be closed during the increasing cases period. All the commercial services should be closed for a certain period until the cases are under control. The essential services which are operative should comply with the highest hygiene and precautionary practices. The workers should adhere to the government guidelines to ensure their and the customers' safety. They should practice touchless service and provide sanitizer at each corner.

UAE should continue to conduct the mass testing approach, for free of cost or minimum price, so that every individual can afford the test.

The healthcare institution should focus on improving the capacity in terms of healthcare staff, beds, and life-supporting devices. This approach can help UAE to face and manage any unexpected or sudden rise in COVID-19 positive patients.

Support from global consultants in sharing the evidence-based practices, guidelines for social measures, healthcare affordable measures, equipment and protective are also valuable. Certainly, these measures were effective in controlling the situation.

Lastly the acknowledgement and affection to the healthcare professionals for their constant support and development must have prevailed. This, in turn, will thank their valuable contribution for the betterment of society from pandemic.

References

1. Rossi, R., Socci, V., Talevi, D., Mensi, S., Niolu, C., Pacitti, F., Di Lorenzo, G.: COVID-19 pandemic and lockdown measures impact on mental health among the general population in Italy. Front. Psych. **11**, 790 (2020)
2. Hsiang, S., Allen, D., Annan-Phan, S., Bell, K., Bolliger, I., Chong, T., Lau, P.: The effect of large-scale anti-contagion policies on the COVID-19 pandemic. Nature **584**(7820), 262–267 (2020)

3. Cheikh Ismail, L., Osaili, T.M., Mohamad, M.N., Al Marzouqi, A., Jarrar, A.H., Abu Jamous, D.O., AlMarzooqi, L.M.: Eating habits and lifestyle during COVID-19 lockdown in the United Arab Emirates: a cross-sectional study. Nutrients **12**(11), 3314 (2020)
4. Hassan, S., Saviour, M., Perkar, S., Augustine, T., Ahmed, S.U., Radhadevi, B., Blelepsma, M.: The impacts of home confinement due to coronavirus (COVID-19) on children: a cross sectional survey study, mediclinic city hospital, Dubai. UAE. Am. J. Pediatr. **6**(4), 408–420 (2020)
5. Masoud, N., Bohra, O.P.: Challenges and opportunities of distance learning during Covid-19 in UAE. Acad. Account. Financ. Stud. J. **24**, 1–12 (2020)
6. Sher, L.: The impact of the COVID-19 pandemic on suicide rates. QJM Int. J. Med. **113**(10), 707–712 (2020)
7. Sharifi, A., Khavarian-Garmsir, A.R.: The COVID-19 pandemic: impacts on cities and major lessons for urban planning, design, and management. Sci. Total Environ. 142391 (2020)
8. Rodríguez-Rey, R., Garrido-Hernansaiz, H., Collado, S.: Psychological impact and associated factors during the initial stage of the coronavirus (COVID-19) pandemic among the general population in Spain. Front. Psychol. **11**, 1540 (2020)
9. Dwivedi, Y.K., Hughes, D.L., Coombs, C., Constantiou, I., Duan, Y., Edwards, J.S., Raman, R.: Impact of COVID-19 pandemic on information management research and practice: transforming education, work and life. Int. J. Inf. Manage. **55**, 102211 (2020)
10. Niankara, I., Muqattash, R., Niankara, A., Traoret, R.I.: COVID-19 vaccine development in a quadruple helix innovation system: uncovering the preferences of the fourth helix in the UAE. J. Open Innov. Technol. Market Complexity **6**(4), 132 (2020)
11. Grimes, D.A., Schultz, K.F.: Descriptive studies: what they can and cannot do. The Lancet **359**, 145–149 (2002)
12. Atallah, B., El Lababidi, R., Jesse, W., Noor, L., Al Mahmeed, W.: Adapting to the COVID-19 pandemic at a quaternary care hospital in the middle East Gulf Region. Heart Views **21**(3), 133 (2020)
13. Lin, C.Y., Cheng, C.H., Lu, P.L., Shih, D.C., Hung, C.T., Lo, H.H., Tsai, M.J., Hung, J.Y.: Active surveillance for suspected COVID-19 cases in inpatients with information technology. J. Hosp. Infect. **105**(2), 197 (2020)
14. Repici, A., Maselli, R., Colombo, M., Gabbiadini, R., Spadaccini, M., Anderloni, A., Lagioia, M.: Coronavirus (COVID-19) outbreak: what the department of endoscopy should know. Gastrointest. Endosc. **92**(1), 192–197 (2020)
15. Wei, W.E., Li, Z., Chiew, C.J., Yong, S.E., Toh, M.P., Lee, V.J.: Presymptomatic transmission of SARS-CoV-2—Singapore. Morb. Mortal. Wkly Rep. **69**(14), 411 (2020)

Millennial Perspective of the Influence of Social Media and Tourism Destination Image on Visiting Decisions

Indah Fatmawati, Prisna Rahmawati, Agneta Prihastuti Lestarisiwi, Nurul Aini, and Fadhilah Tsani

Abstract This study analyzes how millennials use E-WOM on Instagram and destination images on visiting tourist destinations. To test the hypothesis, we performed a causal survey design. The study took place in a natural tourist destination in Yogyakarta, Indonesia. The sampling method uses non-probability sampling, specifically purposive sampling. Data analysis uses Path Analysis, and the results demonstrate that E-WOM has a favorable effect on the image of destinations. Image destination itself positively affects the decision to visit. E-WOM also surprisingly significantly influences the decision to visit. At the same time, our study also showed that image destination mediates the relationship of E-WOM on visiting decisions. Our research revealed that millennials have strong preferences and rely heavily on social media reviews to make purchasing decisions.

Keywords Social media · Destination image · Tourism visit decision

1 Introduction

Communication and information technology development is currently at a very high speed. It enables colleagues to share their experiences openly and reliably through comments, descriptive posts, collaborative discussions, and surveys [1, 2]. According to the Indonesia Digital Report released by Hootsuite (We are social) 2021, 93.8% of social media users use Facebook and Instagram as the most popular social media platforms. Different from Facebook, on Instagram, users directly engage with specific brands they follow. This issue influences marketers to virtualize certain businesses,

I. Fatmawati (✉)
Doctoral of Management, Postgraduate Program, Universitas Muhammadiyah Yogyakarta, Yogyakarta, Indonesia
e-mail: indahfatmawati@umy.ac.id

P. Rahmawati · A. P. Lestarisiwi · N. Aini · F. Tsani
Department of Management, Faculty of Economics and Business, Universitas Muhammadiyah Yogyakarta, Yogyakarta, Indonesia

A. Hamdan et al. (eds.), *The Implementation of Smart Technologies for Business Success and Sustainability*, Studies in Systems, Decision and Control 216,
https://doi.org/10.1007/978-3-031-10212-7_76

strengthening Instagram as a marketing channel and becoming common commercial material.

Instagram is the favorite social media for tourist destination providers to advertise their tourist attractions. Users can judge a destination from online comments among users, which refers to the Electronic Word of Mouth. According to Kotler and Keller [3], E-WOM is used to share the benefits and experiences of buying or using products or services. Consumers can now easily access information, and they can also influence many other consumers by sharing their own experiences [4].

The phrase destination image refers to a grouping of thoughts and impressions established through time due to information processing from multiple sources, resulting in a mental representation of the features and benefits sought of a destination [5]. Echtner and Ritchie [6], as cited by Bruwer and Lesschaeve [7], state that destination image has two functional and psychological components. Functional factors (i.e., tourism infrastructure, prices) are directly observable and measurable, while psychological aspects (i.e., atmosphere, friendliness of local people) are unobservable. Kotler and Keller [3] stated that brand image is the assessment given by consumers to a product or company based on experience and information obtained, which creates positive or negative value.

The concept of a decision to visit could refer to the idea of purchasing decisions. According to Peter and Olson [7], buying decision is an evaluative integration process that combines knowledge of two or more alternatives to choose one of them in fulfilling consumers' needs and wants. A survey of Compete Inc [8], cited by Jalilvand and Samiei [9], stated one-third of online travel customers, before making a travel decision, believe that online reviews will help them make their decision. Ozer [9] and Rebollo [10] stated that opinions and comments surrounding them on social media would influence tourist spending decisions, which results in consumers visiting after comparing destinations and evaluating what suits them. Travel reviews are beneficial in the tourism industry to make travel decisions that provide an indirect experience to tourists [11]. This study looks at the effects of E-WOM and brand image on customer decisions to visit tourism places.

2 Theoretical Framework

2.1 Persuasive Communication

Persuasive communication is necessary for building a good relationship between individuals in society. Anyone who has an audience can use social media to influence that audience [12]. Similarly, marketing a product or service is critical to inform products and services effectively. Persuasive communication is designated to influence, inform listeners to persuade, invite, or influence verbally and non-verbally [13]. The objective of persuasive communication is changing attitudes, beliefs, and opinions according to the communicator's wishes.

2.2 Electronic Word of Mouth

E-WOM is communication between word of mouth through the internet. It can be in a positive and negative statement. Positive or negative comments are formed through consumers' assessment of their experience with a product. Litvin et al. [14] discuss the influence of online interpersonal, or E-WOM, as a potentially cost-effective means of marketing hospitality and tourism and some of the emerging technological and ethical challenges that marketers must overcome as they seek to adopt new E-WOM technologies.

2.3 Destination Image

Destination image in this study refers to brand image theory. The name of a tourist destination will have a particular image like a brand and refers to the term tourist destinations [15–18]. Jalilvand and Samiei [19] stated that understanding the factors influencing image will aid in identifying target markets. Schiffman and Kanuk [20] mention seven factors that can shape a brand image: quality or quality, trustworthiness or reliability, usability or benefits, service, risk, price, and brand image. Someone who wishes to visit a tourist attraction will use social media to find out more information. This information will impact their perception of the destination, influencing their decision to visit.

2.4 Decision to Visit

Travel decisions in tourism pertain to purchase decisions. Referring to Kotler and Keller [3], purchasing decisions are buying decision behavior that refers to the final purchase decision of consumers. Peter and Olson [7] stated that purchasing decisions are an integrated evaluative process that combines knowledge of two or more alternatives to choose one of them. Murray [21] discovered that service consumers trusted personal sources of information more and that personal information had a more significant influence on service purchase decisions. Sukhdial et al. [22], in research Sokolova and Kefi [23], stated that millennials are more likely to purchase a product endorsed by someone who resembles their self-image.

3 Previous Research

Bao and Chang [24] investigate the influence of products and time E-WOM in viral marketing. Shin et al. [25] also describes how much negative impact online oral communication has as a promotion and prevention media provided by consumers. Cheung et al. [26] significantly influence consumers when providing positive information in using the product. Albarq [22] measured the impact of word of mouth on the behavior and intentions of the consumer to visit tourist destinations. This study shows that the three variables used are; E-WOM, tourist attitudes, and interest in seeing, which have a significant effect. Goldsmith and Horowitz [27] show that consumer motivation and information significantly influence E-WOM.

Brand image and equity significantly affect product purchases [28]. Jalilvand and Samiei [19] E-WOM and brand image significantly affect visiting interest. Kanwel et al. [29] brand image significantly affects E-WOM and buying interest, but E-WOM does not directly affect purchase interest. Torlak et al. [30] this study confirms a positive relationship between all variables, including destination image, E-WOM, tourist satisfaction, loyalty, and intention to visit. Park et al. [31] revealed that the quality of online consumer reviews positively affects consumer buying interest. Amble and Bui [32] show that brand reputation significantly increases product reviews given by consumers through social media.

4 Hypothesis Development

4.1 The Effect of EWOM on the Image of the Destination

Jalilvand and Samiei [19] explained that the advancement of internet technology today makes more and more people access information about a product or company. Shimp [33] explains that brand image is an association that appears in consumers when remembering a particular brand. According to Kotler and Keller [4], the more often consumers seek information or provide information about a brand, the stronger the brand image. Hence, when online oral communication occurs positively, it will create a positive brand image of a product.

H1: E-WOM positively affects the destination's image.

4.2 The Influence of Destination Image on Visiting Decisions

According to Peter and Olson [7], purchasing decisions are an integration process that integrates knowledge to evaluate and select one of two or more alternative behaviors. If buyers have no prior experience with a product, they will trust a brand they enjoy or a well-known brand [20]. Based on Rio et al. [33], brand image significantly affects consumer behavior in determining the decision to buy. Mun et al. [34] also argue that customer perceptions and destination image are the basis for discussing how marketing strategies shape customer perceptions and expectations in tourism management. Continuous brand image placement in the minds of consumers creates a strong and positive brand image, which can encourage consumers to buy.

H2: The destination's image has a significant positive effect on the decision to visit.

4.3 The Effect of EWOM on Visiting Decisions

Chang et al. [24] show an influence between E-WOM and consumer decision. Another study from Themba-Monica [35] also showed the effect of E-WOM on purchasing decisions. Kotler and Keller [3] argue that E-WOM is an effective promotion method because satisfied customers convey it with a product to other customers. It can be a medium to advertise specific products or in visiting decisions.

H3: E-WOM has a significant positive effect on visiting decisions.

4.4 The Mediating Effects of Destination Image on the Influence of EWOM on the Decision to Visit

According to Kotler and Keller [3], consumers' brand image assesses products based on experiences that create positive or negative values. Based on this theory, E-WOM influences purchasing decisions indirectly through brand image. These results indicate that consumers who make purchasing decisions initially get information about products and services through E-WOM will form judgments and beliefs that create positive or negative values that consumers will use as purchasing decisions.

H4: E-WOM significantly influences the decision to visit indirectly through the destination's image.

5 Research Model

The discussion in the previous section guides us to develop the following research model, as shown in Fig. 1.

6 Method

Our study uses a quantitative causal survey to explain causal relationships among our research constructs. The setting is Tebing Breksi, an award-winning tourist destination as the most popular new tourist destination in the Special Region of Yogyakarta.

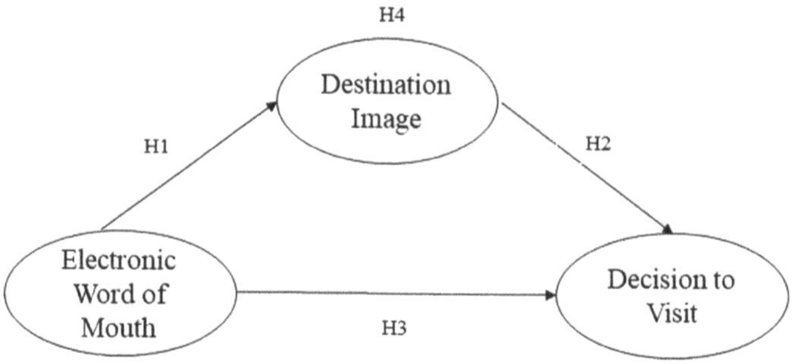

Fig. 1 Research framework

Initially, this tourist spot was a former mining area; in 2014, the government closed the mining activity, and the local community redecorated it into a tourist destination. In May 2015, the Sultan, the Governor of the Special Region of Yogyakarta, inaugurated this site as a new tourist attraction in Jogja named Tebing Breksi (Breksi Hill).

Sampling used a non-probability sampling method with a purposive sampling approach. According to Sekaran and Bougie [36], purposive sampling is a sampling process by selecting certain types of people to provide the desired information or fulfill several criteria determined by the researcher. Young adults who had visited the site and sought input from Instagram comments were our sample criteria. The final sample size in our study was 175 samples. According to Hair et al. [37], the sample size for the complex model is close to 200 or 5–10 times the number of indicators in the questionnaire. Our research data collection using an online questionnaire. We adapted the measurement of variables from several sources such as; Hennig-Thurau et al. [38] for E-WOM, from Kotler and Keller [3] for brand image, Schiffman and Kanuk [25], and Peter and Olson [7] for visiting decisions.

7 Results

The description of respondents there is four characteristics, gender, age, job, and information search sources. Most of the respondents are male. The respondents' age ranges from 17-to 28 years old, and 92 percent are less than 25 years old. This phenomenon implies that respondents are active users of Instagram within this age range, and they are more interested in visiting popular tourist destinations based on social media pages. Most of the respondents are students. This result shows millennials' preference to enjoy natural tourism destinations is exists. The most frequent searching information source used by respondents is the Instagram @explorejogja account via the review/tag of others.

Results of the validity test using Factor Analysis showed that all indicators are valid to measure the variables in this study. Instrument reliability testing uses Cronbach Alpha with the criteria of $\geq$0.6. Test results of the reliability test showed the following scores of Cronbach Alpha: E-WOM (0.689), destination image (0.772), and visiting decision (0.813). As a result, all variables are reliable in terms of stability when retested and consistent amongst items to evaluate the research topic.

We analyzed the data using path analysis. To test the first hypothesis about the effect of E-WOM on destination image, we performed a simple regression analysis. The simple regression model determines whether the independent variable E-WOM (X) affects the destination image (Z). The results shown using path analysis are the coefficient of direct influence of E-WOM on the destination image (p_1) of 0.223 (positive) and the significance of t (Sig.t) of 0.005, which means Sig.t < 0.05 and for t_{stat} is 2867. Thus, H1 is supported; this shows that the better E-WOM occurs, the higher the impact on the destination's image.

In the next step, we perform multiple regression to test the effect of destination image on visiting decisions (H2) and the effect of E-WOM on visiting decisions (H3). The results shown are the path coefficient of the direct influence of destination image on visiting decisions (p_2) of 0.444 (positive), and the probability of significance (Sig.t) of 0.000 and t_{stat} is 7151. Since Sig.t < 0.05, H2 is supported at the 0.05 level of significance. These results mean a positive and significant direct influence on the destination's image on the decision to visit. This finding shows that the better the destination's image, the higher the tourist visits.

The path coefficient of the direct influence of E-WOM on visiting decisions is 0.329 (positive), and the significance (Sig.t) is 0.000. Because Sig.t < 0.05, then H3 is supported at a significance level of 0.05. This finding means a significant direct effect of E-WOM on visiting decisions. This shows that the better E-WOM occurs positively, the higher the tourist visits.

We performed a path analysis to test the mediating effect of destination image on E-WOM and visiting decisions relationship. Hypothesis 4 in our study is that E-WOM (X) influences visiting decisions (Y) indirectly through destination image (Z). The indicator of hypothesis testing is the significance of indirect influence coefficients ($p_1 p_2$) with Z_{stat} based on simple regression models and multiple regressions estimates. The result showed that E-WOM indirectly influences the decision to visit through the destination's image of 0.099. The direct influence was gained by 0.329, so the total effect of E-WOM on visiting decisions was 0.329 + 0.099 = 0.428.

Testing the significance of the indirect influence of E-WOM on visiting decisions through destination image with Z statistic (Z_{stat}) can be calculated by formula, Standard error p_1 (S_{p1}) = p_1/t_{p1}. Where p_1 is 0.223. The value of t_{p1} is the t_{stat} *for* p_1 2867 so that $Sp_1 = 0.223/2867 = 0.0777816$. Standard error p_2 (Sp_2) = p_2/t_{p2}. Where p_2 is 0444. The value of t_{p2} is 7,151, so $Sp_2 = 0.444/7{,}151 = 0.0620892$. Based on the given values, the Z statistic is:

$$\frac{0.099}{\sqrt{(0.444)^2(0.07778)^2 + (0.223)^2(0.06208)^2 - (0.07778)^2(0.06208)^2}} = 2.75$$

Because the value $Z_{stat} = 2.75 > 1.96$ shows significant results at a significance level of 0.05. Hence, it indicated that E-WOM indirectly influences visiting decisions through destination image.

8 Conclusion

After conducting the data analysis, we arrived at the following conclusions. Results of hypothesis 1 testing showed, E-WOM positively affects the tourist destination image. This result supported the findings of Jalilvand and Samiei [19]. Comments and ideas of other visitors via online platforms significantly shaped the tourist destination image in consumers' minds. Thus, positive E-WOM will show that a tourist destination has a good image of the destination and vice versa. The results of hypothesis 2 testing show a significant direct effect of destination image on visits to tourist destinations. This result supported the findings of Peter and Olson [7]. Thus, when a tourist destination has a good image in the consumers' minds of tourists, it will be permanently embedded in the minds of tourists and allow them to visit. Based on the results of hypothesis 3 testing using multiple regression, there is a significant direct influence between online oral communication on the decision to visit a tourist destination. This finding supports the opinion of Kotler and Keller [3]. Hence, the higher the E-WOM, the higher the tourists determine tourist visits. Analysis of hypothesis 4 shows that online communication indirectly affects the decision to visit through the destination's image. These findings support the ideas of Kotler and Keller [3]. First-time consumers indicated that they got information about products and services through E-WOM and will build judgment and self-confidence. Then it will create positive or negative values that consumers will use as purchasing decisions.

References

1. Rizky, R.M., Kusdi, R., Yusri, A.: The impact of e-WOM on destination image, attitude toward destination and travel intention. Russ. J. Agric. Socio-Economic Sci. **61**(1), 94–104 (2017)
2. Al-gasawneh, J., Al-adamat, A.: The relationship between perceived destination image, social media interaction and travel intentions relating to neom city. Acad. Strateg. Manag. J. **19**(2), 1–12 (2020)
3. Kotler, P., Keller, K.L., Manceau, D., Dubois, B.: Marketing Management, 15th edn. Pearson Education, New Jersey (2016)
4. Ward, J.C., Ostrom, A.L.: The Internet as information minefield an analysis of the source and content of brand information yielded by net searches. J. Bus. Res. **56**(11), 907–914 (2003). https://doi.org/10.1016/S0148-2963(01)00277-6
5. Uysal, M., Fesenmaier, D.: Communication and Channel Systems in Tourism Marketing. Routledge, New York (2012)
6. Bruwer, J., Lesschaeve, I.: Wine tourists' destination region brand image perception and antecedents: conceptualization of a winescape framework. J. Travel Tour. Mark. **29**(7), 611–628 (2012). https://doi.org/10.1080/10548408.2012.719819

7. Peter, J., Olson, J.: Consumer Behavior and Marketing Strategy, vol. 1. Salemba Empat, Jakarta (2013)
8. Compete, I.: Consumer generated content: Learning from travel innovators. Compete, Inc. (2007). https://media.competeinc.com/med/uploads/files/traveltrends_consumer_ generated_travel_content.htmlS
9. Ozer, S.: The Effect of Social Media on Consumer Buying Decision Process. National College of Ireland (2012)
10. Rebollo, H.P.M.: A structural model of millennial tourist behavior towards tourism in Davao Region. J. Adv. Humanit. Soc. Sci. **4**(1), 26–36 (2018). https://doi.org/10.20474/jahss-4.1.3
11. Park, D.H., Lee, J., Han, I.: The effect of online consumer reviews on consumer purchasing intention: the moderating role of involvement. Int. J. Electron. Commer. **11**(4), 125–148 (2007). https://doi.org/10.2753/JEC1086-4415110405
12. Chatzigeorgiou, C.: Modelling the impact of social media influencers on behavioural intentions of millennials: the case of tourism in rural areas in Greece. J. Tour. Herit. Serv. Mark. **3**(2), 25–29 (2017). https://doi.org/10.5281/zenodo.1209125
13. Larson, C.U.: Persuasion: Reception and Responsibility, 12th edn. Wadsworth, USA (2010)
14. Litvin, S.W., Goldsmith, R.E., Pan, B.: Electronic word-of-mouth in hospitality and tourism management. Tour. Manag. **29**(3), 458–468 (2008). https://doi.org/10.1016/j.tourman.2007.05.011
15. Gallarza, M.G., Saura, I.G., CalderónGarcía, H.: Destination image towards a conceptual framework. Ann. Tour. Res. **29**(1), 56–78 (2002). https://doi.org/10.1016/S0160-7383(01)00031-7
16. Baloglu, S., Mangaloglu, M.: Tourism destination images of Turkey, Egypt, Greece, and Italy as perceived by US-based tour operators and travel agents. Tour. Manag. **22**(1), 1–9 (2001). https://doi.org/10.1016/S0261-5177(00)00030-3
17. Ko, D., Park, S.: Five aspects of tourism image: a review. Int. J. Tour. Sci. **1**(1), 79–92 (2000). https://doi.org/10.1080/15980634.2000.11434520
18. Hahm, J., Upchurch, R., Wang, Y.: Millennial students, movies and tourism. Tour. Anal. **13**(2), 189–205 (2008). https://doi.org/10.3727/108354208785664229
19. Reza, J.M., Samiei, N.: The effect of electronic word of mouth on brand image and purchase intention (an empirical study in the automobile industry in Iran). J. Mark. Intell. Plan. **30**(4)
20. Schiffman, L.G., Kanuk, L.L., Kumar, S.R., Wisenblit, J.: Consumer Behavior. Pearson Educational, Inc, (2010)
21. Murray, K.B.: A test of services marketing theory: consumer information. J. Mark. **55**(1), 10–25 (1991). 10.1177%2F002224299105500102
22. Albarq, A.N.: Measuring the impacts of online word-of-mouth on tourists' attitude and intentions to visit Jordan: an empirical study. Int. Bus. Res. 7(1) (2014)
23. Sokolova, K., Kefi, H.: Instagram and YouTube bloggers promote it, why should I buy ? How credibility and parasocial interaction influence purchase intentions. J. Retail. Consum. Serv. **53** (2020). https://doi.org/10.1016/j.jretconser.2019.01.011
24. Bao, T.T., Tung-lung Steven, C.: The product and timing effects of eWOM in viral marketing. Int. J. Business. 2(2) (2016)
25. Shin Dongwoo, J.H., Song, A.B.: Electronic word-of-mouth (eWOM) generation in new media platforms: the role of regulatory focus and collective dissonance. Mark. Lett. **25**(2) (2014)
26. Yee, C.M., Chuan Luo, C.L.S., Huaping, C.: The credibility of electronic word-of-mouth: informational and normative determinants of online consumer recommendations. Int. J. Electron. Commer. **13**(4) (2009)
27. Goldsmith, R.E., Horowitz, D.: Measuring motivations for online opinion seeking. J. Interact. Advert. **6**(2) (2006)
28. Del, R., Belen, A., Vazquez, R., Iglesias, V.: The effects of brand associations on consumer response. J. Consum. Mark. **18**(5) (2001)
29. Torlak, O., Behcet Yalin Ozkara, M.: The effect of electronic word of mouth on brand image and purchase intention: an application concerning cell phone brands for youth consumers in Turkey. J. Mark. Dev. Compet. 8(2) (2014)

30. Kanwel, S., Lingqiang, Z., Asif, M., Hwang, J.: The influence of destination image on tourist loyalty and intention to visit: testing a multiple mediation approach. Sustain 11(22) (2019). https://doi.org/10.3390/su11226401
31. Park, D.-H., Lee, J., Han, I.: The effect of online consumer reviews on consumer purchasing intention: the moderating role of involvement. Int. J. Electron. Commer. 11(4) (2007)
32. Amblee, N., Bui, T.: Can brand reputation improve the odds of being reviewed online? Int. J. Electron. Commer. **12**(3) (2008)
33. Shimp, T.: Advertising Promotion Additional aspects of Integrated Marketing Communication. Erlangga, Jakarta (2000)
34. Mun, Y.W., Aziz, Y.A., Bojei, J.: Mediating impact of perceived destination image on intention to recommend Malaysia as a key educational hub. Int. J. Acad. Res. Bus. Soc. Sci. **8**(16), 1–14 (2018)
35. Themba, G., Mulala, M.: Brand related eWOM and its effects on purchase decisions: an empirical study of botswana students. Int. J. Bus. Manag. **8**(8), 31–40 (2013)
36. Sekaran, U., Bougie, R.: Research Methods for Business: Research Methodology For Business Research. Publisher of Salemba Empat, Jakarta (2012)
37. Hair, J.F., Babin, B.J., Anderson, R.E., Black, W.C.: Multivariate Data Analysis, 8th ed. Cengage India (2018)
38. Hennig-Thurau, T., Kevin, G., Gianfranco, G., Dwayne, W.: Electronic word of mouth via consumer-opinion platforms: what motivates consumers to articulate themselves on the internet? J. Interact. Mark. **18**(1) (2004)

Effects of COVID-19 on Domestic Tourism in the West Bank, Palestine: A Case Study of Nablus City

Safaa Hamada and Ahmed Ghodieh

Abstract This study aims to investigate the effects of the COVID-19 pandemic on domestic tourism in the Nablus City located in the West Bank, Palestine. The data for this study are collected using the questionnaire method as a research instrument. Three sets of questionnaires were designed to target three different groups of people: tourists or visitors, employees of tourism facilities, and owners of tourism facilities. These three questionnaires were then distributed online to the participants. The responses from the participants were then collected through email and a descriptive analysis was carried out for a total of 263 valid questionnaires based on the responses. The findings of this study have shown that domestic tourism in the West Bank was negatively affect-ed because of the COVID-19 pandemic. This study also revealed that the average number of employees in the tourism facilities decreased significantly during the pandemic to 22%. The aver-age number of tourists also decreased to around 5%, while 20% of tourists' facilities were closed. Also, the lifestyle of locals has changed significantly. Based on the findings, this study also presented a few suggestions to assist the recovery of domestic tourism in Nablus city. These suggestions include creating a program led by the government and the private sector to support small and medium tourism enterprises.

Keywords Coronavirus (COVID-19) pandemic · Tourism · Palestine · West Bank · Nablus

---

S. Hamada (✉) · A. Ghodieh
An-Najah National University, Nablus 44388, Palestine
e-mail: s.hamada@najah.edu

A. Ghodieh
e-mail: ahmed@najah.edu

A. Hamdan et al. (eds.), *The Implementation of Smart Technologies for Business Success and Sustainability*, Studies in Systems, Decision and Control 216,
https://doi.org/10.1007/978-3-031-10212-7_77

1 Introduction

In recent decades, the tourism sector has achieved high growth rates, and its contribution to GDP has increased at the global level to represent about 10.4% during 2019 [1]. Thus, it becomes one of the most important sectors that contribute to achieving economic growth, increasing employment levels, reducing poverty, reducing inequality in income distribution, increasing levels of demand for goods and services, and strengthening official reserves and foreign exchange receipts [1]. The tourism and travel sector has been the most affected by the continuous closures all over the world, to avoid the spread of the virus [2]. The influences of COVID-19 towards tourism might be different depending on area and time. Aside from the human tool, estimates display a tremendous and global economic effect: international tourist arrivals are predicted to drop by 78% [3], leading to a loss of US$ 1.2 trillion in export sales from tourism and a 120 million direct tourism job cuts, representing seven times the effect of September 11, and the largest economic decline in the history. The tourism industry contains among the maximum amount of essential worldwide employers (1 in 10 jobs directly related to tourism [4]. The coronavirus pandemic has radically reshaped city life: the relationship between work and home, safety in transportation (public and private) essentially, enjoyment of public space, and recreational activities participation [5, 6]. These impacts of COVID-19 are therefore known to be negative in the urban sector [7].

During the COVID-19 pandemic, knowledge and awareness of health risks were established through media channels (especially television) and scientific publications on the subject were published 12 months after the start of the pandemic in China.

However, given the lack of studies that have been done and the successive confinement and closure of borders between countries, its impact on the tourism experience remains unknown. Eventually, the tourism market is expected to experience a prolonged downturn [8]. Therefore, tourists have chosen the neighboring areas to their places of residence [9], which increases domestic tourism compared to the pre-pandemic figures. Moreover, people show a preference for exploring less populated areas where they can be more active, with more natural environments, as evidenced by increased tourism in rural areas [6, 10]. As a result, tourists must now consider various factors (such as infectious disease outbreaks, quarantine status, hygiene, and safety) more carefully when choosing tourist destinations than before the outbreak of COVID-19 [6].

Although no studies have examined the impact of COVID-19 on the tourism sector in Nablus city in Palestine, several studies have addressed this subject and contributed to the discussion about the impact of COVID-19 on the tourism sector in some areas of the world. This includes research by Isaac and Abuaita [11]. Their study helps to fulfill intriguing gaps in the literature about the effects of COVID-19 on Bethlehem. In their study, they applied both qualitative and quantitative methods. They conducted interviews consist of open-ended questions and these interviews were then analyzed. Fifty-two respondents who are tourism business owners and major stakeholders in the tourism sector in Bethlehem participated in the study and were interviewed.

The results showed that the pandemic caused huge job losses in Bethlehem, as the tourism sector in Bethlehem implemented pay cuts and layoffs to mitigate the loss of income caused by the outbreak of the COVID-19 pandemic.

On the other hand, Bitok [12], examines how sustainable tourism has contributed to Kenya's economic growth. The study used Eviews 10 software to analyze the time-series data and based on data from 1995 to 2020, Johansen's approach to co-integration, causation, and regression was used. In conclusion, the study suggests that in the new normal of the post-COVID-19 period, policymakers in the field of tourism should focus on creating policies to promote sustainable tourism and must be managed according to the three pillars of sustainability. Next research is by Riadil [13], which studied the perspectives of Indonesian tourism employees who are working in the hospitality and tourism industry to see the impact of coronavirus on their careers. The researcher used a qualitative study (questionnaire) for this research, and the research data were collected from tourism employees in Indonesia. Results of the study show that the pandemic have negative impact on Indonesian tourism and the hospitality industry and is considered the most affected sector in Indonesia. Alternatively, Persson-Fischer and Shuangqi [14] conducted a literature review of the COVID-19 literature in tourism in 2020. They analyzed a total of 87 articles, in 17 journals, from 4 databases to explore how current scholars perceive COVID-19 and tourism, in light of sustainability perspectives. This study found six major themes in COVID-19 and tourism and provided valuable information with descriptive statistical analyzes of their distributions by theory, methodology, and field of study. This research emphasized two urgent aspects: There is an urgent need to establish a benchmark for assessing the performance of tourist destinations and setting a priority to ensure that other crises are limited (e.g., global warming, wildfires, species extinctions, etc.). Lopes et al. [7] investigated the sensitivity of tourists to crises when making decisions about the daily routine, as well as future travel plans in the presence of a serious health problem. In this article, the researchers examine the perceptions and behaviors of tourists before and during COVID-19 in the municipality of Porto, the main area of the Porto Metropolitan Region, Portugal. As a result of the study, there have been significant changes in the use of public spaces and the way tourists visits are managed: the visiting time (a shorter than usual visit to some tourists attractions), visiting areas with limited space, and the ability to attract regular tourists from some countries where stricter rules are applied to prevent foreclosure.

This present study is hoped to contribute to the existing literature by exploring the asymmetric impact of COVID-19 pandemic through the questionnaires on inbound Palestinian tourists to Nablus city.

2 Materials and Methods

2.1 Study Area

The study area consists of the occupied West Bank of the River Jordan, including East Jerusalem and the northwestern part of the Dead Sea [15, 16]. The West Bank lies between the longitudes 34°53′ and 35°31′ E, and between the latitudes 31°20′ and 32°38′ N [17, 18] The surface area of the West Bank is around 5850 km^2 including around 200 km^2 of the Dead Sea. This area represents around 22% of the area of historical Palestine.

Physically, the West Bank consists of four main geomorphological regions: the Semi-Coastal region, the Mountains region, the Eastern Slopes region, and the Jordan Valley region [19–21].

The population of the West Bank in 2017 was 2,856,691 and estimated to be 3,120,448 in mid. 2021 [22].

Nablus city lies in the mountain's region; between Mount Ebal in the north and Mount Gerizim in thc south, on the longitude 35.25° and latitude 32.24° [23].

Nablus city is a historic city that was established around 5000 years ago by the Canaanites who later named it Shakim. During the Canaanite time, the Romans destroyed it, in addition to nine other cities, on the day they occupied the Levant [24].

The total area of Nablus City is approximately 33 km^2 according to the new domestic councils' borders which was defined by the Palestinian Ministry of Domestic Government [24]. The built-up areas of the city are now connected with the built-up areas of the surrounding villages and domesticities as a result of the normal city growth and expansion, especially from the west and east directions. Israeli settlements which are situated on the tops of hills and mountains around the city stopped the city's expansion towards the north and the south directions.

2.2 Tourism in Palestine in 2019 and Before

Palestine witnessed remarkable developments and improvements in the tourism sector in the last few years. Statistics show that that there are 38 thousand employees in the tourism sector in Palestine, constituting 4% of the total employees from all sectors. There are about 29 thousand employees in the West Bank and 9 thousand employees in Gaza Strip. During the first half of 2019, tourist sites in the West Bank witnessed an active movement of inbound visitors, resulting in an increase of 17% [25] compared to the same period of the previous year. The number of hotel guests in the West Bank reached 363,385 in the first half of 2019, where they spent 1,021,567 nights [26].

Nablus city is one of the oldest and largest cities in the West Bank and is considered to be the most important city in the northern part of the study area. It connects the northern domesticicities with the southern ones and the eastern domesticities with the western ones. Hence, people visit the city from all parts of the West Bank [27]. Furthermore, the Palestinians in Palestine 1948 prefer to visit Nablus for shopping and eating at the restaurants not only because the goods and foods are much cheaper than those in Israel, but they also fit their traditions and lifestyle better. Nablus is known for its olive oil soap, talented goldsmiths, and traditional sweets, Nablus is considered the major center for commercial, industrial and agricultural in the northern region of the West Bank [27]. In terms of recreational tourism, Nablus is classified among the centers of tourist attractions in Palestine due to its picturesque mountainous nature and its mild climate in summer and winter. The markets and shops, restaurants, hotels, Turkish and Shami paths constitute the main attraction for shoppers from different parts of historical Palestine [26].

3 Results and Discussion

It is important to evaluate the impact of the coronavirus pandemic on the domestic tourism sectors in Palestine in terms of the potential impact on the sector's output and on employees, visitors, and owners of tourist facilities as well as the expected period for the recovery of the tourism sector.

The analysis and results were interpreted in three sections: the questionnaire of tourists, the questionnaire of employees in tourist facilities, and the questionnaire of owners of tourist facilities.

3.1 Tourists' Questionnaires

This section consists of subjects related to the behavior of tourists in Nablus city towards the crisis of COVID-19 during the summer of 2020. This section will also reveal the demographic characteristics (e.g., gender, age, education background, and income level) of the tourists.

Based on the survey conducted, it is shown that 48.6% of the tourists are male and 51.4% are female. It is also found that 70% of the tourists are from Nablus and around 20% from Palestine 1948 (Israel), while the rest 10% of tourists are from different parts of the West Bank. The study also revealed that the ages of more than half of the tourists are between 20 and 29 years old, while the ages of the lowest percentage of tourists are in the categories of 50–59 years old and 60 years old and above.

Next, most tourists who participated in this study are highly educated. Around 79.3% of tourists hold a bachelor's degree in different disciplines, 6.5% hold Master's and Ph.D. degrees. The rest hold a general secondary education certificate.

As for the tourists' income level, the study showed that about 17% of them earn less than $500 per month, while the percentage of those who earn between $500 and $1100 is about 53%, and about 30% earn more than $1100 per month.

The study also showed that about 19% of the tourists do not allocate any monthly amount of money for tourism purposes, 23% of them allocate less than 30 dollars, 12% allocate 30–60 dollars, and 27% allocate 60–150 dollars, while about 19% of tourists allocate more than 150 dollars for tourism purposes per month.

For the tourists' frequency of visits to the tourist facilities, the study showed that in a month, about 69% of tourists visit them once, 15% make two visits, and 16% make three visits. About 53% of tourists visit the tourist facilities with their families, 36% visit them with their friends, and 11% visit the tourist facilities alone.

The study also indicated that the percentage of those who depend on the delivery service to get their needs from shops and restaurants increased from about 30% to about 45% after the coronavirus pandemic.

The study showed that about 60% of the tourists prefer to spend the weekend going on trips, while about 24% prefer to spend it at home, and about 13% prefer to spend the weekend in cafes, and the rest prefer to spend it working in agriculture, with friends or with relatives.

This study also revealed the different types of activities done by the tourists during the outbreak of COVID-19. About 36% of them practice different kinds of sports, 34% prefer to study, 15% of them spend the time to watch television, and the rest work mainly in agriculture.

In the questionnaires, the tourists were also asked to what extent do the tourist facilities adhere to the preventive measures during the coronavirus pandemic, and their answers were as follows: about 59% of the facilities adhere to hygiene guidelines, about 64% of the facilities have defined the signs according to safety procedures, and about 61% of the facilities are spacious and well-ventilated. Next, the tourists' adherence to COVID-19 preventive measures will also be discussed in this section. The survey stated that 52% of the tourists are able to practice social distancing in the tourist facilities and around 76% of the tourists feel safe while traveling and visiting tourist facilities. In regards to the use of face masks, 73% of them stated that they are committed to wearing face masks inside the tourist facilities, while 29% of them wear face masks only in public places. In the matter of social distancing, 26% of tourists always keep enough distance between them and others, 52% of the tourists occasionally keep that distance, and around 22% of the tourists do not keep that distance at all.

The study also showed that only 35% of tourists use sanitizers consistently, and 56% use sanitizers occasionally, while 9% do not use sanitizers at all.

Next, the following are the suggestions by the tourists in Nablus city to improve the tourism in the city while overcoming the spreading of COVID-19:

- Tourists' facilities should be supervised better.
- The public should all be vaccinated.
- Encouragement of domestic tourism.
- Rehabilitation and development of domestic tourism facilities.

- Implementation of tour guide service.
- Enhancement of various services in tourist facilities such as create a cable car.
- Increase of natural areas and public facilities such as create a zoo.
- Rehabilitation of mountain climbing areas.
- Rehabilitation of skiing areas.

3.2 Tourism Facilities Employees' Questionnaire

This section consists of subjects related to the behavior of employees in the tourism facilities towards the crisis of COVID-19 during the summer of 2020. This section will also reveal the demographic characteristics of the employees. The study showed that during the day, 70% of the employees are male, and 30% are female, while at night 96.7% are male and 3.3% are female. About 90% of employees are from Nablus and the rest 10% are from neighboring cities, villages, and Palestinian refugee camps. The study also revealed that around 94% of the employees are between 20 and 39 years old, while the rest 6% of the employees are between 40 and 49 years old. None of the employees are less than 20 years or more than 50 years old.

Most employees who participated in this study are highly educated. Around 65.6% of the employees hold a bachelor's degree in different disciplines, 28.1% of the employees hold. The general secondary education certificate, while 6.3% of the employees do not hold the general secondary education certificate.

As for the employees' income, the study showed that about 20% of them earn less than $600 per month, while the percentage of those who earn between $600 and $900 is about 53.3%, and about 26.7% of the employees earn between $900–1200 per month.

During the closure of tourists' facilities, the employees struggled financially, because of the loss of income. Only 13.3% of the employees received inadequate financial aid from the government and the owners of tourists facilities which totaling to around $300.

As for the working experience in tourism sector for the employees, half of the employees have experience of 2–5 years, and only around 3% of them have experience of more than 15 years.

According to employees who working in restaurants and cafés were around 67%, and the rest work in handcrafts shops, hotels, and travel and tourism companies. About 70% of them did not change their place of work because of the coronavirus pandemic, while 30% of the employees change their place of work.

Following are the attitudes of the employees towards working in the tourism sector before and after the coronavirus pandemic.

Table 1 show that there is a major change in the attitude of the employees towards their work in the tourism sector because of the coronavirus pandemic. Before the pandemic; 72% of the employees were satisfied with their work and had

Table 1 Employees' attitudes towards several issues before and after COVID-19 outbreak

Issue	Employees attitude before COVID-19 pandemic		Employees attitude after Corona pandemic	
	Positive %	Negative %	Positive %	Negative %
Nature of work	63.4	36.6	32.0	68.0
Tourists	88.3	11.7	48.3	51.7
Contract of work	56.8	43.2	56.7	43.3
Owner of facility	68.0	32.0	25.1	74.9
Personal safety level	83.5	16.5	63.4	36.6
Average	72.0	28.0	45.1	54.9

a positive attitude towards their work in the tourism sector. However, after the pandemic, this percentage was decreased to 45.1%. These figures prove that the pandemic affected the employees in the tourism sector in Palestine negatively.

To improve the tourism sector in the city, the employees suggested the following:

- Reduction in prices in tourist facilities.
- Owners of tourists facilities should adhere to the minimum wage.
- Owners of tourists facilities should comply with labor laws.
- The government should pay more attention to tourist guides.
- Owners of tourists facilities should improve the expertise of employees through training courses.

3.3 Owners of Tourist Facilities Questionnaire

This section consists of subjects related to the behavior of owners of tourists' facilities towards the crisis of COVID-19 during the summer of 2020. The demographic characteristics were also investigated. The study showed that 93.3% of tourists are male, while 6.7% are female of which 93.3% are from Nablus and around 6.7% are from neighboring areas. The study revealed that 26.7% of the owners are between 25 and 34 years old, 40% of the owners are between 35 and 44 years old, and 33.35% of the owners are between 45 and 55 years old.

Most owners who participated in this study are highly educated. Around 66.7% of them hold a bachelor's degree in different disciplines, 13.3% hold a Diploma degree, and the rest 20.0% hold a general secondary education certificate.

The study showed that 53.3% of the owners own the of tourist facilities, while 46.7% of them are renters. The value of the annual rent varies between less than $4000 and more than $14,000. The geographic site of the tourists' facility plays the most important role in determining the value of the annual rent in the city of Nablus.

Owners of tourists' facilities confirmed that the number of employees in their facilities has been reduced significantly. In 2019 and before the COVID-19 outbreak, the average number of employees in the tourism facilities was 23, but after the COVID-19 outbreak, the average number of employees dropped to 5 employees, and 20% of the tourism facilities were closed.

According to the owners' opinions, the workforce reduction is due to the government policy of frequent and long closures, the lack of tourists' (arrivals, the reduction of income, and the significant decrease in outbound tourists' arrivals.

The daily average of visitors or tourists who visited the tourism facilities was 468 in 2019 or before the COVID-19 outbreak, then dropped to 23 in 2020 or after the outbreak.

Findings of this study also revealed the dissatisfaction of owners of tourist facilities about the absence of the government's role in supporting them, the high value of taxes imposed on them, and the failure of banks to offer them loans with low interest. The owners of tourist facilities also complained about the Israeli occupation measures at the crossings and borders. They reported that the Israeli occupation always warns foreign tourists against entering the Palestinian Authority areas, and this subsequently affects the number of outbound tourists who visit the West Bank.

3.4 Proposed Policies to Support the Recovery of the Tourism Sector in Nablus

According to the estimation of the Arab Tourism Organization and the Arab Transport Federation, the recovery period for the tourism and travel sector will take up to six years. The length of the expected recovery period depends on several factors: the speed of global economic recovery, the extent of coordination between countries regarding the adoption of harmonious health measures, the expected period for the vaccine arrival, and the restoration of tourists' confidence.

Currently, domestic tourism has been recovering and is helping to mitigate the negative impact of coronavirus pandemic on jobs and businesses in some destinations. However, actual recovery will only be possible when domestic and international tourism returns. This requires governmental and private sectors cooperation and evidence-based solutions, so that travel restriction can be safely lifted. Although the governments have taken steps to cushion the blow to tourism, reduce job losses while building recovery in 2021 and beyond, additional efforts are needed and in a more coordinated manner

Top policy priorities include:

- Encouraging and promoting domestic tourism. In this context, the tourism sector recovery plan must include the promotion of domestic tourism through marketing plans that encourage citizens to spend holidays in the cities. It is also necessary to work on participating and organizing exhibitions of national products to highlight and shed light on the distinguished tourism industries in Nablus. This constitutes the Palestinian identity and makes it a distinctive tourist attraction for Nablus.

- Program to support small and medium tourism enterprises. The tourism sector is one of the most important sectors that attract many employees, as it contributes to creating direct job opportunities in tourism activities (hotels, restaurants, coffee shops, travel and tourism companies, etc.). It also contributes to creating indirect job opportunities in economic activities related to or supportive of the tourism sector (food and beverage industry, transportation, and other activities), most of which are small and medium enterprises which represent 80% of the tourism sector. Supporting small and medium enterprises is one of the most important economic recovery policies for the tourism sector in Palestine, because of its impact on the levels of domestic product and employment. Furthermore, the intervention of the Ministry of Tourism and the concerned authorities in Palestine should focus on formulating plans that would maintain employment opportunities in small and medium tourism enterprises by providing a large number of incentives. Central banks must also intervene in financing small and medium enterprises by reducing the risk weighted assets associated with these loans within the frame of capital adequacy.
- Restoring traveller confidence. Restoring the confidence of tourists is among the most important factors supporting the recovery of the tourism sector. It focuses on making health arrangements for safe tourism following health measures and precautionary safety standards.
- Building more resilient, sustainable tourism. Sustainability may become more prominent in tourism options, due to increased awareness of the coronavirus pandemic, and the negative effects of tourism. Natural areas and domestic destinations are expected to promote recovery, and shorter travel distance may have a less environmental impact on tourism. More flexible systems are also needed as they are capable of adapting more quickly to changes in policy orientation. Effective crisis management and implementation of health and safety policy issues are also significant in building sustainable tourism.

References

1. Ismail, M., Jamaal, Q.: The impact of the tourism sector on economic growth in the Arab countries (2020) https://www.amf.org.ae/ar/content
2. Sigala, M.: Tourism and COVID-19: impacts and implications for advancing and resetting industry and research. J. Bus. Res. **117**(September), 312–321 (2020). https://doi.org/10.1016/j.jbusres.2020.06.015
3. UNWTO, World Tourism Barometer.: Global Tourism Statistics (2021a). https://www.unwto.org/unwto-world-tourism-barometer-data

4. UNWTO, World Tourism Barometer and Statistical Annex.: UNWTO World Tourism Barometer (English Version) **18**(7), 1–36 (2020). https://doi.org/10.18111/wtobarometereng.2020.18.1.7
5. Ghosh, S.: Asymmetric impact of COVID-19 induced uncertainty on inbound Chinese tourists in Australia: insights from nonlinear ARDL model. Quant. Finan. Econ. **4**(2), 343–364 (2020). https://doi.org/10.3934/QFE.2020016
6. Jeon, C.-Y., Hee-Won, Y.: The structural changes of a local tourism network: comparison of before and after COVID-19. Curr. Issues Tour. **0**(0), 1–15 (2021). https://doi.org/10.1080/13683500.2021.1874890
7. Lopes, H.D.S., Remoaldo, P., Ribeiro, V., Martín, V.G.: Effects of the COVID-19 pandemic on tourist risk perceptions—the case study of Porto, 2021. Sustainability **13**(11), 6399. https://doi.org/10.3390/su13116399
8. UNWTO, World Tourism Barometer and Statistical Annex.: UNWTO World Tourism Barometer (English Version) **19**(1), 1–42 (2021b). https://doi.org/10.18111/wtobarometereng.2021.19.1.1
9. Everingham, P., Natasha, C.H.: Post COVID-19 ecological and social reset: moving away from capitalist growth models towards tourism as Buen Vivir. Tour. Geogr. **22**(3), 555–566 (2020). https://doi.org/10.1080/14616688.2020.1762119
10. Zhu, H., Fumin, D.: How to influence rural tourism intention by risk knowledge during COVID-19 containment in china: mediating role of risk perception and attitude. Int. J. Environ. Res. Public Health **17**(10), 3514 (2020). https://doi.org/10.3390/ijerph17103514
11. Isaac, R.K., Wahba, A.: Can you imagine Bethlehem without tourism: the impacts of Covid-19 on Bethlehem, Palestine. Curr. Issues Tour. 1–17 (2021). https://doi.org/10.1080/13683500.2021.1883559
12. Bitok, K.: Sustainable tourism and economic growth nexus in Kenya: policy implications for post-Covid-19. J. Sustain. Tour. Entrepreneurship **1**(2), 123–38 (2019). https://doi.org/10.35912/joste.v1i2.209
13. Riadil, I.G.: Tourism industry crisis and its impacts: investigating the Indonesian tourism employees perspectives in the pandemic of COVID-19. Jurnal Kepariwisataan: Destinasi, Hospitalitas Dan Perjalanan **4**(2), 98–108 (2020)
14. Persson, F.U., Shuangqi, L.: The impact of a global crisis on areas and topics of tourism research. Sustainability **13**(2), 906 (2021). https://doi.org/10.3390/su13020906
15. Ghodieh, A.: Urban Built-Up Area Estimation and Change Detection of the Occupied West Bank, Palestine, Using Multi-Temporal Aerial Photographs and Satellite Images (2020). https://pubag.nal.usda.gov/catalog/6833394
16. Hamada, S., Ghodieh, A.: Mapping of solar energy potential in the west bank, Palestine using geographic information systems. Pap. Appl. Geogr. 1–18 (2021). https://doi.org/10.1080/23754931.2020.1870540
17. Hamada, S., Vieira, A., Ghodieh, A.: Natural characteristics of the Palestinian region of the Jordan Valley. Geo-Working Papers **0**(25), 5–46 (2017).
18. Hamada, S.: Towards the development of regional and urban planning in the Palestinian region of the Jordan Valley (2017). http://repositorium.sdum.uminho.pt/
19. Hamada, S.D., Vieira, A., Ghodieh, A.: Israeli Planning Policies in the Jordan Valley (2015). http://repositorium.sdum.uminho.pt/
20. PCBS.: Projected mid-year population for Nablus governorate by locality 2017–2026 (2020). https://www.pcbs.gov.ps/statisticsIndicatorsTables.aspx?lang=en&table_id=698
21. Google Earth.: "Google Earth." (2021). https://www.google.com/earth/download/gep/agree.html?hl=en-GB
22. ARIJ.: Locality Profiles and Needs Assessment in Nablus Governorate (2014). http://www.arij.org/publications/books-atlases/224-2014-publications/706-locality-profiles-and-needs-assessment-in-nablus-governorate.html
23. PCBS.: Palestine Census Bureau of Statistics. Census Final Results—Summary (Population, Buildings, Housing, Establishments) (2009). https://www.pcbs.gov.ps/Downloads/book2364-1.pdf

24. PCBS, MOTA.: Press Release on World Tourism Day 27/09/2019 (2019). https://www.pcbs.gov.ps/site/512/default.aspx?lang=en&ItemID=3555&fbclid=IwAR0WGU95wMX2P60hQHnNc5HqVnHUs7PXzQbkv7Pa_OcPPh_VHN_qwGYL-C0
25. Ghodieh, A., Abahre, J., Abahre, H., Huang, P.: Challenges facing the tourism sector in palestine: a chinese perspective. J. Tour. Hospitality Manag. **7**(June), 2372–5133 (2019). https://doi.org/10.15640/jthm.v7n1a2
26. Ministry of Tourism. Ministry of Tourism and Antiquities —Nablus (2020). http://www.travelpalestine.ps/en/category/28/1/Nablus
27. Nablus Municipality.: 2021 .السياحة اصالة وحداثة. http://nablus.org/index.php/ar/links/index.php?option=com_content&view=article&id=379

Segmentation Analysis of MSMEs in Brunei Darussalam Based on IR4.0 Technology Adoption

Nurul Izzati Naqibah Zaidi, Mohamed Saleem Nazmudeen, and Fadzliwati Mohiddin

Abstract The adoption of IR4.0 technologies is crucial for businesses across all sectors to be more competitive and to maximize their performance. Furthermore, there has been an increase in the adoption of IR4.0 technologies in Micro, Small and Medium Enterprises (MSME) as they provide a significant contribution to every country's economy. In Brunei Darussalam, the government has recently begun to emphasize the importance of IR4.0 technologies adoption to increase the performance of MSMEs in the country due to the recent low economic diversification rate in the country. This study aims to analyse the current level of adoption of IR4.0 in Brunei's MSMEs. The K-means machine learning algorithm was applied to segment the MSMEs' adoption level across sectors and to understand the reasons behind their level of adoption. According to the findings of this study, MSMEs in Brunei Darussalam are currently at the level where they are adopting mostly basic and intermediate technologies and less adoption of advanced IR4.0 technologies. This research used a quantitative method, using primary data from 201 MSMEs in Brunei Darussalam. Implications of the study are outlined in the research and portray a path for future research for researchers, businesses and policymakers.

Keywords Digitalization · Industry 4.0 · MSME · Technology readiness · Technology adoption · K-means · Segmentation · Machine learning

---

N. I. N. Zaidi (✉) · M. S. Nazmudeen · F. Mohiddin
University Technology Brunei, Bandar Seri Begawan, Brunei
e-mail: p20191008@student.utb.edu.bn

M. S. Nazmudeen
e-mail: mohamed.saleem@utb.edu.bn

F. Mohiddin
e-mail: fadzliwati.mohiddin@utb.edu.bn

A. Hamdan et al. (eds.), *The Implementation of Smart Technologies for Business Success and Sustainability*, Studies in Systems, Decision and Control 216,
https://doi.org/10.1007/978-3-031-10212-7_78

1 Introduction

Industry 4.0 is where Information Technology (IT) systems were well-integrated with network and human skills for an organization to work smarter, be more connected, automated, and increased efficiency and effectiveness to achieve a competitive advantage. According to Lasi et al. [1], IR4.0 is the integration of Cyber-Physical systems in industrial manufacturing with the vision to be connected, smarter and automated to increase value in the industry. Before the 4th Industrial Revolution, from the 1st until the 3rd industrial revolution, technologies evolved from mechanical to electricity and information technology (IT) [2]. Technologies are getting more advanced as well as less expensive which leads to easy access by both large and small businesses. Technology innovation with mobile computing, social media, Internet of Things (IoT), sensors and robotics, analytics pushed the businesses to adopt more technologies as things get more advance to maintain their relationship with customers [3].

MSMEs are one of the vital parts of economies throughout the world especially when MSMEs have special behavior and needs in the challenging economic world [4]. Using IR4.0 technologies can help MSMEs to be more sustainable in supply chains, and this is one of the key factors of success. MSMEs with IR4.0 technologies can overcome challenges and yield more benefits with fast changes and value-added [5]. Industry 4.0 readiness refers to the level at which organizations were implementing IR4.0 technologies [6] and how ready are the businesses to step in digitalization and implement IR4.0 technologies [7, 8]. There are different categories and levels of implementing technologies, it depends on the level of complexity and the result is used for marking the usage of IR4.0 technologies by the MSMEs [9] Not all IR4.0 maturity models or assessments fit for businesses, especially for SMEs. Some models can fit most sectors but some models aren't for which there is a need to customize according to the type of industry and area of expertise [10]. Therefore, SMEs need better precision when it comes to identifying the level of maturity, especially considering the study in different regions of the world for example the maturity level of SMEs between Europe and Southeast Asia. Those two regions have a distinctive level of maturity, especially when considering the same type of business. Although there has been an increasing number of studies in this area, there is a lack of empirical validity and ready-to-use documentation [11]. There are also big gaps in the literature in business sustainability and IR4.0 concepts in manufacturing [12, 13]. More investigation on MSMEs is needed in IR4.0 usage as they need a different approach compared to large businesses [5, 13–15]. Developing countries are mostly made up of MSMEs with different structures and support systems [16]. The findings in this paper are based on the study conducted in Brunei Darussalam, which is a developing country although the economic contributors are mainly medium and big businesses from oil and gas industries. This research focuses on identifying IR4.0 technology adoption level using our suitable categories of technologies and segmenting approach to determine the readiness on IR4.0 technology adoption in Brunei Darussalam in all industries as the country have a small number of manufacturing businesses but high in oil and gas services, followed by retailing and food and beverages industries. The remainder of the paper is structured as follows: In the next section, we will be explaining the categories of IR4.0 technologies that we used particularly for categorizing the level of IR4.0 technologies in MSMEs. In Sect. 3, we discuss our methodology and samplings, including data preprocessing to data segmentation using clustering analysis. Section 4 elaborates the results and discussion which is followed by the conclusion.

2 Categories of IR4.0 Technologies

Before going into segmentation analysis, it is important to know what type of technologies are being implemented in an organization. There are different technologies of IR4.0, which might vary according to different works of literature. In this research, the categories for IR4.0 technologies are integration between Nine Pillars of Technological Advancement [17] and study from MSMEs Participation in the Digital Economy in ASEAN [18]. By combining the two sources, we came out with these categories that are well-suited towards ASEAN countries, at the same time still within the fundamental area of IR4.0 technologies. There is no fixed concept or meaning of IR4.0 and its technology however, the fundamentals of the technologies in IR4.0 are the integration of hardware, software and network interacted with a human touch to produce effectiveness and efficiencies of an organization. Technologies in the basic category are usually a low-cost investment, easily adapted and only require basic IT skills for the employees to handle. These basic technologies are highly necessary for the communication process and further reaching the end-user or customers. In this research, basic technologies are business management systems, mobile communication apps, e-commerce systems, smart devices, internet networks, wireless devices. In the intermediate category, most of the technologies are a little higher cost of implementation and needs higher skills to handle the system. Training and knowledge transfer to make this a successful implementation is important. These technologies are mostly used for important decision-making for quality products and more efficient communication and operation processes. Technologies in this category are big data analytics, cloud systems and cybersecurity. In the advanced category, technologies are expensive to implement which need experts or skilled human resources to handle the systems. Those are artificial intelligence, robots and advanced machinery, augmented reality, sensors and networks and cyber-physical systems. However, this is crucial for decision-making and to boost operation performance, makes use of complex algorithm systems. These types of technologies are usually adopted in medium to large-sized manufacturing or assembly industries.

3 Research Design and Methodology

3.1 Research Objective

The research objective is to assess the current level of adoption of IR4.0 technologies among MSMEs in Brunei Darussalam across different sectors.

3.2 Data Collection

Brunei Darussalam's MSMEs are defined according to the number of employees in the businesses. For Micro businesses employees are less than 4 people, small businesses consist of 4–19 employees whereas medium businesses are between 20 and 99 employees [19]. Electronic questionnaire surveys were disseminated to owners of the businesses. The result shows that most respondents are female (64%), majority of them are between 31 and 50 years old followed by 19–30 years old (40%) this shows that business owners answering the survey are mostly in the youth category in Brunei Darussalam. Most of the respondents are having undergraduate qualifications (36%). The sectors of industries covered in this survey are majorly from other services. Further looking into this data reveals these are mostly small businesses under professional service providers such as photography services, professional make-up artists, event services, event decorator specialists and art and creative services. The highest number of respondents are from micro-sized businesses (68%) which in recent years has grown more with the encouragement from the government of Brunei Darussalam.

3.3 Data Preprocessing

The raw data collected from online survey questionnaires were coded and subgrouped to different areas such as demographic and type of technologies implemented (number of basic, intermediate, and advanced technologies implemented).

3.4 Data Segmentation

After the coding has been done and incomplete rows were removed, segmentation analysis was performed using the k-means clustering algorithm. The processed data from the survey is unlabeled and the objective is to find the natural grouping of the respondents with respect to the usage of the IR4.0 technologies. K-means is an unsupervised machine learning algorithm that is ideally suitable for this scenario. The software used for segmentation analysis was Jupyter notebook with python

Fig. 1 K-means algorithm

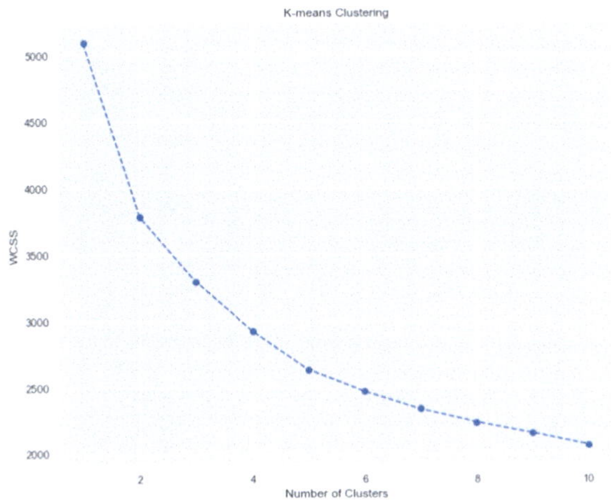

programming language. We used the elbow method to determine the number of clusters and found that it is 2 which is shown in Fig. 1. Once the number of clusters was identified then the k-means algorithm was run on the data with a ‘k’ value of 2. The results are discussed in the following section.

4 Results and Discussions

The mean cluster values for each of the attributes are shown in Tables 1, 2 and 3. The range of the mean values varies from 1 to 4 where 1 is indicative of a particular technology not being used and 4 is indicative of heavy usage. Value 1 has been assigned to non-usage of the listed technologies to avoid confusion with missing data input by the respondents which are assigned a value of 0. According to the results from the K-means process, only two clusters were formed. The cluster output has been split into three tables for the sake of clarity. The first cluster (Segment K-means = 1) has an average mean of 3.37 for basic technologies and 2.44 for intermediate technologies. This shows the businesses in the cluster are using more basic technologies and even slightly more for intermediate technologies. However, the mean average for advanced technologies in this cluster is 1.45, which is very minimal that it is at the mean average of 1. Therefore, this cluster represents the Moderate Category. The second cluster (Segment K-means = 0) has an average mean of 2.56 for basic technologies and 1.18 for intermediate technologies. This shows the businesses in the cluster are using more basic technologies, but a little less compared to Cluster 0 (Moderate Category) with regards to intermediate technologies. As for advanced technologies the mean average for this cluster is 1.14, which is even lower than the Moderate Category. Hence, this cluster represents the Modest Category.

Table 1 Mean average of basic technologies implementation

Segment	Basic technologies									
K-means	Mob. comm. apps	Bus. Mgt Sys	System and software	E-Comm. platform	E-Payment system	Website	Internet network	Wireless device	Smart devices	Mean
0 (Modest)	3.76	2.08	1.81	1.78	2.68	1.49	3.33	2.98	3.16	2.56
1 (Moderate)	3.91	3.05	3.20	2.76	3.40	2.93	3.77	3.71	3.65	3.37

Table 2 Mean average of intermediate technologies implementation

Segment K-means	Intermediate technologies			
	Cloud comp	Big Data analytics	Cyber security systems	Mean
Modest	1.33	1.12	1.09	1.18
Moderate	2.84	2.17	2.32	2.44

Table 3 Overall mean average of technologies implemented

Segment K-means	Overall mean average for technologies implemented		
	Basic technologies	Intermediate technologies	Advanced technologies
Modest	2.56	1.18	1.14
Moderate	3.37	2.44	1.45

Referring to Table 3, to compare between Modest and Moderate category, the mean value of Moderate category is 0.81 more than that of Modest category in implementation of basic technologies. This shows that Moderate categories are using more basic technologies than the Modest categories.

Also, we can see that the businesses in the Moderate category (2.44) are implementing a higher level of intermediate technologies with a mean difference of 1.26 compared to the Modest category (1.18).

Both Modest and Moderate categories have very minimal mean average in implementing advanced technologies. Referring to Table 4, in the Modest category, the mean average of implementing Advanced technologies is 1.14 whereas the Moderate category has a little higher value which is 1.45. However, both mean averages are very low as it is closer to 1, which is equivalent to "None" for the level of implementation. Hence there is no Modernized category in this result. In other words, MSMEs in Brunei Darussalam have a very low rate in terms of implementing advanced technologies.

Table 4 Mean average of advanced technologies implementation

Segment K-means	Advanced technologies					
	AI	Cyber-physical systems	Augmented reality	Robotics and advanced machineries	Sensors and network	Mean
Modest	1.32	1.13	1.14	1.06	1.05	1.14
Moderate	1.59	1.37	1.44	1.35	1.49	1.45

4.1 Type of MSMEs Industry in Modest and Moderate Category

4.1.1 MSMEs Industries in Modest Category

From the 201 MSMEs, a total of 113 businesses falls under the Modest category. The largest group of industry in this category is Wholesale and Retail Trading industry sector (31 companies) while 27 businesses are from Other Services which are mostly in professional and specialized services e.g., Entertainment, healthcare, and beauty, wedding services, catering etc., Printing Services (3 companies), Transport and storage and Agriculture (4 companies), Construction (6 companies), Professional Make-Up Services (8 companies), Food and Beverages Supply (10 companies) and one each from Engineering Services, Fishery, Maintenance Services, Education and Learning Services, Marine Services, Creative Art Services, Mining, Construction, Recycling Waste Management, Government-related service, Photography Services, Finance service, Professional Wedding Services and Communication.

Most of the industry sectors in this category are businesses that supply basic products such as foods, health and living, as well as lifestyles. Only a few of the sectors are coming from highly specialized businesses such as engineering services, maintenance, finance, mining, and construction field of businesses. This indicates that the businesses in this category are needing only basic technologies with low intermediate technologies usage. Looking at the sectors as well, most of them are not businesses that provide complex and sophisticated products or services. Hence, for them basic technologies and a hint of intermediate technologies are sufficient. According to Papachashvili 2018, Internet can be very beneficial to increase a firm's potential to produce competitive products and win new markets. The Internet can help to be more cost-efficient, communicate and get more information as well as to organize with the use of emails, websites, platforms and online marketplace to enter the international market [20]. Looking at their size and nature of businesses, basic technologies are mainly on the usage of internet and communication as well as e-commerce platforms which are well-suited with the current situation of the businesses.

4.1.2 MSMEs Industries in Moderate Category

From the survey, a total of 88 businesses falls under the Moderate category. Highest of all is in Wholesale and Retail trade industry (30 companies), Other Services (16 companies), Mining (7 companies), Food and Beverage Supplies (6 companies), Photography services (5 companies), Constructions (4 companies), Transport and Storage (3 companies), Professional Wedding services (3 companies), Finance services (2 companies) and one each from the rest of the industries such as 3D Modelling Services, Agriculture, Communication, Construction, Education and Learning Services, Event Services, Integrated Facility Management, Marketing

Services, Software Developer and Traditional Healthcare services. The result shows a similar ratio of Wholesale and retail trade sectors in the Modest category (30 businesses) compared to the Moderate category (31 businesses). This illustrates highest sectors in MSMEs are under Wholesale and Retail trade and this shows that such grouping is quite advanced compared to other sectors in both categories. However, here there are more highly specialized businesses compared to the Modest category. This shows industry sectors in this group are more than just daily supplies but more on an industrial level and skilled services which is why they required a more advanced system compared to the businesses in the Modest category. The possible reason for the Moderate category in implementing more basic and advanced technologies is because of the availability of resources which can be associated with the cost factor. For example from a study in India on small retail shops, it was found that customers and retailers didn't prefer to get cashless transactions because of the potential costs, however, suppliers are willing to pay and prefer low fixed transaction cost rather than percentage value [21]. Usually, suppliers are medium businesses that have more resources. Large businesses have advantages over MSMEs because of scale, but IR4.0 technologies help to push MSMEs to be better, especially with the internet of things (IoT), cloud-based manufacturing technology, and big data analytics [14]. From this, it can be said that businesses that fall under the Moderate category are ready in using slightly more basic and intermediate technologies which boost extra performance to the business compared to the Modest category.

5 Conclusion

Most of the businesses in the Modest category are those that are supplying or providing basic food supplies, healthy products, lifestyle products and any other supplies that require a less technical process. In such category, they are most keen on using basic technologies more than intermediate, or even advanced technologies. However, in the Moderate category, on the other hand, are mostly businesses that are more technical or highly specialized services that require more advanced skills and expertise such as mining businesses, education, construction services, engineering, and maintenance services. Most businesses in the Moderate category are Wholesale and Retail Services which shows that this type of business can operate well when using technologies considering the nature of the business which is quite basic in supplying and providing daily necessities. Overall, this shows that MSMEs in Brunei Darussalam are mostly in the Modest category where they focus on basic technologies, whereas the rest of the businesses are in the Moderate category, which shows they are going forward in using intermediate technologies more. There is no Modernized category at the moment, which is a grouping where businesses use advanced technologies. This can be explained because there are less advanced industries businesses and fewer industrial manufacturer or assembly sectors in Brunei Darussalam, hence the lack of using of advanced machinery and technologies at this point. This research introduced important categorization of MSMEs when it comes to adopting

IR4.0 technologies. Researchers will be able to benefit from the research results and hence they could further introduce a more structured adoption level using the three categories of technologies mainly Modest, Moderate and Modernized and identify the impact level of technology adoption towards firm performance. Researchers could replicate this study from the context of their own countries and the results could be compared with Brunei Darussalam's studies. This approach could help to identify extensive validation and may lead to more generalization across countries. Businesses can use this information to strengthen their IR4.0 adoption level and gain more benefits. Policymakers could use these results to develop more focused policies which can encourage MSMEs to adopt more technologies in their business. However, the low or high adoption level of technologies is often influenced by several factors in an organization whether it's external or internal factors. In future research, we plan to study the factors that contribute to the level of adoption for IR4.0 technologies amongst MSMEs in Brunei Darussalam and to do empirical research on adoption levels based on each focused industry to see the level of adoption for IR4.0 technologies.

References

1. Lasi, H., Fettke, P., Kemper, H.G., Feld, T., Hoffmann, M.: Industry 4.0. Bus. Inf. Syst. Eng. **6**(4), 239–242 (2014). https://doi.org/10.1007/s12599-014-0334-4
2. Veza, I., Mladineo, M., Gjeldum, N.: Managing innovative production network of smart factories. IFAC-PapersOnLine **48**(3), 555–560 (2015). https://doi.org/10.1016/j.ifacol.2015.06.139
3. Toanca, L.: Empirical research regarding the importance of digital transformation for Romanian SMEs. Manag. Econ. Rev. **1**(2), 92–108 (2016)
4. Ghafoorpoor Yazdi, P., Azizi, A., Hashemipour, M.: A hybrid methodology for validation of optimization solutions effects on manufacturing sustainability with time study and simulation approach for SMEs. Sustainability **11**(5), 1454 (2019)
5. Birkel, H.S., Veile, J.W., Müller, J.M., Hartmann, E., Voigt, K.-I.: Development of a risk framework for Industry 4.0 in the context of sustainability for established manufacturers. Sustainability **11**(2), 384 (2019)
6. Stentoft, J., Jensen, K.W., Philipsen, K., Haug, A.: Drivers and barriers for Industry 4.0 readiness and practice: a SME perspective with empirical evidence. In: Proceedings 52nd Hawaii International Conference on System Sciences, vol. 6, pp. 5155–5164 (2019). https://doi.org/10.24251/hicss.2019.619
7. Schwab, K.: The fourth industrial revolution. Encyclopedia Britannica, Chicago, IL, USA, 3 (2017) [Online]. Available: https://www.britannica.com/topic/The-Fourth-Industrial-Revolution-2119734
8. Vazire, S.: Implications of the credibility revolution for productivity, creativity, and progress. Perspect. Psychol. Sci. **13**(4), 411–417 (2018). https://doi.org/10.1177/1745691617751884
9. Kagermann, H., Wahlster, W., Helbig, J.: Recommendations for implementing the strategic initiative INDUSTRIE 4.0. Final Rep. Ind. 4.0 WG, no. April (2013)
10. Hizam-Hanafiah, M., Soomro, M.A., Abdullah, N.L.: Industry 4.0 readiness models: a systematic literature review of model dimensions. Inf. **11**(7), 1–13 (2020). https://doi.org/10.3390/info11070364

11. Poeppelbuss, J., Niehaves, B., Simons, A., Becker, J.: Maturity models in information systems research: literature search and analysis. Commun. Assoc. Inf. Syst. **29** (2011). https://doi.org/10.17705/1cais.02927
12. Luthra, S., Kumar, A., Zavadskas, E.K., Mangla, S.K., Garza-Reyes, J.A.: Industry 4.0 as an enabler of sustainability diffusion in supply chain: an analysis of influential strength of drivers in an emerging economy. Int. J. Prod. Res. **58**(5), 1505–1521 (2020)
13. Müller, J.M., Voigt, K.-I.: Sustainable industrial value creation in SMEs: a comparison between industry 4.0 and made in China 2025. Int. J. Precis. Eng. Manuf. Technol. **5**(5), 659–670 (2018)
14. Khanzode, A.G., Sarma, P.R.S., Mangla, S.K., Yuan, H.: Modeling the Industry 4.0 adoption for sustainable production in Micro, Small and Medium Enterprises. J. Clean. Prod. **279**, 123489 (2021)
15. Mue, N.E., Mulongo, N.Y., Kholopane, P.A.: Analysis of the status-quo of industrial 4.0 in the SMMEs: a critical review
16. Jové-Llopis, E., Segarra-Blasco, A.: Eco-efficiency actions and firm growth in European SMEs. Sustainability **10**(1), 281 (2018)
17. Russmann, M., et al.: Industry 4.0: World Economic Forum, pp. 1–20. Boston Consulting Group (2015)
18. ERIA.: Study on MSMEs Participation in the Digital Economy in ASEAN: Nurturing ASEAN MSMEs to Embrace Digital Adoption, pp. 1–28 (2018) [Online]. Available: https://asean.org/storage/2012/05/Study-on-MSME-Participation-in-the-Digital-Economy-in-ASEAN.pdf
19. Department of Statistics, Department of Economic Planning and Statistics, Ministry of Finance and Economy Brunei Darussalam: Annual Census of Enterprises 2019, Report of Summary Findings, p.23 (2018)
20. Papachashvili, N.: Industry 4.0 and its impact on the international trade. PMC Res. Cent. **10**(9), 4–37 (2018). http://ir.kneu.edu.ua/bitstream/2010/24244/1/444-453.pdf
21. Seethamraju, R., Diatha, K.S.: Digitalization of small retail stores—challenges in digital payments. In: Proceedings of the Annual Hawaii International Conference on System Sciences, vol. 2019, Jan, pp. 5175–5184. https://doi.org/10.24251/hicss.2019.621

Perceived Covid-19 Risk and E-Wallet Adoption: An Empirical Evidence MSEs of Indonesia

Laila Maya Riska, Muamar Nur Kholid, and Selfira Salsabilla

Abstract Along with the global growth of Covid-19 infections, transactions involving physical money pose a significant danger of viral transmission. As a result, customers and Indonesian MSEs increasingly utilize E-Wallets to conduct cashless transactions. Given the lack of studies linking Covid-19 and E-Wallets, The goal of this study is to analyze how Indonesian micro and small enterprises (MSEs) adopted E-Wallets during Covid-19. Specifically, this study examines the effect of perceived usefulness, perceived Covid-19 risk, perceived financial cost, and perceived ease of use on Indonesian MSEs' intention to adopt E-Wallet. The data for this study were gathered via a questionnaire-based survey. Respondents in this study were either owners or managers of Indonesian MSEs. The model is tested using SmartPLS 3.0, based on the PLS-SEM approach. The analysis reveals that Perceived Covid-19 Risk significantly affects the perceived usefulness and intention of Indonesian MSEs to adopt E-Wallets. Surprisingly, the perceived financial cost does not affect the intention of Indonesian MSEs to use E-Wallet.

Keywords E-Wallet · Perceived Covid-19 risk · Perceived usefulness · Adoption

1 Introduction

The Gross Domestic Product (GDP) of Indonesia has dropped due to Covid-19 [1]. Micro and Small Enterprises (MSEs) in Indonesia have been a primary driver of GDP growth [2]. The significant contribution of MSEs in Indonesia is predicted to rehabilitate the Indonesian economy and boost the country's GDP [3]. Due to the government's restriction on face-to-face interactions during Covid-19, MSEs in Indonesia have innovated their business practices, one of which is the adoption of E-Wallet in various transactions with buyers and suppliers. This innovation in the

L. M. Riska · M. N. Kholid · S. Salsabilla (✉)
Department of Accounting, Faculty of Business and Economics, Universitas Islam Indonesia, Yogyakarta, Indonesia
e-mail: selfirasalsabilla@uii.ac.id

A. Hamdan et al. (eds.), *The Implementation of Smart Technologies for Business Success and Sustainability*, Studies in Systems, Decision and Control 216,
https://doi.org/10.1007/978-3-031-10212-7_79

usage of E-Wallets is predicted to boost the performance of MSEs in Indonesia, hence reviving the country's economy.

The Covid-19 virus is spread widely by droplets. According to health experts, The Covid-19 virus is spread by frequently touched surfaces, such as physical money [4]. Health experts urge avoiding physical money in transactions in this regard, as the Covid-19 virus can be spread by physical money [5]. Since the information was distributed, E-Wallet usage by Indonesian MSEs has continued to grow, and competition for E-Wallet service providers in Indonesia has grown increasingly heated [6]. Understanding the factors that influence Indonesian MSEs' adoption of E-Wallets would aid E-Wallet service providers in developing customers acquisition and retention strategies. Regarding this issue, it is essential to understand the constructs that influence intention of Indonesian MSEs to adopt E-Wallets.

Numerous studies on e-Wallets have been conducted. Perceived Usefulness (PUS), Perceived Ease of Use (PES), and Trust all had a significant effect on customers' intents to use E-Wallets in the UAE [7]. In Spain and the United States of America, researchers discovered that PUS, attitude, and subjective norm all had a significant effect on the intention to utilize E-Wallets, and PES had a significant effect on PUS [8]. In India, research has discovered that PUS, trust, and attitude affect intention to use an e-Wallet [9]. PES also significantly affected PUS and University Students' intention to use E-Wallets, according to research conducted with a sample of Indian university students [10]. Although earlier research on E-Wallets has been undertaken, some of these studies have not considered the Technology Acceptance Model (TAM) and the risk of distributing Covid-19 in the context of E-Wallet research. There have been studies relating Perceived Covid-19 Risk (PR) and intention to use E-Wallets, but they did not include the TAM component in its whole [11]. In contrast to past research, this study focuses exclusively on Indonesian MSEs, which have been neglected by several earlier studies. Due to the widespread use of E-Wallets in Indonesia, E-Wallet service providers charge users various fees. Additionally, this study attempts to consider the Perceived Financial Cost (PFC). Numerous prior studies demonstrate the essential nature of PFC in utilizing information technology for financial transactions [12, 13].

This paper contributes to the body of knowledge by investigating the effect of Perceived Usefulness (PUS), Perceived Ease of Use (PES), Perceived Financial Cost (PFC), and Perceived Covid-19 Risk (PR) on the intention of Indonesian MSEs to accept E-Wallet for business transactions.

2 Literature Review

2.1 Perceived Usefulness, E-Wallet Adoption, and Perceived Ease of Use

Perceived Usefulness (PUS) and Perceived Ease of Use (PES) are two fundamental variables in the Technology Acceptance Model (TAM) framework used to explain information technology (IT) adoption [14]. PUS is the extent to which a person believes that adopting a specific IT would improve his or her performance on the job [14]. According to TAM, PUS has a positive effect on adopting IT [14]. Numerous studies on various forms of IT adoption have revealed a positive impact of PUS on intentions, including mobile shopping [15], paid mobile media service [16], mobile banking [17], mobile application [18], and mobile accounting app [19].

H1. PUS has a positive effect on Indonesian MSEs' intentions to adopt E-Wallets.

PES is a term that refers to a person's belief that utilizing specific IT will be effortless or straightforward to manage [14]. As with PUS, TAM states that PES will positively impact Indonesian MSEs' intention to use E-Wallets [14]. If Indonesian MSEs perceive that utilizing an E-Wallet is effortless, they will strongly intend to use it. Numerous studies have discovered a positive influence of PES on particular IT adoption, such as mobile tourism [20], mobile learning [21], Islamic Fintech [22], and mobile accounting app [23].

H2. PES has a positive effect on Indonesian MSEs' intentions to adopt E-Wallets.

According to TAM, PES has a positive effect on PUS and a positive impact on adoption intentions [14]. MSEs in Indonesia who believe that using E-Wallet is effortless believe it is useful. Numerous prior research conducted in a variety of IT contexts has found that PES has effect on PUS [15, 16, 23–25].

H3. PES has a positive effect on PUS.

2.2 Perceived Covid-19 Risk, E-Wallet Adoption, and Perceived Usefulness

Perceived risk (PR) is frequently defined as the uncertainty surrounding the possibility of meeting an adverse impact of a service or product [26]. Other scholars state that perceived risk refers to the degree of uncertainty in a purchase transaction [27]. Perceived risk is a multi-faceted concept. Perceived risk comprises several components, including risks associated with time, security, social interaction, and financial [28]. Numerous studies on technology acceptance have identified perceived risk as a significant construct in understanding an individual's intention to utilize a technology [15]. In online transactions, disease risk is also regarded as an essential risk component during Covid-19 [11]. In this study, perceived risk is defined as a state of uncertainty among clients regarding the existence of fresh Covid-19 droplets on cash or physical money [11].

Numerous research has incorporated perceived risk into TAM and found that perceived risk has a significantly negative effect on IT adoption intentions [29–31]. However, in this study's context, perceived risk has a different influence than earlier studies. Indonesian SMEs who believe that there is a lot of Covid-19 virus in physical money, the greater the intention to use E-Wallet to make transactions with customers.

H4. PR has a positive effect on Indonesian MSEs' intentions to adopt E-Wallets.

While the PUS and PES may not encompass all of the user's reasons for embracing IT adoption, there are other constructs that do [32]. According to TAM, External Variables might affect PUS and PES [14]. Several studies have confirmed that PR has a significant effect on PUS [16, 30–32]. In the context of this research, Indonesian SMEs believe that there is a high concentration of the Covid-19 virus in physical money, then his or her has believed that using an E-Wallet is useful.

H5. PR has a positive effect on PUS.

2.3 Perceived Financial Cost and E-Wallet Adoption

The term "perceived financial cost (PFC)" refers to the magnitude to which an IT users perceives that utilizing specific IT would cost them money [13]. Outcomes and motivations of Economic are frequently the focus of adoption studies of IT [13]. MSEs in Indonesia that utilize E-Wallets may be subject to a variety of costs. The more cost that must be paid, the less inclined Indonesian MSEs are to adopt E-Wallets. Numerous research have demonstrated the negative impact of PFC on intention to adopt IT [12, 13].

H6. PFC has a negative effect on Indonesian MSEs' intentions to adopt E-Wallets.

3 Methodology

This paper is a quantitative study conducted using a questionnaire-based survey method. This study uses *purposive sampling*. The sample for this paper consists of Indonesian MSEs that have employed E-wallets in their commercial operations. There were 169 respondents, but 17 respondents did not wholly complete the questionnaire. The demographic features of the survey respondents are shown in Table 1. The response population is predominantly female and between the ages of 20 and 25. Most respondents are from the island of Java and conduct business exclusively online. Most respondents are fashion goods merchants, with most respondents having been in business for less than five years.

Table 1 Respondents profile

Characteristics	Frequency	Percentage (%)	Characteristics	Frequency	Percentage
Age of respondents (years)			*Business media*		
<20	8	5.30	Online	73	48.00
20–25	90	59.00	Offline	17	11.20
26–30	15	9.90	Online and offline	62	40.80
31–35	10	6.60	*Business category*		
36–40	8	5.30	Culinary	29	19.10
40–45	5	3.30	Fashion	47	30.90
46–55	14	9.20	Automotive	5	3.00
>55	2	1.30	Tour and travel	8	5.30
Gender			Agribusiness	7	4.60
Female	78	51.30	Creative products	11	7.20
Male	74	48.70	Information technology	7	4.60
Island			Beauty	24	15.80
Jawa	113	74.30	Others	14	9.20
Sulawesi	8	4.60	*Business age (years)*		
Bali	8	6.60	<5	124	81.60
Sumatra	7	5.30	5–10	20	13.00
Papua	3	5.30	10–15	2	1.30
Nusa Tenggara	3	2.00	15–20	3	2.00
Kalimantan	10	2.00	>20	3	2.00

The study's measurement of each construct was based on prior research. PUS, PES, PFC, and Intention measurements were adapted from [13]. PES, PFC, and Intention are each measured using four questions, while PUS is measured using five questions. Meanwhile, PR is measured using four questions adapted from [11]. The questionnaire uses a six-point Likert scale to measure each question's level of agreement. The data acquired from the questionnaire were analyzed using the PLS-SEM technique with the assistance of SmartPLS 3.0. The test entails assessing the measurement model in order to determine the reliability and validity of each indicator and construct, as well as the structural model in order to ascertain the relationships between construct [33].

4 Discussion and Results

4.1 Test of Measurement Model

The measurement model is tested using many statistical measures, including outer loading, composite reliability (CRE), Cronbach alpha (CAP), average variance extracted (AVE), and the square root of AVE [33]. As indicated in Table 2, all construct indicators have values greater than 0.7, and the AVE value for each construct is greater than 0.5. This result suggests that this study does not have a problem with convergent validity. Meanwhile, when the CRE and CAP values are considered, all constructs have a value more than 0.7, indicating no reliability issue in this study. This result is consistent with the [33] requirements, which require CRE and CAP values to be more than 0.7. Meanwhile, the square root of the AVE value for each variable is greater than the correlation between the variable. This result shows that this study does not have a problem with discriminant validity. The findings of discriminant validity testing utilizing the Fornel-Larcker Criteria are showed in Table 3.

4.2 Test of Structural Model

The structural model test findings indicate that the determinant coefficient value (R^2) for the intention of Indonesian MSEs to adopt E-wallet is 33.10%. This result shows that the model used in this investigation remains weak. Meanwhile, perceived usefulness has a determinant coefficient (R^2) of 30.00%. As illustrated in Table 4, there are five acceptable hypotheses and one rejected hypothesis. PUS has a significant positive effect on MSEs' intentions to utilize E-Wallets in Indonesia ($\beta = 0.334$; $\rho = 0.01$). This result demonstrates that H1 has been accepted. Additionally, test results indicate that PES has a positive significant effect on PUS ($\beta = 0.464$; $\rho = 0.01$) and intention ($\beta = 0.201$; $\rho = 0.01$). These findings corroborate H2 and H3. Meanwhile, PR has a significant positive effect on PUS ($\beta = 0.201$; $\rho = 0.01$) and Intention ($\beta = 0.208$; $\rho = 0.01$). These findings imply that neither H4 nor H5 are rejected. Finally, PFC has no significance influence on intention ($\beta = 0.0.056$; $\rho =$ not sig). Additionally, these findings demonstrate that H6 is not accepted.

4.3 Discussion

The data analysis reveals that PUS is still the most influential factor influencing Indonesian MSEs' intention to adopt E-Wallets. This result demonstrates that Indonesian MSEs prioritize usefulness while deciding whether to use an E-Wallet. These findings corroborate prior studies indicating that PUS is the most influential factor determining IT adoption [11, 22, 32, 34].

Table 2 Constructs reliability and convergent validity results

Constructs/Indicators	Outer loading	CRE	CAP	AVE
Intention (IN)		0.827	0.721	0.544
IN1	0.738			
IN2	0.700			
IN3	0.748			
IN4	0.771			
Perceived ease of use (PES)		0.914	0.874	0.726
PES1	0.871			
PES2	0.864			
PES3	0.861			
PES4	0.812			
Perceived financial cost (PFC)		0.918	0.889	0.737
PFC1	0.852			
PFC2	0.864			
PFC3	0.809			
PFC4	0.907			
Perceived Covid-19 risk (PR)		0.917	0.880	0.736
PR1	0.849			
PR2	0.838			
PR3	0.909			
PR4	0.833			
Perceived usefulness (PUS)		0.902	0.864	0.649
PUS1	0.796			
PUS2	0.846			
PUS3	0.755			
PUS4	0.832			
PUS5	0.795			

Table 3 Discriminant validity results

	IN	PES	PFC	PR	PUS
IN	**0.738**				
PES	0.446	**0.852**			
PFC	−0.131	−0.201	**0.859**		
PR	0.372	0.284	0.071	**0.858**	
PUS	0.516	0.521	−0.147	0.333	**0.806**

The bold number indicates the square root of AVE

Table 4 Model testing results

Hypothesis	Coeff	T-Stat	*P*-values	Results
PES → IN	0.201	2.621	0.009	Accepted
PES → PU	0.464	6.587	0.000	Accepted
PFC → IN	−0.056	0.858	0.391	Rejected
PR → IN	0.208	3.263	0.001	Accepted
PR → PU	0.201	3.021	0.003	Accepted
PUS → IN	0.334	3.607	0.000	Accepted

Other findings indicate that PR has a significant positive effect on Indonesian MSEs' intention to utilize E-Wallet. This significant effect of PR on intentions corroborates prior research findings [29–31]. These findings suggest that the risk of Covid-19 transmission via physical money can encourage Indonesian MSEs to use E-Wallets as a payment channel for customer interactions. This result is because the use of E-Wallets enables merchants and purchasers to avoid direct physical touch, which can minimize the spread of the Covid-19 virus. Given that most respondents in this study perform business transactions online, it is reasonable to conclude that face-to-face business transactions are too risky in terms of Covid-19 viral proliferation. This result demonstrates that the usage of E-Wallets is motivated by a fear of spreading Covid-19 through face-to-face business interactions.

Regrettably, the data reveals that PFC has no effect on Indonesian MSEs' intention to utilize E-Wallet. These findings complement prior research that found no significant effect of PFC on intention [35]. This result is likely due to the intense competition for E-Wallet providers in Indonesia, where providers charge relatively similar fees [36]. Additionally, Indonesian MSEs are confronted with the Covid-19 situation, which encourages Indonesian MSEs to adopt various types of IT to mitigate the spread of Covid-19.

5 Conclusion

This study found that several factors influence the intention of Indonesian MSEs to adopt E-Wallets. Some of these factors include PUS, PES, and PR. Regrettably, PFC has had no significant effect on the intention of Indonesian MSEs to adopt E-Wallets. PES and PR, on the other hand, have a significant impact on PUS. While this research model effectively explains the intention of Indonesian MSEs to use E-Wallet, it remains a weak model. Further research might be conducted in this area to determine additional relevant constructs that possibly influence Indonesian MSEs to use E-Wallet. Additionally, this study has few weaknesses that can be addressed by future study. Several of these limitations include respondents who are predominantly from the island of Java, respondents who are predominantly from the millennial group, and the business's youth. Furthermore, this study approach disregards the possibility of a moderating or mediating variable in the interaction between variables.

References

1. Jayani, D.H.: PDP per Kapita Indonesia Turun 3.7% Akibat Pandemi Covid-19. https://databoks.katadata.co.id/datapublish/2021/02/05/pdb-per-kapita-indonesia-turun-37-akibat-pandemi-covid-19#:~:text=PDB. Last accessed 29 Jan 2022
2. Rahim, A.: Digitalization shows Indonesian MSMEs the way to recovery. https://www.thejakartapost.com/academia/2021/10/17/digitalization-shows-indonesian-msmes-the-way-to-recovery.html. Last accessed 29 Jan 2022
3. Satriaji, S.: Digitalisasi UMKM: Penggerak Perekonomian Bangsa saat Pandemi COVID-19 (2020)
4. Anggraini, A.P.: Bisakah Covid-19 Menular Lewat Uang Tunai? https://health.kompas.com/read/2020/04/13/080000368/bisakah-covid-19-menular-lewat-uang-tunai-?page=all. Last accessed 29 Jan 2022
5. Maggirwar, S.: CBS news—paper money shunned due to Coronavirus. https://smhs.gwu.edu/news/cbs-news-paper-money-shunned-due-coronavirus. Last accessed 29 Jan 2022
6. Octaviano, A.: Persaingan semakin ketat, pemain dompet digital siapkan strategi. https://keuangan.kontan.co.id/news/persaingan-semakin-ketat-pemain-dompet-digital-siapkan-strategi. Last accessed 29 Jan 2022
7. Alshurideh, M.T., Al Kurdi, B., Masa'deh, R., Salloum, S.A.: The moderation effect of gender on accepting electronic payment technology: a study on United Arab Emirates consumers. Rev. Int. Bus. Strateg. **31**, 375–396 (2021). https://doi.org/10.1108/RIBS-08-2020-0102
8. Flavian, C., Guinaliu, M., Lu, Y.: Mobile payments adoption—introducing mindfulness to better understand consumer behavior. Int. J. Bank Mark. **38**, 1575–1599 (2020). https://doi.org/10.1108/IJBM-01-2020-0039
9. Chawla, D., Joshi, H.: Consumer attitude and intention to adopt mobile wallet in India—an empirical study. Int. J. Bank Mark. **37**, 1590–1618 (2019). https://doi.org/10.1108/IJBM-09-2018-0256
10. Sarmah, R., Dhiman, N., Kanojia, H.: Understanding intentions and actual use of mobile wallets by millennial: an extended TAM model perspective. J. Indian Bus. Res. **13**, 361–381 (2020). https://doi.org/10.1108/JIBR-06-2020-0214
11. Aji, H.M., Berakon, I., Husin, M.: COVID-19 and e-wallet usage intention: a multigroup analysis between Indonesia and Malaysia. Cogent Bus. Manag. **7** (2020). https://doi.org/10.1080/23311975.2020.1804181
12. Kholid, M.N.: Determinants of intention to use Islamic mobile banking: evidence from millennial generation. J. Ekononomi Keuang. Islam. **5**, 53–62 (2019). https://doi.org/10.20885/JEKI.vol5.iss2.art2
13. Luarn, P., Lin, H.H.: Toward an understanding of the behavioral intention to use mobile banking. Comput. Human Behav. **21**, 873–891 (2005). https://doi.org/10.1016/j.chb.2004.03.003
14. Davis, F.D.: Perceived usefulness, perceived east of use, and user acceptance of information technology. MIS Q. **13**, 319–340 (1989). https://doi.org/10.1016/S0305-0483(98)00028-0
15. Natarajan, T., Balasubramanian, S.A., Kasilingam, D.L.: Understanding the intention to use mobile shopping applications and its influence on price sensitivity. J. Retail. Consum. Serv. **37**, 8–22 (2017). https://doi.org/10.1016/j.jretconser.2017.02.010
16. Youn, S. yi, Lee, K.H.: Proposing value-based technology acceptance model: testing on paid mobile media service. Fash. Text. **6**, 1–16 (2019). https://doi.org/10.1186/s40691-018-0163-z
17. Sharma, S.K.: Integrating cognitive antecedents into TAM to explain mobile banking behavioral intention: a SEM-neural network modeling. Inf. Syst. Front. **21**, 815–827 (2019). https://doi.org/10.1007/s10796-017-9775-x

18. Huang, Y.C., Chang, L.L., Yu, C.P., Chen, J.: Examining an extended technology acceptance model with experience construct on hotel consumers' adoption of mobile applications. J. Hosp. Mark. Manag. **28**, 957–980 (2019). https://doi.org/10.1080/19368623.2019.1580172
19. Kholid, M.N., Alvian, S., Yumewang, Y.K., Tumewang, Y.K.: Understanding determinants mobile accounting app adoption by micro, small, and medium enterprise. J. Account. Strateg. Fin. **3**, 52–70 (2020). https://doi.org/10.33005/jasf.v3i1.74
20. Chen, C.C., Tsai, J.L.: Determinants of behavioral intention to use the personalized location-based mobile tourism application: an empirical study by integrating TAM with ISSM. Futur. Gener. Comput. Syst. **96**, 628–638 (2019). https://doi.org/10.1016/j.future.2017.02.028
21. Almaiah, M.A., Alismaiel, O.A.: Examination of factors influencing the use of mobile learning system: an empirical study. Educ. Inf. Technol. **24**, 885–909 (2019). https://doi.org/10.1007/s10639-018-9810-7
22. Shaikh, I.M., Qureshi, M.A., Noordin, K., Shaikh, J.M., Khan, A., Shahbaz, M.S.: Acceptance of Islamic financial technology (FinTech) banking services by Malaysian users: an extension of technology acceptance model. Foresight **22**, 367–383 (2020). https://doi.org/10.1108/FS-12-2019-0105
23. Kholid, M.N., Asri, N.: The effect of external variables on mobile accounting app adoption by student entrepreneurs. J. Small Bus. Strateg. **31**, 38–49 (2021). https://doi.org/10.53703/001c.29816
24. Rai, R.S., Selnes, F.: Conceptualizing task-technology fit and the effect on adoption—a case study of a digital textbook service. Inf. Manag. **56** (2019). https://doi.org/10.1016/j.im.2019.04.004
25. Su, C.Y., Chiu, C.H.: Perceived enjoyment and attractiveness influence Taiwanese elementary school students' intention to use interactive video learning. Int. J. Hum. Comput. Interact. **37**, 574–583 (2021). https://doi.org/10.1080/10447318.2020.1841423
26. Featherman, M., Pavlou, P.A.: Predicting e-services adoption: a perceived risk facets perspective. Int. J. Hum. Comput. Stud. **59**, 451–474 (2003). https://doi.org/10.1016/S1071-5819(03)00111-3
27. Im, I., Kim, Y., Han, H.-J.: The effects of perceived risk and technology type on users' acceptance of technologies. Inf. Manag. **45**, 1–9 (2008). https://doi.org/10.1016/j.im.2007.03.005
28. Lee, M.-C.: Factors influencing the adoption of internet banking: an integration of TAM and TPB with perceived risk and perceived benefit. Electron. Commer. Res. Appl. **8**, 130–141 (2009). https://doi.org/10.1016/j.elerap.2008.11.006
29. Nguyen, T.D., Hyunh, P.A.: The roles of perceived risk and trust on e-payment adoption. Econom. Financ. Appl. **760** (2018). https://doi.org/10.1007/978-3-319-73150-6
30. Hanafizadeh, P., Behboudi, M., Koshksaray, A.A., Tabar, M.J.S.: Mobile-banking adoption by Iranian bank clients. Telemat. Informatics. **31**, 62–78 (2012). https://doi.org/10.1016/j.tele.2012.11.001
31. Wang, Y., Wang, S., Wang, J., Wei, J., Wang, C.: An empirical study of consumers' intention to use ride-sharing services: using an extended technology acceptance model. Transportation (Amst). **47**, 397–415 (2020). https://doi.org/10.1007/s11116-018-9893-4
32. Mutahar, A.M., Daud, N.M., Ramayah, T., Isaac, O., Aldholay, A.H.: The effect of awareness and perceived risk on the technology acceptance model (TAM): mobile banking in Yemen. Int. J. Serv. Stand. **12**, 180–204 (2018). https://doi.org/10.1504/IJSS.2018.091840
33. Hair, J.F., Hult, G.T.M., Ringle, C.M., Sarstedt, M.: A Primer on Partial Least Squares Structural Equation MODELING (PLS-SEM). Sage Publications, Los Angeles (2017)
34. Abdullah, F., Ward, R., Ahmed, E.: Investigating the influence of the most commonly used external variables of TAM on students' perceived ease of use (PEOU) and perceived usefulness (PU) of e-portfolios. Comput. Human Behav. **63**, 75–90 (2016). https://doi.org/10.1016/j.chb.2016.05.014

35. Ahmad, S.Z., Khalid, K.: The adoption of M-government services from the user's perspectives: empirical evidence from the United Arab Emirates. Int. J. Inf. Manage. **37**, 367–379 (2017). https://doi.org/10.1016/j.ijinfomgt.2017.03.008
36. Sutianto, F.D.: Peta Persaingan Dompet Digital di Indonesia Setelah Masuknya Hary Tanoe ke Arena. https://kumparan.com/kumparanbisnis/peta-persaingan-dompet-digital-di-indonesia-setelah-masuknya-hary-tanoe-ke-arena-1vspT6lMeLh/full. Last accessed 29 Jan 2021

Consumer Perceived Severity and Application Based Platform Purchase Behavior Amid COVID 19 in Society 5.0

Dhea Radika Putri, Sari Lestari Zainal Ridho, Habsah Binti Haji Mohamad Sabli, Kartini Binti Che Ibrahim, Mohammad Fardillah Bin Wahi, Suhaimi Bin Sibir, Hadi Jauhari, Alditia Detmuliati, Alfitriani, Fernando Africano, Agung Anggoro Seto, Keti Purnamasari, Beni Wijaya, and Jovan Febriantoko

Abstract Society 5.0 is a design where humans live side by side with technology. This design has accelerated implementation in the midst of the COVID-19 pandemic, including in the behavior of buying one of tourism industry, i.e., food and or beverage business. Food and beverages business offered their services in vary platform. One of which is in the form of application-based platform food and beverage service providers. Many factors influence purchasing behavior during the pandemic, and it is believed to have an effect on perceived severity. However, there is still a research gap regarding the effect of perceived severity on purchasing behavior. Therefore, this study aims to evaluate the effect of perceived severity along with several other variables, hygiene awareness, quality, and interface on attitudes and shopping behavior in the community 5.0. The method used is quantitative analysis with Partial Least Square-Structural Equation Modeling. Using 112 samples collected through a questionnaire of the second semester of 2021 in Indonesia, this study found that perceived severity along with several other variables, hygiene awareness, quality, and interface effect the shopping attitudes and behavior in the community 5.0.

Keywords Food and beverage business · Society · Purchase behavior

D. R. Putri · S. L. Z. Ridho (✉) · H. Jauhari · A. Detmuliati · Alfitriani · F. Africano · A. A. Seto · K. Purnamasari · B. Wijaya · J. Febriantoko
Politeknik Negeri Sriwijaya, Palembang, Sumatera Selatan 30139, Indonesia
e-mail: sarilestari@polsri.ac.id

H. B. H. M. Sabli · K. B. C. Ibrahim · M. F. B. Wahi · S. B. Sibir
Politeknik Mukah, KM 7.5 Jalan Oya, 96400 Mukah, Sarawak, Malaysia

A. Hamdan et al. (eds.), *The Implementation of Smart Technologies for Business Success and Sustainability*, Studies in Systems, Decision and Control 216,
https://doi.org/10.1007/978-3-031-10212-7_80

1 Introduction

Society 5.0 is equated with a smart society which is a design where humans and technology coexist as an effort to build social equality and sustainable development with the aim of solving economic and social problems [3, 9, 13]. The coexistence and harmony between humans and technology has accelerated in the midst of the Covid-19 pandemic. More intense interaction between technology and humans occurs because technology really helps solve many problems that arise in the midst of the Covid-19 pandemic. The Indonesian Internet Service Providers Association (APJII), based on a survey they conducted, stated that there was an increase of 25.5 million new users in the second quarter of 2020 [1]. Furthermore, data shows that based on research results in 2019, 58% of the population ordered food through food delivery applications [15, 18, 19]. Based on survey data from the Central Statistics Agency in April 2020, public consumption patterns during the pandemic showed that the level of spending on ready-to-eat food was higher [10]

Provision of food and beverage business or food and beverage service industry (providers of food and beverage services) is one of the businesses included in the tourism and creative economy sectors [11]. Service providers in the tourism industry must be aware of the different motivations for using services in different segments, this is based on the fact that each segment has special needs that may differ from those of other tourists [2]. Society 5.0 is a society with a certain lifestyle, where the lifestyle of society 5.0 is one segment which is a community group with the characteristics of spending time interacting more intensely with technology. One of the triggers for the increase in purchases using applications, apart from being familiar with technology which is the lifestyle of the community, based on previous research, perceived severity during the Covid-19 pandemic affected shopping routines using applications [12], because the pandemic caused several government policies, including restrictions on community activities. Hence, one way to enjoy the food provided by the restaurant is through a food delivery application. Not only the severity of the epidemic, the choice of food delivery service and the food used is believed to be influenced by awareness regarding hygiene [17], and the interface of the application provider [16] as well as the quality of the restaurant providing food and/or beverage services [4].

On the contrary, other previous studies on the one hand found that the perceived severity factor had no significant impact on intention to use online food delivery services during the Covid-19 pandemic [12], although on the other hand there were other studies which stated a significant effect on perceived severity perceived against the use of food and beverage services with an application-based platform [7]. Because there is still a gap regarding the effect of perceived severity on shopping attitudes and behavior, this study aims to evaluate the effect of perceived severity along with several other variables, hygiene awareness, quality, and interface on shopping attitudes and behavior in the community. This paper has been divided into four sections. The first section deals with the introduction, the second section deals with the methodology, the third section deal with the result, and the last section deals with the conclusion.

2 Method

This study uses a quantitative approach method in testing the relationship between variables. This research was using the Partial Least Square-Structural Equation Modeling (PLS-SEM) technique. The SEM was using in data processing, since SEM is an analytical technique to confirm further in order to determine whether a particular model is valid or not. The PLS-SEM model evaluation is consist of 2 evaluations, namely by evaluating the model (evaluating the *outer model*) and evaluating the structural model (evaluating the *inner model*). First, the evaluation of *outer model* includes the value of *outer loading*, *average variance extracted* (AVE), and *composite reliability* (CR), and second, the evaluation of *inner model* includes the value of *path coefficients*, and *R-squre* (R^2) [5, 6, 8].

The population of this study were users of the consumer of food and beverage application based platform in Palembang, Indonesia and the sampling for this research, refer to statistical requirement, the appropriate sample size in the study can be between 30 and 500 [14]. Based on that requirement, in this study, the sample size used was 112 respondents, which was collected successfully through efforts to distribute as many questionnaires as possible during the Covid-19 pandemic, along the second semester of 2021.

3 Result

3.1 Descriptive Statistics

The following tables show the output in the form of a Frequency Distribution Table. Table 1 refers to descriptive statistic of gender and frequency of using food delivery application for the research framework (model) 1:

Regarding the gender of the respondents, the results of observations described that there were 84 female respondents (75%), and the remaining 28 respondents were male (25%). It can be concluded that almost three quarters of the respondents are women. The majority of respondents used food delivery application more than

Table 1 Gender and use frequency

Gender	Frequency	Percentage	Use	Frequency	Percentage
Male	28	25.0	Once	0	0.0
Female	84	75.0	Twice	14	12.5
			Three times	16	14.3
			More than three times	82	73.2
Total	112	100.0	Total	112	100.0

Source Data processing results, 2022

Table 2 Age

Age	Frequency	Percentage
18	23	20.5
19	38	33.9
20	21	18.8
21	7	6.3
22	12	10.7
23	6	5.4
24	3	2.7
25	1	0.9
29	1	0.9
Total	112	100.0

Source Data processing results, 2022

3 times as many as 82 people (73.2%). It can be concluded that three quarters of respondents use food delivery application more than 3 times.

Table 2 refers to the respondent's age category. In the respondent's age category, it was concluded that the respondents' ages spread following a right-sloping normal distribution with ages ranging from 18 to 29. The most respondents aged 19 years were 38 people (33.9%).

3.2 Validity and Reliability Test

Table 3 refers to validity and reliability test. Validity testing was conducted to determine the correlation between each question item (indicator) and the total score of all questions (factor). The indicator is said to be valid when the correlation (r) > 0.50. Besides, to test the reliability, the reliability coefficient used is Cronbach's Alpha. Cronbach's Alpha is a reliability coefficient calculated by the Alpha formula made by Cronbach. This test was conducted to determine the reliability (level of confidence) between the question items (all indicators) to the total score of all questions (factors). All indicators are said to be reliable when Cronbach's Alpha Value > 0.60. From the output results, it is found that the correlation between the research variables and their respective indicators is given in the Table 3.

From these results, the correlation coefficient of all indicators is greater than 0.50, hence the validity test can be concluded that all indicators are said to be valid (accurate). In addition, all indicators have a correlation greater than 0.70, so all indicators in this case can be said to be ideal. In addition, Cronbach's alpha values for Interface Issues (IF), Hygiene Consciousness (HC), Perceived Severity (PS), Quality Issues (QL), Attitude (AT), and Purchase Routine (PR) were 0.919, 0.742, respectively: 0.920, 0.814, 0.855, and 0.811 hence this reliability test can be concluded that

Table 3 Validity and reliability test

Variable	Indicator	Correlation	Cronbach's Alpha
Interface issues (IF)	IF1	0.974	$\alpha = 0.919$
	IF2	0.960	
Hygiene consciousness (HC)	HC1	0.889	$\alpha = 0.742$
	HC2	0.894	
Purchase routine (PR)	SR1	0.924	$\alpha = 0.811$
	SR2	0.911	
Perceived severity (PS)	PS1	0.863	$\alpha = 0.920$
	PS2	0.847	
	PS3	0.890	
	PS4	0.882	
	PS5	0.853	
	PS6	0.776	
Quality issues (QL)	QL1	0.821	$\alpha = 0.814$
	QL2	0.898	
	QL3	0.850	
Attitude (AT)	AT1	0.914	$\alpha = 0.855$
	AT2	0.934	
	AT3	0.792	

Source Data processing results, 2022

all indicators are said to be reliable (can be trusted). Thus, testing the validity and reliability of the constructs of all indicators is concluded to be valid and reliable.

3.3 *Goodness of Fit*

Table 4 refers to the goodness of fit. The goodness-of-fit index results are acceptable. The results are shown in Table 4.

From the overall suitability test of the model, all test results that show the model are of good value, so it can be concluded that the SEM model is good. Furthermore, Table 5 refers to the factor loading, VE, CR, and Cronbach's Alpha values are very satisfactory for the modified SEM model. All factors loading is still above the cut-off value (0.50). The reliability obtained shows good results because all variance extracted (VE) values are greater than 0.50, construct reliability (CR) is above 0.70, and Cronbach's Alpha is above 0.60 (cut-off values). Table 5 provides summary results of score information from factor loading, VE, CR, and Cronbach's Alpha for the SEM model. This test was held to test discriminant validity, with good results

Table 4 Goodness of fit

Goodness of fit	Result	Model requirement	Decision
RMSEA	0.071	RMSEA < 0.08 is a *good fit* RMSEA < 0.05 is a *close fit*	Good fit
GFI	0.858	GFI ≥ 0.90 is a *good fit* 0.80 ≤ GFI ≤ 0.90 is a *marginal fit*	Marginal fit
AGFI	0.790	0 ≤ AGFI ≤ 1 PGFI bigger, better	Good fit
CFI	0.954	CFI > 0.90 is a *good fit* 0.80 ≤ CFI ≤ 0.90 is a *marginal fit*	Good fit
TLI	0.938	TLI > 0.90 is a *good fit* 0.80 ≤ TLI ≤ 0.90 is a *marginal fit*	Good fit
CMIN/DF	1.556	CMIN/DF < 2.00 is a *good fit*	Good fit

Source Data processing results, 2022

on CR. Table 5 refer to the value of factor loading, CR, and Cronbach's alpha for construct investigations.

Table 5 The validity and reliability measurements for all constructs

Item description	Factor loading	VE	CR	Cronbach's Alpha
IF1	0.921	0.875	0.950	0.919
IF2	0.950			
HC1	0.707	0.575	0.729	0.742
HC2	0.806			
PR1	0.851	0.660	0.795	0.811
PR2	0.772			
PS1	0.809	0.679	0.927	0.920
PS2	0.772			
PS3	0.878			
PS4	0.894			
PS5	0.808			
PS6	0.775			
QL1	0.867	0.736	0.892	0.814
QL2	0.727			
QL3	0.963			
AT1	0.867	0.676	0.860	0.855
AT2	0.911			
AT3	0.668			

Source Data processing results, 2022

Table 6 The relationship path among the constructs

Hypothesized paths	Estimate	C.R	*P*	Result	Information
H1: IF → AT	0.009	0.182	0.855	Not significant	Direct effect
H2: HC → AT	0.437	4.068	<0.001***	Significant	Direct effect
H3: PS → AT	0.535	3.986	<0.001***	Significant	Direct effect
H4: QL → AT	0.031	0.692	0.489	Not significant	Direct effect
H5: IF → PR	0.145	1.939	0.052*	Significant	Direct effect
H6: AT → PR	0.369	2.863	0.004***	Significant	Direct effect
H7: QL → PR	0.140	1.822	0.068*	Significant	Direct effect

Note ***, **, and * denote the two-tail statistical significance at 1%, 5%, and 10% respectively
Source Data processing results, 2022

3.4 Interpretation of Path Coefficients in SEM Model

The results of the hypothesis shown in Table 6 refer to the results of hypothesis testing. The table shows the results of hypothesis testing in a direct relationship of several constructs. Of the 3 hypotheses that were formulated, all hypotheses were significant with *p*-value <0.05. In addition, the largest estimate (0.535) is given between Perceived Severity (PS) and Attitude (AT).

$$AT = 0.009IF + 0.437HC + 0.535PS + 0.031QL + e_1 \quad (1)$$

$$PR = 0.14IF + 0.39AT + 0.140QL + e_2 \quad (2)$$

The results of the hypothesis shown in Table 6 refer to the results of hypothesis testing. The table shows the results of hypothesis testing in a direct relationship of several constructs. Of the 3 hypotheses that were formulated, all hypotheses were significant with *p*-value <0.05. In addition, the largest estimate (0.535) is given between Perceived Severity (PS) and Attitude (AT) (Fig. 1; Table 7).

On the structural equations and decomposition effects, the test obtained the following results and conclusions:

1. There is no significant direct effect of Interface Issues (IF) on Attitude (AT). The higher/lower the Interface Issues value, it does not affect the Attitude value.
2. Hygiene Consciousness (HC) has a significant direct effect on Attitude (AT) of 0.437. The higher the Hygiene Consciousness value, the higher the Attitude value. Conversely, the lower the Hygiene Consciousness value, the lower the Attitude value.
3. The direct effect of Perceived Severity (PS) is obtained which is significant on Attitude (AT) of 0.535. The higher the Perceived Severity value, the higher the Attitude value. Conversely, the lower the Perceived Severity value, the lower the Attitude value.

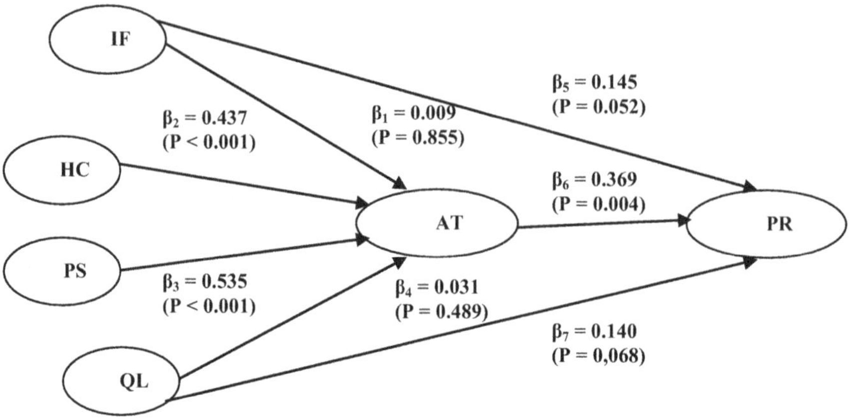

Fig. 1 Path coefficient. *Source* Data processing results, 2022

Table 7 The decomposition effect among the constructs

Hypothesized paths	Passed paths	Estimate	Calculate	Information	Calculate	Information
H8: IF → PR	IF → SR	0.145	–	Direct effect	0.145 + 0.003 = 0.148*	Total effect
	IF → AT AT → PR	0.009 0.369	0.009 × 0.369 = 0.003	Indirect effect		
H9: HC → PR	HC → AT AT → PR	0.437 0.369	–	Indirect effect	0.437 × 0.369 = 0.161*	Indirect effect
H10: PS → PR	PS → AT AT → PR	0.535 0.369	–	Indirect effect	0.535 × 0.369 = 0.197*	Indirect effect
H11: QL → PR	QL → PR	0.140	–	Direct effect	0.140 + 0.011 = 0.151*	Total effect
	QL → AT AT → PR	0.031 0.369	0.031 × 0.369 = 0.011	Indirect effect		

Note *denote significance because the value is not close to zero
Source Data processing results, 2022

4. There is no significant direct effect of Quality Issues (QL) on Attitude (AT). The higher/lower the Quality Issues value, it does not affect the Attitude value.
5. The direct effect of Interface Issues (IF) is obtained which is significant on the Purchase Routine (PR) of 0.145. The higher the Interface Issues value, the higher the Purchase Routine value; conversely, the lower the Interface Issues value, the lower the Purchase Routine value.
6. Attitude (AT) has a significant direct effect on Purchase Routine (PR) of 0.369. The higher the Attitude value, the higher the Purchase Routine value; conversely, the lower the Interface Issues value, the lower the Purchase Routine value.
7. The direct influence of Quality Issues (QL) is obtained which is significant on the Purchase Routine (PR) of 0.140. The higher the Quality Issues value, the higher the Purchase Routine value; conversely, the lower the Quality Issues value, the lower the Purchase Routine value.
8. The direct effect of total Interface Issues (IF) which is significant on Purchase Routine (PR) is 0.148. This occurs because the value of the direct influence of Interface Issues (IF) on the Purchase Routine (PR) is significant, although the direct effect of Interface Issues (IF) on Attitude (AT) is not significant.
9. There is a significant indirect effect of Hygiene Consciousness (HC) on Purchase Routine (PR) of 0.161. This happens because the value of the direct influence of Hygiene Consciousness (HC) on Attitude (AT) is significant, and there is a direct effect of Attitude (AT) on Purchase Routine (PR) is also significant.
10. There is a significant indirect effect of Perceived Severity (PS) on Purchase Routine (PR) of 0.197. This happens because the value of the direct influence of Perceived Severity (PS) on Attitude (AT) is significant, and there is a direct effect of Attitude (AT) on Purchase Routine (PR) is also significant.
11. The direct effect of total Quality Issues (QL) which is significant on Purchase Routine (PR) is 0.151. This happens because the value of the direct influence of Quality Issues (QL) on the Purchase Routine (PR) is significant, although the direct influence of Quality Issues (QL) on Attitude (AT) is not significant.

These findings support some of the previous theories or researches, i.e., regarding the effect of perceived severity, that asserted the perceived severity affect the application based platform purchase behavior [12]. Furthermore, this research in line with the previous researches in term of hygiene consciousness on attitude [17]; a direct influence of quality issues, on purchase behavior [4]; a significant indirect effect of hygiene consciousness on purchase routines and a direct influence of attitude on purchase routines which is also significant; a significant indirect effect of perceived severity on the purchase routine as well as a significant direct effect of attitude on the purchase routine; the effect of total interface issues on the purchase routine; a direct effect of total quality issues that are significant to the purchase routine [12].

4 Conclusion

The purpose of this study was to evaluate the effect of perceived severity along with several other variables, hygiene awareness, quality, and interface on attitudes and shopping behavior in the community 5.0. Based on the results of data processing that has been shown previously, it was found that there was a direct effect of perceived severity, hygiene consciousness on attitude; a direct influence of quality issues, on purchase behavior; a significant indirect effect of hygiene consciousness on purchase routines and a direct influence of attitude on purchase routines which is also significant; a significant indirect effect of perceived severity on the purchase routine as well as a significant direct effect of attitude on the purchase routine; the effect of total interface issues on the purchase routine; a direct effect of total quality issues that are significant to the purchase routine. However, there was no direct influence of significant interface issues on attitude and no direct influence of significant quality issues on attitude was obtained. Hence, it can be said that the finding of this research in line with the previous research that asserted the perceived severity affect the application based platform purchase behavior.

The implication of these findings, that it is important for the food and beverages business to improve even more to enhance the quality and the interface issues. In order to gain the trust of the consumer since the consumer, the society 5.0, considering the hygiene, the quality and more over the severity they perceived amid COVID 19. Furthermore, this research has some limitations; one of it is the scope of the research, in term of the location only in Indonesia. Hence for the future agenda it is recommended to expand the scope of the research, for instance to add some survey location or to add some more variables that hypothetically affect the behavior of the customer.

References

1. APJII, A.P.J.I.I.: Survei Pengguna Internet APJII 2019-Q2 2020: Ada Kenaikan 25,5 Juta Pengguna Internet Baru di RI. https://apjii.or.id/content/read/104/503/BULETIN-APJII-EDISI-74---November-2020 (2020)
2. Fajriyati, I., et al.: Generic and Islamic attributes for non-Muslim majority destinations: application of the three-factor theory of customer satisfaction. Heliyon. **6**, 6 (2020). https://doi.org/10.1016/j.heliyon.2020.e04324
3. Foresti, R., et al.: Smart society and artificial intelligence: Big Data scheduling and the global standard method applied to smart maintenance. Engineering **6**(7), 835–846 (2020). https://doi.org/10.1016/j.eng.2019.11.014
4. Kaur, P., et al.: The value proposition of food delivery apps from the perspective of theory of consumption value. Int. J. Contemp. Hosp. Manag. **33**(4), 1129–1159 (2020). https://doi.org/10.1108/IJCHM-05-2020-0477
5. Latan, H.: Model Persamaan Struktural: Teori dan Implementasi AMOS 21.0. ALFABETA, Bandung (2013)
6. Latan, H., Ghozali, I.: Partial least square: Konsep, Teknik dan Aplikasi Menggunakan Program SmartPLS 2.0 M3. Badan Penerbit Universitas Dipenogoro, Semarang (2012)

7. Li, J.-B., et al.: Self-control moderates the association between perceived severity of Coronavirus Disease 2019 (COVID-19) and mental health problems among the Chinese public. Int. J. Environ. Res. Public Heal. **17**(13), 4820 (2020). https://doi.org/10.1159/000513546
8. Mokhtarian, P.L., Ory, D.T.: Structural equations models. Int. Encycl. Hum. Geogr., 10–17 (2009). https://doi.org/10.1016/B978-008044910-4.00541-1
9. Nagy, K., Hajrizi, E.: Building pillars for adapting society 5.0 in post-conflict countries. IFAC-PapersOnLine **52**(25), 40–45 (2019). https://doi.org/10.1016/j.ifacol.2019.12.443
10. Putranto, W.P.A., Larasaty, P., Kurniasih, A., Pratiwi, A.I., Saputri, V.G., Meilianingsih, T.: Hasil Survei Sosial Demografi Dampak Covid-19 2020. Badan Pusat Statistik (2020). https://doi.org/10.16309/j.cnki.issn.1007-1776.2003.03.004
11. Republik Indonesia: Undang-Undang Republik Indonesia Nomor 10 Tahun 2009 tentang Kepariwisataan. Kementrian Kebudayaan dan Pariwisata, Jakarta (2009)
12. Sharma, R., et al.: Over-ordering and food waste: the use of food delivery apps during a pandemic. Int. J. Hosp. Manag. **96**(April), 102977 (2021). https://doi.org/10.1016/j.ijhm.2021.102977
13. Sugiono, S.: Digital content industry in society 5.0 perspective. J. Ilmu Pengetah. dan Teknol. Komun. **22**(2), 175–191 (2020)
14. Sugiyono: Metode Penelitian Kuantitatif. ALFABETA, Bandung (2018)
15. Susanti, I.: 58% Masyarakat Indonesia Beli Makanan Siap Santap Lewat Aplikasi Smartphone. https://ekbis.sindonews.com/berita/1441214/34/58-masyarakat-indonesia-beli-makanan-siap-santap-lewat-aplikasi-smartphone (2019)
16. Talwar, S., et al.: Barriers and paradoxical recommendation behaviour in online to offline (O2O) services. A convergent mixed-method study. J. Bus. Res. **131**(April), 25–39 (2021). https://doi.org/10.1016/j.jbusres.2021.03.049
17. Talwar, S., et al.: Why do people purchase from online travel agencies (OTAs)? A consumption values perspective. Int. J. Hosp. Manag. **88**(Feb), 102534 (2020). https://doi.org/10.1016/j.ijhm.2020.102534
18. Thomas, V.F.: Nielsen: 58% Masyarakat Pesan Makanan via Aplikasi Pesan-Antar. https://tirto.id/nielsen-58-masyarakat-pesan-makanan-via-aplikasi-pesan-antar-eil7 (2019)
19. Triwijanarkao, R.: Riset: 58% Konsumen Pesan Makanan Via Aplikasi Online. https://www.marketeers.com/58-konsumen-pesan-makanan-via-aplikasi-online/ (2019)

A Systematic Literature Review of the Performance Measurement System in the Hospitals Sector

Ayman Mohammad Aburumman, Rosliza Mat Zin, and Norazlina Ilias

Abstract The goal of this research is to provide a complete review of the literature on Performance Measurement Systems (PMSs) in hospitals. One research question was clarified: which performance measurement dimensions related to the hospital's sector can be observed in the existing studies? The process involved a systematic literature review of 48 studies formed on the basis of rigor, clarity, and trustworthiness. For the first time, we systematically reviewed the 8 Performance Measures (PMs) for the dimensions of the Performance Measurement System Framework (PMSF). The PMSF allows academics to obtain a basic understanding and conduct additional future studies in the area of PMSs in hospitals. Regarding the results of this research, it has been determined that a more thorough approach to PMS in hospitals is required. This research is also useful in that it offers knowledge to decision-makers who are considering using a PMS.

Keywords Systematic literature review · Performance measurement system · Performance dimensions · Hospital performance

1 Introduction

PMS has caught the interest of practitioners and scholars from a variety of disciplines, sectors, and organizations over the years [1, 2]. This interest is reflected in the growth of PMS research literature [3]. The PMS is intended to give meaningful data to aid strategic decision-making in order to meet the organization's goals and objectives [4, 5].

Scholars in various fields have paid close attention to the hospital sector [6]. Health care services are insufficiently presented, reducing service quality in terms of cost, timeliness, efficiency, and delivery quality [7], despite hospitals' crucial role in the presentation of health care services. In terms of business growth and performance,

A. M. Aburumman (✉) · R. M. Zin · N. Ilias
Faculty of Business, Economics and Social Development, Universiti Malaysia Terengganu, Kuala Terengganu, Malaysia
e-mail: aburumman.ayman@yahoo.com

A. Hamdan et al. (eds.), *The Implementation of Smart Technologies for Business Success and Sustainability*, Studies in Systems, Decision and Control 216,
https://doi.org/10.1007/978-3-031-10212-7_81

the hospital sector in developing nations is still lagging behind, and as a result, they are not maximizing their potential to improve the quality of health care services for all stakeholders [8, 9]. Furthermore, according to Van Van Elten et al. [10], the hospital's low performance may be due to its failure to apply effective PMSs. The lack of a comprehensive and internationally acknowledged PMSF for evaluating hospital performance [11] is the reason for this.

In regards to the usual purposes for recognizing PMS to be important for organizations [12], there are some specific issues that apply to the hospital sector, such as weak organizations, lack of performance, service quality, rising health costs [13, 14]. As a result, a broader view of hospital performance, and the creation of dependable PMS, is required.

This paper's goal is to give a comprehensive overview of the literature on PMSs in hospitals. A comprehensive literature review of studies' step-by-step processes as part of the approach. The second section summarizes the literature review of PMS and provides an overview of the hospital industry. Section 3 focuses on research methodology and highlights aspects of Arksey and O'Malley's five-step guidelines [15], describing the performance measures dimensions into 8 categories. Section 4, includes concluding as well as research topics for the future.

2 Literature Review

2.1 Performance Measurement System

The terms and definitions are helpful to each regular follow-up of awareness in a field of study [3, 16]. Define PMS as the process of identifying activities that motivate people to improve their productivity, flexibility, and customer happiness [17]. According to Neely et al. [18], a PMS is a set of PMs used to determine the efficiency and effectiveness of activities.

According to a prior study, the PMS began in the mid-eighteenth century and evolved in stages [19, 20]. Furthermore, in the 1980s, the development of PMSs began, which primarily included criticism of financial/accounting performance measures [12], and the focus shifted from traditional PMSs to new non-financial performance measures, which include various measures [21, 22]. Since then, the PMS revolution has led to suggested PMFs and possible solutions that include both financial and non-financial measures [23–25]. More specifically, recent advances in PMS have emphasized effectiveness, efficiency, flexibility, and long-term viability [7, 26, 27].

2.2 *Hospital Sector: An Overview*

The hospital sector is regarded among the most significant sectors for any country and a significant component of its growth [28]. In both developed and developing countries, the hospital sector is one of the fastest-growing segments of the service economy [29]. In comparison to other industries, hospitals contribute directly to the delivery of high-quality healthcare to the community and patients [30].

Over the last three decades, there has been a growing worry about the efficiency of the hospital sector in both developed and developing countries. Traditional financial indicators have always been used to evaluate hospital performance [31]. As a result, financial performance measures may fail to capture other important hospital characteristics [6]. Finally, the healthcare sector's PMS is still an unanswered problem [26]. As a result, PMF should be used and designed to match the requirements and circumstances of hospitals [10].

3 Research Methodology

First of all, a systematic literature review aims to locate all relevant material, assess it, and synthesize it in order to answer questions about a given topic [32–34]. In addition, in 2022, a literature review was undertaken for this study with the goal of reducing systematic errors or bias. The five-step guidelines of Arksey and O'Malley [15] were used to define the hospital's Performance Measures Dimensions (PMDs). Meanwhile, comprehensive literature reviews have received little attention as a research approach in the field of PMF for hospital sector research. There has been no extensive evaluation of the PMS in the hospital sector to yet. This section covers the implementation of each component of the review process in great depth (see Fig. 1).

3.1 *Selecting Research Question*

The first stage in doing a systematic literature review is to determine the research question that will be addressed, since this will influence the development of search techniques [15]. Creating precise questions to choose the area of study is crucial in various fields [35, 36]. As a result, the research question is: "which performance measures dimensions related to the hospital's sector can be observed in the existing studies?".

Fig. 1 The steps of a systematic literature review

3.2 Search Relevant Studies

Using search databases, the second phase was to find primary papers and reviews that were relevant to answering the research questions [15, 34, 37]. To reduce bias, this study used an approach that included looking for research evidence through a variety of sources, including electronic databases, reference lists, hand-searching of significant journals, existing networks, and conferences at this stage of the scoping review. To assure that the study was not repeated, no systematic reviews on the topic were discovered.

3.3 Study Selection

To maintain consistency in decision-making, systematic review methods define inclusion and exclusion criteria based on a specific research question. Only papers mentioning the PMF or dimensions for performance measures were included in this study. To put it another way, from our initial analysis that the references that did not address our main research question were excluded from the study. The inclusion and exclusion criteria are listed in Table 1.

522 papers with possibly relevant investigations were located through database searches. All duplicates and articles that did not fulfill the admissions criteria were removed, leaving 260 papers for the next stage of the process. As a result of their lack of relation to the review questions, 90 papers were deleted. In the final screening,

Table 1 The exclusionary criteria

Inclusion criteria	Research published in academic journals Full access to full-text Published in the English language Research that discusses PMS and dimensions
Exclusion criteria	Research that does not discuss PMS and dimensions Foreign language Articles such as book reviews, commentaries, and opinion articles were excluded and journals that did not have a precise review process Papers that mention performance measures but do not discuss it

research publications that did not provide relevant information were excluded. There are a total of 48 studies that discuss performance measures of PMS.

3.4 *Analysis and Synthesis*

The findings and analysis from publications that addressed the research question identified throughout the evaluation process are highlighted in this section. The performance measures dimensions of the PMS are identified in this study. The 48 papers that appeared between 1990 and 2021 are examined in depth.

The numbers of publications published in various academic journals, as well as the methodology used in these articles, are summarized in Table 2. Among academic publications, the International Journal of Production Research published the most papers. Table 2 shows the most commonly used research methods, which include Conceptual Research (CR) and Empirical Research (ER), Case Studies (CS), and Literature Reviews (LR).

Categorization of Performance Measures Dimensions for Hospital Sector

For the majority of the research, most of the studies had their own measurement models and unique performance measures for the PMS. As a result, based on the findings of the literature review, no consensus on a PMF for the hospital sector has been reached. This study combined and consolidated diverse PM categorizing them into 8 categories in Table 3.

This study evaluated and discussed the chosen research journals, as well as describing the performance measures dimensions in the hospital sector, as follows:

(1) Patient/employee safety is primarily concerned with the avoidance, mitigation, and amelioration of negative outcomes or injuries resulting from healthcare. International health organizations and the World Alliance have paid attention to patient safety [49–51]. However, patient safety should be prioritized in the expectations of healthcare organizations and, performance measures [52]. Hence, one of the primary dimensions of the PMS is patient safety [11, 39].

Table 2 Papers published in academic

Journal	No. of papers	Frequency (%)	LR	CR	ER	CS
International Journal of Operations and Production Management	5	10.4	2	1	2	
Accounting, Organizations and Society	4	8.3			4	
Management Accounting Research	3	6.7	1		2	
International Journal for Quality in Health Care	3	6.7	3			3
International Journal of Health Care Quality Assurance	2	4.2			1	1
BMC Health Services Research	2	4.2	1			1
The Journal of Business Performance Management	2	4.2		2	2	
Measuring Business Excellence	2	4.2	1		1	
Journal of Management Accounting Research		2.1	1			
Others	24	50.0	8	1	5	10
Total	48	100	14	4	15	15

Table 3 Categorizations of performance measures dimensions

	Performance measures	Source
1	Patient/employee safety	[11, 38–40]
2	Patient/employee satisfaction	[11, 38]
3	Responsive governance	[39–41]
4	Financial	[11, 42–45]
5	Efficiency	[11, 26, 40, 46]
6	Effectiveness	[11, 26, 46, 47]
7	Flexibility	[7, 26]
8	Sustainability	[39, 47, 48]

(2) Patient/employee satisfaction measures are widely used and concerned in order for policymakers to understand better patients' requirements [53–56]. The concept of satisfaction is complex, and it has been extensively explored and evaluated as a separate phenomenon [57]. However, Liu and Itoh [38] suggest that measuring patient satisfaction, as well as employee satisfaction, could be a useful method for obtaining hospital performance. In particular, Gu and Itoh [11]

and Heidegger et al. [58], emphasize that patient/employee satisfaction involves patient satisfaction, patient complaints, and employee satisfaction measures.

(3) Financial measures: Kaplan [21], refers to the fact that the financial dimension is the most widely used and popular management accounting instrument. The financial dimension aimed at providing value that allows hospital managers to make the best decisions, establish a competitive environment for supplying products and services, and define service unit costs [27, 43].

(4) The efficiency dimension necessitates allowing hospital management to use available resources or finances to produce results or outputs [26]. Previous study has suggested that efficiency be included as one of the major performance measures when selecting a PMS [18, 26, 27]. Consequently, Basu et al. [41], and Simou et al. [40], argue that the efficiency dimension should focus on bed occupancy rate, rate of utilization technologies, length of stay.

(5) Effectiveness dimension refers to the degree of achieving desirable outcomes, also the intended goals and objectives, given the correct provide services to all who are likely to benefit most [41, 46]. Effectiveness is considered a major performance measure when designing PMF used in hospitals [26, 59, 60], to assess and improve hospital performance [41]. Moreover, Arah et al. [60] indicated that the effectiveness measures include fair access, timeliness, and appropriateness.

(6) The flexibility dimension refers to the flexibility as part of the healthcare delivery process, due to the necessary response for every patient in the clinical service process [14]. The flexibility dimension requires hospitals to measure a system's ability to respond to diversity or change [61]. The present performance measurement systems lack the flexibility measures to measure hospital performance [26, 62].

(7) Sustainability dimension refers to meeting the requirements of the present generation without compromising the ability of future generations to meet their own requirements [63]. However, there is little published research on evaluating the healthcare sector's sustainability performance [47]. Consequently, there is the need to incorporate dimensions of sustainability in a performance measurement framework [48, 64].

(8) Responsive governance dimension refers to the extent to which a hospital responds to community needs, promotes health, and provides care to all citizens [40, 65]. Thus, there is little research has addressed the Responsive measures in hospitals [11, 38]. Several authors have suggested the measurement of Responsive should consider important aspects of hospital performance [38, 40].

4 Conclusion

The uniqueness of purpose of this literature review is to provide a better conceptual framework for these key research subjects in the hospital sector by focusing on performance measures. This research contributes to the analysis of PMS by using a systematic literature review that covers studies conducted over a 31-year period (1990–2021), giving researchers the most up-to-date knowledge of relevant studies. This analysis of performance measures can assist hospital managers to get a better understanding of the prerequisites for utilizing a PMS. According to literature analysis, performance measures should be used by hospitals if they want to develop and improve their performance. To validate PMS with practical evidence, more rigorous and exploratory empirical research is required.

Finally, this research has certain limitations. Because there is a scarcity of literature providing empirical evidence of PMS in the hospital sector, this study's analyses and contributions are based only on 48 academic journal papers. The articles selected for review and analysis are limited to peer-reviewed academic journal articles, other types of texts, such as conference papers and book chapters, are ignored; these sources might offer a deeper understanding of this topic.

References

1. Aidemark, L.G., Funck, E.K.: Measurement and health care management. Fin. Account. Manage. **25**(2), 253–276 (2009)
2. Ferreira, P.S., Shamsuzzoha, A.H.M., Toscano, C., Cunha, P.: Framework for performance measurement and management in a collaborative business environment. Int. J. Prod. Perform. Manage. **61**(6), 672–690 (2012)
3. Choong, K.K.: Has this large number of performance measurement publications contributed to its better understanding? A systematic review for research and applications. Int. J. Prod. Res. **52**(14), 4174–4197 (2013)
4. Franco-Santos, M., Kennerley, M., Micheli, P., Martinez, V., Mason, S., Marr, B., …, Neely, A.: Towards a definition of a business performance measurement system. Int. J. Oper. Prod. Manage. **27**(8), 784–801 (2007)
5. Munir, R., Baird, K.: Influence of institutional pressures on performance measurement systems. J. Account. Organ. Chang. **12**(2), 106–128 (2016)
6. Nuti, S., Noto, G., Vola, F., Vainieri, M.: Let's play the patients music: a new generation of performance measurement systems in healthcare. Manag. Decis. **56**(10), 2252–2272 (2018)
7. Chahal, H., Gupta, M., Lonial, S.: Operational flexibility in hospitals: scale development and validation. Int. J. Prod. Res. **56**(10), 3733–3755 (2018)
8. Hammad, S.A., Jusoh, R., Ghozali, I.: Decentralization, perceived environmental uncertainty, managerial performance and management accounting system information in Egyptian hospitals. Int. J. Account. Inf. Manag. **2**(4), 314–330 (2013)
9. El-Jardali, F., Fadlallah, R.: A review of national policies and strategies to improve quality of health care and patient safety: a case study from Lebanon and Jordan. BMC Health Serv. Res. **17**(1), 568 (2017)
10. Van Elten, H.J., van der Kolk, B., Sülz, S.: Do different uses of performance measurement systems in hospitals yield different outcomes? Health Care Manage. Rev. **46**(3), 217 (2021)
11. Gu, X., Itoh, K.: Performance indicators: healthcare professionals' views. Int. J. Health Care Qual. Assur. **29**(7), 801–815 (2016)
12. Neely, A.: The evolution of performance measurement research. Int. J. Oper. Prod. Manag. **25**(12), 1264–1277 (2005)
13. Arah, O.A., Westert, G.P., Hurst, J., Klazinga, N.S.: A conceptual framework for the OECD health care quality indicators project. Int. J. Qual. Health Care **18**(suppl_1), 5–13 (2006)

14. Alolayyan, M.N., Ali, K.A.M., Idris, F., Alolayyan, M.N., Ali, K.A.M., Idris, F.: The impact of operational flexibility on hospital performance in Jordanian hospitals: some empirical evidences. J. Glob. Manag. **2**(1), 39–54 (2011)
15. Arksey, H., O'Malley, L.: Scoping studies: towards a methodological framework. Int. J. Soc. Res. Methodol. **8**(1), 19–32 (2005)
16. Gerba, Y., Viswanadham, P.: Performance measurement of small scale enterprises: review of theoretical and empirical literature. Int. J. Appl. Res. **2**(3), 531–535 (2016)
17. Lynch, R.L., Cross, K.F.: Measure up! The essential guide to measuring business performance. Mandarin
18. Neely, A., Gregory, M., Platts, K.: Performance measurement system design: a literature review and research agenda. Int. J. Oper. Prod. Manag. **25**(12), 1228–1263 (2005)
19. Kaplan, R.S., Norton, D.P.: Putting the balanced scorecard to work. In: Performance Measurement. Management, and Appraisal Sourcebook, vol. 66, pp. 17511 (1995)
20. Neely, A., Gregory, M., Platts, K.: Performance measurement system design: a literature review and research agenda. Int. J. Oper. Prod. Manag. **15**(4), 80–116 (1995)
21. Kaplan, R.S.: The evolution of management accounting. Read. Account. Manage. Control **49**(3), 586–621 (1984)
22. Slack, N.: The flexibility of manufacturing systems. Int. J. Oper. Prod. Manag. **7**(4), 35–45 (1987)
23. Fitzgerald, L., Johnston, R., Brignall, T.J., Silvestro, R., Voss, C.: Performance Measurement in Service Businesses. Chartered Institute of Management Accountants, London (1991)
24. Kaplan, R., Norton, D.: The balanced scorecard measures that drive performance. Harv. Bus. Rev. **70**(1), 71–79 (1992)
25. Micheli, P., Mura, M.: Executing strategy through comprehensive performance measurement systems. Int. J. Oper. Prod. Manag. **37**(4), 423–443 (2017)
26. Purbey, S., Mukherjee, K., Bhar, C.: Performance measurement system for healthcare processes. Int. J. Product. Perform. Manag. **56**(3), 241–251 (2007)
27. Pourmohammadi, K., Hatam, N., Shojaei, P., Bastani, P.: A comprehensive map of the evidence on the performance evaluation indicators of public hospitals: a scoping study and best fit framework synthesis. Cost Effectiveness Resour. Alloc. **16**(1), 64 (2018)
28. Rodríguez-Labajos, L., Thomson, C., O'Brien, G.: Performance measurement for the strategic management of health-care estates. J. Facil. Manag. **16**(2), 217–232 (2018)
29. Catuogno, S., Arena, C., Saggese, S., Sarto, F.: Balanced performance measurement in research hospitals: the participative case study of a haematology department. BMC Health Serv. Res. **17**(1), 1–11 (2017)
30. Smith, P.C., Mossialos, E., Papanicolas, I., Leatherman, S. (Eds.).: Performance measurement for health system improvement: experiences, challenges and prospects. Cambridge University Press (2008)
31. Naranjo-Gil, D.: Strategic performance in hospitals: the use of the balanced scorecard by nurse managers. Health Care Manage. Rev. **34**(2), 161–170 (2009)
32. Briner, R.B., Denyer, D.: Systematic review and evidence synthesis as a practice and scholarship tool. In: Handbook of evidence-based management: companies, classrooms and Research, pp. 112–129 (2012)
33. Petticrew, M., Roberts, H.: Systematic Reviews in the Social Sciences: A Practical Guide. John Wiley & Sons (2008)
34. Denyer, D., Tranfield, D.: Producing a systematic review. In: Buchanan, D., Bryman, A. (eds.) The Sage Handbook of Organizational Research Methods, pp. 671–689. Sage, London (2009)
35. Shbail, M., Shbail, A.: Organizational climate, organizational citizenship behaviour and turnover intention: evidence from Jordan. Manage. Sci. Lett. **10**(16), 3749–3756 (2020)
36. Lim, S., Pettit, S., Abouarghoub, W., Beresford, A.: Port sustainability and performance: a systematic literature review. Transp. Res. Part D: Transp. Environ. **72**, 47–64 (2019)
37. Schryen, G., Wagner, G., Benlian, A.: Theory of knowledge for literature reviews: an epistemological model, taxonomy and empirical analysis of IS literature (2015)

38. Liu, H.C., Itoh, K.: Conceptual framework for holistic dialysis management based on key performance indicators. Ther. Apher. Dial. **17**(5), 532–550 (2013)
39. Braithwaite, J., Hibbert, P., Blakely, B., Plumb, J., Hannaford, N., Long, J.C., Marks, D.: Health system frameworks and performance indicators in eight countries: a comparative international analysis. SAGE Open Med. **5**, 2050312116686516 (2017)
40. Simou, E., Pliatsika, P., Koutsogeorgou, E., Roumeliotou, A.: Developing a national framework of quality indicators for public hospitals. Int. J. Health Plann. Manage. **29**(3), e187–e206 (2014)
41. Veillard, J., Champagne, F., Klazinga, N., Kazandjian, V., Arah, O.A., Guisset, A.L.: A performance assessment framework for hospitals: the WHO regional office for Europe PATH project. Int. J. Qual. Health Care **17**(6), 487–496 (2005)
42. Elg, M., Broryd, K.P., Kollberg, B.: Performance measurement to drive improvements in healthcare practice. Int. J. Oper. Prod. Manag. **33**(11/12), 1623–1651 (2013)
43. Grigoroudis, E., Orfanoudaki, E., Zopounidis, C.: Strategic performance measurement in a healthcare organisation: a multiple criteria approach based on balanced scorecard. Omega **40**(1), 104–119 (2012)
44. Buathong, S., Bangchokdee, S.: The use of the performance measures in Thai public hospitals. Asian Rev. Account. **25**(4), 472–485 (2017)
45. Parkinson, J., Tsasis, P., Porporato, M.: A critical review of financial measures as reported in the Ontario hospital balanced scorecard. J. Health Care Fin. **34**(2), 48–56 (2007)
46. Khalifa, M., Khalid, P.: Developing strategic health care key performance indicators: a case study on a tertiary care hospital. Procedia Comp. Sci. **63**, 459–466 (2015)
47. Wong, K.L., Chong, K.E., Chew, B.C., Tay, C.C., Mohamed, S.B.: Key performance indicators for measuring sustainability in health care industry in Malaysia. J. Fund. Appl. Sci. **10**(1S), 646–657 (2018)
48. Zhou, L., Keivani, R., Kurul, E.: Sustainability performance measurement framework for PFI projects in the UK. J. Financ. Manag. Prop. Constr. **18**(3), 232–250 (2013)
49. Gholamzadeh, N.R., Jabbari, B.H., Jannati, A., Asghari, J.M.: Selecting hospital's key performance indicators, using analytic hierarchy process technique (2013)
50. World Alliance For Patient Safety Drafting Group, Sherman, H., Castro, G., Fletcher, M., Hatlie, M., Hibbert, P., …, Virtanen, M.: Towards an international classification for patient safety: the conceptual framework. Int. J. Qual. Health Care **21**(1), 2–8 (2009)
51. Aburumman, O., Salleh, A., Omar, K., Abadi, M.: The impact of human resource management practices and career satisfaction on employee's turnover intention. Manage. Sci. Lett. **10**(3), 641–652 (2020)
52. Vincent, C.: Patient Safety. John Wiley & Sons (2011)
53. Batbaatar, E., Dorjdagva, J., Luvsannyam, A., Savino, M.M., Amenta, P.: Determinants of patient satisfaction: a systematic review. Perspect. Public Health **137**(2), 89–101 (2017)
54. Obeid, M., Salleh, Z., Nor, M.N.M.: The mediating effect of job satisfaction on the relationship between personality traits and premature sign-off. Acad. Account. Fin. Stud. J. **21**(2), 1–17 (2017)
55. Salleh, A.M.M., Omar, K., Aburumman, O.J., Mat, N.H.N., Almhairat, M.A.: The impact of career planning and career satisfaction on employee's turnover intention. Entrepreneur. Sustain. Issues **8**(1), 218 (2020)
56. Zamanan, M., Alkhaldi, M., Almajroub, A., Alajmi, A., Alshammari, J., Aburumman, O.: The influence of HRM practices and employees' satisfaction on intention to leave. Manage. Sci. Lett. **10**(8), 1887–1894 (2020)
57. Heidegger, T., Saal, D., Nuebling, M.: Patient satisfaction with anaesthesia care: what is patient satisfaction, how should it be measured, and what is the evidence for assuring high patient satisfaction? Best Pract. Res. Clin. Anaesthesiol. **20**(2), 331–346 (2006)
58. Boyer, L., Francois, P., Doutre, E., Weil, G., Labarere, J.: Perception and use of the results of patient satisfaction surveys by care providers in a French teaching hospital. Int. J. Qual. Health Care **18**(5), 359–364 (2006)
59. Davis, P., Milne, B., Parker, K., Hider, P., Lay-Yee, R., Cumming, J., Graham, P.: Efficiency, effectiveness, equity (E3). Evaluating hospital performance in three dimensions. Health Policy **112**(1–2), 19–27 (2013)

60. Arah, O.A., Klazinga, N.S., Delnoij, D.M., Asbroek, A.T., Custers, T.: Conceptual frameworks for health systems performance: a quest for effectiveness, quality, and improvement. Int. J. Qual. Health Care **15**(5), 377–398 (2003)
61. Sawhney, R.: Interplay between uncertainty and flexibility across the value-chain: towards a transformation model of manufacturing flexibility. J. Oper. Manag. **24**(5), 476–493 (2006)
62. Ward, M.J., Ferrand, Y.B., Laker, L.F., Froehle, C.M., Vogus, T.J., Dittus, R.S., Kripalani, S., Pines, J.M.: The nature and necessity of operational flexibility in the emergency department. Ann. Emerg. Med. **65**(2), 156–161 (2015)
63. WCED, S.W.S.: World commission on environment and development. Our Common Fut. **17**, 1–91 (1987)
64. Pedersen, E.R.G., Sudzina, F.: Which firms use measures? Internal and external factors shaping the adoption of performance measurement systems in Danish firms. Int. J. Oper. Prod. Manag. **32**(1), 4–27 (2012)
65. McKee, M., Healy, J. (eds.): Hospitals in a Changing Europe, vol. 3. Open University Press, Buckingham (2002)

Institutional Response Strategies of Health Crises Through Public's Lens

Amer Qasem and Donya Janem

Abstract This descriptive study employs the Social Mediated Crisis Communication Model [SMCC] to primarily identify the crisis response strategies adopted by the Palestinian Ministry of Health in facing the Corona Covid-19 pandemic. The extent to which the public accepted these strategies approved by the ministry following the form of the messages directed to the audience is consequently examined. The study, at the same vein, attempts to detect the audiences' emotional responses that have been shaped following the selected form of messages under analysis. Traditional media, social media or offline word-of-mouth entail the form of message. A content analysis was conducted on the ministry's communications during the study period. To triangulate the data, a questionnaire has been designed and electronically distributed on a sample of 370 students enrolled at Hebron University, Palestine. The analysis has shown that the supportive strategies are the most acceptable responses to the audience. They create a positive emotional reaction following the type of message source applied. The study also concludes that—in health crises—social media and offline word-of-mouth are more convenient as a form of message at the outbreak and during the pandemic. The form of message is highly influential on the development of internal and external dependent-attributed emotions. The current study argues that it is essential to adopt social media and offline word-of-mouth as a communication form during health crises.

Keywords SMCC · Coronavirus crisis · Response strategies · Crisis communication

1 Introduction

Life is progressively becoming virtually-oriented. The Social Mediated Crisis Communication Model [SMCC] was developed as a theoretical framework to manage crisis communication in the light of that significant shift towards social media.

A. Qasem (✉) · D. Janem
An-Najah National University, Nablus 282615, Palestine
e-mail: amer.q@najah.edu

A. Hamdan et al. (eds.), *The Implementation of Smart Technologies for Business Success and Sustainability*, Studies in Systems, Decision and Control 216,
https://doi.org/10.1007/978-3-031-10212-7_82

It describes the way in which the form and source of crisis information influence the audience's acceptance and emotional responses. This model examines the organization's options for crisis response strategies from the public's point of view.

The world is witnessing various consecutive and complex crises. The Coronavirus Crisis has recently emerged on a worldwide scale as a result of the rapid spread of Corona Covid-19 virus. The Palestinian Ministry of Health applied a set of preventive measures determined by the World Health Organization [WHO] to confront the pandemic. Lee [1] argues that there is a need to shape a better understanding on how individuals understand the organizational crisis and interact with the course of related actions.

It is essential to detect the public interpretations and interactions of communication process during crises. When the public realize the seriousness of a crisis situation, their participation in the communication process increases. Their attribution of responsibility is activated, as well [2]. The severity of the crisis may, therefore, lead to negative emotional reactions towards the organization. In other words, the public relations practitioners must exert greater efforts to enhance the credibility of message form and source. In so doing, they motivate the audience to follow directions, trust in the source and form, and determine the degree of message acceptance [3].

2 Crisis Communication: A Public Viewpoint

According to Adi [4], previous studies in crisis communication focus more on examining communication theories used by organizations in times of crisis. There is, however, a dearth of research that investigate the crisis communication from public's point of view. In the same vein, literature review has shown a lack of organizational communication research that consider the aspects of stakeholders/public affected by a crisis. Austin and Jin [5] argue, relatedly, the need for applying the SMCC Model to health crises in particular. McDonald et al. [6] agree on the inevitability of hindrance that might be formulated through public's negative emotions against institutional response strategies during a crisis. They see an urgent need to explore a variety of crisis emotions and their impact on public responses and attributions, therefore.

Yang et al. [7] called for conducting more research on the impact of sources credibility on the public's acceptance of and participation in strategies to confront crises. Liu et al. [8] also recommend to broaden a better understanding of how crisis communication can affect the emotional responses of audience.

In her study, Abdullah [9] has shown the effect of messages of a humanitarian nature on the positive response of the public to the crisis when adopting the strategy of denial in particular. She recommends conducting studies on the level of public acceptance of response strategies through social media. According to Mak and Song [10], emotional responses play key role in how the public participates in a crisis. The findings show that organizations, social media creators, and influencers can influence the feelings of social media followers. However, these emotional changes also depend on the content of crisis information introduced through different types of influencers in various forms. Liu [11] found that the public accepts crisis responses via traditional media in the largest proportion, followed by social media and oral communication, respectively.

Austin et al. [12] argue that there is an impact of the third party in communication during crises. They see also a need to integrate traditional media and social media in responding to crises. Park and Avery [13] concluded that in a public health crisis, people desire a variety of opinions and perspectives and turn to social media, especially during health crises.

3 Social-Mediated Crisis Communication Model [SMCC]

Scholars have developed SMCC model as a practical framework for managing crisis communications via social media in the modern media landscape. It serves as the first theoretical framework to describe the relationships between organizations, online and offline audiences, social media, traditional media, and offline word-of-mouth before, during and after crises [12, 14].

The form of information can influence the communication behaviors of the public. Coombs [15] argues that the form of information could be more significant than the actual crisis response messages. Scholars in crisis communication, meanwhile, pointed out the impact of the source of crisis information as it plays a key role in building trust and credibility. Thus, the level of acceptance of crisis response strategies will be affected accordingly [3]. The source of information was classified into two categories: from the organization and from a third party outside the organization [such as media outlets].

Five factors influence the institutional crisis communication of institutions were identified [11]. The origin of the crisis [whether the crisis was caused by an internal organizational problem, or by external factors. The attribution of responsibility, and response strategies are affected on this regard. Secondly, the type of crisis [victim, accident, intentional]. Thirdly, the infrastructure that overlaps the level of the institution's response to the crisis. Finally, the source and the form of the message detailed earlier.

This model hypothesizes that the degree of acceptance of the strategy is affected by the source of the crisis information [first party: the institution itself, or a third party: a source from outside the institution, such as the media]. For instance, the degree of acceptance of a particular strategy by the institution itself may differ from receiving it from a media outlet. Consistent with what the model postulates, the form and source of the message also influence the emotions/feelings that the audience may form when responding to a crisis [5].

The SMCC model argues that the public turns to social media during a crisis for emotional support and the desire of emotional venting. Hence, the type of emotional support that the public receives through information sources affects their perceptions of crisis response strategies. According to Jin [16] the higher the responsibility of a crisis attributed to the institution, the greater the negative feelings of the public are. It is, therefore, significant to investigate to which extent the form of crisis information shapes the feelings of the public experiencing a crisis [11].

Pertaining to institutional response strategies during a crisis, the model suggests twofold question [11]: what are the audience's feelings and emotions shaped following the selection of form and source of message; and to which extent they accept these strategies following the form and source of message? Fig. 1 visualizes the overlapping among variables within the suggested model. It shows ten response strategies introduced by Coombs [2]. These strategies were reclassified into four major groups [15]. The SMCC variables are applied to these strategies, which creates the dimensions as presented in the figure.

This paper solely employs the assumptions related to the form of message as a theoretical thrust. It examines the response strategies adopted by the Palestinian Ministry of Health to confront the Coronavirus Crisis, then investigates the effects of the form of messages on the public's degree of acceptance and in shaping their emotions and feelings towards these messages.

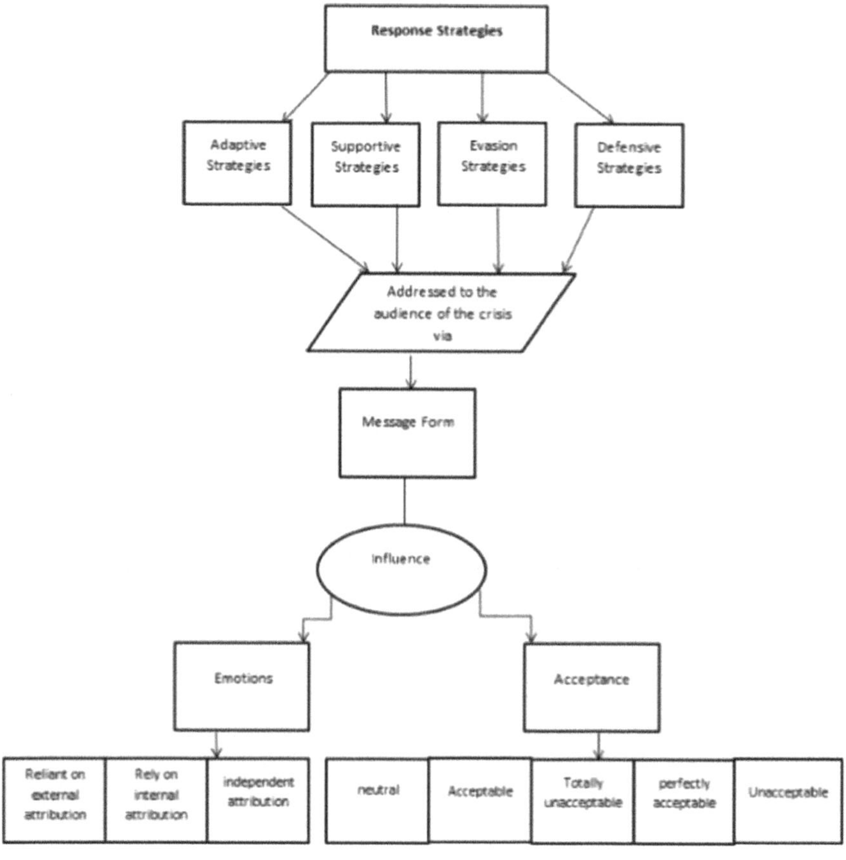

Fig. 1 Theoretical framework of the study

4 Methodology

In this descriptive study, a mixed methodology is adopted. Both quantitative and qualitative tools are used to achieve the research objectives. First, a coding sheet and coding book was designed to identify the institutional response strategies employed by the Palestinian Ministry of Health during the crisis under scrutiny. Following that, a questionnaire was distributed on a random sample of the Palestinian public to examine their degree of acceptance and identify the feelings the shape when they were exposed to the concerned strategies.

The students of Hebron University, West Bank, represented the study population as the Coronavirus Crisis erupted in Hebron then spread out to the rest of the West Bank territories. A random sample of 370 students received the questionnaire electronically. The questionnaire comprises 42 items distributed on three dimensions. The first is communicated with the respondents' demographic variables. The second and third dimensions measure the respondents' degree of acceptance and the type of emotions developed following the form of information [social media, traditional media, offline WOM], respectively. The questionnaire's items related to the type of response strategy are designed following Coombs [15] classification. That is adaptive strategies, supportive strategies, evasive strategies and defensive strategies.

The Cronbach's Alpha mean was calculated to measure the stability of the test, where the total stability of tool was [81.4]. This is deemed a high stability coefficient and satisfies the purposes of scientific research. The arithmetic mean, percentages and standard deviation were calculated to identify answer the research questions.

A content analysis, as well, was conducted on a total of 85 message directed from the ministry to the public during the study period as follows: 50 written posts, 35 videos. The videos included 20 television interviews, 6 press conferences and 9 other videos posted by the ministry on its Facebook page. The study period started from the beginning of March 2020 through the end of September 2020.

Cronbach's Alpha was calculated to measure the reliability and internal consistency of the analytical tool. The alpha coefficient of the related tool was 81.4, which reflects a high stability and satisfies the purposes of scientific research. The percentages and frequencies of the strategies used according to the classification of Coombs [15] were calculated, too. The arithmetic mean, percentages and standard deviation were also taken into account to examine the level of audience's acceptance of the strategies under scrutiny.

5 Analysis and Findings

The analysis have shown that the adaptive strategies are the most prominent. In contrast, the supportive strategies are the least (see Table 1). The adaptive strategies appear to be more acceptable to the public when using social media and/or offline WOM as a form of message. More importantly, this overlapping creates negative emotions against the corporation. This result is consistent with what Liu et al. [11] concluded. The public accepts adaptive crisis responses when issued by the organization itself through offline WOM.

The results have also illustrated that the supportive strategies are more acceptable to the audience when using traditional media and offline WOM. In the same token, Liu et al. [11] suggest that the audience is more receptive to supportive strategies transmitted through traditional media instead of social media. The emotions shaped when employing supportive strategies were positive, particularly if social media and traditional media are used as form of the message.

The evasive strategies were acceptable to the public to a moderate degree following the applied source and form of the message. They are more acceptable to the audience when using social media. This result differs with what Liu et al. [11] found. Their analysis has shown the audience in a more receptive manner to evasive responses communicated via traditional media rather than social one.

When social media is used to communicate strategies to evade crisis responsibility, the audience creates emotions of external attribution. The analysis, as well, has shown that offline WOM messages used to communicate evasive strategies create independent-attributed emotions. However, the defensive strategies are more acceptable to the public when using offline WOM. This result is inconsistent with Liu's et al. [11] findings. They argue that the audience is more receptive to defensive strategies communicated via traditional media. The emotions developed when applying defensive strategies are positive, particularly if offline WOM and traditional media are used as a communication tool.

The results show that choosing the appropriate form of crisis information has an impact on the emotional responses of audience. When adopting offline WOM, the dominant emotions are independently attributed. However, these emotions appear at a lower rate when the crisis information are communicated via traditional media.

The public accepts crisis responses through social media and WOM in a greater proportion than traditional media. Social media and offline WOM have not noticeably contributed in creating a positive emotional responses following the source of message. This result is in line with Yang's et al. [7] findings. They contend that the credibility of source has a slight effect on audience participation in the crisis communication process.

The results also reveal that the public is more receptive of crisis responses via offline WOM, followed by social media, and finally traditional media. This order of

Table 1 Crisis response strategies of Palestinian MOH

Type of response strategy	Frequency	Percentage
Defensive	19	12.5
Adaptive	86	56.6
Evasive	37	24.3
Supportive	10	6.6
Total	152	100.0

message form differs from what Liu et al. [11] suggests. They found that traditional media comes first, followed by social media, then offline WOM.

The results have illustrated that when crisis information are disseminated via social media and offline WOM, the dominant emotions created are independently attributed. When supporting strategies are used, the audience will also form independently attributed emotions in a greater proportion than if other strategies were employed [adaptive strategies, evasive strategies, and defensive strategies].

By using defensive strategies, the audience will form emotions dependent on external attribution in a greater proportion than if other strategies were used. The results have shown that emotions dependent on external attribution recorded the highest percentage when using defensive strategies compared to other strategies. It is also worthy to mention that emotions dependent on internal attribution are the least developed regardless of the type of response strategies. The form of message has greater impact than message source on developing emotions dependent on external and internal attributions.

6 Discussion and Conclusion

Institutional response strategies vary during a crisis. According to Traintafillidou and Yannas [17], corporations experience an urgent need to apply social media as a platform to communicate during crises. Monitoring audience's feedback, communicating with them, and responding to their needs are more facilitated via new media. This will resultantly drive crises managers to build trust with their public. A study on American public revealed that the majority consider the internet as their favorite and most trustworthy medium to receive news [15].

The direct and immediate nature of crisis responses maximize the challenges on PR practitioners to produce effective content. Social media may constitute a serious risk on effective crisis communication if messages are not designed accordingly [18].

Social motivation is deemed the significant reason behind using social platforms by audience. They tend to express themselves, search for information, operate social monitoring and fulfill their political needs [19]. Responding to a crisis is considered the most perilous period in the institution's lifecycle [20]. Therefore, social media should be included in the crisis management plan to reduce its negative effects. Even after the crisis ends, communicating with audience continues.

Park and Kim [21] argue that individuals tend to trust peers more than corporations. Institutions are seen as untrustworthy source of information by the public. They lack credibility and transparency while communicating with their audiences. Social media could offer unique information during crises [4]. However, this paper recommends employing traditional and offline WOM to communicate response strategies when a crisis erupts. In later periods, social media could be considered an effective form of message.

The current study was limited to one segment of audience, namely, university students. It recommends expanding the sample for future research to include further segments of the public. Moreover, It agrees with Xu's [22] recommendation to explore the way of how social media might influence the institutional reputation after a crisis ends. This study solely focused on Facebook as a representative medium of social platform. Other social platforms are recommended to be under scrutiny, therefore.

References

1. Lee, B.: Audience oriented approach to crisis communication: a study of Hong Kong consumers' evaluation of an organizational crisis. Commun. Res. **31**(5), 600–618 (2004)
2. Coombs, T.: Protecting organization reputations during a crisis: the development and application of situational crisis communication theory. Corp. Reput. Rev. **10**(3), 163–177 (2007)
3. Avery, E.: The role of source and the factors audience rely on in evaluating credibility of health information. Public Relations Rev. **36**(1), 81–83 (2010)
4. Adi, W.: Simplified social mediated crisis communication model during crisis in Indonesia: a case study on how customer of Indonesia commuter Line Train Company seek information on a train delay due to the Double Track Trial on April 12, 2019. In: International Conference on Humanities, Education and Sciences, pp. 290–301 (2020)
5. Austin, L., Jin, Y.: Social media and crisis communication: Explicating the social-mediated crisis communication model. In: Due, A., Kahlor, L. (Eds.), Strategic Communication: New Agendas in Communication, pp. 163–186. Routledge, New York (2016)
6. McDonald, L., Sparks, B., Glendon, A.: Stakeholder reaction to company crisis communication and causes. Public Relation Rev. **36**(3), 263–271 (2010)
7. Yang, S., Kang, M., Johnson, P.: Effects of narratives, openness to dialogic communication, and credibility on engagement in crisis communication through organizational blogs. Commun. Res. **37**(4), 473–497 (2010)
8. Liu, B., Fraustino, J., Jin, Y.: Social media use during disasters: how information form and source influence intended behavioral responses. Commun. Res. **43**(5), 626–646 (2016)
9. Abdullah, D.: The effect of negative communication through social media during the product crisis on the behavior of the Egyptian public towards boycotting its purchase: a case study on the Cairo Food Industries Company (Heinz Egypt). Egyptian J. Media Res. **4**, 229–305 (2017)
10. Mak, A., Song, A.: Revisiting social-mediated crisis communication model: the Lancôme regenerative crisis after the Hong Kong umbrella movement. Public Relations Rev. **45**(4), 1–15 (2019)
11. Liu, B., Austin, L., Jin, Y.: How publics respond to crisis communication strategies: the interplay of information form and source. Public Relations Rev. **37**(4), 345–353 (2011)
12. Austin, L., Fisher Liu, B., Jin, Y.: How audiences seek out crisis information: exploring the social-Mediated crisis communication model. J. Appl. Commun. Res. **40**(2), 188–207 (2012)
13. Park, S., Avery, E.: Effect of media channel, crisis type and demographics on audience intent to follow instructing information during crisis. J. Contingencies Crisis Manage. **26**(1), 69–78 (2018)
14. Cheng, Y.: The social-mediated crisis communication research: revisiting dialogue between organizations and publics in crises of China. Public Relations Rev. **46**(1), 1–11 (2020)
15. Coombs, T.: Ongoing Crisis Communication: Planning, Managing, and Responding, 3rd edn. Sage, Thousand Oaks, CA (2012)
16. Jin, Y.: The effects of public's cognitive appraisal of emotions in crises on crisis coping and strategy assessment. Public Relations Rev. **35**(3), 310–313 (2009)
17. Traintafillidou, A., Yannas, P.: Social media crisis communication in racially charged crises: exploring the effects of social media and image restoration strategies. Comput. Hum. Behav. **106**, 106269 (2020)

18. Kapuściński, G., Zhang, N., Zeng, L., Cao, A.: Effects of crisis response tone and spokesperson's gender on employer attractiveness. Int. J. Hosp. Manag. **94**, 102884 (2021)
19. Liu, B., Austin, L., Jin, Y.: How publics use social media to communicate during crises: proposing the social-mediated crisis communication model. In: Proceedings of the Public Relations Society of America Educators Academy, Washington, DC, pp. 142–157 (2010).
20. Graham, M.W., Avery, E.J., Park, S.: The role of social media in local government crisis communications. Public Relations Rev. **41**(3), 386–394 (2015)
21. Park, H., Kim, Y.: Is there still a PR problem online? Exploring the effects of different sources and crisis response strategies in online crisis communication via social media. Corp. Reput. Rev. **20**(1), 76–104 (2017)
22. Xu, J.: Does the medium matter? A meta-analysis on using social media vs. traditional media in crisis communication. Public Relations Rev **46**(4), 101947 (2020).

SPRINGER NATURE

GPSR Compliance

The European Union's (EU) General Product Safety Regulation (GPSR) is a set of rules that requires consumer products to be safe and our obligations to ensure this.

If you have any concerns about our products, you can contact us on ProductSafety@springernature.com

In case Publisher is established outside the EU, the EU authorized representative is:

Springer Nature Customer Service Center GmbH
Europaplatz 3
69115 Heidelberg, Germany

The manufacturer's authorised representative in the EU is Springer Nature Customer Service Centre GmbH, Europaplatz 3, 69115 Heidelberg, Germany. If you have any concerns regarding our products, please contact ProductSafety@springernature.com

Printed and bound by CPI Group (UK) Ltd, Croydon, CR0 4YY
30/09/2026
02225511-0002